CRYSTALLIZATION – SCIENCE AND TECHNOLOGY

Edited by **Marcello Rubens Barsi Andreeta**

INTECHOPEN.COM

Crystallization – Science and Technology
Edited by Marcello Rubens Barsi Andreeta

CBS, Edition 2016

Published by InTech
Janeza Trdine 9, 51000 Rijeka, Croatia

Publishing Process Manager Vana Persen

Technical Editor Damir Jovic

Cover Designer InTech Design Team

Additional hard copies can be obtained from orders@intechopen.com

Crystallization – Science and Technology, Edited by Marcello Rubens Barsi Andreeta
p. cm.
ISBN 978-953-51-0757-6

INTECH

open science | open minds

Contents

Preface

Crystallization is one of the first items of technological knowledge acquired by mankind. In fact, the use of crystallization processes can be considered as ancient as writing. One example of this is a Chinese document dating back to 2700 BC, which describes the process of artificial evaporation for the production of salt [1]. Crystallization is also one of the most interdisciplinary topics of research, ranging from inorganic to organic compounds, and crystals can be produced from melts, liquid solutions, vapors or even in the solid state. The nature of the crystallization process has been described by researchers with diverse scientific backgrounds, each one presenting a different facet of the theme, be it from the standpoint of Chemistry, Physics, Biology, Geology or Engineering. Despite its inherently high complexity, the crystallization process is part of our everyday lives: from ice-making in our homes or falling snow to the most state-of-the-art chemical and electronic industry. This book offers the reader an overview of the science and technology involved in different crystallization processes, compiled by researchers with different scientific backgrounds.

In the first part of the book, entitled *Fundamentals and Theoretical Aspects*, the reader is introduced to various different perspectives of the crystallization process. The first two chapters review the crystallization processes of glasses. The first chapter describes crystallization using the chemical bond approach. This new approach, as presented by Chechetkina, is based on the hypothesis of *initial reorientation*, which is considered a specific pre-nucleation stage, and on the *bond wave* model, considered a carrier for collective processes in glass formers. Chapter two describes the crystallization of chalcogenide glasses, with Singh discussing the fundamentals of nucleation, growth and crystallization processes in amorphous glassy materials based on the classical ideas of crystallization. The next two chapters deal with aspects of the crystallization process using mathematical models. In chapter three, which contains a numerical model of the crystallization of steel, ductile cast iron and ceramic corundo-baddeleyite material, Kavicka and collaborators describe the optimization of the production and properties of these materials after casting. Chapter four consists of a theoretical modeling of the growth process of a single crystal in the shape of a cylindrical tube, obtained by the edge-defined film-fed growth (EFG) technique. Tanasie and Balint use this model to demonstrate that it can be employed in determining the pulling rate, and the thermal and capillary conditions of an actual experiment. In chapter five, Cooper and collaborators present a thermodynamic discussion about crystallization in microemulsions. The authors show that the critical nucleus can be estimated with good accuracy under conditions of thermodynamic control and that the stable form of a material can be identified and readily produced under ambient conditions. The authors of the last two chapters describe the recrystallization process. Umalik focuses his analysis on the chemical, physicochemical and crystal-chemical aspects of the recrystallization process, discussing its effects on the cocrystallization coefficient, $D_{2/1}$. Stinger and Liebenberg, in turn, review the aspects of recrystallization of active pharmaceutical ingredients (API). They draw attention to the possibility of the process creating undesirable polymorphic modifications, and the reason why some of these forms have gone undetected for decades.

The second part of the book, *Applications, Techniques and Mineral Formation,* is dedicated to the practical application of crystallization processes and concepts. This section begins with Chapter 8, which describes the crystallization process under hydrothermal conditions. Rendón-Angeles and collaborators have contributed not only with an interesting review of the hydrothermal method, but have also presented results of synthesis of different perovskite oxides and mineral replacement reactions as a new approach to prepare inorganic materials.

Chapters 9 and 10 describe crystallization in Fe-based compounds. Minić and collaborators use a technique to rapid-quench an iron-based compound, producing an amorphous alloy. They demonstrate that under the correct heat treatment, these alloys can be used as precursors in the preparation of nanocomposite materials composed of nanocrystals in the amorphous matrix. In chapter 10, Selivanov and Gulyaeva report on a study of the composition of metallurgical slag and the influence of iron on the different metals it contains. They show that iron oxidation and the ratio of calcium oxide to silicon oxide play a major role in determining the types of non-ferrous metals in crystallized slag.

The next four chapters introduce techniques and discussions about aspects of organic and large molecule crystallization. In chapter 11, Silva and co-authors describe a technique to monitor the formation of adipic acid crystals in real time, which can be applied industrially in place of off-line measurements. They demonstrate that the off-line measurements used in industry to determine the onset of crystallization present a much higher error than the real-time analysis they have developed. In chapter 12, Moynihan and Kelly present a study of the crystallization of phenacetin. They discuss the influence of two different cooling procedures on the control of crystal size distribution. In the next work, in chapter 13, Françoise presents a review of macromolecular crystallization. He exemplifies this process by using urate oxidase, and shows that pH, salt and polymers play a crucial role when crystallization is controlled by colloidal interactions. In chapter 14, Giulietti and Bernardo describe their work on the sugar crystallization process using an antisolvent approach. They demonstrate that the addition of alcohol or ketone, combined with the use of a suitable cooling system, can improve sugar crystallization. The authors show that this decreases the system's viscosity and also causes the organic antisolvent and the solute to compete for water, thereby segregating the solute from the solution and promoting crystallization.

Thin films are the theme of the next five chapters. We begin with a review of the sputtering technique for thin film deposition. In chapter 15, Alfonso and co-workers discuss the physicochemical parameters involved in the formation of well crystallized thin films, using the sputtering technique. Arciniega and co-authors, in chapter 16, discuss the crystallization of Ge:Sb:Te thin films and the viability of their use as phase change memory devices. In chapter 17, Mohamad and Mamidipudi review the work on metal-induced crystallization in silicon thin films. The authors describe the possible mechanisms involved in this process and discuss Sn and Cr-induced crystallization of silicon thin films. The use of an Excimer laser in the copper-induced crystallization of silicon thin films is discussed by Bidin and Razak in chapter 18. The authors used an Excimer laser to promote surface annealing of the film, creating the condition necessary for the silicon thin film to recrystallize. They also demonstrate that this procedure allows for the production of large grain silicon thin films for potential use in ultra-large-scale integrated circuit devices. In chapter 19, the authors also discuss the use of an Excimer laser in silicon thin film crystallization. Kitahara and Hara investigate the lateral growth orientation of silicon thin films on glass substrates, and report that relatively small defect densities are observed in the case of pure Si films.

This book concludes with a discussion about the formation of minerals through a natural crystallization process. Ramasamy describes the carbonatite complex of Tiruppattur, Tamil Nadu, India. The author discusses the crystallization, fractionation and solidification of a co-magmatic alkaline series emplaced sequentially in the carbonatite complex. He shows that, as the first crystallization occurs, the composition of the parent magma evolves to become poor in alumina and silica, and acts as a primary magma in a closed magmatic chamber under a volatile enriched condition. Later on, during the early crystallizations of calcium-rich clinopyroxenes, the residual magma becomes impoverished in silica and enriched in volatile constituents such as H_2O and CO_2. With the low viscousness and high temperature of shonkinite magma, clinopyroxenes are crystallized with enrichment of Ca, Mg, Fe, Ti and Al and depletion of Si, Al, Na and K in low pressure zones towards the top of the magmatic column. On the other hand, from the top of the magmatic column, sinking clinopyroxenes release Ca, Mg, Fe, Ti and Al and accumulate Na and Al, with crystallization of subsolidus aegirine.

Finally, I would like to thank the authors for their hard work and patience. Not only did they work on the formatting and language review of their chapters, but also did a fantastic job in following the proposed editorial guidelines. Their dedication has enabled us to produce a book with new insights, containing crucial and very useful information for researchers working in this field, while simultaneously creating a comprehensive text about crystallization processes that may serve as a starting point for readers with different backgrounds.

Dr. Marcello Rubens Barsi Andreeta
Crystal Growth and Ceramic Materials Group,
Department of Physics and Material Science,
Physics Institute of São Carlos,
University of São Paulo, São Carlos, SP,
Brazil

Reference

Schoen, H.M., C.S. Grove, and J.A. Palermo, *The early history of crystallization.* Journal of Chemical Education, 1956. 33(8): p. 373.

Section 1

Fundamentals and Theoretical Aspects

1

Crystallization in Glass Forming Substances: The Chemical Bond Approach

Elena A. Chechetkina
Institute of General and Inorganic Chemistry of Russian Academy of Sciences, Moscow, Russia

1. Introduction

Glassy materials are strongly connected with crystallization or - more strictly - with the ability to *avoid* crystallization when cooling a melt. The more stable a supercooled liquid against crystallization, the higher its glass forming ability. It should be noted that almost every substance can be preapared in the form of amorphous solid by means of special methods of fast melt cooling (in the form of ribbons), evaporating (films), deposition (layers) and sol-gel technique (initially porous samples), etc. [1]. The resulting materials are often named "glass" (e.g., "glassy metals"), but they are actually outside the scope of this chapter. Here, we consider only typical inorganic glasses, such as SiO_2 and Se, which correspond to an understanding of glass as a *bulk non-crystalline solid prepared by melt cooling.* "Bulk" means a 3D sample having a size of 1 cm and larger, a condition that implies a cooling rate of about 10 K/s and lower. Despite the "technological" character of this definition, it is the most objective one, being free from the declared or hidden speculations about the nature of glass.

The theoretical background is given in Section 2, beginning with classical notions about crystallization in ordinary melts and the concept of the *critical cooling rate* (CCR), understood as the minimal cooling rate that provides the solidification of a melt without its crystallization. As far as any crystallization ability is reciprocal to a glass forming ability, one can evaluate a *glass forming ability* in terms of the CCR, using both crystallization theory and experiments. It will demonstrate a principal inapplicability of classical crystallization theory in the case of glass forming substances. As an alternative, we have developed a new approach based on the hypothesis of *initial reorientation,* considered as a specific pre-nucleation stage in a non-crystalline network. The classical model of a *continuous random network* consisting of common covalent bonds (two-centre two-electron, 2c-2e) is modified in two respects. First, we introduce *hypervalent bonds* (HVB) in addition to covalent bonds (CB); and besides, such bonds can be transformed into one another (CB↔HVB). Second, the elementary acts of bond exchange are in spatiotemporal correlation, thus providing a *bond wave* by means of which the collective processes - including initial reorientation and further crystallization - may occur. A bond wave also means a non-crystalline order of hierarchical character: from the well known short-range order (it changes in the vicinity of HVB), through the well known but poorly-understood medium-range order (in the limits of the wavefronts populated with HVB), to the non-crystalline long-range order generated by bond waves. Insofar as bond waves change their parameters and direction during melt cooling, being frozen in solid glass, the non-crystalline long-range order depends not only on the substance under consideration but also on the sample prehistory. This is a principal distinction from the ordinary long-range order in crystals, which is strictly determined for a given substance at a given temperature. As such, the problem of crystallization in glass can be considered as a competition between two types of order, and hypervalent bonds and their self-organization in the form of a bond wave play a central role in this process.

In Section 3 and Section 4, original crystallization experiments based on the above approach are described. In Section 3, we consider crystallization in **solid Se-X** glasses, in which additions of a different chemical nature (X = Cl, S, Te, As, Ge) are used. A set of properties, including the abilities of nucleation and surface crystallization, was investigated based upon the composition for each series. A strong *non-linearity* was found in the region of small additions $N<N^*$ (concentration N^* depends on the nature of X), which is discussed in terms of the bond wave interaction with foreign atoms also existing in a hypervalent state. Finally, in Section 4, original experiments on the crystallization of **softening Se-X** glasses in an **ultrasonic field** are presented. In softening glass, bond waves are refrozen and an ultrasonic field can act as an *information field* for them; in our case, a US-field can give a predominant direction for the crystallization process. Actually, the resultant glasses become *anisotropic*, the anisotropy also being non-linear in respect to the composition.

Thus, starting from the classical approach operating with a 3-step process of crystallization (sub-critical unstable nucleation, stable nucleation, crystal growth), fluctuations, atomic jumps, atomic/geometrical structure, one-type bonding and one-type long-range order, we pass to a strange picture of mixed bonding, bond waves, "wavy" long-range order and glass as a self-organizing system.

The text is intended for a wide audience. The readers from the field of glass are invited to meet with new experiments and new ideas. For beginners - let us now enter into the mysterious world of glass!

2. Crystallization in supercooled glass forming liquids ($T>T_g$)

2.1 Critical cooling rate

The idea of the CCR as a measure of a glass forming ability seems to be trivial, but it remained in the shadows up until 1968, when *Sarjeant & Roy* [2] emphasized "the concept of '*critical quenching rate*'… defined as the cooling rate below which detectable crystalline phases are obtained from the melt" and proposed the first theoretical expression for the CCR in the form of:

$$Q_c = 2.0*10^{-6}*(T_m)^2*R/V*\eta \quad (1)$$

Here T_m is the melting point, η (poises) is the viscosity at T_m and V is the volume of the "diffusing species". The main problem is the evaluation of V, which needs "an extreme simplification and interaction of many parameters involved" [2]. Nevertheless, it is seen in **Table 1** that the values thus obtained are in agreement with the later estimations of the CCR [3-5] using the equations of classical crystallization kinetics.

Strictly specking, it is easy to fit into the "classical" gap [3-5] that extends for orders in magnitude. More interesting is the tendency for this gap, $\Delta|\lg Q_c|$, to be especially wide just for glass-formers (G). It is seen in **Fig.1** that in glass forming substances ($Q_c<10^2$ K/s by definition) the difference $\Delta|\lg Q_c|$ increases when Q_c decreases, i.e., the larger the gap, the greater the glass forming ability. Moreover, and to the contrary, for non-glass forming substances (Q_c>10^2 K/s), there is no relation between Q_c and its variation, the latter actually being the same for all non-glass formers.

Substance	T_m, K	lgQ_c(K/s)				$\lvert \Delta lgQ_c \rvert$
	[4]	[2]	[3]	[4]	[5]	
SiO_2 (G)	1993	+0,6	-3,7	-1	-2,9	4,3 (G)
B_2O_3 (G)	733		-1,2	-16	-7,3	8,7 (G)
GeO_2 (G)	1359			-11	-2,9	9,8 (G)
P_2O_5 (G)	853			-23	-5,9	17,1 (G)
As_2O_3	1070			+7	+4,1	2,9
H_2O	273	+7,2	+7	+12	+10,4	5,0
Salol (G)	316.6		+1,7	-15		13,3 (G)
Ethanol	159			-2	+1	3,0
Glycerin (G)	293			-40	-7,9	32,1 (G)
CCl_4	250,2			0	+5,2	5,2
C_6H_6 (G?)	278,4			-14	+4,9	18,9 (G?)
BeF_2 (G)	1070			-3	-4,0	1,0 (G)
$ZnCl_2$ (G)	548			-21	-4,4	16,6 (G)
$BiCl_3$	505			+7	+5,5	1,5
$PbBr_2$	643			+3	+6,3	3,3
NaCl	1073	+9,1		+10	+8,4	1,6
$NaNO_3$	580			+5	+7,0	2,0
$NaCO_3$	1124			+3	+5,0	2,0
$NaSO_4$	973			+10	+7,4	2,6
$NaWO_4$	1162			+7	+8,4	1,6
Ag	1234	+8,0	+10	+8	+9,3	1,3
Cu	1356			+9	+8,5	0,5
Sn	505,7			+2	+7,9	5,9
Pb	600,7			+10	+8,7	1,3

Table 1. The calculated CCRs among the groups of oxides, substances with hydrogen bonding and/or organic glass formers, halides, sodium salts and metals. The "G" in brackets indicates a well-recognized glass former.

One should note the absence of benzene (see C_6H_6 in Table 1) in Fig.1. This is a result of the outstanding disagreement in the evaluations of its CCR: benzene looks like an excellent glass former after [4], but after [5] it is a non-glass former. When lgQ_c=+4,9 after [5], the point for difference $\Delta\lvert lgQ_c\rvert$=18,9 is dramatically lifted above the "amorphous line"; however, if one takes lgQ_c=−14 after [4], the point falls far below the correlation line for glass formers.

The Qc interval in Fig.1 extends for 19 orders in magnitude. The practical interval is far narrower. For glasses, there are three useful CCR points: the glass definition limit Q_c=10^2 K/s, which corresponds roughly to a quenching of a 1 cm^3 sample in water with ice, the rate of 10 K/s corresponding to quenching in the air, and the rate of 1-10^{-2} K/s corresponding to free cooling in a furnace. Therefore, the extremely large negative values of lgQ_c(K/s) in Table 1 - such as -16 for B_2O_3 or -23 for P_2O_5 [4] - only display an extremely high glass forming ability, i.e., the ability for very slow cooling in a furnace, a feature which provides the opportunity to obtain very massive articles, such as lenses for telescopes. For amorphous materials (Q_c>10[2] K/s), the experimental limit is about Q_c=10[6] K/s, which corresponds with the preparation of amorphous ribbons, namely the 2D case. The calculated value of Q_c=10[8]-10[10] K/s displays only an extra low amorphization ability of a substance, which cannot be obtained by melt cooling at all. This is really the 1D case, when the amorphous film is prepared by evaporation onto a cold substrate.

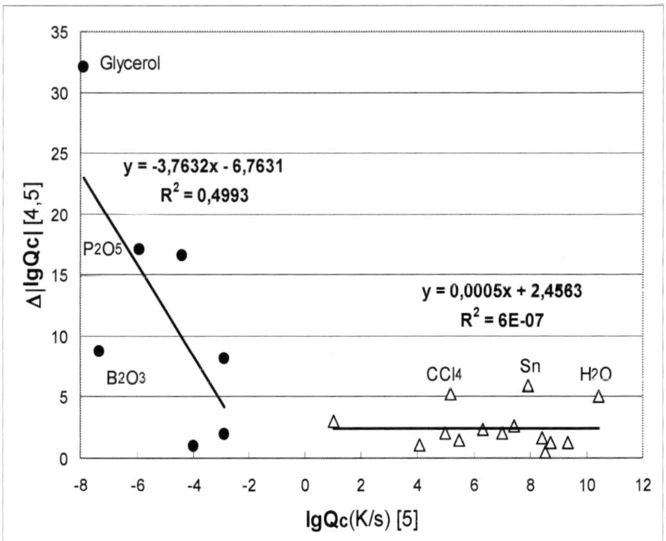

Fig. 1. The absolute difference between the calculated values of the CCRs from [4,5] as a function of the CCR value after [5]. The data is from Table 1.

Based on the tendency in Fig.1, one can conclude that the greater the glass forming ability, the more problematic the application of the classical theory of crystallization. In reality, in the case of a lithium disilicate glass it was shown that "not only do all forms of classical theories predict nucleation rates many orders in magnitude smaller than those observed, but also the temperature dependence of the theoretical rate is quite different from that observed" [6]. Thus, crystallization in glasses needs special methods, both theoretical and experimental.

Although there are many experimental works on the crystallization of glass and glass forming liquids, only a few of them concern the CCR. The known evaluations of Q_c in oxides [7-11] and chalcogenides [12-16] are made by either a direct method covering a relatively small lgQ_c interval (e.g., from -2 to +4 after [7-10]), or indirect methods. By 'indirect' is meant that the crystallization data corresponding to some extent of crystallinity α~0,01-0,1 is recalculated to a much lower α corresponding to Q_c (usually to α=10[-6] [15-16]) using equations of the *Kolmogorov-Avrami* type. However, the applicability of such equations in an extra low α region is also under question. Thus, it is not surprising that interest in the investigation of the CCR is low at present. The reader can find a critical review of the "golden age" of the CCR in the last third of the Twentieth Century in our monograph [17]. The main result of this period seems to be the formulation of new problems concerning glass forming ability.

The general question is what is meant by the demand of melt cooling "without crystallization"? After *Vreeswijk et al.* [4] glass is considered as being partially crystallized even one critical nucleus is formed; this is a **theoretical** limit for estimation of the Q_c value. Later, *Uhlmann* [3] proposed the crystalline fraction $\alpha=10^{-6}$ for the **experimentally** detected limit, and *Ruscenstein & Ihm* [5] proposed $\alpha=10^{-4}$ for a more convenient **practical** limit. Obviously, the lower α, the higher Q_c for a given substance. However, when comparing the calculation results in Table 1 - we see that there is no a relation between the crystalline fraction permitted (α increases from [4] to [5]) and the corresponding Q_c values. Thus, classical crystallization theory needs serious reconsideration for glass forming substances.

2.2 Induction period, transient effects and initial reorientation

The critical nucleus - arising when overcoming the "thermodynamic" barrier W^* with the following steady-state nucleation of the $I(T)$ frequency and the steady-state crystal growth of the $u(T)$ rate - forms the basis of classical crystallization theory. The critical nucleus is considered as the cause of the *induction period* observed in the process of crystallization. It is unclear, however, to what extent the classical notions are true for glass forming substances. In order to test the assumption about steady-state nucleation, *Kelton & Greer* [18] have analysed the data for lithium disilicate (a typical glass former) and two metallic "glasses" ('amorphous metal' would be a more appropriate term). When the authors introduced the time-dependent nucleation frequency $I(\tau,T)$, they revealed the so-called *transient effects* which occurred to be most significant only for typical glass. Transient effects, together with the fact that *"even small uncertainties of material parameters can introduce uncertainties of several orders in magnitude in the calculated nucleation frequency"* [18], make the application of classical crystallization theory problematic in the case of glass forming substances.

On the other hand, from the beginning of the 1980s onwards, we have developed an alternative approach for the prediction of glass forming ability in relation to crystallization ability. This approach is based on *Dembovsky's* "empirical theory of glass formation" [19], by which one can calculate the dimensionless glass forming ability, which was transformed further into an energetic barrier for crystallization E_{cr} [20]. The next stage was the transformation of the semi-empirical E_{cr} (enthalpy) to the barrier $\Delta G^{\#}$ (free energy), and finally $\Delta G^{\#}$ to the CCR measure V_m [21,22]. The reader can find a complete calculation scheme for V_m in [17; p.145] or [23; p.105]. Two points of this CCR are of special interest.

First, in order to calculate V_m, one needs *no kinetic parameters*, such as a coefficient of diffusion and/or viscosity. Instead, we use only the phase diagram of a system (in order to determine the liquidus temperature for a related unit: an element, a compound, an eutectic or else an intermediate), the first coordination number (presumably, in the liquid state at around T_m) and the averaged valence electron concentration (e.g., 6 for Se, 5.6 for As_2Se_3, 5.33 for SiO_2, etc.).

Second, the barrier $\Delta G^{\#}$ was attributed to a special pre-crystalline ordering named the *initial reorientation* (IReO), which is postulated as the stage after which crystal nucleation and growth is possible [21]. IReO means that neither nucleation nor crystal growth can proceed in a non-oriented - i.e., a truly "amorphous" - medium. Since IReO is a simple activation process with the $\Delta G^{\#}$ barrier, the reoriented fraction increases exponentially with time, as the dotted line in Fig.2 shows.

States:

(1) $A \rightarrow A_0 \rightarrow A_{Nu}$

(2) $A_0 \rightarrow A_{Cr}$

(3) $A \rightarrow A_0 \rightarrow A_{Cr}$

Barriers:

(1) $\Delta G^{\#}$; $(W^* + \Delta G^{\#})$

(2) $\Delta G' \rightarrow \Delta G''$

(3) $\Delta G^{\#}$; $\Delta G''$

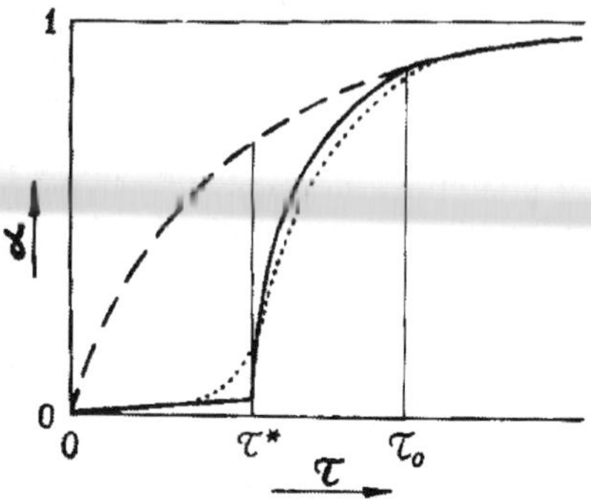

Fig. 2. Isothermal crystallization after [21]: the dotted curve describes the development of the reoriented non-crystalline fraction; the solid line is the model temporal dependence of the extent of crystallinity α; the pointed curve is a real crystallization process.
Atomic states: **A** is the initial truly "amorphous" state for a species in a supercooled liquid or glass; $\mathbf{A_0}$ is the reoriented state; $\mathbf{A_{Nu}}$ is the state of the crystal nucleus; $\mathbf{A_{Cr}}$ is the state of the growing crystal.
Barriers: $\Delta G^{\#}$ is the initial orientation; W^* is the critical nucleus; $\Delta G'$ is the steady-state nucleation; $\Delta G''$ is the crystal growth.

The three intervals in Fig.2 correspond to an *induction period* at $0<\tau<\tau^*$, a *transient period* at $\tau^*<\tau<\tau_0$ and a *saturation* at $\tau>\tau_0$. At first, the reoriented areas are relatively small and/or disconnected, such that only nucleation can proceed. Although nucleation is permitted, the extent of the crystalline fraction (the solid line) is negligible when compared with the reoriented fraction (the dotted line). At the moment τ^*, the reoriented areas become large enough and/or percolated so as to permit not only nucleation but also crystal growth, using both the arising and newly existing portions of the reoriented species. In this *transient* region $\tau^*<\tau<\tau_0$, crystallization slows gradually and at τ_0 it becomes limited by the process of initial reorientation when each new portion of the reoriented species joins directly to the growing crystals.

One should note in Fig.2 the well known barriers G', G'' and W^*. Fortunately, the problems with these barriers (e.g., varying G' after [18] or an anomalous increase of W^* with a decreasing temperature at about T_g [24]) are outside of our method for the estimation of the CCR since V_m relates only to the initial reorientation stage with the $\Delta G^{\#}$ barrier which determines the boundary between glassy and crystalline states. Thus, V_m is the *upper* theoretical limit for the CCR and in this sense V_m satisfies the condition set by *Vreeswijk et al.* [4] concerning the absence of even one critical nucleus when determining the CCR.

By 'initial reorientation' is meant some kind of order. Interestingly, the notions concerning pre-crystalline ordering in glasses appear consistently in the literature. One can see the IReO in a real SEM image, which is interpreted by the authors as "a pre-crystallization stage, in which glass matrix becomes inhomogeneous, forming nano-sized volumes" [25]. Indirect evidence of the IReO is seen in the conclusion that "glass-transition kinetics can be treated as pre-crystallization kinetics" [26] as well as in the models of "dynamic heterogeneity" in an amorphous matrix before its crystallization [27-29]. The problem is that both "dynamic heterogeneity" and "initial reorientation" are the only terms that need decoding.

2.3 Glass structure, hypervalent bonds and bond waves

The theory of glass structure was actually formed at 1932 when, beginning with the words *"It must be frankly admitted that we know practically nothing about the atomic arrangement in glasses,"* the young scientist *Zachariasen* proposed his famous model of a **continuous random network** (CRN) [30]. This network consists of covalent bonds, - the same as that in a related crystal - and the only difference is the "random" arrangement of the bonds in CRN in contrast with their regular arrangement in a crystalline network. As far as the atomic arrangement in CRN is determined by the arrangement of chemical bonds, one might suppose that only chemists - who deal with chemical bonds - would play a leading role in the further development of the CRN model. However, the development was directed in another "physical" way. As a result, the chemical bond is present in contemporary glass theory in the form of "frustration", "elastic force" and rigid "sticks" connecting atoms-balls, etc., but not as a real object with its own specificity. We think that a pure "physical approach" opens a rich field for theoretical speculation, rather than clarifying the nature of glass. For example, to understand glass transition, the analogues with spin glasses, granular systems and colloids are used (e.g., see the excellent review [31]), although any similarity in behaviour does not mean that the similarity is the reason behind that behaviour.

One can find, however, deviations from the main "physical" stream of glass science, even among physicians. As a fresh example, we may cite that: *"It is widely believed that crystallization in three dimensions is primary controlled by positional ordering, and not by bond orientational ordering. In other words, bond orientational ordering is usually considered to be merely a consequence of positional ordering and thus has often been ignored. ... Here we proposed that bond orientational ordering can play a key role in (i) crystallization, (ii) the ordering to quasi-crystal and (iii) vitrification"* [29]. Although this is a rare case, it may be a sign that the need for a chemical approach is now stronger.

Our goal was to return the chemical bond - as chemists know it - to the theory of glass. As far back as in 1981, *Dembovsky* [32] has connected the experimental fact of an *increased* coordination number in glass forming melts with the model of *quasimolecular defects* (QMD) following *Popov* [33] for the creation of the chemical-bond basis for glass theory. The principal feature of a QMD is its hyper-coordinated nature, a property that provides for the connectivity of a covalent network even at melt temperature, thus resulting in high melt viscosity, which is a characteristic property of a glass forming melt. QMDs in a non-crystalline network can also explain general features of glass [34-38]. The problem is that the concentration of QMDs should be high enough, especially in a glass forming liquid, to provide these features. Therefore, we have quickly substituted the initial term QMD (quasimolecular *defect*) for TCB (three-centre *bond*) and then, being based on a special quantum-chemical study (see the reviews [39] for chalcogenide glasses and [40] for oxide ones), TCB for HVB, **hypervalent bond.** TCB is only a particular case of HVB, which is rarely realizes in glass. For example, in the simplest case of Se not TCB, having two three-coordinated atoms (-Se<) and one "normal" two-coordinated atom (-Se-) [33], but a four-coordinated atom (>Se<)was revealed [41]. Thus, we have introduced *alternative bonds* into CRN, the bonds whose concentration is commensurable with that of ordinary covalent bonds constituting classical CRN.

The term "hypervalent", which was introduced by *Muscher* in 1969 [42], has a long and controversial history, beginning from the 1920s up until the present (see [43] for an introduction). Currently, a large number of hypervalent molecules is known, and various methods of their theoretical description exist. For a long time, one of the most popular was the *Pimentel's* model [44] of the electron-rich three-centre four-electron (3c-4e) bond; this model was used by *Popov* to construct his "quasimolecular defect" in a covalent network of Se glass [33]. The principal step made by *Dembovsky* was in the understanding of QMD not as a "defect" but as the second type of bonding in a glassy network, the first being a common two-centre two-electron (2c-2e) covalent bond. The next step was made by means of *ab initio* quantum-chemical modelling, which reveals various metastable hypervalent configurations - configurations in which a central atom has more bonds with its neighbours than the "normal" surrounding covalent species - in glasses. We use the term *hypervalent bonds* in order to emphasize the additional bonding state in non-classical CRN, and so the additional structure possibilities in it. This network is not "random" now.

Three types of order can be realized in such a mixed network. When the diffraction pattern of glass is transformed into a radial distribution function one can observe the so-called *short-range order* (SRO) extending to the limits of at least two coordination spheres around an arbitrary atom. The peak's position relates to the distances between the nearest neighbours and the peak square to the number of these neighbours. The SRO in glass is close (but not coinciding completely!) with the SRO in a counterpart crystal, with both SROs being determined by a bond length (first distance), valence (first coordination number) and valence angles (second coordination sphere). The SRO is the basis for a conventional continuous random network, but the problem is that a CRN consisting of covalent bonds cannot exist because of the rigid and directional character of the covalent bond. When moving from the first atom, the stress due to bond distortion accumulates rapidly up to a critical value above which the covalent bond should be destroyed (the covalent bond limits are known in the chemistry of related compounds). This means that a real CRN should contain either internal fractures (however, glass is known to be an optically transparent and mechanical stable material) or additional "soft" regions for relaxation. These regions are provided by HVB and are soft and flexible when compared with covalent bonds.

The next step of order which surely exists in a non-crystalline state is the so called *medium-range order* (MRO), which is observed directly in the diffraction pattern of glass or glass forming liquids in the form of the well known *first sharp diffraction peak* (FSDP). It is a very narrow (for an amorphous state) peak located at about Q_1=1-1,5Å^{-1}; the peak position and intensity depend upon the chemical composition, temperature and pressure (see [45,46] for a review). The two generally accepted parameters of the MRO are the correlation length $d=2\pi/Q_1\approx$6-4Å and the coherence length $L= 2\pi/\Delta Q_1\approx$10-20Å, where ΔQ_1 is the FSDP half-width.

The nature of the FSDP/MRO is debatable at present. Our interpretation of the MRO is based on hypervalent bonds and their collective behaviour in the form of a *bond wave*, which is illustrated in **Fig.3**.

A bond wave is the spatiotemporal correlation of elementary acts of the reversible transformation of a covalent bond and a hypervalent bond: CB↔HVB. A snapshot of a bond wave, which spreads in the Se-like network, is shown in the left part of Fig.3. The wavefronts represent equidistant layers populated with hypervalent bonds (they are modelled by TCB in Fig.3), and so the correlation length $d=2\pi/Q_1$ roughly corresponds with the HVB length. The layers give a true Bragg reflex at $Q_1=2\pi/d$, so the FSDP intensity and the FSDP width depend upon the number of these reflecting layers and their reflection ability, with both depending on the temperature.

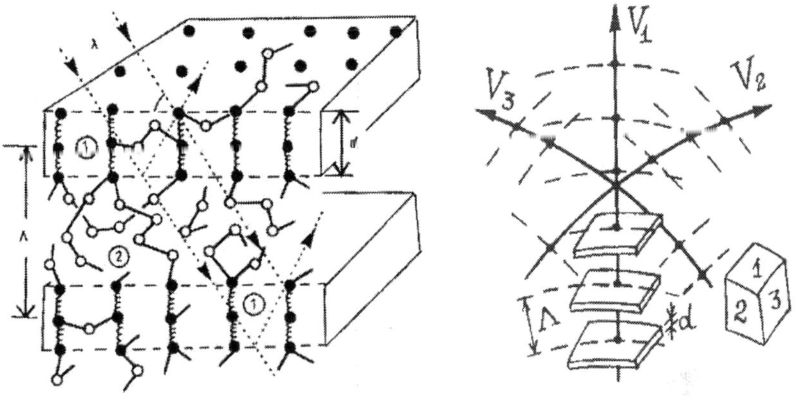

Fig. 3. Bond waves after [46]. On the left: two adjacent wavefronts – layers "1" populated with a HVB (here modelled by a TCB: black atoms linked by spring) together with a CRN "2" between the layers (here an Se-like CRN, which consists of white atoms, each having two covalent bonds as shown by the lines). On the right: the intersection of three bond waves, which gives the 1-2-3 elementary cell.

A bond wave with a Λ wavelength represents a totality of d-layers, and the totality represents a periodic structure of the Λ period. Thus, there appears a Λ-lattice and a corresponding *long-range order* (LRO) in glass. This order is, in some respects, similar to the smectic-type LRO; however, in glass - in contrast with liquid crystal - there are two types of bonding species in the layer and the d-layers themselves are punctuated with thick pieces of the CRN. Such long-range periodicity is "invisible" to ordinary X-ray analysis, by which one can see a CRN (in the form of the peaks in the radial distribution function) and d-layers (in the form of a FSDP in an initial diffraction pattern), but not the Λ-lattice. The reflex from the Λ-lattice should be disposed in the 0.01-0.001$Å^{-1}$ range, which is inaccessible to the usual X-ray techniques. Thus, special techniques (e.g., those using synchrotron radiation) are needed to see a LRO in glass.

One can note that the layer structure shown in the left side of Fig.3 is *anisotropic*, while glass is known to be an *isotropic* medium. Note, however, that anisotropic glasses can also be prepared (our own experiments of this sort will be considered below). The generally observed isotropic behaviour of glass means the *solitonic* behaviour of bond waves, i.e., the waves ability to intersect each other without distortion, as is shown on the right in Fig.3. One wave propagating through a CRN with the velocity V_1 forms the layer structure, two waves (V_1+V_2) create the columnar structure, and three waves (V_1+V_2+V_3) correspond to the cellular structure, which is isotropic on the macroscopic scale. The factors which govern bond waves and, consequently, the structure that they form, can be divided into internal (chemical composition, temperature, pressure) and external (flows of energy and/or information) categories.

As far as the elementary act CB↔HVB is considered to be a thermally activated process, both the wave frequency and the HVB concentration increase with temperature, while the wavelength Λ decreases [47]. Thus, when approaching the glass transition temperature T_g, the interlayer distance Λ becomes so large that the correlation between the layers/wavefronts becomes impossible, although the intimated HVBs within the layers continue to "feel" each other. This means that a 3D bond wave (Fig.3, right side) stops its propagation through the structure and the 2D bond wave within the stopped d-layers (Fig.3, left side) remains mobile. Thus, the glass transition can be considered as a 3D→2D bond wave transition [48].

As far HVBs represent active sites in the mixed CB/HVB network, one should distinguish between the processes above and below T_g. More specifically, the abovementioned pre-nucleation stage in the form of initial reorientation (IReO) proceeds within the stopped d-layers below T_g and only after the reconstruction of the layers can it penetrate into the CRN: the process is slow and has an induction period (see $\tau<\tau_0$ in Fig.2). Above T_g, where the 3D bond waves exist, the d-layers pass through every structure element: the induction period is short, if it even exists, and the process is much faster and homogeneous.

Crystallization is not the first property considered by means of the bond wave model. Earlier, this model was applied successively for the interpretation of the thermodynamic features of glass forming substances [49], characteristic glass fractures [50], the first sharp diffraction peaks [46] and the temperature dependence of viscosity [51]. Unfortunately, there are only interpretations and so far there is no direct evidence of bond waves at present, i.e., the direct observation of a Λ-lattice in a structural experiment. Instead, we performed a computational experiment [52], presented in **Fig.4**, in which we intended to investigate the ability of the HVB for association - this property alone is a necessary requirement for the existence of a bond wave.

In model clusters like those shown in Fig.4, a single HVB looks like a "defect" embedded into an ordinary continuous random network (CRN). A single HVB was shown to be a low-energy "defect" (compare 0,3 eV for Se_4^0 [41], with 2 eV needed to generate a broken bond $2Se_2 \rightarrow 2Se_1$ or with 1 eV proposed for the so called "valence alternation pair" $Se_3^+Se_1^-$ after [53], the index below is the coordination and the index above is the charge). Nevertheless, even such low-energy defects cannot ensure the above-mentioned over-coordination in glass (there needs to be ~1% for four-coordinated atoms) and especially so in a melt (ten%) [37]. From Fig.4, it follows that the associated HVBs are much more stable, so that even *negative energy* regions can arise in a CRN for a definite HVB arrangement (SS).

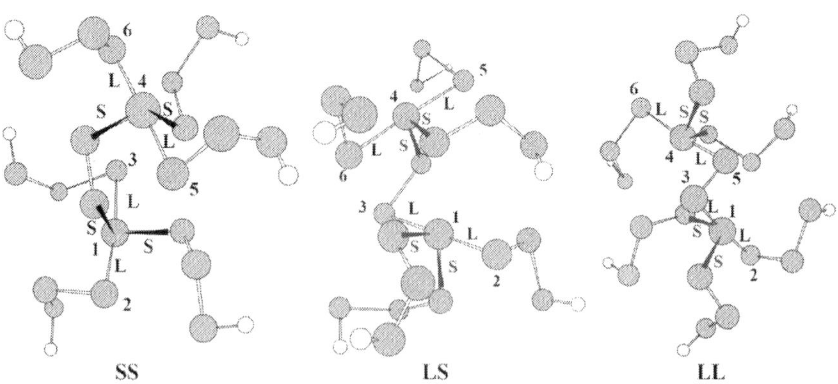

Fig. 4. Quantum-chemical modelling of the HVB interaction in Se after *Zyubin & Dembovsky* [52]. The cluster energies are **−0.16 eV** (SS), +0.08 eV (LS) and +0.09 eV (LL); the notations correspond with the mutual orientation of short (S) and long (L) pairs of bonds surrounding each Se_4^0 atom (1 or 4). The white spheres are the hydrogen atoms terminating the clusters.

In this way, not only is a principal ability to form a bond wave justified, so is the connection of a *structural order* due to a Λ-lattice with an *"energetic order"*, which is based on the intuitive belief that the more structurally ordered substance has the lower energy (with other conditions being equal). As such, the processes of the crystallization of glass, both below and above T_g, can be considered as the competition of two types of order: a crystalline long-range order and a specific non-crystalline long-range order, provided by hypervalent bonds and bond waves. In what follows, these notions are tested by means of special crystallization experiments.

3. Crystallization in solid glass ($T<T_g$)

3.1 The composition dependent rate of nucleation

For these experiments, we chose selenium as the simplest one-element glass; the additions, with various valence abilities, were introduced into the Se matrix, giving five series of Se-X (X = S, Te, As, Ge, Cl) with a varied but relatively low concentration of the second component. The as-prepared samples were cylinders measuring 25 mm in diameter and 15 mm in height; the two cylinder ends were polished. The optical transmission and ultrasonic velocity were measured through the ends; the X-ray fluorescent spectra were measured from the end surface.

The as-prepared samples were quite transparent, actually having the same value of transmission at 1000 cm^{-1} (this value, which corresponds with entry into the so-called "window of transparency" for selenide glasses, we shall call *transparency*) of around 60%, a value that is typical for chalcogenide glasses of high quality for the given thickness of 15 mm. The two other properties (V and r) investigated in the fresh glasses, however, were strongly dependent upon composition, as is seen in **Fig.5**. Note that the both properties are *macroscopic* in character. The ultrasound velocity, V, characterizes the elastic ability of the Se-based network. The relative intensity of the X-ray fluorescence, $r=S_{val}/S_{char}$ (S_{val} and S_{char} are the integral intensities of the $K_{\beta 2}$ and $K_{\beta 2}$ emission lines for Se, the first corresponding to the 4p→1s transition from a valence band and the second to the characteristic 3p→1s inner transition for Se), belongs to the totality of selenium atoms, reflecting the average valence state for Se.

We see a strong non-linear character for all of the dependencies in Fig.5. The two upper cases of Se-Te and Se-S seem to be the most surprising because S and Te belong to the same VI group of the Periodic Table, having the same number of covalently bonded neighbours per atom. Additionally, the Se-Te phase diagram is a simple "fish", which corresponds with the discontinuous series of liquid and solid solutions. However, the *metastable phase diagram* (remember that glass is formed far below the melting point from a metastable liquid) has a "two-fish" form, with "two series of solid solutions meeting at 96.8 at%Se [i.e., 3.2 at%Te - ECh] and temperature 180°C" [55].

Note that the Se-Te glasses display extrema in the 1-4%Te range for the other composition-property dependences, e.g., for electrical and crystallization properties [56] and for the glass transition temperature [57]. *Dembovsky et al.* [58] have also revealed non-linearity when investigating crystallization kinetics in Se-Te glasses in the 0-5% range. In **Fig.6**, the data for CCRs calculated with the use of the data obtained is shown; one can see not only the conditional character of the CCR, the composition dependence of which depends on the chosen α, but also the non-linearity of $Q(N)$ for every chosen extent of crystallinity with the extrema located in the region 1-5%Te. Note that our calculation of the CCR in the form of V_m

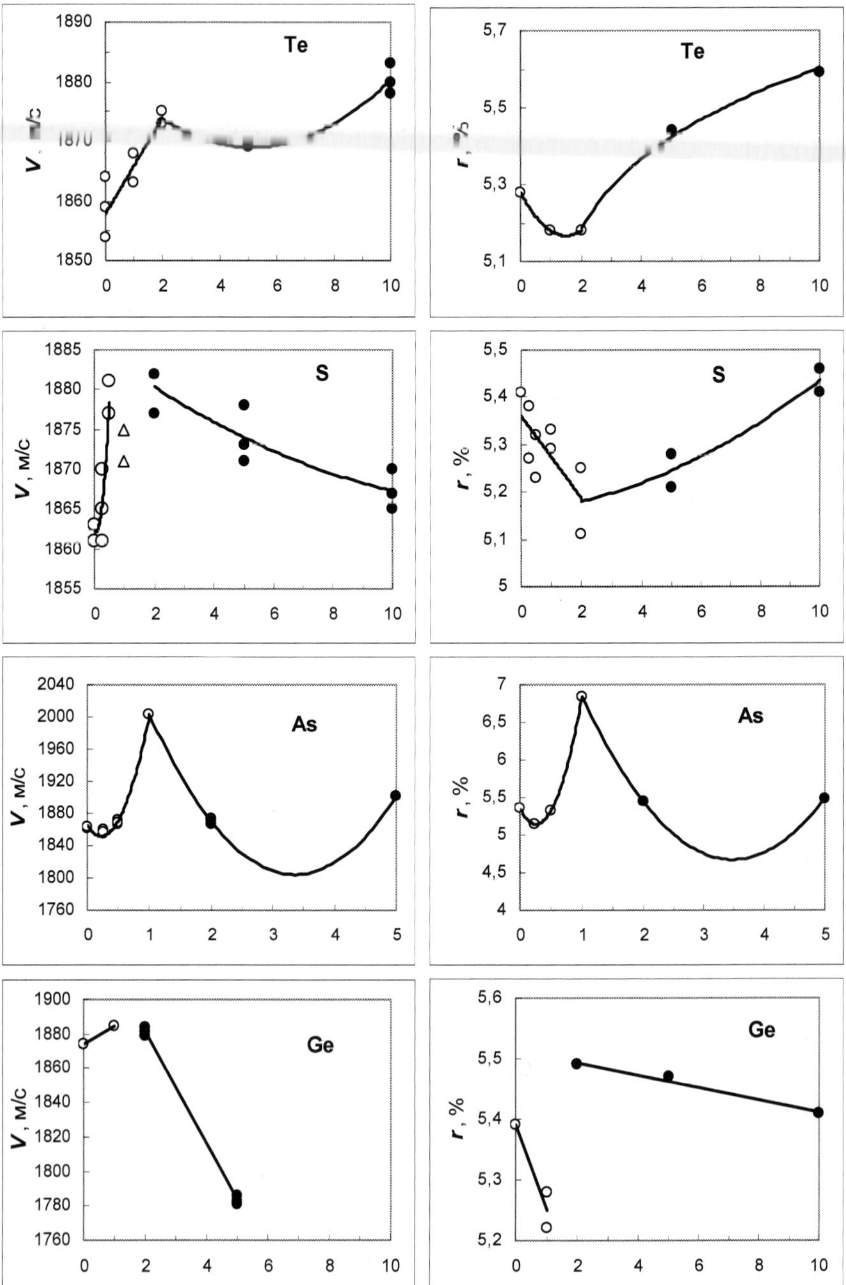
Te
V, м/с
Te
r, %
S
V, м/с
S
r, %
As
V, м/с
As
r, %
Ge
V, м/с
Ge
r, %

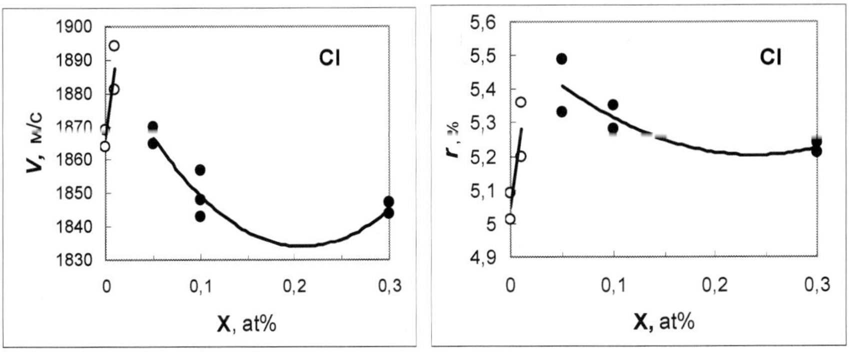

Fig. 5. Longitudinal ultrasound velocity (V) and relative intensity of X-ray fluorescence from the Se valence band (r) in the as-prepared Se-X glasses after [54].

[21,22], which corresponds to the pre-crystallization IReO stage, really represents the upper limit for the CCR, but only outside the non-linearity range. Remember, however, that when calculating V_m we only take into account the *equilibrium* fish-like phase diagram, whereas real glass relates more to metastable diagram(s) with peculiarities at about 5%Te.

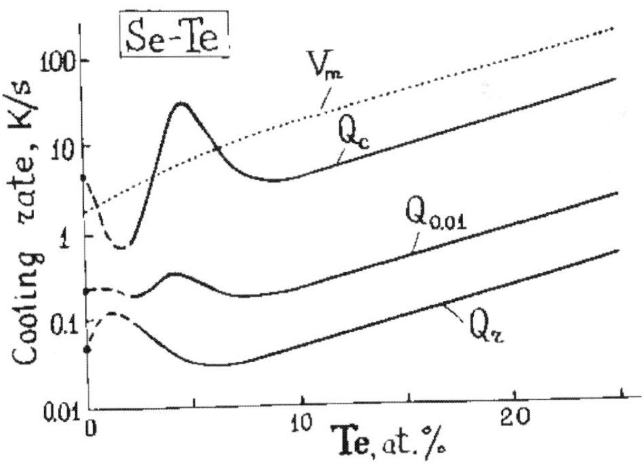

Fig. 6. Critical cooling rates (CCRs) in the Se-Te system after [58]. Q_r, $Q_{0.01}$ and Q_c correspond to the crystalline fraction α=0,63; 0,01 and 10^{-6} respectively. V_m is our calculation after [21,22], without the use of crystallization data.

The Se-S series in Fig.5 looks similar to Se-Te, although one can propose a more complex behaviour below 2%S, which can be compared with a more complex equilibrium Se-S phase diagram [59] in contrast with the simple "fish" for Se-Te.

In the Se-As and Se-Ge series, the additions belong to other groups of Periodic Table (IV for Ge and V for As), so one can expect another property-composition behaviour. Accordingly, the *V(N)* and *r(N)* dependencies change in the same way in Se-As glasses (maximum on the both), in contrast with the behaviour in Se-Te and Se-S glasses, which display a maximum on *V(N)* and a minimum on *r(N)*. Although Se-Ge glasses are similar to Se-Te as concerns the maximum on *V(N)* at 2%Ge and the minimum on *r(N)* at about 1%Ge, the extrema in Se-Ge are much sharper and there is different post-extreme behaviour in Se-Ge as compared with Se-S and Se-Te.

The Se-Cl glasses are of a special interest because, in contrast with the above four additions which belong to the well recognized glass forming Se-X systems, chlorine in *not* a glass forming addition for Se. It is usually assumed that Cl can only break the Se-Se bonds, thereby decreasing viscosity and thus the glass forming ability. In fact, Se-Cl glasses can be only prepared in the 0-0.3%Cl range, but in other respects the Se-Cl series demonstrates the same behaviour: the as-prepared Se-Cl glasses are quite transparent (about 60%) and their properties are strongly non-linear (see Fig 5, at the bottom).

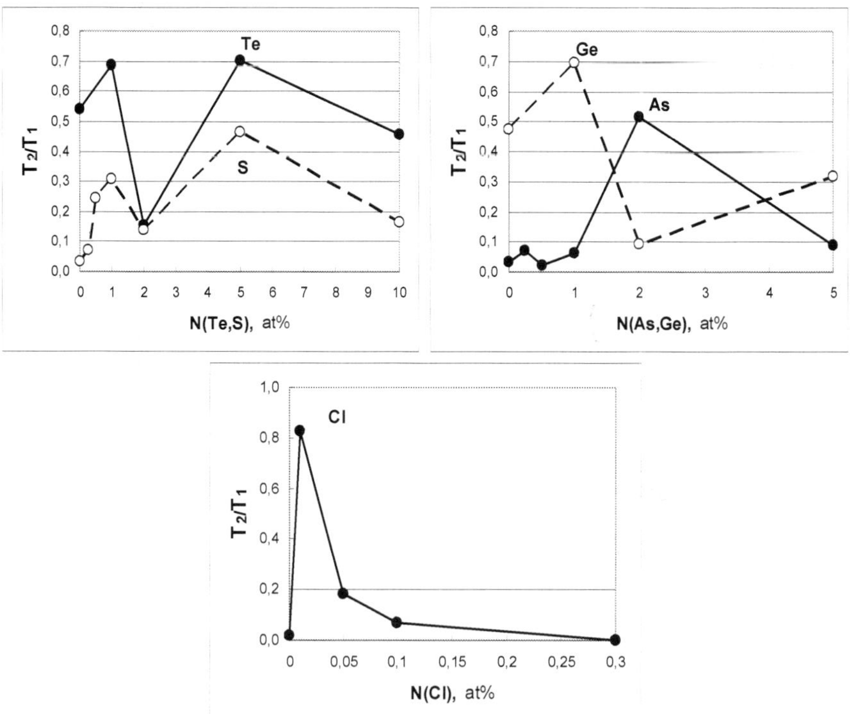

Fig. 7. Darkening in Se-X glasses after 5-year ageing at room temperature [60]. Here, *T* is the transparency (optical transmission at 1000 cm^{-1}); index "1" refers to the as-prepared glass and "2" to the aged glass. The unchanged transparency, i.e., the absence of darkening, corresponds to $(T_2/T_1)=1$ and complete darkening to $(T_2/T_1)=0$.

Next, the five series of glasses were stored for 5 years at room temperature, which is somewhat below the glass transition temperature ($T_g\approx35°C$ for Se). In accordance with the expected crystallization, the **aged glasses** became less transparent, but not to the same extent. It can be seen in **Fig 7** that the transparency shows a non-linear character, like that shown in Fig.5 for ultrasound velocity and electron emission from the valence Se band. Thus, one can suppose that even in fresh Se-X glasses the special regions of initial reorientation (IReO) for further crystallization have been developed. These regions manifested themselves in *V(N)* and *r(N)* in Fig.5, but not in the transparency because the IReO regions were congruently inserted into a glassy matrix. During the process of ageing, the IReO regions are transformed into crystalline regions and so become visible due to light scattering from new crystal/glass boundaries.

The question is: what stage of crystallization do we observe in the aged glasses? The X-ray pattern of the aged samples usually looked like a wide hill of a low intensity; only the darkest samples display very weak crystalline peaks on the hill. Thus, because of the low extent of crystallinity (usually lower than 1%) together with the considerable darkening, we can conclude that by the optical transmission method we can observe the *nucleation stage* of crystallization. Next, from Fig.7 one can conclude that compositions of 1%Te and 5%Te, 5%S, 1%Ge, 2%As and (surprisingly!) 0.01%Cl are the most stable against nucleation.

3.2 The composition dependent extent of heterogeneity

Nucleation in glass is known to be usually *heterogeneous*, i.e., the nuclei tend to appear on the surface of a sample rather than distribute evenly in the volume. Note that only the case of homogeneous nucleation is considered in classical crystallization kinetics, a fact that creates additional problems when comparing theory with experiment for such non-classical objects like glass.

In order to evaluate heterogeneity in a numerical way, we have elaborated a simple method which includes the removal of the surface layer [61]. The main effect of the removal is the resection of the surface nuclei with the consequent rise of transparency. Next, the extent of heterogeneity *G* can be evaluated as:

$$G = (T_3 - T_2)/T_1\,, \tag{2}$$

where T_1 is transparency (i.e., the optical transmission at 1000 cm^{-1}) for the as-prepared glass (of 15 mm thickness), T_2 is the transparency after ageing and T_3 is the transparency of the aged sample after the removal of the surface layer (0.5 mm from each side of the cylinder, a procedure that gives a sample of 14 mm thickness). The method is illustrated on the Se-Te series as an example in the left-top and left-bottom (Te) of Fig.8.

From Fig.8, it can be seen that the distribution of crystalline nuclei is most homogeneous in the Se-Ge system having G=0-0.2. The negative *G* for the 5%Ge sample may indicate the "inverse heterogeneity", a situation whereby nuclei prefer to burn within the volume rather than at the surface. The opposite deviation G>1, which is observed in the Se-Cl series for the 0.05%Cl sample when the volume becomes more transparent after ageing ($T_3 > T_1$), may be a result of the sample thinning. Another possibility, of *non-crystalline ordering* during ageing, seems to be more interesting. The reason for this ordering may be the development of the IReO regions in the glassy matrix (see the dotted line in Fig.2) without nucleation.

One can note a remarkable similarity in the *r(N)* graphs in Fig.5 for the as-prepared glasses and the *G(N)* graphs in Fig.8 for the aged glasses, and not only in the extrema location, which coincides for every Se-X series, but also in the dependence shape which is qualitatively the same for the Se-Te, Se-As and Se-Cl series. For the Se-S and Se-Ge series, the shapes are somewhat different. One should note, however, that the Se-S glass is a special case because of the introduction of sulphur with T_g= -25°C, which would decrease the glass transition temperature in the series. Accordingly, Se-S glasses are especially inclined to crystallization when ageing at room temperature.

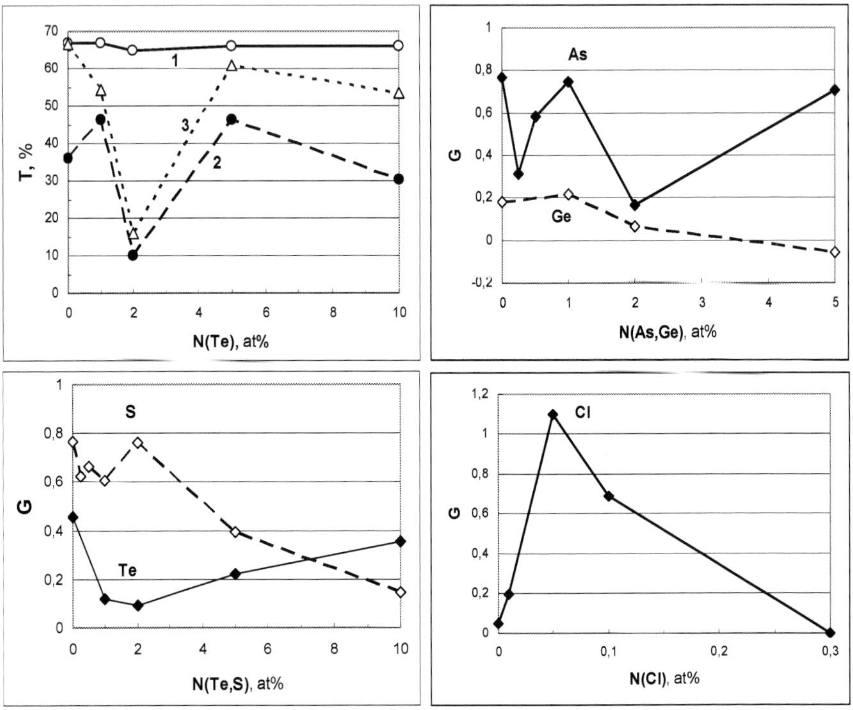

Fig. 8. The extent of heterogeneity, *G* by eq.(2), for spontaneous nucleation in Se-X glasses. T_i is the optical transmission at 1000 cm^{-1} for the as-prepared glass (1), the aged glass (2) and the aged glass without a surface layer (3).

In spite of these deviations, a non-trivial relation between the inclination to heterogeneous nucleation, *G(N)*, and a change in chemical bonding, reflecting by *r(N)*, is observed. Note that by means of the X-ray fluorescence method we probe only the *surface* layer of the sample; therefore, there exists a direct connection between *nucleation ability* and the *density of valence electrons*, with which the value of *r(N)* can be related. This is a direct route to hypervalent electron-rich bonds, which were discussed in section 2.3: the higher the electron density, the easier the formation of such HVBs representing active centres in covalent network, the centres which provide for processes including crystallization. The concentration of HVBs may be high owing to the decrease of energy when HVBs are associated (see Fig.4), probably in the form of bond waves. Thus, foreign atoms can interact strongly and non-linearly with the bond waves which spread in non-crystalline network during the glass preparation (3D waves) and ageing (2D waves). What is a mechanism for such an interaction?

3.3 Non-linearity and self-organization in glass

From a phenomenological point of view, one can distinguish between "*impurity*" (non linear) and "*component*" (smooth) concentration regions [54]. The boundary concentration, after which non-linearity begins to vanish, depends upon the system; it is about 2% for Se-S, Se-Te and Se-Ge glasses, about 1% for Se-As glasses, and about 0.05% for Se-Cl glasses (Figs.5-8). Let us discuss a possible nature of non-linearity in a low-concentration "impurity" region.

Starting from classical CRN [30], one cannot wait for any singularities except for those stipulated by a phase diagram, and it is generally accepted that foreign atoms enter into a CRN with "normal" valence (e.g., Te forms the -Se-Se-**Te**-Se-Se- chains), thus causing a proportional change in glass properties. The above-demonstrated non-linearity means that this is not the case and, hence, that the valence might be unusual. We have a model of *hypervalent bonds* - HVBs - after Dembovsky (section 2.3) for just this situation. It is clear, however, that the HVBs in a concentration of about 1% cannot explain the *macroscopic* effects observed. This is the *bond wave* model for the case of the collective interaction between "normal" bonds and HVBs. Next, the question about the nature of non-linearity should be reformulated: how do selenium bond waves interact with foreign atoms in a selenium network? Let us begin with the quantum-chemical modelling of the simplest case of Se-Te.

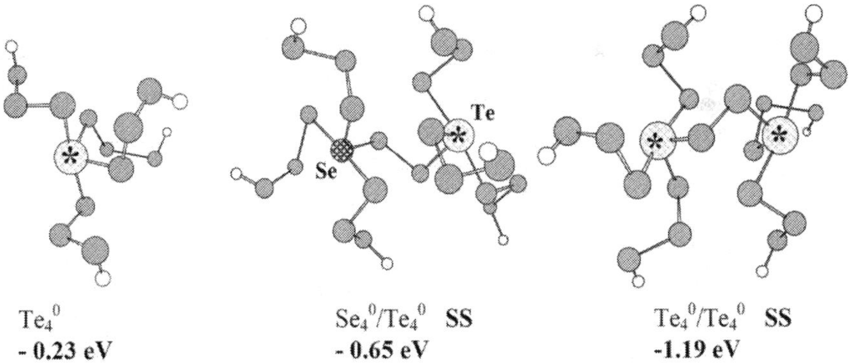

Fig. 9. A quantum-chemical modelling of HVBs in a selenium clusters containing Te atoms after [62]. The energies below are compared with the selenium CRN of the same size.

It can be seen in **Fig.9** that even a simple hypervalent Te_4^0 decreases the energy of the surrounding network (-0.23 eV). When the selenium wavefront (see d-layer in Fig.3, on the left) moves closer, a more stable $Se_4^0Te_4^0$ (-0.65 eV) is formed. Hence, a hypervalent Te atom represents an energetic trap for the selenium bond wave; however, this is so for only that local part of the wavefront, a part that interacts with the atom. When the Te concentration is high enough, the atoms can form hypervalent associates with the additional decrease in energy - see dimer $2(Te_4^0)$ in Fig.9.

The result of such an interaction is currently unclear. Given only the traps, foreign HVBs - such as Te_4^0 - should slow down or even destroy a bond wave, most probably to the extent that it is proportional to their concentration. However, we observe both maxima and minima in concentration dependencies in different Se-X series (Figs.5-8); hence, the interaction is not simple and depends on the concentration of foreign atoms. The reasons may be both in their bonding state and/or their arrangement – random or ordered. An ordered state can be realized by a directed diffusion of impurity under the action of any bond waves spreading through the melt, before the waves are frozen in solid glass.

The nature of the added atoms is also significant for their interaction with the bond wave. For example, a similar quantum chemical study for the Se-Cl system [63] leads to the same qualitative result as for with Se [52] and Se-Te [62], namely, the aggregation of HVBs with the participation of foreign atoms which leads to a lowering of energy - see **Fig.10** (chlorine atoms are chequered).

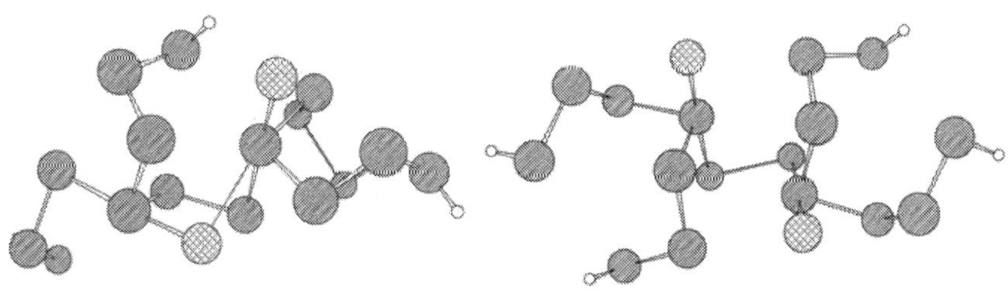

E= −0,57 eV E= −0,37 eV

Fig. 10. Two of the lowest energy complex HVBs in Se-Cl after [63].

The configurations and energies in the Se-Cl case differ from those of Se-Te (Fig.9); in this way, the specificity of foreign atoms is revealed. Note that in the left configuration of the lowest energy in **Fig.10**, the Cl_2Se_5 fragment has not only over-coordinated Cl (a two-coordinated instead of the usual one-coordinated state) but also five-coordinated Se, i.e. chlorine induces the most coordinated Se state obtained by us so far (see four-coordinated Se in Fig.4, Fig.9 and Fig.10, as well as the configurations with three-coordinated Se in other chalcogenide glasses [39]). One should note, however, that the relation between coordination and valence is not strictly determined, especially in the hypervalent case [64]. Indeed, by use of only the geometric arrangement, it is impossible to conclude does a chemical bonding exist between the atoms under consideration - this is the question to the quantum-chemistry study of a concrete configuration.

The method of adding foreign atoms in a simple non-crystalline matrix with the further comparison of experimental features with quantum-chemical models seems to be a fruitful way for understanding what is the nature of chemical bonding in glass formers and how one can creates various bonding states and/or structures. As to the structures, let us consider an additional feature that we found when investigating Se-X glasses by means of IR spectroscopy.

In **Fig.11** one can see a *narrow* absorption band at 792 cm^{-1} on the high-frequency side of the third overtone for the 245 cm^{-1} band of Se: the line half width is one-third of that for the overtone. Note that the 792 cm^{-1} band develops only for a definite concentration of chlorine, existing in the sample 0.01%Cl, which is most persistent for nucleation (Fig.7), and in the sample 0.05%Cl, which has the greatest capacity for heterogeneous nucleation (Fig.8). Interestingly, the same 792 cm^{-1} band also observed in the Se-As series for the 0.25%As sample (compared with the minimal heterogeneity in Fig.8) and unexpectedly at 5%As [65] (one can also see the extrema at the same concentrations of 0.01-0.05%Cl and 0.025%As in other properties for fresh glasses in Fig.5). Hence, the *ordering* which reveals itself as the 792 cm^{-1} narrow band can be related to the selenium matrix, the structure of which is modulated by means of the impurity introduced into it. Interestingly, the ordering can manifest in properties in the opposite manner, depending upon the impurity, as it is seen in Fig.5 for the extrema at 0.01-0.05%Cl and 0.025%As .

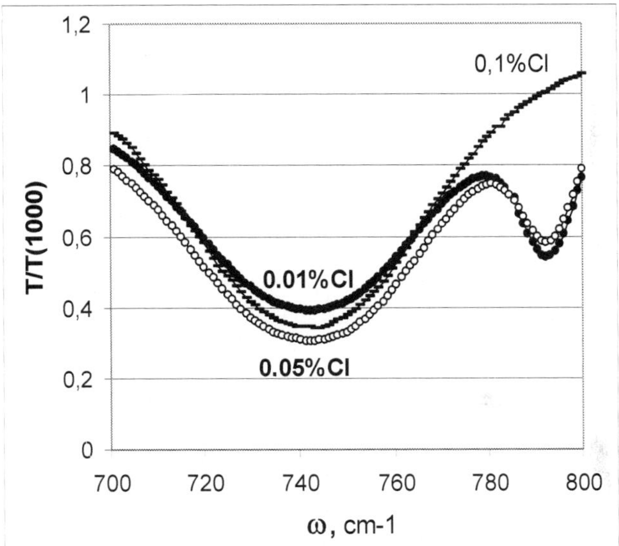

Fig. 11. IR spectra of Se-Cl glasses within the "phonon" range (400-1000 cm^{-1}) after [60]. T is the transmission at a given frequency, T(1000) is the transmission at 1000 cm^{-1}, i.e., the *transparency* (see Fig.7 and Fig.8).

When analysing the properties, a thoughtful reader can find out the disagreement in the $r(0)$ values in Fig.5 and in the T_2/T_1 values in Fig.7 for pure selenium glass (X=0) in different Se-X systems, a fact that may devaluate the data itself. The explanation is rather simple: different series were formed in different regimes: the first regime was applied for Se-Te and Se-Ge, the second one for Se-As and Se-S, and the third one for Se-Cl. Indeed, Se glasses have similar properties in the series of the same preparation. This is the *memory effect*, which is well known for anyone who works with glasses.

The non-linearity, spontaneous ordering and the memory effect discussed above are signs of a *self-organizing* system, a system that possesses various scenarios for evolution, depending upon the system's nature and the information provided by external medium [66]. When glass is considered as a self-organizing system with bond waves as the basis for self-organization, one acquires an appropriate tool not only for the explanation of the above-described phenomena but also for the planning of special experiments. In the following experiments, we use an external *information field* for governing the development of a glass structure by means of the directed spreading of bond waves in the region where they are most effective, i.e., at $T>T_g$ where 3D bond waves exist. Note that the above-described experiments concern nucleation in solid glass ($T<T_g$), a process that is provided by 2D bond waves spreading in the limits of frozen wavefronts. Thus, let us defreeze them.

4. The crystallization of softening glass in an ultrasonic field

4.1 Experimental conditions

The same Se-X series of glasses is used and optical transmission is also used as the general method for the observation of the crystallization process. The distinctions from the above experiments are as follows. First, the *temperature* range of the treatment 50-72°C is somewhat *above* the glass transition temperature T_g (which is about 35°C for Se), although the samples retain their form (in general). Second, the *time* of the treatment was 5-10 min instead of 5 years, as in the previous case. Third, the *ultrasonic field* is applied during treatment. Finally, the optical transmission was measured in *two directions* perpendicular to one another, before and after treatment.

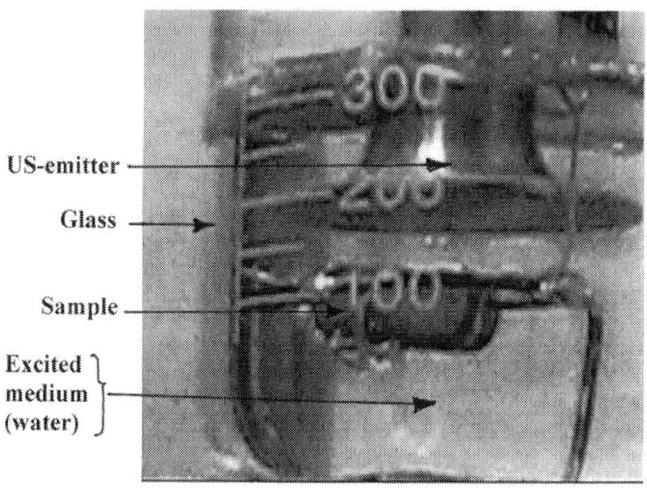

Fig. 12. The treatment of a Se-X sample in an ultrasonic field in a cavitation regime. Note the glass with water, the sample in the holder near the 100 ml mark and the end of a US-emitter at the 200 ml mark, as well as the large bubbles at the water's surface and near the sample.

In the framework of our understanding of glass as a self-organizing system, owing to the collective behaviour of hypervalent bonds in a covalent network, these conditions mean, first, the existence of 3D bond waves which activate *all of the volume* of a substance. Second, the crystallization process *accelerates* owing to not only the increase of the bond wave dimensionality but also of the bond wave velocity. Third, an ultrasonic field can play the role of an *information field* that can orient bond waves. Finally, *anisotropy* can arise after the treatment.

The experimental equipment shown in **Fig.12** is rather simple: a glass with water (the excited medium for ultrasonic cavitation), in which the sample is placed in the holder. The sample has two pairs of perpendicular grains - A-A and B-B - the distance between like grains being equal. $d_{A-A}=d_{B-B}$. The US-emitter (its own frequency is 24 kHz and its vibration amplitude is 3 μ) is disposed at 10 mm above the upper grain A. The temperature varied from 50°C to 72°C and the time of treatment from 10 to 5 min. Note that the field is *weak*, having an intensity of about 0.2-0.3 W/cm² (the cavitation threshold is 0.1 W/cm²). The vibration frequency in the excited medium is within the range of 1 kHz - 1 GHz, with a maximum at 5-10 MHz. At the end of treatment, the US-input is turned off, the emitter is lifted and the sample together with the holder is taken out of the glass with further cooling in the air down to room temperature. Then all the samples of a given Se-X series are subjected to optical investigation, which is performed at the same day.

The optical measurements were made in two directions: in the A-A direction along the axis of the US-emitter (i.e., through two "frontal" A grains, which are parallel to the emitter end in Fig.12) and in the B-B direction perpendicular to the emitter axis (one can see one of the lateral grains B facets in face in Fig.12). These directions are easily distinguished in the samples owing to the curve sides remaining after transformation of initial cylinders/tablets, which were used in the previous experiments at $T<T_g$, into the blocks using in the present experiments; the A-A grains correspond to the ends of the initial cylinders. The spectra were measured within the 400-5000 cm^{-1} range and the data up to 1000 cm^{-1} are discussed here. It may be desirable to catch the eye of future investigators to the 1000-5000 cm^{-1} interval: although we do not discuss the related data, it is a quite sensitive area for the observation of the crystallization process in the scale from 10μ to 2μ, respectively.

4.2 Basic experimental features

This study is currently in progress, although two articles describing experiments on the Se-As series [65] and Se-Te series [67] have just become available in English. Here, only basic experimental features are discussed; for additional information the reader is invited to look at the originals. Note that we use the optical transmission method for watching the crystallization process, which is again attributed primarily to the nucleation stage, a process that decreases transmission due to light scattering from a well-developed internal glass/crystal surface.

The first feature is a very fast darkening (for minutes) as compared with the previously considered darkening due to ageing (years). Thus, the first question is: what causes such rapid darkening? Temperature or cavitation?

It can be seen in **Fig.13** that even at 72°C - the highest temperature which was applied - the temperature itself has only a low effectiveness (compare the spectra 12 to 13) and only the US treatment strongly accelerates nucleation in softening glass (see spectrum 5 of a very low intensity).

The second feature is optical *anisotropy*, which is very weak in the initial glasses but can develop after US treatment. The anisotropy value in a given series depends on composition, as can be seen in **Fig.14** with Se-As as an example. Despite the series investigated, anisotropy always arises as a result of a strong darkening in the A-A direction together with a negligible change in transparency in the B-B direction. As to the anisotropy peak position, a special concentration of 0.25%As in Fig.14 corresponds with the above-mentioned appearance of a narrow 792 cm^{-1} band - the same as that which is shown in Fig.11 for the Se-Cl series. This fact establishes a connection between the ability of pre-crystalline ordering in an as-prepared glass and the ability for the development of gigantic optical anisotropy in the glass after its further high-temperature treatment in an ultrasonic field.

In the framework of the bond wave model, anisotropy arises due to the definite location of an excitation source (see Fig.12): although one can suppose a nearly spherical distribution for the dissipation of the input energy in a cavitating medium, the input itself violates the symmetry. The excited medium, in addition, has a temperature above the glass transition temperature for the Se-X glasses investigated. Therefore, when an initially isotropic glass is placed into the cell (Fig.12), the bond waves - which were initially frozen in different directions, as in the case shown in the right-hand part of Fig.3 - begin to refreeze (2D→3D), and the refrozen bond wave will move in a definite direction, remaining this direction after glass cooling.

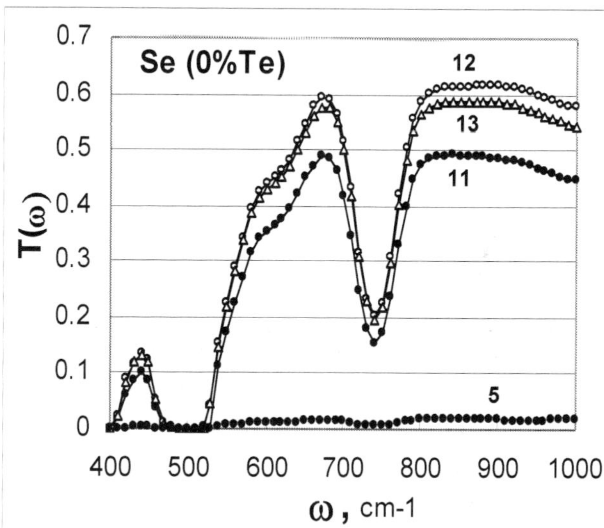

Fig. 13. Optical transmission spectra of the aged Se glass (11), the aged glass after polishing (12) and the polished glass after a 5 min treatment at 72°C (13). Curve 5 corresponds with the polished sample (an analogue of 12) after a 5 min treatment in a cavitation US field at the same temperature, 72°C. The measurements are made in the A-A direction.

The question is: in what direction will the refrozen bond waves move? It seems likely that the layers/wavefronts will be oriented in the A-A direction, parallel to the vibrating end of the US-emitter (see Fig.13). Insofar as crystallization begins within the layers containing active hypervalent bonds, then one obtains a system of partially-crystallized layers divided by periodic glass-like regions. This system is frozen in the sample after its removal from the cell with further quenching in the air. The sample reveals optical anisotropy (see line 5 in Fig.14) because in such a layered structure a measuring beam should scatter strongly when it probes in the A-A direction (the beam falls onto the nucleated layers) and scatter much less in the B-B direction (the beam spreads within transparent glass-like regions between the layers, as like along the optical wire).

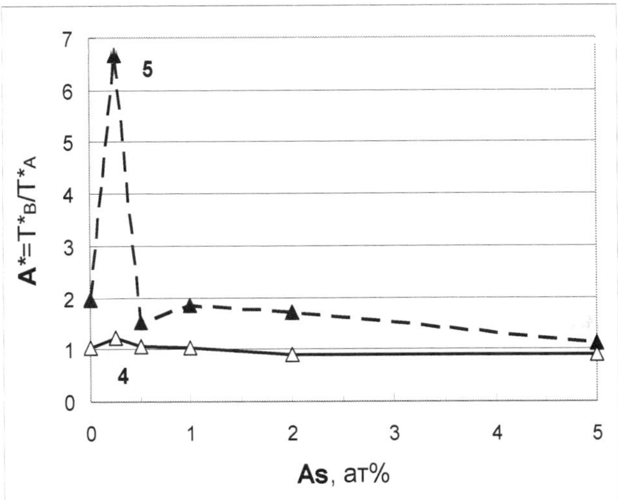

Fig. 14. The anisotropy in transparency (transmission at 1000 cm^{-1} measured in the A-A and B-B directions) for the Se-As series. The two lines correspond with the measurements before (4) and after (5) US treatment (72^{0}C, 5 min).

In an attempt to see the proposed internal structure, we used electron microscopy; the two images shown in **Fig.15** correspond to fresh fractures made in the two perpendicular directions.

Fig. 15. A SEM image of the fracture made parallel to the A grain (on the left) and the B grain (on the right) for a selenium sample after US treatment at 72°C.

In order to exclude the effects of many-component crystallization, we made the fractures using pure Se glass subjected to US treatment. This is that glass belonging to the Se-As series with the induced anisotropy of A*=2 (see 0% in Fig.14): after the US treatment, the transparency in the A-A direction was decreased doubly, while the transparency in the B-B direction was unchanged. In **Fig.15**, we present two images for the two fractures. The left image probably corresponds to the crystallized *d*-layer, along which a fracture developed. The right image looks like ordinary glass, probably corresponding to the glass-like region between the two adjacent *d*-layers.

For the crystallized layer (Fig.15, on the left) it is interesting to note the unusual *needle* form of nucleation in the selenium, which is known to be inclined to form the spheric-like crystallites. Of course, a much more intensive SEM investigation, as well as a wide set of other methods, is needed for a more adequate characterization of the materials obtained and, consequently, a deeper understanding of the processes involved in their formation. We hope that the described experiments of crystallization in glass under the action of impurities and/or ultrasonic field will help the coming investigators of glass and glass-ceramics in obtaining the new materials using non-traditional ways.

5. Conclusions

The structure of glass, considered from the chemical bond point of view using the bond wave model, acquires specific elements of order which develop during cooling (open system) owing to the alternation of chemical bonds (two-stable element) in the form of a bond wave (collective feedback between the elements). In the brackets, there are three general conditions for self-organization; they are naturally fulfilled in glass forming substances owing to their ability to form alternative hypervalent bonds. The processes in a self-organizing system proceed in a specific manner, and this is the reason why classic notions about crystallization works poorly in glass formers. On the other hand, self-organization opens new possibilities for both the understanding and management of related materials. One should note that although "self-organization in glass" becomes a hot topic (see [68] as an example), a real chemical bond is actually absent in the related works. We hope that our approach, connecting contemporary notions about chemical bonding - from the one side - and the self-organization theory (synergetics) - from the other side - will help to fill a significant gap in glass theory, including the understanding of the mechanism of glass-crystal transition and, in glass practice, the elaboration of new principles for the formation of glass-ceramic materials.

6. Acknowledgements

Dedicated to Prof. S. A. Dembovsky (1932-2010); his ideas and our long-term collaboration actually make him the co-author of this work.

This work was supported by the Russian Foundation of Basic Research (Grant No. 09-03-01158).

7. References

[1] J. Zarzycki. Glasses and the vitreous state. Cambridge University Press, Cambridge-New-York-Melbourne, 1991.

[2] P.T. Sarjeant, R. Roy. Mater. Res. Bull., 3, 265 (1968).
[3] D.R. Uhlmann. J. Non-Cryst. Solids, 7, 337 (1972).
[4] J.C.A. Vreeswijk, R.G. Gossink, J.M. Stevels. J. Non-Cryst. Solids, 16, 15 (1974).
[5] E. Ruckenstein, S.K. Ihm. J. Chem. Soc. Faraday Trans. I, 72, 764 (1976).
[6] G.F. Neilson, M.C. Weinberg. J. Non-Cryst. Solids, 34, 137 (1979).
[7] R.J.H. Gelsing, H.N. Stein, J.M. Stevels. Phys. Chem. Glasses, 7, 185 (1966).
[8] J.C.Th.G.M. Van der Wielen, H.N. Stein, J.M. Stevels. J. Non-Cryst. Solids, 1, 18 (1968).
[9] A.C.J. Havermans, H.N. Stein, J.M. Stevels. J. Non-Cryst. Solids, 5, 66 (1970).
[10] M.H.C. Baeten, H.N. Stein, J.M. Stevels. Silicates Industriels, 37, 33 (1972).
[11] G. Whichard, D.E. Day. J. Non-Cryst. Solids, 66, 477 (1984).
[12] W. Huang, C.S. Ray, D.E. Day. J. Non-Cryst. Solids, 86, 204 (1986).
[13] M.D. Mikhailov, A.C. Tver'janovich. Glass Phys. Chem., 6, No.5 (1980).
[14] M.D. Mikhailov, A.C. Tver'janovich. Glass Phys. Chem., 12, No.3 (1986).
[15] S.A. Dembovsky, L.M. Ilizarov, E.A. Chechetkina, A.Yu. Khar'kovsky. In: Proc. "Amorphous Semiconductors - 84". Gabrovo, Bulgaria, 1984. p.39.
[16] S.A. Dembovsky, L.M. Ilizarov, A.Yu. Khar'kovsky. Mater. Res. Bull. 21, 1277 (1986).
[17] S.A. Dembovsky, E.A. Chechetkina. Glass Formation (in Russ.). Nauka, Moscow, 1990.
[18] K.F. Kelton, A.L. Greer. J. Non-Cryst. Solids, 79, 295 (1986).
[19] S.A. Dembovsky. Zh. Neorg. Khim. (Russ. J. Inorg. Chem.), 22, 3187 (1977).
[20] S.A. Dembovsky. Proc. "Amorphous Semiconductors - 80". Kishinev, USSR, 1980. p.22.
[21] S.A. Dembovsky, E.A. Chechetkina. Mater. Res. Bull., 16, 606 (1981).
[22] S.A. Dembovsky, E.A. Chechetkina. Mater. Res. Bull., 16, 723 (1981).
[23] S.A. Dembovsky, E.A. Chechetkina. J. Non-Cryst. Solids 64, 95 (1984).
[24] V.M. Fokin, E.D. Zanotto, W.P. Schmelzer, O.V. Potapov. J. Non-Cryst. Solids, 351, 1491 (2005).
[25] J.L. Nowinski, M. Mroczkowska, J.E. Garbarczyk, M. Wasiucionek. Materials Science-Poland 24, 161 (2006).
[26] A.H. Moharram, A.A. Abu-sehly et al. Physica B, 324, 344 (2002).
[27] L. Berthier. Physics 4, 42 (2011).
[28] P.N. Pusley, E. Zaccarelli et al. Phil. Trans. R. Soc. A, 367, 4993 (2009).
[29] H. Tanaka. J. Phys.: Condens. Matter, 23, 284115 (2011).
[30] W.H. Zachariasen. J. Amer. Ceram. Soc., 54, 3841(1932).
[31] G. Biroli. Sèminaire Poincarè XII, 37 (2009).
[32] S.A. Dembovsky. Mater. Res. Bull., 16, 1331 (1981).
[33] N.A. Popov. JETP Lett., 31, 409 (1980).
[34] S.A. Dembovsky, E.A. Chechetkina. J. Non-Cryst. Solids, 85, 346 (1986).
[35] S.A. Dembovsky, E.A. Chechetkina. Philos. Mag., B53, 367 (1986).
[36] S.A. Dembovsky, E.A. Chechetkina. J. Optoel. Adv. Mater., 3, 3 (2001).
[37] S.A. Dembovsky. J. Non-Cryst. Solids, 353, 2944 (2007).
[38] E.A. Chechetkina. J. Optoel. Adv. Mater., 13, 1385 (2011).
[39] S.A. Dembovsky, A.S. Zyubin. Russ. J. Inorg. Chem., 46, 121 (2001).
[40] S.A. Dembovsky, A.S. Zyubin, O.A. Kondakova. Mendeleev Chemistry Journal (Zhurnal Ross. Khim. Ob-va im.D.I.Mendeleeva), 45, 92 (2001).
[41] S. Zyubin, F.V. Grigoriev, S.A. Dembovsky. Russ. J. Inorg. Chem., 46, 1350 (2001).
[42] J.L. Muscher. Angew. Chem. Int. Ed., 8, 54 (1969).
[43] W.B. Jensen. J. Chem. Educ., 83, 1751 (2006).

[44] G. Pimentel. J. Chem. Phys., 19, 446 (1981).

[45] S.C. Moss, D.L. Price. In: Physics of Disordered Materials (Eds. D. Adler, H. Fritzsche, S.R. Ovshinsky), Plenum, New York, 1985. p.77.

[46] E.A. Chechetkina. J. Phys.: Condes. Matter, 7, 3099 (1995).

[47] E.A. Chechetkina. Solid State Commun., 87, 171 (1993).

[48] E.A. Chechetkina. In: Proc XVII Inter. Congr. Glass, Beijing, 1995, vol 2, p 285

[49] E.A. Chechetkina. J. Non-Cryst. Solids, 128, 30 (1991).

[50] E.A. Chechetkina. In: Fractal Concepts in Materials Science. MRS Proc., vol.367 (1995).

[51] E.A. Chechetkina. J. Non-Cryst. Solids, 201, 146 (1996).

[52] S.A. Dembovsky, A.S. Zyubin. Russ. J. Inorg. Chem. 54, 455 (2009).

[53] M. Kastner, D. Adler, H. Fritzsche. Phys. Rev. Lett. 37, 1504 (1976).

[54] S.A. Dembovsky, E.A. Chechetkina, T.A. Kupriyanova. Materialovedenie (Mater. Sci. Trans.), No.4, 37 (2004).

[55] M.F. Kotkata, E.A. Mahmoud, M.K. El-Mously. Acta Phys. Acad. Sci. Hung., 50, 61 (1981).

[56] M.F. Kotkata, M.K. El-Mously. Acta Phys. Hung., 54, 303 (1983).

[57] G. Parthasarathy, K.J. Rao, E.S.R. Gopal. Philos. Mag., B50, 335 (1984).

[58] S.A. Dembovsky, L.M. Ilizarov, A.Yu. Khar'kovsky. Mater. Res. Bull., 21, 1277 (1986).

[59] M. Hansen, K. Anderko. Constitution of Binary Alloys. McGraw-Hill, 1958.

[60] E.A. Chechetkina, A.B. Vargunin, V.A. Kuznetzov, S.A. Dembovsky. Materialovedenie (Mater. Sci. Trans.), No.7, 28 (2006).

[61] E.A. Chechetkina, E.B. Kryukova, A.B. Vargunin. Materialovedenie (Mater. Sci. Trans.), No.6, 29 (2007).

[62] A.S. Zyubin, E.A. Chechetkina, S.A. Dembovsky. Russ. J. Inorg. Chem., 57, 913 (2012).

[63] S. Zyubin, S.A. Dembovsky. Russ. J. Inorg. Chem., 56, 329 (2011).

[64] D.W. Smith. J. Chem. Educ., 82, 1202 (2005).

[65] E.A. Chechetkina, E.V. Kisterev, E.B. Kryukova, A.I. Vargunin. Inorg. Mater.: Appl. Res., 2, 360 (2011).

[66] H. Haken. Information and Self-Organization. A macroscopic Approach to Complex Systems. Springer, 1988, 2000, 2006.

[67] E.A. Chechetkina, E.V. Kisterev, E.B. Kryukova, A.I. Vargunin. J. Optoel. Adv. Mater., 11, 2034 (2009).

[68] P. Boolchand, G. Lucovsky, J.C. Phillips, M.F. Thorpe. Philos. Mag., 85, 3823 (2005).

2

Crystallization Kinetics of Chalcogenide Glasses

Abhay Kumar Singh
Department of Physics, Banaras Hindu University, Varanasi, India

1. Introduction

1.1 Background of chalcogenides

Chalcogenide glasses are disordered non crystalline materials which have pronounced tendency their atoms to link together to form link chain. Chalcogenide glasses can be obtained by mixing the chalcogen elements, viz, S, Se and Te with elements of the periodic table such as Ga, In, Si, Ge, Sn, As, Sb and Bi, Ag, Cd, Zn etc. In these glasses, short-range inter-atomic forces are predominantly covalent: strong in magnitude and highly directional, whereas weak van der Waals' forces contribute significantly to the medium-range order. The atomic bonding structure is, in general more rigid than that of organic polymers and more flexible than that of oxide glasses. Accordingly, the glass-transition temperatures and elastic properties lay in between those of these materials. Some metallic element containing chalcogenide glasses behave as (super) ionic conductors. These glasses also behave as semiconductors or, more strictly, they are a kind of amorphous semi-conductors with band gap energies of 1±3eV (Fritzsche, 1971). Commonly, chalcogenide glasses have much lower mechanical strength and thermal stability as compared to existing oxide glasses, but they have higher thermal expansion, refractive index, larger range of infrared transparency and higher order of optical non-linearity.

It is difficult to define with accuracy when mankind first fabricated its own glass but sources demonstrate that it discovered 10,000 years back in time. It is also difficult to point in time, when the field of chalcogenide glasses started. The vast majority of time the vitreous glassy state was limited to oxygen compounds and their derivatives. Schulz-Sellack was the first to report data on oxygen-free glass in 1870 (Sellack, 1870). Investigation of chalcogenide glasses as optoelectronics materials in infra-red systems began with the rediscovery of arsenic trisulfide glass (Frerichs, 1950, 1953) when R. Frerichs was reported his work. Development of the glasses as a practical optoelectronic materials were continued by W. A. Fraser and J. Jerger in 1953 (Fraser et. al., 1953). During the 1950-1970 periods (Hilton, 2010) the glasses ware made in ton quantities by several companies and it frequently used in commercial devices. As an example, devices were made to detect the overheated bearings in the railroad cars. Hot objects could be detected by the radiation transmitted through the 3- to 5 μm atmospheric window, for this transparent arsenic trisulfide glass was used. While, to make chalcogenide glass compositions which capable in transmitting longer wavelengths arose the concept of passive thermal optical systems was adopted.

Jerger, Billian, and Sherwood (Hilton, 1966 & 2010) extended their investigation on arsenic glasses containing selenium and tellurium and later adding germanium as a third constituent. The goal was to use chalcogen elements heavier than sulfur to extend long-wavelength transmission to cover the 8- to 12 µm window with improve physical properties. A subsequent work also in Ioffee Institute, in Lenin-grad under the direction of Boris Kolomiets was also reported in 1959 (Hilton, 2010). Work along the same line was begun in the United Kingdom by Nielsen and Savage (Nielsen, 1962, Savage et. al., 1964, Savage et. al., 1966) as well as work also at Texas Instruments (TI) began as an outgrowth of the thermoelectric materials program. The glass forming region for the silicon-arsenic-tellurium system was planed by Hilton and Brau (Hilton et. al., 1963). This development led to an exploratory DARPA- ONR program from 1962 to 1965 (Hilton, 2010). The ultimate goal of the program was to find infrared transmitting chalcogenide glasses with physical properties comparable to those of oxide optical glasses and a softening point of 500^{0}C. Futher, Hilton (Hilton, 1974) also worked on sulfur-based glasses in 1973 to 1974. The exploratory programs resulted the eight chalcogenide glass U.S. patents and a number of research paper published in an international journal detailing the results (Hilton et. al., 1966). After that scientific community made serious effort in development of chalcogenide glasses and organized symposia and meeting in All-Union Symposium on the Vitreous Chalcogenide Semiconductors held in May 1967 (Kokorina, 1966) in Leningrad (now called St. Petersburg). In that symposium Stanley Ovshinsky (founder of Electron Energy Conversion Devices in Troy, Michigan) was presented his first paper dealing switching devices based on the electronic properties of chalcogenide glasses. Similar work was reported by A. D. Pearson (Pearson, 1962) in United States, thus started a great world wide effort to investigate chalcogenide glasses and their electronic properties. The purpose was to pursue a new family of inexpensive electronic devices based on amorphous semiconductors. The effort in this field far exceeded the effort directed toward optoelectronics applications. Some of the results of the efforts in the United States were reported in a symposium (Doremus, 1969) and in another symposium (Cohen et. al.,1971). In this order Robert Patterson mapped out the glass forming region for the germanium-antimony-selenium system and granted a U.S. patent (Patterson, 1966-67) covering the best composition selection.

In 1967 Harold Hafner was made many important contributions including a glass casting process and a glass tempering process in Semiconductor Production Division under the direction of Charlie Jones. There work concentrated the efforts on a glass from the germanium-arsenic-selenium system and outcomes (Jones et. al., 1968) agreed with the conclusions of the Russian, U.K. Alternatively, Servo group efforts that the germanium-arsenic-selenium system produced the best glasses for infrared system applications. Don Weirauch (Hilton, 2010) was conducted a crystallization study on the germanium-arsenic selenium family of glasses and identified a composition in which crystallites would not form. In 1972 a commercial group was successfully cast (12 in 24 in 0.5 in) a window which flat polished, parallel and antireflection-coated (Hafner, 1972). In the late 1960s and early 1970s, passive 8- to 12µ m systems began to be produced in small numbers mostly for the defense uses.

In 1968, Ovshinsky and his co-workers was discovered (Stocker, 1969) the some chalcogenide glasses exhibited memory and switching effects. After this discovery it became clear that the electric pulses could be switch the phases in chalcogenide glasses back and forth between amorphous and crystalline state.Around the same period in 1970's, Sir N. F. Mott (a former Noble Prices winner in Physics-1977) and E.A. Davis were developed the theory on the electronic processes in non-crystalline chalcogenide glasses (Mott et. al., 1979), and Kawamura (Kawamura et. al., 1983) was discovered xerography. Applications of solar cells were developed by Ciureanu and Middehoek (Ciureanu et. al., 1992) and Robert and his coworkers (Robert et.al., 1998). Infrared optics applications were studied by Quiroga and Leng and their coworkers (Quiroga et al., 1996, Leng et. al., 2000). The switching device applications were introduced by Dicerono and Ovshinsky (Dicerono et. al.,1985) and Ovshinsky (Ovshinsky, 1994). P. Boolchand and his coworkers (Boolchand et.al., 2001) was discovered intermediate phase in chalcogenide glasses. In this order several investigators have been also reported that the useful optoelectronics applications in infrared transmission and detection, threshold and memory switching (Selvaraju et al., 2003), optical fibers (Bowden et al., 2009, Shportko et al., 2008, Milliron et al.,2007) functional elements in integrated-optic circuits (Pelusi et al.2009) non-linear optics (Dudley et al., 2009), holographic & memory storage media (Vassilev et al., 2009, Wuttig et al.,2007), chemical and bio-sensors (Anne et al.,2009, Schubert et at., 2001), infrared photovoltaics (Sargent, 2009), microsphere laser (Elliott, 2010), active plasmonics (Samson,2010), microlenses in inkjet printing (Sanchez, 2011) and other photonics (Eggleton, 2011) applications. In this respect, the analysis of the composition dependence of their thermal properties was an important aspect for the study (Singh et al., 2009, 2010, 2011).

Subsequently, several review books were published on chalcogenide glasses e. g. "The Chemistry of Glasses" by A Paul in 1982, "The Physics of Amorphous Solids" by R.Zallen in 1983 and "Physics of Amorphous Materials" by S.R.Elliott in 1983. However, first book entirely dedicated to chalcogenide glassy materials entitled "Chalcogenide Semiconducting Glasses" was published in 1983 by Z.U.Borisova. In this order, G.Z. Vinogradova was published her monograph "Glass formation and Phase Equilibrium in Chalcogenide Systems" in 1984. M.A. Andriesh dedicates a book to some specific applications of chalcogenide glasses entitled "Glassy Semiconductors in Photo-electric Systems for Optical Recording of Information". M.A. Popescu gave large and detailed account on physical and technological aspect of chalcogenide systems in his book "Non-Crystalline Chalcogenides". The compendium of monographs on the subject of photo-induced processes in chalcogenide glasses entitled "Photo-induced Metastability in Amorphous Semiconductors" was compiled by A. Colobov-2003. Robert Fairman and Boris Ushkov-2004 described physical properties in "Semiconducting Chalcogenide Glass I: Glass formation, structure, and simulated transformations in Chalcogenide Glass". Finally, A. Zakery and S.R. Elliott demonstrated the "Optical Nonlinearities in Chalcogenide Glasses and their Applications" in 2007.

1.1.1 Binary chalcogenides

Structure of chalcogenide glasses have been extensively studied in binary compositions considering both the bulk and thin film forms. Chalcogens can form alloys together, Se-S, Se-Te (Gill, 1973) and S-Te amorphous (Sarrach et. al, 1976, Hawes, 1963) compounds were identified; however the scientific community seems to have, for the moment at least, left these glasses aside. Many binary compounds can be synthesized by associating one of the chalcogen with another element of the periodic table like, indium, antimony, copper, germanium, phosphorus, silicon and tin. A few other compounds based on heavy or light elements and alkali atoms have also been investigated. Abrikosov and his co-workers in 1969 (Lopez, 2004) were first reported the molecular structures of most extensively studied As-S, As-Se binary chalcogenide alloys in their monograph, the phase diagrams for the As- S and As-Se systems. As-S alloys can be formed with an As content up to 46%, while in As-Se this maximum content can be raised to almost 60%. Glasses with low As content can easily crystallize (e.g. for a content of 6% As the glass crystallizes at room temperature in one day) in the range 5-16 weight %, in a couple of days at 60°C while it takes 30 days for As S at 280°C (Lopez, 2004). As-Se alloys can crystallize along the all composition range, however this was to be done under pressure and at elevated temperatures. The typical As_2S_3 structure has usually pictured as an assembly of six AsS pyramids (the As atom being the top of the pyramid while three S atoms form the base). Goriunova and Kolomiets in 1958 (Lopez, 2004) were pointed out that the importance of covalent bonding in chalcogenide glasses as the most important property to make stability of these glasses. As opposed to metallic bonding, covalent bonding ensures easier preparation of the glasses. Thus, the crosslinking initiated by the As atoms should reduce the freedom for disorder in which bonds are covalent. Further Vaipolin and Porai-Koshits reported X-ray studies in beginning of the 1960's (Lopez, 2004), for the vitreous $As_2 S_3$ and As_2Se_3 and a number of binary glass compositions based on these two compounds. These glasses were shown to contain corrugated layers, which deformed with increasing size of the chalcogens and arsenic atoms became octohedrally coordinated. The character of the bonds was also found become more ionic when at equimolecular compositions. At the beginning of the 1980's, Tanaka was also characterized the chalcogenide glasses as a phase change materials and demonstrated that they structurally rigid and not having long-range ordering.

Alternatively, Se-S, Se-Te and S-Te (Hamada et. al., 1968, Bohmer & Angell, 1993) were extensively studied binary chalcogens alloys in which Se and S taken as host material. Amorphous selenium and sulfer molecular structure become the mixture of chain and rings which bridging the gap between molecular glasses and polymers. They covalently bonded with two coordination number. The most stable trigonal structural phase a-Se consists of parallel helical chains and two monoclinic bonds forms the composed of rings of eight atoms. These polymorphs distinguished by the correlation between neighboring dihedral angles. The amorphous selenium has relatively low molecular weight polymer with low concentration of rings (Bichara & pellegatti, 1993, Caprion & Schober, 2000 & 2002, Echeveria et. al., 2003, Malek et. al., 2009, etc).

Particularly Se-In binary chalcogenide compositions were getting much attention due to their versatile technical applications. VI- III family compounds Se- In form layered structures with strong covalent bonds. Basically VI-III group Se- In compounds have hexagonal symmetry structure. It consists of two layers which separated by tetrahedrally or pentagonally coordinated Se and In (JablÇonska et. al., 2001 & Pena et al., 2004). Amorphous Se-In compounds contains a-In_2Se_3, a-InSe, and a-In_4Se_3 binary phases. The number of In-Se nearest neighbor heteropolar bonds considerably larger than the homopolar bonds. The In-In and Se-Se homopolar bonds contribute mainly to the left- and right-hand side of the first peak in the radial distribution function, but they do not influence original position. The number of nearest-neighbor Se-Se bonds in a-InSe and a-In_4Se_3 structures is generally negligible (Kohary et. al., 2005).

1.1.2 Ternary chalcogenides

Ternary chalcogenide glasses also broadly studied from more than three decades. Ternary chalcogenides can be prepared by introducing a suitable additive element in well known or new binary matrix. Most extensively studied ternary As-S-Se system was shown a very wide glass-forming region (Flaschen et. al., 1959). The solid solutions can be formed along the line $As_2 S_3$ -$As_2 Se_3$ which proved via IR spectra and X-ray analysis by Velinov and his coworkers (Velinov et. al., 1997). The Covalent Random Network (CRN) and the Chemically Ordered Network (CON) models both satisfy the 8-N rule under the distribution of bond types in a covalent network with multi elements. As- rich glasses can be formed As–As, As–Se, and As–S bonds; thus Se-rich glasses have As–Se, As–S, and Se–Se bonds and S-rich glasses As–Se, As–S, and S–S bonds. The relative weight of each of the above units is expected to be proportionate to the overall composition of the glass itself (Yang et. al., 1989).

In recent years Zn containing ternary chalcogenide glasses attracted much attention due to higher melting point, metallic nature and advanced scientific interest (Boo et. al., 2007). Crystalline state zinc has hexagonal close-packed crystal structure with average coordination number four. While, in amorphous structure it is expected to metallic Zn dissolve in Se chains and makes homopolar and heteropolar bonds. Addition of third element concentration in binary alloy affects the chemical equilibrium of exiting bonds, therefore newly form ternary glass stoichiometry would heavily cross-linked, and makes homopolar and heteropolar bonds in respect of alloying elements. Specifically, Se-Zn-In ternary chalcogenide glasses can form Se-In heteronuclear bonds with strong fixed metallic Zn-In, Zn-Se bonds. Incorporation of indium concentration as-cost of selenium amount, the Se-Zn-In became heavily cross-linked results the steric hindrance increases at the threshold compositional concentration and beyond the threshold concentration a drastic change in physical properties has been observed.

1.1.3 Multicomponent chalcogenides

Addition of more than three elements in chalcogenide alloys refers as multicomponent alloys. In recent years there is an intensive interest made on study of new multi-component chalcogenide glasses to make sophisticated device technology as well as from the point of view of basic physics. Although Se rich binary and ternary chalcogenide glasses exhibit high resistivity, greater hardness, lower aging effect, enhanced electrical and optical properties with good working performance. But ternary glasses have certain drawbacks which implying the limitation in applications. It is worth then to add more than two components into selenium matrix can produce considerable changes in the properties complex glasses. Predominantly, metal and semimetal containing multi-component amorphous semiconductors promising materials to investigations such as; Ge-Bi-Se-Te, Al-(Ge-Se-Y), Ge–As–Se–Te, Cd (Zn)- Ge(As), $GeSe_2$– Sb_2Se_3–PbSe, $Cu_2ZnSnSe_4$ etc. (Thingamajig et. al., 2000, Petkov, 2002, Vassilev, 2006, Wibowo et. al. 2007). More specifically Se-Zn-Te-In multi-component chalcogenide glasses make Se-In heteronuclear bonds with other possible bonds Zn-In and Te-In. Due to the addition of Indium in quaternary glassy matrix, the structures become heavily cross-linked and steric hindrance increases. Therefore, at the expanse of Se chains and replacement of weak Se-Se bonds by Se-In bonds results the increase and decrease in their associative physical properties of Se-Zn-Te-In glasses.

Therefore, the thermal, electrical and optical properties of chalcogenide glasses widely depend on alloying concentration and intrinsic structural changes make them a chemical threshold a particular concentration of alloy. In view of these basic property several past research work in chalcogenide glasses were reported on binary, ternary and very few on multicomponent systems (Tonchev et. al., 1999, Wagner et. al., 1998, Mehta et. al., 2008, Patial et. al., 2011, Malek et. al., 2003, Soltan et. al., 2003, Song et. al., 1997, Usuki et. al., 2001, Fayek et. al., 2001, Wang et. al., 2007, Vassilev et. al., 2007, Eggleton et. al., 2011, Prashanth, et. al., 2008, Vassilev et. al.,2007, Othman et. al., 2006, Zhang et. al.,2004, Hegab et. al., 2007, Narayanan et. al., 2001, etc). Scientific and technological drawbacks, like low thermal stability, low crystallization temperature and aging effects (Liao et. al., 2007, Boycheva et. al., 2002, Ivanova et. al., 2003, Vassilev et. al., 2005, Xu et. al., 2008, Troles et. al., 2008) of non-metallic binary and ternary alloys motivates to investigators to make metallic multicomponent chalcogen alloys to achieve high thermal stability and harder chalcogenide glasses (Pungor, 1997, Demarco et. al., 1999, Kobelke et. al.,1999, Zhang et. al., 2005, Singh, 2011).

Extensive research on metal containing multicomponent chalcogenide alloys was begin nearly end of nineties when Kikineshy and Sterr 1989 & 1990 (Kokenyesi, et. al.,2007 & Ivan & Kikineshi, 2002) were demonstrated the multilayer of chalcogen alloys simply nanostructures materials which can be rather easily produced with controlled geometrical parameters. In this order Ionov and his coworkers (Ionov et. al.,1991) were demonstrated the electrical and electrophotographic properties of selenium based metal containing multicomponent chalcogenide glass and outlined these materials would be useful for electrophotographic and laser printer photoreceptors. Saleh and his coworkers (Saleh et. al., 1993) were studied the nuclear magnetic resonance relaxation of Cu containing chalcogenide glasses. Carthy & Kanatzidis (Carthy & Kanatzidis, 1996) were introduced the bismuth and antimony containing new class of multicomponent chalcogenide glasses. In the same year Natale and his coworkers (Natale et. al.,1996) were demonstrated the heavy metal multicomponent glasses useful for array sensors. Further, Nesheva and his coworkers (Nesheva et. al., 1997) were studied the amorphous pure and allying selenium based multilayers and demonstrated the photoreceptor properties at room temperature unaltered throughout in a year. Efimov (Efimov,1999) was described the mechanism of formation of the vibrational spectra of glasses such as quasi-molecular model, central force model and its recent refinements (model of phonon localization regions) and deduce the trends in the IR and Raman band assignments in inorganic systems. Goetzberger & Hebling (Goetzberger & Hebling, 2000) were commented on the present, past and future of photovoltaic materials. In the same year Naumis (Naumis, 2000) demonstrated the jump of the heat capacity in chalcogenide glasses during glass transition and show change in glass fragility and excess thermal expansivity is a function of average coordination number. While Mortensen and his coworkers (Mortensen et. al., 2000) were used heavy metals based chalcogenides sensors in detection of flow injection. Mourizina and his coworkers (Mourizina et. al., 2001) were demonstrated the ion selective light addressable poentiometric senor based on metal containing chalcogenide glass film. In the same year Rau and his coworkers (Rau et. al., 2001) were studied the effect of the mixed cation in chalcogenide glasses and reported that the non-linear structural changes in Raman and infrared spectra. Further, Messaddeq and his coworkers (Messaddeq et. al., 2001) were demonstrated the light induced volume expansion in chalcogenide glasses under the irradiation UV light. Moreover, Hsu and Narayanan and their coworkers (Hsu et. al., 2001, Narayanan et. al., 2001) were studied the near field microscopic properties of electronic & photonic materials and devices and large switching fields in metal containing chalcogenide glasses owing to chemical disordering. Salmon & Xin (Salmon & Xin, 2002) were studied the effect of high modifier content corresponding to coordination number and demonstrate the structural motifs change in such materials. Subsequently, Agarwal & Sanghera (Agarwal & Sanghera, 2002) were discussed the development and application of chalcogenide glass optical fibers in near scanning field microscopy/spectroscopy and Jackson & Srinivas (Jackson & Srinivas, 2002) demonstrated the modeling of metallic chalcogenide glasses using density function theory calculations. Tanaka (Tanaka, 2003) reviewed the nanoscale structures of chalcogenide glasses and inspect surface modifications at nanometer resolution and Micoulaut & Phillips (Micoulaut & Phillips ,2003) were shown the three elastic phases of covalent networks ((I) floppy, (II) isostatically rigid, and (III) stressed-rigid) depend on the degree freedom of material. They were also suggested that the ring factor is responsible for high crystallization temperature in metallic/ semi-metallic chalcogenide glasses. Lezal and his coworkers (Lezal et. al., 2004) were reviewed the chalcogenide glasses for optical and photonics applications.

Vassilev & Boycheva (Vassilev & Boycheva, 2005) were critically reviewed the achievements in application of chalcogenide glasses as membrane materials. They were also demonstrated that the advantages and disadvantages in analytical performance and compared with the corresponding polycrystalline analogous. Emin (Emin, 2006) was explained the polaron conduction machines in amorphous semiconductors and Kokenyesi (Kokenyesi, 2006) reviewed the amorphous chalcogenide nano-multilayers: research and developments. While Phillips (Phillips, 2006) demonstrated the, ideally glassy materials have hydrogen-bonded networks. Bosch and his coworkers (Bosch et. al. 2007) were critically reviewed the last decade developments in optical fibers in bio sensing and Vassilev and his coworkers (Vassilev et. al.,2007) introduced the new Se- based multicomponent chalcogenide glasses and studied their composition dependence physical properties. Furthermore, Wachter and Taeed their coworkers (Wachter et. al., 2007, Taeed et. al., 2007) were demonstrated the composition dependence reversible and tunable glass-crystal-glass phase transition properties in new class of multicomponent chalcogenide glasses and show chalcogenide glasses useful for all-optical signal processing devices due to their large ultrafast third-order nonlinearities, low two-photon absorption and the absence of free carrier absorption in a photosensitive medium. Dahshan and Lousteau and their coworkers (Dahshan et. al.,2008, Lousteau et. al.,2008) were demonstrated the thermal stability and activation energy of some Cu doped chalcogenide glasses and the fabrication of heavy metal fluoride glass to explore the optical planar waveguides by hot-spin casting. Further, Klokishner (Klokishner et. al., 2008) were studied the concentration effects on the photoluminescence band centers in multicomponent metallic chalcogenide glasses and Ielmini and his coworkers (Ielmini et. al., 2008) demonstrated the threshold switching mechanism by high-field energy gain in the hopping transport of chalcogenide glasses. Mehta and his coworkers (Mehta et. al., 2009) were studied the effect of metallic and non metallic additive elements on Se-Te based chalcogenide glasses. In the same year Turek and Anne their coworkers (Turek et. al.,2009, Anne et. al.,2009) were demonstrated the artificial intelligence/fuzzy logic method for analysis of combined signals from heavy metal chemical sensors and commented on, due to the remarkable properties of chalcogenide glasses can be used as a biosensor which can collect the information on whole metabolism alterations rapidly. Further, Khan and his coworkers (Khan et. al., 2009) were demonstrated the composition dependence electrical transport and optical properties of metallic element doped Se based chalcogenide glasses. In order to this Snopatin and his coworkers (Snopatin et. al., 2009) were demonstrated the some high purity multicomponent chalcogenide glasses for fiber optics. Kumar and his coworkers (Kumar et. al., 2010) were demonstrated the calorimetric studies of Se-based metal containing multicomponent chalcogenide glasses. Peng & Liu (Peng & Liu ,2010) were reviewed the advances and achievements in SPM-based data storage in viewpoint of recording techniques including electrical bistability, photoelectrochemical conversion, field-induced charge storage, atomic manipulation or deposition, local oxidation, magneto-optical or magnetic recording, thermally induced physical deformation or phase change, and so forth as well as achievements in design and synthesis of organic charge-transfer (CT) complexes towards thermochemical-hole-burning memory, the correlation between hole-burning performances and physicochemical properties of CT complexes.

Story of the investigations will be remain continue in field of metallic chalcogenides (not limited to above outlined the major events) to deduce the new future prospective multicomponent chalcogenide glassy alloys. Amorphous chalcogenide alloys which full fill the essential requirement of modern optoelectronics. So, it can be outlined potential field of optoelectronics and advanced material is rapidly growing owing to their possible uses. Therefore, it is important to have an understanding regarding on crystallization process of chalcogenide glasses (predominately in metal containing alloys).

1.2 Crystallization

Crystallization is a natural process of formation of solid crystals from a solution/- melt. Crystallization of a substance can also achieve from the chemical solid-liquid separation technique, in which mass transfer from the liquid solution to a pure solid crystalline phase. The crystallization process of a substance mainly consists of two major events nucleation and crystal growth. In nucleation process the molten molecules dispersed in solid solution and begin to formation of clusters at the nanometer scale. Crystal growth is the subsequent growth of the nuclei which develop critical size of the formed clusters (because size of clusters plays an important role in the application of the material). Hence, the nucleation and growth are the continuous process which occurs simultaneously when supercooling exists in a system. Thus, system supercooling state acts as a driving force for the crystallization process. The supercooling driving force depending upon the conditions, either nucleation or growth may be predominant over to other and outcomes can be formed crystals with different sizes and shapes. Once the supercooling is established in a solid-liquid system and reached at equilibrium then crystallization process is completed (Mersmann, 2001).

In general supercooled materials/ or alloys have ability to crystallize with different crystal structures, this process is known as polymorphism. Each polymorph is in fact a different thermodynamic solid state and crystal polymorphs of the same alloy/-compound which exhibited different physical properties, such as dissolution rate, shape, melting point, etc. Thus the crystallization process of a substance is governed by both thermodynamic and kinetic factors which highly variable and difficult to control. Factors those affect the crystallization process of a substance/alloy are the impurity level, mixing regime, vessel design, and cooling profile and shape of crystals. Usually in those materials crystallization process occurs at lower temperatures, in supercooling situation they obey the law of thermodynamics. Its literal meaning a crystal can be more easily destroyed than it is formed.

Subsequently it is easier to a perfect crystal in a molten solid than to grow again a good crystal from the resulting solution. Hence the nucleation and growth process of a crystal are well control under the thermodynamic kinetic.

1.3 Nucleation and growth

1.3.1 Nucleation

Nucleation of the substance reflects the initiation of a phase change in a small region cause the formation of a solid crystal from a liquid solution. It is a consequence of rapid local fluctuations on a molecular scale in a homogeneous phase which define as a metastable equilibrium state. The whole nucleation process of a substance is the sum of heterogeneous (nucleation that occurs in the absence of a second phase) and heterogeneous (nucleation that occurs in the presence of a second, foreign phase) category of nucleation. Homogeneous nucleation due to clustering of molecules (embryos) in a supersaturated environment, in which a process began, combines two or more than molecules. In the reversible clustering process a few molecules grew at the same time and others dissolving. Once embryos attained a certain critical size then it decrease its total free energy by growing and becomes stable (Reid et. al., 1970). But, in practice it is difficult to find complete homogeneous nucleation owing to presence of insoluble amounts of matter even in pure material. Therefore, heterogeneous nucleation is always associated with homogeneous nucleation due to presence of second phase in bulk molten material. The heterogeneous nucleation occurs in a random fashion at various sites in a matter.

1.3.2 Crystal growth

As earlier mentioned crystal growth is the successive process of nucleation in which the critical nuclei of microscopic size form a crystal. Crystal growth in crystallization process takes place by fusion and re-solidification of the material. In this process within a solid material constituent of molecules (embryos) are arranged in an orderly repeating pattern extending in all three spatial dimensions. Crystal growth is a major stage of a crystallization process which consists the addition of new molecules (embryos) strings into the characteristic arrangement of a crystalline lattice. The growth typically follows an initial stage of either homogeneous or heterogeneous nucleation. The crystal growth process yields a crystalline solid whose molecules are typically close packed with fixed positions in space relative to each other. In general crystalline solids are typically formed by cooling and solidification from the molten (or liquid) state. As per the Ehrenfest classification it is first-order phase transitions with a discontinuous change in volume (and thus a discontinuity in the slope or first derivative with respect to temperature, dV/dT) at the melting point. Hence, the crystal and melt are distinct phases with an interfacial discontinuity having a surface of tension with a positive surface energy. Thus, a metastable parent phase represents it always stable with respect to the nucleation of small embryos from a daughter phase with a positive surface of tension. Hence, crystal growth process is first-order transitions consist advancement of an interfacial region whose structure and properties vary discontinuously from the parent phase. In the crystal growth process stiochiometry of glass compositions do not undergo in compositional changes during crystallization, mean, no need to long-range diffusion (Swanson, 1977) for crystal growth in chalcogenide glasses; thus, interfacial rearrangements are likely to control the crystal growth process. This (melt quenched) type of crystal growth is generally described from these three basic standard models: (i) the screw dislocation model; (ii) the normal or continuous growth model; and (iii) the two-dimensional surface nucleation growth.

Theoretically, nucleation and crystal growth process of molten solids first reported by Volmer and Weber (Volmer and Weber, 1925) and later as it explained in a large amount of literature. Tammann (1925) discussed the theory of nucleation and crystal growth and outlined the various parameters involved in terms of a probabilistic model involving a functional relation to pressure, temperature, and time. Many later studies deal with the nucleation and growth of crystals on an atomistic level. Nucleation of crystals from a melt is the mobility of atoms and molecules in the melt as measured by the diffusion coefficient. In glass-forming systems, liquid diffusion coefficients drop markedly with decreasing temperature (Towers and Chipman, 1957). When temperature drops below the liquidus, nucleation will increase from zero to a maximum at some undercooling. Diffusion rates then very low and nucleation decreases with further decrease in temperature. This explains the pattern of nucleation in liquids where the liquid diffusivity decreases with temperature and finally it grow a crystal. Rate of growth is thus a function of mobility of crystal-forming species within the melt. Mobility, can be measured by the diffusion coefficient, as drops with decreasing temperature and growth rate, like nucleation.

Thermodynamics of crystal growth process in molten solids can be expressed as (Warghese, 2010); the thermodynamical equilibrium between solid and liquid phases occur when the free energy of the two phases are equal

$$G_L = G_S \tag{1}$$

here G_L, and G_S are representing solid and liquid phases free energies

Free energy, internal energy and entropy of a system can be related from the Gibbs equation

$$G = H - TS \tag{2}$$

here G is the Gibbs free energy, H is the enthalpy, S is the entropy and T is the temperature.

Formation of a crystal can be considered as a controlled change of phase to the solid state. Therefore, the driving force for crystallization comes from the lowering of the free energy of the system during this phase transformation. Change in free energy in such transition can be related as;

$$\Delta G = \Delta H - T\Delta S \tag{3}$$

Where $\Delta H = H_L - H_S$

$$\Delta S = S_L - S_S$$

$$\Delta G = G_L - G_S$$

For the equilibrium $\Delta G = 0$

$$\Delta H = T_e . \Delta S$$

Where, T_e is the equilibrium temperature

$$\Delta G = \Delta H.\ \Delta T / T_e \quad (4)$$

Where $\Delta T = T_e - T$

ΔG is positive when $T_e > T$ and it depends on the latent heat of transition. The change in free energy can also represent as the product of change in entropy and super cooling temperature ΔT.

$$\Delta G = \Delta S.\ \Delta T \quad (5)$$

Although equation (5) representing melt growth, in which one may depend on concentration rather than supercooling for solution growth and vapour growth. Equation (5) in more convent form can be expressed as;

$$\Delta G \sim RT \ln (C/C_0) \quad (6)$$

$$\Delta G \sim RT \ln (P/P_0) \quad (7)$$

In general form

$$\Delta G \sim RT \ln S \quad (8)$$

where R is the Rydberg constant, C, C_0 are the concentration of solid solution and concentration at critical transition, P, P_0 are the pressers in vapour phase and S is the supercooling ratio.

Thus the equations (4) and (6) explain how the free energy changes depend on the supercooling parameters which are decisive in the process of crystallization. The rate of growth of a crystal can be expressed as a monotonically increasing function of Gibbs free energy, when other parameters remain the same.

Since in nucleation process in the supercooled solution forms the small clusters of molecules (Joseph, 2010), therefore, free energy change between the solid and liquid can be expressed as ΔG_v and Gibbs equation can be written as;

$$\Delta G = 4\pi r^2 \sigma - 4/3\ \pi r^3\ \Delta G_v \quad (9)$$

where σ- is the interfacial energy and r is the spherical radius of the molecule

The surface energy increase in term of r^2 and volume energy decreases in term of r^3. The critical size of nucleus can be obtained by the equation (9)

$$r^* = 2\ \sigma / \ \Delta G_v \quad (10)$$

Hence the critical size **(r^*)** of the nucleus decreases with increasing cooling rate.

Further, in formation of critical nucleus size free energy **(ΔG^*)** change can be calculated as;

$$\Delta G^* = 16\pi\sigma^3 / \ 3\Delta G_v^2 \quad (11)$$

Equation (11) in terms of Gibbs thermodynamical relation can be written as;

$$\Delta G^* = 16\pi\sigma^3\Omega / \ 3(kT \ln.S)^2 \quad (12)$$

where Ω is the molecular volume

The rate of nucleation **(J)** can be expressed as;

$$J = J_0 \exp [-\Delta G^*/ kT] \quad (13)$$

or in terms of themodynamical parameters

$$J = J_0 \exp [-16\pi\sigma^3\Omega^2/ 3k^3T^3 (\ln S)^2] \quad (14)$$

where J_0 is the pre-exponential factor

For critical supercooling condition J = 1, so that ln.J = 0, then expression can be expressed as;

$$S_{cri} = \exp [16\pi\sigma^3\Omega^2/ 3k^3T^3 \ln J_0]^{1/2} \quad (15)$$

1.4 Crystallization kinetics

Study of the crystallization of the amorphous materials with respect to time and temperature is called crystallization kinetics. Crystallization kinetic study of the materials can be performed in either isothermal or non-isothermal mode of Differential Scanning Calorimetry (DSC). In the isothermal method, the sample is brought near to the crystallization temperature very quickly and the physical quantities, which change drastically are measured as function of time. In the non-isothermal method the sample is heated at a fixed rate and physical parameters recorded as a function of temperature. Investigators (Sbirrazzuoli, 1999) preferred to perform DSC measurements in non-isothermal mode. Owing to fact, it is not possible to ensure the homogeneity (or constant) of DSC furnace temperature in isothermal mode during the injection of material sample.

Crystallization kinetics parameters of the materials generally interpreted at glass transition temperature (T_g), crystallization temperature (T_c) and peak crystallization temperature (T_p) with help well defined statistical approximations (such as Hurby, Ozawa, Augis and Bentt, Moynihan and Kissinger) (Hruby,1972, Ozawa, 1970, Augis & Bennett, 1978, Moynihan et al., 1974, Kissinger,1957). All the existing approximations described on the basis of JMA (Johnson,1939, Avrami, 1939& 1940) model statics, although in recent years investigators also reported the (Sanchez-Jimenez, 2009) JMA model not a universal model to explain the crystallization kinetics of the materials, because it has few limitations. Despite of this majority view of investigators toward to kinetic methods based on JMA model are more reliable to explain the crystallization of chalcogenide glasses, polymers, metallic and oxide glasses.

1.5 Theoretical basis of crystallization kinetics

In isothermal phase transformation the extent of crystallization (α) of a certain material can be represent from the Avrami's equation (Moynihan et al., 1974, Kissinger,1957, Johnson,1939)

$$\alpha(t)=1-\exp[(-Kt)^n] \quad (16)$$

where K is the crystallization rate constant and n is the order parameter which depends upon the mechanism of crystal growth.

In general the value of crystallization rate constant K increases exponentially with temperature. The temperature dependence behaviour of K indicates that the crystallization of amorphous or amorphous glassy materials is a thermally activated process. Mathematically it can be expressed as

$$K = K_0 \exp(-E_c / RT) \tag{17}$$

Here E_c is the activation energy of crystallization, K_0 is the pre-exponential factor, R is the universal gas constant and T is the temperature. For Isothermal condition parameters E_c and K_0 in Eq. (17) can be assumed practically independent of the temperature (at least in the temperature interval accessible in the calorimetric measurements).

In non-isothermal crystallization, it is assumed that the constant heating rate during the experiment. The relation between the sample temperature T and the heating rate β can be written as

$$T = T_i + \beta t \tag{18}$$

Here T_i is the initial temperature. The crystallization rate is obtained by taking the derivative of expression (1) with respect to time, keeping in mind that the reaction rate constant is a time function which represents Arrhenius temperature dependence.

$$(d\alpha/dt) = n\,(Kt)^{n-1}[K+(dK / dt)\,t]\,(1-\alpha) \tag{19}$$

The derivative of K with respect to time can be obtained from Eqs. (17) and (18), which follows as:

$$(dK/dt) = (dK / dT)\,(dT/dt) = (\beta E_c/RT^2)\,K \tag{20}$$

From Eq. (19) we obtained

$$(d\alpha/dt) = nK^n t^{n-1}[1+at].(1-\alpha) \tag{21}$$

where $a = (\beta E_c/RT^2)$.

Using Eq. (21) Augis and Bennett (Augis & Bennett, 1978) have developed a crystallization kinetic method. They have taken proper account of the temperature dependence crystallization reaction rate. Their approximation results have also been verified the linear relation between ln $(T_c - T_i)/\beta$ versus $1/T_c$ (here T_c is the onset crystallization or critical transition temperature). This can be deduced by substituting the u for Kt in Eq. (21), accordingly the rate of reaction can be expressed as

$$(d\alpha/dt) = n\,(du/dt)\,u^{(n-1)}\,(1-\alpha) \tag{22}$$

Where

$$(du/dt) = u\,[(1/t) + a] \tag{23}$$

Second derivatives of Eqs. (22) and (23) are given as:

$$(d^2\alpha/dt^2) = [(d^2u/dt^2)\,u - (du/dt)^2 \times (nu^2 - n + 1)]\,nu^{(n-2)}\,(1-\alpha) = 0 \tag{24}$$

$$(d^2u/dt^2) = (du/dt)\,[(1/t) + a] + u\,[(-1/t^2)] + (da/dt) \tag{25}$$

In Eq. (25) substituting for $(da/dt) = -(2\beta/T)a$, then it can be written as

$$(d^2u/dt^2) = u\,[a^2 + (2a\,T_i/\,tT)] \tag{26}$$

The last term in the above equation was omitted in the original derivation of Augis and Bennett [33] (T_i T) and resulted in the simple form:

$$(d^2u/dt^2) = a^2 u \tag{27}$$

Substitution of (du/dt) and (d^2u/dt^2) in Eqs. (23), (27), and (24) gives the following expression:

$$(nu^n - n + 1) = [at/(1+at)]^2. \tag{28}$$

For $E/RT >> 1$, the right-hand bracket approaches its maximum limit and consequently u (at the peak) = 1, or

$$u = (Kt)_c = K_0 \exp(-E_c / RT_c)\,[\,(T_c - T_i) / \beta] \approx 1 \tag{29}$$

In logarithm form, for $T_i << T_c$

$$\ln(\beta/T_c) \approx (-E_c / RT_c) + \ln K_0 \tag{30}$$

Value of E_c and K(T) can be obatined from the equation (30) by using the plots of ln β/T_c against $1/T_c$. Further, by using the Eq (17) Hu et.al. (Sbirrazzuoli, 1999) have introduced the crystallization rate constant stability criterion corresponding to T_c.

$$K(T_c) = K_0 \exp(-E_c/RT_c) \tag{31}$$

1.6 Differential Scanning Calorimetry (DSC) thermograms

Endothermic and exothermic peaks in amorphous glassy materials arise due to thermal relaxation from a state of higher enthalpy toward to metastable equilibrium states of lower enthalpy. The process of the thermal relaxation depends on temperature and may quite fast near the glass-transition temperature. The glass transition peak in DSC measurement represents the abrupt change in specific heat and decrease in viscosity (Matusita, 1984), while the crystallization peak demonstrate to the production of excess free-volume, and endothermic peak at T_m reflects the amount of energy which liberate owing to complete destroy the solid phase structure cause braking of all type existing bonds in solid alloy. Hence, the materials crystallizations temperatures as well as mode of crystallizations extensively depend on the compositions of alloys.

In general DSC thermograms of amorphous glassy (i.e. chalcogenide glasses) materials have exhibited a considerable shifts in endothermic glass-transition and exothermic crystallization temperatures with increasing heating rates. But in recent investigations (Singh & Singh, 2009) investigators have also been reported vary small or negligible endothermic glass transitions shifts in metal, semi-metal and non metal containing multicomponent chalcogenide glasses. In order to this, we have performed the DSC measurements on recent developed $Se_{93-x}Zn_2Te_5In_x$ ($0 \leq x \leq 10$) chalcogenide glasses.

These materials could be prepared by the well known most convenient melt quenched method. The high purity elements Selenium, Zinc, Tellurium and Indium were used. The suitable amounts of elements were weighed by electronic balance and put into clean quartz ampoules (length of ampoules 8 cm and diameter 14 mm). All the ampoules were evacuated and sealed under at a vacuum of 10^{-5} Torr to avoid the reaction of glasses with oxygen at high temperature. A bunch of sealed ampoules was heated in electric furnace up to 1173K at a rate of 5-6 K/min and held at that temperature for 10-11 h. During the melting process ampoules were frequently rocked to ensure the homogeneity of molten materials. After achieving desired melting time, the ampoules with molten materials were frequently quenched into ice cooled water. Finally ingots of glassy materials were obtained by breaking the ampoules. The preparation and characterizations technique of the test materials also outlined in our past [Singh & Singh, 2010] research work.

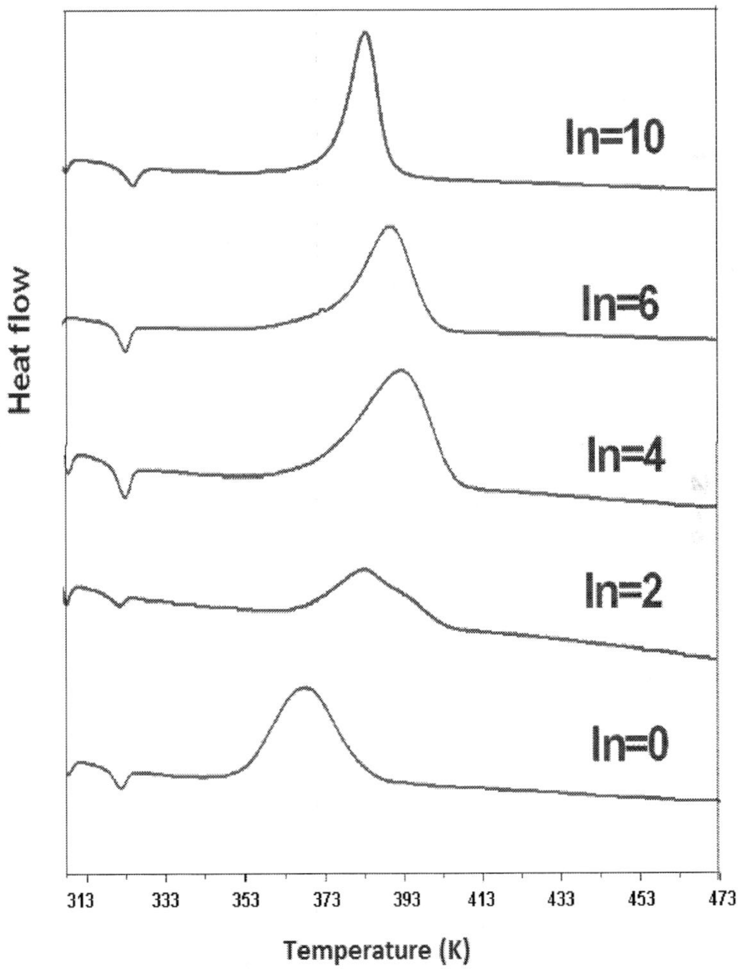

Fig. 1. DSC patterns of Se_{93-x}-Zn_2-Te_5-In_x ($0\leq X\leq 10$) chalcogenide glasses at heating rate 5 K/min

DSC patterns of $Se_{93-x}Zn_2Te_5In_x$ ($0 \leq x \leq 10$) glasses at heating rate of 5 K $(min)^{-1}$ is given in Figure.1. DSC traces clearly show the endothermic and exothermic phase reversal peaks at the glass transitions and crystallizations temperatures. Obtained values of the glass transitions temperatures (T_g), onset crystallizations temperatures (T_c), peak crystallizations temperatures (T_p) and melting temperatures (T_m) at heating rates of 5, 10, 15 and 20 K $(min)^{-1}$ is listed in Table 1.

Se_{93-x}-Zn_2-Te_5-In_x (X=0, 2, 4, 6 and 10)					
	Heating rate	T_g (K)	T_c (K)	T_p (K)	T_m(K)
Se_{93}-Zn_2-Te_5	5	318	354	368	501
	10	320	359	374	502
	15	322	363	378	503
	20	323	367	381	503
Se_{91}-Zn_2-Te_5-In_2	5	319	370	385	502
	10	322	376	392	503
	15	323	382	398	504
	20	324	385	402	504
Se_{89}-Zn_2-Te_5-In_4	5	319	372	393	503
	10	322	381	402	504
	15	324	385	407	504
	20	325	389	411	504
Se_{87}-Zn_2-Te_5-In_6	5	320	375	396	504
	10	323	383	405	504
	15	325	389	412	506
	20	327	392	415	506
Se_{83}-Zn_2-Te_5-In_{10}	5	319	370	386	504
	10	322	378	394	504
	15	323	383	400	504
	20	324	386	405	505

Table. 1 Obatined values of T_g, T_c, T_p and T_m at heating rates 5, 10, 15 and 20 K/min

Outcome demonstrates, a very small glass-transitions temperatures shifts and a considerable shifts in onset and peak crystallizations temperatures, the corresponding T_c-T_g result is given in Figure.2 and their values listed in Table.2. This result revealed the values of T_c and T_p crystallizations temperatures increases upto 6 at. wt. % indium and beyond this decreased for 10 atomic percentage composition glass.

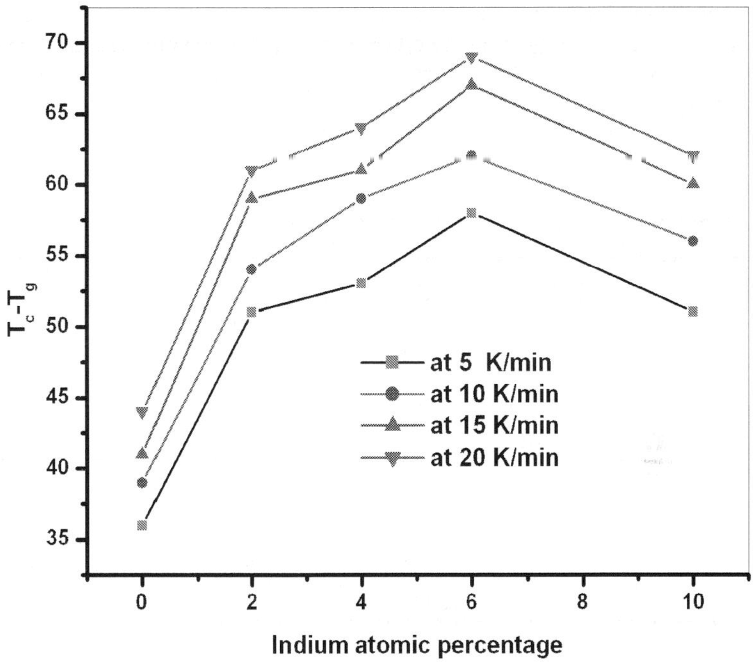

Fig. 2. Variations of T_c-T_g with Indium atomic percentage at 5, 10, 15 and 20 K/min heating rates

	Heating rate	$Se_{93}Zn_2Te_5$	$Se_{91}Zn_2Te_5In_2$	$Se_{89}Zn_2Te_5In_4$	$Se_{87}Zn_2Te_5In_6$	$Se_{83}Zn_2Te_5In_{10}$
T_c-T_g	5	36	51	53	55	51
	10	39	54	59	60	56
	15	41	59	61	64	60
	20	44	61	64	65	62
T_p-T_c	5	14	15	21	21	16
	10	15	16	21	22	16
	15	15	16	22	23	17
	20	14	17	22	23	19
T_m-T_c	5	147	132	131	126	134
	10	143	127	123	119	126
	15	140	122	119	114	121
	20	136	119	115	110	119

Table 2. Evaluated values of T_c-T_g, T_p-T_c and T_m-T_c at different heating rates

1.7 Glass forming ability (GFA)

GFA of material describes the relative ability to a set of compound adopt the amorphous structure (Mehta et al., 2006, Jain et al., 2009). In practice criteria to establish the GFA of vitreous materials are based on DSC measurements. Usually, unstable glass has show a crystallization peak near to the glass transition temperature while stable glass peak close to melting temperature. GFA can be evaluated by mean of the difference between the crystallization temperature (peak temperature T_p and/or onset temperature T_c) and the glass transition temperature (T_g). This difference varies with alloys concentrations and higher and lower for a certain composition. To evaluate the GFA of glassy alloys several quantitative methods have been introduced from the investigators. Most of the methods (Saad & Poulin,1987, Dietzel,1968) based on characteristics temperatures of glassy alloys. Dietzel (Dietzel,1968) has introduced the first GFA criterion $D_T = T_c - T_g$. Further Hruby (Hruby,1972) developed the H_R GFA criterion [$H_R = T_c-T_g / T_m-T_c$].This method has additional advantage to describe the thermal stability of amorphous materials.

To applying the GFA criterion method, obtained critical characteristics temperatures difference T_c-T_g, T_p-T_c and T_m-T_c (Here T_g is glass transition temperature, T_p is peak crystallization temperature and T_m is the melting temperature) values of under examine materials is listed in Table 2. Using these values H_R parameter of GFA can be described as:

$$H_R = \left(\frac{T_c - T_g}{T_m - T_c} \right) \tag{32}$$

GFA variation with indium atomic weight percentage at heating rates 5, 10, 15 and 20 is given in Figure. 3 and their corresponding average values listed in Table 3. The higher GFA value is obtained for threshold indium concentration glass. High GFA value of threshold composition also reflects their high order thermal stability as compare to other glasses of this series.

1.8 Activation energy

Activation energy reflects the involvement of molecular motions and rearrangements of the atoms around the critical transitions temperatures (Suri et al., 2006). In DSC measurement atoms undergo infrequent transitions between the local (or metastable state) potential minima which separated from different energy barriers in the configuration space, where each local minima represent a different structure. The most stable configuration has local minima structure in glassy region. This literal meaning a glass atoms possessing minimum activation energy have a higher probability to jump in metastable state of lower internal energy configuration. This local minima configuration occurs at particular composition of alloy which refers as a most stable glass (Imran et al., 2001). The activations energies of chalcogenide glasses at the critical temperatures can be interpreted in these words: the glass transition activation energy (E_g), onset crystallization activation energy (E_c) and peak crystallization activation energy (E_p) are the amount of energies which absorbed by a group of atoms for a jump from one metastable state to another state (Imran et al., 2001, Agarwal et al., 1991

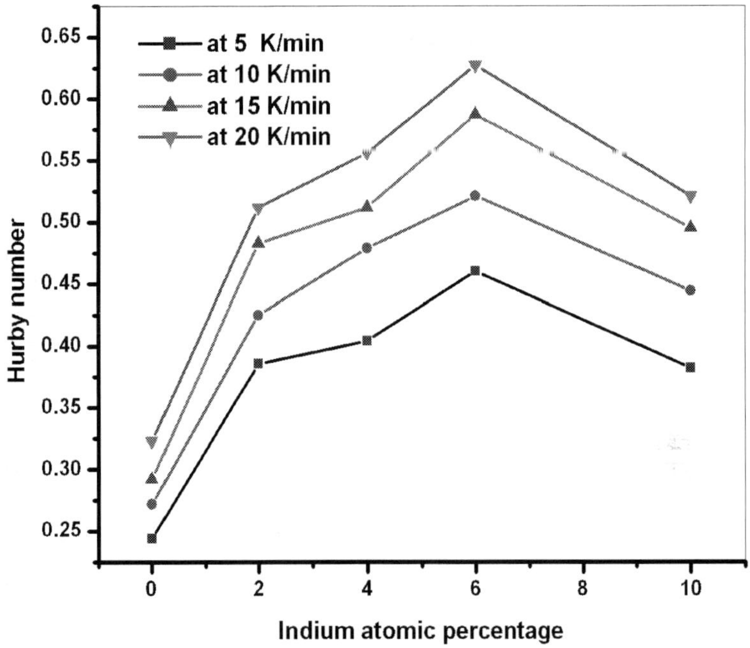

Fig. 3. Plots of GFA parameter with indium atomic percentage at 5, 10, 15 and 20 K/min heating rates

1.8.1 Glass transition activation energy

The glass transition activation reflects the endothermic energy of the material which produced due to unsaturated or hydrogen like bond braking at the pre-crystallization critical temperature. Glass transitions activations energies of the under test glasses can be defined by using Ozawa method (Ozawa, 1970).

$$\ln\beta = -\left(\frac{E_g}{RT_g}\right) + C \tag{33}$$

where β is the heating rate, E_g is the glass transition activation energy, C is the constant in usual meaning. Obtained Ozawa plots, ln β vs $1000/T_g$, for these materials is given in Figure.4 and their corresponding E_g values is listed in Table 3.

Outcomes revel the E_g values have a maxima and a minima for 0 and 6 percentage indium compositions glasses. Thus the E_g values have very small increasing and decreasing trend in T_g values (see Table 3&Table 1) with increasing DSC heating rates in these metal, semi-metal and non-metal containing multicomponent chalcogenide glasses. The high activations energies at pre-crystallizations reflect the materials rigidity. While, normally reported T_g values for non-metallic chalcogenides compositions has show a considerable shifts with

DSC heating rates. Hence the obtained T_g values results in increasing DSC heating rates for metal, semi-metal and non-metal containing multicomponent chalcogenide glasses are not in good agreement with previous reported non metallic compositions. Deviations in the results arise due to existence relatively hard metallic, semi-metallic characters unsaturated bonds with hydrogen like week bonds in the alloys stoichiometrics. Further, it is quite possible to large amount of metallic, semi-metallic characters unsaturated bonds sustain over to T_g critical transition temperature of the materials owing to requirement greater amount of energy to bark the heteropolar unsaturated bonds. As consequence the metal,

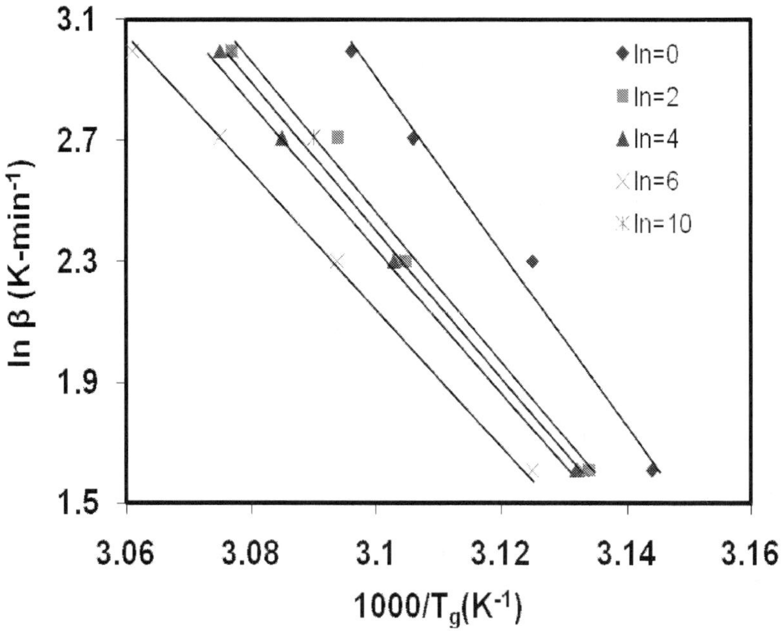

Fig. 4. Ozwa polts of Se_{93-x}-Zn_2-Te_5-In_x ($0 \leq X \leq 10$) glasses to obtain E_g

Se_{93-x}-Zn_2-Te_5-In_x (X=0, 2, 4, 6 and 10)				
Alloy compositions	Glass transition activation energy E_g (KJ/mol)	Onset crystallization activation energy E_c (KJ/mol)	Peak crystallization activation energy E_p (KJ/mol)	Average (GFA)
$Se_{93}Zn_2Te_5$	229.94	115.64	118.23	0.282
$Se_{91}Zn_2Te_5In_2$	205.02	106.11	96.40	0.451
$Se_{89}Zn_2Te_5In_4$	195.47	99.02	91.09	0.487
$Se_{87}Zn_2Te_5In_6$	174.23	93.66	82.77	0.548
$Se_{83}Zn_2Te_5In_{10}$	199.88	101.84	92.70	0.460

Table 3. E_g, E_c, E_p and GFA values of Se_{93-x}-Zn_2-Te_5-In_x($0 \leq x \leq 10$) chalcogenide glasses

semi-metal and non-metal containing multicomponent chalcogenide glasses have exhibited either very small or negligible glass transitions temperatures shifts with increasing DSC heating rates.

1.8.2 Onset crystallization activation energy

Onset crystallization activation energy (E_c) is the amount of thermal energy which requires to begin the phase transformation from glassy to crystallization state. Quantitive knowledge of onset crystallization activation energy at T_c defines the heat/energy storage capability of the material which useful for different physical applications. The exothermic onset crystallization activation energy at T_c arises due to barking of existing covalent bonds in glassy configuration. In case of complex metallic multicomponent chalcogenide glasses high energy homopolar and heteropolar covalent bonds formed as compare to metallic binary and ternary compositions. Due to this the critical onset crystallization temperatures of the complex metallic glasses (see Table 1&Table 3) increases and their corresponding activation energies tend to be decrease upto threshold composition then visa-verse direction. While in case of non-metallic binary, ternary and multicomponet chalcogenide alloys reports demonstrated they have lower values of onset crystallizations temperatures owing to existence of week homopolar and heteropolar bonds in glassy configuration.

Onset crystallizations activations energies of under examine complex metallic multicomponent chalcogenide glasses described by employing the Ozawa method (Ozawa, 1970).

$$\ln\beta = -\left(\frac{E_c}{RT_c}\right) + C \tag{34}$$

Here symbols (β is heating rate, E_c is the onset crystallization activation energy, T_c is the onset crystallization temperature and R & C are the constant) are in usual meaning. Obtained Ozawa plots ln β vs $1000/T_c$ is given in Figure. 5 and their corresponding E_c values is listed in Table 3. Outcomes show a phase reversal in E_c values which have a maxima and minima corresponding to 0 and 6 atomic weight percentage of indium glasses.

1.8.3 Peak crystallization activation energy

Peak crystallization activation energy (E_P) of a glass expresses the amount of heat energy which requires for utmost crystallization. By mean at peak crystallization point almost all the existing heteropolar covalent bonds have to be broken and material achieve to maximum crystallization i.e. a glassy phase material completely transform to crystalline phase and relax toward to original state. The E_P values of examined materials can be described by using the Kissinger method (Kissinger,1957).

$$\ln\left(\frac{\beta}{T_P^2}\right) = -\frac{E_P}{RT_P} + C \tag{35}$$

Here symbols (β is heating rate, E_P is the peak crystallization activation energy, T_P is the paek crystallization temperature and R & C are the constant) are in usual meaning. Obtained E_P values from Kissinger plots (see Figure. 6) are listed in Table.3.

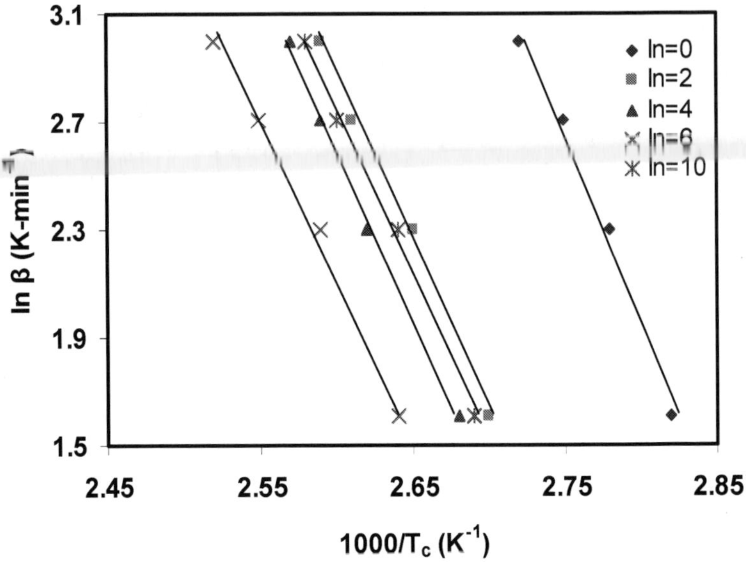

Fig. 5. Ozwa polts of Se_{93-x}-Zn_2-Te_5-In_x ($0\leq X \leq 10$) glasses to obtain E_c

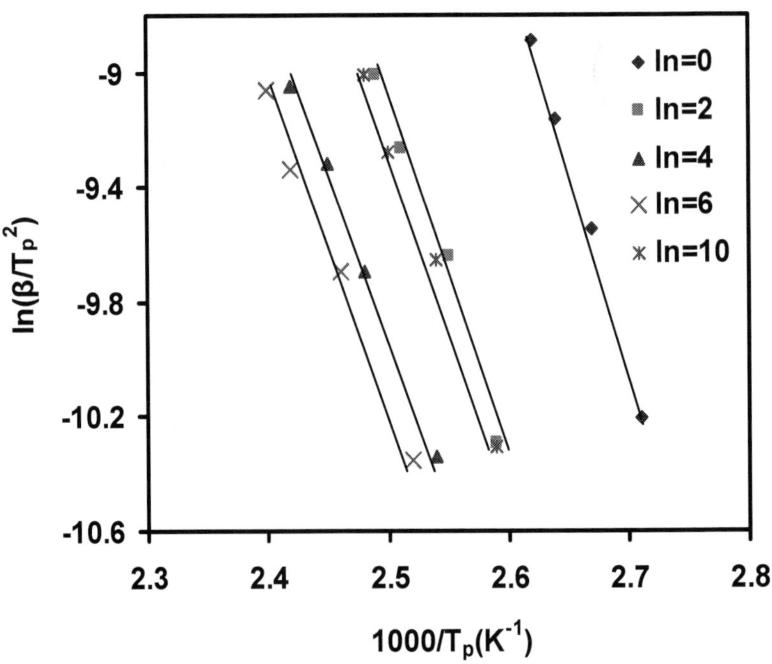

Fig. 6. Kissinger polts of Se_{93-x}-Zn_2-Te_5-In_x ($0\leq X \leq 10$) glasses to obtain E_p

Values of E_p also show a phase reversal with alloying compositions and have a maxima and minima respectively for 0 and 6 atomic weight of indium. Commonly, in metal, semimetal and non-metallic elements containing multicomponent chalcogenides show a sharp and continuous crystallization process (exception is also reported in few composition of chalcogenide glasses) with lower E_p values. The sharp crystallization prevail between T_c(where crystallization began) and T_p (where crystallization completed) owing to continuous braking of rigid heteropolar bonds cause generation of greater amount of heat energy in the specimen.

1.9 Melting temperature (T_m)

Melting temperature of amorphous glassy materials defines as; temperature at which solid state materials destroy all the existing homopolar and heteropolar bonds and alloying elements separated. Melting temperatures of amorphous glassy materials extensively depend on the constituent of the alloys. Technologically kinetics at T_m have less impotence, therefore investigators interest to provide only introductory information regarding to phase transformation at T_m in amorphous glassy materials.

2. Discussions

Crystallizations kinetics variations in under test metal, semi-metal and non-metal containing multicomponent chalcogenide glasses can be interpreted in term of bond formation in solids. It is expected to Zn and Te dissolved in Se chains and makes Zn-Zn, Te-Te, Se-Se, Se-Zn, Se-Te, Se-Zn-Te homopolar and heteropolar bonds. Essentially ternary Se-Zn-Te glass can forms cross-link heteropolar metastable state structure. The heteropolar bonds will be produced the defects in density of localized state owing to existence of dangling bonds in alloy configuration (Maharjan et al., 2000, Saffarini, 2002, Abdel Latif, 1998). Further incorporation of foreign element Te concentration in ternary configuration transforms the whole stoichiometry into quaternary or multicomponent system. The metal, semi-metal and non-metal multicomponent glassy configuration possibly makes them dominating Se-In heteropolar bonds with other metallic character bonds Zn-In, Te-In. The Se-In heteropolar bonds play an important role in crystallization kinetics variations due to fixed amounts of Zn and Te. Addition of additional indium concentration has produced the heavily cross-linked structure in which steric hindrance increases. Therefore the expanse of Se chains and replacement of weak Se-Se bonds by Se-In bonds results the increase and decrease in associative activations energies. A chemical threshold has established at critical composition (6 at wt % of In). At this concentration glassy structure become more chemically ordered and contains large number of Se-In bonds (Singh, 2011). As consequence a significant change is appeared in crystallization parameters of threshold composition alloy.

Furthermore, incorporation of indium concentration beyond the threshold composition reduced the Se-In bonds and increases the In-In bond strength in glassy configuration. The increase and decrease bonds strengths of Se-In and In-In influenced the defects / dangling bonds concentrations in the glassy stoichiometry. Owing to alternation in dangling bonds densities the GFA, activations energies E_g, E_c and E_p of the corresponding glass show a significant change in kinetic parameters.

3. Summary

In summary, in this work an effort is made to present the fundamentals (in short form) of nucleation and growth processes and crystallization in amorphous glassy (chalcogenide glasses) materials. Further, project a clear view on natural crystallization and process of non–isothermal crystallization kinetics of chalcogenide glasses. Subsequently a concrete explanation on origin of endothermic and exothermic peaks in DSC measurements is also discussed. Glass forming ability and crystallizations kinetics of recent developed $Se_{93-x}Zn_2Te_5In_x$ ($0 \leq x \leq 10$) metal, semi-metal and non-metal containing multicomponent chalcogenide glasses have also been taken under discussion. Outcomes revealed such combinations glasses have high order GFA and thermal stability and low E_c, E_p activations energies as compare to previous reported non-metallic compositions. In subsequent it has found these materials critical kinetic parameters extensively depend on their alloys constituents and have a maxima and a minima in respective manor for threshold composition glass. It is also concluded that the crystallizations kinetics variations in these glasses can be occur owing to fluctuations in solid state bonds densities in localized states. In case of metallic multicomponent chalcogenide glasses heteropolar unsaturated and covalent bonds may play an important role in crystallization kinetics variations.

Indeed, chalcogenide glasses are the potential materials which used in various optoelectronics applications, but still have plenty of room for their applications in different areas which have either less studied or undiscovered. A few thrust areas of these materials are outlined here in which they have (or may have) potential applications. Surface plasmon resonance (SPR) is a very versatile and accurate technique to determining small changes in optoelectronics parameters like, refractive index at the interface of a metal layer and the adjacent dielectric medium. The SPR detection mechanism has secured a very important place among several sensing techniques due to its better performance and reliable procedure. Chalcogenide materials can also perform as infrared sensing for an efficient, non-destructive and highly selective technique in detection of organic and biological species. This technique has combined benefits of ATR spectroscopy with the flexibility of using a fiber as the transmission line of the optical signal, which allows for remote analysis during field measurements or in clinical environments. Sensing mechanism based on absorption of the evanescent electric field, which propagates outside the surface of the fiber and interacts with any absorbing species at the fiber interface. However, their efficiency controlled multilayer optical filters by periodically switching and evaporation angle, leading to periodic dielectric structure makes them a potential candidate for chemical sensing application. Further, their larger refractive index amongst the glasses makes them to made chalcogenide based ultra-low loss waveguides devices. Chalcogenide materials have most promising applications in area of phase change memory (PCM) and in photonics. Their unique physical characteristic the reversible amorphous to crystalline phase change which can be induced by controlled thermal cycling (through laser absorption or current flow) in certain chalcogenide alloys. Phase-change materials have always technological importance to make read–write storage device (commercially rewritable CD/DVD), because they can be switched (in nanoseconds) rapidly back and forth between amorphous and crystalline phases by applying appropriate laser heat pulses. Although optical phase-change storage is a widespread and successful technology, further advances in areal densities will be very challenging. Moreover, chalcogenide (in glassy or nano embedded) based photovoltaic cells applications are also identified as a prominent alternative of conventional energy which can provides terawatts capacity at cheaper cost. But improvement in low efficiency chalcogenide based photovoltaics is challenging in future. Thus, in view of author glassy and nano embedded (glassy) chalcogenides would be matter of future research to undersatnd the molecular /- or nano-phase photonics of the materials, particularly for thrust areas in PCM memory and photovoltaics applications. In author opinion chalcogenide glasses (bulk or nano phase) has a bright future and it is still open for further inventions.

4. Acknowledgment

Author thankful to Dr.Kedar Singh and senior faculty members of Department of Physics, Banaras Hindu University, Varanasi-221005, India, for their kind support to carry out this work.

5. References

Abdel Latif, R. M. (1998). DC Electrical measurements on evaporated thin films of vanadium pentoxide. *Physica B,*Vol. 254,pp. 273-276, ISSN 0921-4526

Agarwal, P., Goel, S., Rai, J. S. P., Kumar, A. (1991). Calorimetric studies in glassy $Se_{80-x}Te_{20}In_x$, *Phys Stat Sol (a),* Vol. 127, pp.363-369,ISSN 1862-6319

Aggarwal, I. D. & Sanghera, J. S. (2002). Development and applications of chalcogenide glass optical Fibers at NRL. *Journal of Optoelectronics and Advanced Materials,* Vol. 4, pp.665 - 678, ISSN 2066-0049

Anne, M.L., Keirsse, J., Nazabal, V., Hyodo, K., Inoue, S., Pledel, C.B., Lhermite, H., Charrier, J., Yanakata, K., Loreal, O., Person, J. L., Colas, F., Compère, C. & Bureau, B. (2009). Chalcogenide Glass Optical Waveguides for Infrared. Biosensing. *Sensors,*Vol. 9, pp.7398-7411,ISSN1424-8220

Anne, M.L., Keirsse, J., Nazabal, V., Hyodo, K., Inoue, S., Pledel, C. B., Lhermite, H., Charrier, J., Yanakata, K., Loreal, O., Person, J. L., Colas, F., Compère, C.& Bureau, B. (2009). Chalcogenide Glass Optical Waveguides for Infrared Biosensing. *Sensors,* Vol. 9, pp.7398-7411, ISSN 1424-8220

Aravinda Narayanan, R., Asokan, S. & Kumar, A. (2001). Influence of chemical disorder on electrical switching in chalcogenide glasses, *Physical Review B,* Vol.63, pp.092203-1-092203-4,ISSN 0163-1829

Augis, J. A.& Bennett, J. E. (1978). Calculation of the Avrami parameters for heterogeneous solid state reactions using a modification of the Kissinger method. *Journal of Thermal Analysis and Calorimetry,*Vol. 13, No.2, pp. 283-292, ISSN 1572-8943

Avrami, M . (1939). Kinetics of Phase Change. I General Theory. *Journal of Chemical Physics,* Vol. 7,pp. 1103-1112,ISSN.1089-769

Avrami, M. (1940). Kinetics of Phase Change. II Transformation-Time Relations for Random Distribution of Nuclei, *Journal of Chemical Physics,* Vol. 8,pp. 212-224, ISSN.1089-769

Bicerono, J. & Ovshinsky, S.R. (1985). Chemical Bond Approach to the Structure of chalcogenide glasses with Reversible Switching Properties, *Journal of Non-Crystalline Solids,* Vol.74, pp. 75-84,ISSN 0022-3093

Bichara, C. & pellegatti, A. (1993). Chain structure of amorphous selenium by tight-binding Mont Carlo simulation. *Physical Review B*, Vol.49, pp.6581-6586, ISSN 1943-2879

Bohmer, R. & Angell, C.A. (1993). Elastic and viscoelastic properties of amorphous and identification of phase transition between ring and chain structures. *Physical Review B*, Vol.48, pp.5857-5864,ISSN 1943-2879

Boo, B. H., Chu, H. & Kang, D. E. (2007). Ab initio and DFT investigation of structures and energies of low-lying isomers of Zn_xSe_x (x = 1–4) clusters. *Journal of Molecular Structure: THEOCHEM*, 806, 77-83, ISSN: 0166-1280

Boolchand, P., Georgiev, D. G. & Goodmana, B. (2001). Discovery of The Intermediate Phase in Chalcogenide Glasses, *Journal of Optoelectronics and Advanced Materials*, Vol.3, pp.703-720,ISSN 1454-4164

Bosch,V., Sánchez, A. J. R., Rojas, F. S. & Ojeda, C. B. (2007). Recent Development in Optical Fiber Biosensors. *Sensors*, Vol. 7, pp.797-859, ISSN 1424-8220

Bowden, B. F.& Harrington, J. A. (2009). Fabrication and characterization of chalcogenide glass for hollow Bragg fibers. *Applied Optics*,Vol. 48, pp. 3050-3054, ISSN 2155-3165

Boycheva, S. & Vassilev, V. (2002). Electrode-limited conductivity of amorphous chalcogenide thin films from the $GeSe_2$-Sb_2Se_3-ZnSe system , *Journal of Optoelectronics and Advanced Materials*, Vol. 4, pp. 33-40, ISSN 1454 - 4164

Caprion, D. & Schober, H. R. (2000). Structure and relaxation in liquid and amorphous selenium. *Physical Review B*, Vol.62, pp.3709-3716, ISSN 1943-2879

Caprion, D. & Schober, H. R. (2002). Influence of the quench rate and the pressure on the glass transition temperature in selenium. *Physical Review B*, Vol.117, pp.2814-2818, ISSN 1943-2879

Carthy, T.M. & Kanatzidis, M.G. (1996). Synthesis in molten alkali metal polythiophosphate fluxes. The new quaternary bismuth and antimony thiophosphates $ABiP_2S_7$ (A = K, Rb), $A_3M(PS_4)_2$ (A = K, Rb, Cs; M = Sb, Bi), $Cs_3Bi_2(PS_4)_3$, and $Na_{0.16}$ $Bi_{1.28}P_2S_6$. *Journal of Alloys and Compounds*, Vol.236, pp.70-85, ISSN 0925- 8388

Ciureanu, P. & Middelhoek, S. (1992). Thin Film Resistive Sensors London, *IOP Publishing*

Cohen, M. H. & Lucovsky, G. (1971). Amorphous and Liquid Semiconductors, North Holland Publishers, Amsterdam.

Dahshan, A., Aly, K.A. & Dessouky, M.T. (2008). Thermal stability and activation energy of some compositions of Ge-Te-Cu chalcogenide system. *Philosophical Magazine*, Vol. 88, pp.2399–2410, ISSN ISSN 1478-6443

Demarco, P. & Pejcic, B. (1999). Electrochemical Impedance Spectroscopy and X-ray Photoelectron Spectroscopy Study of the Response Mechanism of the Chalcogenide Glass Membrane Iron(III) Ion-Selective Electrode in Saline Media, *Analytical Chemistry*, Vol. 72, pp. 669-679, ISSN 0003-2700

Dietzel, A. (1968). Glass Structure and Glass Properties, I, *Glass Science and Technology*, Vol.22,pp. 41-50, ISSN 0017-1085.

Doremus, W. (1969). Semiconductor Effects in Amorphous Solids, North Holland Publishers, Amsterdam

Dudley, J. M. & Taylor, J. R. (2009). Ten years of nonlinear optics in photonic crystal fibre.*Nature Photonics*, Vol. 3, pp. 85-90, ISSN 1749-4885

Echeveria, I., Kolek, P. L., Plazek, D. J. & Simon, S. L. (2003). Enthalpy recovery, creep and creep-recovery measurements during physical aging of amorphous selenium. *Journal of Non-Crystalline Solids*, Vol.324, pp.242-255, ISSN 0022-3093

Efimov, A. M. (1999). Vibrational spectra, related properties, and structure of inorganic glasses. *Journal of Non-Crystalline Solids*. Vol. 253,pp. 95-118, ISSN 0022-3093

Eggleton, B. J., Davies, B. L. & Richardson, K. (2011). Chalcogenide photonics, *Nature Photonics* ,Vol. 5,pp.141-148, ISSN

Eggleton, B.J., Davies, B. L. & Richardson, K. (2011). Chalcogenide photonics, Nature Photonics, Vol. 5, pp.141-148, ISSN 1749-4885

Elliott, G. R., Senthil Murugan, G., Wilkinson, J. S., Zervas, M. N.& Hewak, D. W. (2010) Chalcogenide glass microsphere laser. *Optical Express*,Vol. 18,pp. 26720-26727,ISSN 1094-4087

Emin, D. (2006). Current-driven threshold switching of a small polaron semiconductor to a metastable conductor. *Physical Review B* 74, 035206-10, ISSN 1943-2879

Reid, R. C., Botsaris, G. D., Margolis, G., Kirwan, D.J., Denk, E. G., Ersan, G.S., Tester, J. & Wong, F. (1970). Crystallization—Part I - Transport Phenomena of Nucleation and Crystal Growth. *Industrial & Engineering Chemistry*, Vol. *62*, pp 52–67, ISSN 0888-5885

Fayek, S. A., El Sayed, S. M., Mehana, A. & Hamza, A. M. (2001). Glass formation, optical properties and local atomic arrangement of chalcogenide systems GeTe-Cu and GeTe-In, *Journal of Materials Science*, Vol. 36, pp.2061-2066,ISSN 0022-2461

Flaschen, S.S., Pearson, A.D., Northover, W.R. (1959). Low-Melting Inorganic Glasses with High Melt Fluidities Below 400°C. *Journal of the American Ceramic Society*, Vol.42, pp.450, ISSN 1551-2916

Fraser, W. A. & Jerger, J. (1953). New Optical Glasses with Good Transparency in the Infrared. *Journal of Optical Society of America*, Vol. 43, pp.332A, ISSN 1084-7529

Frerichs, R. (1950). "New optical glasses transparent in the Infra-red up to 12 ,u". *Physical Review*, Vol.78, pp.643- 643, ISSN 1050-2947

Frerichs, R. (1953). New optical glasses with good transparency in the infrared. *Optical Society ofAmerica*, Vol.43, pp.1153- 1157, ISSN 1084-7529

Fritzsche, H. (1971). Optical and electrical energy gaps in amorphous semiconductors. *Journal of Non-Crystalline* Solids, Vol.6, pp.49-71,ISSN 0022-3093

Gill, W.D. & Street, G.B. (1973). Drift mobility in amorphous selenium-sulfur alloys. *Journal of Non-Crystalline Solids*, Vol.13, pp.120,ISSN ISSN 0022-3093

Goetzberger, A. & Hebling, C. (2000). Photovoltaic materials, past, present, future. *Solar Energy Materials & Solar Cells*, Vol. 62, pp.1-19, ISSN 0927-0248

Guo, H., Zhai, Y., Haizheng, T., Yueqiu, G. & Zhao, X. (2007). Synthesis and properties of GeS_2–Ga_2S_3–PbI_2 chalcohalide glasses, *Materials Research Bulletin*, Vol. 42, pp.1111-1118, ISSN0025-5408.

Hafner, H. C. (1972) Final Technical Report, AFML-TR-72-54

Hamada, S., Sato, T. & Shirai, T. (1968). Glass Transition Temperature and Isothermal Volume Change of Selinium. *Bulletin of the Chemical Society of Japan*, Vol.41, pp.135-139, 1348-0634

Hawes, L. (1963). Sulphur-Selenium and Sulphur-Tellurium Cyclic Interchalcogen Compounds.*Nature*, Vol. 198, pp.1267-1270, ISSN0028-0836

Hegab, N. A. & Atyia, H. E. (2007). Dielectric Studies Of Amorphous $As_{45}Te_{33}Ge_{10}$ Si_{12} Films, *Journal of Ovonic Research*, Vol.3, pp.93-102, ISSN 1584 - 9953

Hilton, A. R. & Brau, M. (1963). New high temperature infrared transmitting glasses. *Infrared Physics*, Vol 3, pp 69-72, ISSN 1350-4495

Hilton, A. R. (1966). Nonoxide Chalcogenide Glasses as Infrared Optical Materials. Applied Optics, Vol. 5, pp. 1877-1882, ISSN 2155-3165

Hilton, A. R. (2010). Chalcogenide Glasses for Infrared Optics, Chapter-2,*The McGraw-Hill companies*, pp.17-59, ISBN 978-0-07-159698-5

Hilton, A. R. (June 1974). Defense Advanced Research Projects Agency (DARPA), Contract No. N00014-73-C-0367.

Hruby, A. (1972). Evaluation of glass-forming tendency by means of DTA. *Czechoslovak Journal of Physics*, Vol.22, No.11, pp. 1187-1193, ISSN1572-9486

Hsu, J. W.P. (2001). Near-field scanning optical microscopy studies of electronic and photonic materials and devices. *Materials Science and Engineering. R:Reports,* Vol. 33, pp.1-50, ISSN 0927-796X

Ielmini, D. (2008). Threshold switching mechanism by high-field energy gain in the hopping transport of chalcogenide glasses. *Physical Review B,* Vol.78, pp.035308 -8, ISSN 0163-1829

Imran, M. M. A., Bhandari, D.& Saxena, N. S. (2001). Enthalpy recovery during structural relaxation of $Se_{96}In_4$ chalcogenide glass. *Physica B,*Vol.294,pp.394-401, ISSN 0921-4526

Ionov, R., Nesheva, D. & Arsova, D. (1991). Electrical and electrophotographic properties of CdSe/SeTe and CdSe/Se multilayers, *Journal of Non-Crystalline Solids.* Vol. 137,pp. 1151-1154, ISSN 0022-3093

Ivan, I.& Kikineshi, A. (2002).Stimulated interdiffusion and expansion in amorphous chalcogenide multilayers. *Journal of Optoelectronics and Advanced Materials,*Vol.4, pp.743 – 746, ISSN 2066-0049

Ivanova, Z.G., Cernoskova, E., Vassilev, V.S. & Boycheva, S.V. (2003). Thermomechanical and structural characterization of $GeSe_2$–Sb_2Se_3–ZnSe glasses, *Materials Letters,* Vol. 57, pp. 1025-1028, ISSN 0167-577X

JablÇonska, A., Burian, A., Burian, A.M., Lecante, P. & Mosset, A., (2001). Modelling studies of amorphous In–Se films. *Journal of Alloys and Compounds*, Vol.328, pp.214-217, ISSN: 0925-8388

Jackson, K. & Srinivas, S. (2002). Modeling the ^{119}Sn Mössbauer spectra of chalcogenide glasses using density-functional theory calculations. *Physical Review B,* Vol. 65, pp.214201-8, ISSN 1943-2879

Jain, P. K., *Deepika & Saxena, N. S.,* (2009).Glass transition, thermal stability and glass-forming ability of $Se_{90}In_{10-x}Sb_x$ (x = 0, 2, 4, 6, 8, 10) chalcogenide glasses, *Philosophical Magazine,*Vol. 89,pp. 641-650, ISSN 1478-6443

Johnson, W. A. & Mehl, R. F. (1939). Reaction kinetics in processes of nucleation and growth. *Transactions of the American Institute of Mining, Metallurgical, Engineers* ,Vol.135,pp. 416-458, ISSN 0096-4778

Jones, C. & Hafner, H. (1968) Final Technical Report, Contract No. AF 33 (615)-3963

Joseph, C . (2010). Crystal Growth: Theory and Techniques. Chapter-1 (*shodhganga.inflibnet.ac.in/bitstream/10603/265/8/08_chapter1.pdf*)

Kawamura,T.,Yamamoto, N., & Nakayama, Y. (1983). Electrophtographic Application of Amorphous Semiconductors, Amorphous Semiconductor Technologies and Devices, *JARECT Hamakawa, Y. North Holland OHMSHA*,Vol.6, pp.325-336

Khan, M.A.M., Khan, M.W., Husain, M. & Zulfequar, M. (2009). Electrical transport and optical properties of Zn doped Bi–Se chalcogenide glasses. *Journal of Alloys and Compounds.* Vol.486, pp.876– 880, ISSN 0925- 8388

Kissinger, H. E. (1957).Reaction Kinetics in Differential Thermal Analysis. *Analytical Chemistry*, Vol. 29, pp.1702-1706, ISSN1520-6882

Klokishner, S.I., Kulikova, O.V., Kulyuk, L.L., Nateprov, A.A., Nateprov, A.N., Ostrovsky, S.M., Palii, A.V., Reu, O.S. & Siminel, A.V. (2008). Concentration effects in the photoluminescence spectra of $ZnAl_{2(1-x)}Cr_{2x}S_4$. *Optical Materials,* Vol. 31, pp.284-290, ISSN 0925-3467

Kobelke, J. , Kirshhof, J., Scheffer, S. & Schuwuchow, A. (1999). Chalcogenide glass single mode fibres - preparation and properties, *Journal of Non-Crystalline Solids*, Vol.256-257, pp. 226-231, ISSN 0022-3093

Kohary, K., Burlakov, V. M., Pettifor, D. G. & Manh, D. N. (2005). Modeling In-Se amorphous alloys. *Physical Review B*, Vol.71, pp.184203-1- 184203-7, ISSN 0022-3093

Kokenyesi, S. (2006). Amorphous chalcogenide nano-multilayers: research and development. *Journal of Optoelectronics and Advanced Materials,* Vol. 8, pp.2093 - 2096, ISSN 2066-0049

Kokenyesi, S., Takats, V., Ivan, I., Csik, A., Szabo, I., Beke, D., Nemec, P., Sangunni, K., & Shiplyak, M. (2007). Amorphous chalcogenide nano-multilayers: research and development. *Acta Physica Debreceniensis*, Vol. XLI, pp. 51-57, ISSN 1789-6088

Kokorina, V. (1996). Glasses for Infrared Optics, *CRC Press, Boca Raton, Florida,* ISBN: 0-8493-3785-2

Kumar, S., Singh, K. & Mehta, N. (2010). Calorimetric studies of crystallisation kinetics of $Se_{75}Te_{15-x}Cd_{10}In_x$ multi-component chalcogenide glasses using non-isothermal DSC. *Philosophical Magazine Letters,* Vol.90, pp.547–557, ISSN 1362-3036

Lenz, G., Zimmermann, J.K. & Katsufuji, T. (2000). Large kerr effect in bulk Se-based chalcogenide glasses, *Optical Letter*,Vol. 25, pp. 254 -256,ISSN 1539-4794

Lezal, D., Pedlikov, J. & Zavadil, J. (2004). Chalcogenide glasses for optical and photonics applications. *Chalcogenide Letteres,* Vol. 1, pp.11 - 15, ISSN ISSN 1584-8663

Lopez, C., (2004).Evaluation of the photo-induced structural Mechanisms in chalcogenide glass materials, Thesis (Ph.D.) *University of Central Florida,* pp.1-213, Publication Number: AAI3163612; ISBN: 9780496978441

Lousteau, J., Furniss, D., Arrand, H.F., Benson, T.M., Sewell, P.& Seddon, A.B. (2008). Fabrication of heavy metal fluoride glass, optical planar waveguides by hot-spin casting. *Journal of Non-Crystalline Solids,* Vol. 354, pp.3877–3886, ISSN 0022-3093

Maharjan, N.B., Bhandari, D., Saxena, N.S., Paudyal, D.D.& Husain, M. (2000). Kinetic Studies of Bulk $Se_{85-x}Te_{15}Sb_x$ Glasses with x = 0, 2, 4, 6, 8 and 10 . *physica status solidi (a),* Vol.178, pp.663-670,ISSN 1862-6319

Malek, J., Pustkova, P. & Shanilova, J. (2003). Kinetic Phenomena In Non-Crystalline Materials Studied By Thermal Analysis, *Journal of Thermal Analysis and Calorimetery,*Vol.72, pp.289-297, ISSN 1388-6150

Malek, J., Svoboda, R., Pustkova, P. & Cicmanec, P. (2009). Volume and enthalpy relaxation of a-Se in the glass transition region. *Journal of Non-Crystalline Solids,* Vol.355,pp.[illegible], ISSN 0022-3093

Matusita, K., Komatsu, T.& Yokota, R. (1984).Kinetics of non-isothermal crystallization process and activation energy for crystal growth in amorphous materials. *Journal Of Materials Science,* Vol.19, No.1, pp. 291-296, ISSN 1573-4803

Mehta, N. & Kumar, A. (2007). Comparative Analysis Of Calorimetric Studies In $Se_{90}M_{10}$ (M=In, Te, Sb) Chalcogenide Glasses, *Journal of Thermal Analysis and Calorimetery,* Vol.87, pp.345-150,ISSN 1388-6150

Mehta, N., Singh, K. & Kumar, S. (2009).Effect of Sb and Sn additives on the activation energies of glass transition and crystallization in binary $Se_{85}Te_{15}$ alloy. *Phase Transitions,* Vol. 82, pp.43–51, ISSN 0141-1594

Mehta, N., Tiwari, R.S. & Kumar, A. (2006). Glass forming ability and thermal stability of some Se–Sb glassy alloys, *Materials Research Bulletin,* Vol. 41,pp. 1664-1672,ESSN0025-5408

Mersmann, A. (2001). *Crystallization Technology Handbook,* CRC; 2nd ed. ISBN 0-8247-0528-9

Messaddeq, S. H., Tikhomirov, V. K., Messaddeq, Y., Lezal, D.& Siu Li, M. (2001). Light-induced relief gratings and a mechanism of metastable light-induced expansion in chalcogenide glasses. *Physical Review B,* Vol.63, pp.224203-5, ISSN 1943-2879

Micoulaut, M. & Phillips, J. C. (2003). Rings and rigidity transitions in network glasses, *Physical Review B,* Vol. 67, pp.104204 -9, ISSN 1943-2879

Milliron, D. J., Raoux, S., Shelby, R. M. & Sweet, J. J. (2007). Solution-phase deposition and nanopatterning of GeSbSe phase-change materials.*Nature Materials* Vol.6, pp.352, ISSN 1476-4660

Mortensen, J., Legin, A., Ipatov, A., Rudnitskaya, A., Vlasov, Y. & Hjuler, K. (2000). A flow injection system based on chalcogenide glass sensors for the determination of heavy metals. *Analytica Chimica Acta,* Vol. 403, pp.273–277, ISSN 0003-2670

Mott, N.F. & Davis, E.A. (1979). Electronic Processes in the Non-Crystalline Materials, *Oxford University Press, Second Edition*

Mourizina, Y., Yoshinobu, T., Schubert, J., Luth, H., Iwasaki, H.& Schoning, M. J. (2001). Ion-selective light-addressable potentiometric sensor (LAPS) with chalcogenide thin film prepared by pulsed laser deposition. *Sensor and Actuators B,* Vol. 80, pp.136-140, ISSN 0925-4005

Moynihan, C.T. , Easteal, A.J., Wilder, J., Tucker, J. (1974). Dependence of the glass transition temperature on heating and cooling rates. *Journal of Physical Chemistry,* Vol.78,No. 26,pp.2673-2677, ISSN 1520-5207

Narayanan, R. A., Asokan, S. & Kumar, A. (2001). Influence of chemical disorder on electrical switching in chalcogenide glasses. *Physical Review B,* Vol. 63,pp. 092203 -4, ISSN 1943-2879

Natale, C. D., Davide, F., Brunink, J. A.J., Amico, A. D., Vlasov, Y. G., Legin, A. V. & Rudnitskaya, A. M. (1996). Multicomponent analysis of heavy metal cations and inorganic anions in liquids by a non-selective chalcogenide glass sensor array. *Sensors and Actuators B* Vol.34, pp.539-542, ISSN 0925-4005

Naumis, G. G. (2000). Contribution of floppy modes to the heat capacity jump and fragility in chalcogenide glasses. *Physical Review B-Rapid Communications,* Vol. 61, pp.9205-9208, ISSN 1089-4896

Neoheva, D., Arsova, D.& Valeva, E. (1997). Electrophotographic photoreceptors including selenium-based multilayers. *Semiconductor Science and Technology,* Vol.12, pp.595-599, ISSN 1361-6641

Nielsen, S. (1962). Note on the preparation and properties of glasses containing germanium disulphide. *Infrared Physics,* Vol. 2, pp.117-119,ISSN 1350-4495

Othman, A.A., Aly, K.A. & Abousehly, A.M. (2006). Crystallization kinetics in new $Sb_{14}As_{29}Se_{52}Te_5$ amorphous glass, *Solid State Communications,*Vol.138, pp.184-189,ISSN 0038- 1098

Ovshinsky, S.R.(1994). An History of Phase Change Technology, *Memories Optics Systems,*Vol. 127, pp. 65, ISSN

Ozawa, T. (1970). Kinetic analysis of derivative curves in thermal analysis. *Journal of Thermal Analysis and Calorimetry,*Vol.2, No.3, pp. 301-324, ISSN 1572-8943

Part 1, Hilton, A. R., Jones, C .E. & Brau, M., Part 2, Hilton, A. R. & Jones, C. E., Part 3, Hilton, A. R., Jones, C. E., Dobrott, R. D., Klein, H. M., Bryant, A. M. & George, T. D. (1966). Non-oxide I V A-V A-VIA chalcogenide glasses. *Physics and Chemistry of Glasses,* Vol. 7, pp.105-126, ISSN 0031-9090

Patial, B. S., Thakur, N. & Tripathi, S.K. (2011). On the crystallization kinetics of In additive Se–Te chalcogenide glasses, *Thermochimica Acta,*Vol. 513, pp.1-8, ISSN 0040-6031

Patterson, R. J. (1966)15th National Infrared Information Symposium (IRIS) at Ft Monmouth N.J., (1967) U.S. Patent 3,360,649.

Pearson, A. D. (1962). Electrochemical Society Meeting, *Los Angeles, California-121st Meeting*

Pelusi, M., Luan, F., Vo, T. D., Lamont, M. R. E., Madden, S. J., Bulla, D. A., Choi, D. Y., Davies, B. L.& Eggleton, B. J. (2009). Photonic-chip-based radio-frequency spectrum analyser with terahertz bandwidth. *Nature Photonics,* Vol. 3, pp.139-143, ISSN 1749-4885

Pena, E.Y., Mejia, M., Reyes, J.A., Valladares, R.M., Alvarez, F. & Valladares, A.A. (2004). Amorphous alloys of $C_{0.5}Si_{0.5}$, $Si_{0.5}Ge_{0.5}$ and $In_{0.5}Se_{0.5}$:atomic topology, *Journal of Non-Crystalline Solids,* Vol.338, pp.258-261, ISSN 0022-3093

Peng, H. & Liu, Z. (2010). Organic charge-transfer complexes for STM-based thermochemical-hole-burning memory. *Coordination Chemistry Review.* Vol.254, pp.1151–1168, ISSN 0010-8545

Phillips, J. C. (2006). Ideally glassy hydrogen-bonded networks. *Physical Review B,* Vol. 73, pp.024210 -10, ISSN 1943-2879

Prashanth, S.B. B. & Asokan, S. (2008). Composition dependent electrical switching in $Ge_xSe_{35-x}Te_{65}$ ($18 \leq x \leq 25$) glasses – the influence of network rigidity and thermal properties, *Solid State Communications,* Vol. 147, pp.452-456,ISSN 0038- 1098

Pungor, E. (1997) . Ion-selective electrodes – istory and conclusions, *Journal of Analytical Chemistry,* Vol. 357, pp. 184-188, ISSN 1061-9348

Quiroga, I., Corredor, C., Bellido, F., Vazquez, J., Villares, P. & Garay, R. J. (1996). Infrared studies of Ge-Sb-Se glassy semiconductor, *Journal of Non- Crystalline Solids* Vol.196, pp.183-186,ISSN0022-3093

Rau, C., Armand, P., Pradel, A., Varsamis, C. P. E., Kamitsos, E. I., Granier, D., Ibanez, A. & Philippot, E. (2001). Mixed cation effect in chalcogenide glasses Rb_2S-Ag_2S-GeS_2. *Physical Review B*, Vol. 63, pp.184204-9, ISSN 1943-2879

Robert, E.J., Kasap, S.O., Rowlands, J. & Polischuk, B. (1998). Metallic electric contacts to stabilized amorphous selenium for use in X-ray image detectors, *Journal of Non-Crystalline Solids*,Vol. 2,pp. 227-240, 1359-1362, ISSN 0022-3093

Saad, M.& Poulin, M. (1987). Glass Forming Ability Criterion.*Materials Science Forum*,Vol. 19–20,pp. 11-18,ISSN 1662-9752.

Saffarini, G. (2002) The effect of compositional variations on the glass-transition and crystallisation temperatures in Ge-Se-In glasses. *Applied Physics A*,Vol.74, No.2, 283-285,ISSN 1432-0630

Saleh, Z.M., Williams, G.A.& Taylor, P.C. (1993). Nuclear-magnetic-resonance relaxation in glassy Cu-As-Se and Cu-As-S. *Physical Review B*, Vol. 47, pp.4990-5001, ISSN 1943-2879

Salmon, P. S. & Xin, S. (2002). The effect of covalent versus ionic bonding in chalcohalide glasses:$(CuI)_{0.6}(Sb_2Se_3)_{0.4}$. *Physical Review B*, Vol. 65, pp.064202-4, ISSN 1943-2879

Samson, Z. L., Yen, S. C., MacDonald, K. F., Knight, K., Li , S., Hewak, D. W., Tsai, D.P. & Zheludev, N. I. (2010). Chalcogenide glasses in active plasmonics. *Physica Status Solidi RRL*, Vol.4,pp. 274–276, ISSN 1862-6270

Sanchez, E. A., Waldmann, M.& Arnold, C. B. (2011). Chalcogenide glass microlenses by inkjet printing," *Applied Optics*, Vol. 50,pp. 1974-1978 ISSN 2155-3165

Sanchez-Jimenez, P.E. et al., (2009).Combined kinetic analysis of thermal degradation of polymeric materials under any thermal pathway. *Polymer Degradation and Stability*, Vol. 94, pp. 2079-2085,ISSN 0141-3910

Sargent, E. H. (2009). Infrared photovoltaics made by solution processing.*Nature Photonics*,Vol. 3,pp. 325-331, ISSN 1749-4885

Sarrach, J., de Neufville, J.P. & Haworth, W.L. (1976).Studies of amorphous Ge□Se□Te alloys (I): Preparation and calorimetric observations. *Journal of Non-Crystalline Solids*, Vol. 22, PP.245, ISSN ISSN 0022-3093

Savage, J. A. & Nielsen, S. (1964). Preparation of glasses transmitting in the infra-red between 8 and 15 microns. *Physics and Chemistry of Glasses*, Vol.5, pp.82-86, ISSN 0031-9090

Savage, J. A.& Nielsen, S. (1966). The infra-red transmission of telluride glasses. *Physics and Chemistry of Glasses*, Vol.7,pp. 56-59, ISSN 0031-9090

Sbirrazzuoli, N. (1999). Isothermal and Non-Isothermal Kinetics When Mechanistic Information Available.*Journal of Thermal Analysis and Calorimetry*,Vol.56, No.2, pp. 783-792, ISSN 1572-8943

Schubert, J., Schoning, M. J., Mourzina, Y.G., Legin, A.V., Vlasov, Y.G., Zander, W.& Luth, H. (2001). Multicomponent thin films for electrochemical sensor applications prepared by pulsed laser deposition. *Sensor and Actuators B*,Vol. 76, pp.327-330, ISSN 0925-4005

Sellack, C. S. (1870). An early, but only qualitative, observation of good transmission of A_2S_3 in the infrared. *Ann. Physik,* Vol. 215, pp. 182–187, ISSN 1521-3889.

Selvaraju, V.C., Asokan, S. & Srinivasan, V. (2003). Electrical switching studies on $As_{40}Te_{60-x}Se_x$ and $As_{35}Te_{65-x}Se_x$ glasse. *Applied Physics A,* Vol. 77, No.1, pp. 149-153, ISSN 1432-0630

Shportko, K., Kremers, S., Woda, M., Lencer, D., Robertson, J. & Wuttig, M. (2008). Resonant bonding in crystalline phase-change materials.*Nature Materials,*Vol. 7,pp. 653-658, ISSN 1476-4660

Singh, A. K. & Singh, K.(2010). Observation of Meyer Neldel rule and crystallization rate constant stability for $Se_{93-x}Zn_2Te_5In_x$ chalcogenide glasses. *The European Physical Journal - Applied Physics,*Vol.51, pp.30301 (5pp) ISSN 1286-004

Singh, A. K.& Singh, K. (2009). Crystallization kinetics and thermal stability of $Se_{98-x}Zn_2In_x$ chalcogenide glasses, *Philosophical Magazine,* Vol. 89, pp.1457-1472, ISSN 1478-6443

Singh, A.K & Singh, K. (2011). Localized structural growth and kinetics of $Se_{98-x}Zn_2In_x$ ($0\leq x\leq 10$) amorphous alloys. *Physica Scripta,*Vol.83, No.2, pp. 025605 (6pp),ISSN1402-4896

Singh, A.K, Mehta,N.&Singh, K. (2009). Correlation between Meyer–Neldel rule and phase separation in $Se_{98-x}Zn_2In_x$ chalcogenide glasses.*Current Applied Physics,* Vol.9, pp.807-811,ISSN1567-1739

Singh, A.K, Mehta,N.&Singh, K. (2010). Effect of indium additive on glass-forming ability and thermal stability of Se-Zn-Te chalcogenide glasses. *Philosophical Magazine Letters,* Vol.90,pp.201-208,ISSN 1362-3036

Singh, A.K. (2011). Effect Of Indium Additive on Heat Capacities Of Se-Zn-Te Multicomponent Chalcogenide Glasses.*Chalcogenide Letters,*Vol.8, No.2,pp.123-128, ISSN 1584-8663

Singh, A.K. (2011). Effect of indium additive on the heat capacity of Se-Zn chalcogenide glasses, *The European Physical Journal Applied Physics,* Vol. 55, pp. 11103-1-11103-4, ISSN 1286-0042

Snopatin, G. E., Shiryaev, V. S., Plotnichenko, V. G., Dianov, E. M. & Churbanov, M. F. (2009). High-purity chalcogenide glasses for fiber optics. *Inorganic Materials,* Vol.45, pp.1439–1460, ISSN 0020-1685

Soltan, A.S., Abu EL-Oyoun, M., Abu-Sehly, A.A. & Abdel-Latief, A.Y. (2003). Thermal annealing dependence of the structural, optical and electrical properties of selenium –tellerium films, *Materials Chemistry and Physics,* Vol. 82,pp. 101-106,ISSN 0254-0584

Song, S.M., Choi, S.Y. & Yong-Keun, L. (1997). Crystallization property effects in $Ge_{30}Se_{60}Te_{10}$ glass, *Journal of Non-Crystalline Solids,* Vol.217,pp.79-82, ISSN 0022-3093

Stocker, H. J. (1969). Bulk and thin film switching and memory effects in semiconducting chalcogenide glasses .*Applied Physics Letters,* Vol.15, pp. 55-57, ISSN1432-0630

Suri, N., Bindra, K.S., Kumar, P., Kamboj, M.S.& Thangaraj, R. (2006). Thermal Investigations In Bulk $Se_{80-x}Te_{20}Bi_x$ Chalcogenide Glass. *Journal of Ovonic Research,* Vol. 2, No. 6, pp. 111 – 118,ISSN 1584 - 9953

Swanson, S.E. (1977). Relation of nucleation and crystal-growth rate to the development of granitic textures. *American Mineralogist,* Vol.62, pp. 966-978, ISSN 0003-004X

Taeed, V. G., Baker, N. J., Fu, L., Finsterbusch, K., Lamont, M. R.E., Moss, D. J., Nguyen, H. C., Eggleton, B. J., Yong Choi, D., Madden, S. & Davies, B. L. (2007). Ultrafast all-optical chalcogenide glass photonic circuits. *Optics Express,* Vol.15, pp.9205- 9221, ISSN 1094-4087

Tanaka, K. (2003). Nanostructured chalcogenide glasses. *Journal of Non-Crystalline Solids,*Vol. [illegible],pp.[illegible], ISSN 0022-3093

Thingamajig, B., Ganesan, R., Asha Bhat, N., Sangunni, K. S. & Gopal, E. S. R. (2000). Determination of thermal diffusion length in bismuth doped chalcogenide glasses, by photoacoustic technique. *Journal of Optoelectronics and Advanced Materials,* Vol. 2, pp.91-94, ISSN 1454 - 4164

Tikhomirov, V.K., Furniss, D., Seddon, A.B., Savage, J.A., Mason, P.D.,Orchard, D.A. & Lewis, K.L. (2004). Glass formation in the Te-enriched part of the quaternary Ge–As–Se–Te system and its implication for mid-infrared optical fibres. *Infrared Physics & Technology,* Vol.45, pp.115-123, SN: 1350-4495

Tonchev, D. & Kasap, S.O. (1999). Thermal properties of Sb_xSe_{100-x} glasses studied by modulated temperature di.erential scanning calorimetry, *Journal of Non-Crystalline Solids,*Vol. 248,pp. 28-36,ISSN 0022-3093

Towers, H. and Chipman, J. (1957). Diffusion of calcium and silicon in a lime-alumina-silica slag. *Transactions of the American Institute of Mining, Metallurgical, Engineers,* Vol. 209, pp.769-773, ISSN 0096-4778

Troles, J. , Niu, Y., Arfuso, D.C. , Smektala, F. , Brilland, L., Nazabal, V., Moizan, V., Desevedavy, F. & Houizot, P. (2008).Synthesis and characterization of chalcogenide glasses from the system Ga-Ge-Sb-S and preparation of a single-mode fiber at 1.55 μm , *Materials Research Bulletin,*Vol. 43, pp. 976-982

Turek, M., Heiden, W., Riesen, A., Chhabda, T.A., Schubert, J., Zander, W., Kruger, P., Keusgen, M. & Schoning, M.J. (2009). Artificial intelligence/fuzzy logic method for analysis of combined signals from heavy metal chemical sensors. *Electrochimica Acta,* Vol. 54, pp.6082–6088, ISSN 0013-4686

Usuki, T., Uemura, O., Konno, S., Kameda, Y. & Sakurai, M. (2001). Structural and physical properties of Ag-As-Te glasses, *Journal of Non-Crystalline Solids,* Vol. 293, pp.799-805, ISSN 0022-3093

Vassilev, V. (2006). Multicomponent Cd (Zn)-containing Ge(As)-chalcogenide glasses. *Journal of the University of Chemical Technology and Metallurgy,* Vol.41, pp.257-276,ISSN 1311-7629

Vassilev, V., Parvanov, S., Vasileva, T. H., Aljihmani, L., Vachkov, V. & Evtimova, T. V. (2007). Glass formation in the As_2Te_3–As_2Se_3–SnTe system, *Materials Letters,* Vol.61,pp. 3676-3578,ISSN 0167-577X

Vassilev, V., Parvanov, S., Vasileva, T. H., Parvanova, V. & Ranova, D. (2007). Glass formation in the As-Te-Sb system, *Materials Chemistry and Physics,*Vol.105, pp.53-57, ISSN 0254-0584

Vassilev, V., Tomova, K., Parvanova, V. & Parvanov, S. (2007). New chalcogenide glasses in the $GeSe_2$–Sb_2Se_3–PbSe system. *Materials Chemistry and Physics,* Vol. 103, pp.312–317, ISSN0254-0584

Vassilev, V., Tomova, K., Parvanova, V.& Boycheva, S. (2009). Glass-formation in the $GeSe_2$–Sb_2Se_3–SnSe system. *Journal of Alloys and Compounds*,Vol. 485, pp.569-572, ISSN 0925-8388

Vassilev, V.S. & Boycheva, S.V. (2005). Chemical sensors with chalcogenide glassy membranes. *Talanta,* Vol. 67,pp. 20–27, ISSN 0039-9140

Vassilev, V.S., Hadjinikolova, S.H. & Boycheva, S.V. (2005). Zn(II)-ion-selective electrodes based on $GeSe_2$–Sb_2Se_3–ZnSe glasses , *Sensors Actuators B* , Vol. 106, pp. 401-406, ISSN 0925-4005

Velinov, T., Gateshki, M., Arsova, D. & Vateva, E. (1997). Thermal diffusivity of Ge-As-Se(S) glasses, *Physical Review B,* Vol.55, pp.11014-4, ISSN 1943-2879

Volmer, M. and Weber, A. (1925). *Z. Phys. Chem.,* Vol.119, pp.227, ISSN Tammann, G. (1925). The States of Aggregation. *Van Nostrand, New York.*

Wachter, J.B., Chrissafis, K., Petkov, V., Malliakas, C.D., Bilc, D., Kyratsia, Th., Paraskevopoulos, K.M., Mahanti, S.D., Torbrugge, T., Eckert, H. & Kanatzidis, M.G. (2007). Local structure and influence of bonding on the phase-change behavior of the chalcogenide compounds $K_{1-x}Rb_xSb_5S_8$. *Journal of Solid State Chemistry,* Vol. 180, pp.420–431, ISSN 0022-4596

Wagner, T., Kasap, S.O., Vlcek, M., Sklenar, A. & Stronski, A. (1998). The structure of $As_x S_{100-x}$ glasses studied by temperature-modulated differential scanning calorimetry and Raman spectroscopy, *Journal of Non-Crystalline Solids*,Vol. 2227,pp. 752-756,,ISSN 0022-3093

Wang, R. P., Zha, C. J., Rode, A. V. Madden, S. J. & Davies, B. L. (2007). Thermal characterization of Ge–As–Se glasses by differential scanning calorimetry, *Journal of Materials Science: Materials in Electronics,* Vol. 18, pp.419-422,ISSN 0957-4522

Warghese, G. (2010). Crystal Growth. Chapter -1 (*shodhganga.inflibnet.ac.in/bitstream/10603/.../08_chapter%201.pdf*).

Wibowo, R. A., Kim, W. S., Lee, E. S., Munir, B. & Kim, K. H. (2007). Single step preparation of quaternary $Cu_2ZnSnSe_4$ thin films by RF magnetron sputtering from binary chalcogenide targets. *Journal of Physics and Chemistry of Solids,* Vol. 68, pp.1908-1913, ISSN: 0022-3697

Wuttig, M .& Yamada, N. (2007). Phase-change materials for rewriteable data storage.*Nature Materials*,Vol. 6, pp.824-832, ISSN 1476-4660

Xu, Y., Zhang, Q., Wang, W., Zeng, H., Xu, L. & Chen, G. (2008). Large optical Kerr effect in bulk $GeSe_2$–In_2Se_3–CsI chalcohalide glasses , *Chemical Physics Letters,* Vol.462, pp. 69-71, ISSN 0009-2614

Yang, C.Y., Paesler, M.A. &Sayers, D.E. (1989).Chemical order in the glassy As_xS_{1-x} system: An x-ray-absorption spectroscopy study, *Physical Review B,* Vol.39, pp.10342-11, ISSN 1943-2879

Zhang, G., Gu, D., Jiang, X., Chen, Q. & Gan, F. (2005). Femtosecond laser pulse irradiation of Sb-rich AgInSbTe films: Scanning electron microscopy and atomic force microscopy investigations, *Applied Physics A Materials Science & Processing,* Vol. 80, pp. 1039-1043, ISSN 0947-8396

Zhang, X. H., Ma, H., & Lucas, J. (2004). Evaluation of glass fibers from the Ga-Ge-Sb-Se system for infrared applications, *Optical Materials*, Vol. 25, pp.85-89, ISSN 0925-3467

3

Numerical Models of Crystallization and Its Direction for Metal and Ceramic Materials in Technical Application

Frantisek Kavicka[1], Karel Stransky[1], Jana Dobrovska[2],
Bohumil Sekanina[1], Josef Stetina[1] and Jaromir Heger[1]
[1]Brno University of Technology,
[2]VSB TU Ostrava
Czech Republic

1. Introduction

Structure of metallic and also majority of ceramic alloys is one of the factors, which significantly influence their physical and mechanical properties. Formation of structure is strongly affected by production technology, casting and solidification of these alloys. Solidification is a critical factor in the materials industry, e.g. (Chvorinov, 1954). Solute segregation either on the macro- or micro-scale is sometimes the cause of unacceptable products due to poor mechanical properties of the resulting non-equilibrium phases. In the areas of more important solute segregation there occurs weakening of bonds between atoms and mechanical properties of material degrade. Heterogeneity of distribution of components is a function of solubility in solid and liquid phases. During solidification a solute can concentrate in inter-dendritic areas above the value of its maximum solubility in solid phase. Solute diffusion in solid phase is a limiting factor for this process, since diffusion coefficient in solid phase is lower by three up to five orders than in the melt (Smrha, 1983). When analysing solidification of these alloys so far no unified theoretical model was created, which would describe this complex heterogeneous process as a whole. During the last fifty years many approaches with more or less limiting assumptions were developed. Solidification models and simulations have been carried out for both macroscopic and microscopic scales. The most elaborate numerical models can predict micro-segregation with comparatively high precision. The main limiting factor of all existing mathematical micro-segregation models consists in lack of available thermodynamic and kinetic data, especially for systems of higher orders. There is also little experimental data to check the models (Kraft & Chang, 1997).

Many authors deal with issues related to modelling of a non-equilibrium crystallisation of alloys. However, majority of the presented works concentrates mainly on investigation of modelling of micro-segregation of binary alloys, or on segregation of elements for special cases of crystallisation - directional solidification, zonal melting, one-dimensional thermal field, etc. Moreover these models work with highly limiting assumption concerning phase diagrams (constant distribution coefficients) and development of dendritic morphology (mostly one-dimensional models of dendrites); e.g. overview works (Boettingen, 2000; Rappaz, 1989; Stefanescu, 1995). Comprehensive studies of solidification for higher order real alloys are rarer. Nevertheless, there is a strong industrial need to investigate and simulate more complex alloys because nearly all current commercial alloys have many components often exceeding ten elements. Moreover, computer simulation have shown that even minute amounts of alloying elements can significantly influence microstructure and micro-segregation and cannot be neglected (Kraft & Chang, 1997). Dendritic crystallisation is general form of crystallisation of salts, metals and alloys. At crystallisation of salts from solutions a dendritic growth of crystals occurs at high crystallisation rate, which requires high degree of over-saturation. Findings acquired at investigation of crystallisation of salts were confirmed at investigation of crystallisation of metals. If negative temperature gradient is present in the melt before the solidification front, this leads to a disintegration of the crystallisation front and to formation of dendritic crystallisation (Davies, 1973). High crystallisation rate is characteristic feature of dendritic crystallisation. Solutes have principal influence on the crystallisation character, as they are the cause of melt supercooling before the crystallisation front and formation of the negative temperature gradient. This kind of supercooling is called constitutional supercooling. For example a layer of supercooled melt is formed in a steel ingot in the immediate vicinity of the interface melt-solid, in principle at the very beginning of crystallisation as a result of segregation of solutes, which causes decrease in solidification temperature of this enriched steel. Increased concentration of solutes creates soon a broad zone of constitutionally supercooled steel, in which the crystallisation rate is high. During subsequent solidification, when the crystallisation rate is low, the value of temperature gradient is also low, which means that conditions for dendritic crystallisation are fulfilled again (Šmrha, 1983). More detailed information on dendritic crystallisation - see classical works (Chalmers, 1940), (Flemings, 1973), (Kurz &Fisher, 1986). According Chvorinov (1954), Smrha (1983) and others metallic alloys and also majority of ceramic alloys in technical application are always characterised by their dendritic crystallisation. It is therefore of utmost importance that their final desirable dendritic structure has appropriate properties that can be used in technical practice. These properties depend of the kind of practical use and they comprise flexibility, elasticity, tensile strength, hardness, but also for example toughness. In the case of ceramic materials the properties of importance are brittleness, fragility and very often also refractoriness or resistance to wear. This chapter presents numerical models of crystallisation of steel, ductile cast iron and ceramics EUCOR aimed at optimisation of their production and properties after casting.

2. Numerical models of the temperature field and heterogeneity

Crystallization and dendritic segregation of constitutive elements and admixtures in solidifying (crystallising) and cooling gravitationally cast casting or continuously cast blank (shortly concast blank) is directly dependent on character of formation of its temperature field. Especially rate and duration of the running crystallisation at any place of the blank, so called local solidification time, is important. Solidification and cooling-down of a gravitationally cast casting as well as the simultaneous heating of a metal or non-metal mould is a rather complex problem of transient heat and mass transfer. This process in a system casting- mould-ambient can be described by the Fourier's equation (1) of 3D transient conduction of heat.

$$\frac{\partial T}{\partial \tau} = \frac{\lambda}{\rho c}\left(\frac{\partial^2 T}{\partial x^2} + \frac{\partial^2 T}{\partial y^2} + \frac{\partial^2 T}{\partial z^2}\right) + \frac{Q_{SOURCE}}{\rho c} \tag{1}$$

In equation (1) are T temperature [K], τ time [s], λ heat conductivity [$W.m^{-1}.K^{-1}$], ρ density [$kg.m^{-3}$], c specific heat capacity [$J.kg^{-1}.K^{-1}$], Q_{SOURCE} latent heat of the phase or structural change [$W.m^{-3}$], x,y,z axes in given directions [m].

Implementation of continuously casting (shortly concasting), which considerably increased rate of melt cooling between temperatures of liquidus and solidus brought about time necessary for crystallic structure homogenisation. 3D transient temperature field of the of the system of concast blank-mould or concast blank-ambient is described by Fourier-Kirchoff's equation (2).

$$\frac{\partial T}{\partial \tau} = \frac{\lambda}{\rho c}\left(\frac{\partial^2 T}{\partial x^2} + \frac{\partial^2 T}{\partial y^2} + \frac{\partial^2 T}{\partial z^2}\right) + \left(w_x \frac{\partial T}{\partial x} + w_y \frac{\partial T}{\partial y} + w_z \frac{\partial T}{\partial z}\right) + \frac{Q_{SOURCE}}{\rho c} \tag{2}$$

In equation (2) w_x, w_y, w_z are velocity in given directions [$m.s^{-1}$]

These equations are solvable only by means of modern numerical methods. Therefore original models of the transient temperature field (models **A**) of both systems of gravitational casting or continuous casting were developed. Both models are based on the 1st and 2nd Fourier's laws on transient heat conduction, and the 1st and 2nd law of thermodynamics. They are based on the numerical method of finite differences with explicit formula for the unknown temperature of the mesh node in the next time step, which is a function of temperatures of the same node and six adjacent nodes in Cartesian coordinate system in the previous time step. Models take into account non-linearity of the task, it means dependence of thermo-physical properties of all materials of the systems on temperature and dependence of heat transfer coefficients on temperature of all external surfaces. Models are equipped with and interactive graphical environment for automatic generation of a mesh, and for evaluation of results, it means by so called pre-processing and post-processing.

Another model, which has also been already mastered is model of chemical heterogeneity of chemical elements (model **B**), enables description and measurement of dendritic segregation of constitutive elements and admixtures in crystallising and cooling blank (casting or concasting). This model is based on the 1st and 2nd Fick's laws of diffusion and it comprises implicitly also the law of conservation of mass. The solution itself is based on the Nernst distribution law, which quantifies at crystallisation distribution of chemical elements between liquid and solid phases of currently crystallising material in the so called mushy zone (i.e. in the zone lying between the temperature of liquidus and solidus). Majority of parameters necessary for application of the models **A** is known, but parameters necessary for use of the model **B** had to be determined by measurements on the work itself, i.e. on suitably chosen samples from continuously crystallised blanks.

Measurement was realised in the following manner: at selected segments of the cast blank concentration of main constitutive, additive and admixture elements was determined from the samples taken in regular steps. In dependence on chemical heterogeneity and structure of a blank the segments with length of 500 to 1000 μm were selected, and total number of steps, in which concentration of elements was measured, was set to 101. Measurement of concentration of elements was performed by methods of quantitative energy dispersive analysis (EDA) and wave dispersive analysis (WDA) of X-ray spectral microanalysis, for which special software and special measurement device was developed for use in combination with the analytical complex JEOL JXA 8600/KEVEX.

After completion of measurement the sample surface was etched in order to make visible the original contamination of surface by electron beam, and the measured traces were photographically documented, including the mean distance of dendrites axes within the measured segment. It was verified that the basic set of measured concentration data of elements (8 to 11 elements) makes it possible to obtain a semi-quantitative to quantitative information on chemical heterogeneity of the blank, and that it is possible to apply at the same time for evaluation of distribution of elements in the blank structure the methods of mathematical- statistical analysis. It is possible to determine the distribution curve of the element concentration in the measured segment of the analysed blank and their effective distribution coefficient between the solid and liquid phase during crystallisation. In this way the crucial verified data necessary for creation of the conjugated model (**AB**) of crystallising, solidifying and cooling down blank were obtained. It was verified that re-distribution of constitutive, additive elements and admixtures can be described by effective distribution coefficient, which had been derived for parabolic growth by Brody and Flemings (Brody&Flemings, 1966). At the moment of completed crystallisation, at surpassing of an isosolidic curve in the blank, it is possible to express the ratio of concentration of dendritically segregating element C_S to the mean concentration of the same element at the given point of the blank C_0 by the relation

$$C_S/C_0 = k_{ef}[1-(1-2\alpha k_{ef})g_S]^{(k_{ef}-1)/(1-2\alpha)} \quad (3)$$

where k_{ef} is effective distribution coefficient, g_S is mass share of the solidified phase, and α is dimensionless Fourier's number of the 2nd kind for mass transfer. This number is given by the relation

$$\alpha = D_S\,\theta_{lS}/L^2 \quad (4)$$

in which D_S is diffusion coefficient of the segregating element in solid phase, θ_{lS} is local time of crystallisation (i.e. time of persistence of the assumed dendrite between the temperature of liquidus and solidus) and L is mean half distance of dendritic axes (namely of axes of secondary dendrites). In the next step it is necessary to express the ratio of concentrations C_S/C_0 express as a function of concrete heterogeneity index I_H and of statistical distribution of the measured element, expressed by distributive curve of crystallization segregation. In this manner the following equation is available for each measured element:

$$C_S/C_0 = I_H \quad (5)$$

which expresses by concrete numbers the parameters, defined by the equation (3). By solving equations (3, 4) it is then possible to determine for each analysed element (i.e. for its measured index of dendritic heterogenity, effective distribution coefficient and distribution curve of dendritic segregation, i.e. for the established statistic character of distribution of the analysed element in structure of the blank) certain values of dimensionless criterion α. Afterwards on the basis of semi-empiric relations and rates of movement of the crystallisation front, calculated from the thermal field model in confrontation with the results of experimental metallographic analysis, the mean value of distances between branches of secondary dendrites L was determined for 9 samples of the blank. The values θ_{S} and L for the criterion α, are calculated from the model for each sample, which were determined for individual measured elements in each sample of the blank. It is possible to make from the equation (4) an estimate of the diffusion coefficient of each analysed element in individual samples of the blank. At the moment, when temperature of any point of the mesh drops below the liquidus temperature, it is valid that the share of the forming solid phase g_s grows till its limit value $g_s = 1$ (i.e. in solid phase). In this case segregation of the investigated element achieves in the residual inter-dendritic melt its maximum.

The combination of mentioned models and methodology of chemical heterogeneity investigation are presented on following technical applications.

3. Gravitational casting

3.1 Solidification of massive casting of ductile cast iron

The quality of a massive casting of cast iron with spheroidal graphite is determined by all the parameters and factors that affect the metallographic process and also others. This means the factors from sorting, melting in, modification and inoculation, casting, solidification and cooling inside the mould and heat treatment. The centre of focus were not only the purely practical questions relating to metallurgy and foundry technology, but mainly the verification of the possibility of applying two original models - the 3D model of transient solidification and the cooling of a massive cast-iron casting and the model of chemical and structural heterogeneity.

3.1.1 Calculation and meassurement of the temperature field

The application of the 3D numerical model on a transient temperature field requires systematic experimentation, including the relevant measurement of the operational parameters directly in the foundry. A real 500×1000×500 mm ductile cast-iron block had been used for the numerical calculation and the experiment. They were cast into sand moulds with various arrangements of steel chills of cylindrical shape. The dimensions of the selectid casting, the mould, the chills and their arrangements are illustrated in Figure 1. The

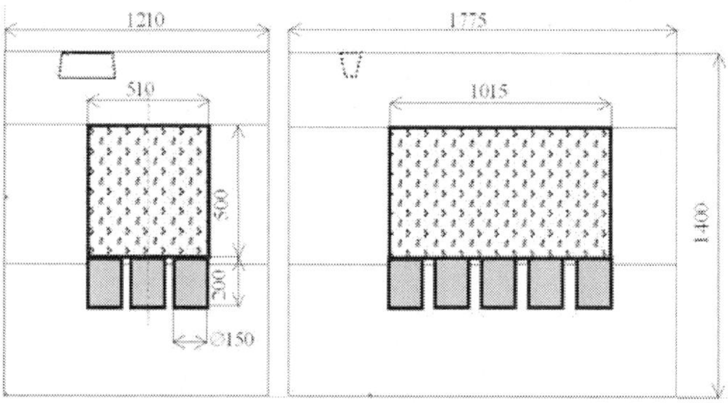

Fig. 1. The forming of casting no. 1 with chills on one side

courses of the temperatures on casting No. 1 were measured for 19 hours 11 min after pouring. The iso-zones, calculated in castings No.1 and in the chills in the total solidification time after casting, are illustrated in Figure 2 (Dobrovska et al., 2010).

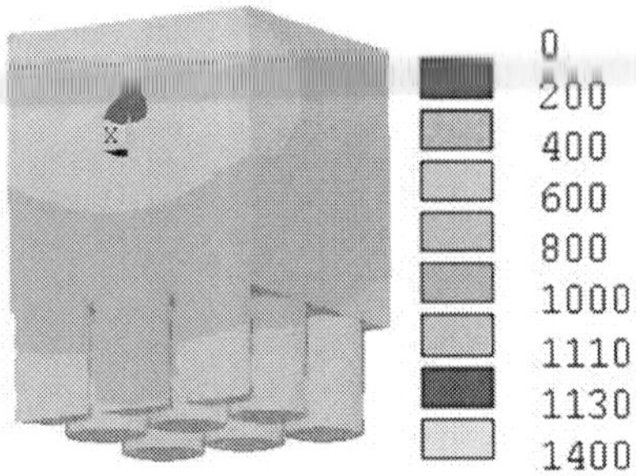

Fig. 2. The calculated iso-zones in casting No. 1 (5 hours)

3.1.2 The relation between the model of the temperature field and the model of structural and chemical heterogeneity

The 3D numerical model of the temperature field of a system comprising the casting-mold-ambient is based on the numerical finite-element method. The software ANSYS had been chosen for this computation because it enables the application of the most convenient method of numerical simulation of the release of latent heats of phase and structural changes using the thermodynamic enthalpy function. The software also considers the non-linearity of the task, i.e. the dependence of the thermophysical properties (of all materials entering the system) on the temperature, and the dependence of the heat-transfer coefficients (on all boundaries of the system) on the temperature of the surface—of the casting and mold. The original numerical model had been developed and used for estimation of structural and chemical heterogeneity. Initial and boundary conditions had been defined by means of theory of similarity. The verifying numerical calculation of the local solidification times θ - conducted according to the 3D model proved that, along the height, width and length of these massive castings, there are various points with differences in the solidification times of up to two orders. The aim was to verify the extent to which the revealed differences in the local solidification times affect the following parameters (Stransky et al., 2010):

a. The average size of the spheroidal graphite particles;
b. The average density of the spheroidal graphite particles;
c. The average dimensions of the graphite cells, and
d. The chemical heterogeneity of the elements in the cross-sections of individual graphite cells.

The relationships - among the given four parameters and the corresponding local solidification times - were determined in the series of samples that had been selected from defined positions of the massive casting. The bottom part of its sand mould was lined with (a total number of) 15 cylindrical chills of a diameter of 150 mm and a height of 200 mm. The upper part of the mould was not lined with any chills. The average chemical composition of the cast-iron before casting is given in Table 1.

Element	C	Mn	Si	P	S	Ti	Al	Cr	Ni	Mg
wt.%	3.75	0.12	2.15	0.039	0.004	0.01	0.013	0.07	0.03	0.045

Table 1. Chemical composition of ductile cast-iron (casting No. 1)

A 500×500×40 mm plate had been mechanically cut out of the middle of the length by two parallel transversal cuts. Then, further samples were taken from exactly defined points and tested in terms of their structural parameters and chemical heterogeneity. Samples in the form of testing test-samples for ductility testing, with threaded ends, were taken from the bottom part of the casting (A), from the middle part (C) and from the upper part (G). The 15 mm in diameter and 12 mm high cylindrical samples served the actual measurements in order to determine the structural parameters and chemical heterogeneity. In the points of the defined positions of the samples prepared in this way, the quantitative metallographic analysis was used to establish the structural parameters of cast-iron, the in-line point analysis to establish the chemical composition of selected elements and numerical calculation using the 3D model to establish the local solidification time. Quantitative analyses of the basic micro-structural parameters in the samples have been the subject of a special study. On each sample a total number of 49 views were evaluated. On the basis of average values of these results the structural parameters of graphite, i.e. the radius of the spheroids of graphite - R_g, the distances between the edges of graphite particles - L_g and the radius of the graphite cells - R_c have been determined for each sample. The concentration of selected elements in each of the samples was measured on the line of L_g between the edges of two particles of spheroidal graphite. The actual measurements of concentrations of ten elements - Mg, Al, Si, P, S, Ti, Cr, Mn, Fe, Ni - was carried out. On each of the samples, the concentrations of all ten elements had been measured in three intervals with each individual step being 3 μm. Before the actual measurement, the regions were selected on an unetched part of the surface and marked with a micro-hardness tester. After the micro-analysis, the samples were etched with nitric acid in alcohol in order to make the contamination of the ground surface visible using an electron beam. Then, using a Neophot light microscope, the interval within which the concentrations were measured was documented. The method of selection of the measurement points is illustrated in Figures 3a,b.

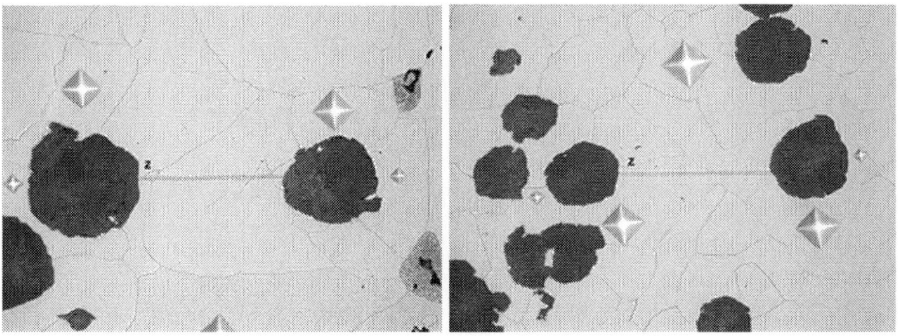

(a) (b)

Fig. 3. An example of the chemical micro-heterogeneity measurement of ductile cast-iron a) L_g = 165 μm, b) L_g = 167 μm (in Fig. $L_g \equiv$ z)

The results of the measurements of the chemical heterogeneity were evaluated statistically. The element heterogeneity index $I_{H(i)}$ is defined by the quotient of standard deviation of element concentration $\sigma_{c(i)}$ and average element concentration C_i^{av} in the analysed area, i.e. $I_{H(i)} = \sigma_{c(i)}/C_i^{av}$. The element segregation index $I_{S(i)}$ is defined by the quotient of element maximum concentration C_i^{max} and average element concentration C_i^{av} in the analysed area, i.e. $I_{S(i)} = C_i^{max}/C_i^{av}$. The local solidification times of the selected samples of known coordinates within the massive experimental casting were calculated by the 3D model. The calculation of the temperature of the liquidus and solidus for a melt with a composition according to the data in Table 1 was performed using special software with the values: 1130 °C and 1110 °C (the liquidus and solidus temperature). If the local solidification time is known, then it is also possible to determine the average rate of cooling w of the mushy zone as a quotient of the temperature interval and the local solidification time $w = \Delta T/\theta$ [°C/s]. It is obvious from the results that in the vertical direction y from the bottom of the massive casting (sample A: y = 50 mm) to the top (gradually samples C: y = 210 mm and G: y = 450 mm) the characteristic and significant relations are as follows:

a. The average size of the spheroids of graphite R_g, the average size of the cells of graphite R_c and also the average distance between the individual particles of the graphite L_g are all increasing. This relation was confirmed by the quantitative metallographic analysis.
b. The chemical heterogeneity within the individual graphite cells is also changing. The increase in the chemical heterogeneity is reflected most significantly in the increase in the indexes of segregation I_S for titanium which are increasing in the direction from the bottom of the massive casting to the top in the following order: $I_{S(Ti)}$ = 5.79-to-9.39-to-11.62
c. The local solidification time increases very significantly - from the bottom of the casting to the top - from 48 s more than 50× (near the centre of the casting) and 95× (at the top of the massive casting).

The relationships between the structural characteristics of graphite in the casting and the local solidification time were expressed quantitatively using a semi-logarithmic dependence:

$$R_g = 19.08 + 2.274 \ln\theta \,,\ R_c = 61.33 + 5.567 \ln\theta \,,\ L_g = 84.50 + 6.586 \ln\theta \ \ [\mu m, s] \qquad (6, 7, 8)$$

As far as chemical heterogenity of the measured elements is concerned, the analogous relation was established only for the dependence of the segregation index of titanium on the local solidification time, which has a steadily increasing course from the bottom of the casting all the way up to the top. The relevant relation was expressed in the form of a logarithmic equation

$$\ln I_S^{Ti} = 1.201 + 0.1410 \ln \theta_{ls}, \quad [s] \qquad (9)$$

The local solidification time θ naturally affects the mechanical properties of cast-iron however with regard to the dimensions of the test pieces; it is not possible to assign the entire body a single local solidification time. To assess relationship among structural parameters, chemical microheterogenity and mechanical properties of analysed cast-iron casting, the selected mechanical properties have been measured. The samples for testing of the tensile strength were taken from the test-sample of the experimental casting in such a way that one had been taken from under the metallographic sample and the second was taken from above. The testing indicates that the local solidification time θ has significant influence on the ductility A_5. The relationship between the ductility and the local solidification time (equation 10) indicates that the reduction in the ductility of cast-iron in the state immediately after pouring is - in the first approximation - directly proportional to the square of the local solidification time.

$$A_5 = 23.399\text{-}8.17030\theta^2 \qquad [\%, hr] \qquad (10)$$

It can be seen from previous experimentation and the evaluations of the results that - in the general case of the solidification of ductile cast-iron - there could be a dependence of the size of the spheroids of graphite, the size of the graphite cells and therefore even the distance among the graphite particles on the local solidification time. The described connection with the 3D model of a transient temperature field, which makes it possible to determine the local solidification time, seems to be the means via which it is possible to estimate the differences in structural characteristics of graphite in cast-iron and also the effect of the local solidification time on ductility in the poured casting. The main economic goal observed is the saving of liquid material, moulding and insulation materials, the saving of energy and the already mentioned optimization of pouring and the properties of the cast product.

3.2 Casting of corundo-baddeleyit ceramic material

The corundo-baddeleyit material (CBM) belongs to the not too well known system of the Al_2O_3-SiO_2-ZrO_2 oxide ceramics. Throughout the world, it is produced only in several plants, in the Czech Republic under the name of EUCOR. This production process entails solely the utilisation of waste material from relined furnaces from glass-manufacturing plants. EUCOR is heat resistant, wear resistant even at extreme temperatures and it is also resistant to corrosion. It was shown that from the foundry property viewpoint, EUCOR has certain characteristics that are similar to the behavior of cast metal materials, especially steel for castings (Chvorinov, 1954; Smrha, 1983 and others).

3.2.1 Measurements and computation results (the original riser)

The assignment focussed on the investigation of the transient 3D temperature field of a system comprising a casting-and-riser, the mold and ambient, using a numerical model (Heger et al., 2002; Kavicka et. al., 2010) . The dimensions of the casting – the so-called "stone" – were 400 x 350 x 200 mm (Figure 4). The results attained from the numerical analysis of the temperature field of a solidifying casting and the heating of the mold represent only one quadrant of the system in question. Figure 5 shows the 3D temperature field of the casting with the original riser and the mold at two times after pouring. The riser-mold interface is an interesting place for monitoring. Once this point solidifies, the riser can no longer affect the process inside the casting. The initial temperature of the mold was 20°C. The pouring temperature of the melt was 1800°C. That was approximately 300°C higher, when compared with, for example, the steel pouring temperature. The temperature of the liquidus was 1775°C and the solidus 1765°C. The temperature field was symmetrical along the axes, i.e. it was sufficient for the investigation of the temperature field of a single quadrant only.

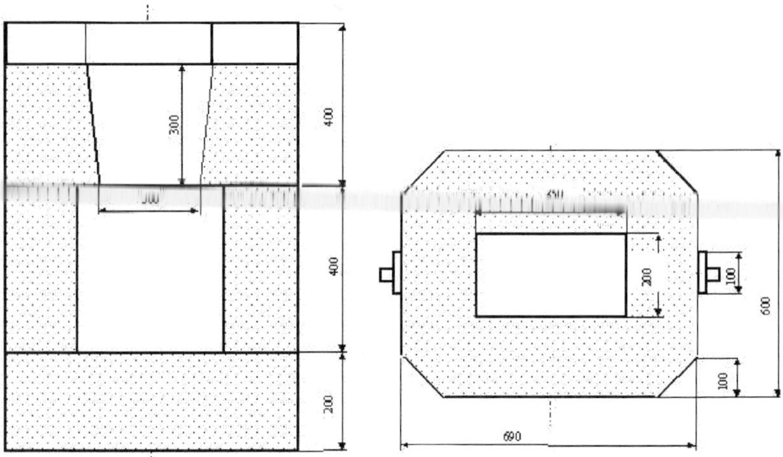

Fig. 4. The casting-riser-mold system

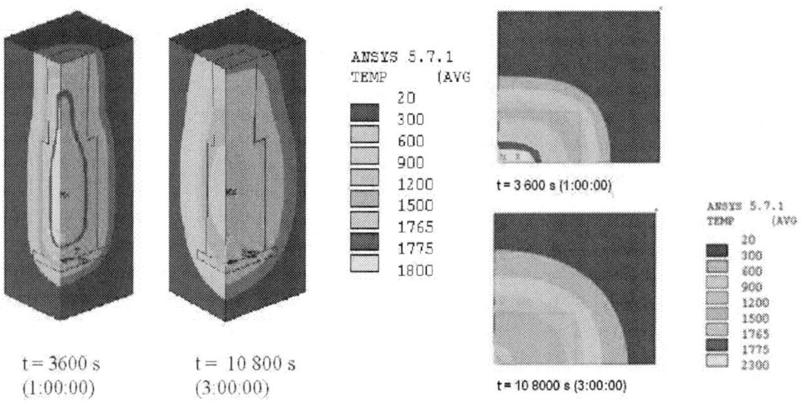

(a) (b)

Fig. 5. a) The 3D teperature field of one quadrant of the casting wirh riser- mold, b) The 2D temperature field on the riser mold and riser casting interferace

3.2.2 The model of the chemical heterogeneity and its application

The concentration distribution of individual oxides, making up the composition of the ceramic material EUCOR, was determined using an original method (Dobrovska, et al., 2009) and applied in the process of measuring the macro- and micro-heterogeneity of elements within ferrous alloys. This method was initially modified with respect to the differences occurring during solidification of the ceramic material, when compared to ferrous alloys. It was presumed that within EUCOR, the elements had been already distributed, together with oxygen, at the stoichiometric ratio (i.e. the chemical equation), which characterized the resulting composition of the oxides of individual elements after solidification. The preconditions for the application of the model of chemical heterogeneity on the EUCOR material are as follows:

If the analytically expressed distribution of micro-heterogeneity of the oxides of the ceramic material is available, if their effective distribution coefficient is known, and if it is assumed that it is possible to describe the solidification of the ceramic material via analogical models as with the solidification of metal alloys, then it is possible to conduct the experiment on the mutual combination of the calculation of the temperature field of a solidifying ceramic casting with the model describing the chemical heterogeneity of the oxides.

If the Brody-Flemings Model (Brody & Flemings, 1966) is applied for the description of the segregation of oxides of the solidifying ceramic material and if an analogy with metal alloys is assumed, then it is possible to express the relationship between the heterogeneity index I_H of the relevant oxide, its effective distribution coefficient k_{ef} and the dimensionless parameter a using the equation

$$[\ln(2\alpha k_{ef})]/(1 - 2\alpha k_{ef}) = \{\ln[(1 + nI_H^{(m)})/k_{ef}]\}/(k_{ef} - 1) \quad (11)$$

the right-hand side of which $\{ln[(1+ nI_H^{(m)})/k_{ef}]\}/(k_{ef} - 1)$, based on the measurement of micro-heterogeneity, is already known and through whose solution it is possible to determine the parameter a, which is also on the right-hand side of the equation in $2ak_{ef} = X$. The quantity n has a statistical nature and expresses what percentage of the measured values could be found within the interval $x_s \pm ns_x$ (where x_s is the arithmetic mean and s_x is the standard deviation of the set of values of the measured quantity). If $n = 2$, then 95% of all measured values can be found within this interval. If the dimensionless parameter a is known for each oxide, then a key to the clarification of the relationship exists between the local EUCOR solidification time θ, the diffusion coefficient D of the relevant oxide within the solidifying phase and the structure parameter L, which characterizes the distances between individual dendrites in metalic and ceremic alloys (Figure 6). The equation of the dimensionless parameter a is

$$\alpha = D\theta/L^2 \qquad [\text{-}], [m^2s^{-1}, s, m] \quad (12)$$

It is possible to take the dimension of a structure cell as the structure parameter for the EUCOR material. The verification of the possibility of combining both methods was conducted on samples taken from the EUCOR blocks – from the edge (sample B) – and from the centre underneath the riser (sample C). Both the measured and the computed parameters of chemical micro-heterogeneity and the computed parameters of the local solidification time θ (according to the temperature-field model) were calculated. The local solidification time of the sample B was $\theta_B = 112.18$ s and of the sample C was $\theta_C = 283.30$ s. The computed values of parameter a and the local solidification time θ determine, via their ratio, the quotient of the diffusion coefficient D and the square of the structure parameter L, which means that the following relation applies:

$$\alpha/\theta = D/L^2 \qquad [s^{-1}] \quad (13)$$

The calculated values of relation (13) for oxides of the samples B and C are arranged in Table 2 together with the parameters a

Oxide	Sample B: α	$\alpha/\theta_B \cdot 10^4$ [1/s]	Sample C: α	$\alpha/\theta_C \cdot 10^4$ [1/s]
Na_2O	0.0732	6.53	0.0691	2.44
Al_2O_3	0.0674	6.01	0.0662	2.34
SiO_2	0.0741	6.61	0.0663	2.34
ZrO_2	0.00035	0.0312	0.00008	0.0028
K_2O	0.0721	6.43	0.0665	2.35
CaO	0.075	[illegible]	0.0703	2.48
TiO_2	0.0759	6.77	0.0757	2.67
Fe_2O_3	0.0732	6.53	0.0711	2.51
HfO_2	0.0165	1.47	0.00017	0.006

Table 2. Calculated values of the equation (13)

It comes as a surprise that the values of the parameter $a/\theta = D/L^2$ of the oxides of elements Na, Al, Si, K, Ca, Ti, and Fe differed by as much as an order from the value of the same parameter of the oxide of zirconium and hafnium. This could be explained by the fact that zirconium contains hafnium as an additive and, therefore, they segregate together and the forming oxides of zirconium and hafnium show the highest melting temperatures. From the melt, both oxides segregated first, already in their solid states. Further redistribution of the oxides of both elements ran on the interface of the remaining melt and the successive segregation of other oxides only to a very limited extent. It was therefore possible to count on the fact that the real diffusion coefficients of zirconium and hafnium in the successively forming crystallites were very small (i.e. $D_{Zr} \rightarrow 0$ and $D_{Hf} \rightarrow 0$). On the other hand, the very close values of the parameters $a/\theta = D/L^2$ of the remaining seven analyzed oxides:

$$D/L_B^2 = (6.51 \pm 0.25).10^{-4} \text{ and } D/L_C^2 = (2.45 \pm 0.12).10^{-4} \quad [s^{-1}] \qquad (14,15)$$

indicated that the redistribution of these oxides between the melt and the solid state ran in a way, similar to that within metal alloys, namely steels.

It would be possible to count - in the first approximation - the diffusion coefficients of the oxides in the slag having the temperatures of 1765°C (solidus) and 1775°C (liquidus), the average value of $D = (2.07 \pm 0.11) \times 10^{-6}$ cm²/s (the data referred to the diffusion of aluminum in the slag of a composition of 39% CaO-20% Al_2O_3-41% SiO_2). For these cases, and using Equation (13), it was possible to get the magnitude of the structure parameters that governed the chemical heterogeneity of the values:

$$L_B = \sqrt{\left[\left(2.07 \times 10^{-6}\right)/\left(6.51 \times 10^{-4}\right)\right]} = 0.05639$$
$$L_C = \sqrt{\left[\left(2{,}07 \times 10^{-6}\right)/\left(2.45 \times 10^{-4}\right)\right]} = 0.09192 \quad [cm] \qquad (16,17)$$

It corresponded to 564 μm in the sample B (which was taken from the edge of the casting block) and 919 μm in the sample C (which was taken from underneath the riser of the same casting block). The comparison of the micro-structures of the analyses samples B and C (Figures 6a,b) has clearly shown that the sample B micro-structure (L_B) was significantly finer than the micro-structure of the sample C (L_C), which semi-quantitatively corresponded to the qualified estimate of the structure parameters L, conducted on the basis of calculations using the data obtained from both models.

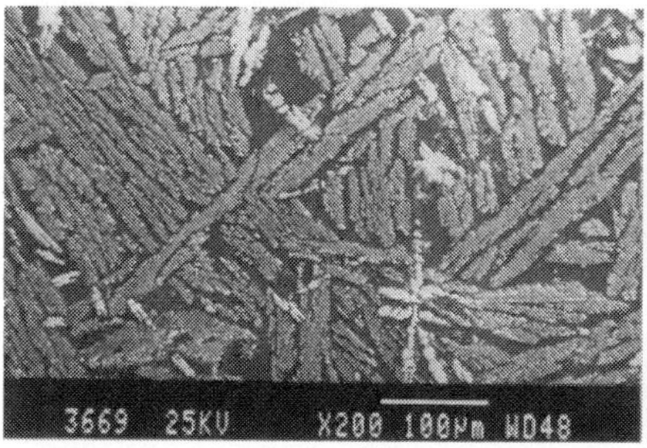

(a)

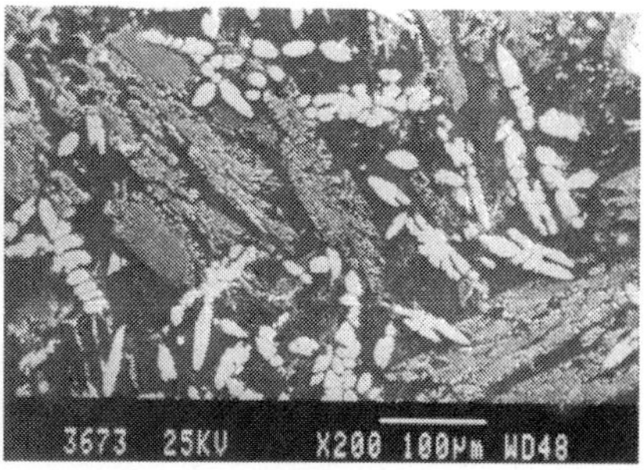

(b)

Fig. 6. a)The structure of the sample B (L_b = 564 μm), b) The structure of the ample C (L_c = 919 μm)

4. Continuously casting

4.1 Chemical microheterogeneity of continuously cast steel slab

Structure of metallic alloys is one of the factors, which significantly influence their physical and mechanical properties. Formation of structure is strongly affected by production technology, casting and solidification of these alloys. Solidification is a critical factor in the materials industry (Kavicka et al., 2007). Solute segregation either on the macro- or micro-scale is sometimes the cause of unacceptable products due to poor mechanical properties of the resulting non-equilibrium phases. In the areas of more important solute segregation there occurs weakening of bonds between atoms and mechanical properties of material degrade. Heterogeneity of distribution of components is a function of solubility in solid and liquid phases. During solidification a solute can concentrate in inter-dendritic areas above the value of its maximum solubility in solid phase. Solute diffusion in solid phase is a limiting factor for this process, since diffusion coefficient in solid phase is lower by three up to five orders than in the melt. When analysing solidification of steel so far no unified theoretical model was created, which would describe this complex heterogeneous process as a whole. During the last fifty years many approaches with more or less limiting assumptions were developed. Solidification models and simulations have been carried out for both macroscopic and microscopic scales. The most elaborate numerical models can predict micro-segregation with comparatively high precision. The main limiting factor of all existing mathematical micro-segregation models consists in lack of available thermodynamic and kinetic data, especially for systems of higher orders. There is also little experimental data to check the models. Many authors deal with issues related to modelling of a non-equilibrium crystallisation of alloys. However, majority of the presented works concentrates mainly on investigation of modelling of micro-segregation of binary alloys, or on segregation of elements for special cases of crystallisation – directional solidification, zonal melting, one-dimensional thermal field, etc. Moreover these models work with highly limiting assumption concerning phase diagrams (constant distribution coefficients) and development of dendritic morphology (mostly one-dimensional models of dendrites. Comprehensive studies of solidification for higher order real alloys are rarer. Nevertheless, there is a strong industrial need to investigate and simulate more complex alloys because nearly all current commercial alloys have many components often exceeding ten elements. Moreover, computer simulation have shown that even minute amounts of alloying elements can significantly influence microstructure and micro-segregation and cannot be neglected.

4.1.1 Methodology of chemical heterogeneity investigation

Original approach to determination of chemical heterogeneity in structure of poly-component system is based on experimental measurements made on samples taken from characteristic places of the casting, which were specified in advance. Next procedure is based on statistical processing of concentration data sets and application of the original mathematical model for determination of distribution curves of dendritic segregation of elements, characterising the most probable distribution of concentration of element in the frame of dendrite (Dobrovska et al., 2009), and the original mathematical model for determination of effective distribution coefficients of these elements in the analysed alloy.

4.1.2 Application of methodology of chemical heterogeneity investigation – investigation into chemical micro-heterogeneity of CC steel slab

A continuously cast steel slab (CC steel slab, Figure 7) with dimensions 1530x250 mm was chosen for presentation of results, with the following chemical composition in (wt. %): 0.14C; 0.75Mn; 0.23Si; 0.016P; 0.010S; 0.10Cr; 0.050Cu; 0.033Al_{total}.

After solidification and cooling of the cast slab a transversal band was cut out, which was then axially divided into halves. Nine samples were taken from one half for determination of chemical heterogeneity according to the diagram in Figure 8. The samples had a form of a cube with an edge of approx. 20 mm, with recorded orientation of its original position in the CC slab. Figure 9 shows an example of microstructure of the analysed slab. On each sample a concentration of seven elements (*aluminium, silicon, phosphor, sulphur, titanium, chromium and manganese*) were measured along the line segment long 1000 µm. The distance between the measured points was 10 µm.

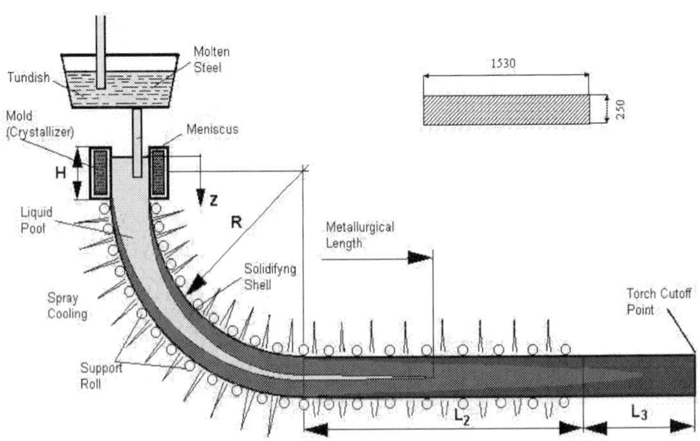

Fig. 7. The steel slab caster

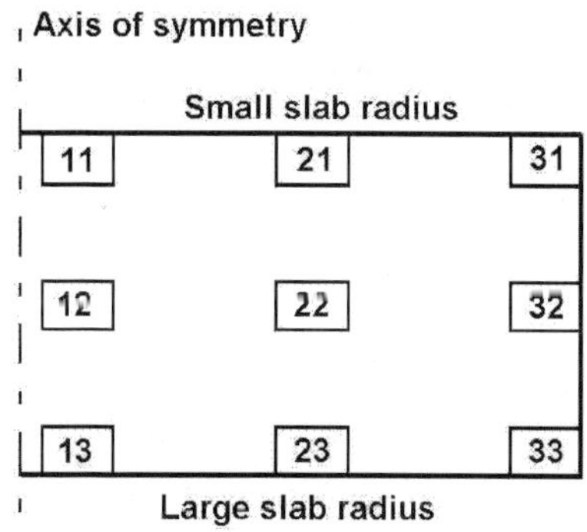

Fig. 8. Scheme of sampling from a slab and marking of samples

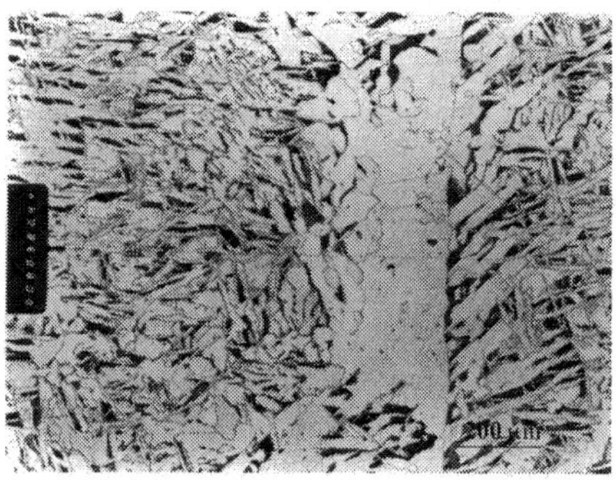

Fig. 9. Example of structure of the sample 21 with a microscopic trace of 1000 μm long

Analytical complex unit JEOL JXA 8600/KEVEX Delta V Sesame was used for determination of concentration distribution of elements, and concentration was determined by method of energy dispersive X-ray spectral micro-analysis. As an example, Figures 10 a,b present t*he basic concentration spectrum* of Mn, Si, P and S.

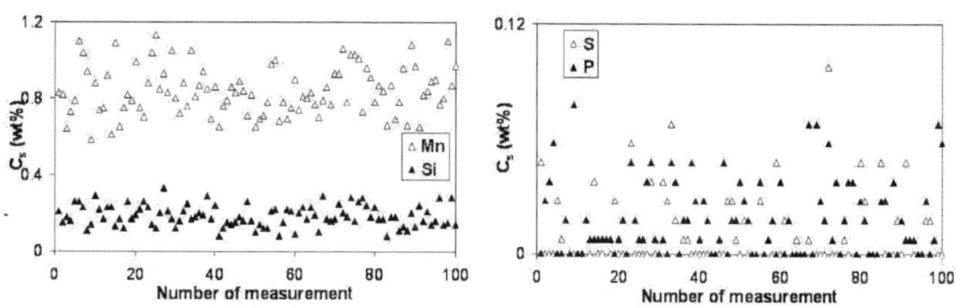

Fig. 10. Sample 21. Basic concentration spectrum a) of Mn and Si, b) of S and P

Chemical micro-heterogeneity, i.e. segregation of individual elements at distances, order of which is comparable to dendrite arms spacing, can be quantitatively evaluated from the basic statistical parameters of the measured concentrations of elements in individual samples. These parameters comprise: C_x average concentration of element (arithmetic average) in the selected section, s_x standard deviation of the measured concentration of element, C_{min} minimum concentration of element and C_{max} maximum concentration of element measured always on the selected section of the sample. It is possible to calculate from these data moreover *indexes of dendritic heterogeneity* I_H of elements in the measured section of individual samples as ratio of standard deviation s_x and average concentration C_x of the element. Then the element distribution profiles can be plotted according to the Gungor's method (Gungor,1989) from the concentration data sets measured by the method ED along the line segment 1000 μm long. Data plotted as the measured weight percent composition versus number of data (Figures 10 a,b) were put in an ascending or descending order and *x*-axis was converted to the fraction solid ($f_S \approx g_s$ in Equation 3) by dividing each measured data number by total measured data number. The element composition versus fraction solid, i.e. element distribution profile (*distribution curve of dendritic segregation*) was then plotted; Figures 11 a,b represent such dependences for manganese, silicon, phosphor and sulphur. The slope of such curve (ascending or descending) depended on the fact, whether the element in question enriched the dendrite core or the inter-dendritic area in the course of solidification.

From these statistical data it is also possible to determine with use of original mathematical model for each analysed element from the given set of samples the values of *effective distribution coefficients* k_{ef}. The procedure of the effective distribution coefficient calculation will be outlined here as follows:

The sequence of such arranged concentrations (Figures 11 a,b) was seen as a distribution of concentrations of the measured element in the direction from the axis ($f_S = 0$) to the boundary ($f_S = 1$) of one average dendrite.

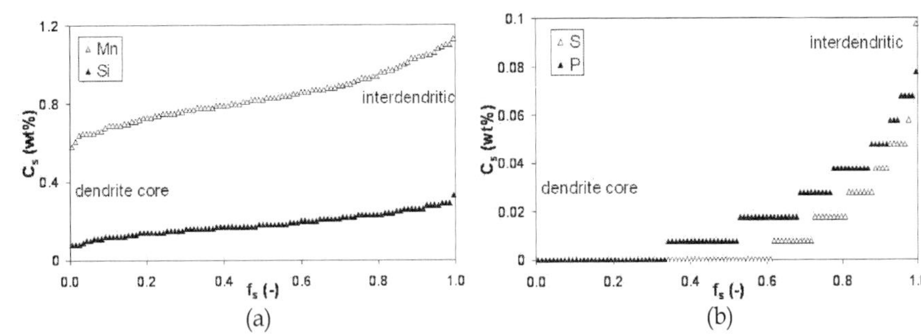

Fig. 11. Experimentally determined distribution curve of dendritic segregation (sample 21) a) for Mn and Si ,b) for P and S

The effective distribution coefficient k_{ef} was in this case defined by the relation

$$k_{ef}(f_s) = C_s(f_s) / C_L(f_s) \tag{18}$$

where C_S is the solute concentration in the solidus and C_L is its concentration in liquidus and argument (f_S) expressed the dependence of both concentrations on the fraction solid. A perfect mixing of an element in the interdendritic melt was then assumed (this assumption.

It was therefore possible to substitute the equation (18) by the formula

$$k_{ef}(i) = C_i / C_R(i) \tag{19}$$

where C_i is the concentration in i-th point of the sequence (i.e. in the i-th point of the curve in Figures 11 a, b) and $C_R(i)$ is the average concentration of the element in the residual part of the curve (i.e. for $f_S \in \langle i, 1\rangle$), expressed by the relation:

$$C_R(i) = [1/(n-i+1)]\sum_{j=1}^{n} C_j \tag{20}$$

where n was the number of the measured points. In this way it was possible to determine the values of effective distribution coefficients for all $i \in \langle 1, n\rangle$, i.e. for the entire curve characterising the segregation during solidification. The effective distribution coefficients of all the analysed elements were calculated by this original method. The average values of determined effective distribution coefficients are listed in Table 3. No segregation occurs when $k_{ef} = 1$; the higher is the deviation from the number 1, the higher is the segregation ability.

The effective distribution coefficients calculated in this way inherently include in themselves both the effect of segregation in the course of alloy solidification and the effect of homogenisation, occurring during the solidification as well as during the cooling of alloy.

Average values of measured and calculated quantities in the set of samples are in Table 4.

Sample		Element						
		Al	Si	P	S	Ti	Cr	Mn
11	I_H	1.24	0.28	1.22	1.45	0.30	0.22	0.14
	k_{ef}	0.32	0.78	0.33	0.26	0.76	0.83	0.88
12	I_H	1.54	0.30	1.12	1.74	0.29	0.27	0.15
	k_{ef}	0.24	0.77	0.36	0.20	0.78	0.79	0.88
13	I_H	1.44	0.30	1.25	1.48	0.30	0.29	0.15
	k_{ef}	0.27	0.78	0.32	0.26	0.77	0.78	0.88
21	I_H	1.33	0.29	1.58	1.49	0.31	0.24	0.13
	k_{ef}	0.29	0.78	0.24	0.25	0.76	0.81	0.89
22	I_H	1.14	0.28	1.31	1.41	0.30	0.26	0.14
	k_{ef}	0.35	0.78	0.30	0.27	0.77	0.80	0.88
23	I_H	1.56	0.29	1.34	1.86	0.26	0.28	0.13
	k_{ef}	0.24	0.78	0.29	0.18	0.80	0.78	0.89
31	I_H	1.11	0.28	1.22	2.34	0.31	0.23	0.16
	k_{ef}	0.37	0.78	0.33	0.18	0.76	0.82	0.87
32	I_H	1.44	0.27	1.16	1.49	0.34	0.25	0.14
	k_{ef}	0.27	0.79	0.34	0.25	0.74	0.80	0.88
33	I_H	1.32	0.29	1.24	1.64	0.35	0.26	0.13
	k_{ef}	0.30	0.78	0.32	0.22	0.74	0.80	0.89

Table 3. The average values of the heterogeneity index I_H and the effective distribution coefficient k_{ef} of elements in the individual samples

	c_x ± s_x	I_H ± s_I	k_{ef} ± s_k	$k^{(ref)}$ according to Dobrovska et al., 2009
Al	0.0136 0.0029	1.352 0.162	0.294 0.046	0.12 -0.92
Si	0.1910 0.0060	0.285 0.011	0.781 0.005	0.66 -0.91
P	0.0141 0.0023	1.270 0.133	0.314 0.035	0.06 -0.50
S	0.0136 0.0030	1.657 0.297	0.232 0.035	0.02 -0.10
Ti	0.0951 0.0032	0.306 0.027	0.765 0.019	0.05 -0.60
Cr	0.1758 0.0076	0.255 0.023	0.799 0.017	0.30 -0.97
Mn	0.8232 0.0169	0.143 0.009	0.873 0.033	0.72 -0.90

Table 4. Average values of the measured and calculated quantities in the set of all samples

Data represented in Table 3 and Table 4 make it possible to evaluate dendritic heterogeneity (micro-heterogeneity) of elements, as well as their effective distribution coefficients in individual samples, and also in the frame of the whole analysed half of the slab cross-section. It is obvious from these tables that dendritic heterogeneity of accompanying elements and impurities is comparatively high. This is demonstrated by the index of dendritic heterogeneity I_H. It follows from Table 3, that distinct differences exist between micro-heterogeneity of individual elements. Figures 12 a, b show distribution of indexes of micro-heterogeneity of sulphur (the most segregating element) and manganese (the least segregating element) on slab cross-section.

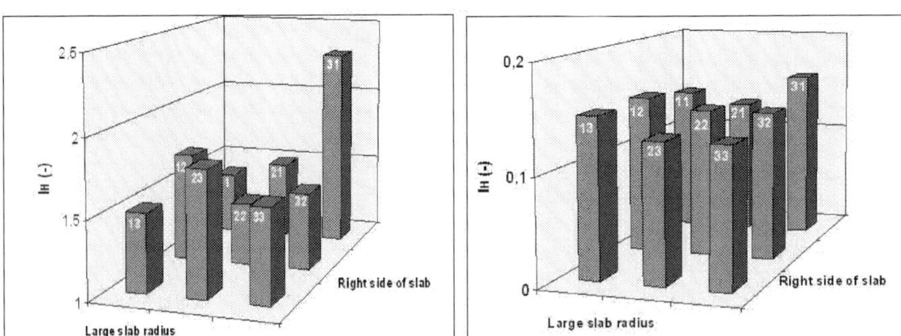

(a) (b)

Fig. 12. a) Differences in sulphur micro-heterogeneity in samples taken from one-half of slab cross-section. b) Differences in manganese micro-heterogeneity in samples taken from one-half of slab cross-section.

Average value of this coefficient for all the analysed elements and the whole set of nine samples is given in Table 4. It follows from this table that dendritic heterogeneity of slab decreases in this order of elements: sulphur, aluminium, phosphor, titanium, silicon, chromium and manganese, which has the lowest index of heterogeneity. Dendritic heterogeneity of the analysed elements is expressed also by the values of their effective distribution coefficients, arranged for individual samples in Table 3 and for the set of samples in Table 4.

It is obvious from the tables that pair values of the index of dendritic heterogeneity and effective distribution coefficient for the same element do mutually correspond. The higher the value of the heterogeneity index, the lower the value of effective distribution coefficient and vice versa. The lowest value of the effective distribution coefficient is found in sulphur and the highest value is found in manganese. It follows from the Table 4, that effective distribution coefficient increases in this order of elements: sulphur, aluminium, phosphor, titanium, silicon, chromium and manganese. All the analysed elements segregate during solidification into an inter-dendritic melt, and their distribution coefficient is smaller than one. For comparison, the Table 4 contains also the values of distribution coefficients found in literature. It is obvious that our values of effective distribution coefficients, calculated according to the original model, are in good agreement with the data from literature, only with the exception of sulphur (and titanium). The reason for this difference is probably the means of calculation of the effective distribution coefficient - the value of this parameter is calculated from concentration data set measured on solidified and cooled casting. Consequently, the effective distribution coefficients calculated in this way inherently include in themselves both the effect of segregation in the course of alloy solidification and the effect of homogenisation, occurring during the solidification as well as during the cooling of alloy.

4.2 Effect of elelectromagnetic stirring on the dendritic structure of steel billets

Currently, casters use rotating stators of electromagnetic melt-stirring systems. These stators create a rotating magnetic induction field with an induction of **B**, which induces eddy-current **J** in a direction perpendicular to **B**, whose velocity is **v**. Induction **B** and current **J** create an electromagnetic force, which works on every unit of volume of steel and brings about a stirring motion in the melt. The vector product ($\mathbf{v} \times \mathbf{B}$) demonstrates a connection between the electromagnetic field and the flow of the melt. The speeds of the liquid steel caused by the elelectromagnetic stirring is somewhere from 0.1-to-1.0 m/s. The stirring parameters are within a broad range of values, depending on the construction and technological application of the stirrer. The power output is mostly between 100 and 800 kW, the electric current between 300 and 1000 A, the voltage up to 400 V and with billet casting the frequency from 5 to 50 Hz.

The elelectromagnetic stirring applied on the steel caster is basically a magneto-hydraulic process together with crystallisation processes and solidification of billet steel. The complexity of the entire process is enhanced further by the fact that the temperatures are higher than the casting temperatures of concast steel. The temperature of the billet gradually decreases as it passes through the caster down to a temperature lying far below the solidus temperature. From the viewpoint of physics and chemistry, the course of the process is co-determined by a number of relevant material, physical and thermokinetic characteristics of the concast steel and also electrical and magnetic quantities. There is also a wide range of construction and function parameters pertaining to the caster and elelectromagnetic stirring as well as parameters relating to their mutual arrangement and synchronisation. Numerous works from recent years relate that exact mathematical modelling of elelectromagnetic stirring on a caster is still unsolvable (Stransky et al., 2009).

The basic elelectromagnetic stirring experiment was conducted on a continuously steel billet caster where two individual mixers were working (Figure 13). The first stirrer, entitled MEMS (Mould Electromagnetic Stirring), is mounted directly on the mould and the second stirrer, entitled SEMS (Strand Electromagnetic Stirring), is mounted at the beginning of the flow directly after the first cooling zones but in the secondary-cooling zone. Here the outer structure of the billet is already created by a compact layer of crystallites, however, in the centre of the billet there is still a significant amount of melt that is mixed by the SEMS.

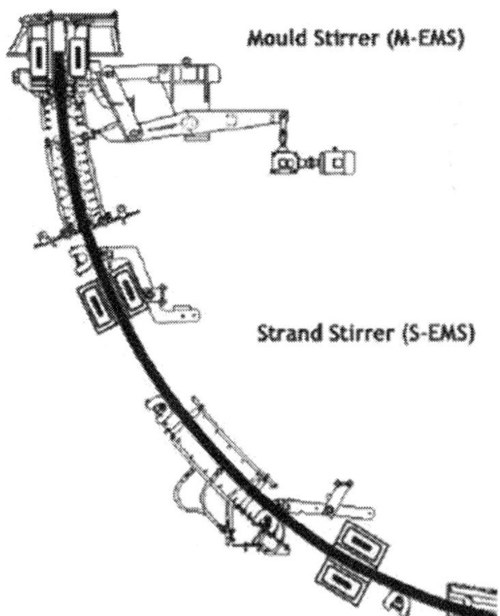

Fig. 13. The steel billet caster of 150x150 mm. The positions of the MEMS and SEMS stirrers

4.2.1 The temperature field of a billet

The temperature field of the billet of 150x150 mm computed via original numerical model (Stransky et al., 2009,2011) is in Figures 14-15.

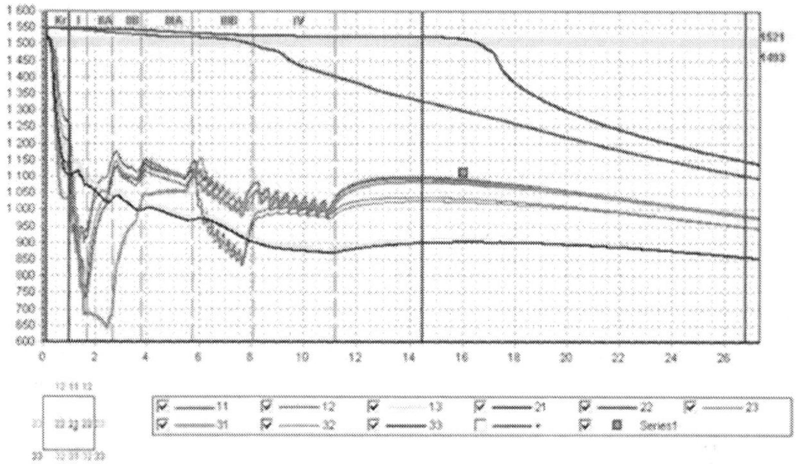

Fig. 14. The temperature history of marked points of the cross-section of the steel billet 150x150 mm

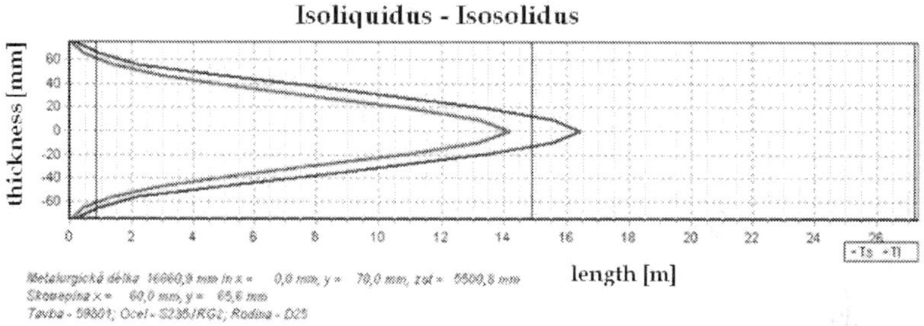

Fig. 15. Computed iso-solidus and iso-liquidus curves in the axial longitudinal section

4.2.2 The experiment

The first stirrer (MEMS) stirs the melt still in the mould while the billet is undergoing crystallization and solidification. The second stirrer (SEMS) works at a time when the melt is already enclosed by a shell of crystallites around the perimeter of the billet and inside the billet there is less melt than above in the active zone of the first stirrer. When both stirrers were switched off, the crystallisation and solidification continued in the normal way, i.e. the solidifying melt did not undergo a forced rotational movement. Samples were taken throughout the course of the experiment – from parts of the billet cast using the MEMS and SEMS and without and also using either one. The samples were taken in the form of cross-sections (i.e. perpendicular to the billet axis). The samples were fine-ground and etched with the aim of making visible the dendritic structure which is characteristic for individual variants of the solidification of the billet. The verification of the influence of MEMS and SEMS on the macrostructure of the billet was carried out on two melts of almost the same chemical composition (Table 5).

Melt	C	Mn	Si	P	S	Cu	Cr	Ni	Al	Ti
A	0.14	0.31	0.22	0.014	0.009	0.03	0.05	0.02	0.02	0.002
B	0.13	0.32	0.22	0.018	0.012	0.09	0.06	0.04	0.02	0.002

Table 5. Chemical composition of experimental melts [wt.%]

The timing of the concasting process of the billets - without the involvement of the stirrers and with the working of the elelectromagnetic stirring of individual variants of stirrers (MEMS and SEMS) - is given in Table 6. The speed of the concasting (i.e. the movement, the proceeding of the billet through the mould) of the billet was maintained constant during the experimentation at a value of 2.7 m/min. Table 6 shows that as many as nine concasting variants were verified. The lengths of individual experimental billets - from which samples had been taken - were always a multiple of the metallurgical length. The average superheating of the steel above the liquidus was 32.8 ± 3.1 °C in melt A and 28.0 ± 4.6 °C in melt B, which lies within the standard deviation of the temperature measurements.

Melt	Concasting mode - sampling	Superheating of steel above liquidus [°C]	MEMS stirring [Amperes]	SEMS stirring [Amperes]	Fig.
A	1A	37	210	0	
	2A	31	0	0	Fig. 16a
	3A	33	0	29	
	4A	30	210	57	Fig. 16b
B	1B	35	210	0	
	2B	30	0	0	
	3B	27	0	57	
	4B	24	210	57	
	5B	24	210	29	

Table 6. The billet concasting modes and sampling

Note: Detailed records of the experimental verification of the effects of MEMS and SEMS during concasting on the relevant device pertain to Table 6. The data are appended with a time history of the MEMS and SEMS connection and with information relating to the lengths of individual billets and the points from which the actual samples had been taken (i.e. the cross-sections from which the dendritic structures had been created). Evaluation of all nine variants of concasting (Table 6) indicates that the arrangement of dendrites in the cross-section follow the same tendency in the first phase of crystallization. The structure is created by columnar crystals - dendrites - perpendicular to the walls of the billet (Figure 16a).

In the billets that were not stirred the dendrites gradually touch one another on the diagonals of the cross-section. Here their growth either ceases, or the dendrites bend in the directions of the diagonals and their growth continues all the way to the centre of the billet. The columnar dendrites that grow from the middle part of the surface maintain their basic orientation - perpendicular to the surface - almost all the way to the centre of the billet. In the central part of the cross-section there is an obvious hollow on all nine macroscopic images. This is most probably a shrinkage.

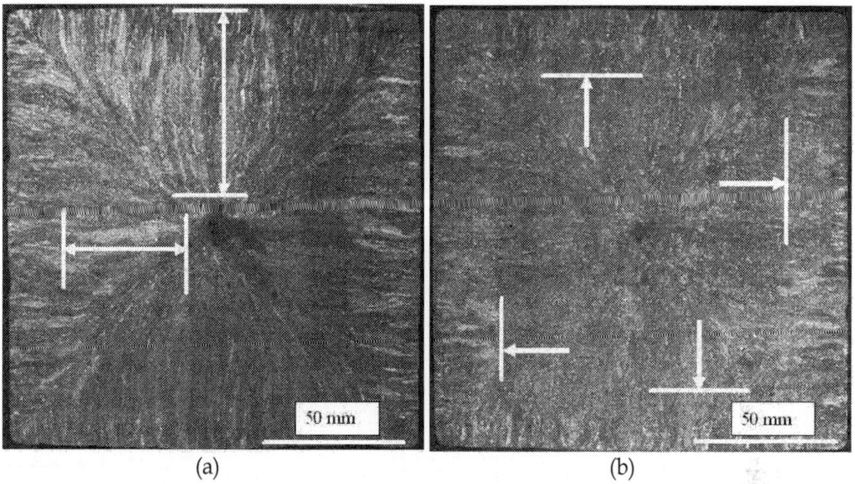

Fig. 16. a) Dendrite growth in the concasting structure without elelectromagnetic stirring - mode 2A, b) The growth of dendrites in the billet structure using the MEMS and SEMS - mode 4A (Table 6)

The above-described mechanism of dendrite growth during concasting without stirring is frequently the object of interest (Figure 16a). Inside the billets, when using the MEMS stirrer (or both MEMS and SEMS), the kinetics of solidification and dendrite growth is initially the same as without stirring. This also creates columnar dendrites which touch along the diagonals, however, soon their growth ceases still near the surface. Dendrites, which are called equiaxed dendrites continue to grow - their orientation is more random and only partly directed towards the centre of the billet (Figure 16b). It appears that this dendrite growth mechanism manifests itself the most when both stirrers are working simultaneously (Table 6: 4A, 4B and 5B). If MEMS and SEMS are working simultaneously, the stirring effect significantly destroys the formation of columnar crystals. If only MEMS is working and SEMS is switched off (1A and 1B), then the desctruction of columnar crystals is less evident. The working mode of SEMS alone (modes 3A and 3B) cannot be clearly differentiated from the changes in the dendritic structure in relation to the structure formed without stirring (2A and 2B).

Figure 16b (the macro-ground dendritic structure) shows the depth of the columnar band of dendrites in the direction away from the surface of the billet (Figure 16b - see arrows) and its value, which (with the simultaneous stirring of MEMS and SEMS) is 23.4±1.8 mm. The same qualified guess was made for ordinary billet casting (i.e. without stirring). Here, the depth of the dendrites can be guessed almost all the way to the central shrinkage at 70 mm (Figure 16a - see arrows). It is known that additives and impurities during solidification are often concentrated in points of contact of the growing dendrites, where the maximum of segregated additives and impurities and the greatest probability of technological defects occurs. In the given case, this undesirable effect can be expected along the diagonals which have a length of up to 100-to-103 mm towards the central shrinkage. This point of contact of the dendrites during the simultaneous working of SEMS and MEMS is only 29.8±1.9 mm, i.e. 3.4× less. The central area of the billet containing a hollow as a result of a shrinkage is then filled with dendrites growing into a vacuum (i.e. underpressure) (Figure 17).

Fig. 17. Dendrites in the centre of the billet

Under the assumption that the maximum of defects (i.e. vacancies, impurities, additives and micro-shrinkages) are formed along the diagonals it is possible to expect that in the areas of the corners - specifically on the edges - the nucleation of cracks will be higher than on the walls of the billet. If the first approximation of the fracture toughness of the relevant billet made from low-carbon steel is $K_{IC} \sim 75.0$ MPa.m$^{1/2}$, then in the ordinary concasting process it can be assumed that the length of the contact of columnar dendrites along the diagonal will be approximately $\Delta l_{normal} \approx 101.5$ mm (Figure 16a). On the other hand, if both electromagnetic stirrers (MEMS and SEMS) are engaged simultaneously, the contact length of the columnar dendrites along the diagonal decreases to $\Delta l_{el.magnl} \approx 29.8$ mm (Figure 16b). Along these lengths (i.e. the areas) it could be expected that during concasting the concentration of the primary defects will increase.

A comparison of limit stresses and strains in the area of the edges of the billets during concasting without elelectromagnetic stirring and if both MEMS and SEMS stirrers are engaged indicates that the billets (otherwise cast under the same conditions) cast without stirring are almost twice as susceptible to cracking along the edges as billets cast using both stirrers. A similar assumption can be made even in the case of assessing the effect of columnar dendrites in the central part of the surface of the billet where, without stirring, their length grows from the surface of the wall all the way to the central shrinkage (Figure 16a), while with the stirrers the dendrites are significantly shorter. The boundaries of the dendrites are however much less damaged by technological defects (vacancies, etc.) than the areas of their touching - of the peaks along the diagonals. Long-term statistical monitoring of the quality of 150×150 mm billets and the chemical composition has proven that the application of elelectromagnetic stirring has significantly reduced the occurrence of defects (in this case cracks).

5. Conclusion

Progressive creation of numeric model of unsteady thermal field A connected with the model of chemical heterogeneity B, leads to a completely novel conjugated numeric model, which necessarily requires respecting reality of poly-component crystallising metallic system, formed usually by eight to eleven constitutive elements. Constitutive elements forming conjugated model have during crystallisation completely different physical-chemical properties in dependence of temperature. Their redistribution in the volume of crystallising tangible macroscopic system is governed by the 2nd Fick's law. Mutual functional connection of both models A and B into one mutually cooperating conjugated model AB represents a completely new step resting on real crystallising poly-component system. This connection of two models AB necessarily requires large amount of consistent concentration data of constitutive elements forming real crystallising tangible macroscopic poly-component system. These data concern alloying elements (e.g. Ni, Cr), basic tramp elements (Mn, Si, Ti, V, Mo), data on admixture elements and impurities (S, P), as well as data on de-oxidising elements (Al, Ca, etc.).

The most complicated conjugated model will be the model for continuous casting. The authors have prepared for its creation 50,000 experimentally verified and mutually consistent data on elements. These data, make it possible to express concentrations C, effective distribution coefficient of elements between melt and crystallising solid phase k_{ef}, diffusion coefficients D_s in the melt of all segregating elements of dendritically crystallising system in the sense of the equation (3), express also the degree of heterogeneity and shares of solidified phase g_s. It contains also the equation (4), which is dimensionless Fourier's number α of the second kind for mass transfer, which contains implicitly, apart from the diffusion coefficient also share of local solidification time and squares of half distance of dendritic axes θ_{lS}/L^2. Equation (5) postulates by share of concentrations at the interface of the melt and solid phase (dendrite) the degree of heterogeneity - these mutually consistent and already verified data on elements form the basic starting point for progressive functional creation of the above mentioned conjugated model AB.

6. Acknowledgment

This research was conducted using a program devised within the framework of the GA CR projects GA CR projects No. 106/08/0606, 106/09/0370, 106/09/0940, 106/09/0969, P107/11/1566 and MSMT CR- MSM6198910013.

7. References

Boettinger, W.J. et al.(2000). Solidification microstructures: recent development, future directions, *Acta Mater.*, 48 pp. 43-70 ISSN 1359-6457

Brody, H.D. & Flemings, M.C. (1966). Solute redistribution in dendritic solidification, *Trans. AIME*, 236 (1966), pp. 615-624

Chalmers, B. (1964). *Principles of Solidification*. John Wiley & Sons, New York.

Chvorinov, N. (1954). *Krystalisace a nestejnorodost oceli [Crystallization and heterogeneity of steel]*. Nakladatelstvi CSAV, Praha

Davies, G.J. (1973). *Solidification and Casting*. Applied Science Pub., London.

Dobrovska, J., Stransky, K., Dobrovska, V. & Kavicka, F. (2009).Characterization of Continuously Cast Steel Slab Solidification by Means of Chemical Micro-heterogeneity Assessment, *Hutnicke listy* No.5, LXII , pp. 4-9, ISSN 0018-8069

Dobrovska, J., Kavicka F., Stransky, K., et al. (2010). Numerical Optimization of the Method of Cooling of a Massive Casting of Ductile Cast-Iron , *NUMIFORM 2010*, Vols 1 and

2 - Dedicated to professor O. C. Zienkiewicz (1921-2009) Book Series: AIP Conference Proceedings Vol. 1252 , pp. 578-585 , ISBN 978-0-7354-0799-2

Gungor, M. N. (1989) A statistically significant experimental technique for investigating micro-segregation in cast alloys, *Metall. Trans. A*, 20A, pp. 2529-2538.

Heger, J., Stetina, J., Kavicka, F., et al. (2002). Pilot calculation of the temperature field of the ceramic material EUCOR, *Advanced Computational Methods in Heat Transer VII* Book Series: Computational Studies, Vol. 4, pp. 223-232 ISBN 1-85312-906-2

Flemings, M. C. (1974). *Solidification Processing*. McGraw-Hill, New York.

Kavicka, F., Stetina, J., Sekanina, B., et al. (2007). The optimization of a concasting technology by two numerical models , *Journal of Materials Processing Technology* , Vol. 185, Issue 1-3, Sp. Iss. SI, pp. 152-159 , ISSN 0924-0136

Kavicka, F, Dobrovska, J., Sekanina, B., et al. (2010). A Numerical Model of the Temperature Fi.eld of the Cast and Solidified Ceramic Material , *NUMIFORM 2010*, Vols 1 and 2 - Dedicated to professor O. C. Zienkiewicz (1921-2009) Book Series: AIP Conference Proceedings , Vol. 1252 , pp. 571-577 ISBN 978-0-7354-0799-2

Kraft, T., Chang, Y. A. (1997) Predicting microstructure and microsegregation in multicomponent alloys, *JOM*, pp. 20-28.

Kurz, W., Fisher, D.J. (1986). *Fundamentals of Solidification*. Trans Tech Publications, Switzerland.

Rappaz, M. (1989). Modelling of microstructure formation in solidification processes, *Int. Mater. Rev.*, 34, pp. 93-123 ISSN 0950-6608

Smrha, L.(1983). *Tuhnutí a krystalizace ocelových ingotů* [*Solidification and crystallisation of steel ingots*] . SNTL, Praha

Stefanescu, D. M.(1995). Methodologies for modelling of solidification microstructure and their capabilities, *ISIJ Int.*, 35, pp. 637-650 ISSN 0915-1559

Stransky, K., Kavicka, F. , Sekanina, B. et al. (2009). Electromagnetic stirring of the melt of concast billets and its importance, *METAL 2009*, Hradec nad Moravici, ISBN 978-80-87294-04-8

Stransky, K., Dobrovska, J., Kavicka, F., et al. (2010). Two numerical models of the solidification structure of massive ductile cast-iron casting , *Materiali in Tehnologije / Materials and Technology* , Vol. 44, Issue 2, (MAR-APR 2010), pp. 93-98 , ISSN 1580-2949

Stransky, K., Kavicka, F., Sekanina, B. et al. (2011).The effect of electromagnetic stirring on the crystallization of concast billets, *Materiali in Tehnologije /Materials and Technology,* Vol. 45 , Issue 2, pp. 163–166, ISSN 1580-2949

4

A Mathematical Model for Single Crystal Cylindrical Tube Growth by the Edge-Defined Film-Fed Growth (EFG) Technique

Loredana Tanasie and Stefan Balint
West University of Timisoara
Romania

1. Introduction

1.1 Crystal growth from the melt by E.F.G. technique

Modern engineering does not only need crystals of arbitrary shapes but also plate, rod and tube-shaped crystals, i.e., crystals of shapes that allow their use as final products without additional machining. Therefore, the growth of crystals of specified sizes and shapes with controlled defect and impurity structures are required. In the case of crystals grown from the melt, this problem appears to be solved by profiled-container crystallization as in the case of casting. However, this solution is not always possible, for example growing very thin plate-shaped crystals from the melt (to say nothing of more complicated shapes), excludes container application completely.[Tatarchenko, 1993]

The techniques which allow the shaping of the lateral crystal surface without contact with the container walls are appropriate for the above purpose. In the case of these techniques the shapes and the dimensions of the grown crystals are controlled by the interface and meniscus-shaping capillary force and by the heat- and mass-exchange conditions in the crystal-melt system. The edge-defined film-fed growth (EFG) technique is of this type. Whenever the E.F.G. technique is employed, a shaping device is used (Fig. 1). In the device a capillary channel is manufactured (Fig. 1) in which the melt raises and feeds the growth process. Frequently, a wettable solid body is used to raise the melt column above the shaper, where a thin film is formed. When a wettable body is in contact with the melt, an equilibrium liquid column embracing the surface of the body is formed. The column formation is caused by the capillary forces being present. Such liquid configuration is usually called a meniscus (Fig. 1) and in the E.F.G. technique, its lower boundary (Fig. 1 - point C) is attached to the sharp edge of the shaper.

Let be the temperature of the meniscus upper horizontal section (Fig. 1 - $\overline{AB}$) the temperature of the liquid crystallization. So, above the plane of this section, the melt transforms in solid phase. Now set the liquid phase into upward motion with the constant rate, v, keeping the position of the phase-transition plane invariable by selection of the heat conditions. When the motion starts, the crystallized position of the meniscus will continuously form a solid upward or downward tapering body. In the particular case when the line tangent at the triple point B to the liquid meniscus surface makes a specific angle (angle of growth) with the vertical, the lateral wall of the crystal will be vertical. Thus, the initial body, called the seed, serves to form a meniscus which later on determines the form of the crystallized product, the phase transition position being fixed.

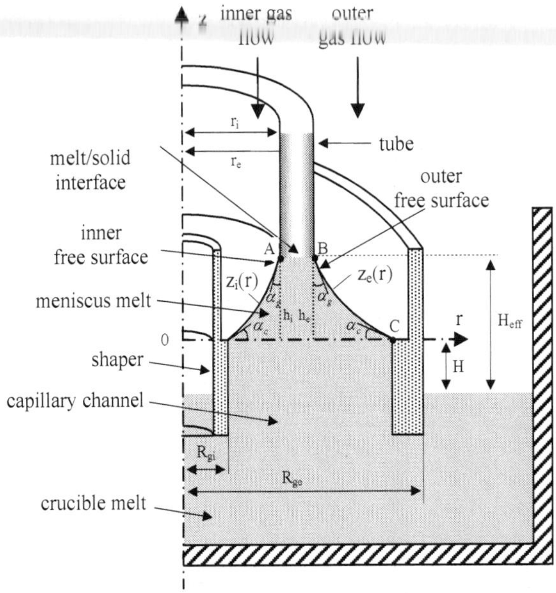

Fig. 1. Prototype tubular crystal growth by E.F.G. method

Based on this description, a conclusion can be drawn that the dimensions and shapes of the specimens being pulled by the E.F.G. technique depend upon the following factors: (i) the shaper geometry; (ii) the pressure of feeding the melt to the shaper; (iii) the crystallization front position; (iv) the seed's shape. The seed's shape is only important for stationary pulling; in this case its cross-section should coincide with the desired product's cross-section. Frequently, especially when complicated profiles are grown, the pulling process is carried out under unstationary conditions by lowering the crystallization surface, which then enhances the dependence of the shapes on the crystal cross-section. With such an approach applied to the pulling process, the dimensions and the shape of the grown crystal are determined by the above-mentioned factors and by the pulling rate-to-crystallization front displacement ratio.[Tatarchenko, 1993].

1.2 Background history of tube growth from the melt by E.F.G. method

The technology of growing tubes can have a significant impact for example on the solar cell technology. The growth of silicon tubes by E.F.G. process was first reported by Erris et al. [Erris et al.,1980]. Tubes were grown with a diameter of 95×10^{-4} [m], wall thickness in the range of $5\times10^{-5}-1\times10^{-3}$ [m] at rates up to 2×10^{-3} [m/s]. In [Erris et al.,1980] a theory of tube growth by the E.F.G. process is developed to show the dependence of the tube wall thickness on the growth variables. The theory concerns the calculations of the shape of the liquid-vapor interface (or meniscus) and of the heat flow in the system. The inner and outer meniscus shapes, (Fig.1), are both calculated from Laplace's capillary equation, in which the pressure difference Δp across a point on meniscus is considered to be $\Delta p = \rho \cdot g \cdot H_{eff}$ = constant, where H_{eff} represents the effective height of the growth interface above the horizontal liquid level in the crucible (Fig.1). According to [Surek et al.,1977], [Swartz et al., 1975], it includes the effects of the viscous flow of the melt in the shaper capillary and in the meniscus film, as well as that of the hydrostatic head. The above approximation for Δp is valid for silicon ribbon growth [Surek et al.,1977], [Kalejs et al., 1990], when $H_{eff} >> h$, where h is the height of the growth interface above the shaper top (i.e. the meniscus height). Another approximation used in [Erris et al.,1980], concerning the meniscus, is that the inner and outer meniscus shapes are approximated by circular segments. With these relatively tight tolerances concerning the menisci in conjunction with the heat flow calculation in the system, the predictive model developed in [Erris et al.,1980] has been shown to be a useful tool in understanding the feasible limits of wall thickness control. A more precise predictive model would require an increase of the acceptable tolerance range introduced by approximation.

Later, this process was scaled up by Kaljes et al. [Kalejs et al., 1990] to grow 15×10^{-2} [m] diameter silicon tubes, and the stress behavior in the grown tube was investigated. It has been realized that numerical investigations are necessary for the improvement of the technology. Since the growth system consists of a small die tip (1×10^{-3} m width) and a thin tube (order of 200×10^{-6} [m] wall thickness) the width of the melt/solid interface and meniscus are accordingly very small. Therefore, it is essential to obtain an accurate solution for the temperature and interface position in this tiny region.

In [Rajendran et al., 1993] an axisymmetric finite element model of magnetic and thermal field was presented for an inductively heated furnace. Later the same model was used to determine the critical parameters controlling silicon carbide precipitation on the die wall [Rajendran et al., 1994]. Rajendran et al. also developed a three dimensional magnetic induction model for an octagonal E.F.G, system. Recently, in [Roy et al., 2000a], [Roy et al., 2000b], a generic numerical model for an inductively heated large diameter Si tube growth system was reported. In [Sun et al., 2004] a numerical model based on multi-block method and multi-grid technique is developed for induction heating and thermal transport in an E.F.G. system. The model is applied to investigate the growth of large octagon silicon tubes of up to 50×10^{-2} m diameter. A 3D dynamic stress model for the growth of hollow silicon polygons is reported in [Behnken et al., 2005]. In [Mackintosh et al., 2006] the challenges fixed in bringing E.F.G. technology into large-scale manufacturing, and ongoing development of furnace designs for growth of tubes for larger wafer production using hexagons with 1×10^{-2} m face widths, and wall thicknesses in he range $250\times10^{-6}-300\times10^{-6}$ m is described. In [Kasjanow et al., 2010] the authors present a 3D coupled electromagnetic and thermal modeling of E.F.G. silicon tube growth, successfully validated by experimental tests with industrial installations.

The state of the art at 1993-1994 concerning the calculation of the meniscus shape in general in the case of the growth by E.F.G. method is summarized in [Tatarchenko, 1993]. According to [Tatarchenko, 1993], for the general equation describing the surface of a liquid meniscus possessing axial symmetry, there is no complete analysis and solution. For the general equation only numerical integration was carried out for a number of process parameter values that are of practical interest at the moment. The authors of papers [Borodin&Borodin&Sidorov&Petkov, 1999],[Borodin&Borodin&Zhdanov, 1999] consider automated crystal growth processes based on weight sensors and computers. They give an expression for the weight of the meniscus, contacted with a crystal and shaper of arbitrary shape, in which there are two terms related to the hydrodynamic factor. In [Rosolenko et al., 2001] it is shown that the hydrodynamic factor is too small to be considered in the automated crystal growth and it is not clear what equation (of non Laplace type) was considered for the meniscus surface. Finally, in [Yang et al., 2006] the authors present theoretical and numerical study of meniscus dynamics under symmetric and asymmetric configurations. A meniscus dynamics model is developed to consider meniscus shape and its dynamics, heat and mass transfer around the die-top and meniscus. Analysis reveals the correlations between tube thickness, effective melt height, pull-rate, die-top temperature and crystal environmental temperature.

The purpose of this chapter is the mathematical description of the growth process of a single crystal cylindrical tube grown by the edge-defined film-fed growth (EFG) technique. The mathematical model defined by a set of three differential equations governing the evolution of the outer radius and the inner radius of the tube and of the crystallization front level is the one considered in [Tatarchenko, 1993]. This system contains two functions which represent the angle made by the tangent line to the outer (inner) meniscus surface at the three-phase point with the horizontal. The meniscus surface is described mathematically by the solution of the axi-symmetric Young-Laplace differential equation. The analysis of the dependence of solutions of the Young-Laplace differential equation on the pressure difference across the free surface, reveals necessary or sufficient conditions for the existence of solutions which represent convex or concave outer or inner free surfaces of a meniscus. These conditions are expressed in terms of inequalities which are used for the choice of the pressure difference, in order to obtain a single-crystal cylindrical tube with specified sizes.

A numerical procedure for determining the functions appearing in the system of differential equations governing the evolution is presented.

Finally, a procedure is presented for setting the pulling rate, capillary and thermal conditions to grow a cylindrical tube with prior established inner and outer radius. The right hand terms of the system of differential equations serve as tools for setting the above parameters. At the end a numerical simulation of the growth process is presented.

The results presented in this chapter were obtained by the authors and have never been included in a book concerning this topic.

Since the calculus and simulation in this model can be made by a P.C., the information obtained in this way is less expressive than an experiment and can be useful for experiment planing.

2. The system of differential equations which governs the evolution of the tube's inner radius r_i, outer radius r_e and the level of the crystallization front h

According to [Tatarchenko, 1993] the system of differential equations which governs the evolution of the tube's inner radius r_i, the outer radius r_e and the level of the crystallization front h is:

$$\begin{cases} \dfrac{dr_e}{dt} = -v \cdot \tan\left[\overline{\alpha}_e(r_e,h,p_e) - \left(\dfrac{\pi}{2} - \alpha_g\right)\right] \\ \dfrac{dr_i}{dt} = v \cdot \tan\left[\overline{\alpha}_i(r_i,h,p_i) - \left(\dfrac{\pi}{2} - \alpha_g\right)\right] \\ \dfrac{dh}{dt} = v - \dfrac{1}{\Lambda \cdot \rho_1} \cdot \left[\lambda_1 \cdot G_1(r_e,r_i,h) - \lambda_2 \cdot G_2(r_e,r_i,h)\right] \end{cases} \tag{1}$$

In equations $(1)_1$ and $(1)_2$: v is the pulling rate, $\overline{\alpha}_e(r_e,h,p_e)$ ($\overline{\alpha}_i(r_i,h,p_i)$) is the angle between the tangent line to the outer (inner) meniscus at the three phase point of coordinates (r_e,h) ((r_i,h)) and the horizontal Or axis (Fig.1 b), α_g is the growth angle (Fig. 1), p_e (p_i) is the controllable part of the pressure difference across the free surface given by:

$$p_e = p_m - p_g^e - \rho_1 \cdot g \cdot H_e \quad (p_i = p_m - p_g^i - \rho_1 \cdot g \cdot H_i) \tag{2}$$

where p_m is the hydrodynamic pressure in the melt under the free surface, which can be neglected in general, with respect to the hydrostatic pressure $\rho_1 \cdot g \cdot H_e$ ($\rho_1 \cdot g \cdot H_i$); p_g^e (p_g^i) is the pressure of the gas flow, introduced in order to release the heat from the outer (inner) wall of the tube; H_e (H_i) is the melt column height between the horizontal crucible melt level and the shaper outer (inner) top level (Fig. 1a); ρ_1 is the melt density; g is the gravity acceleration.

The angle $\overline{\alpha}_e(r_e,h,p_e)$ ($\overline{\alpha}_i(r_i,h,p_i)$) fluctuates due to the fluctuations of: the outer (inner) radius r_e (r_i), the level h of the crystallization front and the outer (inner) pressure p_e (p_i)

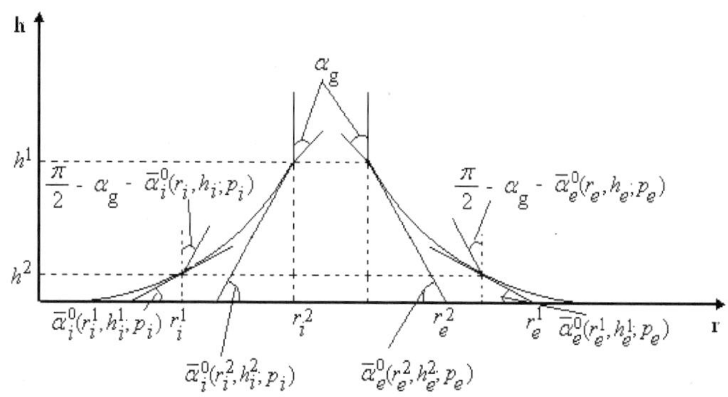

Fig. 2. Fluctuations at the triple point

In the equation $(1)_3$: Λ is the latent melting heat; λ_1, λ_2 are the thermal conductivity coefficients in the melt and the crystal respectively; G_1^j, G_2^j are the temperature gradients at the interface in the melt *(i=1)* and in the crystal *(i=2)* respectively, given by the formulas:

$$G_1^j(r_e,r_i,h)=\frac{1}{SINH(\beta_{1j}\cdot h)}\left[\left(T_0-T_{en}(0)-\frac{v\cdot k}{(F_j)^2\cdot Bi\cdot\chi_1}\right)\cdot\left(-\beta_{1j}\cdot e^{\delta_1\cdot h}\right)+\right.$$
$$\left.\left(T_m-T_{en}(0)+k\cdot h-\frac{v\cdot k}{(F_j)^2\cdot Bi\cdot\chi_1}\right)\cdot\left(\delta_1\cdot SINH(\beta_{1j}\cdot h)+\beta_{1j}\cdot COSH(\beta_{1j}\cdot h)\right)\right]-k \tag{3}$$

$$G_2^j(r_e,r_i,h)=\frac{1}{SINH(\beta_{2j}\cdot L)}\left[\left(T_m-T_{en}(0)+k\cdot h-\frac{v\cdot k}{(F_j)^2\cdot Bi\cdot\chi_2}\right)\cdot\right.$$
$$\left.\left(\delta_2\cdot SINH(\beta_{2j}\cdot L)-\beta_{2j}\cdot COSH(\beta_{2j}\cdot L)\right)-\frac{v\cdot k}{(F_j)^2\cdot Bi\cdot\chi_2}\cdot\beta_{2j}\cdot e^{-\delta_2\cdot L}\right]-k \tag{4}$$

where χ_i - the thermal diffusivity coefficient equal to $\frac{\lambda_i}{\rho_i\cdot c_i}$, ρ_i - the density, c_i - the heat capacity, Bi - the Biot number equal to $\frac{\mu_i\cdot r_e}{\lambda_i}$ (*i=1* - the melt, *i=2* - the crystal*)*, $\mu_i=\mu_i^k+\mu_i^r$ - the coefficient of the heat-exchange with environment (μ_i^k - the convective heat-exchange coefficient and μ_i^r - the linearized radiation heat-exchange coefficient), F_j ($j=1,2,3$) the crystal (meniscus) cross – section perimeter – to – its area ratio: $j=1$ and $F_1=\frac{2}{r_e}$, for a thick-walled tube with small inner radius, for which heat is removed from the external surface only, $j=2$ and $F_2=\frac{2\cdot r_e}{r_e^2-r_i^2}$ for a tube of not to large inner radius for which heat is removed from the external surface only, $j=3$ and $F_3=\frac{2}{r_e-r_i}$ for a tube for which heat is removed from both the outer and inner surfaces([Tatarchenko, 1993], pp. 39-40, 146). T_0 - the melt temperature at the meniscus basis, T_m - melting temperature, $T_{en}(0)$ - the environment temperature at $z=0$, k - the vertical temperature gradient in the furnace, r_e - the outer radius of the tube equal to the upper radius of the outer meniscus, r_i - the inner radius of the tube equal to the upper radius of the inner meniscus, L - the tube length and $\delta_i=\frac{v}{2\chi_i}$, $\beta_{ij}=\sqrt{\frac{v^2}{4\chi_i^2}+(F_j)^2\cdot Bi}$, $i=1,2, j=1,2,3$, $SINH$ and $COSH$ are the hyperbolic sine and hyperbolic cosine functions.

Due to the supercooling in this gradients it is assumed that: $T_0<T_m$, $k>0$, $T_{en}(0)<T_m$.

In the following sections we will show in which way $\bar{\alpha}_e(r_e,h,p_e)$ and $\bar{\alpha}_i(r_i,h,p_i)$ can be found starting from the Young-Laplace equation of a capillary surface in equilibrium.

3. The choice of the pressure of the gas flow and the melt level in silicon tube growth

In a single crystal tube growth by edge-defined film fed growth (E.F.G.) technique, in hydrostatic approximation, the free surface of a static meniscus is described by the Young-Laplace capillary equation [Finn, 1986]:

$$\gamma\cdot\left(\frac{1}{R_1}+\frac{1}{R_2}\right)=\rho\cdot g\cdot z-p \tag{5}$$

Here γ is the melt surface tension, ρ denotes the melt density, g is the gravity acceleration, $1/R_1, 1/R_2$ denote the mean normal curvatures of the free surface at a point M of the free surface, z is the coordinate of M with respect to the Oz axis, directed vertically upwards, p is the pressure difference across the free surface. To calculate the outer and inner free surface shape of the static meniscus it is convenient to employ the Young-Laplace eq.(5) in its differential form. This form of the eq.(5) can be obtained as a necessary condition for the minimum of the free energy of the melt column [Finn, 1986].For a tube of outer radius $r_e\in\left(\frac{R_{gi}+R_{ge}}{2},R_{ge}\right)$ and inner radius $r_i\in\left(R_{gi},\frac{R_{gi}+R_{ge}}{2}\right)$, the axi-symmetric differential equation of the outer free surface is given by:

$$z''=\frac{\rho\cdot g\cdot z-p_e}{\gamma}\left[1+(z')^2\right]^{3/2}-\frac{1}{r}\cdot\left[1+(z')^2\right]\cdot z' \quad \text{for } r\in[r_e,R_{ge}] \tag{6}$$

which is the Euler equation for the energy functional

$$\begin{gathered}I_e(z)=\int_{r_e}^{R_{ge}}\left\{\gamma\cdot\left[1+(z')^2\right]^{1/2}+\frac{1}{2}\cdot\rho\cdot g\cdot z^2-p_e\cdot z\right\}\cdot r\cdot dr,\quad z(r_e)=z_e(r_e),\\ z(R_{ge})=z_e(R_{ge})=0\end{gathered} \tag{7}$$

The axi-symmetric differential equation of the inner free surface is given by:

$$z''=\frac{\rho\cdot g\cdot z-p_i}{\gamma}\left[1+(z')^2\right]^{3/2}-\frac{1}{r}\cdot\left[1+(z')^2\right]\cdot z' \quad \text{for } r\in\left[R_{gi},r_i\right] \tag{8}$$

which is the Euler equation for the energy functional:

$$I_i(z)=\int_{R_{gi}}^{r_i}\left\{\gamma\cdot\left[1+(z')^2\right]^{1/2}+\frac{1}{2}\cdot\rho\cdot g\cdot z^2-p\cdot z\right\}\cdot r\cdot dr,\quad z(R_{gi})=z_i(R_{gi})=0,\quad z(r_i)=z_i(r_i) \tag{9}$$

In papers [Balint & Balint, 2009b], [Balint&Balint&Tanasie, 2008], [Balint & Tanasie, 2008] , Balint, Tanasie, 2011] some mathematical theorems and corollaries have been rigorously proven regarding the existence of an appropriate meniscus. These results are presented in Appendixes. In the following we will shown in which way the inequalities can be used for creation of the appropriate meniscus.

5.1 Convex free surface creation

In this section, it will be shown in which way the inequalities presented in Appendix 1 can be used for the creation of an appropriate static convex meniscus by the choice of p_e and p_i [Balint, Tanasie, 2011].

Inequalities (A.1.1) establish the range where the pressure difference p_e has to be chosen in order to obtain a static meniscus with convex outer free surface, appropriate for the growth of a tube of outer radius equal to $\frac{R_{ge}}{n}$.

If the pressure difference satisfies (A.1.2), then a static meniscus with convex outer free surface is obtained which is appropriate for the growth of a tube of outer radius $r_e \in \left[\frac{R_{ge}}{n}, R_{ge}\right]$.

If the pressure difference satisfies inequality (A.1.4) and the value of p_e is close to the value of the right hand member of the inequality (A.1.4) then a static meniscus with convex outer free surface is obtained which is appropriate for the growth of a tube of outer radius equal to $\frac{R_{ge} + R_{gi}}{2}$.

If the pressure difference satisfies inequality (A.1.5), then a static meniscus with convex outer free surface is obtained which is appropriate for the growth of a tube of outer radius in the range $\left[\frac{R_{ge}}{n}, \frac{R_{ge}}{n'}\right]$.

Theorem 5 (Appendix 1) shows that a static meniscus having a convex outer free surface, appropriate for the growth of a tube of outer radius r_e situated in the range $\left[\frac{R_{ge}}{n}, \frac{R_{ge}}{n'}\right]$, is stable.

Inequalities (A.1.6) establish the range where the pressure difference p_i has to be chosen in order to obtain a static meniscus with convex inner free surface appropriate for the growth of a tube of inner radius equal to $m \cdot R_{gi}$.

If the pressure difference p_i satisfies (A.1.7) then a static meniscus with convex inner free surface is obtained which is appropriate for the growth of a tube of inner radius $r_i < m \cdot R_{gi}$.

If the pressure difference p_i satisfies the inequality (A.1.9) and the value of p_i is close to the value of the right hand term of the inequality (A.1.9) then a static meniscus with convex inner free surface is obtained which is appropriate for the growth of a tube of inner radius equal to $\frac{R_{ge}+R_{gi}}{2}$.

If the pressure difference p_i satisfies inequality (A.1.10) then a static meniscus with convex inner free surface is obtained which is appropriate for the growth of a tube of inner radius which is in the range $\left[m' \cdot R_{gi}, m \cdot R_{gi}\right]$.

Theorem 10 (Appendix 1) shows that a static meniscus having a covex inner free surface appropriate for the growth of a tube of inner radius r_i situated in the range $\left[R_{gi}, \frac{R_{gi}+R_{ge}}{2}\right]$ is stable.

For numerical illustrations, the inner radius of the shaper was taken $R_{gi} = 4.2 \times 10^{-3}$ [m] and outer radius of the shaper was chosen $R_{ge} = 4.8 \times 10^{-3}$ [m] [Eriss, 1980]. Computations were performed in **MathCAD 14**. and for **Si** the following numerical values were considered: $\rho = 2.5 \times 10^3$ [kg/m³]; $\gamma = 7.2 \times 10^{-1}$ [N/m]; α_c =30°; α_g =11°; g=9.81[m/s²].

To create a convex meniscus appropriate for the growth of a tube having the outer radius r_e^1 equal to $r_e^1 = 4.65 \times 10^{-3}$ [m] ($n_1 = 1.03226$), according to the **Theorem 1** (Appendix 1), p_e has to be chosen in the range: $[-3480.07, -612.35]$ [Pa]. According to the **Corollary 3** (Appendix 1), from this range for the values of p_e smaller than -1702.52 [Pa] the point r_e where $z_e'(r_e) = -\tan\left(\pi/2 - \alpha_g\right)$ is close to 4.5×10^{-3} [m]. Hence, we have to find for $p_e = -1702.52$ [Pa] the point r_e for which the above condition is satisfied. This can be made by integrating numerically the following system for $z(R_{ge}) = 0$, $\alpha(R_{ge}) = \alpha_c$ and $p_e = -1702.52$ [Pa] (see Fig. 3):

$$\begin{cases} \frac{dz_e}{dr} = -\tan(\alpha_e) \\ \frac{d\alpha_e}{dr} = \frac{p_e - \rho \cdot g \cdot z_e}{\gamma} \cdot \frac{1}{\cos\alpha_e} - \frac{1}{r} \cdot \tan\alpha_e \end{cases} \tag{10}$$

Since the obtained r_e is $r_e = 4.609 \times 10^{-3}$ [m], and it is smaller than the desired value $r_e^1 = 4.65 \times 10^{-3}$ [m], the value of p_e has to be chosen in the range $[-3480.07, -1702.52]$ [Pa].

The results of the integrations of the system (10) for $z(R_{ge}) = 0$, $\alpha(R_{ge}) = \alpha_c$ and different values of p_e in this range, are presented in Fig. 4. This figure shows that the outer radius $r_e^1 = 4.65 \times 10^{-3}$ [m] is obtained for $p_e^1 = -2198$ [Pa].

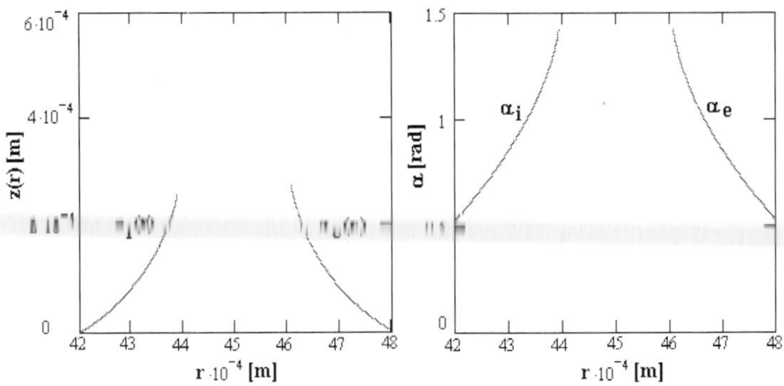

Fig. 3. The results of the integration of systems (10) and (11) for $p_e = -1702.52$ [Pa] and $p_i = -1945.80$ [Pa]

Taking $p_m \approx 0$ [Eriss et al., 1980], [Rossolenko et al., 2001], [Yang et al., 2006], the melt column height in this case is $H_e^1 = \frac{1}{\rho \cdot g}\left(2198 - p_g^e\right)$, where $p_g^e \geq 0$ is the pressure of the gas flow (introduced in the furnace for release, the heat from the outer wall of the tube). When $p_g^e = 0$, then $H_e^1 = 8.96 \times 10^{-2}$ [m] i.e. the shaper's outer top level has to be with $H_e^1 = 8.96 \times 10^{-2}$ [m] above the crucible melt level.

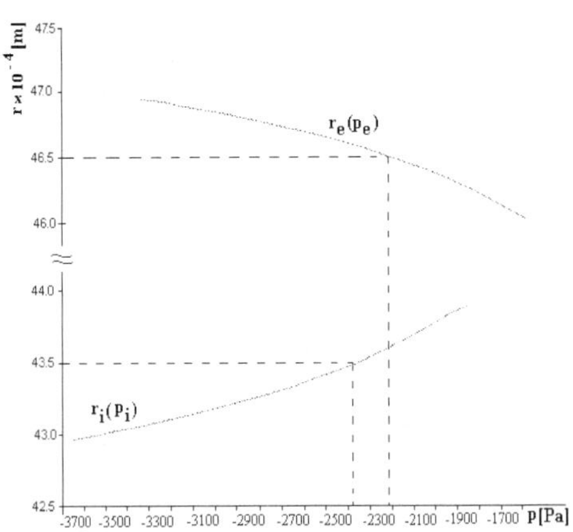

Fig. 4. The tube outer radius and inner radius versus p_e and p_i

To create a convex meniscus appropriate for the growth of a tube having the inner radius $r_i^1 = 4.35 \times 10^{-3}$ [m] ($m_1 = 1.03571$), according to the **Theorem 6** (Appendix 1), p_i has to be chosen in the range: $[-3723.32, -847.10]$ [Pa]. According to the **Corollary 8** (Appendix 1), from this range for the values of p_i smaller than -1945.80 [Pa], the point r_i where the condition $z_i'(r_i) = \tan\left(\pi/2 - \alpha_g\right)$ is satisfied is close to 4.5×10^{-3} [m]. Therefore, we have to find now for $p_i = -1945.80$ [Pa] the point r_i where the above condition is achieved. This can be made by integrating numerically the system:

$$\begin{cases} \dfrac{dz_i}{dr} = \tan\alpha_i \\ \dfrac{d\alpha_i}{dr} = \dfrac{\rho \cdot g \cdot z_i - p_i}{\gamma} \cdot \dfrac{1}{\cos\alpha_i} - \dfrac{1}{r} \cdot \tan\alpha_i \end{cases} \tag{11}$$

for $z(R_{gi}) = 0$, $\alpha(R_{gi}) = \alpha_c$ and $p_i = -1945.80$ [Pa]. (see Fig. 3).

Since the obtained r_i is $r_i = 4.390 \times 10^{-3}$ [m] and it is higher than the desired value $r_i^1 = 4.35 \times 10^{-3}$ [m], we have to choose the value of p_i in the range $[-3723.32 - 1945.80]$ [Pa].

The results of the integrations of the system (11) for $z(R_{gi}) = 0$, $\alpha(R_{gi}) = \alpha_c$ and for different values of p_i in this range are represented in Fig. 4.

This figure shows that the inner radius $r_i^1 = 4.35 \times 10^{-3}$ [m] is obtained for $p_i^1 = -2434$ [Pa].

Taking $p_m \approx 0$ [Eriss et al., 1980], [Rossolenko et al., 2001], [Yang et al., 2006], the melt column height is $H_i^1 = \dfrac{1}{\rho \cdot g}\left(2434 - p_g^i\right)$, where $p_g^i \geq 0$ is the pressure of the gas flow (introduced in the furnace for releasing the heat from the inner wall of the tube). When $p_g^i = 0$, then $H_i^1 = 9.92 \times 10^{-2}$ [m], i.e. the shaper's inner top level has to be with $H_i^1 = 9.92 \times 10^{-2}$ [m] above the crucible melt level. When $p_g^i \geq 2434$, then H_i^1 is negative, i.e. the crucible melt level has to be above the shaper's inner top level.

To create a convex meniscus appropriate for the growth of a tube with the outer radius $r_e^1 = 4.65 \times 10^{-3}$ [m] and inner radius $r_i^1 = 4.35 \times 10^{-3}$ [m], when the shaper's inner top is at the same level as the shaper's outer top, we have to take: $\dfrac{1}{\rho \cdot g}\left(2198 - p_g^e\right) = \dfrac{1}{\rho \cdot g}\left(2434 - p_g^i\right)$.

It follows that the pressure of the gas flow, introduced in the furnace for releasing the heat from the inner wall of the tube has to be higher than the pressure of the gas flow, introduced in the furnace for releasing the heat from the outer wall of the tube and we have to take: $p_g^i - p_g^e = 236$ [Pa].

3.2 Concave free surface creation

In this section, it will be shown in which way the inequalities presented in Appendix 2 can be used for the creation of an appropriate static concave meniscus by the choice of p_e and p_i [Balint&Balint, 2009a].

Inequalities (A.2.1) establish the range in which the pressure difference p_e has to be chosen in order to obtain a static meniscus with concave outer free surface appropriate for the growth of a tube of outer radius equal to $\frac{R_{ge}}{n}$.

If the pressure difference satisfies inequality (A.2.2) then a static meniscus with concave outer free surface is obtained which is appropriate for the growth of a tube of outer radius in the range $\left[\frac{R_{ge}}{n}, \frac{R_{ge}}{n'}\right]$.

Theorem 13 (Appendix 2) shows that a static meniscus having a concave outer free surface appropriate for the growth of a tube of outer radius $r_e \in \left[\frac{R_{gi}+R_{ge}}{2}, R_{ge}\right]$ is stable.

Inequalities (A.2.3) establish the range in which the pressure difference p_i has to be chosen in order to obtain a static meniscus with convex inner free surface appropriate for the growth of a tube of inner radius equal to $m \cdot R_{gi}$.

If the pressure difference p_i satisfies inequality (A.2.4) then a static meniscus with concave inner free surface is obtained which is appropriate for the growth of a tube of inner radius in the range $\left[m' \cdot R_{gi}, m \cdot R_{gi}\right]$.

Theorem 16 (Appendix 2) shows that a static meniscus having a concave inner free surface appropriate for the growth of a tube of inner radius $r_i \in \left[R_{gi}, \frac{R_{gi}+R_{ge}}{2}\right]$ is stable.

Computations were performed for an InSb tube growth: $\alpha_c = 63.8^0$; $\alpha_g = 28.9^0$; $\rho = 6582\left[kg/m^3\right]$; $\gamma = 4.2 \times 10^{-1}[N/m]$.

If there exists a concave outer free surface, appropriate for the growth of a tube of outer radius $r_e^1 = 4.65 \times 10^{-3}[m]$ ($n_1 = 1.03226$), then according to the **Theorem 11** (Appendix 2) this can be obtained for a value of p_e in the range $(134.85; 164.49)[Pa]$.

Taking into account the above fact, in order to create a concave outer free surface, appropriate for the growth of a tube of which outer radius is equal to $r_e^1 = 4.65 \times 10^{-3}[m]$ we have solved the i.v.p. (A.1.3) for different values of p_e in the range $(134.85; 164.49)[Pa]$.

More precisely, we have integrated the system (10) for $z_e(R_{ge}) = 0$, $z_e'(R_{ge}) = -\tan\alpha_c$ and different p_e. The obtained outer radii r_e versus p_e are represented in Fig.5, which shows that the desired outer radius $r_e^1 = 4.65 \times 10^{-3}[m]$ is obtained for $p_e' = 149.7[Pa]$.

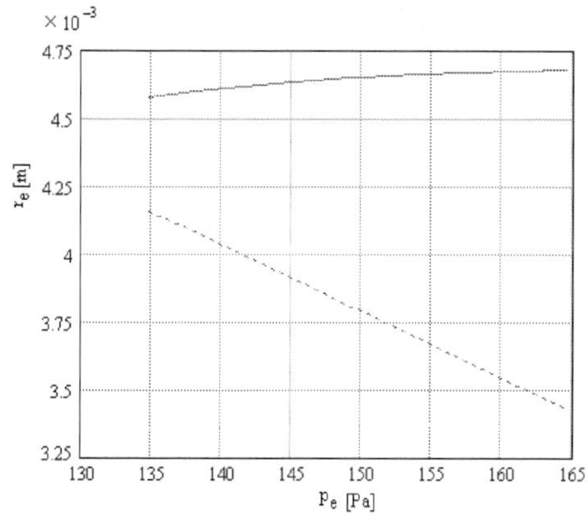

Fig. 5. Outer radii r_e versus p_e in the range $(134.85; 164.49)[Pa]$.

Actually, as it can be seen in the same figure, for $p_e{}'=149.7[Pa]$ we can also obtain a second outer radius $r_e^2 = 3.8\times10^{-3}[m]$, which is not in the desired range $\left(\frac{R_{gi}+R_{ge}}{2}, R_{ge}\right)$. Moreover, the outer free surface of this meniscus is not globally concave; it is a convex-concave meniscus (Fig.6).

Taking into account $p_m \approx 0$ [Eriss, 1980], [Rossolenko, 2001], [Yang, 2006], the melt column height in this case is $H_e{}' = -\frac{1}{\rho\cdot g}\cdot[p_e{}'+p_g^e]$, where $p_g^e \geq 0$ is the pressure of the gas flow (introduced in the furnace for releasing the heat from the outer side of the tube wall). When $p_g^e = 0$, then $H_e{}'$ is negative, $H_e{}' = -2.31\times10^{-3}[m]$, i.e. the crucible melt level has to be with $-H_e{}' = 2.31\times10^{-3}[m]$ above the shaper top level.

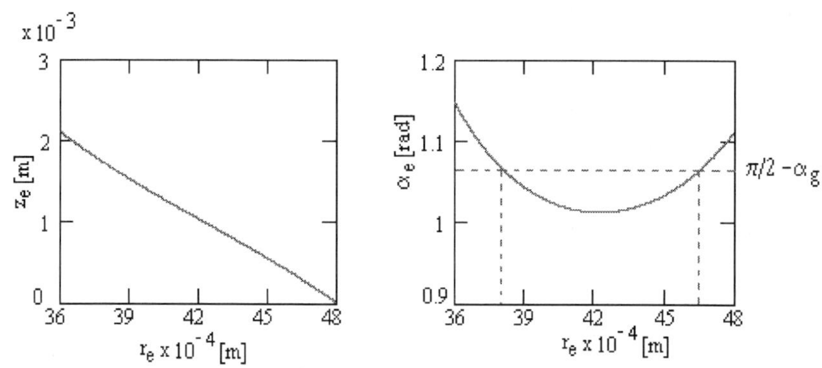

Fig. 6. Non globally concave outer free surface obtained for $p_e{}'=149.7[Pa]$.

If there exists a concave inner free surface for the growth of a tube of inner radius

$r_i' = 4.35 \times 10^{-3} [m]$ ($m_1 = 1.03571$), then according to the **Theorem 14** (Appendix 2), this can be obtained for a value of p_i which is in the range $(-31.46, -1.07)[Pa]$.

Taking into account the above fact, in order to create a concave inner free surface, appropriate for the growth of a tube whose inner radius is equal to $r_i' = 4.35 \times 10^{-3} [m]$, we have solved the i.v.p. (A.1.8) for different values of p_i in the range $(-31.46, -1.07)[Pa]$. More precisely, we have integrated the system (11) for $z_i(R_{gi}) = 0$, $z_i'(R_{gi}) = \tan\alpha_c$ and different p_i. The obtained inner radii r_i versus p_i are represented in Fig.7 which shows that the desired inner radius $r_i^1 = 4.35 \times 10^{-3} [m]$ is obtained for $p_i' = -16.2[Pa]$.

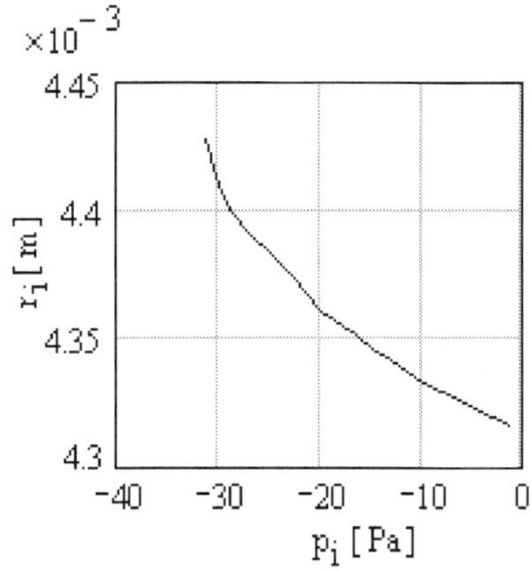

Fig. 7. Inner radii r_i versus p_i in the range $(-31.46, -1.07)[Pa]$.

Taking $p_m \approx 0$ [Eriss et al., 1980], [Rossolenko et al., 2001], [Yang et al. 2006], the melt column height in this case is $H_i' = -\frac{1}{\rho \cdot g} \cdot [p_i' + p_g^i]$, where $p_g^i \geq 0$ is the pressure of the gas flow (introduced in the furnace for releasing the heat from the inner side of the tube wall). When $p_g^i = 0$, then H_i' is positive, $H_i' = 0.25 \times 10^{-3} [m]$, i.e. the crucible melt level has to be with $H_i' = 0.25 \times 10^{-3} [m]$ under the shaper top level. To create a concave meniscus, appropriate for the growth of a tube with outer radius $r_e' = 4.65 \times 10^{-3}$ [m] and inner radius $r_i' = 4.35 \times 10^{-3}$ [m] the melt column heights (with respect to the crucible melt level) have to be $H_i^1 = -\frac{1}{\rho \cdot g} \cdot \left[-16.2 + p_g^i\right]$ and $H_e^1 = -\frac{1}{\rho \cdot g} \cdot \left[149.7 + p_g^e\right]$. When the shaper's outer top is at the same level as the shaper's inner top, with respect to the crucible melt level, then the relation $H_e^1 = H_i^1$ holds . It follows that the pressure of the gas flow, introduced in the furnace for releasing the heat from the inner wall of the tube, p_g^i has to be higher than the pressure of the gas flow introduced in the furnace for releasing the heat from the outer wall of the tube, p_g^e; $p_g^i - p_g^e = 149.7 + 16.2 = 165.9[Pa]$.

4. The angles $\bar{\alpha}_e(r_e, h_e, p_e)$ and $\bar{\alpha}_i(r_i, h_i, p_i)$ which appear in system (1) describing the dynamics of the outer and inner radius of a tube, grown by the E.F.G. technique

4.1 The procedure for the determination of the angles $\bar{\alpha}_e(r_e, h_e, p_e)$ and $\bar{\alpha}_i(r_i, h_i, p_i)$ for a convex free surface

The angles $\bar{\alpha}_e(r_e, h_e, p_e)$ ($\bar{\alpha}_i(r_i, h_i, p_i)$) represent the angles between the tangent line to the outer free surface (inner free surface) of the meniscus at the three phase point, of coordinates (r_e, h_e) ((r_i, h_i)) and the horizontal axis Or. These angles can fluctuate during the growth. The deviation of the tangent to the crystal outer (inner) free surface at the triple point from the vertical is the difference $\bar{\alpha}_e(r_e, h_e, p_e) - \left(\frac{\pi}{2} - \alpha_g\right)$, $\left(\bar{\alpha}_i(r_i, h_i, p_i) - \left(\frac{\pi}{2} - \alpha_g\right)\right)$ (Fig. 2), where α_g is the growth angle. The deviation can fluctuate also and the outer (inner) radius r_e (r_i) is constant when the deviation is constant equal to zero

The angles $\bar{\alpha}_e(r_e, h_e, p_e)$ and $\bar{\alpha}_i(r_i, h_i, p_i)$ cannot be obtained directly from the Young-Laplace equation. For this reason for this equation the following strategy is adopted: two conditions are imposed at the outer radius R_{ge} of the shaper (inner radius R_{gi} of the shaper) $z_e(R_{ge}) = 0$; $z_e'(R_{ge}) = -\tan(\alpha_c^e)$ ($z_i(R_{gi}) = 0$, $z_i'(R_{gi}) = \tan(\alpha_c^i)$). In the last condition α_c^e (α_c^i) is a parameter which can fluctuate in a certain range, during the growth. For different values of α_c^e (α_c^i) in a given range the solution $z_e(r; \alpha_c^e, p_e)$ ($z_i(r; \alpha_c^i, p_i)$) of the Young-Laplace equation which satisfies the conditions $z_e(R_{ge}) = 0$; $z_e'(R_{ge}) = -\tan(\alpha_c^e)$ at R_{ge} (($z_i(R_{gi}) = 0$, $z_i'(R_{gi}) = \tan(\alpha_c^i)$) at R_{gi}) is found.

With $z_e(r; \alpha_c^e, p_e)$ ($z_i(r; \alpha_c^i, p_i)$) the function $\alpha_e(r_e; \alpha_c^e, p_e) = -\arctan z_e'(r_e; \alpha_c^e, p_e)$ ($\alpha_i(r_i; \alpha_c^i, p_i) = \arctan z_i'(r_i; \alpha_c^i, p_i)$) is constructed. After that, from $h_e = z_e(r; \alpha_c^e, p_e)$ ($h_i = z_i(r; \alpha_c^i, p_i)$) α_c^e (α_c^i) is expressed as function of r_e, h_e and p_e (r_i, h_i and p_i).

$\alpha_c^e = \alpha_c^e(r_e; h_e, p_e)$ ($\alpha_c^i = \alpha_c^i(r_i; h_e, p_i)$) is introduced in $\alpha_e(r_e; \alpha_c^e, p_e)$ ($\alpha_i(r_i; \alpha_c^i, p_i)$) obtaining the function

$$\bar{\alpha}_e(r_e, h_e, p_e) = \alpha_e(r_e; \alpha_c^e(r_e; h_e, p_e), p_e) \qquad (\bar{\alpha}_i(r_i, h_i, p_i) = \alpha_i(r_i; \alpha_c^i(r_i; h_i, p_i), p_i)).$$

To the best of our knowledge, there is no algorithm in the literature concerning the construction of $\bar{\alpha}_e(r_e, h_e, p_e)$ ($\bar{\alpha}_i(r_i, h_i, p_i)$) at the level of generality presented here.

Due to the nonlinearity, the above described procedure can't be realized analytically. This is the reason why for the construction of the function $\overline{\alpha}_e(r_e;\alpha_c,p_e)$ in [Balint&Tanasie, 2010] the following numerical procedure was conceived:

Step 1. For a given α_c^{e0}; $\alpha_c^{e0} \in \left(0, \pi/2 - \alpha_g\right)$ and $n = \dfrac{2 \cdot R_{ge}}{R_{ge} + R_{gi}}$ an $n' \in (1,n)$ is found such that $E_1^e\left(n', \alpha_c^{e0}\right) < E_2^e\left(n, \alpha_c^{e0}\right)$ where:

$$E_1^e\left(n',\alpha_c^{e0}\right) = -\gamma \cdot \frac{\pi/2 - \left(\alpha_c^{e0} + \alpha_g\right)}{R_{ge}} \cdot \frac{n'}{n'-1} \cdot \sin\alpha_g + \rho_1 \cdot g \cdot R_{ge} \cdot \frac{n'-1}{n'} \cdot \tan\left(\pi/2 - \alpha_g\right) + \frac{\gamma}{R_{ge}} \cdot n' \cdot \cos\alpha_g \tag{12}$$

$$E_2^e(n,\alpha_c^{e0}) = -\gamma \cdot \frac{\pi/2 - \left(\alpha_c^{e0} + \alpha_g\right)}{R_{ge}} \cdot \frac{n}{n-1} \cdot \cos\alpha_c^{e0} + \frac{\gamma}{R_{ge}} \cdot \sin\alpha_c^{e0}$$

Step 2. For α_c^e a range $\left[\underline{\alpha}_c^e, \overline{\alpha}_c^e\right]$ is determined such that $0 < \underline{\alpha}_c^e < \alpha_c^{e0} < \overline{\alpha}_c^e < \pi/2 - \alpha_g$ and for every $\alpha_c^e \in \left[\underline{\alpha}_c^e, \overline{\alpha}_c^e\right]$ the inequality $E_1^e\left(n', \alpha_c^e\right) < E_2^e\left(n, \alpha_c^e\right)$ holds.

Step 3. For p_e the range $\left[\underline{p}_e, \overline{p}_e\right]$ defined by:

$$\underline{p}_e = \sup_{\alpha_c^e \in \left[\underline{\alpha}_c^e, \overline{\alpha}_c^e\right]} E_1^e\left(n', \alpha_c^e\right) \qquad \overline{p}_e = \inf_{\alpha_c^e \in \left[\underline{\alpha}_c^e, \overline{\alpha}_c^e\right]} E_2^e\left(n, \alpha_c^e\right) \tag{13}$$

is considered.

Step 4. In the range $\left[\underline{\alpha}_c^e, \overline{\alpha}_c^e\right]$ a set of l different values of α_c^e is chosen.

Step 5. In the range $\left[\underline{p}_e, \overline{p}_e\right]$ a set of m different values of p_e is chosen.

Step 6. In a given range $\left[\underline{\alpha}_e, \overline{\alpha}_e\right]$ possessing the property $\overline{\alpha}_c^e < \underline{\alpha}_e < \pi/2 - \alpha_g < \overline{\alpha}_e < \pi/2$ a set of j values of α_e is chosen: $\underline{\alpha}_e = \alpha_e^1 < \alpha_e^2 < \ldots < \alpha_e^j = \overline{\alpha}_e$.

Step 7. For a given p_e^k, $k = \overline{1,m}$ and α_c^{eq}, $q = \overline{1,l}$ the solution of the system (10) corresponding to the conditions: $z_e(R_{ge}) = 0, \alpha_e(R_{ge}) = \alpha_c^{eq}$ is determined numerically obtaining the functions (profiles curves Refs [Tatarchenko, 1993]): $z_e = z_e(r;\alpha_c^{eq}, p_e^k)$ and $\alpha_e = \alpha_e(r;\alpha_c^{eq}, p_e^k)$.

Step 8. The values $r_{k\,q\,s}^e$ for which $\alpha_e(r_{k\,q\,s}^e;\alpha_c^{eq}, p_e^k) = \alpha_e^s \in \left[\underline{\alpha}_e, \overline{\alpha}_e\right]$, $k = \overline{1,m}$, $q = \overline{1,l}$ are determined.

Step 9. The values $h_{kqs}^{e}=z_{e}(r_{kqs}^{e};\alpha_{c}^{eq},p_{e}^{k})$ are found.

Step 10. Fitting the data r_{kqs}^{e} h_{kqs}^{e} p_{e}^{k} and α_{e}^{s}, the function $\bar{\alpha}_{e}(r_{e},h_{e},p_{e})$ is found.

For the same reason as in the case of $\bar{\alpha}_{e}(r_{e},h_{e},p_{e})$ for the construction of $\bar{\alpha}_{i}=\bar{\alpha}_{i}(r_{i},h_{i},p_{i})$ the following numerical procedure was conceived:

Step 1. For $\alpha_{c}^{i0}=\alpha_{c}^{e0}$ and $m\in\left(1,\dfrac{R_{ge}+R_{gi}}{2\cdot R_{gi}}\right]$ an $m'\in(1,m)$ is determined such that $E_{1}^{i}\left(m',\alpha_{c}^{i0}\right)<E_{2}^{i}\left(m,\alpha_{c}^{i0}\right)$ where:

$$E_{1}^{i}\left(m',\alpha_{c}^{i0}\right)=-\gamma\cdot\frac{\pi/2-\left(\alpha_{c}^{i0}+\alpha_{g}\right)}{(m'-1)\cdot R_{gi}}\cdot\sin\alpha_{g}+\rho_{1}\cdot g\cdot R_{gi}\cdot(m'-1)\cdot\tan\left(\pi/2-\alpha_{g}\right)-$$
$$-\frac{\gamma}{m'\cdot R_{gi}}\cdot\sin\alpha_{c}^{i0} \tag{14}$$
$$E_{2}^{i}(m,\alpha_{c}^{i0})=-\gamma\cdot\frac{\pi/2-\left(\alpha_{c}^{i0}+\alpha_{g}\right)}{(m-1)\cdot R_{gi}}\cdot\cos\alpha_{c}^{i0}-\frac{\gamma}{R_{gi}}\cdot\cos\alpha_{g}$$

Step 2. For α_{c}^{i} a range $\left[\underline{\alpha}_{c}^{i},\bar{\alpha}_{c}^{i}\right]$ is determined such that $0<\underline{\alpha}_{c}^{i}<\alpha_{c}^{i}<\bar{\alpha}_{c}^{i}<\pi/2-\alpha_{g}$ and for every $\alpha_{c}^{i}\in\left[\underline{\alpha}_{c}^{i},\bar{\alpha}_{c}^{i}\right]$, the inequality $E_{1}^{i}\left(m',\alpha_{c}^{i}\right)<E_{2}^{i}\left(m,\alpha_{c}^{i}\right)$ holds.

Step 3. For p_{i} the range $\left[\underline{p}_{i},\bar{p}_{i}\right]$ defined by:

$$\underline{p}_{i}=\sup_{\alpha_{c}^{i}\in\left[\underline{\alpha}_{c}^{i},\bar{\alpha}_{c}^{i}\right]}E_{1}^{i}\left(m',\alpha_{c}^{i}\right)\qquad\bar{p}_{i}=\inf_{\alpha_{c}^{i}\in\left[\underline{\alpha}_{c}^{i},\bar{\alpha}_{c}^{i}\right]}E_{2}^{i}\left(m,\alpha_{c}^{i}\right) \tag{15}$$

is considered.

Step 4. In the range $\left[\underline{\alpha}_{c}^{i},\bar{\alpha}_{c}^{i}\right]$ a set of l different values of α_{c}^{i} are chosen.

Step 5. In the range $\left[\underline{p}_{i},\bar{p}_{i}\right]$ a set of n different values of p_{i} are chosen.

Step 6. In a given range $\left[\underline{\alpha}_{i},\bar{\alpha}_{i}\right]$, possessing the property $\bar{\alpha}_{c}^{i}<\underline{\alpha}_{i}<\pi/2-\alpha_{g}<\bar{\alpha}_{i}<\pi/2$, a set of j values of α_{i} are chosen: $\underline{\alpha}_{i}=\alpha_{i}^{1}<\alpha_{i}^{2}<\ldots<\alpha_{i}^{j}=\bar{\alpha}_{i}$.

Step 7. For a given p_{i}^{k}, $k=\overline{1,n}$ and α_{c}^{iq}, $q=\overline{1,l}$ the solution of the system (11) which satisfies the conditions: $z_{i}(R_{gi})=0,\ \alpha_{i}(R_{gi})=\alpha_{c}^{iq}$ is found numerically obtaining the functions (profiles curves Ref. [Tatarchenko, 1993]): $z_{i}=z_{i}(r;\alpha_{c}^{iq},p_{i}^{k})$ and $\alpha_{i}=\alpha_{i}(r;\alpha_{c}^{iq},p_{i}^{k})$.

Step 8. The values r^{i}_{kqs} for which $\alpha_i(r^{i}_{kqs};\alpha^{iq}_{c},p^{k}_{i})=\alpha^{s}_{i}\in\left[\underline{\alpha}_{e},\overline{\alpha}_{e}\right], k=\overline{1,n},\ q=\overline{1,l}$ are determined.

Step 9. The values $h^{i}_{kqs}=z_i(r^{i}_{kqs};\alpha^{iq}_{c},p^{k}_{i})$ are found.

Step 10. The function $\bar{\alpha}_i(r_i,h_i,p_i)$ is found by fitting the data r^{i}_{kqs} h^{i}_{kqs} p^{k}_{i} and α^{s}_{i}.

For the case of a silicon tube and the outer free surface the function

$$\bar{\alpha}_e(r_e,h_e,p_e)=\frac{a_1(p_e)+a_2(p_e)\cdot r_e+a_3(p_e)\cdot r_e^2+a_4(p_e)\cdot h_e+a_5(p_e)\cdot h_e^2+a_6(p_e)\cdot h_e^3}{1+a_7(p_e)\cdot r_e+a_8(p_e)\cdot h_e+a_9(p_e)\cdot h_e^2+a_{10}(p_e)\cdot h_e^3}$$

with: $a_1(p_e)=0.400314546\text{-}0.00494101\cdot \mathrm{p_e}$ $\quad a_2(p_e)=\text{-}374.789455+2.353132518\cdot \mathrm{p_e}$

$a_3(p_e)=56161.27414\text{-}244.013259\cdot \mathrm{p_e}$ $\quad a_4(p_e)=393.2446055+0.010192507\cdot \mathrm{p_e}$

$a_5(p_e)=91143.39025\text{-}555.405995\cdot \mathrm{p_e}$ $\quad a_6(p_e)=\text{-}1097500000\text{-}1168100\cdot \mathrm{p_e}$

$a_7(p_e)=\text{-}200.489376+0.001998751\cdot \mathrm{p_e}$ $\quad a_8(p_e)=114.0391478+0.002922594\cdot \mathrm{p_e}$

$a_9(p_e)=83485.77784\text{-}477.770821\cdot \mathrm{p_e}$ $\quad a_{10}(p_e)=415910000+476372.6048\cdot \mathrm{p_e}$

was obtained and for the inner free surface the function:

$$\bar{\alpha}_i(r_i;h_i,p_i)=\frac{b_1(p_i)+b_2(p_i)\cdot r_i+b_3(p_i)\cdot r_i^2+b_4(p_i)\cdot h_i+b_5(p_i)\cdot h_i^2+b_6(p_i)\cdot h_i^3}{1+b_7(p_i)\cdot r_i+b_8(p_i)\cdot h_i+b_9(p_i)\cdot h_i^2+b_{10}(p_i)\cdot h_i^3}$$

with $b_1(p_i)=0.102617985+0.004931999\cdot \mathrm{p_i}$ $\quad b_2(p_i)=\text{-}172.7251\text{-}2.34898717\cdot \mathrm{p_i}$

$b_3(p_i)=29313.52855+277.5275727\cdot \mathrm{p_i}$ $\quad b_4(p_i)=444.327355+0.367481455\cdot \mathrm{p_i}$

$b_5(p_i)=\text{-}2687700\text{-}817.408936\cdot \mathrm{p_i}$ $\quad b_6(p_i)=5760710000+3810000\cdot \mathrm{p_i}$

$b_7(p_i)=\text{-}263.39243\text{-}0.00872161\cdot \mathrm{p_i}$ $\quad b_8(p_i)=707.770183+0.334375383\cdot \mathrm{p_i}$

$b_9(p_i)=\text{-}2709500\text{-}696.727273\cdot \mathrm{p_i}$ $\quad b_{10}(p_i)=3090180000+1348660\cdot \mathrm{p_i}$

was obtained. For $p_e=-2000[Pa]$ and $p_i=-2242[Pa]$ the functions $\bar{\alpha}_e(r_e,h_e,p_e)$ and $\bar{\alpha}_i(r_i,h_i,p_i)$ are represented in Fig. 8.:

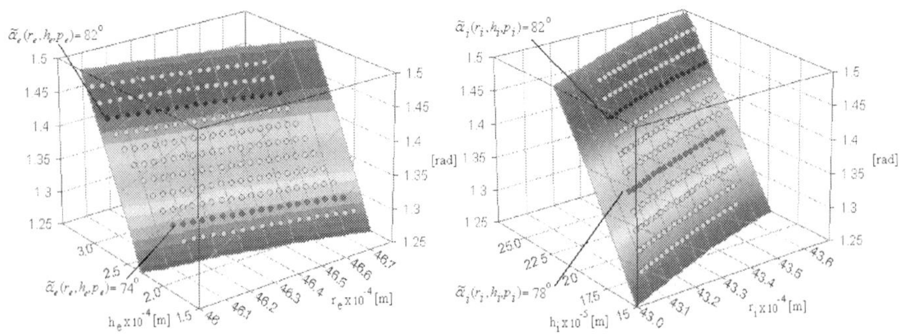

Fig. 8. The graphics of $\bar{\alpha}_e(r_e,h_e,p_e)$ ($p_e=-2000\,[Pa]$) and $\bar{\alpha}(r_i,h_i,p_i)$ ($p_i=-2242\,[Pa]$)

4.2 The procedure for the determination of the angles $\bar{\alpha}_e(r_e,h_e,p_e)$ and $\bar{\alpha}(r_i,h_i,p_i)$ for a concave free surface

The numerical procedure for the construction of the function $\bar{\alpha}_e(r_e,h_e,p_e)$ for a concave free surface (when $\alpha_c > \frac{\pi}{2} - \alpha_g$) is similar to those applied for a convex free surface. Only **Step 1-Step 3**. present the some differences. For the outer free surface we have to consider:

Step 1. For a given α_c^{e0}; $\alpha_c^{e0} \in \left(\pi/2 - \alpha_g, \pi/2\right)$ and $n = \dfrac{2 \cdot R_{ge}}{R_{ge} + R_{gi}}$ an $n' \in (1,n)$ is found such that $E_1^e\left(n,\alpha_c^{e0}\right) < E_2^e\left(n',\alpha_c^{e0}\right)$ where:

$$\begin{aligned} E_1^e\left(n,\alpha_c^{e0}\right) &= \gamma \cdot \frac{\alpha_c^{e0} + \alpha_g - \pi/2}{R_{ge}} \cdot \frac{n}{n-1} \cdot \sin\alpha_g + \rho_1 \cdot g \cdot R_{ge} \cdot \frac{n-1}{n} \cdot \tan\alpha_c^{e0} + \frac{\gamma}{R_{ge}} \cdot n \cdot \sin\alpha_c^{e0} \\ E_2^e(n',\alpha_c^{e0}) &= \gamma \cdot \frac{\alpha_c^{e0} + \alpha_g - \pi/2}{R_{ge}} \cdot \frac{n'}{n'-1} \cdot \cos\alpha_c^{e0} + \frac{\gamma}{R_{ge}} \cdot \cos\alpha_g \end{aligned} \tag{16}$$

Step 2. For α_c^e a range $\left[\underline{\alpha}_c^e, \bar{\alpha}_c^e\right]$ is determined such that $\pi/2 - \alpha_g < \underline{\alpha}_c^e < \alpha_c^{e0} < \bar{\alpha}_c^e < \pi/2$ and the inequality $\sup\limits_{\alpha_c^e \in \left[\underline{\alpha}_c^e, \bar{\alpha}_c^e\right]} E_1^e\left(n,\alpha_c^e\right) < \inf\limits_{\alpha_c^e \in \left[\underline{\alpha}_c^e, \bar{\alpha}_c^e\right]} E_2^e\left(n',\alpha_c^e\right)$ holds.

Step 3. For p_e the range $\left[\underline{p}_e, \bar{p}_e\right]$ defined by:

$$\underline{p}_e = \sup_{\alpha_c^e \in \left[\underline{\alpha}_c^e, \bar{\alpha}_c^e\right]} E_1^e\left(n,\alpha_c^e\right) \qquad \bar{p}_e = \inf_{\alpha_c^e \in \left[\underline{\alpha}_c^e, \bar{\alpha}_c^e\right]} E_2^e\left(n',\alpha_c^e\right) \tag{17}$$

is considered.

For the inner free surface we have to make:

Step 1. For $\alpha_c^{i0} = \alpha_c^{e0}$ and $m \in \left(1, \dfrac{R_{ge} + R_{gi}}{2 \cdot R_{gi}}\right]$ an $m' \in (1,m)$ is determined such that $E_1^i\left(m,\alpha_c^{i0}\right) < E_2^i\left(m',\alpha_c^{i0}\right)$ where:

$$\begin{aligned} E_1^i\left(m,\alpha_c^{i0}\right) &= \gamma \cdot \frac{\alpha_c^{i0} + \alpha_g - \pi/2}{(m-1) \cdot R_{ge}} \cdot \sin\alpha_g + \rho_1 \cdot g \cdot R_{gi} \cdot (m-1) \cdot \tan\alpha_c^{i0} - \frac{\gamma}{m \cdot R_{gi}} \cdot \cos\alpha_g \\ E_2^i(m',\alpha_c^{i0}) &= \gamma \cdot \frac{\alpha_c^{i0} + \alpha_g - \pi/2}{(m'-1) \cdot R_{gi}} \cdot \cos\alpha_c^{i0} + \frac{\gamma}{R_{gi}} \cdot \sin\alpha_c^{i0} \end{aligned} \tag{18}$$

Step 2. For α_c^i a range $\left[\underline{\alpha}_c^i, \bar{\alpha}_c^i\right]$ is determined such that $\pi/2 - \alpha_g < \underline{\alpha}_c^i < \alpha_c^{i0} < \bar{\alpha}_c^i < \pi/2$ and the inequality $\sup\limits_{\alpha_c^i \in \left[\underline{\alpha}_c^i, \bar{\alpha}_c^i\right]} E_1^i\left(m,\alpha_c^i\right) < \inf\limits_{\alpha_c^i \in \left[\underline{\alpha}_c^i, \bar{\alpha}_c^i\right]} E_2^i\left(m',\alpha_c^i\right)$ holds.

Step 3. For p_i the range $\left[\underline{p}_i, \overline{p}_i\right]$ defined by:

$$\underline{p}_i = \sup_{\alpha_c^i \in \left[\underline{\alpha}_c^i, \overline{\alpha}_c^i\right]} E_1^i\left(m, \alpha_c^i\right) \qquad \overline{p}_i = \inf_{\alpha_c^i \in \left[\underline{\alpha}_c^i, \overline{\alpha}_c^i\right]} E_2^i\left(m', \alpha_c^i\right) \tag{19}$$

is considered.

In the case of of the InSb tube considered in section 2.2, for the outer free surface the function:

$$\overline{\alpha}_e(r_e, h_e, p_e) = \frac{a_1(p_e) + a_2(p_e)\cdot r_e + a_3(p_e)\cdot r_e^2 + a_4(p_e)\cdot h_e + a_5(p_e)\cdot h_e^2 + a_6(p_e)\cdot h_e^3}{1 + a_7(p_e)\cdot r_e + a_8(p_e)\cdot h_e + a_9(p_e)\cdot h_e^2 + a_{10}(p_e)\cdot h_e^3}$$

with
$a_1(p_e) = 16.95005004 - 0.1803412 \cdot p_e$ $\quad a_2(p_e) = \text{-}6991.91018 + 76.11771837 \cdot p_e$
$a_3(p_e) = 721251.8226 - 8031.90955 \cdot p_e$ $\quad a_4(p_e) = 82.08229042 - 1.41357516 \cdot p_e$
$a_5(p_e) = -820223.151 + 11947.58366 \cdot p_e$ $\quad a_6(p_e) = 75949100 - 3370800 \cdot p_e$
$a_7(p_e) = -207.085329 - 0.00600066 \cdot p_e$ $\quad a_8(p_e) = 0.10447655 - 0.82800845 \cdot p_e$
$a_9(p_e) = -544970.373 + 8774.44 \cdot p_e$ $\quad a_{10}(p_e) = -237870000 - 663676.426 \cdot p_e$

was obtained. For the inner free surface the function:

$$\overline{\alpha}_i(r_i; h_i, p_i) = \frac{b_1(p_i) + b_2(p_i)\cdot r_i + b_3(p_i)\cdot h_i + b_4(p_i)\cdot h_i^2 + b_5(p_i)\cdot h_i^3}{1 + b_6(p_i)\cdot r_i + b_7(p_i)\cdot r_i^2 + b_8(p_i)\cdot h_i + b_9(p_i)\cdot h_i^2}$$

with $b_1(p_i) = \text{-}0.03323515 + 0.0000618675 \cdot \mathrm{p_i}$ $\quad b_2(p_i) = 7.907785322\text{-}0.01471419 \cdot \mathrm{p_i}$
$b_3(p_i) = \text{-}10.9920301 + 0.019065783 \cdot \mathrm{p_i}$ $\quad b_4(p_i) = 81321.53642 + 10.42303935 \cdot \mathrm{p_i}$
$b_5(p_i) = \text{-}6164200\text{-}109065.464 \cdot \mathrm{p_i}$ $\quad b_6(p_i) = \text{-}478.279159 + 0.004407571 \cdot \mathrm{p_i}$
$b_7(p_i) = 57185.4894\text{-}1.04591888 \cdot \mathrm{p_i}$ $\quad b_8(p_i) = \text{-}7.40209299 + 0.012781953 \cdot \mathrm{p_i}$
$b_9(p_i) = 57826.05599 + 9.103117035 \cdot \mathrm{p_i}$

was obtained. For $p_e = 180[Pa]$ and $p_i = 290[Pa]$ the functions $\overline{\alpha}_e(r_e, h_e, p_e)$ and $\overline{\alpha}_i(r_i, h_i, p_i)$ are represented in Fig. 9:

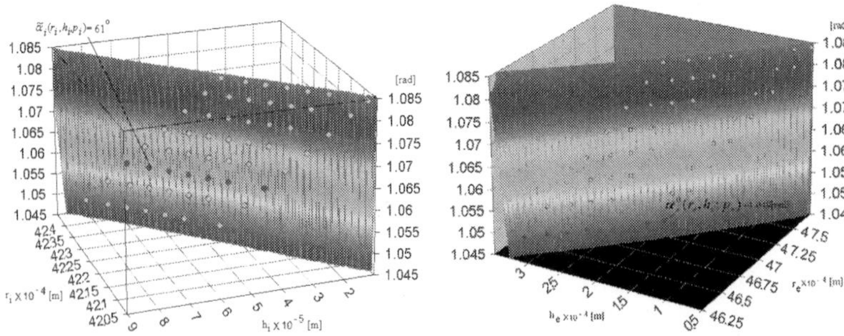

Fig. 9. The graphics of $\overline{\alpha}_i(r_i, h_i, p_i)$ $(p_i = 290[Pa])$ and $\overline{\alpha}_i(r_i, h_i, p_i)$ $(p_e = 180[Pa])$

5. Setting the pulling rate, the thermal and capillary conditions

In this section it will be shown that the results presented in the above sections can be used for setting the pulling rate, the thermal and capillary conditions in view of an experiment [Tanasie&Balint, 2010].

According to [Tatarchenko, 1993] at the level of the crystallization front h the crystallization rate v_c^j is given by:

$$v_c^j = \frac{1}{\Lambda \cdot \rho_1} \cdot \left[\lambda_1 \cdot G_1^j(r_e, r_i, h) - \lambda_2 \cdot G_2^j(r_e, r_i, h) \right], j = 1,2,3. \tag{20}$$

The difference between the pulling rate v and the crystallization rate v_c^j is equal to the crystallization front displacement rate $\frac{dh^j}{dt}$, $j = 1,2,3$.

In order to keep the crystallization front level h^j constant, the pulling rate and the thermal conditions have to satisfy the following conditions:

$$v - \frac{1}{\Lambda \cdot \rho_1} \cdot \left[\lambda_1 \cdot G_1^j(r_e, r_i, h) - \lambda_2 \cdot G_2^j(r_e, r_i, h) \right] = 0, j = 1,2,3 \tag{21}$$

When the radii r_e, r_i and the length L of the tube, which has to be grown, are prior given, and h is known, then the condition (21) can be regarded as an equation in which the pulling rate v is unknown. If this equation has a positive solution v, it depends on the following parameters: h, $T_{en}(0)$, T_0, and k. The setting of the pulling rate, thermal conditions means the choice of v, $T_{en}(0)$, T_0 and k such that the following conditions be satisfied:

- $300 < T_{en}(0) < T_m < T_0$; $0 < k < \frac{T_{en}(0) - 300}{L}$;
- equation (21) has a positive solution v in an acceptable range.
- v is practically the same for every L': $L_0 \le L' \le L$ (L_0 = the seed length).

The setting of the capillary condition means to take the tube radii r_e, r_i (prior given) and the crystallization front level h_c, determined form (21) (for the above chosen v, $T_{en}(0)$, T_0, k) and find the pressures p_e, p_i solving the followings equations:

$$\overline{\alpha}_e(r_e, h_c, p_e) = \pi/2 - \alpha_g \text{ and } \overline{\alpha}_i(r_i, h_c, p_i) = \pi/2 - \alpha_g \tag{22}$$

If the solutions p_e, p_i of this equations are in the range for which $\overline{\alpha}_e$, $\overline{\alpha}_i$ was build up, then the values p_e, p_i will be used to set p_g^e, p_g^i, H_e, H_i using (2) with $p_m = 0$ or

$$H_e = -\frac{p_g^e + p_e}{\rho_1 \cdot g}, \quad H_i = -\frac{p_g^i + p_i}{\rho_1 \cdot g} \tag{23}$$

For the growth of a silicon tube with convex profile curves the following numerical data will be used: $\rho_1 = 2.5 \times 10^3$ [kg/m³]; $\rho_2 = 2.3 \times 10^3$ [kg/m³]; $T_m = 1683$ [K]; $\lambda_1 = 60$ [W/m·K]; $\lambda_2 = 21.6$ [W/m·K]; $\Lambda = 1.81 \times 10^6$ [J/kg]; $\chi_1 = \dfrac{\lambda_1}{c_1 \cdot \rho_1}$; $\chi_2 = \dfrac{\lambda_2}{c_2 \cdot \rho_2}$; $c_1 = 913$ [J/kg·K]; $c_2 = 703$ [J/kg·K]; $\mu_1 = 7300.42$ [K]; $\mu_2 = 2822.58$ [K]; $R_{gi} = 4.2 \times 10^{-3}$ [m]; $R_{ge} = 4.8 \times 10^{-3}$ [m];

$R_i^c = 4.339 \times 10^{-3}$ [m]; $R_e^c = 4.66 \times 10^{-3}$ [m]; $L_1 = 0.4$ [m]; $L_2 = 0.2$ [m]; $L_3 = 0.1$ [m].

Step 1. A stable static outer meniscus is chosen, whose characteristic parameters r_e, h, p_e are in the range where $\overline{\alpha}_e(r_e, h, p_e)$ is valid and for which r_e is close to r_e^c. In the case considered here such a static meniscus is obtained for $p_e = -1980$ [Pa] and its characteristic parameters are: $r_e = 4.660112250074 \times 10^{-3}$ [m] and $h = 2.14370857185 \times 10^{-4}$ [m].

Step 2. An initial input for T_0, $T_{en}(0)$ and k has to be chosen. For T_0, the start can be $T_0 = T_m + 1$. Concerning $T_{en}(0)$ and k the start can be $T_{en}(0) = T_m - 1$ and $k = \dfrac{T_{en}(0) - 300}{L}$.

Using this input and the values r_e, h, r_i^c the value of the pulling rates $v_1, v_2, \ldots, v_{40}$ given by the equation (21) have to be found. If all these values are positive, then: if the average $\overline{v}$ and standard deviation σ of the set of values of v, are acceptable, then the average pulling rate $\overline{v}$ and the initial input thermal conditions can be set, else the initial input thermal conditions have to be reset lowering in general $T_{en}(0)$ and/or increasing T_0.

Step 3. Consider $\overline{v}$, $T_{en}(0)$, T_0, k obtained above and solve equation (22) for these values choosing $r_e = r_e^c$ and $r_i = r_i^c$ (the desired radii) and h unknown. Denote by h_c the obtained solution. Replace r_e^c, r_i^c, h_c in equation (23) and solve this equations finding p_e, p_i.

Step 4. Using p_e, p_i find $H_i - H_e$, for $p_g^e = p_g^i$ (in the case of an open crucible) or find $p_g^i - p_g^e$ for $H_i - H_e = 0$ (in the case of a closed crucible).

Following the above steps for the considered silicon tube growth, some of the computed possible settings, are presented in Table 1.

For the above settings the growth process stability analysis is made through the system of nonlinear ordinary differential equations (1) which governs the evolution of r_e, r_i, h for the established settings. It means to verify first of all that the desired r_e^c, r_i^c and the obtained h_c is a steady state of (1).

Furthermore to verify if at the start r_e^c, r_i^c, h_c are perturbed (i.e. the seed sizes are different from r_e^c, r_i^c) after a period of transition the values r_e^c, r_i^c, h_c are recovered. In other words, to verify if the steady state (r_e^c, r_i^c, h_c) is asymptotically stable. This last requirement is satisfied if the Hurwitz conditions are satisfied [Tatarchenko, 1993] i.e.:

$$\begin{gathered} -a_{11}-a_{22}-a_{33}>0,\; -a_{11}\cdot a_{22}\cdot a_{33}+a_{31}\cdot a_{13}\cdot a_{22}>0 \\ (-a_{11}-a_{22}-a_{33})\cdot(-a_{31}\cdot a_{13}+a_{11}\cdot a_{22}+a_{22}\cdot a_{33}+a_{11}\cdot a_{33})-(-a_{11}\cdot a_{22}\cdot a_{33}+a_{31}\cdot a_{13}\cdot a_{22})>0 \end{gathered} \tag{24}$$

$$\begin{gathered} a_{11}=-v\cdot\frac{\partial\bar{\alpha}_e(r_e^c,h_c,p_e)}{\partial R_e}\quad a_{12}=-v\cdot\frac{\partial\bar{\alpha}_e(r_e^c,h_c,p_e)}{\partial R_i}=0\quad a_{13}=-v\cdot\frac{\partial\bar{\alpha}_e(r_e^c,h_c,p_e)}{\partial h} \\ a_{21}=v\cdot\frac{\partial\bar{\alpha}_i(r_i^c,h_c,p_i)}{\partial R_e}=0\quad a_{22}=v\cdot\frac{\partial\bar{\alpha}_i(r_i^c,h_c,p_i)}{\partial R_i}\quad a_{23}=v\cdot\frac{\partial\bar{\alpha}_i(r_i^c,h_c,p_i)}{\partial h} \\ a_{31}=\frac{\partial S(r_e^c,r_i^c,h_c)}{\partial R_e}\quad a_{32}=\frac{\partial S(r_e^c,r_i^c,h_c)}{\partial R_i}\quad a_{33}=\frac{\partial S(r_e^c,r_i^c,h_c)}{\partial h} \end{gathered} \tag{25}$$

$$S(r_e,r_i,h)=\bar{v}-\frac{1}{\Lambda\cdot\rho_1}\cdot\left[\lambda_1\cdot G_1(r_e,r_i,h)-\lambda_2\cdot G_2(r_e,r_i,h)\right]$$

	F_1	F_2	F_3		F_1	F_2	F_3
$\bar{v}$	$8.32993\cdot10$	$8.29566\times1($	$8.29107\ \times1($	H_i-H_e	$1.10328\times10^{-}$	$1.10320\times10^{-}$	$1.10324\times10^{-}$
T_0	1716.46	2110.76	2933.32	$p_g^i-p_g^e$	-270.58	-270.56	-270.57
$T_{en}(0)$	400.00	400.00	400.00	h_c	$2.14380\times10^{-}$	$2.14324\times10^{-}$	$2.14371\times10^{-}$
k	250.00	500.00	1000.00	r_e^c	$4.66000\times10^{-}$	$4.65990\times10^{-}$	$4.65990\times10^{-}$
p_e	-1981.44	-1982.93	-1981.61	r_i^c	$4.33900\times10^{-}$	$4.33900\times10^{-}$	$4.33899\times10^{-}$
p_i	-2252.02	-2253.49	-2252.18				

Table 1. Possible settings of $\bar{v}$, T_0, $T_{en}(0)$, p_e, p_i, $p_g^i-p_g^e$

	a_{11}	a_{13}	a_{22}	a_{23}	a_{31}	a_{32}	a_{33}
F_1	-0.20278	-0.29136	-0.1984	0.285445	121.8176	0	-2534.0535
F_2	-0.20191	-0.29033	-0.19755	0.284434	27285.12	-27215.77	-42171.2316
F_3	-0.20183	-0.29002	-0.19747	0.284131	88368.31	-88368.31	-132203.96

Table 2. The coefficients a_{ij} of the linearized system in the steady states

The values of the numbers a_{ij} in the considered cases are given in Table 2. It is easy to verify that in all cases the Hurwitz condition are satisfied.

In Fig. 10 simulations of the silicon tube growth is presented when the seed length is 10^{-2} and the radii of the seed are $R_i=4.3389\times10^{-3}$ [m]; $R_e=4.659\times10^{-3}$ [m]. The meniscus height at the start is $h=2.14838\times10^{-4}$ [m].

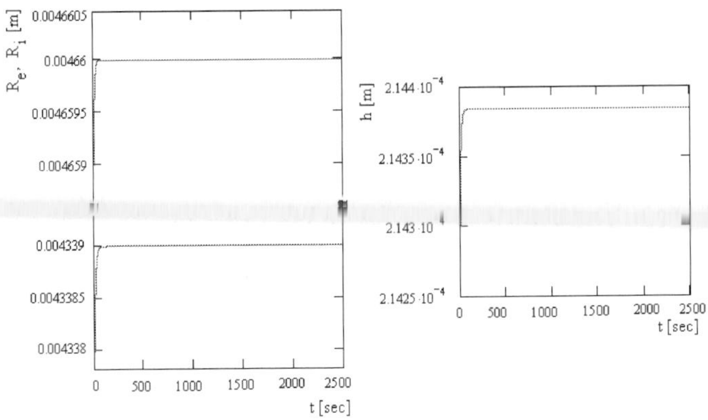

Fig. 10. The evolution of the outer radius, the inner radius of the tube and the meniscus height obtained integrating numerically the system (1) for $F_1 = 2/R_e$, $\bar{v} = 1.6585917365 \cdot 10^{-4}$ [m/s], $T_0 = 1714.81$ [K] and $T_{en}(0) = 400[K]$

6. Conclusions

Knowing the material constants (density, heat conductivity, etc), the size of the single crystal tube which will be grown from that material, the size of the shaper which will be used and the cooling gas temperature at the entrance, it is possible to predict values of pulling rate, temperature at the meniscus basis, cooling gas temperature at the exit, vertical temperature gradient in the furnace, inner and outer walls cooling gas pressure differences, melt column height differences, crystallization front level, which can be used for a stable growth.

According to the model the predicted values are not unique i.e. there are several possibility to obtain a tube with prior given size from a given material using the same shaper. So, even if in our computation the material and the size of the shaper and tube is the same as in [Eriss] experiment, the computed data given in Table 1 can be different from that used in the real experiment. For this reason our purpose is not rely to compare the computed results with the experimental data. Moreover we want to reveal that a tube of prior given size can be obtained by different settings and the model permit to compute such settings. The choice of a specific setting is the practical crystal grower decision. The model provide possible settings and can be helpful in a new experiment planning.

Concerning the limits of the model it is clear that it is limited in applicability, as all models. The main limits are those introduced by approximations made in equations defining the model.

7. Appendix 1. Inequalities for single crystal tube growth by E.F.G. technique - Convex outer and inner free surface

Consider the differential equation (6) for $\frac{R_{gi} + R_{ge}}{2} \le r_e \le R_{ge}$, α_c, α_g such that $\alpha_g \in \left(0, \pi/2\right)$, $0 < \alpha_c < \pi/2 - \alpha_g$.

Definition 1. A solution $z = z(x)$ of the eq. (6) describes the outer free surface of a static meniscus on the interval $\left[r_e, R_{ge}\right]$ if it possesses the following properties: $z\left(R_{ge}\right) = 0$, $z'(r_e) = -\tan\left(\pi/2 - \alpha_g\right)$, $z'\left(R_{ge}\right) = -\tan\alpha_c$ and $z(r)$ is strictly decreasing on $[r_e, R_{ge}]$. The described outer free surface is convex on $[r_e, R_{ge}]$ if $z''(r) > 0 \quad \forall r \in [r_e, R_{ge}]$.

Theorem 1. If there exists a solution of the eq. (6), which describes a convex outer free surface of a static meniscus on the closed interval $[r_e, R_{ge}]$, then for $n = \dfrac{R_{ge}}{r_e}$, p_e satisfy:

$$\begin{aligned}
&-\gamma \cdot \frac{\pi/2 - (\alpha_c + \alpha_g)}{R_{ge}} \cdot \frac{n}{n-1} \cdot \cos\alpha_c + \frac{\gamma}{R_{ge}} \cdot \sin\alpha_c \le p_e \\
&\le -\gamma \cdot \frac{\pi/2 - (\alpha_c + \alpha_g)}{R_{ge}} \cdot \frac{n}{n-1} \cdot \sin\alpha_g + \frac{\rho \cdot g \cdot R_{ge} \cdot (n-1)}{n} \cdot \tan\left(\pi/2 - \alpha_g\right) + \\
&+ \frac{\gamma}{R_{ge}} \cdot n \cdot \cos\alpha_g
\end{aligned} \tag{A.1.1}$$

Theorem 2. Let be n such that $1 < n < \dfrac{2 \cdot R_{ge}}{R_{gi} + R_{ge}}$. If p_e satisfies the inequality:

$$p_e < -\gamma \cdot \frac{\pi/2 - (\alpha_c + \alpha_g)}{R_{ge}} \cdot \frac{n}{n-1} \cdot \cos\alpha_c + \frac{\gamma}{R_{ge}} \cdot \sin\alpha_c \tag{A.1.2}$$

then there exists $r_e \in \left[\dfrac{R_{ge}}{n}, R_{ge}\right]$ such that the solution of the initial value problem:

$$\begin{cases} z'' = \dfrac{\rho \cdot g \cdot z - p_e}{\gamma} \cdot \left[1 + (z')^2\right]^{3/2} - \dfrac{1}{r} \cdot \left[1 + (z')^2\right] \cdot z' \quad \text{for } \dfrac{R_{gi} + R_{ge}}{2} < r \le R_{ge} \\ z\left(R_{ge}\right) = 0, \; z'\left(R_{ge}\right) = -\tan\alpha_c \end{cases} \tag{A.1.3}$$

on the interval $[r_e, R_{ge}]$ describes the convex outer free surface of a static meniscus.

Corollary 3. If for p_e the following inequality holds:

$$p_e < -2 \cdot \gamma \cdot \frac{\pi/2 - (\alpha_c + \alpha_g)}{R_{ge} - R_{gi}} \cdot \cos\alpha_c + \frac{\gamma}{R_{ge}} \cdot \sin\alpha_c \tag{A.1.4}$$

then there exists $r_e \in \left(\dfrac{R_{ge} + R_{gi}}{2}, R_{ge}\right)$ (close to $\dfrac{R_{ge} + R_{gi}}{2}$) such that the solution of the i.v.p. (12) on the interval $[r_e, R_{ge}]$ describes a convex outer free surface of a static meniscus.

Corollary 4. If for $1 < n' < n < \frac{2 \cdot R_{ge}}{R_{gi} + R_{ge}}$ the following inequalities holds:

$$-\gamma \cdot \frac{\pi/2 - (\alpha_c + \alpha_g)}{R_{ge}} \cdot \frac{n'}{n'-1} \cdot \sin\alpha_g + \rho \cdot g \cdot R_{ge} \cdot \frac{n'-1}{n'} \cdot \tan\left(\pi/2 - \alpha_g\right) + \frac{\gamma}{R_{ge}} \cdot n' \cos\alpha_g$$
$$< p_e < -\gamma \cdot \frac{\pi/2 - (\alpha_c + \alpha_g)}{R_{ge}} \cdot \frac{n}{n-1} \cdot \cos\alpha_c + \frac{\gamma}{R_{ge}} \cdot \sin\alpha_c \qquad \text{(A.1.5)}$$

then there exists $r_e \in \left[\frac{R_{ge}}{n}, \frac{R_{ge}}{n'}\right]$ such that the solution of the i.v.p. (A.1.3) on the interval $[r_e, R_{ge}]$ describes a convex outer free surface of a static meniscus.

Theorem 5. If a solution $z_1 = z_1(r)$ of the eq. (6) describes a convex outer free surface of a static meniscus on the interval $[r_e, R_{ge}]$, then it is a weak minimum for the energy functional of the melt column (7).

Definition 2. A solution $z = z(x)$ of the eq.(8) describes the inner free surface of a static meniscus on the interval $\left[R_{gi}, r_i\right]$, $\left(R_{gi} < r_i < \frac{R_{gi} + R_{ge}}{2}\right)$ if it possesses the following properties: $z'(R_{gi}) = \tan\alpha_c$, $z'(r_i) = \tan\left(\pi/2 - \alpha_g\right)$, $z(R_{gi}) = 0$ and $z(r)$ is strictly increasing on $[R_{gi}, r_i]$. The described inner free surface is convex on $[R_{gi}, r_i]$ if $z''(r) > 0$, $\forall r \in [R_{gi}, r_i]$.

Theorem 6. If there exists a solution of the eq. (8), which describes a convex inner free surface of a static meniscus on the closed interval $[R_{gi}, r_i]$ and $r_i = m \cdot R_{gi}$ with $1 < m < \frac{R_{gi} + R_{ge}}{2 \cdot R_{gi}}$, then the following inequalities hold:

$$-\gamma \cdot \frac{\pi/2 - (\alpha_c + \alpha_g)}{(m-1) \cdot R_{gi}} \cdot \cos\alpha_c - \frac{\gamma}{R_{gi}} \cdot \cos\alpha_g \le p_i$$
$$\le -\gamma \cdot \frac{\pi/2 - (\alpha_c + \alpha_g)}{(m-1) \cdot R_{gi}} \cdot \sin\alpha_g + \rho \cdot g \cdot R_{gi} \cdot (m-1) \cdot \tan\left(\pi/2 - \alpha_g\right) - \frac{\gamma}{m \cdot R_{gi}} \cdot \sin\alpha_c \qquad \text{(A.1.6)}$$

Theorem 7. Let m be such that $1 < m < \frac{R_{gi} + R_{ge}}{2 \cdot R_{gi}}$. If p_i satisfies the inequality:

$$p_i < -\gamma \cdot \frac{\pi/2 - (\alpha_c + \alpha_g)}{(m-1) \cdot R_{gi}} \cdot \cos\alpha_c + \frac{\gamma}{R_{gi}} \cdot \cos\alpha_g \qquad \text{(A.1.7)}$$

then there exists $r_i \in [R_{gi}, m \cdot R_{gi}]$, such that the solution of the initial value problem:

$$\begin{cases} z'' = \dfrac{\rho \cdot g \cdot z - p_i}{\gamma} \cdot \left[1 + (z')^2\right]^{3/2} - \dfrac{1}{r} \cdot \left[1 + (z')^2\right] \cdot z' \text{ for } R_{gi} < r \le \dfrac{R_{gi} + R_{ge}}{2} \\ z(R_{gi}) = 0, \; z'(R_{gi}) = \tan \alpha_c \end{cases} \quad (A.1.8)$$

on the interval $[R_{gi}, r_i]$ describes the convex inner free surface of a static meniscus.

Corollary 8. If for p_i the following inequality holds,

$$p_i < -2 \cdot \gamma \cdot \frac{\pi/2 - (\alpha_c + \alpha_g)}{R_{ge} - R_{gi}} \cdot \cos \alpha_c - \frac{\gamma}{R_{gi}} \cdot \cos \alpha_g \quad (A.1.9)$$

then there exists $r_i \in \left(R_{gi}, \dfrac{R_{gi} + R_{ge}}{2} \right)$ (close to $\dfrac{R_{gi} + R_{ge}}{2}$) such that the solution of the i.v.p. (A.1.8) on the interval $[R_{gi}, r_i]$ describes a convex inner free surface of a static meniscus.

Corollary 9. If for $1 < m' < m < \dfrac{R_{gi} + R_{ge}}{2 \cdot R_{gi}}$ the following inequalities hold

$$-\gamma \cdot \frac{\pi/2 - (\alpha_c + \alpha_g)}{(m'-1) \cdot R_{gi}} \cdot \sin \alpha_g + \rho \cdot g \cdot R_{gi} \cdot (m'-1) \cdot \tan\left(\pi/2 - \alpha_g\right) - \frac{\gamma}{m' \cdot R_{gi}} \cdot \sin \alpha_c \quad (A.1.10)$$

$$< p_i < -\gamma \cdot \frac{\pi/2 - (\alpha_c + \alpha_g)}{(m-1) \cdot R_{gi}} \cdot \cos \alpha_c - \frac{\gamma}{R_{gi}} \cdot \cos \alpha_g$$

then there exists r_i in the interval $[m' \cdot R_{gi}, m \cdot R_{gi}]$ such that the solution of the i.v.p. (A.1.8) on the interval $[R_{gi}, r_i]$ describes a convex inner free surface of a static meniscus.

Theorem 10. If a solution $z_1 = z_1(r)$ of the eq. (8) describes a convex inner free surface of a static meniscus on the interval $[R_{gi}, r_i]$, then it is a weak minimum for the energy functional of the melt column (9).

8. Appendix 2. Inequalities for single crystal tube growth by E.F.G. technique - Concave outer and inner free surface

Consider the equation (6) for $0 < R_{gi} < \dfrac{R_{gi} + R_{ge}}{2} \le r_e < R_{ge}$, α_c, α_g such that $\alpha_g \in \left(0, \pi/2\right)$, $0 < \pi/2 - \alpha_g < \alpha_c < \pi/2$.

Definition 3. The outer free surface is concave on $[r_e, R_{ge}]$ if $z''(r) < 0$, $\forall r \in [r_e, R_{ge}]$.

Theorem 11. If there exists a concave solution $z_e = z_e(r)$ of the equation (6) then $n = \frac{R_{ge}}{r_e}$ and p_e satisfy the following inequalities:

$$\frac{n}{n-1}\cdot\gamma\cdot\frac{\alpha_c+\alpha_g-\pi/2}{R_{ge}}\cdot\cos\alpha_c+\frac{\gamma}{R_{ge}}\cdot\cos\alpha_g \le p_e \le \frac{n}{n-1}\cdot\gamma\cdot\frac{\alpha_c+\alpha_g-\pi/2}{R_{ge}}\cdot\sin\alpha_g + \frac{n-1}{n}\cdot\rho\cdot g\cdot\tan\alpha_c+\frac{n\cdot\gamma}{R_{ge}}\cdot\sin\alpha_c \tag{A.2.1}$$

Theorem 12. If for $1 < n' < n < \frac{2\cdot R_{ge}}{R_{ge}+R_{gi}}$ and p_e the inequalities hold:

$$\frac{n}{n-1}\cdot\gamma\cdot\frac{\alpha_c+\alpha_g-\pi/2}{R_{ge}}\cdot\sin\alpha_g+\frac{n-1}{n}\cdot\rho\cdot g\cdot R_{ge}\cdot\tan\alpha_c+\frac{n\cdot\gamma}{R_{ge}}\cdot\sin\alpha_c < p_e < \frac{n'}{n'-1}\cdot\gamma\cdot\frac{\alpha_c+\alpha_g-\pi/2}{R_{ge}}\cdot\cos\alpha_c+\frac{\gamma}{R_{ge}}\cdot\cos\alpha_g. \tag{A.2.2}$$

then there exists $r_e \in \left[\frac{R_{ge}}{n}, \frac{R_{ge}}{n'}\right]$ and a concave solution of the equation (6).

Theorem 13. A concave solution $z_e(r)$ of the equation (6) is a weak minimum of the free energy functional of the melt column (7).

Consider now the differential equation (8) for $0 < R_{gi} < r_i < \frac{R_{gi}+R_{ge}}{2} < R_{ge}$ and α_c, α_g such that $0 < \pi/2 - \alpha_g < \alpha_c < \pi/2$, $\alpha_g \in (0, \pi/2)$.

Theorem 14. If there exists a concave solution $z_i = z_i(r)$ of the equation (8) then $m = \frac{r_i}{R_{gi}}$ and p_i satisfies the following inequalities:

$$\frac{1}{m-1}\cdot\gamma\cdot\frac{\alpha_c+\alpha_g-\pi/2}{R_{gi}}\cdot\cos\alpha_c-\frac{\gamma}{R_{gi}}\cdot\sin\alpha_c \le p_i \le \frac{1}{m-1}\cdot\gamma\cdot\frac{\alpha_c+\alpha_g-\pi/2}{R_{gi}}\cdot\sin\alpha_g + (m-1)\cdot\rho\cdot g\cdot R_{gi}\cdot\tan\alpha_c-\frac{\gamma}{m\cdot R_{gi}}\cdot\cos\alpha_g \tag{A.2.3}$$

Theorem 15. If for $1 < m' < m < \frac{2\cdot R_{ge}}{R_{ge}+R_{gi}}$ and for p_i the following inequalities hold:

$$\frac{1}{m-1}\cdot\gamma\cdot\frac{\alpha_c+\alpha_g-\pi/2}{R_{ge}}\cdot\sin\alpha_g+(m-1)\cdot\rho\cdot g\cdot R_{gi}\tan\alpha_c-$$
$$\frac{\gamma}{m\cdot R_{gi}}\cdot\cos\alpha_g<p_i<\frac{1}{m'-1}\cdot\gamma\cdot\frac{\alpha_c+\alpha_g-\pi/2}{R_{gi}}\cdot\cos\alpha_c+\frac{\gamma}{R_{gi}}\cdot\sin\alpha_c \tag{A.2.4}$$

then there exists r_i in the interval $\left[m'\cdot R_{gi}, m\cdot R_{gi}\right]$ and a concave solution of the eq. (8).

Theorem 16. A concave solution $z_i(r)$ of the equation (8) is a weak minimum of the free energy functional of the melt column (9).

9. References

St. Balint, A.M. Balint (2009), *On the creation of the stable drop-like static meniscus, appropriate for the growth of a single crystal tube with prior specified inner and outer radii*, Mathematical Problems in Engineering, vol. 2009, Article ID:348538 (2009), pp 1-22

St. Balint, A.M. Balint (2009), *Inequalities for single crystal tube growth by edge-defined film-fed (E.F.G.) technique* , Journal of Inequalities and Applications, vol.2009, Article ID: 732106, pp.1-28

St.Balint, A.M.Balint, L.Tanasie (2008) - *The effect of the pressure on the static meniscus shape in the case of tube growth by edge-defined film-fed growth (E.F.G.) method*, Journal of Crystal Growth, Vol. 310, pp.382-390

St. Balint, L.Tanasie (2008), *Nonlinear boundary value problems for second order differential* equations describing concave equilibrium capillary surfaces, Nonlinear Studies 15, Vol. 4, pp.277-296.

St.Balint, L.Tanasie(2011), *Some problems concerning the evaluation of the shape and size of the meniscus occurring in silicon tube growth* - Mathematics in Engineering, Science and Aerospace Vol. 2, pp. 53-70.

St.Balint, L.Tanasie (2010), *A procedure for the determination of the angles* $\tilde{\alpha}_e^0(r_e,h_e;p_e)$ *and* $\tilde{\alpha}_i^0(r_i,h_i;p_i)$ *which appears in the nonlinear system of differential equations describing the dynamics of the outer and inner radius of a tube, grown by the edge-defined film-fed growth (EFG) technique*, Nonlinear Analysis: Real World Applications, Vol. 11(Issue 5), pp 4043-4053

St.Balint, L.Tanasie (2011), *The choice of the pressure of the gas flow and the melt level in silicon tube growth*, Mathematics in Engineering, Science and Aerospace, Vol. 4, pp.

H.Behnken, A.Seidl and D.Franke (2005), *A 3 D dynamic stress model for the growth of hollow silicon polygons*, Journal of .Crystal Growth, Vol 275, pp. e375-e380.

A.V.Borodin, V.A.Borodin, V.V.Sidorov and I.S.Petkov (1999), *Influence of growth process parameters on weight sensor readings in the Stepanov (EFG) technique*, Journal of .Crystal Growth, Vol. 198/199, pp.215-219.

A.V.Borodin, V.A.Borodin and A.V.Zhdanov (1999), *Simulation of the pressure distribution in the melt for sapphire ribbon growth by the Stepanov (EFG) technique*, Journal of .Crystal Growth, Vol. 198/199, pp.220-226.

L.Erris, R.W.Stormont, T.Surek, A.S.Taylor (1980), *The growth of silicon tubes by the EFG process*, Journal of .Crystal Growth, Vol. 50, pp.200-211.

R. Finn, *Equilibrium capillary surfaces* (1986), Vol. 284, Grundlehren der mathematischen Wissenschaften, Springer, New York, NY, USA.

I.P.Kalejs, A.A.Menna, R.W.Stormont and I.W.Hudrinson (1990), *Stress in thin hollow silicon cylinders grown by the edge-defined film-fed growth technique*, Journal of .Crystal Growth, Vol 104, pp.14-19.

H.Kasjanow, A.Nikanorov, B.Nacke, H.Behnken, D.Franke and A.Seidl (2007), 3Dcoupled electromagnetic and thermal modeling of EFG silicon tube growth, Journal of .Crystal Growth, Vol. 303, pp.175-179.

B.Mackintosh, A.Seidl, M.Quellette, B.Bathey, D.Yates and J.Kalejs (2006) *Large silicon crystal hollow-tube growth by the edge-defined film-fed growth (EFG) method*, Journal of .Crystal Growth, Vol 287, pp.428-432.

S.Rajendram, M.Larousse, B.R. Bathey and J.P.Kalejs (1993), *Silicon carbide control in the EFG system*, Journal of .Crystal Growth, Vol. 128, pp.338-342.

S.Rajendram, K.Holmes and A.Menna (1994), *Three-dimensional magnetic induction model of an octagonal edge-defined film-fed growth system*, Journal of .Crystal Growth, Vol. 137(No 1-2), pp.77-81.

S.N.Rossolenko (2001), *Menisci masses and weights in Stepanov (EFG) technique: ribbon, rod, tube*, Journal of .Crystal Growth, vol. 231, pp.306-315.

A.Roy, B.Mackintosh, J.P.Kalejs, Q.S.Chen, H.Zhang and V.Prasad(2000), *A numerical model for inductively heated cylindrical silicon tube growth system*, Journal of Crystal Growth , Vol 211, pp.365-371.

A.Roy, H.Zhang, V.Prasad, B.Mackintosh, M.Quellette and J.Kalejs (2000), *Growth of large diameter silicon tube by EFG technique: modeling and experiment*, Journal of .Crystal Growth , Vol 230, pp.224-231.

D.Sun, Ch.Wang, H.Zhang, B.Mackintosh, D.Yates and J.Kalejs (2004), *A multi-block method and multi-grid technique for large diameter EFG silicon tube growth*, Journal of .Crystal Growth, Vol 266, pp. 167-174

T.Surek, B.Chalmers and A.I.Mlavsky (1977), *The edge film-fed growth of controlled shape crystals*, Journal of .Crystal Growth, Vol. 42, pp.453-457

J.C.Swartz, T.Surek and B.Chalmers (1975), *The EFG process applied to the growth of silicon ribbons*, J.Electron.Mater, Vol. 4, pp.255-279.

L.Tanasie, St.Balint (2010) , *Model based, pulling rate, thermal and capillary conditions setting for silicon tube growth*, Journal of Crystal Growth, vol. 312, pp. 3549-3554

V.A.Tatarchenko (1993), *Shaped crystal growth*, Kluwer Academic Publishers, Dordrecht.

B.Yang, L.L.Zheng, B.MacKintosh, D.Yates and J.Kalejs (2006), *Meniscus dynamics and melt solidification in the EFG silicon tube growth process*, Journal of .Crystal Growth, Vol. 293, pp.509-516.

Crystallization in Microemulsions: A Generic Route to Thermodynamic Control and the Estimation of Critical Nucleus Size

Sharon Cooper, Oliver Cook and Natasha Loines
Durham University
UK

1. Introduction

Crystallization is ubiquitous. It is evident in natural processes such as biomineralization and gem formation, and is important in industrial processes both as a purification step and in the production of materials with specific properties, including drug polymorphs, cocrystals, mesocrystals, quasicrystals, quantum dots and other inorganic nanocrystals. Consequently it is essential to gain greater understanding of the process to be able to elicit more control over its outcome.

Crystallization occurs from melts that are supercooled, i.e. cooled below their equilibrium melting temperatures, T_{eq}. For crystallization from solution, the solutions must be supersaturated, i.e. have solute concentrations above their saturation values, c_{eq}, defined as the solute concentration in equilibrium with the macroscopic crystal. The supersaturation is the driving force for crystallization, being the difference in chemical potential, $\Delta\mu$, between the parent (melt or solution) and daughter (new crystal) phases. For crystallization from the melt, $\Delta\mu = \Delta_{fus}H\Delta T / T_{eq}$, where $\Delta_{fus}H$ is the enthalpy of fusion and $\Delta T = T_{eq} - T$, is the supercooling with T denoting the temperature. Here it is assumed that $\Delta_{fus}H$ is invariant between T and T_{eq}. For an ideal solution, the supersaturation is $\Delta\mu = kT\ln\left(c / c_{eq}\right)$, where k is the Boltzmann constant, and c/c_{eq}, is the ratio of the solute concentration compared to its saturation value, which is known as the supersaturation ratio.

The formation of any new phase from a bulk parent phase requires the creation of an interface between the two phases, which requires work. Hence there exists an energy barrier to the formation of the new phase. The process of overcoming this energy barrier is known as nucleation. In crystallization, once nucleation has occurred, crystal growth onto the nuclei proceeds until the supersaturation is relieved. Owing to this nucleation stage, crystallization from the bulk melt or solution is typically under kinetic control, with metastable forms often crystallizing initially in accordance with Ostwald's rule of stages (Ostwald, 1897). In contrast, microemulsions have the unique ability to generically exert thermodynamic control over the crystallization process. This provides significant advantages; the size of the critical nucleus can be estimated with good accuracy under thermodynamic control conditions and importantly, the stable form of a material can be identified and readily produced under ambient conditions. In this chapter we discuss the scientific rationale for this thermodynamic control and provide practical examples.

2. Theoretical considerations

2.1 Classical Nucleation Theory (CNT)

Nucleation can be modelled most simply using classical nucleation theory (CNT). Gibbs thermodynamic treatment of liquid nucleation from a vapour (Gibbs, 1876, 1878) shows that the free energy change, ΔF, involved in producing a spherical liquid nucleus from the vapour is given by:

$$\Delta F = -n\Delta\mu + \gamma A = -\frac{4\pi r^3 \Delta\mu}{3v_c} + 4\pi r^2 \gamma \quad (1)$$

where n is the number of molecules in the nucleus, $\Delta\mu$ denotes the chemical potential difference between the vapour and the liquid nucleus which defines the supersaturation, γ denotes the surface tension between the nucleus and the surrounding vapour, A denotes the surface area of the nucleus, r denotes its radius and v_c denotes the molecular volume of the new condensed phase, i.e. the liquid. This same thermodynamic treatment is often used for crystallization. For crystallization from bulk solutions at constant pressure, the relevant free energy to use is the Gibbs free energy. For crystallization from microemulsions, however, there is a very small Laplace pressure difference across the microemulsion droplet interface and so the Helmholtz free energy for constant volume systems is the appropriate free energy to employ.

Equation (1) clearly shows that the favourable formation of the bulk new phase in any supersaturated system, given by the first term $-n\Delta\mu$, is offset by the unfavourable surface free energy term, γA, that necessarily arises from creating the new interface. The surface free energy term dominates at smaller nucleus sizes and leads to the nucleation energy barrier (see Figure 1a). In particular, differentiating equation (1) with respect to r leads to a maximum at

$$r^* = 2\gamma v_c / \Delta\mu , \quad (2)$$

i.e. the well-known Gibbs-Thomson equation, with the nucleation barrier given by:

$$\Delta F^* = \frac{16}{3\Delta\mu^2}\pi\gamma^3 v_c^{\,2} . \quad (3)$$

The r^* nucleus is termed the critical nucleus. It is of pivotal importance in CNT because it determines the size above which it is favourable for the new phase to grow. Nuclei smaller than r^* have a greater tendency to dissolve (or melt) than grow, whilst nuclei larger than r^* will tend to grow. The r^* nucleus has an equal probability of growing or dissolving (melting) and is in an unstable equilibrium with the surrounding solution (or melt). At larger r, a stable r_0 nucleus with $\Delta F = 0$ occurs (see Figure 1a).

Equation (3) gives the magnitude of the nucleation barrier when the new phase forms within the bulk parent phase, which is known as homogeneous nucleation. If the new phase forms on a foreign surface, however, heterogeneous nucleation occurs. The corresponding heterogeneous nucleation barrier, ΔF_{het} that arises from forming a cap-shaped critical nucleus with a contact angle, θ, on the foreign surface is given by:

$$\Delta F^*_{het} = \frac{16\pi\gamma^3 v_c^2}{3\Delta\mu^2} f(\theta) = \Delta F^*_{hom} \frac{v^*_{het}}{v^*_{hom}} \tag{4}$$

where: ΔF^*_{hom} denotes the homogeneous nucleation barrier of the system given by equation (3), $f(\theta) = (2 - 3\cos\theta + \cos^3\theta)/4$ and v^*_{het} and v^*_{hom} denote the volumes of the heterogeneous and corresponding homogeneous critical nuclei, respectively. Note that equation (4) ignores the entropic free energy contribution arising from the number of surface sites upon which the nucleus may form, as this factor is incorporated in the pre-exponential factor, Ω, instead.

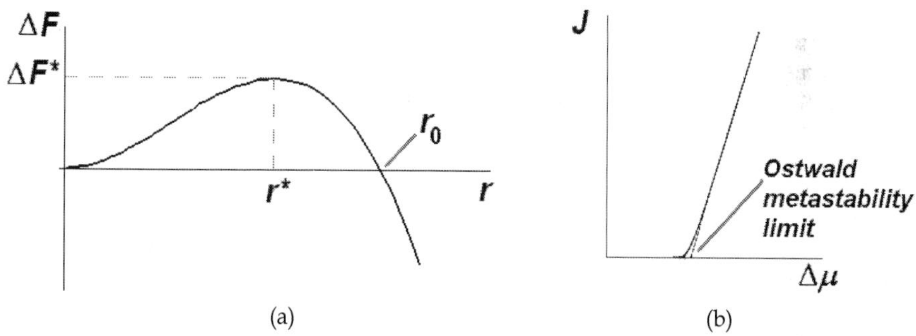

Fig. 1. (a) Schematic graph of free energy change, ΔF vs. nucleus size, r, for a homogeneously nucleating crystal showing the critical nucleus, r^*, and stable nucleus, r_0, sizes. (b) Schematic graph of nucleation rate, J, vs. supersaturation, $\Delta\mu$, showing the Ostwald metastability limit which gives the onset crystallization temperature, T_c.

The kinetic theory of CNT (Volmer & Weber, 1926; Becker & Döring, 1935) used the nucleation barrier, ΔF^*, in an Arrhenius-type equation to derive the nucleation rate, J as:

$$J = \Omega \exp(-\Delta F^*/kT) \tag{5}$$

where Ω is the pre-exponential factor accounting for the rate at which the molecules impinge upon, and are incorporated into, the critical nuclei. The form of equation (5) is such that J remains negligibly small until the supersaturation reaches a critical value, the Ostwald metastability limit, at which point J suddenly and dramatically increases (see Figure 1b). Hence, an onset crystallization temperature, T_c, can be identified with this metastability limit, and the corresponding nucleation rate can be set, with little loss in accuracy, to a suitable detection limit for the technique monitoring the crystallization.

CNT is widely adopted because of its simplicity. However this simplicity limits its ability to model real systems. Two main assumptions are: firstly that it considers the nuclei to be spherical with uniform density and a structure equivalent to the bulk phase, and secondly that the nuclei interfaces are infinitely sharp and have the same interfacial tensions as found at the corresponding planar interfaces. A recent review by Erdemir et al., 2009, details all the assumptions of CNT, and its applicability to different systems. Notably, given that CNT stems from the condensation of a liquid from its vapour, it cannot model two stage nucleation (Vekilov, 2010), where solute molecules organize initially into an amorphous cluster, from which long range crystal order then emerges on cluster rearrangement. Despite these many limitations, CNT is useful to benchmark crystallization experiments because this approach has been so widely adopted. More importantly here, it allows useful insights into the crystallization process that are readily apparent due to its simplicity. In particular, the use of CNT has enabled us to establish that thermodynamic control of crystallization is possible in 3D nanoconfined volumes, as shown below.

2.1.1 Adoption of CNT to curved interfaces

For crystallization in nanodroplets, the planar substrate of CNT's heterogeneous nucleation formulation is replaced by a highly curved concave substrate. For such curved substrates, the free energy becomes (Cooper et al., 2008; Fletcher, 1958):

$$\Delta F_{het} = -\frac{4}{3v_c}\pi r^3 \Delta\mu[(f(\theta+\phi) - (R/r)^3 f(\phi)] + 2[1-\cos(\theta+\phi)]\pi r^2\gamma - 2\cos\theta(1-\cos\phi)\pi R^2\gamma \quad (6)$$

where θ is the contact angle between the nucleus and the spherical substrate, ϕ is the angle between the spherical substrate and the plane connecting the nucleus edge and $f(\alpha) = 0.25\left(2 - 3\cos\alpha + \cos^3\alpha\right)$ (see Figure 2a). Note that for concave surfaces, corresponding to crystallization within the curved substrate, R and ϕ are assigned negative values.

The maximum in ΔF_{het} gives the barrier to nucleation, ΔF^*_{het}, and again this condition is satisfied when $r^* = 2\gamma v_c / \Delta\mu$ to give:

$$\Delta F^*_{het} = \frac{\Delta F^*_{\text{hom}}}{2}\left\{1 - 3x^2\cos\theta + 2x^3 + y(1 + x\cos\theta - 2x^2)\right\} = \Delta F^*_{\text{hom}} f(\theta_p) \quad (7)$$

where $x = R/r^*$, $y = \pm\left(x^2 - 2x\cos\theta + 1\right)^{0.5}$ with the positive and negative roots applying to a nucleus on a convex and concave surface, respectively and $\cos\theta_p = x - y$. θ_p is the angle between the corresponding planar critical nucleus and the plane tangential to the curved substrate surface, as shown in Figure 2a.

Equation (7) shows that at a given temperature, and hence constant ΔF^*_{hom} value, ΔF^*_{het} depends only upon the θ_p value. Consequently in Figure 2b, all the spherical substrates depicted result in the same ΔF^*_{het} value. This can be rationalized as follows. For nucleation on a concave surface, the critical nucleus volume, v^*, is reduced compared to the planar case, and hence fewer molecules need to cluster together to form the critical nucleus. However this effect is negated by the greater contact angle, θ, compared to θ_p, which means that more work is required to create unit area of the nucleus-substrate interface, and so the mean energy increase on addition of a molecule to the sub-critical nucleus is larger. In contrast on a convex substrate, v^* is increased compared to the planar case, but θ is decreased.

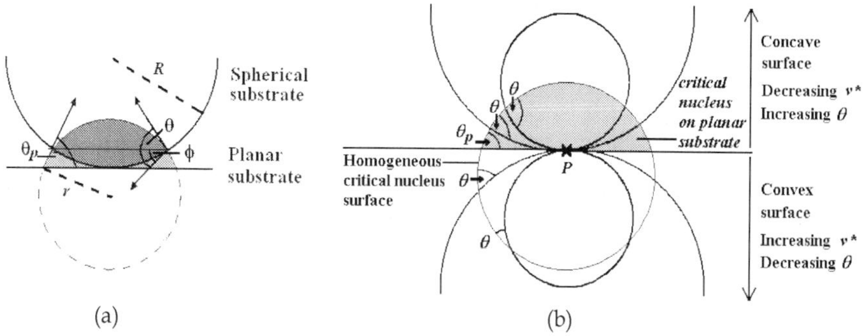

Fig. 2. (a) Schematic diagram describing nucleation upon a concave substrate of radius, R. The dark grey regions depict the nucleus on the concave surface. (b) Schematic diagram showing that for a given supersaturation, and hence critical nucleus radius, r^*, all surfaces through point P that cross the homogeneous critical nucleus surface produce the same ΔF^*_{het} value for nucleation, since they all have the same θ_p value (Cooper et al, 2008).

The onset temperature for the phase transition, e.g. the highest temperature, T_{trans}, at which crystallization should be observable, can then be found by setting the nucleation rate, J_{trans}, at T_{trans} to a suitable detection limit, where $J_{trans} = \Omega \exp(-\Delta F^*/kT_{trans})$. The pre-exponential factor, Ω, can be considered constant provided the temperature range is narrow, since ΔF^* has the greater temperature dependence. Using $(\Delta F^*/k) = (T_{eq} - \Delta T_{trans})\ln(\Omega / J_{trans})$, where $\Delta T_{trans} = T_{eq} - T_{trans}$, and substituting in equations (3), (5) and (7), gives:

$$\Delta T_{trans}{}^3 - \Delta T_{trans}{}^2 T_{eq} + \frac{16\pi\gamma^3 v_c{}^2 T_{eq}{}^2 f(\theta_p)}{3k\Delta_{fus}H^2 \ln(\Omega / J_{trans})} = 0 \,. \tag{8}$$

This equation has three roots corresponding to (1) The onset crystallization temperature T_c, (2) the expected onset melting temperature, T_m, for a nucleation-based melting transition, which would be required if surface melting did not occur, and (3) a non-physical root $T_c = (2T_{eq}/3)\{1 - \cos W\}$ close to 0 K corresponding to the case where the critical nucleus contains only one molecule and the energy barrier is vanishingly small. T_c and T_m are given by:

$$T_{c,m} = \frac{T_{eq}}{3}\left\{2 + \cos W \mp (3^{0.5} \sin W)\right\} \tag{9}$$

where $W = \frac{1}{3}\arccos\left[1 - \frac{72\pi\gamma^3 v_c{}^2 f(\theta_p)}{k\Delta_{fus}H^2 T_{eq} \ln(\Omega / J_{trans})}\right]$.

Hence T_c and T_m can be found with x, θ, T_{eq}, v_c, γ, $\Delta_{fus}H$ and Ω/J_{trans} as input. r^* and R can then be obtained from the Gibbs-Thomson equation, and $R = r^* x$, respectively.

From equation (7), we find that for a constant contact angle, θ, the energy barrier to nucleation is smaller for a concave surface than a convex one, and that the reduction in energy increases as $|x|$ increases. The onset crystallization temperatures, T_c, obtained from equation (9) are therefore correspondingly higher for concave surfaces. Figure 3 shows the expected T_c as a function of $|R|$ for the case of ice crystallization on a concave surface with a contact angle, θ, between the crystal nucleus and substrate of (a) 180°, i.e. the homogeneous nucleation case and (b) 100°. The onset melting temperatures, T_m, expected for the same systems (i.e. with supplementary contact angles between the molten lens and substrate of (a) 0° and (b) 80°) in the absence of surface melting are also shown in this Figure. This melting is denoted nucleation-melting. The T_m and T_c curves meet at $|R_{min}|$ and at smaller concave radii, the melting curve falls below the crystallization one, which is clearly non-physical. Hence, equation (9) cannot model crystallization in 3D nanoconfinements smaller than $|R_{min}|$. This demonstrates a fundamental limitation of the theory, and shows that a key factor necessary for crystallization has been ignored.

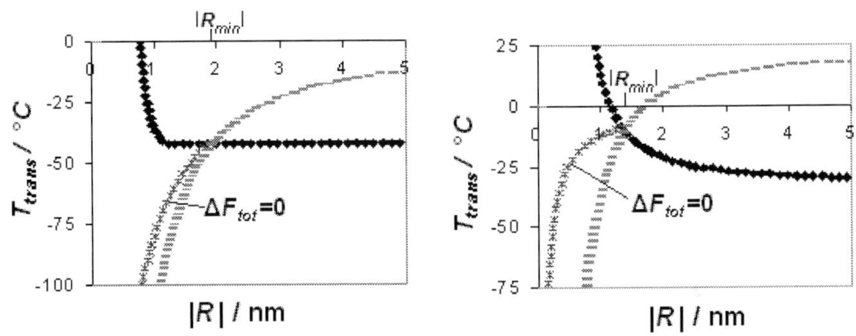

(a) (b)

Fig. 3. The predicted ice onset crystallization temperatures, T_c, (filled diamonds) and nucleation-melting temperatures, T_m, (dashes), which would occur in the absence of surface melting, as a function of substrate radius, $|R|$, for crystallization within a spherical substrate with (a) θ = 180° and (b) θ=100°. The ice T_c have been determined using reasonable values (Pruppacher, 1995; Bartell, 1998; Speedy, 1987) of $v_c = 3.26 \times 10^{-29}$ m^3, γ= 20 mN m^{-1}, $\Delta_{fus}H$ = 4060 J mol^{-1} and $\Omega/J_{trans} = 10^{18}$ for homogeneous nucleation and values of $v_c = 3.26 \times 10^{-29}$ m^3, γ= 22 mN m^{-1}, $\Delta_{fus}H$ = 5000 J mol^{-1} and $\Omega/J_{trans} = 10^{15}$ for the heterogeneous nucleation case. Note that for $|R| < |R_{min}|$, the ice T_c and T_m are both given by the curve labelled ΔF_{tot} = 0, as the transition temperature is now determined by the condition that the nucleus grows to a size r_0.

2.2 Phase transitions within nanoconfined volumes (Cooper et al., 2008)

2.2.1 Crystallization from the melt

The phase transition temperature in equation (9) is determined only by the ability to surmount the nucleation barrier. The thermodynamic feasibility of the transition, i.e. whether the new phase is stable with $\Delta F \leq 0$, however, is not considered. In fact, the crossing point of the crystallization and nucleation-melting curves, occurring at a droplet size denoted R_{min}, corresponds to the system where the total phase transition of the droplet, from all liquid to all crystal, or vice versa, occurs with a free energy change $\Delta F_{tot} = 0$. For droplet sizes smaller than $|R_{min}|$, the nucleation-melting and crystallization curves describe systems where $\Delta F_{tot} > 0$, so the phase transformation would not proceed. This means that for sizes below $|R_{min}|$, the critical nucleus size can be attained as ΔF^*_{het} is surmountable, but there is then insufficient material within the confining substrate for ΔF to decrease to zero through further nucleus growth, e.g. in Figure 1a, the nucleus cannot grow to a size r_0. Consequently, below $|R_{min}|$ the crystallization and nucleation-melting curves must both follow the same curve labelled $\Delta F_{tot} = 0$ in the Figure to ensure a thermodynamically feasible phase transformation occurs. Hence the hysteresis normally observed upon heating and cooling the *same* system would be expected to disappear for phase transformations confined to within volumes with $|R| \leq |R_{min}|$. Unfortunately, there is a difficulty in verifying this lack of hysteresis experimentally, because this would require the r_0 nuclei to be constrained to this size. Typically, however, the r_0 nuclei subsequently grow via collisions with uncrystallized droplets, by oriented attachment of other nuclei, or by Ostwald ripening, and so it is difficult to ensure that subsequent melting and crystallization cycles are indeed performed on the same system.

Using the condition that for droplet sizes with $|R| \leq |R_{min}|$, $\Delta F_{tot}=0$ on complete crystallization of the spherical droplet, we find:

$$R = \frac{3 v_c \gamma}{\Delta \mu} \cos\theta = \frac{3}{2} r^* \cos\theta \,. \qquad (10)$$

Here we have retained the convention that the substrate radii, R, must take negative values for nucleation on a concave surface. Consequently, the critical nucleus size can be obtained from equation (10) if R and θ are known. This is an important finding because determination of r^* usually relies on the Gibbs-Thomson equation and the inappropriate application of bulk interfacial tension values to small nuclei. The number of molecules, n^*, in the critical nucleus is then given by:

$$n^* = \frac{v^*}{v_c} = \frac{4\pi |R|^3}{3 v_c} \left\{ \frac{1}{2} - \frac{4}{27\cos^3\theta} \left(1 - \frac{2 - (3/4)\cos^2\theta + (9/8)\cos^4\theta}{\left(4 - 3\cos^2\theta\right)^{0.5}} \right) \right\} . \qquad (11)$$

The dependence of n^* on θ is relatively weak, so that even if θ can only be estimated to within ~10%, n^* is known with good precision if R can be measured. For homogeneous nucleation (θ = 180°), equation (10) and (11) simplify to $r^* = 2|R|/3$ and $n^* = 32\pi |R|^3 / 81 v_c$, so experimental measurement of R directly gives r^* and n^* provided $|R| \leq |R_{min}|$. So we just need to find the value of $|R_{min}|$.

An empirical determination of the R_{min} value is possible for homogeneous nucleation because the onset crystallization temperature, T_c, should be approximately constant for confinements with sizes above $|R_{min}|$. Note though that nucleation is a stochastic process, so repeated experiments will show some slight variation but an expected homogeneous nucleation temperature should nevertheless be apparent. For instance, the homogeneous nucleation temperature for ice is ≈-40 °C (Wood & Walton, 1970; Clausse et al., 1983).

Consequently, $|R_{min}|$ is readily identifiable as the droplet size at which T_c begins to decrease with decreasing $|R|$, provided the system nucleates homogeneously. For heterogeneous nucleation, θ is also likely to be a function of R, so an empirical determination is more difficult. Instead, the theoretical R_{min} value can be used, which is obtained as follows.

Substituting for r^* in equation (10) using the Gibbs-Thompson equation, $r^* = 2\gamma v_c / \Delta\mu = 2\gamma v_c T_{eq} / \Delta_{fus} H \Delta T_{trans}$, we find that for $|R| \leq |R_{min}|$, T_c and T_m are given by:

$$T_c = T_m = T_{eq}\left[1 + \frac{3\gamma v_c \cos\theta}{|R|\Delta_{fus}H}\right]. \tag{12}$$

Equation (12) has previously been identified (Couchman & Jesser, 1977) as giving the minimum possible melting temperature of a small particle. Our analysis shows that it also gives the maximum possible freezing temperature of a confined object (Vanfleet & Mochel, 1995; Enustun et al., 1990). R_{min} is then readily obtained by substituting in equation (9), since the T_c and T_m curves from equation (9) meet at R_{min}. Hence:

$$R_{min} = \frac{9\gamma v_c \cos\theta}{\Delta_{fus}H\left\{1 - \cos W + \left(3^{0.5}\sin W\right)\right\}} \tag{13}$$

Finding reliable R_{min} values from equation (13) requires knowledge of θ, γ and $\Delta_{fus}H$, which is problematic since the use of bulk θ, γ and $\Delta_{fus}H$ values for such small systems is likely to introduce unquantifiable errors. Consequently, the preferred methodology is that of using homogeneously nucleating systems to identify $|R_{min}|$ from the confinement size below which the onset crystallization temperatures, T_c, start to decrease. Then critical nucleus sizes can be reliably found for all sizes below $|R_{min}|$ using $r^* = 2|R|/3$ and $n^* = 32\pi|R|^3/81v_c$. Fortunately, crystallization in microemulsions often proceeds via homogeneous nucleation, making these the system of choice.

2.2.2 Crystallization from nanoconfined solution

Our crystallization model can be extended to crystallization of solutes from nanoconfined solutions, though here the situation is complicated by the decrease in supersaturation that arises as the nucleus grows. In particular, by adopting a classical homogeneous nucleation approach for crystallization from an ideal solution in a spherical confining volume, V, the free energy change, ΔF, to produce a nuclei containing n molecules would be given by (Cooper et al., 2008; Reguera et al., 2003) :

$$\Delta F = -n\Delta\mu + \gamma A + NkT\left\{\ln\left(1 - \frac{v}{Vv_c c_0}\right) - \frac{1}{v_c c_0}\ln\left(1 - \frac{v}{V}\right)\right\} \tag{14}$$

where $\Delta\mu = kT\ln\left(c/c_{eq}\right)$ denotes the supersaturation at that nucleus size, with $c = (N-n)/(V-v) = c_0\left\{1 - \left(v/Vv_c c_0\right)\right\}\left\{1/\left(1 - v/V\right)\right\}$, γ and A denote the interfacial tension and surface area, respectively, at the nucleus-solution interface, N is the initial number of solute molecules when n = 0, v denotes the nucleus volume, v_c denotes the molecular volume of the crystalline species and c_0 denotes the initial solute concentration when $n = 0$. The first two terms comprise those expected from classical nucleation theory for crystallization from unconfined volumes, whilst the third term provides the correction due to the supersaturation depletion as the nucleus grows. The free energy difference, ΔF, now exhibits a maximum, ΔF^*, corresponding to the critical nucleus radius, r^*, *and a minimum*, ΔF^*_{min}, at a larger nucleus radius, r^*_{min}, owing to the decrease in the supersaturation as the nucleus grows. r^* and r^*_{min} are both given by the usual Gibbs-Thomson equation with ΔF^* and ΔF^*_{min} both given by:

$$\Delta F^* = \frac{16\pi\gamma^3 v_c^2 T_{eq}^2 f(\theta_p)}{3\Delta_{fus}H^2 \Delta T_c^2} + NkT\left\{\ln\left(1 - \frac{v^*}{Vv_c c_0}\right) - \frac{1}{v_c c_0}\ln\left(1 - \frac{v^*}{V}\right)\right\}, \tag{15}$$

where v^* denotes the nucleus volume when $r = r^*$, with the subscript *min* used to distinguish the minimum value from the maximum, and T_{eq} denotes the saturation temperature for the solution at concentration c^* surrounding the critical nucleus, r^*.

As before, the onset crystallization temperature is found from $\left(\Delta F^* / kT_c\right) = \ln(\Omega / J_{trans})$, with Ω assumed constant. This leads to a quartic equation (Cooper et al., 2008):

$$\Delta T_c^4 - T_{eq}\Delta T_c^3 + (X - Y)\Delta T_c + T_{eq}Y = 0 \tag{16}$$

where $X = \dfrac{16\pi\gamma^3 v_c^2 f(\theta_p) T_{eq}^2}{3k\ln(\Omega / J_{trans})\Delta_{fus}H^2}$

and $Y = -\dfrac{32\pi c_0}{3\ln(\Omega / J_{trans})}\left(\dfrac{\gamma v_c T_{eq} x}{\Delta_{fus}H}\right)^3 \left\{\ln\left(1 - \dfrac{v^*}{Vv_c c_0}\right) - \dfrac{1}{v_c c_0}\ln\left(1 - \dfrac{v^*}{V}\right)\right\}.$

Equation (16) reduces to equation (8), the melt crystallization case, when $Y = 0$. Equation (16) is solvable with x, θ, c_0, T_{eq}, v_c, γ, $\Delta_{fus}H$ and Ω/J_{trans} as input. For the typical case where $T_c \leq T_{eq}$, T_c is then given by:

$$T_c = \frac{3T_{eq}}{4} + z_1^{0.5} + z_2^{0.5} - z_3^{0.5} \tag{17}$$

where: $z_1 = \left\{-\left[\dfrac{T_{eq}(X + 3Y)}{48}\right]^{0.5}\left(\cos\left(\dfrac{Z}{3}\right) - 3^{0.5}\sin\left(\dfrac{Z}{3}\right)\right)\right\} + \dfrac{T_{eq}^2}{16},$

$$z_2 = \left\{-\left[\frac{T_{eq}(X + 3Y)}{48}\right]^{0.5}\left(\cos\left(\frac{Z}{3}\right) + 3^{0.5}\sin\left(\frac{Z}{3}\right)\right)\right\} + \frac{T_{eq}^2}{16},$$

and

$$z_3 = \left\{\left[\frac{T_{eq}(X + 3Y)}{12}\right]^{0.5}\cos\left(\frac{Z}{3}\right)\right\} + \frac{T_{eq}^2}{16},$$

where $Z = \arccos\left\{ \dfrac{27^{0.5}\left[T_{eq}^{\ 3}Y + (X-Y)^2\right]}{2\left[T_{eq}(X+3Y)\right]^{1.5}} \right\}$.

r^* and R are then found from the Gibbs-Thomson equation, and $R = r^* x$, respectively.

Of the other three solutions to the quartic equation (16), two are non-physical, as they correspond to either crystallization close to 0 K, or the onset crystallization temperature close to T_{eq} found when the minimum free energy radius, r^*_{min}, is used instead of the maximum value, r^*. The remaining solution provides the onset crystallization temperature for rare cases when $T_c > T_{eq}$, which could in principle arise for sufficiently soluble species when $\theta < 90°$. In this case, positive values of x are used since r is negative as well as R and

$$T_c = \frac{3T_{eq}}{4} + z_1^{\ 0.5} - z_2^{\ 0.5} + z_3^{\ 0.5}. \tag{18}$$

As with the melt crystallization case, equations (17) and (18) are valid until the confinement size decreases to such an extent that there is insufficient crystallizing material present to ensure an energetically feasible phase transformation. For instance, crystallization would not be possible in a 3D-nanoconfined solution having the ΔF vs r curve shown in Figure 4a, since $\Delta F^*_{min} > 0$. For these small nanoconfinements, crystallization becomes feasible when the *minimum energy* ΔF^*_{min} is set to zero so that $r^*_{min} = r_0$, i.e. a stable nucleus can form. We then obtain (Cooper et al., 2008):

$$R = \frac{\gamma f(\theta_{p,min})}{kT_c c_0 x_{min}^{\ 2}\left\{ \ln\left(1 - \dfrac{v^*_{min}}{V v_c c_0}\right) - \dfrac{1}{v_c c_0}\ln\left(1 - \dfrac{v^*_{min}}{V}\right)\right\}} = \frac{2\gamma v_c x_{min}}{kT_c \ln(c^*_{min}/c_{eq})} \tag{19}$$

from which:

$$x_{min} = \frac{R}{r^*_{min}} = \left(\frac{\ln(c^*_{min}/c_{eq}) f(\theta_{p,min})}{2v_c c_0 \ln\left(1 - \dfrac{v^*_{min}}{V v_c c_0}\right) - 2\ln\left(1 - \dfrac{v^*_{min}}{V}\right)} \right)^{\frac{1}{3}} \tag{20}$$

where v^*_{min} denotes the volume of the r^*_{min} nucleus with $\Delta F = 0$, $\partial \Delta F/\partial r = 0$, and $\partial^2 \Delta F/\partial r^2 > 0$, and c^*_{min} denotes the solute concentration surrounding the r^*_{min} nucleus. Equation (20) can be solved iteratively to give $x_{min} = R/r^*_{min}$ with inputted values for c_0, θ, v_c and c_{eq}, but again, crucially not the γ or $\Delta_{fus}H$ values. The x value is then given by:

$$\frac{x}{\ln(c^*/c_{eq})} = \frac{x_{min}}{\ln(c^*_{min}/c_{eq})}. \tag{21}$$

Thus we can work out $r^* = R/x$ just by measuring R and T_c values, when $|R| \leq |R_{\min}|$. The number of molecules, $n^* \ (= v^*/v_c)$ in the critical nucleus is also obtained from R, θ, c_0, v_c, and T_c as input and is given by:

$$n^* = \frac{4\pi R^3}{3v_c x^3}\left\{\frac{1}{2} - \frac{x^3}{2} + \frac{2 - 2x\cos\theta + x^2 - x^2\cos^2\theta - 2x^3\cos\theta + 2x^4}{4y}\right\}, \tag{22}$$

which for homogeneous nucleation (θ = 180°, $y = x + 1$) reduces to the expected $n^* = 4\pi r^{*3}/3v_c$.

Hence provided $|R| \leq |R_{min}|$, we can again determine both n^* and r^* without reliance on macroscopic γ and $\Delta_{fus}H$ values, provided we know or can estimate θ. The experimental onset crystallization temperature, T_c, can then be compared with the values predicted from the Gibbs-Thomson and ideal solubility equations using the experimentally found R, x, and c_{eq} values. This provides a measure of how bulk values of γ and $\Delta_{fus}H$ are likely to be perturbed for solute crystallization in nanosystems. Figure 4a shows a schematic graph of ΔF vs r, whist Figure 4b shows theoretical calculations using equations (17) and (19) to model the homogeneous nucleation of octadecane from dodecane solutions, illustrating again the decrease in T_c that occurs below $|R_{min}|$, from which the critical nucleus size can be estimated.

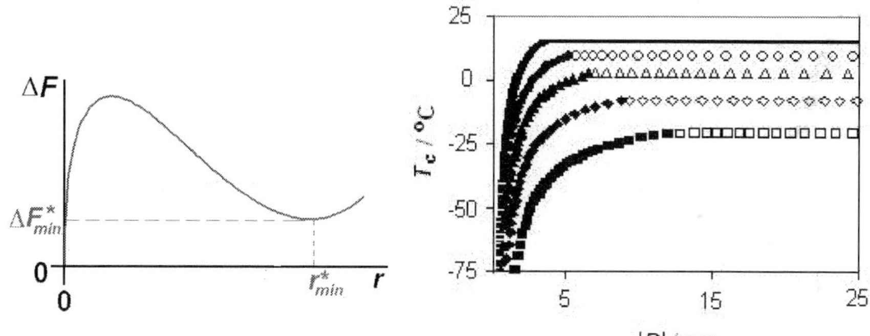

Fig. 4. (a) Schematic graph of ΔF vs r for crystallization in a 3D-nanoconfined solution. (b) Graph showing T_c, with confinement radius, $|R|$, for homogeneous nucleation from solutions of octadecane in dodecane. Open symbols correspond to the regime where $\Delta F^* = kT_c \ln(\Omega / J_{trans})$ gives T_c, filled symbols to the regime where $|R| \leq |R_{min}|$ and T_c is controlled by $\Delta F^*_{\min} = 0$. Squares = 0.1 mole fraction of octadecane in dodecane, diamonds = 0.25 mole fraction, triangles = 0.5 mole fraction and circles = 0.75 mole fraction. The uppermost curve corresponds to the pure octadecane liquid case, with the thicker line portion showing the regime controlled by $\Delta F_{tot} = 0$ (Cooper et al., 2008).

2.2.3 Thermodynamic control of crystallization

The above analysis show that for all confinement sizes below $|R_{min}|$, crystallization is not governed by the ability to surmount the nucleation barrier, ΔF^*, but by the ability to create a thermodynamically feasible new phase , i.e. $\Delta_{tot}F \leq 0$ for melt crystallization or $\Delta F^*_{min} \leq 0$ for solution crystallization. This means that crystallization is under thermodynamic, rather than the usual kinetic, control. This is significant because crystallization can then be directed to generically produce the most stable crystalline phase in 3D nanoconfined solutions and liquids. This finding is particularly important for polymorphic compounds.

3. Polymorphism

Polymorphism occurs when a substance can crystallize into more than one crystal structure. Each polymorph of a substance will have differing physical properties e.g. melting points, solubilities, compaction behaviour etc. In the pharmaceutical industry, it is imperative that a drug does not transform post-marketing, as this can affect its bioavailability, and reduce the drug's effectiveness. An infamous example of this occurred for the anti-HIV drug, Ritonavir (Chemburkar et al., 2000). In 1998, 2 years after marketing the drug in the form of soft gelatine capsules and oral solutions, the drug failed dissolution tests; a less soluble, thermodynamically more stable, polymorph had formed. This resulted in the precipitation of the drug and a marked decrease in the dissolution rate of the marketed formulations, reducing its bioavailability. Consequently Ritonavir had to be withdrawn from the market and reformulated, to the cost of several hundred million dollars.

The Ritonavir case highlights that crystallization is typically under kinetic control, with metastable polymorphs often crystallizing initially in accordance with Ostwald's rule of stages (Ostwald, 1897). For pharmaceutical companies, Ostwald's rule is a nemesis, as it means that their strategy of relying on high throughput screening of different crystallization conditions in order to identify stable polymorphs is scientifically flawed and so may not succeed. Consequently the industry remains vulnerable to another Ritonavir-type crisis. If the crystallization can be conducted from 3D-nanoconfined solutions, however, the crystallization can be exerted under thermodynamic control so that the thermodynamically stable polymorph is crystallized directly. In particular, in Figure 5a it is evident that neither polymorph A in red or polymorph B in blue will crystallize from the nanoconfined solution,

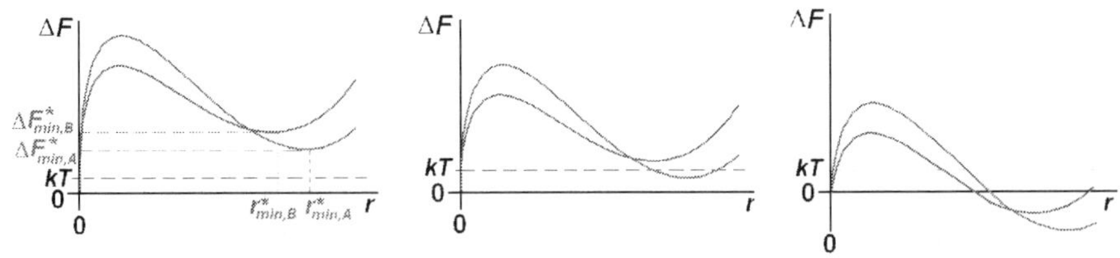

(a) (b) (c)

Fig. 5. Example graphs of free energy change, ΔF *vs.* nucleus size, r for crystallization of a polymorphic system from 3D nanoconfined solutions of monodisperse size and supersaturation. (a) System stabilized due to 3D nanoconfinement, with no observable crystallization. (b) Crystallization is under thermodynamic control with stable polymorph A (red) crystallizing, even though it has the higher nucleation barrier. (c) Crystallization occurs under kinetic control with metastable polymorph B (blue) as the majority product.

since for both $\Delta F^*_{\min} > kT$. In Figure 5b, however, polymorph A can form because it can produce a stable nucleus ($\Delta F^*_{min,A} < 0$) whereas polymorph B cannot. Hence provided the nucleation barrier is surmountable, this system will crystallize under thermodynamic control to directly give the stable polymorph A. In the system shown in Figure 5c, however, both polymorph A and B can form stable nuclei; crystallization will tend to be under kinetic control with polymorph B forming at a faster rate due to its lower nucleation energy barrier. Thus metastable B becomes the majority product in this case.

The thermodynamic arguments stated above are valid in the thermodynamic limit, where the system is so large that fluctuations do not significantly contribute to the equilibrium properties of the system. Though, of course, it must be remembered that it is these very fluctuations that enable critical nuclei to form and hence initiate the phase transformation. Consequently, in a system comprising a limited number of nanoconfined solutions, the fluctuations do need to be included to accurately model the equilibrium properties of the system (Reguera et al., 2003). We neglect this statistical thermodynamic description in our simple model of onset crystallization temperatures because the system in which we apply it, namely microemulsions, typically consists of a sufficiently large number of droplets, ~10^{18} per gram of microemulsion, which makes its additional complexity unwarranted. Moreover our simple model contains the essential features required to show how reliable estimates of critical nucleus sizes and thermodynamic control of crystallization are realizable in microemulsions.

4. Microemulsions

Microemulsions are thermodynamically stable, transparent mixtures of immiscible liquids. Typically they comprise oil droplets in water (an oil-in-water microemulsion) or water droplets in oil (a water-in-oil microemulsion). Bicontinuous microemulsions are also possible, however the absence of a 3D nanoconfined solution in these systems make them less suitable for thermodynamic control of crystallization. In the droplet microemulsions, the droplet size is typically 2-10 nm, with a relatively narrow polydispersity of $\sigma_R/R_{max} \approx$ 0.1-0.2, where σ_R is the Gaussian distribution standard deviation and R_{max} is the modal droplet radius (Eriksson and Ljunggren, 1995). The droplets are stabilized by surfactants, frequently in combination with a co-surfactant, which reside at the droplet interface, reducing the interfacial tension there to ~10^{-3} mN m^{-1}. This ultralow interfacial tension provides the thermodynamic stability of microemulsions, since the small free energy increase involved in creating the droplet interface is more than offset by the increased entropy of the dispersed phase. Note that this ultralow interfacial tension also ensures that the LaPlace pressure difference, $\Delta P = 2\gamma/R$, across the highly curved droplets is small. When the volume fraction of the dispersed phase becomes so low that its properties differ measurably from its usual bulk properties, the terms "swollen micelles", "swollen micellar solutions", "solubilized micellar solutions" or even simply "micellar solutions" can be used instead of microemulsions for oil-in-water systems, whilst for water-in-oil systems, the same terms but with "inverse" or "reverse" inserted before "micelle" or "micellar" may be used. However, because there is, in general, no sharp transition from a microemulsion containing an isotropic core of dispersed phase and a micelle progressively swollen with the dispersed phase, many researchers use the term "microemulsion" to include swollen micelles (or swollen inverse micelles) but not micelles containing no dispersed phase. This is the context in which the term "microemulsion" is used here. In the microemulsions, dissolved solutes may be supersaturated within the discontinuous (dispersed) phase, or the dispersed liquid may be supercooled, so that crystallization in the microemulsions can occur. The solute molecules are assumed to be distributed amongst the microemulsion droplets with a Poisson distribution, so that most droplets will have a supersaturation close to the mean, but a minority will have supersaturations significantly higher than the mean.

Microemulsions are dynamic systems with frequent collisions occurring between the droplets. The most energetic of these collisions cause transient dimers to form, allowing the exchange of interior content between the droplets. This exchange means that microemulsions can act as nanoreactors for creating quantum dots and other inorganic nanoparticles, for example. A recent review (Ganguli et al., 2010) on inorganic nanoparticle formation in microemulsions highlights the progress that has been made in this area from its inception with metal nanoparticle synthesis in 1982 by Boutonnet et al., followed by its use in synthesizing quantum dots (Petit et al. 1990) and metal oxides (see e.g. Zarur & Ying, 2000). The transient dimer mechanism also enables crystallization to proceed in microemulsions whenever a transient dimer forms between a droplet containing a crystal nucleus and a nucleus-free droplet which contains supersaturated solution, since the crystal nucleus can then gain access to this supersaturated solution and thereby grow (see Figure 6). Crystallization of organic compounds from microemulsions was first studied by Füredi-Milhofer et al. in 1999 for the aspartame crystal system, with studies on glycine crystallization (Allen et al., 2002; Yano et al., 2000; Nicholson et al., 2011; Chen et al., 2011) and carbamazepine (Kogan et al., 2007) following. The use of microemulsions in producing both inorganic nanoparticles and macroscopic organic crystals shows that the size of the particulates grown can vary from a few nm to mm, depending upon the nucleation rate, the ability to form stable nuclei, and the extent of surfactant adsorption on the resulting particles. Our interest in microemulsions stems from their intrinsic ability to enable reliable estimates of critical nucleus sizes and to exert generic thermodynamic control over the crystallization process for the first time.

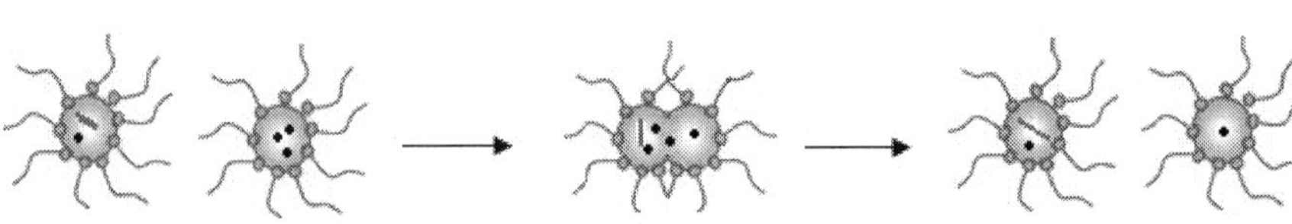

Fig. 6. Schematic diagram illustrating an energetic collision between a microemulsion droplet containing a crystal nucleus and a nucleus-free droplet containing supersaturated solution. The energetic collision results in a transient dimer forming, enabling the nucleus to gain access to more molecules and grow. The crystal nucleus is shown in red and the solute molecules are shown in black. The surfactant molecules stabilizing the microemulsion droplets are depicted as blue circles with tails.

4.1 Measurement of the critical nucleus size in microemulsions

Reliable estimates for the critical nucleus size can be found for homogeneous nucleation in microemulsions. The homogeneous nucleation temperature should be approximately constant until the droplet size decreases to below $|R_{min}|$, and thereafter the temperature should decrease. In the region where T_c decreases, the crystallization is controlled by the requirement that $\Delta_{tot}F \leq 0$, rather than the size of the nucleation energy barrier. Consequently, the critical nucleus size can be estimated by measuring the droplet size $|R|$ and assuming a spherical nucleus so that $r^* = 2|R|/3$ and $n^* = 32\pi|R|^3/81v_c$. Whilst homogeneous nucleation is comparatively rare in bulk systems, in microemulsions it is more prevalent for two main reasons. Firstly, the droplets are too small to contain foreign material onto whose surfaces heterogeneous nucleation could arise. Secondly, the ability of the surfactants to induce heterogeneous nucleation is often reduced in microemulsions compared to that at planar interfaces and in emulsions, particularly for crystallization from solution. This is because nuclei formation on the surfactant layer is disfavoured at this ultra low interfacial tension interface, and the high curvature may also hinder any templating mechanism that operates at more planar interfaces. The reduction in adsorption is readily apparent from Young's equation, $\cos\theta = (\gamma_1 - \gamma_2)/\gamma$ where θ is the contact angle, γ is the interfacial tension between the crystallizing species and the surrounding melt/solution, with γ_1 and γ_2 denoting the interfacial tensions between the surfactant and immiscible phases, and the surfactant and crystallizing species, respectively. The lowering of γ_1 that occurs in going from an emulsion to a microemulsion results in a higher contact angle, θ, and hence reduced adsorption for the crystallizing species.

Given this, we might expect heterogeneous nucleation to be impeded in microemulsions, and indeed other systems with low interfacial tensions, γ_1. Such an effect was observed at the phase inversion of an emulsion using Span 80 and Tween 80 surfactants to induce β-glycine crystallization at the oil-aqueous interface (Nicholson et al., 2005). Similarly, the ability of the nonionic surfactants, Span 80 and Brij 30 to heterogeneously nucleate glycine was negligible in microemulsions containing these mixed surfactants (Chen et al., 2011), whereas they promoted the metastable β-glycine nucleation at planar interfaces and in emulsions (Allen et al., 2002; Nicholson et al., 2005). There are literature examples where nonionic surfactants do induce heterogeneous nucleation in microemulsions, though. For instance, ice nucleation was promoted by adding heptacosanol, a long chain alcohol cosurfactant, to AOT microemulsions (Liu et al., 2007). Long chain alcohols can induce ice nucleation at temperatures of ≈-2 °C at planar air-water interfaces (Popovitz-Biro et al., 1994) and at ≈-8 °C at emulsion interfaces (Jamieson et al., 2005). This nucleating ability was diminished in the microemulsions. Nevertheless, ice crystallization still tended to occur at temperatures in the range of ≈-9 to -30 °C depending upon the heptacosanol concentration in the microemulsion droplets (Liu et al., 2007), i.e. much higher than the homogeneous nucleation temperature of ≈-40 °C (Wood & Walton, 1970; Clausse et al., 1983). For ionic surfactants, like AOT, heterogeneous nucleation in microemulsions can also occur. The longer range electrostatic interactions of ionic surfactants can induce order without requiring direct adsorption onto the surfactant layer. It is important, therefore, to choose surfactants that do not promote crystallization when placed at the planar air-liquid or air-solution interface to ensure that homogeneous nucleation occurs in the microemulsions.

Once a suitable homogeneous nucleating microemulsion system has been found, the onset crystallization temperature, T_c, for microemulsions of varying size, R, can be found by a suitable technique, such as differential scanning calorimetry (DSC). It can be assumed that the exothermic DSC peak arising from the crystallization of the droplets corresponds to the temperature range in which the majority of droplets crystallize, and that the mean droplet size can be used to accurately determine r^* and n^*. The mean droplet size of the microemulsion can be determined from small angle X-ray scattering, or small angle neutron scattering measurements. This methodology allows a simple and accurate measurement of the critical nucleus size, and is particularly useful for crystallization of liquids, or solutes which have a high solubility in the confined phase, so that there is sufficient crystallizable material present within the microemulsion for the exothermic crystallization peak to be observable by DSC. We have recently applied this methodology to ice crystallization in AOT microemulsions (Liu et al., 2007). Figure 7a shows homogeneous ice nucleation in AOT microemulsions, compared to heterogeneous nucleation in Figure 7b where the cosurfactant heptacosanol is added to the AOT microemulsions. The larger error bars in the heterogeneous nucleation case shown in Figure 7b reflect the varying number of heptacosanol molecules in the droplets that cause ice nucleation. For the homogeneous case, the reduction in T_c with $|R|$ occurs at $|R_{min}| \approx 2$ nm in Figure 7a, in good agreement with the theoretical value shown in Figure 3a. From this, the critical nucleus at R_{min} can be estimated to contain ≈280 molecules (Liu et al., 2007).

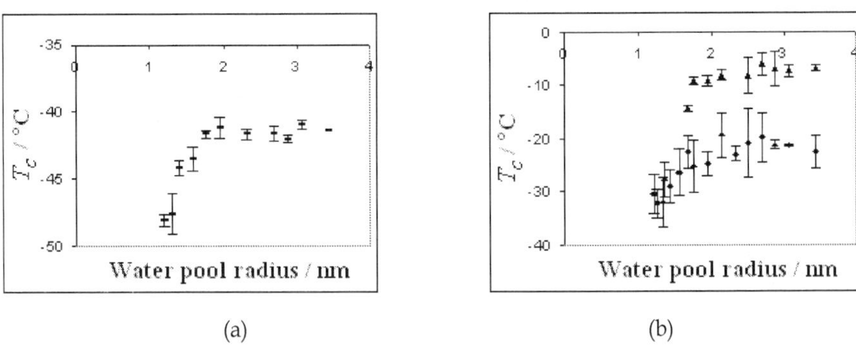

Fig. 7. Variation of observed ice onset crystallization temperatures, T_c, with water pool size for microemulsions with (a) AOT and (b) AOT plus heptacosanol The error bars show the standard deviation from three or more measurements. The nucleation is homogeneous in (a) and heterogeneous in (b). In the AOT microemulsions with added heptacosanol, several crystallization peaks were often evident, due to the droplets having differing numbers of the heptacosanol cosurfactant molecules that help nucleate ice. Consequently in (b) the highest T_c peak (upper curve) and largest T_c peak (lower curve) are just plotted for clarity (Liu et al. 2007).

4.2 Thermodynamic control of crystallization in microemulsions: Leapfrogging Ostwald's rule of stages

As detailed previously, crystallization within 3D nanoconfined solutions differs fundamentally from bulk crystallization because the limited amount of material within a nanoconfined solution results in the supersaturation decreasing substantially as the nucleus grows, leading to a minimum[11,12] in the free energy, $\Delta F^*_{\mathrm{min}}$, at a post-critical nanometre nucleus size, r^*_{min} (see Figure 4a). This fact, in particular, allows stable polymorphs to be solution-crystallized from microemulsions even when they have high nucleation barriers. Hence thermodynamic control of crystallization can be generically achieved for the first time. Note that microemulsions differ in two main ways from the theoretical nanoconfined solutions considered previously in sections 2 and 3. Firstly, transient droplet dimer formation enables the nuclei to grow beyond that dictated by the original droplet size; in fact crystals ranging from nm to mm can be produced. Secondly, microemulsions are polydisperse. There will be a range of droplet sizes and supersaturations in any microemulsion system, which must be considered to enable thermodynamic control of crystallization. An effective strategy is detailed below.

The equilibrium population of the r^*_{min} nuclei in the microemulsion depends upon the Boltzmann factor, $\exp\left(-\Delta F^*_{\mathrm{min}} / kT\right)$. Consequently, if $\Delta F^*_{\mathrm{min}} > kT$, the equilibrium population of the r^*_{min} nuclei will be very low. In contrast, if the r^*_{min} nuclei have free energies, $\Delta F^*_{\mathrm{min}} < kT$, they will have a sizeable equilibrium population. We term such r^*_{min} nuclei, (near) stable nuclei. Crystallization in microemulsions proceeds initially via the 3D nanoconfined nuclei gaining access to more material and growing during the energetic droplet collisions that allow transient dimers to form. *This crystallization process will be severely hindered if the population of such* r^*_{min} *nuclei is so low* as for the case depicted in Figure 5a, that the colliding droplets are highly unlikely to contain nuclei. In this case, the supersaturated system is stabilized due to the 3D nanoconfinement. In contrast, the crystallization can proceed readily via this transient droplet dimer mechanism if $\Delta F^*_{\mathrm{min}} < kT$

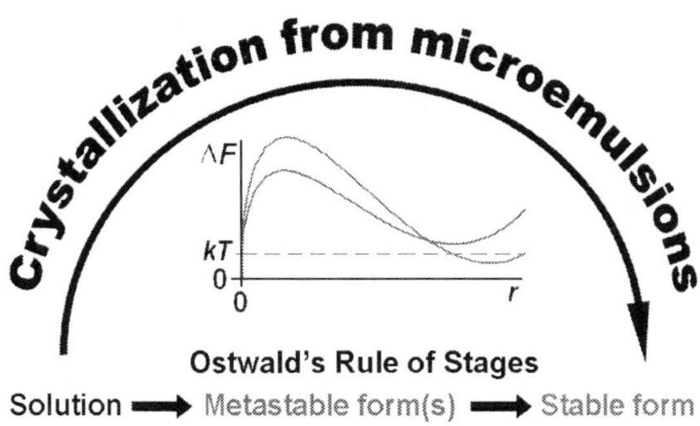

Fig. 8. A schematic diagram showing how Ostwald's rule of stages can be 'leapfrogged' by crystallizing from microemulsions. Stable polymorph A is in red, and metastable B is in blue. The ΔF vs r plot corresponds to the case where crystallization is brought under thermodynamic control so that stable polymorph A crystallizes directly.

because then a sizeable population of droplets will contain a (near) stable nucleus. *Thus, crystallization in microemulsions is governed by the ability to form (near) stable nuclei rather than critical nuclei.* In particular, recalling that there will be a range of droplet sizes and solute concentrations within the microemulsion droplets, thermodynamic control of crystallization can be achieved using the following methodology. The supersaturation in a microemulsion can be increased from one that is stabilized due to 3D nanoconfinement until the following condition is met: only the largest and highest supersaturation droplets, which contain the most material, can form (near) stable nuclei of only the most stable crystal form or polymorph. This situation is exemplified in Figure 8 with $\Delta F^*_{\min,A} \leq kT$ but $\Delta F^*_{\min,B} > kT$. Crystallization is then only just possible, and importantly, only the most stable polymorph will crystallize, as this is the only form for which (near) stable nuclei are likely to exist and so grow during transient droplet dimer formation. This generic strategy allows us to 'leapfrog' Ostwald's rule of stages and crystallize stable polymorphs directly.

The strategy detailed above will work provided $\Delta F^*_{\min,A} < \Delta F^*_{\min,B}$, and (near) equilibrium populations of the $r^*_{\min}$ nuclei are obtained. $\Delta F^*_{\min,A} < \Delta F^*_{\min,B}$ will typically be true because the stable polymorph has the greater bulk stability and it is the least soluble. Hence $r^*_{\min,A} > r^*_{\min,B}$ as stable polymorph A can grow to larger nucleus sizes, with typically lower free energies, than metastable B before its supersaturation is depleted. Equilibrium populations of the $r^*_{\min}$ nuclei will arise provided the $r^*_{\min}$ nuclei formation and dissolution processes are sufficiently rapid. This depends upon the magnitude of the nucleation barriers, ΔF^*, and the dissolution energy barriers, ΔF^* - $\Delta F^*_{\min}$, respectively. It can be ensured that the nucleation barriers to all polymorphs are surmountable by crystallizing from sufficiently small droplets. This is because the substantial supersaturation depletion that arises in a small droplet as the nucleus grows means that very high initial supersaturations with respect to the most stable polymorph are required to enable a (near) stable nuclei to form. Consequently it can reliably be assumed that the initial solutions in these droplets will also be sufficiently supersaturated with respect to all polymorphs that all nucleation barriers are indeed surmountable. The nuclei dissolution barriers, i.e. ΔF^* - $\Delta F^*_{\min}$, will be surmountable when the ΔF^* barriers are surmounted and $\Delta F^*_{\min} \geq 0$ as in Figure 8, since then ΔF^* - $\Delta F^*_{\min} \leq \Delta F^*$. Hence thermodynamic control will indeed be obtainable if the largest and highest supersaturation droplets have free energy curves corresponding to Figure 8. Ostwald's rule will have been 'leapfrogged' because the high initial supersaturations ensure the nucleation barriers are 'leapt over' to directly give the most stable polymorph. In contrast, the analogous unconfined bulk system would crystallize the metastable polymorph initially, in accordance with Ostwald's rule, owing to its lower nucleation barrier.

Note that the ability of microemulsions to exert thermodynamic control over crystallization is independent of the nucleation path; it depends solely upon the ability to form (near) stable nuclei, rather than their formation pathway. Consequently this ability is generic, applying equally to systems that nucleate via a classical one stage mechanism, and to those where two stages are implicated.

Once formed, the (near) stable nuclei can grow via transient droplet dimer formation until the nuclei become larger than the droplets or the supersaturation is relieved. For crystallites larger than the droplets, subsequent growth can then occur via the following processes; energetic collisions with droplets that allow access to the droplets' interior contents, oriented attachment of other nuclei, and impingement from the (typically minuscule) concentration of their molecules in the continuous phase. Thus, the final crystal size can vary from nm to mm, depending upon the concentration of (near) stable nuclei, the supersaturation and the extent to which surfactant adsorption on the crystallites limits their growth rate.

In order to test the ability of microemulsions to exert thermodynamic control over crystallization, we chose three problem systems that had well-documented difficulty in obtaining their stable polymorphs: namely glycine, mefenamic acid and ROY (Nicholson et al., 2011). In each case, it was successfully demonstrated that the stable polymorph crystallized directly from the microemulsions under conditions where crystallization was only just possible. The time-scale at which ~mm sized crystals grew ranged from minutes to months. Solvent-mediated transformations to more stable polymorphs can occur in this timeframe, although such transformations would be expected to have a significantly reduced rate in microemulsions owing to the exceedingly low concentration of the polymorph's molecules in the continuous phase. Accordingly, transmission electron microscopy (TEM) was used to show that the nanocrystals crystallized within the first 24 hours in the microemulsions were also of the stable form, thereby proving that the initial crystallization was indeed in this form. TEM was performed on the microemulsions by dropping small aliquots of the microemulsions onto TEM grids, allowing the drops to (mostly) evaporate, and then washing the grids with the microemulsion continuous phase to remove residual surfactant. This left predominantly just the crystallites grown in the microemulsion droplets on the TEM grids. Figure 9 shows TEM images of stable Form I nanocrystals of mefenamic acid grown within 1 day from DMF microemulsions at 8 °C containing 80 mg/ml of mefenamic acid in the DMF (Nicholson & Cooper, 2011).

With increasing supersaturation, metastable polymorphs also crystallized from the microemulsions. The relative supersaturation increase that led to the emergence of metastable forms was highly system dependant, though. For instance, for glycine the stable γ-polymorph crystallized as the majority polymorph under mean initial supersaturation ratios of 2.0 to 2.3, for mefenamic acid the corresponding range was 4.1 to 5.3, whereas for ROY a much larger range of ≈10 to 24 was found. The small supersaturation range in which the stable γ-glycine polymorph crystallized as the majority form reflected the small relative energy difference of 0.2 kJ mol between the γ- and α-polymorphs and the much faster growth rate (~500 times) of the α-polymorph in aqueous solutions (Chew et al., 2007). Recall that when crystallization is only just possible in the microemulsions, the formation of (near) stable nuclei will be confined to only the largest droplets with the highest supersaturations. Hence the actual initial supersaturations that are required for crystallization to be just possible are always much higher than the mean initial values. An estimate of this actual initial supersaturation was found in the glycine system as follows. Assuming a Poisson distribution of solute molecules amongst the droplets, then for the glycine system, $<10^{-8}$ of the droplets formed (near) stable nuclei under conditions where crystallization was only just possible, since the 0.2 kJ mol^{-1} stability difference between the γ- and α-forms would only lead to thermodynamic preference for the stable γ-form if the (near) stable nuclei contained ~20-30 molecules (Nicholson et al., 2011). This meant initial supersaturation ratios of ~11-15 were necessary for crystallization to be possible in the glycine system. A similar analysis for the mefenamic acid case gives initial supersaturation ratios >15 (Nicholson & Cooper, 2011).

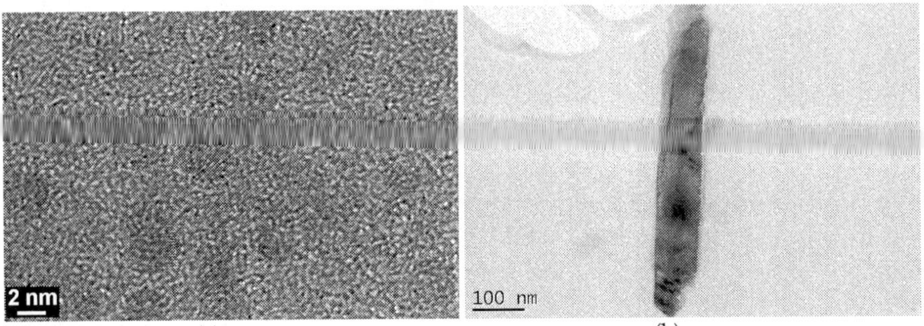

(a) (b)

Fig. 9. TEM images of stable Form I nanocrystals of mefenamic acid grown from DMF microemulsions. (a) ~4 nm nanocrystals grown in 12 hours and (b) a Form I nanocrystal grown in 24 hours.

These very high initial supersaturation ratios highlight two key factors governing the solution crystallization of stable polymorphs from microemulsions. Firstly, the substantial supersaturation decrease as a nucleus grows in a droplet means that very high initial supersaturations are needed for a (near) stable nucleus to form. Secondly, these very high initial supersaturations help ensure that the nucleation barriers to all possible polymorphs are surmountable. Hence, the use of microemulsions is the only way to crystallize a stable polymorph that has a very high nucleation barrier in bulk solution.

4.3 Practical considerations for choosing microemulsion systems

In choosing suitable microemulsion systems for achieving thermodynamic control of crystallization to obtain stable polymorphs, or for obtaining reliable estimates of critical nucleus sizes, the following should be considered.

1. The material to be crystallized should be insoluble, or only sparingly soluble in the microemulsion continuous phase. Since the material will have the same chemical potential at equilibrium in the continuous phase, as in the confined phase, the supersaturated material could potentially crystallize in the continuous phase with attendant loss of thermodynamic control. If the crystallizing material has a very low/negligible solubility in the continuous phase, however, the impingement rate onto a nucleus in this phase is so low that the nucleation rate and subsequent growth of such nuclei will be minimized. Then it can be reliably assumed that the crystallization is initially confined to the dispersed phase so that thermodynamic control is possible.
2. The nucleation should ideally be homogeneous. Choosing surfactants that do not promote crystallization at planar or emulsion interfaces helps ensure this. Homogeneous nucleation enables critical nucleus sizes to be obtained more reliably, since the contact angle of 180° is, of course, invariant with droplet size, R, and the relationships $r^* = 2|R|/3$ and $n^* = 32\pi|R|^3/81v_c$ are valid for droplet sizes

$|R| \leq |R_{\min}|$, i.e. where T_c decreases with $|R|$. Homogeneous nucleation is also preferred when aiming to crystallize stable polymorphs, since heterogeneous nucleation could potentially lead to a metastable polymorph having a lower $\Delta F^*_{\min}$ than the stable one, and thereby crystallizing in preference to the stable form.

3. Given that a metastable polymorph can potentially have the lowest $\Delta F^*_{\min}$ in a microemulsion if, for example, it is heterogeneously nucleated by the surfactant, or it is sufficiently stabilized by the surrounding solvent, or an inversion of stability between polymorphs occurs at nm sizes, then this possibility should be checked. This can be done readily by using a different solvent and/or surfactant. In addition, the supersaturation of the microemulsion should be gradually increased from the point at which crystallization is only just possible, until crystals/nanocrystals of more than one polymorph crystallize. In this way, all low energy polymorphs can be identified.
4. The crystallizable species, or more often the solvent, may be absorbed in the surfactant interfacial layer and so the solute concentration within the interior pool of the microemulsion droplet may differ substantially from the bulk concentration used in making the microemulsion. This possibility must be checked by measuring the solubility of the crystallizing species in the microemulsion, and then determining the mean initial supersaturation ratios accordingly. Accurate solubility measurements require adding powdered material to a microemulsion and leaving for a prolonged period (weeks) to ensure equilibration.
5. Bicontinuous and percolating microemulsions are not suitable systems for determining critical nucleus sizes or obtaining thermodynamic control of crystallization. In bicontinuous microemulsions, the absence of 3D nanoconfinement precludes their use. In percolating microemulsions, the droplets cluster and transiently form much larger droplets in which (near) stable nuclei of metastable polymorphs can form and grow, alongside the (near) stable nuclei of the stable form in the non-clustering droplets. For water-in-oil microemulsions, the absence of percolation and bicontinuous structures can be assumed if the microemulsions show minimal conductivity.
6. Microemulsions are thermodynamically stable and so form spontaneously upon mixing the constituents. Hence shaking by hand and vortexing are suitable preparation methods. Prolonged sonication should be avoided in case this affects the crystallization.
7. When the supersaturation in the microemulsions is achieved via anti-solvent addition or by a chemical reaction, a mixed microemulsion method should generally be implemented whereby two microemulsions are prepared. A different reactant is in each microemulsion, or for the antisolvent crystallization method, one microemulsion contains the undersaturated crystallizable species in a good solvent and the other microemulsion contains the antisolvent. On mixing the two microemulsions, transient dimer formation between droplets containing different reactants and/or solvents enables a relatively rapid equilibration of interior droplet content to take place, on the timescale of ~μs to ms (Ganguli et al., 2010). After this, it can be assumed that the dispersed reactants and solvents are distributed amongst the droplets with a Poisson distribution. Adding the antisolvent or second reactant drop-wise to the microemulsion should be avoided as this can create high concentration fluctuations immediately after the addition (Chen et al., 2011). Alternatively, if one of the reactants is soluble in the continuous phase, a solution of this reactant should be added to the microemulsion

containing the second reactant. Here, it is necessary to ensure the reaction proceeds predominantly in the microemulsion droplet, or at the droplet interface, by making sure that the second reactant has a negligible solubility in the continuous phase, whilst the first reactant must be able to partition into the droplet interface and/or interior.

8. For determining critical nucleus sizes, crystallization of liquids or high concentration solutes are preferred so that crystallization peaks are observable in the DSC. The crystallization peak is then associated with the mean droplet size $|\overline{R}|$, since close to this peak most droplets can form stable nuclei and crystallize. To obtain stable polymorphs, solution-crystallization from microemulsions is preferred, as then the substantial supersaturation decrease as the nuclei grow means that large initial supersaturations are required in order to create (near) stable nuclei and these help ensure the nucleation barriers to all polymorphs are surmountable. To ensure crystallization of stable polymorphs from microemulsions, the crystallization should only just be possible, so that the (near) stable nuclei only form in the largest droplets with the highest supersaturations. Hence the initial supersaturations and number of solute molecules in these droplets will be significantly higher than the mean values.

4.4 Crystallization of inorganic systems in microemulsions

Recently we have extended the thermodynamic control of crystallization methodology to inorganic polymorphic systems. There is much literature detailing inorganic nanoparticle formation via the mixed microemulsion approach (Ganguli et al, 2010). However, often the nanoparticles obtained are amorphous and so require high temperature and/or high pressures to introduce crystallinity. For instance, many literature examples concerning the formation of titania nanoparticles produce the crystallinity via subsequent calcining (e.g. Fernández-Garcı et al., 2007; Fresno et al., 2009) or a combined microemulsion-solvothermal process (e.g. Kong et al., 2011). Many methods also involve continual stirring. This is not necessary for microemulsions since they are thermodynamically stable, and stirring may disrupt any potential thermodynamic control if larger transient droplets are formed from multiple colliding droplets. Our microemulsion methodology enables direct crystallization of the nanocrystals of rutile, the stable form of titania, at room temperature provided the solute concentrations are kept sufficiently low and the crystallization is confined (predominantly) to the dispersed phase. To illustrate this, a microemulsion comprising 1.74 g of cyclohexane, 1-hexanol and Triton X-100 in the volume ratio of 7 : 1.2 : 1.8, and 180 μl of 2M HCl as the dispersed phase, was prepared. To this was added a solution of 180 μl of titanium isopropoxide (TIPO) in 1.74 g of the cyclohexane, 1-hexanol and Triton X-100 surfactant solution. The TIPO molecules reacted with the water predominantly at the droplet interface so that the resulting titanium dioxide resided mainly in the droplets. The use of 2M HCl slowed down the production of titanium dioxide, preventing gellation, and allowing the crystallization to proceed under thermodynamic control to give the stable rutile phase. TEM after 12 hours confirmed that the nanoparticles of size ~4 nm were crystalline rutile (see Figure 10a). After 3 days the nanocrystals were ~100 nm (see Figure 10b).

Similar microemulsion compositions with surfactant:aqueous mass ratios of 2.5:1 or less and ≤9% by volume of TIPO, produced rutile nanoparticles of good crystallinity. Indeed calcining these particles at 450 °C for 18 hours led to only a small increase in crystallinity (see Figure 10c). The microemulsions gradually took on a blue tinge over several days due to scattering from the growing rutile particles. The growth of the nanoparticles could be increased by the subsequent addition of water to swell the droplets so that an emulsion formed. Notably, even if the water addition occurred only a few minutes after mixing the TIPO solution with the aqueous HCl microemulsion, good crystallinity rutile particles were still formed. In contrast, when the reaction was carried out in the bulk phase, poor crystallinity/amorphous titania was obtained, demonstrating that the microemulsion stage was crucial for the formation of seed rutile nanocrystals. This general strategy of slowing the reaction rate via limited reactants and/or an appropriate pH range can be used to help introduce, or increase, crystallinity of inorganic nanoparticles obtained from microemulsions.

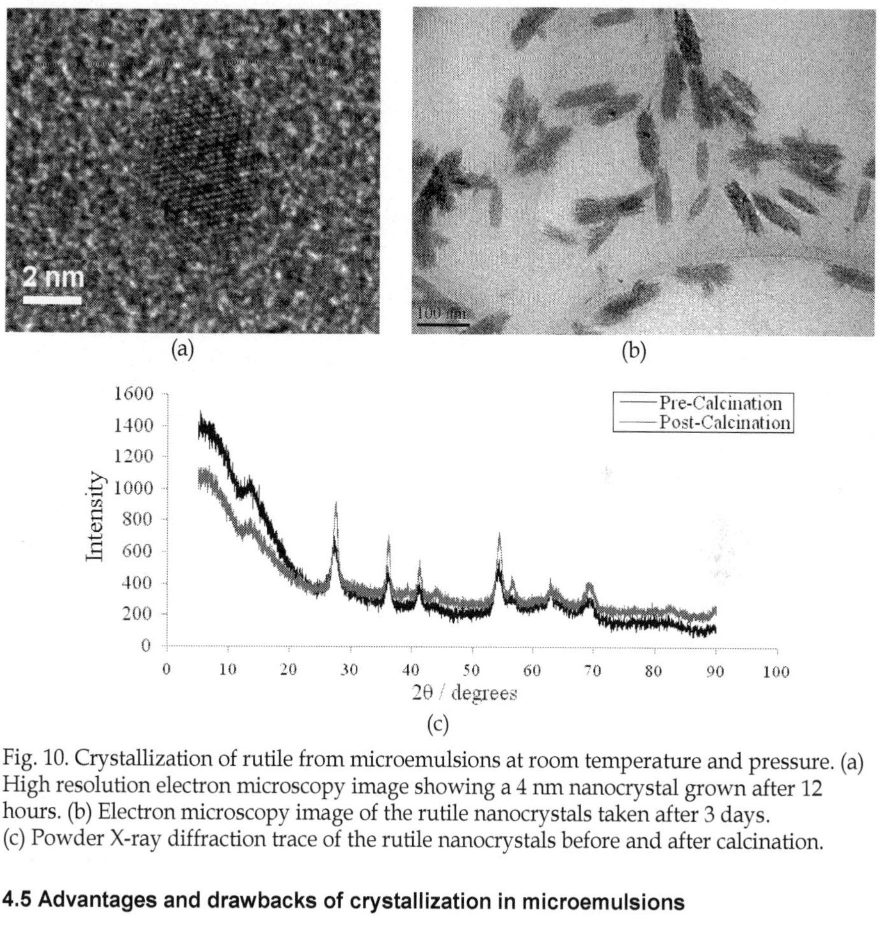

Fig. 10. Crystallization of rutile from microemulsions at room temperature and pressure. (a) High resolution electron microscopy image showing a 4 nm nanocrystal grown after 12 hours. (b) Electron microscopy image of the rutile nanocrystals taken after 3 days. (c) Powder X-ray diffraction trace of the rutile nanocrystals before and after calcination.

4.5 Advantages and drawbacks of crystallization in microemulsions

The use of microemulsions to exert thermodynamic control of crystallization is clearly an advantage whenever stable crystal forms are needed, such as in drug formulations and in obtaining nanocrystals with specific size-dependant properties. However, the much slower growth of crystals in microemulsions, compared to that in bulk solution, may limit industrial applications. A strategy whereby the microemulsion is controllably destabilized once the (near) stable nuclei have formed may circumvent this problem. For instance, the addition of more dispersed phase to swell the droplets into an emulsion may prove advantageous, provided the additional solution results in growth only on the existing (near) stable nuclei and nanocrystals, rather than the nucleation of new crystals, since the latter would be produced under kinetic, rather than thermodynamic, control. An effective approach to ensure this would be to induce the supersaturation of the additional dispersed phase slowly only after the emulsion has formed e.g. by cooling or adding a separate microemulsion containing an antisolvent.

5. Conclusion

Microemulsions present a unique opportunity for both the reliable estimate of critical nucleus sizes and the thermodynamic control of crystallization. The 3D droplet nanoconfinement results in crystallization being limited by the ability to form (near) stable nuclei, rather than critical nuclei, under conditions where crystallization is only just possible. This is a direct consequence of the limited amount of material within a droplet. In solution crystallization there is a substantial supersaturation decrease as a nucleus grows in a nanodroplet. The supersaturation decrease means that very high initial supersaturations are required in a droplet to achieve a (near) stable nucleus, thus enabling nucleation barriers to be readily surmountable. Hence solution crystallization from microemulsions is the only methodology known to-date that can generically crystallize stable polymorphs directly, even when they have insurmountable nucleation barriers in bulk solution. The transient dimer formation in microemulsions provides a mechanism for nuclei growth; we find it is possible to grow crystals ranging from nm to mm in size. Crystallization in microemulsions has already been successfully applied to 'leapfrog' Ostwald's rule of stages and directly crystallize the stable polymorphs of three 'problem' organic systems: glycine, mefenamic acid and ROY. The methodology should be of significant use in the pharmaceutical industry, as it provides the first generic method for finding the most stable polymorph for any given drug, thereby preventing another Ritonavir-type crisis. Microemulsions have also been used to synthesis nanocrystals of rutile without requiring a subsequent calcination step. Other inorganic systems that typically produce amorphous nanoparticulates are also likely to benefit from this approach. Its application to protein crystallization may prove problematic, given the larger size of protein molecules, though droplet clustering to fully encase the protein may occur in these systems alleviating this limitation. Future work will investigate this possibility. A disadvantage of the methodology is that once the (near) stable nuclei are generated, their growth is significantly impeded. Initially this is due to their nanoconfinement, and subsequently, when the nanocrystals grow bigger than these droplets, results from the limited concentration of their molecules in the continuous phase. Controlled microemulsion destabilization strategies, such as adding more of the dispersed phase to form an emulsion, may prove a viable route to circumvent this problem. Finally, given that the ultimate crystal size can vary from nm to mm, depending upon the population of (near) stable nuclei and their subsequent growth rates, there is a significant need for greater understanding of how the growth rates can be tuned. Then the use of microemulsions in crystallization would be truly unrivalled in producing both high crystallinity forms and the desired crystal size.

6. Acknowledgement

We thank EPSRC for funding.

7. References

Allen, K.; Davey, R. J.; Ferrari, E.; Towler, C. & Tiddy, G. (2002). The Crystallization of Glycine Polymorphs from Emulsions, Microemulsions and Lamellar Phases. *Crystal Growth & Design.*, Vol. 2, No. 6 (November, 2002), pp523-527.

Bartell, L. S. (1998). Structure and Transformation: Large Molecular Clusters as Models of Condensed Matter. *Annu. Rev. Phys. Chem.*, Vol. 49, (October 1998), pp43-72.

Becker, R. & Döring, W. (1935). Kinetic Treatment of Germ Formation in Supersaturated Vapour *Ann. Phys.*, Vol. 24, No. 8, (December 1935), pp719-752.

Bowles, R. K.; Reguera D.; Djikaev Y. & Reiss H. (2001). A Theorem for Inhomogeneous Systems: The Generalization of the Nucleation Theorem. *Journal o Chemical Physics*, Vol. 115, (July 2001) pp1853-1866.

Bowles, R. K.; Reguera D.; Djikaev Y. & Reiss H. (2002). Erratum: A Theorem for Inhomogeneous Systems: The Generalization of the Nucleation Theorem. *Journal of Chemical Phyics*, Vol. 116, (February 2002) pp2330.

Chemburkar, S. R.; Bauer, J.; Deming, K.; Spiwek, H.; Patel, K.; Morris, J.; Henry, R.; Spanton, S.; Dziki, W.; Porter, W.; Quick, J.; Bauer, P; Donaubauer, J.; B.; Narayanan, B. A.; Soldani, M.; Riley, D. & McFarland, K. (2000). Dealing with the Impact of Ritonavir Polymorphs on the Late Stages of Bulk Drug Process Development. *Organic Process Research & Development*, Vol. 4, No. 5 (June 2000), pp413-417.

Chen C.; Cook, O.; Nicholson C. E. & Cooper S. J. (2011). Leapfrogging Ostwald's rule of stages: Crystallization of stable γ-glycine directly from microemulsions. *Crystal Growth & Design*, Vol. 11, No. 6 (April 2011), pp2228-2237.

Chew, J. W.; Black, S. N.; Chow, P. S.; Tan, R. B. H. & Carpenter, K. J. (2007). Stable Polymorphs: Difficult to Make and Difficult to Predict. *Crystal Engineering Communications* Vol. 9, No. 2 (January 2007), pp128-130.

Clausse, D.; Babin, L.; Broto, F.; Aguerd, M. & Clausse, M. (1983). Kinetics of Ice Nucleation in Aqueous Emulsions. *Journal of Physical Chemistry* Vol. 87, No. 21 (October 1983), pp4030-4034.

Cooper, S. J.; Nicholson, C. E. & Lui, J. (2008). A simple classical model for predicting onset crystallization temperatures on curved substrates, and its implications for phase transitions in confined volumes. *Journal of Chemical Physics* Vol. 129, No. 12 (September 2008), 124715.

Couchman, P. R. & Jesser, W. A. (1977). Thermodynamic Theory of Size Dependance of Melting Temperature in Metals. *Nature*, Vol. 269, No. 5628, (October 1977), pp481-483.

Denoyel, R. & Pellenq R. J. M. (2002). Simple Phenomenological Models for Phase Transitions in a Confined Geometry. 1: Melting and Solidification in a Cylindrical Pore. *Langmuir*, Vol. 18, No. 7, (April 2002), pp2710-2716.

Erdemir, D.; Lee, A. Y. & Myerson, A. S. (2009). Nucleation of Crystals from Solution: Classical and Two-Step Models. *Accounts of Chemical Research*, Vol. 42, No. 5, (May 2009), pp621-629.

Eriksson, J. C. & Ljunggren, S. (1995). Comments on the Alleged Formation of Bridging Cavities/Bubbles Between Planar Hydrophobic Surfaces. *Langmuir*, Vol. 11, No. 6, (Month 1995), pp1145-1153.

Fletcher, N. H. (1958). Size Effect in Heterogeneous Nucleation. *Journal of Chemical Physics*, Vol. 29, No. 3, (May 1958), pp572-576.

Fernández-Garcıa, M.; Belver, C.; Hanson, J. C.; Wang, X. & Rodriguez, J. A. (2007). Anatase-TiO2 Nanomaterials: Analysis of Key Parameters Controlling Crystallization. *Journal of the American Chemical Society*, Vol. 129, No. 44, (January 2007), pp13604-13612.

Fresno, F.; Tudela, D.; Coronado, J. M. & Soria J. (2009). Synthesis of Ti1_xSnxO2 nanosized photocatalysts in reverse microemulsions. *Catalysis Today*, Vol. 143, No. 3-4 (May 2009), pp230–236.

Füredi-Milhofer, H.; Garti, N. & Kamyshny, A. (1999). Crystallization from microemulsions - a novel method for the preparation of new crystal forms of aspartame. *Journal of Crystal Growth*, Vol. 198, (March 1999), pp1365-1370.

Ganguli, A. K.; Ganguly A. & Vaidya, S. (2010). Microemulsion-based Synthesis of Nanocrystalline Materials. *Chemical Society Reviews*, Vol. 39, No. 2, (January 2010), pp474–485.

Gibbs, J. W. (1876). On the equilibrium of heterogeneous substances. *Trans. Connect. Acad. Sci.*, Vol. 3, pp108–248.

Gibbs, J. W. (1878). On the equilibrium of heterogeneous substances. *Trans. Connect. Acad. Sci.*, Vol. 16, pp343–524.

Kashchiev, D. (1982). On the Relation Between Nucleation Work, Nucleus Size and Nucleation Rate. *Journal of Chemical Physics*, Vol. 76, No. 10, (May 1982), pp5098-5102.

Jamieson, M. J.; Nicholson, C. E. & Cooper, S. J. (2005). First Study on the Effects of Interfacial Curvature and Additive Interfactial Density on Heterogeneous Nucleation. Ice Crystallization in Oil-in-Water Emulsions and Nanoemulsions with Added 1-heptacosanol. *Crystal Growth & Design*, Vol. 5, No. 2, (March 2005), pp451-459.

Kong, W.; Liu, B.; Ye, B.; Yu, Z.; Wang, H.; Qian, G. & Wang, Z. (2011) An Experimental Study on the Shape Changes of TiO2 Nanocrystals Synthesized by Microemulsion-Solvothermal Method, *Journal of Nanomaterials* Volume 2011, Article ID 467083, 6 pages.

Liu, J.; Nicholson, C. E. & Cooper, S. J. (2007). Direct measurement of critical nucleus size in confined volumes. *Langmuir*, Vol. 23, No. 13, (June 2007), pp7286-7292.

Nicholson, C. E.; Chen, C.; Mendis, B. & Cooper, S. J. (2011) 'Stable Polymorphs Crystallized Directly under Thermodynamic Control in Three-Dimensional Nanoconfinement: A Generic Methodology' *Crystal Growth & Design*, Vol 11, No. 2 (February 2011), pp. 363-366.

Nicholson, C. E. & Cooper, S. J. (2011). Crystallization of Mefenamic Acid from DMF Microemulsions: Obtaining Thermodynamic Control through 3D Nanoconfinement. *Crystals* Vol. 1, No. 3, (September 2011), pp. 195-205.

Nicholson, C. E.; Cooper, S. J.; Marcellin, C. & Jamieson, M. J. (2005). The Use of Phase-Inverting Emulsions to Show the Phenomenon of Interfacial Crystallization on both Heating and Cooling. *Journal of the American Chemical Society,* Vol. 127, No. 34, (August 2005), pp11894-11895.

Ostwald, W. Z. (1897). Studies of the Formation and Transformation of solid Substances. *Zeitschrift für Physikalische Chemie,* Vol. 22, (1897), pp289-330.

Oxtoby, D. W. & Kashchiev, D. (1994). A General Relation Between the Nucleation Work and the Size of the Nucleus in Multicomponent Nucleation. *Journal of Chemical Physics,* Vol. 100, No. 10, (May 1994), pp7665-7671.

Petit, C.; Lixon, P. & Pileni, M. P. (1990). Synthesis of Cadmium Sulfide in situ in Reverse Micelles. 2: Influence of the Interface on the Growth of the Particles. *Journal of Physical Chemisty,* Vol. 94, No. 4, (February 1990), pp1598–1603.

Petrov, O. & Furó, I. (2006). Curvature-dependant Metastability of the Solid Phase and the Freezing-Melting Hysteresis in Pores. *Phyical Reiew E,* Vol. 73, No. 1, (January 2006), 001608-1 - 001606-7.

Popovitz-Biro, R.; Wang, J. L.; Majewski, J.; Shavit, E.; Leiserowitz, L.; Lahav, M. (1994). Induced Freezing of Supercooled Water into Ice by Self-Assembled Crystalline Monolayers of Amphiphilic Alcohols at the Air-Water Interface. *Journal of the American Chemical Society,* Vol. 116, No. 4, (February 1994), pp1179-1191.

Pruppacher, H. R. (1995). A New Look at Homogeneous Ice Nucleation in Supercooled Water Drops. *Journal of Atmospheric Science,* Vol. 52, No. 11, (June 1995), pp1924-1933.

Reguera, D.; Bowles, R. K.; Djikaev Y. & Reiss, H. (2003). Phase Transitions In Systems Small Enough to be Clusters. *Journal of Chemical Physics,* Vol. 118, No. 1, (January 2003), pp340-353.

Speedy, R. J. (1987). Thermodynamic Properties of Supercooled Water at 1 atm. *Journal of Phyical Chem*istry, Vol. 91, No. 12, (June 1987), pp3354-3358.

Vanfleet, R. R. & Mochel, J. M. (1995). Thermodynamics of Melting and Freezing in Small Particles. *Surf. Sci.,* Vol. 341, No. 1-2, (November 1995), pp40-50.

Vekilov, P. G. (2010). Nucleation. *Crystal Growth & Design,* Vol. 10, No. 12, (December 2010), pp5007–5019.

Volmer, M, & Weber, A. (1926). Germ Formation in Oversaturated Figures. *Zeitschrift fürPhysikalische Chemie,* Vol. 119, No. 3-4, (March 1926), pp277-301.

Wood, G. R. & Walton, A. G. (1970). Homogeneous Nucleation Kinetics of Ice from Water. *Journal of Applied Physics,* Vol. 41, No. 7 (June 1970), pp3027-3036.

Xu, D. & Johnson, W. L. (2005). Geometric Model for the Critical-Value Problem of Nucleation Phenomena Containing the Size Effect of Nucleating Agent. *Physical Review B,* Vol. 72, No. 5, (August 2005), pp052101.

Yano, J.; Füredi-Milhofer, H.; Wachtel, E. & Garti, N. (2000). Crystallization of Organic Compounds in Reversed Micelles. 1: Solubilization of Amino Acids in Water-

Isooctane-AOT Microemulsions. *Langmuir*, Vol. 16, No. 26, (December 2000), pp10005-10014.

Zarur, A. J. & Ying, J.Y. (2000). Reverse Microemulsion Synthesis of Nanostructured Complex Oxides for Catalytic Combustion. *Nature*, Vol. 403, No. 6765, (January 2000), pp65-67.

Chemical, Physicochemical and Crystal – Chemical Aspects of Crystallization from Aqueous Solutions as a Method of Purification

Marek Smolik
Faculty of Chemistry, Silesian University of Technology, Gliwice
Poland

1. Introduction

This chapter is intended to discuss the effect of chemism of crystallizing and co-crystallizing substances (i.e., their chemical, physicochemical and crystal-chemical properties), as well as some other factors on efficiency of their separation and purification during crystallization from aqueous solutions.

There are three main aims of crystallization (Rojkowski & Synowiec, 1991): creation of the solid phase, forming crystals, purification of substances.

While using crystallization for purification and separation of various substances, as well as for enrichment of trace amounts of new-found radioactive elements, it was established that (in addition to many others) chemical factors strongly affected the mentioned operations. Mechanisms of trace radioactive elements' co-crystallization and the significance of these factors on their enrichment efficiency were reviewed in some works (Przytycka, 1968; Niesmeanov, 1975).

The influence of factors determining the structures of salts of crystal ionic lattices, (salts - considered as ionic coordination compounds) and their ability to isomorphous and isodimorphous mixing on their possibility to crystallization separation was thoroughly discussed by Balarew (1987). Whereas developments concerning inclusions of isomorphous impurities during crystallization from solutions were reviewed by Kirkova et al. (1998).

The discovered settlements were useful in the preliminary assessment of the effectivity of the crystallization method for purification of substances (Kirkova, 1994), for concentration microimpurities (Zolotov & Kuzmin, 1982), for growing of single crystals of specific properties (Byrappa, et al., 1986; Demirskaya, et al. 1989), as well as in explanation of the genesis of some minerals (Borneman-Starinkevich, 1975).

In spite of development of solvent extraction and ionic exchange methods, crystallization is still a very attractive method of purification, particularly in the preparation of numerous high-purity inorganic substances (HPIS). There are two main reasons for that:

- for crystallization purification of a substance only the simplest reagents are necessary (like water or other solvents, sometimes salting out or complexing agends), which can be easily purified to the level suitable for HPIS and readily removed after crystallization;

- in the case of many HPIS (especially crystalline preparations) crystallization is often the final stage of their preparation, which can be simply carried out without incidental contamination.

2. Crystallization as a method of purification

The crystallization should, in principle, yield very significant purification of a substance, but for the phenomenon of transport of accompanying impurities into the crystal, which may happen in the following ways presented in the simplified scheme below.

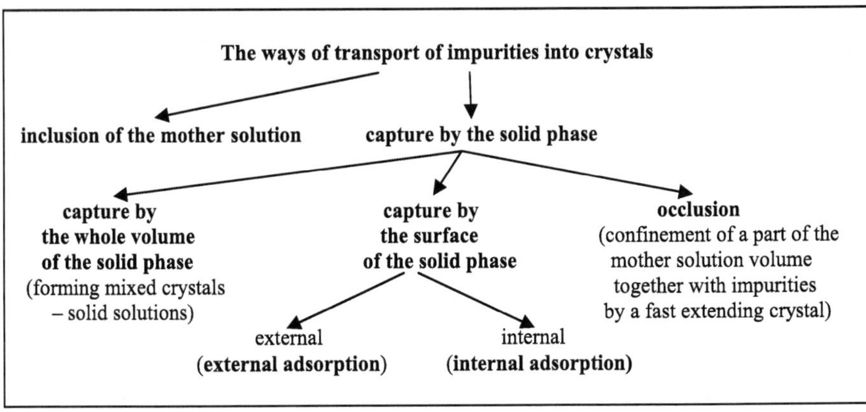

Although a suitable choice of crystallization conditions (supersaturation, rate of crystallization), as well as the ways of separation of crystals from mother solutions (filtration, washing) permits minimizing the capture of impurities derived from the inclusion of the mother solution, occlusion or external adsorption, it is impossible to restrict impurities originated from the capture by the whole volume of the solid phase or internal adsorption.[1]

The highest effect of purification may be expected when impurities are not captured by the solid phase of crystals but get into crystals as a result of the mother solution's residue, which cannot be removed by filtration[2]. In this boundary case the efficiency of crystal purification after its separation from the mother solution (without washing) is defined by the equation (Gorshtein, 1969):

$$\frac{1}{K_k} = \frac{e'_k}{e'_o} = \frac{100 - C_k}{C_k} \cdot \frac{C_r}{100 - C_r} \cdot \frac{1}{1 - \alpha} \tag{1}$$

[1] Internal adsorption takes place when microcomponents cannot form solid solutions with macrocomponent (Niesmieanov,1975)], it is a rather sparsely occurring phenomenon (Przytycka,1968).

[2] During the crystallization without stirring or with not vigorous stirring, big, aggregated crystals (twins, intergrowth) are obtained. The presence of cavities on their surface obstructs the separation of mother solution from these crystals, which results in lowering their purity. Stirring during the crystallization at a considerable concentration of crystals causes rounding of crystals because of abrasion. Then large crystals adopt the form of spheres or ellipsoids, whose separation from mother solution by means of filtration is easier, which leads to higher purity of final product (Matusevich, 1961; Bamforth, 1965).

where:
K_k– crystal purification coefficient (multiplicity of lowering initial microcomponent contents in crystal)
e'_k– relative contents of a microcomponent in the crystal [ppm],
e'_o– initial relative contents of a microcomponent (before crystallization) [ppm],
C_k– contents of the macrocomponent in the crystal [%], (100–C_k) - crystal humidity [%],
C_r– contents of the macrocomponent in the mother solution, (its solubility) [%],
α– degree of crystallization of the macrocomponent.

However, in reality, in numerous cases a microcomponent is captured by the solid phase, mainly by forming mixed crystals. Micro and macrocomponents form real mixed crystals (solid solutions) if they are isomorphous or isodimorphous.

2.1 Co–crystallization coefficients

2.1.1 Homogeneous distribution coefficient $D_{2/1}$ (Henderson– Kraček, Chlopin)

Homogeneous partition takes place in equilibrium conditions between the whole mass of a crystal and the mother solution, and is described by the Chlopin equation:

$$\frac{n\rho_s}{m_s} = K_X \frac{(n_o - n)\rho_r}{m_r} \tag{2}$$

where : n - number of moles of the microcomponent in a crystal, n_o - the whole number of moles of the microcomponent in the system, m_s - the mass of a crystal, m_r - the mass of solution, ρ_s - density of crystal, ρ_r - density of solution, K_X - Chlopin constant.

Taking into account that $\frac{n\rho_s}{m_s} = \frac{n}{\frac{m_s}{\rho_s}} = \frac{n}{V_s} = C_s$ as well as $\frac{(n_o - n)\rho_r}{m_r} = \frac{(n_o - n)}{\frac{m_r}{\rho_r}} = \frac{(n_o - n)}{V_r} = C_r$

where: V_s and V_r - volumes of the solid phase and the solution, C_s and C_r - concentration of a microcomponent in the solid phase and in the solution it is possible to obtain an equation, identical to the well-known Berthelot-Nernst equation describing the partition of a substance between two immiscible solvents $K_x = C_s/C_r$.

During the crystallization from the solution containing two components: macrocomponent (1) and microcomponent (2) the ratio of their partition coefficients defines the equilibrium co-crystallization coefficient:

$$D = D_{2/1} = \frac{K_2}{K_1} = \frac{\left(\frac{C_s}{C_r}\right)_2}{\left(\frac{C_s}{C_r}\right)_1} = \frac{\left(\frac{C_2}{C_1}\right)_s}{\left(\frac{C_2}{C_1}\right)_r} = \frac{C_{2s} \cdot C_{1r}}{C_{1s} \cdot C_{2r}}. \tag{3}$$

Substituting: n_o and n - number of moles of microcomponent in the whole system and in the solid phase, and z_o and z - number of moles of macrocomponent in the whole system and in the solid phase into equation (3) and suitable rearranging, it is possible to obtain a more convenient Henderson & Kraček equation (Niesmieanov, 1975):

$$D_{2/1} = \frac{n(z_o - z)}{z(n_o - n)} \tag{4}$$

Further transformation of this equation gives other, often used practical formulae (Smolik, 2004):

$$D_{2/1} = \frac{\frac{n}{z}}{\frac{(n_o - n)}{(z_o - z)}} = \frac{e'_s}{e'_r} \tag{5}$$

and

$$D_{2/1} = \frac{\frac{n}{n_0}\left(\frac{z_0}{z_0} - \frac{z}{z_0}\right)}{\frac{z}{z_0}\left(\frac{n_0}{n_o} - \frac{n}{n_0}\right)} = \frac{\beta(1-\alpha)}{\alpha(1-\beta)}, \tag{6}$$

where $\alpha = z/z_o$ is the degree of crystallization of macrocomponent, $\beta = n/n_o$ is the degree of co-crystallization of microcomponent, e'_s and e'_r are relative concentrations of microcomponent in the solid phase and in the mother solution, respectively ([ppm] in relation to macrocomponent), $D_{2/1}$ – homogeneous partition coefficient (co-crystallization coefficient).

2.1.2 Heterogeneous (logarithmic) distribution coefficients λ (Doerner–Hoskins)

Logarithmic partition can take place if the equilibrium between the whole mass of crystal does not exist, but only between the surface layer of a crystal and solution. If $D_{2/1} \neq 1$, the concentration of microcomponent in the solution during the crystallization will be changing continuously. So the microcomponent will distribute in the crystal in a stratified manner ("onion" structure). This process for the elementary layer of the crystal may be described by the equation parallel to that of Henderson-Kraček: $\frac{dn}{dz} = \lambda \frac{(n_o - n)}{(z_o - z)}$,

where the meaning of n_o, n, z_o, z is the same as previously described and λ is the heterogeneous (logarithmic) partition coefficient. The integration of this expression yields the known equations (Doerner & Hoskins, 1925):

$$\ln \frac{n_o}{n_o - n} = \lambda \ln \frac{z_o}{z_o - z}, \tag{7}$$

or

$$\lambda = \frac{\log(1-\beta)}{\log(1-\alpha)} \tag{8}$$

Both homogeneous and heterogeneous partitions are boundary cases of distribution of the microcomponent between the solid phase and the mother solution. Experimental study involving which of both coefficients retains constant value with the increase of the degree of crystallization gives information on what partition is actually taking place.

2.2 Homogeneous distribution coefficients $D_{2/1}$ as indicators of crystallization efficiency

Homogeneous partition coefficients $D_{2/1}$ are a convenient measure of crystallization efficiency as a method of purification. In the case of homogeneous partition of microcomponent in the solid phase, the final result of purification (without washing) may be expressed by the formula (Gorshtein, 1969), derived on the basis of the balance of amounts of the microcomponent during the crystallization:

$$\frac{1}{K_k} = \frac{e_k'}{e_o'} = \frac{D_{2/1}'}{\alpha D_{2/1}' + 1 - \alpha} + \frac{100 - C_k}{C_k} \cdot \frac{C_r}{100 - C_r} \cdot \frac{1 - D_{2/1}'}{\alpha D_{2/1}' + 1 - \alpha} \quad (9)$$

After careful washing of the crystals by pure, saturated solution of the macrocomponent, inclusions of the mother solution, as well as microcomponents adsorbed on the surface of the crystal, will be removed. The result of the purification in this case will be:

$$\frac{1}{K_k} = \frac{e_{k(p)}'}{e_o'} = \frac{D_{2/1}}{\alpha D_{2/1} + 1 - \alpha} \quad (10)$$

where: K_K - multiplicity of lowering initial microcomponent contents (e'_0, $e'_{k(p)}$ - initial contents of microcomponent -[ppm] in crystal and in washed crystal after crystallization); $D_{2/1}(D'_{2/1}) = e_s(e'_{k(p)})/e'_r$ (e'_s and e'_r - contents of micomponent - [ppm] in the solid phase and the mother solution); $D_{2/1}$ - isomorphous co-crystallization coefficient of microcomponent, $D_{2/1}'$ - adsorption-isomorphous co-crystallization coefficient of microcomponent; expression $(D'_{2/1} - D_{2/1})/D'_{2/1}$ qualifies a relative importance of adsorption in the capture of microcomponent by the solid phase (Gorshtein, 1969).

Knowing $D_{2/1}(D_{2/1}')$ and using the equations (9) and (10) it is possible to evaluate a number of crystallizations in the conditions of homogeneous partition of the microcomponent (at $D_{2/1}(D_{2/1}')$ =const.) necessary to achieve a desirable degree of purification of crystals. An example of such evaluation for $NiSO_4{\cdot}7H_2O$ is presented in Table 1.

$D_{2/1}'$	$D_{2/1}$	Number of crystallizations (k)	The whole yield of purification $(m_f/m_0)\cdot100\% = (\alpha^k)\cdot100\%$
0.75	0.75	15	0.003
0.50	0.50	6	1.56
0.25	0.25	3	12.5
0.10	0.10	2	25.0
0.05	0.05	1	50.0

Table 1. The number of $NiSO_4{\cdot}7H_2O$ crystallizations necessary to achieve 10-fold lowering of its initial contents of microcomponent for various levels of coefficient $D_{2/1}$ ($D_{2/1'}$) (Smolik, 2004), C_k=98%, C_r=50%, α=0,50 (50%), $m_0(m_f)$ – initial (after k crystallizations) mass of crystals

The data presented in Table 1 show that the level of $D_{2/1}$ ($D_{2/1}'$) is a very important parameter for the evaluation of crystallization efficiency as a method of purification, and therefore, its knowledge is significant in planning the utilization of crystallization in different stages of preparation of high purity substances.

2.3 Practical and equilibrium $D_{2/1}$ coefficients

However, crystallization processes are usually realized at non-equilibrium conditions and the obtained, in this case practical (effective), partition coefficients ($D^{p}_{2/1}$) depend on the ways in which the crystallization is carried out. This dependence may be presented by the following expression (Kirkova et al., 1996):

$$D^{p}_{2/1} = D^{o}_{2/1}\,\Theta(T, \varsigma, \omega, m_2, m_j\,\kappa) \tag{11}$$

where $D^{o}_{2/1}$ - equilibrium co-crystallization coefficient, Θ - imbalance factor, which is a function of temperature(T), supersaturation of the solution (ς), rate of stirring (ω), concentration of microcomponent 2 (m_2), concentration of other microcomponents (m_j) and other factors (κ).

Since the equilibrium co-crystallization coefficient does not depend on crystallization conditions, it may be compared with various crystallization systems. Hence, it is important to trace the determination methods of such coefficients.

2.4 Methods of determination of equilibrium distribution coefficients

The possibility of achieving equilibrium homogeneous partition of microcomponents between solution and crystal solid phase was proved by Chlopin. Subsequent investigations in this area (Gorshtein, 1969; Chlopin, 1957; Zhelnin & Gorshtein, 1971) elaboration upon several methods to accomplish equilibrium partition of microcomponents in the crystal, among which the method of isothermal decreasing of supersaturation and the method of long–time stirring of crushed crystals in their saturated solution are most often used.

2.4.1 The method of isothermal decreasing of supersaturation

It relies on cooling a saturated solution without stirring to the end temperature of crystallization (so that no crystal would appear) and after that on vigorous stirring at the constant end temperature until a complete removal of supersaturation takes place (usually for 3 - 360 h) (Zhelnin & Gorshtein, 1971; Chlopin, 1957). An example of such determination of $D_{2/1}^{eq}$ is presented in Fig. 1

2.4.2 The method of long–time stirring of crushed crystals in their saturated solution

The equilibrium is reached starting either from the initial concentration ratio of a microcomponent in crystal and in solution exceeding the expected value of its equilibrium coefficient ($D^{o}_{MAX}=e'_{ko}$("contaminated" crystal)/e'_{ro}("purified" solution)) - achieving equilibrium "from above" or from this ratio lower than the expected value mentioned above ($D^{o}_{min}=e''_{ko}$("purified crystal)/e''_{ro}("contaminated" solution)) – achieving equilibrium "from below" (Fig. 2).

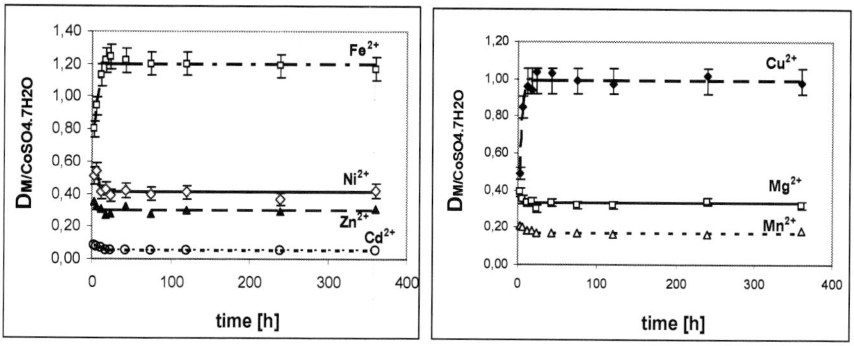

Fig. 1. Changes of co-crystallization coefficients, $D_{2/1}$ of M^{2+} ions as the effect of isothermal levelling of supersaturation during the crystallization of $CoSO_4{\cdot}7H_2O$ at 20 °C (Smolik, 2003)

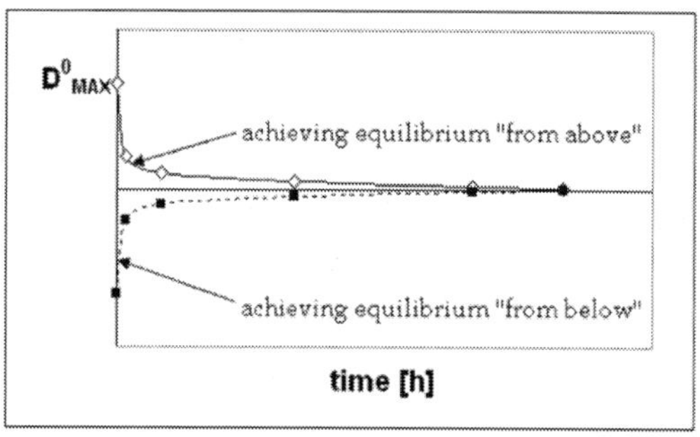

Fig. 2. The principle of the long–time stirring method for the determination of the equilibrium coefficients $D_{2/1}$ (Zhelnin & Gorshtein, 1971; Chlopin, 1957)

When selecting values D^o_{max} and D^o_{min} the highest and the lowest values of $D_{2/1}$ obtained during crystallization by the first method are usually taken into consideration. The experiments are carried out in the following way:

Micro-compo-nent M^{2+}	Initial $D_{2/1}$		Average $D_{2/1}$ after long time stirring		Average equilibrium $D_{2/1}$
			for D^o_{min}	for D^o_{MAX}	
	D^o_{min}	D^o_{MAX}	$\bar{D} \pm t_\alpha \frac{s}{\sqrt{n}}$	$\bar{D} \pm t_\alpha \frac{s}{\sqrt{n}}$	$\bar{D} \pm t_\alpha \frac{s}{\sqrt{n}}$
Ni^{2+}	1.22	1.86	1.52 ± 0.06	1.62 ± 0.09	1.57 ± 0.06
Cu^{2+}	0.08	0.22	0.14 ± 0.02	0.15 ± 0.01	0.14 ± 0.01
Co^{2+}	0.77	1.89	1.18 ± 0.09	1.20 ± 0.05	1.19 ± 0.04
Fe^{2+}	0.42	1.44	0.72 ± 0.05	0.79 ± 0.06	0.76 ± 0.04
Mg^{2+}	0.16	3.47	1.33 ± 0.12	1.40 ± 0.09	1.36 ± 0.07
Mn^{2+}	0.13	0.28	0.17 ± 0.02	0.20 ± 0.02	0.19 ± 0.02

Table 2. Determination of equilibrium $D_{2/1}$ coefficients of M^{2+} ions during the crystallization of $ZnSO_4{\cdot}7H_2O$ at 23°C (Smolik, 2000a)

Crushed "contaminated" crystals (crushed to pass a 0.1mm sieve - ϕ<0.1mm) are introduced into several beakers together with their saturated "purified" solution. Crushed "purified" crystals (crushed to pass a 0.1mm sieve - ϕ<0.1mm) and their "contaminated" saturated solution are introduced to some other beakers. Contents of the beakers are stirred for ~360 h with a magnetic stirrer at constant temperature (**Table 2**).

4. Thermodynamic approach to the calculation of equilibrium $D_{2/1}$ coefficients

The possibility of achieving a thermodynamic equilibrium during crystallization from solutions, as well as melts, as proved by Chlopin (1957), permits introducing a thermodynamic partition coefficient. Substituting concentrations of microcomponent (2) and macrocomponent (1) in equation (3) with their activities (a_{1s}, a_{2s}, a_{1r}, a_{2r}) it is possible to obtain an expression for thermodynamic co-crystallization coefficient $D^o_{2/1}$. (Kirkova et al., 1996; Ratner, 1933).

$$D^o_{2/1} = \frac{a_{2s} \cdot a_{1r}}{a_{2r} \cdot a_{1s}} = \exp\left(-\frac{\Delta\mu^o_{2/1}}{RT}\right) \tag{12}$$

$\Delta\mu^o_{2/1} = \Delta\mu^o_2 - \Delta\mu^o_1$, where $\Delta\mu^o_2$ and $\Delta\mu^o_1$ are the changes of standard molar chemical potential of components (2) and (1) respectively during the transition from the liquid phase (r) into the solid phase (s). Two cases should be distinguished here: 1) substance (2) is not isomorphous with substance (1), i.e., it crystallizes in different crystal systems (different space groups); 2) substance (2) is isomorphous with substance (1).

In the first case the formation of mixed crystals by substance (2) with substance (1) may indicate the existence (besides the basic form [II] of microcomponent [2]) of a polymorphous form (I), metastable in suitable conditions, which is isomorphous with the crystal of the host: macrocomponent - substance (1). The transition of the substance (2) of the structure (II) into its metastable form of the structure (I) is connected with the increase of chemical potential: $\Delta\mu^o_{II\to I} = \mu^{o,I}_{2s} - \mu^{o,II}_{2s}$. Then $\Delta\mu^o_2 = \mu^{o,I}_{2s} - \mu^o_{2r} = \mu^{o,I}_{2s} - \mu^{o,II}_{2s} + \mu^{o,II}_{2s} - \mu^o_{2r} = \Delta\mu^o_{II\to I} + \mu^{o,II}_{2s} - \mu^o_{2r}$, where $\Delta\mu^o_{II\to I} = \mu^{o,I}_{2s} - \mu^{o,II}_{2s}$ – free partial molar enthalpy of phase transition II→I of crystals of microcomponent (2) of structure (II) into the structure (I) proper to that of macrocomponent(1). Therefore (Smolik, 2004):

$$\begin{aligned} D^o_{2,II/1,I} &= \frac{a_{1r}a_{2s}}{a_{1s}a_{2r}} = \frac{a_{1r}a^{I}_{2s}}{a_{1s}a_{2r}} = \exp\left(-\frac{\Delta\mu^o_2 - \Delta\mu^o_1}{RT}\right) = \exp\left(-\frac{\Delta\mu^o_{II\to I} + \mu^{o,II}_{2s} - \mu^o_{2r} + \mu^o_{1r} - \mu^o_{1s}}{RT}\right) = \\ &= \exp\left(\frac{\mu^o_{1s} - \mu^o_{1r}}{RT}\right) \cdot \exp\left(\frac{\mu^o_{2r} - \mu^{o,II}_{2s}}{RT}\right) \cdot \exp\left(-\frac{\Delta\mu^o_{II\to I}}{RT}\right) = \frac{a^o_{1r}}{a^o_{2r}} \cdot \exp\left(-\frac{\Delta\mu^o_{II\to I}}{RT}\right) \end{aligned} \tag{13}$$

If two double salts: $B_bE_eL_l$ (1) i $B_bE_eL_l$ (2) capable of forming solid solutions by the exchange of B ions into B′ones (it is possible to exchange E ions into E′ or L into L′) dissociate into ions in aqueous solution according to the reaction:

$$B_bE_eL_l \leftrightarrows bB^{(z1)+} + eE^{(z2)+} + lL^{(z3)-}$$

and

$$B'_bE_eL_l \leftrightarrows bB'^{(z1)+} + eE^{(z2)+} + lL^{(z3)-}$$

(obviously: $bz_1 + ez_2 - lz_3 = 0$) the following general formula may be derived for thermodynamic co-crystallization coefficient $D^o_{2/1}$ (Balarew, 1987; Smolik & Kowalik, 2010):

$$D_{2/1} = \left(\frac{x_2 \cdot m_1}{x_1 \cdot m_2}\right) = \left(\frac{m_{01} \cdot \gamma_{m01}}{m_{02} \cdot \gamma_{m02}}\right)^{\frac{\nu}{b}} \cdot \left(\frac{\gamma_{m2}}{\gamma_{m1}}\right)^{\frac{\nu}{b}} \cdot \frac{f_1}{f_2} \cdot \exp\left(-\frac{\Delta\mu^o{}_{II\to I}}{bRT}\right) \tag{14}$$

where: $m_{01}(m_{02})$, $\gamma_{m01}(\gamma_{m02})$ – molal solubility ([mol/kg]) of the salt $B_bE_eL_l(B'_bE_eL_l)$) and mean molal activity coefficient of the salt $B_bE_eL_l(B'_bE_eL_l)$ in its binary saturated solution; $m_1(m_2)$, $\gamma_{m1}(\gamma_{m2})$ – molality and mean molal activity coefficient of the salt $B_bE_eL_l(B'_bE_eL_l)$ in the ternary solution being in equilibrium with $B_b(B'_b)E_eL_l$) solid solution; $x_1(x_2)$ – mole fraction of B(B′) ion and $f_1(f_2)$ – activity coefficient of ion B(B′) in this solid solution; $\Delta\mu^o_{II\to I}$ – the partial molar Gibbs free energy of the phase transition of the salt $B'_bE_eL_l$ from its structure (II) into the structure (I) of the salt $B_bE_eL_l$, $\nu = b + e + l$, R – gas constant, T – temperature [K]

In the other case involving isomorphous substances (1) and (2) $\Delta\mu^o_{I\to I} = 0$. Hence:

$$D_{2/1} = \left(\frac{m_{01} \cdot \gamma_{m01}}{m_{02} \cdot \gamma_{m02}}\right)^{\frac{\nu}{b}} \cdot \left(\frac{\gamma_{m2}}{\gamma_{m1}}\right)^{\frac{\nu}{b}} \cdot \frac{f_1}{f_2} \tag{15}$$

Equations (14) and (15) should, in principle, permit calculating exactly the equilibrium partition coefficient $D_{2/1}$, if molal solubilities and all activity coefficients (in the aqueous and the solid phases), as well as the partial molar Gibbs free energy of the phase transition, were known. However, these data (except for molal solubilities) are rarely available. In contrast to mean molal activity coefficients in binary saturated solutions (γ_{m01}, γ_{m02}), as well as those in the ternary solution being in equilibrium with $B_b(B'_b)E_eL_l$) solid solution (γ_{m1}, γ_{m2}) which are sometimes directly accessible or calculable by means of Pitzer equations, activity coefficients in the solid solution (f_1, f_2), as well as the partial molar free energy of the phase transition, are generally unknown (except for very rare individual cases of crystallization systems: macrocomponent(1) – microcomponent(2)).

The attempts to estimate $D_{2/1}$ coefficients by means of simplified equations (taking into account only activity coefficients in the liquid phase) are connected with huge errors, which proves that they result from the lack of the activity coefficients in the solid solution (f_1, f_2) as well as the partial molar Gibbs free energy of the phase transition $\Delta\mu^o_{II\to I}$ (Smolik, 2004).

$$\frac{\Delta[\%]}{100} = \frac{\left|D_{\exp.} - D_{cal.}\right|}{D_{\exp.}} = \left|1 - \frac{D_{cal.}}{D_{\exp.}}\right| = \left|1 - \frac{f_2}{f_1} \cdot \exp\left(\frac{\Delta\mu^0_{II\to I}}{bRT}\right)\right| \tag{16}$$

In the case where coefficients $D_{2/1}$ are independent of mixed crystal composition, the ratio f_2/f_1 remains constant (Balarew, 1987). Assuming the regular solution approximation this ratio may be expressed by the following equation:

$$\frac{f_1}{f_2} = \exp\left(\frac{\Delta\overline{H}_1 - \Delta\overline{H}_2}{RT}\right) \tag{17}$$

where $\Delta\overline{H}_1 - \Delta\overline{H}_2$ is the difference in the partial molar enthalpies of mixing.

According to Balarew (1987) this is the result of the difference in coordination environment around the two substituting ions, affected by ionic size differences (Δr/r), metal - ligand bond energy differences (Δε) with respect to the enthalpy of mixing (Urusov, 1977), as well as the difference in the energy determined by the crystal field (in the case non Jahn–Teller ions):

$$\Delta\overline{H}_1 - \Delta\overline{H}_2 = w_1 \cdot f\left(\frac{\Delta r}{r}\right) + w_2 \cdot \varphi(\Delta\varepsilon) + w_3 \cdot \psi(\Delta s) + ... \tag{18}$$

Hence:

$$D_{2/1} = \left(\frac{c_{01} \cdot \gamma_{c01}}{c_{02} \cdot \gamma_{c02}}\right)^{\frac{\nu}{b}} \cdot \left(\frac{\gamma_{c2}}{\gamma_{c1}}\right)^{\frac{\nu}{b}} \cdot \exp\left(-\frac{\Delta\mu_{II \to I}}{RT}\right) \cdot \exp\frac{w_1 \cdot f\left(\frac{\Delta r}{r}\right) + w_2 \cdot \varphi(\Delta\varepsilon) + w_3 \cdot \psi(\Delta s) + ...}{bRT} \tag{19}$$

where: f, φ, ψ, … - functions sought for, w_1, w_2, w_3, … - estimated coefficients.

To derive an equation for estimating $D_{2/1}$ by finding the functions (f, φ, ψ, …) and coefficients (w_1, w_2, w_3, …), it is necessary to check how $D_{2/1}$ coefficients depend on various factors.

4. The dependence of co–crystallization coefficients, $D_{2/1}$ on chemical, physicochemical and crystal–chemical properties of co–crystallizing salts and ions

Equilibrium co-crystallization coefficients are determined in the conditions ensuring that they do not depend on hydrodynamic and kinetic conditions of crystallization. However, they are affected by several factors both "external" (in relation to the co-crystallizing substances) and "internal" (resulting from chemical, physicochemical and crystal-chemical properties of the co-crystallizing substances).

"External" factors have chemical characteristics (kind and composition of the solvent - the liquid phase, the presence of ions or other foreign substances, the presence of complexing agents, acidity (pH) of solution, from which crystallization takes place) or non-chemical ones (e.g., temperature). "Internal" factors are presented in Table 3.

Chemical, physicochemical and crystal-chemical properties of	
Co-crystallizing salts	Co-crystallizing ions
Solubility in water (m_0)	Charge of cation
Crystal system (CS)	Geometrical factor (ionic radius) (r)
Number of molecules of crystallization water (n)	Character of chemical bond (electronegativity) (ε)
Reciprocal solubility in the solid phase (C^s_{MAX})	Electronic configuration
The volume of one formal molecule of salt (η^3)	Crystal field stabilization energy (CFSE)
	Cation hardness (h)

Table 3. Chemical, physicochemical and crystal-chemical properties of co-crystallizing salts and co-crystallizing ions

To analyse the influence of the above mentioned factors on co-crystallization coefficients $D_{2/1}$, it is convenient to use correlation coefficients (ρ_{xy}) in the case of the properties that can be formulated quantitatively. For other properties (qualitative), mean $D_{2/1}$ values for salts revealing and not revealing may be compared. On the other hand, co-crystallization coefficients may be considered as a measure of mutual solubility of co-crystallizing salts in the solid phase. The famous Latin rule: "Similia similibus solvuntur" (similar substances will dissolve similar substances) may be useful in the prediction of this solubility and the evaluation of $D_{2/1}$ level.

4.1 Chemical, physicochemical and crystal–chemical properties of co-crystallizing salts

4.1.1 Solubility in water (m_0)

This is the most important factor affecting $D_{2/1}$coefficients. For the co-crystallization of isomorphous salts ($\Delta\mu^0_{II\rightarrow I} = 0$) forming ideal solid and liquid solutions $((\gamma_{m01}/\gamma_{m02})^{(\nu/b)} \cdot (\gamma_{m2}/\gamma_{m1})^{(\nu/b)}$ $(f_1/f_2)=1))$ they are expressed by $D_{2/1}=(m_{01}/m_{02})^{\nu/b}$.

However, this equation is proved true only for non-numerous salts fulfilling the additivity rule (Balarew, 1987). This simplified equation is the basis of the Ruff rule (Ruff et al., 1928):

If $m_{01} > m_{02} \rightarrow (m_{01}/m_{02})^{\nu/b} = D_{2/1} > 1$. (During crystallization of two components, the less soluble one grows rich in crystal).

As it can be seen in Table **4**, despite its simplicity and obviousness, this qualitative rule is not always fulfilled.

Kind of co-crystallizing salts	Number of considered crystallization systems	Crystallization systems fulfilling the Ruff rule [%]
$MSO_4{\cdot}nH_2O$	100	62
$MCl_2{\cdot}nH_2O$	23	74
$M(HCOO)_2{\cdot}2H_2O$	37	95
Alums, M(III)	9	100
$M^I_2M^{II}(SO_4)_2{\cdot}6H_2O$ M(I), M(II)	59	96
$M^I_2SO_4$	24	54
MX	16	69
$MClO_3$, $MClO_4$, MNO_3, M_2CrO_4	17	59
mean		**72**

Table 4. The degree of fulfilling the Ruff rule in some crystallization systems: macrocomponent - microcomponent (Smolik, 2004)

Mean molal activity coefficients of some isomorphous double salts, forming ideal solid solutions ($M^I_2M^{II}(SO_4)_2{\cdot}6H_2O$ or $M^IM^{III}(SO_4)_2{\cdot}12H_2O$) in their binary saturated solutions are inversely proportional to the square root of their molal solubility (Hill et al., 1940). Therefore:

$$D_{2/1} = \left(\frac{m_{01}\cdot\gamma_{m01}}{m_{02}\cdot\gamma_{m02}}\right)^{\frac{\nu}{b}} = \left(\frac{m_{01}\cdot\sqrt{m_{02}}}{m_{02}\cdot\sqrt{m_{01}}}\right)^{\frac{\nu}{b}} = \left(\frac{m_{o1}}{m_{o2}}\right)^{\frac{\nu}{2b}} \quad (20)$$

For double salt dissociating: $NiSO_4 \cdot M^I_2SO_4 \cdot 6H_2O \leftrightarrows Ni^{2+} + 2M^+ + 2SO_4^{2-} + 6H_2O$ ($\nu = 5$), it is possible to obtain for M^+ ions (b = 2): $D_{2/1} = (m_{01}/m_{02})^{1,25}$ **(Fig. 3)** and for M^{2+} ions (b = 1): $D_{2/1} = (m_{01}/m_{02})^{2,5}$ **(Fig 4)**.

However, for a similar, but simple salt ($NiSO_4 \cdot 7H_2O$), an analogous dependence does not exist (Fig. 5) ($\rho_{xy} = 0.201$ is insignificant).

This is because of the significant differences in the crystal system of proper sulfate hydrates, while all of the investigated double salts are isomorphous, of the same space group ($P2_1/a$) and of almost identical unit cell parameters (a, b, c, β) (their relative standard deviations do not exceed 0.8%)

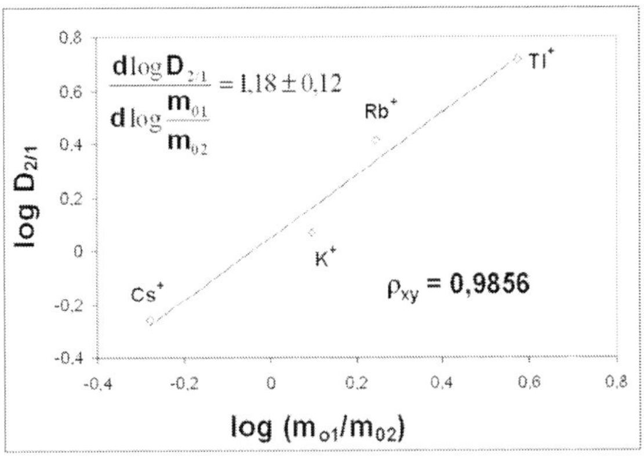

Fig. 3. The dependence of coefficients, $D_{2/1}$ of co-crystalliztion of Cs^+, K^+, Rb^+ and Tl^+ with $NiSO_4 \cdot (NH_4)_2SO_4 \cdot 6H_2O$ on the molality of saturated solutions of suitable salts (Smolik, 1998a)

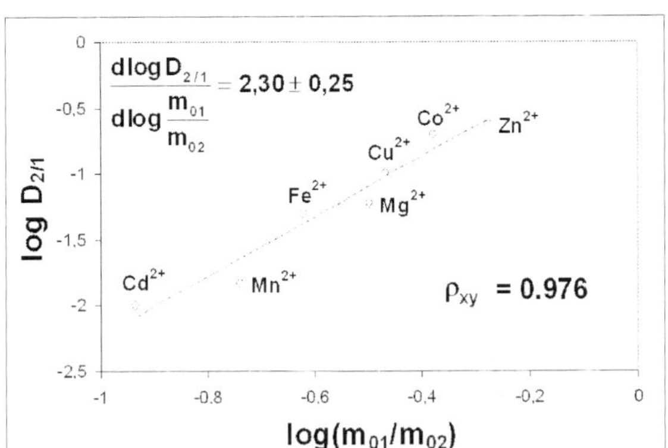

Fig. 4. The dependence of coefficients, $D_{2/1}$ of co-crystallization of Cd^{2+}, Mn^{2+}, Fe^{2+}, Cu^{2+}, Mg^{2+}, Co^{2+} and Zn^{2+} with $NiSO_4 \cdot (NH_4)_2SO_4 \cdot 6H_2O$ on the molality of saturated solutions of suitable salts (Smolik, 2001)

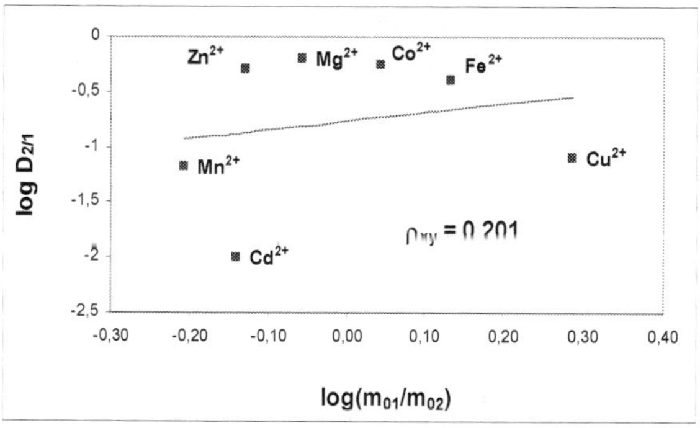

Fig. 5. The dependence of coefficients, $D_{2/1}$ of co-crystallization of Cd^{2+}, Mn^{2+}, Fe^{2+}, Cu^{2+}, Mg^{2+}, Co^{2+} and Zn^{2+} with $NiSO_4{\cdot}7H_2O$ at 20 °C on the molality of saturated solutions of suitable sulfates (Smolik, 2000b]

4.1.2 Crystal system (CS)

The last three examples point to the crystal structure of co-crystallizing salts as a very important factor significantly affecting $D_{2/1}$ coefficients. The dependence of similarity of the crystal structure of macro and microcomponent on mean co-crystallization coefficients in sulfate ($MSO_4{\cdot}nH_2O$) crystallization systems is presented in Fig. 6 and Fig. 7.

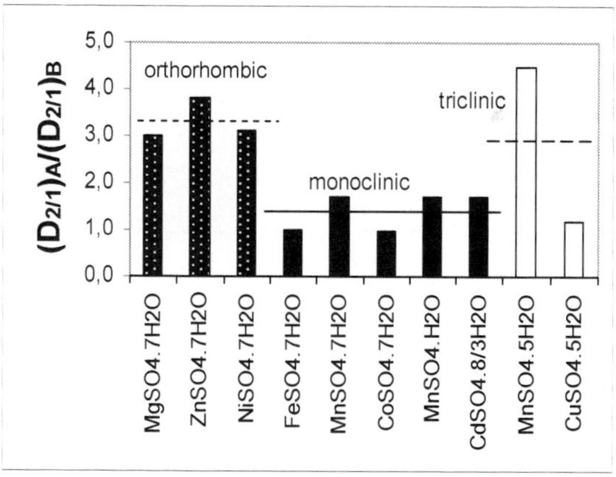

Fig. 6. The dependence of coefficients $D_{2/1}$ on the similarity of the crystal system of macrocomponent (CS_M) and microcomponent (CS_m) $(D_{2/1})_A$ – mean $D_{2/1}$ when $(CS_M) = (CS_m)$ $(D_{2/1})_B$ – mean $D_{2/1}$ when $(CS_M) \neq (CS_m)$ (Smolik, 2002a, 2004)

As it can be seen, mean $D_{2/1}$ coefficients of microcomponents whose hydrates belong to the same crystal system as the macrocomponent are ~ 3 times (for orthorhombic and triclinic macrocomponents) and ~1.5 times (for monoclinic macrocomponents) greater than those whose hydrates belong to a different crystal system than that of macrocomponent) (Fig. 6).

The mean coefficients $D_{2/1}$ of microcomponents belonging to the same crystal system as the macrocomponent are the highest, and they drop as the similarity of their crystal structure and that of the macrocomponent decreases (taking into account the following direction of the increase of crystal systems symmetry: triclinic<monoclinic<orthorhombic) (Fig. 7).

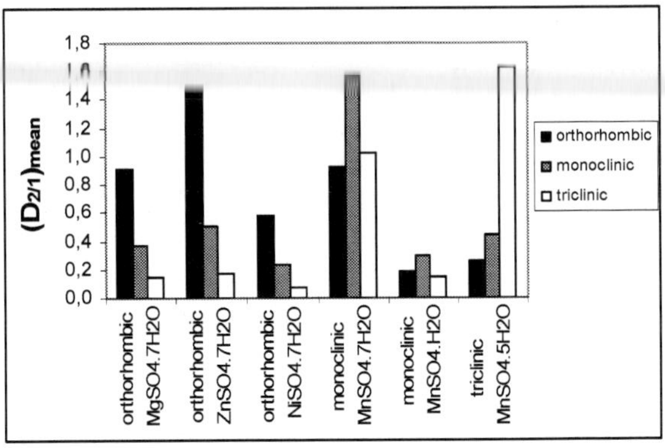

Fig. 7. The dependence of coefficients $D_{2/1}$ on the similarity of crystal systems of macrocomponent (CS_M) and microcomponent (CS_m) $(D_{2/1})_{mean}$ – mean $D_{2/1}$ of microcomponents belonging to the same crystal system (Smolik, 2002a, 2004)

4.1.3 Number of molecules of crystallization water (n)

During the crystallization of hydrates the number of molecules of crystallization water (n) is an additional factor which may influence co-crystallization coefficients $D_{2/1}$. It affects the coordination environment of the metal ion, which in the case of hepta or hexahydrates consists of only water molecules (the linkage of coordination octahedra in these crystals'

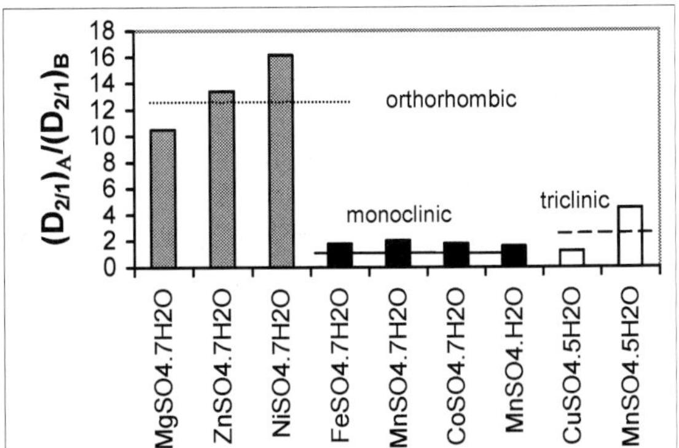

Fig. 8. $D_{2/1} = f(|\Delta n| = |n_M - n_m|)$ $n_M(n_m)$ – number of molecules of crystallization water of macrocomponent (microcomponent) (Smolik, 2002a, 2004),$(D_{2/1})_A$ – mean $D_{2/1}$ when $|\Delta n| = 0$ $(D_{2/1})_B$ – mean $D_{2/1}$ when $|\Delta n| \neq 0$

structures is determined by weak hydrogen bonds), but in lower hydrates oxygen atoms of polyatomic anions enter the coordination environment of the metal ions (causing the formation of chains, closed rings, planar or space networks by vertices-sharing coordination polyhedral) (Balarew, 1987). Generally, mean $D_{2/1}$ values are higher the more similar are n values of macro and microcomponents (Fig. 8 and Fig. 9).

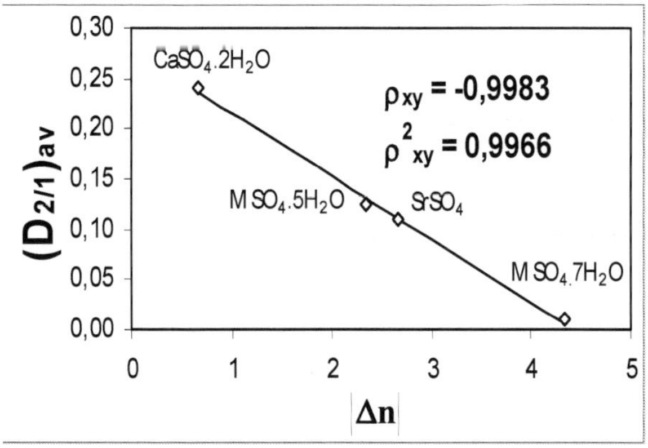

Fig. 9. $(D_{2/1})_{av} = f(\Delta|n| = |n_M - n_m|)$ during the crystallization of $CdSO_4{\cdot}8/3H_2O$ at 20 °C (Smolik, 2002b, 2004), ρ_{xy} – correlation of $(D_{2/1})_{av}$ and $\Delta|n|$

4.1.4 Reciprocal solubility in the solid phase C^s_{MAX}

It is known that the coefficients of co-crystallization of impurities are proportional to their solubilities in the solid phase in the case of crystallization of Ge and Si (Fisher, 1962), as well as several dozen molten metals (Vachobov et al., 1968). However, such regularity has not

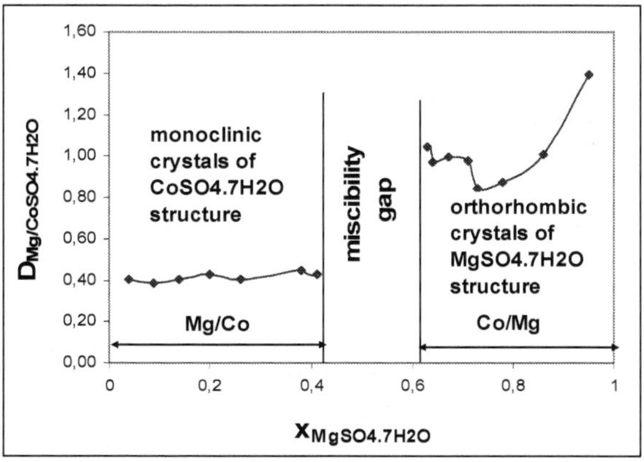

Fig. 10. The dependence of $D_{Mg/CoSO4.7H2O}$ on mole fraction of $MgSO_4{\cdot}7H_2O$ in the solid phase (Oikova et al., 1976) ; Mg/Co – solubility of $MgSO_4{\cdot}7H_2O$ in $CoSO_4{\cdot}7H_2O$, Co/Mg – solubility of $CoSO_4{\cdot}7H_2O$ in $MgSO_4{\cdot}7H_2O$

been found yet during the crystallization of salts from aqueous solutions. The term "solubility in the solid phase" is explained in Fig 10. It is the maximal concentration of a hydrate in another hydrate, which does not cause the change in its structure.

The effect of the solubility in the solid phase (C^s_{MAX}) on $D_{2/1}$ coefficients is presented in Table 5. As it can be seen in most analysed cases, $D_{2/1}$ is proportional to the maximal reciprocal solubility in the solid phase ($C^s_{MAX.}$) which is proved by relatively high and significant correlation coefficients (ρ_{xy}) marked bold.

No	Macrocomponent	Microcomponents	ρ_{xy}
1.	$MgSO_4 \cdot 7H_2O$	Ni^{2+}, Mn^{2+}, Fe^{2+}, Cu^{2+}, Zn^{2+}, Cd^{2+}, Co^{2+}	**0.7635**
2.	$ZnSO_4 \cdot 7H_2O$	Ni^{2+}, Mn^{2+}, Fe^{2+}, Cu^{2+}, Mg^{2+}, Cd^{2+}, Co^{2+}	**0.8608**
3.	$NiSO_4 \cdot 7H_2O$	Zn^{2+}, Fe^{2+}, Cu^{2+}, Mg^{2+}, Cd^{2+}, Co^{2+}	**0.8455**
4.	$CoSO_4 \cdot 7H_2O$	Ni^{2+}, Mn^{2+}, Fe^{2+}, Mg^{2+}, Cd^{2+}, Zn^{2+}	**0.9172**
5.	$MnSO_4 \cdot 5H_2O$	Zn^{2+}, Cu^{2+}, Mg^{2+}	**0.9971**
6.	$FeSO_4 \cdot 7H_2O$	Ni^{2+}, Zn^{2+}, Cu^{2+}, Mg^{2+}, Cd^{2+}, Co^{2+}	**0.7983**
8.	$Ni(NO_3)_2 \cdot 6H_2O$	Mn^{2+},Zn^{2+}, Mg^{2+}, Co^{2+}	0.7022
9.	$Ni(NO_3)_2 \cdot 6H_2O$	Mn^{2+}, Mg^{2+},Co^{2+}	**0.9970**
10.	$NiCl_2 \cdot 6H_2O$	Zn^{2+}, Mn^{2+}, Fe^{2+}, Cu^{2+}, Co^{2+}	**0.9974**
11.	K_2SO_4	Cs^{+}, Tl^{+}, Rb^{+}	0.1041

Table 5. Correlation (ρ_{xy}) of co-crystallization coefficients $D_{2/1}$ and solubility in the solid phase (C^s_{MAX}) (Smolik, 2004)

4.1.5 The volume of one formal molecule

The volume of one formal molecule can be calculated, knowing the molar mass of crystallizing salt (compound) and its density, by the following formula $\eta^3 = 10^{24}\frac{M_x}{d \cdot N}$ [Å³] where: M_x - molar mass [g/mole], d - density [g/cm³], N - Avogadro number = $6{,}022 \cdot 10^{23}$/mole). This is very close to that calculated using unit cell parameters (a, b, c, α, β, γ).

Macro-component	Micro-components	y	x		Ref.
			$\left\|\frac{\Delta\eta^3}{\eta^3}\right\|$	$\left(\frac{\Delta\eta}{\eta}\right)^2$	
			ρ_{xy}		
(1)	Ni^{2+}, Cu^{2+}, Co^{2+}, Zn^{2+}, Mn^{2+}, Cd^{2+}, Ca^{2+} Sr^{2+}	$\ln D_{2/1}$	**-0.8657**		(Smolik, 2008)
	Ni^{2+}, Co^{2+}, Mn^{2+}	$\pi = \ln D_{2/1} - \ln(m_{01}/m_{02})^3$		**-0.9973**	
(2)	Mg^{2+}, Co^{2+}, Ni^{2+}	$\pi = \ln D_{2/1} - \ln (m_{01}/m_{02})^3$		**-0.9987**	(Smolik, 2011)

Table 6. The dependence of $\ln D_{2/1}$ ($\pi = \ln D_{2/1} - \ln (m_{01}/m_{02})^3$) on various functions of η during the crystallization of $Mg(CH_3COO)_2 \cdot 4H_2O$ (1) and $Mn(CH_3COO)_2 \cdot 4H_2O$ (2) at 25 °C

According to Urusov (1977) this parameter is better than ionic radius in the evaluation of the effect of geometric factor on $D_{2/1}$ coefficients, because it takes into account real interatomic distances defined by crystal system and unit cell parameters. However, η is unambiguous only in ionic crystals of high (cubic) symmetry and in the case of complicated heterodesmic structures of low symmetry these distances become equivocal.

The significant effect of this factor has occurred in some acetate crystallization systems (Table 5).

4.2 Chemical, physicochemical and crystal–chemical properties of co-crystallizing ions

4.2.1 Charge of cation

The cation charge is one of the most important factors influencing $D_{2/1}$ coefficients. Taking into account the earlier mentioned Latin rule "Similia similibus solvuntur" it might be expected that microcomponent ions having the same charge as that of macrocomponent ion should co-crystallize in higher degree than those of different ion charge. In many crystallization systems this rule is fulfilled, e.g., in the case of crystallization of $FeSO_4 \cdot 7H_2O$ at 20 °C the mean $D_{2/1}$ of M^{2+} ions (of the same charge as the macrocomponent Fe^{2+}) are ~14 (50) times greater than those of $M^{+}(M^{3+})$ ions (of the charge different from that of macrocomponent) (Fig. 11a), and in the case of $Fe(NH_4)(SO_4)_2 \cdot 12H_2O$ crystallization at 20 °C the mean $D_{2/1}$ of M^{2+} ions (of different charge from macrocomponent ions NH_4^+ and Fe^{3+}) is ~400 - 1000 times lower than those of M^+ and M^{3+} ions (Fig 11b).

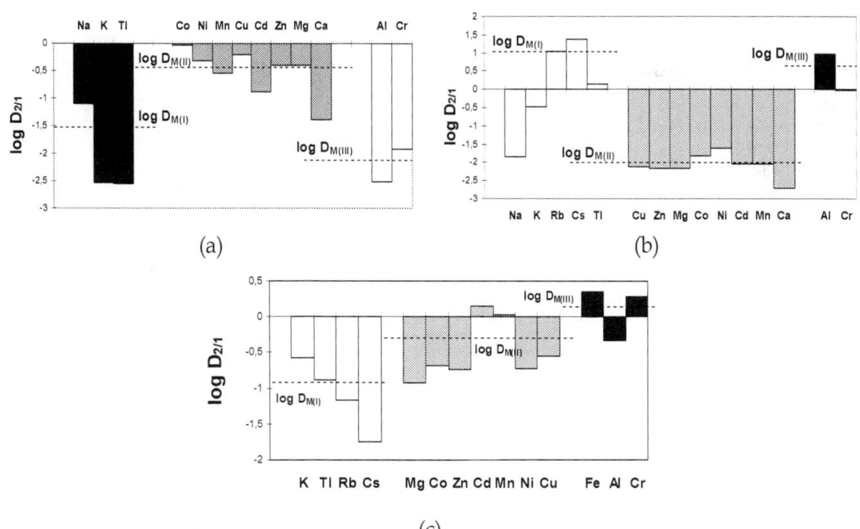

Fig. 11. The effect of ion charge on $D_{2/1}$ during crystallization of: a) $FeSO_4 \cdot 7H_2O$ at 20 °C (Smolik & Lipowska, 1995); b) $Fe(NH_4)$ alum at 20 °C; (Smolik, 1995a); c) Na_2SO_4 at 50 °C; (Smolik, 1998b); log $D_{M(I)(M(II),M(III))}$ – logarithms of mean $D_{2/1}$ for $M^+(M^{2+}, M^{3+})$ ions

However, there are crystallization systems where this simple and evident rule is not fulfilled, e.g., during the crystallization of Na_2SO_4 at 50 °C (Fig. 11c). In this case the mean $D_{2/1}$ of M^+ ions (of the same charge as the macrocomponent Na^+) is the lowest. This is caused by the formation of double, less soluble salts by Na_2SO_4 with M (II) and M (III) sulfates which, because of their structures, are capable of in-build into Na_2SO_4 crystal.

4.2.2 Geometrical factor (Ionic radius (r))

The geometrical factor, determined by the difference in size of mutually substituting ions, has been considered for a long time as one of the most significant factors affecting the existence of isomorphism. Beginning from the empirical Goldschmidt rule postulating the border of 15% relative difference of ionic radii for the occurrence of isomorphic substitution, various values of this border (e.g., 5%) have been given by other authors. In addition it has occurred that they are dependent on other factors. This parameter was not recommended by Urusov (1977), who preferred to take into consideration the differences in interionic distances in the solid phase. In this way he calculated $D_{2/1}$ values for M^+ ions were strictly consistent with those experimental ones during the crystallization of alkali metal halides from melt (Urusov & Kravchuk, 1976). However, as proved by the same author, correlation of interionc distances in the solid phase and $D_{2/1}$ values obtained during the crystallization from aqueous solutions occurred as significantly weaker (Urusov, 1980). Moreover, the results of many investigations [(Smolik, 1993, 1995, 1998a, 2003,2007, 2010) indicate that ionic radius can be a useful parameter in the evaluation of co-crystallization coefficients. Some typical dependences of $D_{2/1}$ coefficients on ionic radii have been presented in Fig. 12 a,b,c,d.

We can observe some types of the dependence of $D_{2/1}$ coefficients on ionic radius: a) monotonic, hyperbolic drop of $D_{2/1}$ with the increase of ionic radius (if the case of Cu^{2+} is ignored, because of the structure of triclinic $CuSO_4 \cdot 5H_2O$ significantly departing from the structures of other sulfates); b) the existence of the maximum of $D_{2/1}$ coefficients for ions, whose radii are closest to the radius of macrocomponent and the monotonic drop of $D_{2/1}$ as the absolute value of the difference in ionic radii of macrocomponent and microcomponent increases; c) similar type like "b", disturbed in the case of (mainly) Cu^{2+} because of the almost identical structure of triclinic $CuSO_4 \cdot 5H_2O$ and $MnSO_4 \cdot 5H_2O$; d) there are two ranges of higher $D_{2/1}$ coefficients corresponding to the values of ionic radii very close to those of two macrocomponent ions NH_4^+ and Fe^{3+}.

Very high correlation coefficients of ln $D_{2/1}$ and $|\Delta r/r_M|$ or $(\Delta r/r_M)^2$ in some crystallization systems (Table 7) indicate that these co-crystallization coefficients strongly depend on the similarity of ionic radii of micro and macrocomponents.

Coefficients of Ca^{2+} co-crystallization with various sulfate hydrates $MSO_4 \cdot nH_2O$, taking into account slight solubility of $CaSO_4 \cdot 2H_2O$, should be very high. However, because of its large radius r_{Ca2+} , Ca^{2+} ion cannot in-build into $MSO_4.nH_2O$ crystals of smaller M^{2+} ions. Therefore, the ionic radius of the macrocomponent is a more important factor than the solubility determining $D_{Ca/MSO4.nH2O}$ values (Fig. 13) (Smolik, 2004).

Thus, ionic radii in the case of crystallization of salts from aqueous solutions can be a convenient and important parameter for the investigation and sometimes evaluation of $D_{2/1}$

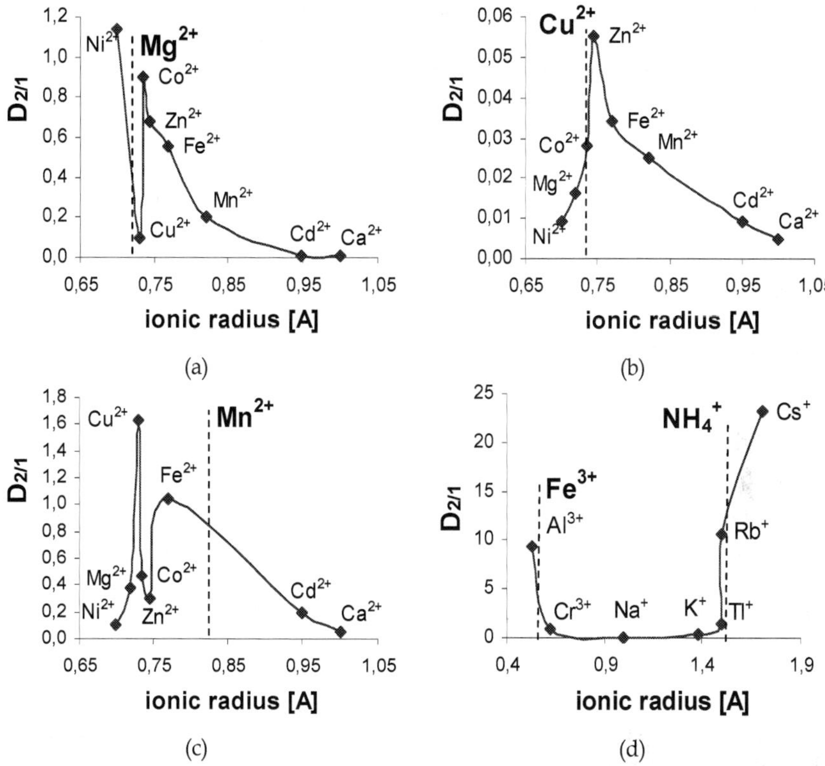

Fig. 12. The effect of ionic radius on coefficients $D_{2/1}$ during the crystallization of : a – $MgSO_4 \cdot 7H_2O$ at 25 °C (Smolik, 1999a), b – $CuSO_4 \cdot 5H_2O$ at 25 °C (Smolik & Zolotajkin, 1993), c – $MnSO_4 \cdot 5H_2O$ at 20 °C (Smolik et al., 1995), d – NH_4Fe alum at 20 °C (Smolik, 1995c)

Macro-component	Microcomponents (M^{2+})	Correlation coefficients ρ_{xy} of ln $D_{2/1}$ and:	
		$\lvert \Delta r/r_M \rvert$	$(\Delta r/r_M)^2$
Orthorhombic $MgSO_4 \cdot 7H_2O$	Ni^{2+}, Cu^{2+}, Co^{2+}, Zn^{2+}, Fe^{2+}, Mn^{2+}, Cd^{2+}, Ca^{2+}	**-0.9061***	
	Co^{2+}, Fe^{2+}, Mn^{2+}, Cd^{2+}, Ca^{2+} (monoclinic)	**-0.9992**	
triclinic $MnSO_4 \cdot 5H_2O$	Ni^{2+}, Cu^{2+}, Co^{2+}, Zn^{2+}, Fe^{2+}, Mn^{2+}, Cd^{2+}, Ca^{2+}		**-0.8353**
	Co^{2+}, Fe^{2+}, Cd^{2+}, Ca^{2+} (monoclinic)		**-0.9948**
triclinic $CuSO_4 \cdot 5H_2O$	Ni^{2+}, Cu^{2+}, Co^{2+}, Zn^{2+}, Fe^{2+}, Mn^{2+}, Cd^{2+}, Ca^{2+}		-0.6659
	Co^{2+}, Fe^{2+}, Mn^{2+}, Cd^{2+}, Ca^{2+} (monoclinic)		**-0.9949**

* -significant ρ_{xy} (for the confidence level of 0.95) are marked bold (Smolik, 2004)

Table 7. Correlation coefficients (ρ_{xy}) of ln $D_{2/1}$ and $(\Delta r/r_M)^2$ (or $\lvert \Delta r/r_M \rvert$) in some sulfate crystallization systems for all ions or those ions whose sulfate hydrates are monoclinic coefficients. The comparison of correlation coefficient (ρ_{xy}) of $\pi = [\ln D_{2/1} - \ln(m_{01}/m_{02})^3][y]$ and $|((r_{Co2+})^3 - (r_{M2+})^3)/(r_{Co2+})^3|[x]$ ($\rho_{xy} = -0.9174$) with analogous correlation coefficient of π [y] and $|((\eta_{Co2+})^3 - (\eta_{M2+})^3)/(\eta_{Co2+})^3|[x]$ ($\rho_{xy} = -0.8590$) indicates that the latter parameter, preferred by Urusov (1977) to estimate $D_{2/1}$ values, is in the case of $Co(CH_3COO)_2 \cdot 4H_2O$ crystallization not better than the first one related to ionic radius (Smolik et al., 2007).

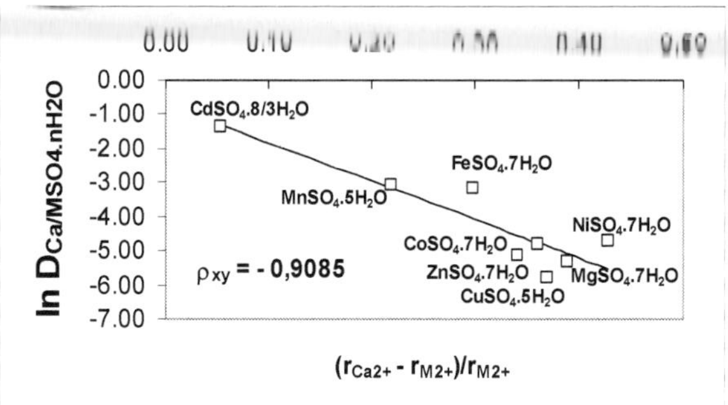

Fig. 13. The dependence of ln $D_{Ca/MSO4.nH2O}$ on ionic radii of macrocomponent ions (M^{2+}) (Smolik, 2004)

4.2.3 Electronegativity of mutually substituting elements Φ(ε)

Apart from the geometric factor, the partition coefficients may be affected by the nature (polarity) of the chemical bond of mutually substituting components. The exact quantitative characteristic of the polarity of the chemical bond is given by the integral of overlapping of atomic orbitals, but such data for many isomorphous systems are not available yet. So some authors (Urusov, 1977; Ringwood, 1955) consider the difference in electronegativity of elements as a measure of polarity.

Spectacular examples of a huge effect of this factor on $D_{2/1}$ are given by (Urusov, 1977) where during the crystallization from the melt in systems NaBr - AgBr, NaCl - AgCl, and NaCl - CuCl, co-crystallization does not take place, although the relative differences of interionic distances of co-crystallizing isomorphous salts are very low or close to zero. The difference in the effective ionic charges of Ag(Cu) and Na in these systems was believed to be responsible for the extremely low miscibility of these systems (Kirkova et al., 1996).

In the case of crystallization of several groups of salts from aqueous solutions, the effect of electronegativity of mutually substituting components on distribution coefficients has been compared with the effect of ionic radius. However, for this comparison function $\pi = \ln D_{2/1} - \ln(m_{01}/m_{02})^{v/b}$ has been taken into consideration (it allows for the solubility ratio of co-crystallizing salts). The results are given in Table **8**.

As it can be seen, higher correlation coefficients $|\rho_{xy}|$ of π and $\Phi(\varepsilon)$ than those of π and $f(r)$ occur in the case of chlorides ($MCl_2 \cdot nH_2O$) and formates ($M(HCOO)_2 \cdot 2H_2O$). Generally, more significant correlation coefficients $|\rho_{xy}|$ occur in the case when in coordination surroundings of appropriate cations M^{n+} in the solid phase, halogen or formate ions appear as well, apart from oxide ligands originating from water or inorganic anions. However, a kind of dependence of $D_{2/1}$ coefficients on electronegativity has not been univocally defined and the kind of function $\Phi(\varepsilon)$ is very important to decide if unilateral isomorphism occurs, in spite of a buffering action of surrounding hydrate mantle (Kirkova et al., 1996).

Groups of considered salts	maximal values of correlation coefficients $\lvert\rho_{xy}\rvert$ of function $\pi = \ln D_{2/1} - \ln(m_{01}/m_{02})^{v/b}$	
	and $f(r)$	and $\Phi(\varepsilon)$
$MSO_4 \cdot nH_2O$	0.7309	0.6377
$M(NO_3)_2 \cdot nH_2O$	0.9421	0.6502
$MCl_2 \cdot nH_2O$	0.6258	0.7328
$M^IM^{III}(SO_4)_2 \cdot 12H_2O$, M(I)	0.9188	non-significant
$M^IM^{III}(SO_4)_2 \cdot 12H_2O$, M(III)	0.7651	non-significant
$M^IM^{II}(SO_4)_2 \cdot 6H_2O$, M(II)	0.4921	0.3358
$M(HCOO)_2 \cdot 2H_2O$	non-significant	0.7604
$M^I_2SO_4 \cdot nH_2O$	0.8005	0.6900
MX	0.8749	0.8745

Table 8. Comparison of absolute maximal values of correlation coefficients of functions $\pi = \ln D_{2/1} - \ln(m_{01}/m_{02})^{v/b}$ and $f(r)$ or $\Phi(\varepsilon)$ in considered groups of salts (Smolik, 2004)

To evaluate if unilateral isomorphism occurs during low temperature crystallization from aqueous solutions (according to the rule known for a long time in geology [Urusov, 1970] that lithophilic elements are substituted in the solid phase by chalcophilic and siderophilic elements and not to the contrary) several criteria may be applied (Smolik, 2004). The strongest of them in the case of co-crystallization of more (M_1) and less (M_2) electronegative ions ($\varepsilon_1>\varepsilon_2$) looks as follows: $\Theta>3$, where $\Theta = \theta_{1/2}/\theta_{2/1}$; $\theta_{1/2} = (D_{1/2})_{exp.}/(D_{1/2})_{cal}$ and $\theta_{2/1} = (D_{2/1})_{exp.}/(D_{2/1})_{cal}$; $(D_{1/2})_{cal} = (m_{02}/m_{01})^{v/b}$; $(D_{2/1})_{cal} = (m_{01}/m_{02})^{v/b}$

So:

$$\ln\Theta = \ln\theta_{1/2} - \ln\theta_{2/1} = [\ln(D_{1/2})_{exp} - \ln(m_{02}/m_{01})^{v/b}] - [\ln(D_{2/1})_{exp} - \ln(m_{01}/m_{02})^{v/b}] > 1.1.$$

The value $\theta_{1/2}$ ($\theta_{2/1}$) is a measure of the extension (or diminution) of the experimental $(D_{1/2})_{exp}$ ($(D_{2/1})_{exp}$) in relation to $(D_{1/2})_{cal}$($(D_{2/1})_{cal}$), which may be brought about by the effect of electronegativity. $\theta_{1/2}$ and $\theta_{2/1}$ coefficients allowing for the solubility of corresponding salts are independent of the structure and the relative difference in interionic distances of co-crystallizing isomorphous salts. Therefore, they are most adequate to evaluate the occurrence of unilateral isomorphism. The results of such an analysis of all isomorphous salts forming hydrates, for which $D_{2/1}$ ($D_{1/2}$) coefficients have been available, show the lack of any example of unilateral isomorphism (Smolik, 2004)

4.2.4 Electronic configuration

Electronic configuration of M^{n+} ions, as well as crystal field stabilization energy in high spin octahedral complexes [ML_6], may influence the crystal structure of co-crystallizing salts. According to their electronic configuration these ions may be divided into two groups:

closed shell ions having the configuration p^6 (Mg^{2+}, Ca^{2+}, Sr^{2+}, Ba^{2+}) or d^{10} (Zn^{2+}, Cd^{2+}), as well as d^5, but only when they are in the high spin state (Mn^{2+}). The crystal field stabilization energy (CFSE) of such ions is zero;

open shell ions having the configuration d^n ($n \neq 0, 5, 10$), where CFSE $\neq 0$.

In the first case the energy of these ionic coordination compounds due to the metal ions would be independent of the spatial orientation of the metal–ligand bonds. For this reason these metal ions permit variations over wide ranges of structural parameters, mainly the structure defining angles (angular deformations) (Balarew, 1987).

In the second case the CFSE depends on the orientation of metal–ligand bonds. Therefore, there are some preferred structures, for which CFSE has a maximum value, and the change in geometry of coordination polyhedron with respect to these preferred structures is related to CFSE losses (Balarew, 1987). However, the amount of the CFSE is only 5 - 10% of the whole bonding energy in the crystals and other factors mentioned previously (ionic radii, their charge, energy of metal - ligand bonds) determine the structure of predominantly ionic crystals (Balarew, 1987). Hence its effect on $D_{2/1}$ coefficients is rarely observable.

Some examples of the direct influence of electron configuration of ion on $D_{2/1}$ coefficients are presented in Tables **9-10**.

Ion M^{2+}	Electron configuration	$D_{2/1}$
Mg^{2+}	$1s^22s^2p^6$	0.009 ± 0.005
Ca^{2+}	$1s^22s^2p^63s^2p^6$	0.022 ± 0.008
Sr^{2+}	$1s^22s^2p^63s^2p^64s^2p^6$	0.013 ± 0.006
Zn^{2+}	$1s^22s^2p^63s^2p^6d^{10}$	0.014 ± 0.005
Cd^{2+}	$1s^22s^2p^63s^2p^6d^{10}4s^2p^6d^{10}$	0.010 ± 0.006
Cu^{2+}	**$1s^22s^2p^63s^2p^6d^9$**	**0.040 ± 0.007**
Mn^{2+}	**$1s^22s^2p^63s^2p^6d^5$**	**0.40 ± 0.02**
Fe^{2+}	**$1s^22s^2p^63s^2p^6d^6$**	**1.70 ± 0.20**
Co^{2+}	**$1s^22s^2p^63s^2p^6d^7$**	**2.60 ± 0.30**

Table 9. The effect the electron configuration on $D_{2/1}$ coefficients during the crystallization of $NiCl_2 \cdot 6H_2O$ at 20 °C (Smolik, 1999b)

As it can be seen, the direct effect of electronic configuration of microcomponent ions on coefficients $D_{2/1}$ is most distinct in the case of chloride, formate and acetate crystallization systems, where mean $(D_{2/1})_{open\ shell}$ coefficients are several times greater than $(D_{2/1})_{closed\ shell}$ ones. This effect is lower in sulfate and nitrate crystallization systems.

The direct effect of the electron configuration of ions depends on the kind of anion of the crystallizing salt. It is slightly perceptible in the case of nitates and sulfates, where, besides water molecules, oxoanions NO_3^- and SO_4^{2-} appear. The valence available between oxygen and metal ion, which is a measure of their anion base strength equals: 0,33 and 0,50, respectively, and is very close to that of water molecules (0,40) (Balarew, 1987). Because of a great excess of water both in the liquid phase and in the solid one (particularly in hepta and hexahydrates), these anions cannot compete with water molecules in the bonding of metal ion. So the environment around both metal cations will be formed mainly by water molecules.

Group of salt	Lewis base strength of anion (x) (Brown, 1981)	Mean coefficients ($D_{2/1}$) for microcomponents (ions)		$\kappa = \frac{D^{o}_{mean}}{D^{C}_{mean}}$ (y)
		closed shell D^{c}_{mean}	open shell D^{o}_{mean}	
$MSO_4 \cdot nH_2O$	0.50	0.28	0.50	1.79
$M(NO_3)_2 \cdot nH_2O$	0.33	0.48	0.75	1.56
$MCl_2 \cdot nH_2O$	1.00	0.02	0.37	18.5
$M(HCOO)_2 \cdot 2H_2O$	0.50	1.43	6.48	4.53
$M(CH_3COO)_2 \cdot nH_2O$	0.55	1.26	8.07	6.40
		$\rho_{xy} = 0.9723$		

Table 10. The effect of electron configuration of microcomponents (ions) on their $D_{2/1}$ coefficients in considered groups of simple salts (Smolik, 2004)

The base strength of anions occurring in chloride, acetate and formate systems is generally higher than that for water molecules and equals: 1,00 for Cl^-, 0,55 for CH_3COO^- and 0,50 for $HCOO^-$ (Balarew, 1987)]. Due to this, as well as because of lower excess of water in relation to CH_3COO^- and $HCOO^-$ in the solid phase, these anions can compete with water molecules in coordination surrounding cations of macro and microcomponents. The presence of both kinds of ligands (water molecule and anions Cl^-, CH_3COO^- or $HCOO^-$) differing in size and charge) causes stronger deformation of the octahedral surrounding of these cations as compared with the presence of one ligand. This deformation depends on the electron configuration of the cation of both the macrocomponent and microcomponent. Therefore, this factor may influence the ability of mutual substitution of those octahedra (deformed to a different degree), whose measure is coefficient $D_{2/1}$.

The dependence of coefficients $D_{2/1}$ on electron configuration is usually connected with their dependence on the crystal field stabilization energy (s), which may be expressed quantitatively for high spin octahedral complexes (most of them occurring in the structures of the considered salts) in kJ/mol or in Dq (where Dq – natural theoretical unit for crystal-field splitting energies (Porterfield, 1993). Thus, it is possible to characterize quantitatively this dependence calculating the correlation coefficients of $\ln D_{2/1}$ or $\pi = \ln D_{2/1} - \ln(m_{01}/m_{02})^{v/b}$ and $\Delta s = s_{MACR} - s_{micr}$, $|\Delta s|$, or $(\Delta s)^2$.

Number of crystallization systems	correlation coefficients (ρ_{xy}) of $\pi = \ln D_{2/1} - \ln (m_{01}/m_{02})^3$ and				
	$f(r)$	$f(\delta)$	Δn	$\Delta\varepsilon$	Δs
37	0.4156	-0.1998	-0.1170	**0.7604**	**0.8486**

Table 11. Comparison of correlation coefficients (ρ_{xy}) of $\pi = \ln D_{2/1} - \ln (m_{01}/m_{02})^3$ and functions of some factors affecting co-crystallization coefficients $D_{2/1}$ in formate crystallization systems (Smolik, 2004)

The direct effect of Δs on $\ln D_{2/1}$ is very slight in most considered groups of salts, but having taken into account the solubility ratio of the co-crystallizing salts, the found correlation coefficients of $\pi = \ln D_{2/1} - \ln (m_{01}/m_{02})^{v/b}$ and Δs are relatively high only for formate crystallization systems (ρ_{xy} = **0.8486**). In this group of salts, this correlation coefficient of π and Δs is the highest as compared to the ones involving all analysed factors (Table **11**).

Significant correlation coefficients of $\pi = \ln D_{2/1} - \ln (m_{01}/m_{02})^{\nu/b}$ and $|\Delta s|$ or $(\Delta s)^2$ occur in none of the considered groups of salts, which means that π does not depend (in them) on the similarity of the CFSE of the macrocomponent and microcomponent ion.

4.2.5 Cation hardness (h)

The concept of cation and anion hardness introduced by Pearson (1963) was utilized by Balarew and co-workers (1984) to solve some crystal chemical problems. Based on a quantitative definition of hardness (Klopman, 1968) and using his procedure they determined hardness of several open shell cations (Mn^{2+}, Fe^{2+}, Co^{2+}, Ni^{2+} i Cu^{2+}), which together with the values given by Klopman for anions and other cations they used for the anticipation of a kind of coordination polyhedra in these compounds and their structure. The hardness of several other cations was given by Tepavičarova et al. (1995).

The hardness of cations affecting their coordination surrounding in the case of several ligands of different anion hardness may change the crystal structure of appropriate hydrates causing the formation of pure simple salts, solid solutions or double salts (Balarew, 1987) and therefore, influencing $D_{2/1}$ coefficients.

Absolute values of the determined correlation coefficients (ρ_{xy}) of $\ln D_{2/1}$ and h are generally low (Smolik, 2004). However, after taking into account the solubility ratio of the co-crystallizing salts (function $\pi = \ln D_{2/1} - \ln (m_{01}/m_{02})^{\nu/b}$), ρ_{xy} values are significant for sulfates ($MSO_4 \cdot nH_2O$ and $M_2SO_4 \cdot xH_2O$), chlorides ($MCl_2 \cdot nH_2O$) and alkali halides MX (X = Cl^-, Br^-, I^-), particularly in the presence of Tl^+ and become the highest ($|(\rho_{xy})_{śr}| > 0.80$) and significant in chloride ($MCl_2 \cdot nH_2O$) and halide (MX) crystallization systems. In each case they are negative, which means that with the increasing similarity of macro and microcomponent (with regard to hardness) π values grow.

Low values of ρ_{xy} occur in groups of salts, where coordination surrounding of cations is homogeneous. It is composed of water molecules, which are hard ligands and anions NO_3^-, SO_4^{2-} and CH_3COO^-, classified as hard bases. In chloride crystallization systems ($MCl_2 \cdot nH_2O$) anions Cl^- occur, whose hardness is less than that of water molecules, and in halogen crystallization systems the cation surrounding in the solid phase consists only of chloride, bromide and iodide anions, which are classified as decidedly soft anions. The greatest effect of hardness on co-crystallization coefficients appears here (particularly in the presence of Tl^+ ions significantly differing in their hardness from alkali ions).

5. Possibility of estimation of $D_{2/1}$ coefficients basing on the determined dependences

The determined correlation coefficients of $D_{2/1}$ or $\pi = \ln D_{2/1} - \ln(m_{01}/m_{02})^{\nu/b}$ and various functions of ionic radii (r), electronegativity (ε), crystal field stabilization energy (s), hardness of cations (h), number of molecules of crystallization water (n), the volume of one formal molecule of hydrate salt (η^3) permit to find a kind of functions (f, φ, ψ, ...) of these parameters and to estimate coefficients (w_1, w_2, w_3, ...) in the general equation (19) for the evaluation of $D_{2/1}$ coefficients. Some particular equations of such a general type for the estimation of $D_{2/1}$ in several groups of crystallization systems (macrocomponent - microcomponents) at average error not exceeding 31% are presented in Table **12**.

Macro-component	Micro-components (M^{n+}) (1)	k	Equation	Δ_{av} [%]
orthorhombic $M'SO_4\cdot 7H_2O$ M' = {Mg, Zn, Ni}	Co^{2+}, Fe^{2+}, Mn^{2+}, Cd^{2+}	21	$D_{2/1} = 0.362 \cdot \left(\frac{m_{01}}{m_{02}}\right)^2 \cdot \exp\left(-25.95 \cdot \left(\frac{r_1 - r_2}{r_1}\right)^2 + 0{,}0072\right)$	21.6
	Cu^{2+}		$D_{2/1} = 0.020 \cdot \left(\frac{m_{01}}{m_{02}}\right)^2 \cdot \exp\left(-25.95 \cdot \left(\frac{r_1 - r_2}{r_2}\right)^2 + 0{,}0072\right)$	
	Ni^{2+}, Mg^{2+}, Zn^{2+}		$D_{2/1} = \left(\frac{m_{01}}{m_{02}}\right)^2 \cdot \exp\left(-25.95 \cdot \left(\frac{r_1 - r_2}{r_1}\right)^2 + 0{,}0072\right)$	
orthorhombic $M'(NO_3)_2\cdot 6H_2O$ M' = {Zn, Mn}	Ni^{2+}, Mg^{2+}, Zn^{2+}, Co^{2+} Mn^{2+}	8	$D_{2/1} = \left(\frac{m_{01}}{m_{02}}\right)^3 \cdot \exp\left(-55.83 \cdot \left(\frac{r_1 - r_2}{r_1}\right)^2 + 0.143\right)$	14.8
$CoCl_2\cdot 6H_2O$	Ni^{2+}, Mg^{2+}, Zn^{2+}, Co^{2+} Cu^{2+}, Mn^{2+}, Cd^{2+}, Ca^{2+}, Sr^{2+}	9	$D_{2/1} = \exp\left(0.0078\eta^3 - 1.42\lvert\Delta h\rvert + 0.32\Delta s - 2.92\right)$	27.0
alums	Al^{3+}, Fe^{3+}, Cr^{3+}	9	$D_{2/1} = \left(\frac{m_{01}}{m_{02}}\right)^2 \cdot \exp(16.85 \cdot \eta - 130.2)$	29.8
M^I_2 $M^{II}(SO_4)_2\cdot 6H_2O$ $M^{I'}$={NH_4^+, Rb^+}	Na^+, K^+, Rb^+, Cs^+, Tl^+,	7	$D_{2/1} = \left(\frac{m_{01}}{m_{02}}\right)^{1.25} \cdot \exp(1.48 \cdot \eta - 10.2)$	9.1
M^I_2 $M^{II}(SO_4)_2\cdot 6H_2O$ $M^{I'}$={Ni, Mg, Cu, Co, Zn, Fe, Mn}	Ni^{2+}, Mg^{2+}, Zn^{2+}, Mn^{2+}, Co^{2+} Cu^{2+}, Cd^{2+}	43	$D_{2/1} = \left(\frac{m_{01}}{m_{02}}\right)^{2.5} \cdot \exp(0.835 \cdot \frac{1}{r_2^2} - 0.082)$	25.9
$M'(HCOO)_2\cdot 2H_2O$ $M^{I'}$={Ni, Mg, Co, Zn, Fe, Mn, Cd}	Ni^{2+}, Mg^{2+}, Zn^{2+}, Mn^{2+}, Co^{2+} Cu^{2+}, Cd^{2+}	37	$D_{2/1} = \left(\frac{m_{01}}{m_{02}}\right)^3 \cdot \exp\left(1.67\frac{\eta_1^3 - \eta_2^3}{\eta_1^3} + 1.37\Delta\varepsilon - 0.057\Delta h + 1.65\Delta s + 0.153\right)$	31.0
$M'_2SO_4\cdot nH_2O$ M'={ Na, K, Tl}	Na^+, K^+, Rb^+, Cs^+, Tl^+,	9	$D_{2/1} = \exp\left(12.88\frac{1}{r_2^2} - 4.829\Delta\varepsilon - 7{,}684\right)$	22.8 (4)
MX X={Cl, Br, I}	Cs^+, Rb^+, K^+	9	$D_{2/1} = \left(\frac{m_{01}}{m_{02}}\right)^2 \cdot \exp\left(-132.8\left(\frac{r_1 - r_2}{r_1}\right)^2 + 4.38\Delta h + 0.072\right)$	29.3 (4)
MNO_3 M' ={K, Rb, Cs}	Cs^+, Rb^+, K^+	5	$D_{2/1} = \exp\left(-16.57\frac{1}{r_2} - 349.4(\Delta\varepsilon)^2 + 9{,}209\right)$	18.9

Table 12. Part 1. Some examples of the estimation of coefficients $D_{2/1}$ (Smolik, 2004)

$MClO_3$ M' ={K, Rb}	Cs^+, Rb^+, K^+	3	$D_{2/1}=\left(\frac{m_{01}}{m_{02}}\right)^2\cdot\exp(-4.282\cdot r_2+4.074)$	5.3
$MClO_4$ M' ={K, Rb, Cs}	Cs^+, Rb^+, K^+	5	$D_{2/1}=\left(\frac{m_{01}}{m_{02}}\right)^2\cdot\exp\left(-74.20\cdot\left(\frac{r_1-r_2}{r_1}\right)^2+ +2.545\Delta\eta - 0.18[illegible]\right)$	26.5
M_2CrO_4 M' ={K, Rb, Cs}	Cs^+, Rb^+, K^+	4	$D_{2/1}=\left(\frac{m_{01}}{m_{02}}\right)^2\cdot\exp\left(-7.967\cdot\left(\frac{r_1-r_2}{r_1}\right)-0.424\right)$	5.1

(1) – M^{n+} is not microcomponent, when M'=M;

(2) – subscripts "1" and "2" relate to macrocomponent and microcomponent respectively;

(3) – $\eta = 10^8\sqrt[3]{\frac{M}{D_x N}}$ [Å], where M – molar mass of salt [g/mol], D_x – its density [g/cm^3],

N – Avogadro number [$6.022\cdot10^{23}$/mol];

(4) – in this case $|(r_1-r_2)/r_1|<0.20$ (for the other cases $D_{2/1} < 0.06$); k – number of crystallizations systems $(\text{macrocomponent})_j$ – $(\text{microcomponent})_{ji}$ in a given group of salt.

Table 12. Part 2. Some examples of the estimation of coefficients $D_{2/1}$ (Smolik, 2004)

6. Methods of lowering $D_{2/1}$ values as the way of increasing purification efficiency of crystallization

From the practical point of view it is very interesting to ascertain how to increase the efficiency of crystallization purification of inorganic substances. This is possible when co-crystallization coefficients $D_{2/1}$ can be lowered. They depend generally, as shown above, on chemical, physicochemical and crystal-chemical properties of co-crystallizing salts and ions. However, there are some previously mentioned "external" factors (such as the kind and composition of the solvent – the liquid phase, the presence of ions or other foreign substances, the presence of complexing agents, acidity [pH] of the solution, from which crystallization takes place, temperature) which may influence these coefficients. Their effect on $D_{2/1}$ coefficients will be discussed below.

6.1 The effect of the kind and composition of the solvent – the liquid phase

The change of composition of the solvent, from which the crystallization takes place, alternates solubilities of co-crystallizing salts, as well as activity coefficients of all components in the liquid phase and indirectly in the solid phase.

Because of decreased water activity, the formed crystal hydrates have a lower number of molecules of crystallization water of different structures, which may influence the similarity of the crystal structure of macro and microcomponents. All these mentioned factors vary in different directions and to a different degree, and therefore, they may finally cause the change of co-crystallization coefficients. Examples are given in Table **13**.

Composition of the mother solution [% v/v]						Co-crystallization coefficients $D_{2/1}$				
H_2O	iso-PrOH	CH_3OH	C_2H_5OH	Et_2O	Me_2CO	Mg^{2+}	Co^{2+}	Fe^{2+}	Mn^{2+}	Cu^{2+}
100						0.50± 0.04	0.96± 0.07	0.60± 0.05	0.22± 0.02	0.21± 0.02
50	20	30				0.42± 0.04	0.80± 0.06	0.31± 0.03	0.10± 0.01	0.14± 0.01
60			38	2		0.39± 0.03	0.96± 0.07	0.46± 0.05	0.14± 0.01	0.17± 0.02
63					37	0.57± 0.05	0.70± 0.06	0.49± 0.04	0.18± 0.02	0.15± 0.02

Table 13. The effect of addition of various organic solvents: iso-propyl alcohol (iso-PrOH), methanol (CH_3OH), ethanol (C_2H_5OH), diethyl ether (Et_2O), acetone (Me_2CO) on coefficients $D_{2/1}$ during the crystallization of $NiSO_4{\cdot}7H_2O$ at 25 °C (Smolik, 1984)

6.2 The effect of the presence of other ions or substances in the liquid phase

Interactions which happen in the aqueous phase may also influence $D_{2/1}$ coefficients. This effect is formally taken into consideration in equation (19) by the mean activity coefficients of both the macrocomponent (γ_{m1}) and microcomponent (γ_{m2}).

If the action of various factors in aqueous solution causes the same changes in both mean activity coefficients, so that $\frac{\gamma_{m1}}{\gamma_{m2}} = const.$, then at unchanged properties of the solid phase $D_{2/1}$ coefficient remains constant (e.g., the addition of HBr during the co-crystallization of Ra^{2+} with $BaBr_2$ does not affect $D_{2/1}$ coefficient, likewise the introduction of weak electrolytes (glucose or $CH_3COONa + CH_3COOH$) having no common ions with micro and macrocomponent ($Ra(NO_3)_2$ i $Ba(NO_3)_2$) and not reacting with them also does not change the $D_{2/1}$ value (Chlopin, 1938).

However, if substances are present in the solution that react in a different way with the macro and microcomponent ions forming slightly dissociated compounds, an essential change of $D_{2/1}$ coefficients takes place (e.g., the addition of $CH_3COONa + CH_3COOH$ in the crystallization system $Pb(NO_3)_2$ – $Ra(NO_3)_2$ – H_2O causes bonding a part of Pb^{2+} ions in slightly dissociated acetate, which leads to the lowering of its mean activity coefficient and finally to the increase of radium co-crystallization coefficient (Chlopin, 1938).

6.3 The effect of the presence of complexing agents

The presence of complexing agents has a significant influence on coefficient, $D_{2/1}$. In the case of forming complexes by both macrocomponent and microcomponent, the relationship between the value of co-crystallization coefficient in the presence of complexing agent $(D_{2/1})^k$ and that in the case of its absence ($D_{2/1}$) is expressed by equation (Mikheev et al., 1962): $D^k_{2/1} = D_{2/1} \cdot \dfrac{1+\frac{[ML]}{[M]}}{1+\frac{\beta}{\beta'}\cdot\frac{[ML]}{[M]}}$, where β (β') – stability constant of the complex of microcomponent (macrocomponent), [ML] and [M] are the concentrations of macrocomponent complexes and its free ions. Hence if $\beta/\beta'>1$ (stability of the complex with microcomponent is higher), then $D^{k}_{2/1} < D_{2/1}$ (microcomponent co-crystallizes to a lower degree and vice versa.

An example of the use of a complexing agent to significantly lower $D_{2/1}$ coefficients is presented in Table 14.

The presence of $EDTA^{4-}$ in stoichiometric amount causes 4 - 100 fold lowering of $D_{2/1}$ coefficients of M^{2+} and M^{3+} ions, because of the formation of $[M(EDTA)]^{2-}$ or $[M(EDTA)]^{-}$ anionic complexes, and in the case of some of them (Cd^{2+}, Mn^{2+}, Ni^{2+}, Al^{3+} and Cr^{3+}) two-fold increase of $EDTA^{4-}$ excess leads to an additional drop of $D_{2/1}$ coefficients.

Ions	Co-crystallization coefficients , $D_{2/1}$		
	Ratio of the number of moles of $EDTA^{4-}$ to the sum of the number of moles of M^{2+} and M^{3+} ions before crystallization		
	0:1	1:1	2:1
Fe^{3+}	2.24	0.02	0.02
Co^{2+}	0.21	0.01	0.01
Zn^{2+}	0.18	<0.01	<0.01
Cd^{2+}	1.40	0.10	<0.01
Mn^{2+}	1.08	0.27	0.07
Ni^{2+}	0.19	0.01	0.01
Cu^{2+}	0.28	0.01	0.01
Al^{3+}	0.46	0.05	0.01
Cr^{3+}	1.92	0.16	0.03

Table 14. The effect of $EDTA^{4-}$ addition on $D_{2/1}$ coefficients of co-crystallization of M^{2+} i M^{3+} ions with Na_2SO_4 at 50°C. (1998b)

6.4 The effect of the acidity (pH) of the solution, from which crystallization takes place

Cations of macrocomponent ($[M(H_2O)_x]^{n+}$) and microcomponent ($[M'(H_2O)_{x'}]^{n+}$) present in the solution, from which crystallization usually takes place, may hydrolyze according to the following equations: $[M(H_2O)_x]^{n+} + H_2O \leftrightarrows [M(H_2O)_{x-1}(OH)]^{(n-1)+} + H_3O^+$ and $[M'(H_2O)_{x'}]^{n+} + H_2O \leftrightarrows [M'(H_2O)_{x'-1}(OH)]^{(n-1)+} + H_3O^+$. The degree of hydrolysis depends on $[M(H_2O)_x]^{n+}$ ($[M'(H_2O)_{x'}]^{n+}$) cation acidic strength, meant as Brönstedt acid

($K_{hi} = \{[M(H_2O)_{x-1}(OH)]^{(n-1)+}\} \cdot \{H_3O^+\} / \{[M(H_2O)_x]^{n+}\}$. If the difference in K_h of both ions is significant (e.g., $K_{h2} >> K_{h1}$), ions ($[M'(H_2O)_{x'-1}(OH)]^{(n-1)+}$) of different charge than those of macrocomponent ($[M(H_2O)_x]^{n+}$) are present in the solution, which in-build into crystals of macrocomponent to a lower degree (e.g., $D_{Fe(III)/NH4Al\ alum} = 0.038 \pm 0.005$ in 0,1 M H_2SO_4 solution and $D_{Fe(III)/NH4Al\ alum} = 0.063 \pm 0.009$ in 1,0 M H_2SO_4 solution) (Smolik, 1995b).

In the case of $NiCl_2 \cdot 6H_2O$ crystallization with increasing HCl concentration, the lowering of co-crystallization coefficients $D_{2/1}$ of Co^{2+}, Mn^{2+}, Cu^{2+} and Fe^{2+} (Table 15) is caused not only by the rise of acidity of the solution, but also by the formation of chloride complexes at higher Cl^- concentrations. A significant decrease of Mn^{2+} coefficient (D_{Mn}) occurs even at 0.5 M HCl, but that of Fe^{2+} and Co^{2+} only at 5M HCl.

Microcomponent	Average $D_{2/1}$ coefficients for HCl concentrations [mol/L]		
	0	0.5	5
Co^{2+}	2.60±0.30	2.30±0.30	1.80±0.10
Mn^{2+}	0.46±0.02	0.20±0.02	0.21±0.01
Cu^{2+}	0.04±0.01	0.04±0.01	0.02±0.01
Fe^{2+}	1.70±0.20	1.20±0.20	0.40±0.08

Table 15. The effect of HCl concentration on $D_{2/1}$ coefficients of some M^{2+} ions during the crystallization of $NiCl_2 \cdot 6H_2O$ at 25 °C (Smolik, 1999b)

6.5 The effect of the change of the oxidation state

In some cases it is possible to change easily the oxidation state of the microcomponent or macrocomponent during or before crystallization. Usually this is accompanied by a significant alteration of $D_{2/1}$ coefficients, which may be utilized for the rise of purification efficiency. Several examples of the change of microcomponent oxidation state are presented in Table 16.

Crystallized salt temperature	Oxidation state of		Factor changing oxidation state	$D_{2/1}$	Ref.
	Macro-component	Micro-component			
$MnSO_4 \cdot 5H_2O$ at 20 °C	Mn^{2+}	Fe^{2+}	H_2O_2	1.04	(Smolik et al., 1995)
		Fe^{3+}		< 0.03	
$CoSO_4 \cdot 7H_2O$ at 20 °C	Co^{2+}	Fe^{2+}	H_2O_2	1.20	(Smolik, 2003)
		Fe^{3+}		< 0.03	
$NH_4Al(SO_4)_2 \cdot 12H_2O$ at 25 °C	Al^{3+}	Fe^{3+}	$NH_2OH.H_2SO_4$	0.04± 0.01	(Smolik, 1995b)
		Fe^{2+}		< 0.01	

Table 16. The effect of the change of oxidation state of microcomponent on $D_{2/1}$ coefficients

The change of the oxidation state of macrocomponent can be used for the purification of iron salts: crystallization of $FeSO_4 \cdot 7H_2O$ at 20 °C permits, with great efficiency, removal of all M^+ and M^{3+} ions (Fig. 12a) and the remaining M^{2+} ions can be easily removed after transferring $FeSO_4 \cdot 7H_2O$ into $NH_4Fe(SO_4)_2 \cdot 12H_2O$ by oxidation and its crystallization at 20 °C (Fig. 12b).

6.6 The effect of temperature

The effect of temperature on $D_{2/1}$ coefficients is very complex. As temperature increases, the solubilities of the macro and microcomponent (m_{01}, m_{02}), the mean activity coefficients in their binary saturated solutions (γ_{m01}, γ_{m02}), the mean activity coefficients in the ternary solution being in equilibrium with their mixed crystal (γ_{m1}, γ_{m2}) (temperature affects dehydration of ions, processes of hydrolysis or complex formation in solution (Kirkova et al., 1996), as well as activity coefficients of both in their solid solution (f_1, f_2) (temperature influences enthalpy of mixing in the solid phase) change in different directions in various crystallization systems. Therefore, both the increase and drop of co-crystallization coefficient may be observed or sometimes the maintenance of its constant value (in the case of compensation of the all mentioned changes). However, the alteration of $D_{2/1}$ runs generally in a continuous manner as long as there are no phase transitions. If a phase transition in the system takes place, a jump change of $D_{2/1}$ appears, connected with the transition from isomorphous co-crystallization into the isodimorphous one. The determination of temperatures at which such a jump change of co-crystallization coefficients takes place, permits finding the temperatures of the phase transitions for many hydrate sulfates (Purkayastha & Das, 1972, 1975), as well as predicting the existence (in specific conditions) of hydrates of some salts, as yet unknown (Purkayastha & Das, 1971).

Some examples of the changes of co-crystallization coefficients $D_{2/1}$ at various temperatures have been presented in Table **17**.

(macro-component)	ion	Temperature [°C]	Kind of hydrate (crystal structure) of Macro-component	Micro-component	$D_{2/1}$	Ref.
$MnSO_4$	Cu^{2+}	20	$MnSO_4 \cdot 5H_2O$ (tcl.)	$CuSeO_4 \cdot 5H_2O$ (tcl.)	1.63	(Smolik, 2004)
		50	$MnSO_4 \cdot H_2O$ (mcl.)	$CuSeO_4 \cdot 5H_2O$ (tcl.)	0.15	
$ZnSeO_4$	Cu^{2+}	25	$ZnSeO_4 \cdot 6H_2O$ (tetr.)	$CuSeO_4 \cdot 5H_2O$ (tcl.)	0.51	(Smolik & Kowalik, 2010, 2011)
		40	$ZnSeO_4 \cdot 5H_2O$ (tcl.)	$CuSeO_4 \cdot 5H_2O$ (tcl.)	1.77	
		50	$ZnSeO_4 \cdot H_2O$ (mcl.)	$CuSeO_4 \cdot 5H_2O$ (tcl.)	0.12	
	Ni^{2+}	25	$ZnSeO_4 \cdot 6H_2O$ (tetr.)	$NiSeO_4 \cdot 6H_2O$ (tetr.)	2.93	
		40	$ZnSeO_4 \cdot 5H_2O$ (tcl.)	$NiSeO_4 \cdot 6H_2O$ (tetr.)	0.21	
$Sr(NO_3)_2$	Pb^{2+}	29	$Sr(NO_3)_2 \cdot 4H_2O$(mcl.)	$Pb(NO_3)_2$ (cub.)	0.66	(Niesmieanov, 1975)
		34	$Sr(NO_3)_2$ (cub.)	$Pb(NO_3)_2$ (cub.)	3.30	
Na_2SO_4	Fe^{3+}	25	Na_2SO_4 (rhomb.)		0.01	(Smolik, 1998b)
		50	$Na_2SO_4 \cdot 10H_2O$(mcl.)		2.24	

Table 17. The effect of temperature on $D_{2/1}$ coefficients

The observation of the alterations of $D_{2/1}$ coefficients with changing temperature sometimes permits finding such ranges of this parameter, where they are low enough that crystallization purification of the macrocomponent from a given microcomponent will be very effective (Purkayastha & Das, 1972; Smolik & Kowalik, 2010, 2011). In such a manner it turned out to be possible to accomplish essential purification of $CoSeO_4$ from almost all M^{2+} ions (most difficult to remove) solely by the crystallization method (Kowalik et al., 2011).

7. Conclusions

Crystallization of substances from solutions seems still to be a convenient method of their purification, particularly in obtaining of high purity inorganic compounds. The effectiveness of this process depends on the kind of both macrocomponent (1) and microcomponent (2) and can be evaluated by means of co-crystallization coefficient, $D_{2/1}$ (Henderson - Kraček, Chlopin). These coefficients are affected by conditions applied in the crystallization process, but those which are equilibrium ones, depend exclusively on "internal" and "external" factors.

"Internal" factors (resulting from chemical, physicochemical and crystal-chemical properties of co-crystallizing salts and ions to a significant degree) determine the level of $D_{2/1}$

coefficients. The investigation involving the effect of these factors is of great importance because it makes it possible to evaluate the usefulness of crystallization in the separation of a given pair of macrocomponent (1) and microcomponent (2).

By means of "external" factors a significant lowering of $D_{2/1}$coefficients is sometimes possible, and thus, the improvement of the efficiency of crystallization purification.

Growing knowledge concerning coefficients $D_{2/1}$ in new crystallization systems, as well as better understanding of the dependences of these coefficients on different factors, permits evaluating in a progressively better way the possibilities of the crystallization method in new crystallization systems and more effective control with "external" conditions to achieve higher yields of crystallization purification, enrichment of trace amounts of rare, scattered elements for preparative or analytical purposes. In addition it also helps in improving the growing of single crystals of specific properties or explaining the genesis of some minerals.

8. References

Balarew, Chr.; Duhlev R. & Spassov D. (1984). Hydrated metal halide structures and the HSAB concept, *Crystal Res. Technol.* Vol. 19(11), pp. 1469

Balarew, Chr. (1987). Mixed crystals and double salts between metal (II) salt hydrates, *Zeitschrift fuer Kristallographie.*, Vol. 181, pp. 35-82

Blamforth, A. W. (1965). *Industrial Crystallization,* Leonard Hill, London 1965, (Great Britain) England

Borneman-Starinkevich, I. D. (1975) Calculation of the crystallochemical formula as one of the methods for the investigation of minerals, In *Izomorfizm Minerallov* (Eds. F. V. Chukhrov, B. E. Borutsky & N. N. Mozgova), pp. 125-33 . Nauka, Moskva

Brown, I. D. (1981). Bond-valence method: an empirical approach to chemical structure and bonding. In *Structure and bonding in crystals.* (Eds. M. O'Keeffe, A. Nawrotsky) Vol. II, p. 1 - 30. New York: Academic Press 1981

Byrappa, K.; Srikantaswamy, S.; Gopalakrishna G. S. & Venkatachalapaty V. (1986) Influence of admixtures on the crystallization and morphology of $AlPO_4$ crystals. *J. Mater.Sci.,* Vol. 21, pp. 2202 - 2206

Chlopin, V. G. (1938). The distribution of electrolytes between solid crystals and liquid phase, *Trudy Rad. Inst. AN SSSR* Vol. 4, pp. 34-79

Chlopin, V. G. (1957). Izbrannye Trudy, (Selected Work) Vol. 6, pp. 173, *Izd. AN SSSR,* Moskva, Russia

Demirskaya, O. V.; Kislomed A.N.; Velikhov Y.N.; Glushova L. V. & Vlasova D. I. (1989) Distribution of impurities during crystallization of potassium dihydrophosphate from aqueous solutions at 25 °C. *Vysokochistye Veshchestva,* Vol. 1, pp. 14-16

Doerner, H. A. & Hoskins, W. M. (1925). Coprecipitation of radium and barium sulfates, *J. Am. Chem. Soc.,*Vol.47, pp 662-75

Fisher, S. (1962). Correlation between maximum solid solubility and distribution coefficient for impurities in Ge and Si, *J. Appl. Phys.*, Vol.33, pp 1615

Gorshtein, G. I. (1969). In *Methods of obtaining of high-purity inorganic substances* (in Russ.), Izd. Chimia, Sanct Petersburg, Russia, (63-125)

Hill, A. E.; Durham, G. S. & Ricci, J .E. (1940). Distribution of isomorphous salts in solubility equilibrium between liquid and solid phases, *J. Am. Chem. Soc.*, Vol. 62, pp 2723-32

Kirkova, E. (1994). *Vysokochisti Vechtestva – Metodi na poluchavane*, Sofia University Press, Sofia

Kirkova, E.; Djarova, M. & Donkova, B. (1996). Inclusion of isomorphous impurities during crystallization from solutions, *Prog. Crystal Growth and Charact.*, Vol.32, pp 111 -134

Klopman, G. (1968). Chemical reactivity and the concept of charge- and frontier-controlled reaction, *J. Am.Chem.Soc.* Vol.90, pp. 223-234

Kowalik, A.; Smolik M. & Maczka K. (2011). The influence of temperature on the values of distribution coefficients $D_{2/1}$ during the crystallization of [illegible]. The 1st international conference on methods and materials for separation processes – *Separation Science – Theory and Practice*, Kudowa Zdrój 2011, Poland

Matusievich, L. M. (1961). Influence of diffusion on the process of joint crystallization of isomorphous salts, *Zh. Neorg. Khim.*, Vol.6, pp. 1020-7

Mikheev, N. B.; Mikheeva, L. M.; Malinin, A. B. & Nikonov, M. D, (1962). Effect of complex formation on the separation of elements by cocrystallization process obeying logarithmic rule, *Zh. Neorg. Khim*, Vol. 7, pp. 2267-2270

Niesmiejanov, A. N. (1975). *Radiochemia,* PWN, Warszawa, Poland

Oikova, T.; Balarew, Chr. & Makarov, L. L. (1976). Thermodynamic study of magnesium sulfate-cobalt sulfate-water and magnesium sulfate-zinc sulfate-water systems at 25 °C, *Zh. Fiz. Khim.*, Vol.50(2), pp. 347-52

Pearson, R . G. (1963). Hard and soft acids and bases, *J. Am. Chem. Soc.* 85(22), 3533-9

Porterfield, W. W. (1993). *Inorganic Chemistry. Unified Approach,* Academic Press, Inc. San Diego, New York, Boston, London, Sydney, Tokyo, Toronto,

Przytycka, R. (1968). Coprecipitation of radioactive elements, (in Polish) *Wiadomości Chemiczne,* Vol. 22, pp. 121-141

Purkayastha, B. C. & Das, N. R. (1971). Transition temperature of heptahydrated copper sulfate *J. Indian Chem. Soc.*, Vol. 48, pp. 70-4

Purkayastha, B. C. & Das, N. R. (1972). Study of transition temperature through mixed crystal formation II. *J. Indian Chem. Soc.*, Vol. 49, pp. 245-50

Purkayastha, B. C. & Das, N. R. (1975). Transition temperature through mixed crystal formation. IV, *J. Radioanal. Chem*, Vol. 25, pp. 35-46

Ratner, A. P. (1933). Theory of the distribution of electrolytes between a solid crystalline and a liquid phase, *J. Chem. Phys.*, Vol.1, pp 789-94

Ringwood, A. E. (1955). Principles governing trace-element distribution during magmatic crystallization. I. Influence of electronegativity. *Geochim. Cosmochim. Acta*, Vol. 7, pp. 139 – 202

Rojkowski, Z. & Synowiec, J.(1991). *Krystalizacja i krystalizatory,* (Crystallization and crystallizers) (in Polish) WNT, ISBN 83-204-1375-3, Warszawa, Poland

Ruff, O.; Ebertt F.& Luft, F. (1928). Röntgenographic methods for determining substances adsorbed on carbon, *Z. anorg. allgem. Chem.*,Vol. 170, pp. 49-61

Smolik, M. (1984). Investigations on the method of preparation of high-purity $NiSO_4 \cdot 7H_2O$, *Internal report of the Department of Inorganic Chemistry and Technology, Silesian University of Technology, Gliwice* unpublished, (in Polish)

Smolik, M. & Zołotajkin, M. (1993). Partition of trace amounts of impurities during the crystallization of $CuSO_4.5H_2O$, *Polish J. Chem.* Vol. 67(3), pp. 383-389

Smolik, M, & Lipowska, B. (1995). Partition of trace amounts of impurities during the crystallization of $FeSO_4.7H_2O$ at 20°C., *Indian J. Chem.*, Vol. 34A, pp.230-4

Smolik, M.; Zołotajkin, M. & Kluczka, J. (1995). Distribution of trace amounts of impurities during manganese (II) sulfate crystallization at 20° and 2°C., *Polish J. Chem.*, Vol.69, pp.1322-7

Smolik, M. (1995a). Cocrystallization of trace amounts of metal ions with $NH_4Fe(SO_4)_2.12H_2O$, (in Polish) - 5th Symposium "*Industrial crystallization*" Rudy, Institute of Inorganic chemistry, Gliwice, pp. XVIII (1-8), Poland

Smolik, M. (1995b). Partition of trace amounts of impurities during the crystallization of $NH_4Al(SO_4)_2 \cdot 12H_2O$, (in Polish), Scientific Meeting of the Polish Chemical Society, Lublin, Poland

Smolik, M. (1998a). Distribution of cocrystallized microamounts of some M^+ ions during $NiSO_4.(NH_4)_2SO_4.6H_2O$ crystallization., *Acta Chem. Scand.*, Vol.52, pp 891-6

Smolik, M. (1998b). Cocrystallization of trace amounts of metal ions with Na_2SO_4 (in Polish), 6th Symposium "*Industrial crystallization*" Rudy 1998 (*in Polish*), Institute of Inorganic chemistry, Gliwice, pp. 91-101

Smolik, M. (1999a). Partition of microamounts of some M^{2+} ions during $MgSO_4 \cdot 7H_2O$ crystallization. *Austr. J. Chem.* Vol. 52, pp. 425-430

Smolik M. (1999b) Distribution of trace amounts of some M^{2+} ions during nickel chloride crystallization coefficients, *Polish J. Chem.* Vol.73, pp. 2027-2033

Smolik, M.; (2000a). Distribution of microamounts of some M^{2+} ions during $ZnSO_4 \cdot 7H_2O$ crystallization., *Canadian J. Chem.*, Vol.78(7), pp. 993-1002

Smolik, M. (2000b). Distribution of trace amounts of M^{2+} ions during crystallization of $NiSO_4 \cdot 7H_2O$, *Polish J. Chem.*, Vol. 74, pp. 1447-1461

Smolik, M. (2001). Distribution of microamounts of M^{2+} during $NiSO_4 \cdot (NH_4)_2SO_4 \cdot 6H_2O$ crystallization, *Separation Science and Technology* Vol. 36(13), pp. 2959-2969

Smolik, M. (2002a). Factors influencing cocrystallization coefficients D of trace amounts of M^{2+} ions in some sulfate crystallization systems ($M'SO_4 \cdot nH_2O$ - H_2O), 10th *International Symposium on Solubility Phenomena*, Varna 2002, Abstract Book, pp.37

Smolik, M. (2002b). Cocrystallization of trace amounts of M^{2+} ions with $CdSO_4 \cdot 8/3H_2O$, 10th *International Symposium on Solubility Phenomena*, Varna 2002, Abstract Book, pp. 93

Smolik, M. (2003). Cocrystallization of low amounts of M^{2+} ions during $CoSO_4 \cdot 7H_2O$ crystallization, *J. Chilean Chem. Soc.*,Vol. 48(3), pp. 13-18

Smolik, M. (2004) (Effect of some chemical, physicochemical and crystal-chemical factors on cocrystallization coefficients $D_{2/1}$ of trace amounts of metal ions during the crystallization of chosen salts from water solutions (in Polish), *Zesz. Nauk. Pol. Sl. Chemia,* No 1617, Vol. 148, pp 1-186

Smolik, M. Jakóbik, A. & Trojanowska J. (2007). Distribution of trace amounts of M^{2+} ions during $Co(CH_3COO)_2 \bullet 4H_2O$ crystallization, *Sep. Pur. Techn.*, Vol. 54, pp. 272-276

Smolik, M. (2008). Effect of chemical, physiochemical and crystal-chemical factors on co-crystallization coefficients of trace amounts of metal ions M2+ during the crystallization of selected acetates M(CH3COO)2 ·nH2O, *Chemik,* Vol.10, pp 526-529

Smolik, M. & Kowalik, A. (2010). Co-crystallization of trace amounts of M^{2+} ions with $ZnSeO_4 \cdot 6H_2O$ at 25°C, *J. Cryst. Growth,* Vol. 312, pp, 611- 616

Smolik, M. & Kowalik, A. (2011a). Equilibrium coefficients of co-crystallization of M2+ ions with MgSeO4·6H2O and their dependences on various physicochemical factors, *Crystal Research and Technology*, Vol. 46, pp. 74-9

Smolik, M. & Kowalik, A. (2011b). Coefficients, $D_{2/1}$ of cocrystallization of M^{2+} ions with $ZnSeO_4 \cdot 5H_2O$ at 40 °C and $ZnSeO_4 \cdot H_2O$ at 50 °C and their dependences on various physicochemical and crystal-chemical factors, *J. Cryst. Growth,* Vol. 337, pp. 46-51

Smolik, M. & Siepietowski, L. (2011). Some factors influencing distribution coefficients of trace amounts of M^{2+} ions during the crystallization of $Mn(CH_3COO)_2 \cdot 4H_2O$, *S. Afr. J. Chem.*, Vol. 64, pp. 34-37

Tepavicharova, S., Balarew, Chr. & Trendafilova S. (1995). Double salts obtained from $Me+X-CuX_2-H_2O$ systems ($Me+ = K^+$, $NH4^+$, Rb^+, Cs^+; $X^- = Cl^-$, Br^-), *J. Solid State Chem.* 114, 385-91

Urusov, V. S. (1970). Energy theory of isovalent isomorphism, *Geochimia,* Vol. 4, pp. 510-24

Urusov, V. S. & Kravchuk I. F. (1976). Energy analysis and calculations of partition coefficients of isovalent isomorphous admixtures during crystallization of melts, *Geochimia,* Vol. 8, pp. 1204-23

Urusov, V. S. (1977). *Theory of the Isomorphous Miscibility,* Nauka, Moskva, Russia

Urusov, V. S. (1980). Energy formulation of the problem of equilibrium cocrystallization from a water solution, *Geochimia,* Vol.54, pp. 627-44

Vachobov, A. V.; Khudaiberdiev, V. G. & Vigdorovich, V. N. (1968). Relation between the distribution coefficient for crystallization purification and the limited solubility of the impurity in the solid state, *Dokl. Tadzh. Akad. Nauk,* Vol.11(8), pp 19-22

Zhelnin, B. I. & Gorshtein G. I. (1971). Investigations on equilibrium in the system $MnSO_4$-$MgSO_4$-H_2O at 25 and 100 °C, *Zh. Neorg. Khim.* Vol. 16, pp. 3146-50

Zolotov, Yu. A. & Kuz'min, N. M. (1990). *Preconcentration of trace elements,* Elsevier, Amsterdam, Oxford, New York, Tokyo

7

Recrystallization of Active Pharmaceutical Ingredients

Nicole Stieger and Wilna Liebenberg
North-West University, Unit for Drug Research & Development
South Africa

1. Introduction

Recrystallization can be described simply as a process whereby a crystalline form of a compound may be obtained from other solid-state forms, being themselves crystalline or amorphous, of the same substance. Recrystallization is the process most often employed for the intermediate separation and last-step purification of solid active pharmaceutical ingredients (APIs) (Shekunov & York, 2000; Tiwary, 2006). Chemical purity is however not the only property of a pharmaceutical active that will affect its performance. Crystal structure (Table 1), crystal habit (Table 2) and particle size all play a part. Polymorphism (different crystal structures of the same substance) affects the physico-chemical properties and stability of an API, whereas crystal habit and particle size mostly affect various indices impacting on dosage form production and performance: particle orientation; flowability; packing and density; surface area; aggregation; compaction; suspension stability; and dissolution (Blagden *et al.*, 2007; Doherty & York, 1988; Tiwary, 2006).

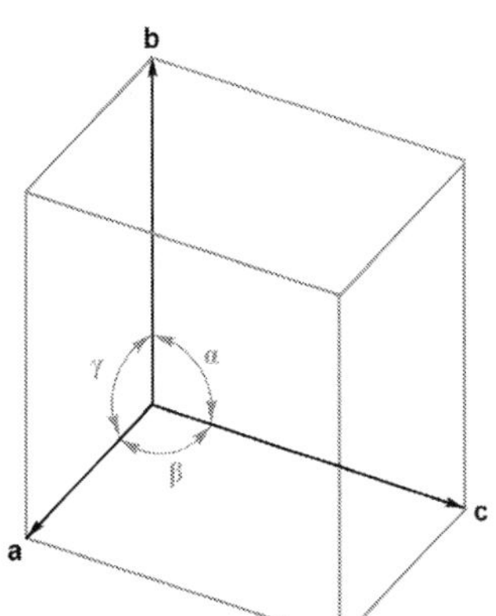

System	Relationship between lattice parameters
Cubic	$a = b = c$ $\alpha = \beta = \gamma = 90°$
Tetragonal	$a = b \neq c$ $\alpha = \beta = \gamma = 90°$
Orthorhombic	$a \neq b \neq c$ $\alpha = \beta = \gamma = 90°$
Rhombohedral (or trigonal)	$a \neq b \neq c$ $\alpha = \beta = \gamma \neq 90°$
Hexagonal	$a = b \neq c$ $\alpha = \beta = 90°, \gamma = 120°$
Monoclinic	$a \neq b \neq c$ $\alpha = \gamma = 90°, \beta \neq 90°$
Triclinic	$a \neq b \neq c$ $\alpha \neq \beta \neq \gamma \neq 90°$

Table 1. Crystal systems and lattice parameters. (Adapted, with permission of Informa Healthcare, from Rodríguez-Homedo *et al.*, 2006.)

Crystal morphology	Description
Acicular	Slender, needle-like crystal
Aggregate	Mass of adhered crystals
Blade	Long, thin, flat crystal
Dendritic	Tiny crystallites forming a tree-like pattern
Equant/cubic	Crystal with similar length, width and thickness
Fiber	Long, thin needle; longer than acicular
Flake/lath	Thin, flat crystal similar in width and length
Plate/platy	Flat crystals with similar width and length but thicker than a flake
Prismatic/bipyramid	Hexagonal crystals with faces parallel to the growth axis; width and thickness greater than acicular and shorter in length
Rod	Cylindrical crystals elongated along one axis
Rosette/spherulite	Sphere composed of needles or rods radiating from a common center
Tablet/tabular	Flat crystal with similar width and length but thicker than a plate

Table 2. Descriptions of common pharmaceutical crystal morphologies. (Adapted, with permission of Informa Healthcare, from Rodríguez-Homedo et al., 2006.)

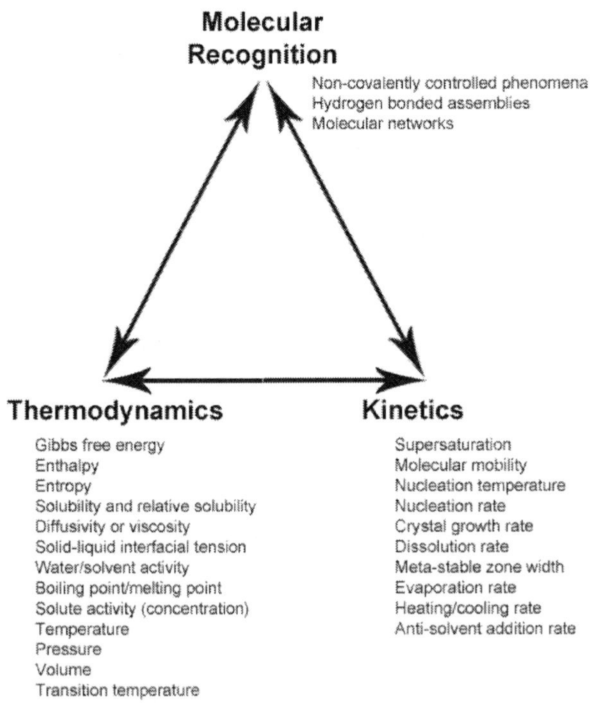

Fig. 1. Diagram showing the phenomena that govern solid-state transformations. Mechanical, thermal and chemical (solvents, additives, impurities, relative humidity) stresses affect the competition (or reinforcement) among these processes. (Adapted, with permission of Informa Healthcare, from Rodríguez-Homedo *et al.*, 2006.)

The number of solid-state forms that an API could exist in relies on the range of non-covalent interactions and molecular assemblies, the order range, and the balance between entropy and enthalpy that defines the free energy states and processes (Figure 1). When an API exists as more than one solid-state form, thermodynamics control their relative stability and the conditions and direction in which a transformation can occur, whilst kinetics determine how long it will take for a transformation to reach equilibrium. Thermodynamics establish the stability domains of the solid-states, but once a metastable domain is encountered, the kinetic pathways will determine which form will be created and for how long it will survive (Rodríguez-Spong *et al.*, 2004).

Small changes in recrystallization procedure can influence the crystallization process and may lead to changes in API crystal structure, crystal habit and particle size with subsequent variability of raw material characteristics and dosage form performance. Herein lies the challenge and the opportunity: manufacturers of pharmaceutical actives go to great lengths to ensure production uniformity, but a pharmaceutical researcher might choose to alter recrystallization conditions to manipulate API characteristics.

Crystallization is generally thought of as the evolution from solution or melt of the crystalline state (Blagden *et al.*, 2007), but a broader definition includes precipitation and solid-state transitions (Shekunov & York, 2000). Crystallization methods can be solvent or non-solvent based and the varied reaction conditions generate different crystal forms (Banga *et al.*, 2004). Non-solvent based methods include sublimation, thermal treatment, desolvation of solvates, grinding and crystallization of a melt (Guillory, 1999). Applied in polymorph screening studies, traditional crystallization approaches most often will not yield all possible polymorphs of a given API. Therefore, there is a continuous search for innovative methods of manipulating the crystallization process (Rouhi, 2003).

2. Solvent based recrystallization

In solution, crystallization is the creation of a crystalline phase by a process (Figure 2) initiated by molecular aggregation, leading to the formation of nuclei (the smallest possible units with defined crystal lattice) and ultimately crystal growth (Banga *et al.*, 2004).

2.1 Parameters that influence nucleation and crystal growth

The sections that follow will address factors that influence solvent recrystallization in general and will not venture into the specifics of specialized solvent-based production methods like spray-drying, freeze-drying, etc. The reader should keep in mind that although different parameters are discussed under separate headings, they are interactive and not independent of each other. For example: changing the composition of a solvent system will change the level of saturation; changes in temperature alter viscosity and level of saturation; agitation may increase temperature; etc.

Each of the parameters discussed can influence polymorphism, and polymorphism in turn affects crystal habit. It is also well-known that each polymorph of an API can exhibit multiple crystal habits, therefore a parameter might influence crystal habit without changing the internal structure of the crystals produced (Tiwary, 2006).

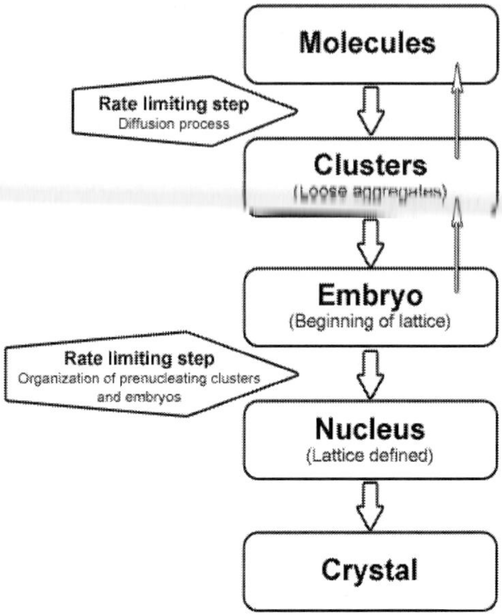

Fig. 2. The course of crystallization and its rate-limiting steps. (Adapted, with permission of Touch Briefings, from Banga *et al.*, 2004.)

2.1.1 Concentration and temperature

The difference in chemical potential between the crystallization solution and the solid phase is the fundamental driving force for crystallization, but it is more convenient to express the driving force in terms of supersaturation (Fujiwara *et al.*, 2005). Supersaturation (the difference between solution concentration and saturation concentration at a specific temperature) leads to the creation of metastable (far from equilibrium) liquid states and crystallization provides a means to reduce the free energy of the system to the most stable state (equilibrium) (Rodríguez-Homedo *et al.*, 2006). The kinetics of nucleation and crystal growth are strongly dependent on supersaturation (Braatz, 2002; Togkalidou *et al.*, 2002). An increase in the degree of supersaturation of a solution leads to a reduction in the size of crystals produced. At high supersaturation nucleation is more rapid than growth, resulting in the precipitation of fine particles (Carstensen *et al.*, 1993; Tiwary, 2006). It is therefore important to control the degree of supersaturation during crystallization, because the size, shape and solid-state form of the crystals produced are all influenced by supersaturation (Fujiwara *et al.*, 2005).

Supersaturation is typically achieved through processes that either increase the solute concentration (evaporation or dissolution of a metastable solid phase with subsequent transformation to the more stable, but less soluble form) or decrease the solubility of the solute (cooling, addition of an antisolvent, pH change or the addition of ions that participate in precipitation of the solute) or through a combination of these strategies (Fujiwara *et al.*, 2005; Rodríguez-Homedo *et al.*, 2006).

The main effect of temperature on crystallization from solution is secondary to its influence on the solubility of the solute, and subsequently the degree of saturation of the solution. Although molecular recognition is required for the formation of molecular clusters, and this is dependent on molecular mobility and collision rates, both of which increase at higher temperatures, the molecular mobility in liquids is too high to be the rate-limiting factor in nucleation and crystal growth (Rodríguez-Spong *et al.*, 2004).

The relationship between nucleation rate and supersaturation is well known and Figure 3 illustrates that the number of nuclei generated by a high rate of cooling increases exponentially with increasing supersaturation. A high number of nuclei at the outset limits their growth potential (Tung *et al.*, 2009).

In cases where the crystallization of a metastable polymorph precedes in-solvent transformation to a more stable polymorph as described by Ostwald's Rule of Stages (Boistelle & Astier, 1988; Ostwald, 1897), the solvent-mediated transformation can be affected by the temperature of the crystallization medium (Stieger *et al.*, 2009), because the process is thermally activated (Beckman, 2000).

Temperature cycling or oscillation is sometimes used to accelerate the effect of post-crystallization Ostwald Ripening in slurries containing just one polymorph of an API. Small particles and rough edges of larger particles dissolve faster during heating periods, followed by their recrystallization onto the existing crystals during cooling. The overall effect is more uniformly shaped particles and a narrower particle size distribution range (Kim *et al.*, 2003; Mullin, 2001).

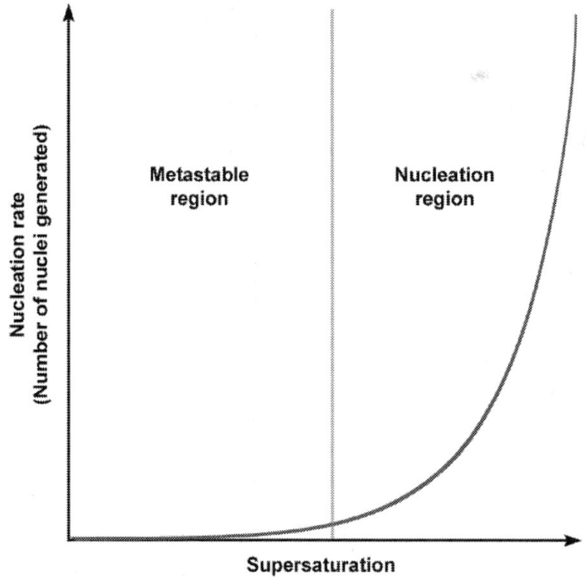

Fig. 3. Nucleation versus supersaturation (Adapted, with permission of John Wiley and Sons, from Tung *et al.*, 2009).

2.1.2 Interfaces and surfaces

Nucleation can be either homogeneous or surface catalyzed (Figure 4). Homogeneous nucleation seldom occurs in volumes greater than 100 µl, because solutions contain random impurities that may induce nucleation. Surface catalyzed nucleation can be promoted by surfaces of the crystallizing solute (secondary nucleation) or a surface/interface of different composition than the solute may induce nucleation (heterogeneous nucleation) by decreasing the energy barrier for the formation of nuclei (Rodríguez-Homedo *et al.*, 2006).

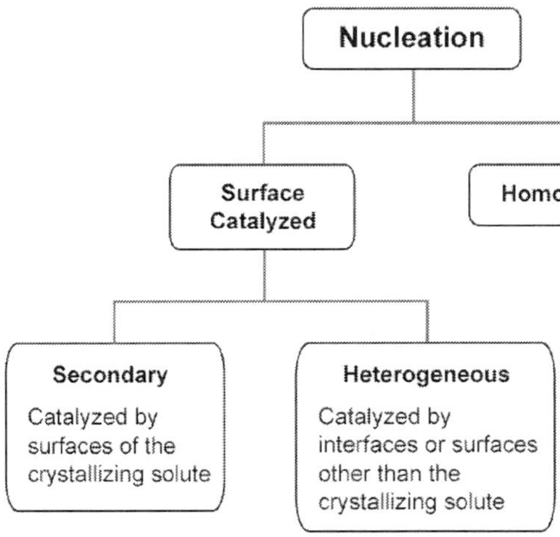

Fig. 4. Mechanisms for crystal nucleation. (Adapted, with permission of Informa Healthcare, from Rodríguez-Homedo *et al.*, 2006.)

Surfaces promoting heterogeneous nucleation may be introduced into the crystallization solution intentionally (as a means of controlling crystal form of the product) (Rodríguez-Spong *et al.*, 2004) or unintentionally (dust and other impurities), or they may be an unavoidable part of the process (crystallization vessel, vessel-solution interface and the solution-air interface) (Florence & Attwood, 2006; Kuzmenko *et al.*, 2001).

The intentional introduction, into crystallization solutions, of surfaces that catalyze nucleation is known as "seeding". Usually, seeding is performed by introducing crystals of the solute that have the preferred crystal structure one wishes to obtain. Seeding can also be performed using isomorphous substances that differ from the solute (Florence & Attwood, 2006). Seeding techniques can be applied to initiate crystallization; to control particle size – usually when larger crystals with uniform size distribution are required; to avoid encrustation through spontaneous nucleation; to control polymorphic form; and to obtain crystals of high purity, high perfection, desired orientation and sufficient size for crystal structure determination by X-ray diffraction (Beckman, 2000).

When the concentration of an API in the crystallization medium is increased past its solubility curve (Figure 5), whether by cooling or by evaporation of solvent, nucleation does not immediately occur. The solution has to reach a certain concentration-temperature point where spontaneous nucleation occurs - this is the border of the metastable zone. For seeding to be successful in producing just the required product, the concentration and temperature of the crystallization medium must be strictly controlled within this zone. If the solubility line is crossed to the left, the seed crystals will dissolve and if the metastable zone limit is crossed to the right, spontaneous nucleation will take place and could result in unwanted crystal forms (Beckman, 2000; Fujiwara *et al.*, 2005).

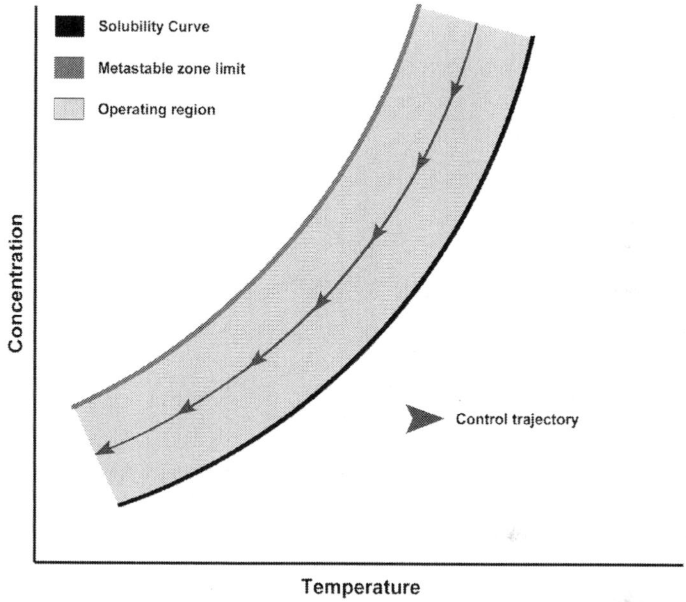

Fig. 5. The operating region of seeded industrial batch crystallization is the metastable zone, which is bound by the solubility curve and the metastable limit for the specific API. The concentration-temperature profile (control trajectory) to be used lies within the metastable zone (operating region). (Adapted, with permission of Elsevier, from Fujiwara *et al.*, 2005.)

In evaporative crystallization, crystals are sometimes observed to form preferentially near the surface of the solution, due to a higher local concentration of solute. The meniscus of a solution can also have geometry favoring higher evaporation rates, with crystals then forming at the contact line with the crystallization vessel (Capes & Cameron, 2007).

2.1.3 Solvent

Dependent on the conditions employed, the crystallization of API polymorphs from a solvent may be under kinetic or thermodynamic control. When crystallization of a dimorphic compound (Figure 6) is conducted sufficiently above or below the transition point, in an area defined by the solubility curves of the two polymorphs, the solvent used is immaterial provided that the API solubility is adequate to allow the prescribed concentrations to be reached. Irrespective of the kinetics, the outcome is under total thermodynamic control. If crystallization takes place outside the area described above (B2, C1, D1, E1 and E2), the choice of solvent may or may not be critical. This will be determined by the relative kinetics of formation, growth, and transformation of the two polymorphs in the various solvents (Threlfall, 2000).

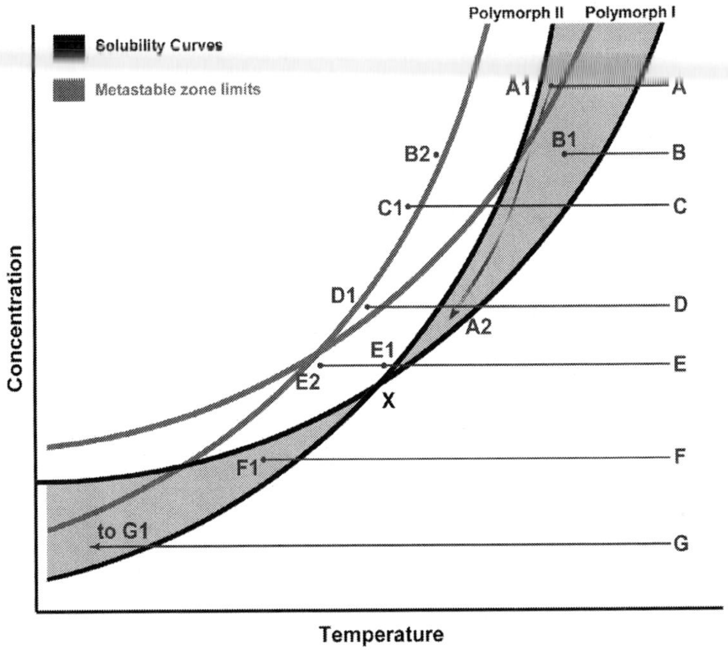

Fig. 6. Polymorphic system of two enantiomorphically related polymorphs I and II. (Transition point X at T_X; A-G, initial state of hot, undersaturated solutions; A1-G1, B2 and E2, state of solution at point of initial crystallization. If B is seeded at B1, it behaves as A1.) (Adapted, with permission of the American Chemical Society, from Threlfall, 2000.)

Recrystallization from solvents often leads to the isolation of solvates, in fact, API solvates are very common (Griesser, 2006). Although they have a recognized potential to improve dissolution kinetics (Brittain & Grant, 1999; Haleblian, 1975; Tros de Ilarduya *et al.*, 1997), they are rarely selected for further development or dosage form formulation - the only exceptions being hydrates. The main reasons for the rarity of marketed API solvates are their solid-state metastability and the relative toxicity of any included solvent (Douillet *et al.*, 2011). It goes without saying that, in the production of solvates, the solvent/s used for recrystallization will be the one/s that could potentially be included in the crystal lattice.

The nature of the crystallization solvent can affect crystal habit, regardless of change in polymorphism (Stieger *et al.*, 2010a). The interaction of the solvent at the different crystal-solution interfaces may lead to altered roundness of growing crystal faces and/or edges, changes in crystal growth kinetics, and enhancement or inhibition of crystal growth at certain faces (Figure 7), thereby changing the crystal habit (Stoica *et al.*, 2004; Tiwary, 2006).

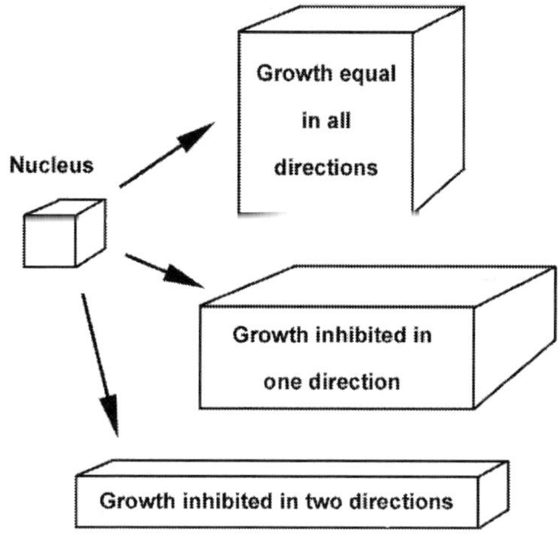

Fig. 7. Crystal habits arising from growth inhibition at crystal faces. (Adapted, with permission of Informa Healthcare, from Carstensen *et al.*, 1993.)

2.1.4 Agitation, mixing and stirring

Mixing of crystallization solutions and crystal slurries is often necessary, especially in industry, to ensure homogeneity (heat transfer, dispersion of additives, uniformity of crystal suspension, avoidance of settling, etc.). Its effects on nucleation, both primary and secondary, and crystal growth are far reaching and complex. An unagitated solution can, in general, be cooled further before onset of nucleation, because the overall result of mixing is a decrease in the width of the metastable zone. In a mixed solution at constant supersaturation, with no crystals yet present, mixing intensity can reduce induction time - the time elapsed before crystals first appear. Induction time decreases up to a critical speed and then remains unchanged (Karpinski & Wey, 2001; Mullin, 2001; Tung *et al.*, 2009).

Mixing actively generates secondary nuclei through crystal-crystal, crystal-vessel and crystal-impeller impacts. A greater number of nuclei generated in agitated systems leads to a decrease in the ultimate crystal size of the product. Mixing also has an intensity dependent effect on the mass transfer rate of growing crystals with a growth limiting outcome (Tung *et al.*, 2009).

It has been found that stirring can reproducibly affect the chiral symmetry of crystallization products. If a substance crystallizes as an equal mixture of dextro and levo crystals when unstirred, its chiral symmetry can be disrupted by stirring (Kondepudi *et al.*, 1990; McBride & Carter, 1991). This phenomenon has been attributed to the effect of stirring on secondary nucleation (Kondepuddi & Sabanayagam, 1994).

2.1.5 Pressure

Almost all polymorph screening studies and industrial crystallization processes are performed under ambient pressure, but low pressure is also often applied to increase the rate of solvent evaporation. As with rapid cooling, rapid evaporation leads to higher nucleation rates.

Only recently have researchers turned their attention to high pressure recrystallization from solution as an added dimension in the search for new pharmaceutical polymorphs and solvates. High pressure encourages denser structures in which molecules must pack more efficiently and this means that changes in relative orientations are likely to occur. This gives rise to different degrees of molecular interactions, the strengths of which are in turn sensitive to distance and therefore the effects of pressure. The interactions between solute and solvent molecules are also modified by pressure, changing the solubility of a given polymorph or solvate. In some instances, the differences in solubility between two forms might also change, thereby encouraging recrystallization of one form at the expense of another (Fabbiani *et al.*, 2004).

2.1.6 Moisture and humidity

When recrystallizing from hygroscopic organic solvents, care should be taken to use only properly dried solvents (Table 3) or newly opened containers from reputable suppliers. If a crystallization process generates a metastable intermediary form, it can absorb water when exposed to moisture and change into a hydrate. Water molecules, because of their small size and multidirectional hydrogen bonding capabilities, are particularly suited to fill structural voids (Manek & Kolling, 2004). If a crystallization system is open to atmospheric conditions, as is likely to be the case with evaporative crystallization, hygroscopic solvents will absorb moisture from the air if present. Should an evaporative crystallization process be sensitive to the presence of moisture, additional measures will have to be put in place to eliminate atmospheric humidity.

Organic Solvent	Drying Agent
Alcohols	Anhydrous potassium carbonate; anhydrous magnesium or calcium sulphate; calcium oxide.
Alkyl halides Aryl halides	Anhydrous calcium chloride; anhydrous sodium, magnesium or calcium sulphate; phosphorous pentoxide.
Saturated and aromatic hydrocarbons Ethers	Anhydrous calcium chloride; anhydrous calcium sulphate; metallic sodium; phosphorous pentoxide.
Aldehydes	Anhydrous sodium, magnesium or calcium sulphate.
Ketones	Anhydrous sodium, magnesium or calcium sulphate; anhydrous potassium carbonate.
Organic bases (amines)	Solid potassium or sodium hydroxide; calcium oxide; barium oxide.
Organic acids	Anhydrous sodium, magnesium or calcium sulphate.
Halogenated solvents*	Calcium hydride; phosphorous pentoxide.

* Never dry halogenated solvents with alkali metals or alkali metal hydrides as this can cause violent explosions.

Table 3. Common drying agents for organic solvents. (Adapted, with permission of Indian Streams Research Journal, from Shinde *et al.*, 2011.)

2.1.7 Addition of salts, polymers or antisolvents

The addition of salts, polymers or antisolvents can be used to create supersaturation of the solution by decreasing the solute solubility in the crystallization medium. Ions, polymeric molecules, or other substances introduced into the crystallization solvent can also act as impurities for growing crystals. These substances may get adsorbed in the crystal lattice of a growing crystal and disturb the regular and repeating arrangements of the crystal, creating defects and leading to polymorphic modifications. Impurities, and surfactants in particular, can also inhibit crystal growth at certain crystal faces, resulting in crystal habit changes (Tiwary, 2006).

When selecting suitable solvents for antisolvent crystallization, one should select a pair that is miscible. The solute must be more soluble in one solvent and less soluble in the other (antisolvent). A water-soluble API will most likely be more soluble in a polar solvent and less soluble in a non-polar solvent (Table 4). The opposite holds true for a poorly water-soluble API. Antisolvent crystallization is performed by dissolving the API in the solvent and then gradually adding an antisolvent. This results in a decrease of solute solubility and an increase in supersaturation - much like cooling crystallization (Nonoyama *et al.*, 2006; Stieger *et al.*, 2010a; Zhou *et al.*, 2006). Reverse addition is a variation of antisolvent recrystallization whereby the API solution is added to the antisolvent. The resulting rapid increase in supersaturation leads to swift nucleation and the precipitation of very fine particles (Tung *et al.*, 2009). The composition of the crystallization medium can affect both the crystal form and crystal habit of the product (Stieger *et al.*, 2010a).

Solvent 1 (more polar)		Solvent 2 (less polar)	
Solvent	Dielectric constant (ε)	Solvent	Dielectric constant (ε)
Water	78.3	Ethanol	24.3
Water	78.3	Acetone	20.7
Methanol	32.6	Dichloromethane	9.08
Ethanol	24.3	Acetone	20.7
Acetone	20.7	Diethyl ether	4.34
Acetone	20.7	Petroleum ether	1.90
Diethyl ether	4.34	Hexane	1.89
Ethyl acetate	6.02	Cyclohexane	1.97
Ethyl acetate	6.02	Petroleum ether	1.90
Dichloromethane	9.08	Petroleum ether	1.90
Toluene	2.38	Petroleum ether	1.90

Table 4. Common solvent-antisolvent pairs. (Adapted, with permission of the author, from Skonieczny, 2009.)

2.2 Novel strategies and new trends in solvent-based crystallization

One of the biggest challenges for pharmaceutical researchers is finding all the solid-state forms of an API that can exist at ambient conditions. There is no single method for producing all conceivable forms and a particular polymorph may go undetected for many years. Scientists are continually searching for novel ways of uncovering hidden solid-states. A few of these methods are briefly discussed below.

2.2.1 Laser-induced crystallization

Non-photochemical laser-induced nucleation (NPLIN) is a crystallization technique that can affect both nucleation rate and the crystal form produced. Laser pulses act predominantly on pre-existing molecular clusters by assisting in the organization of pre-nucleating clusters and embryos into nuclei (Figure 2), leading to dramatically increased nucleation rates for supersaturated solutions. It is believed that the plane-polarized light aligns the pre-nucleating clusters and thereby reduces the entropic barrier to the free energy of activation for critical nucleus formation (Banga *et al.*, 2004; Garetz *et al.*, 1996; Rodríguez-Spong *et al.*, 2004; Zaccaro *et al.*, 2001). This method has not yet been applied to pharmaceuticals, but it should produce similar results as for other organic substances.

2.2.2 Capillary crystallization

In order to access metastable forms of an API, high levels of supersaturation are often required. Capillary tubes as recrystallization vessels are ideal for manipulating the metastable zone width through slow evaporation and because the small volumes of solution isolate heterogeneous nucleants, and reduce turbulence and convection. This technique offers the additional advantage that the crystals need not be removed from the capillary tube prior to characterization by single crystal- or powder X-ray diffraction (Banga *et al.*, 2004; Rodríguez-Spong *et al.*, 2004).

2.2.3 Sonocrystallization

This technique utilizes ultrasound to increase nucleation rate, but it is also an effective means of crystal size reduction that eliminates many of the disadvantages associated with mechanical size reduction. Sonic waves give rise to a phenomenon called cavitation - the formation of bubbles that decrease in size until a critical size is reached, leading to collapse and the formation of cavities. Cavitation provides energy that accelerates the nucleation process (Banga *et al.*, 2004; Kim *et al*, 2003).

3. Non-solvent based recrystallization

Much emphasis has been placed on the importance of solvent based recrystallization in the production of different solid-state forms of APIs, but non-solvent based recrystallization is equally important to industry and researchers alike. Recrystallization through solid-state transitions affects not only the production of APIs, but also the stability of the final product. Additionally, physical vapor deposition (PVD) recrystallization offers yet more opportunities for the preparation of polymorphs.

3.1 Recrystallization through solid-state transitions

Polymorphs may be obtained through solid-state transition by recrystallization of metastable polymorphs and amorphous forms, or through desolvation of solvates (including hydrates). No solid-state transition will take place unless it is thermodynamically favored and solids will always tend to transform to their lowest energy state - the most stable form. Stability is of major concern where APIs are concerned, but it is also true that stable forms are less soluble than their metastable counterparts and solubility of an API often affects its bioavailability, and therefore also its efficacy (Lipinsky *et al.*, 1997). This necessitates a compromise between stability and solubility: we search for a metastable form with improved solubility that is still relatively stable enough to withstand the rigors of pharmaceutical processing and will yield a product with acceptable shelf-life.

3.1.1 Transformation of polymorphs

Polymorphism, occurring in a single-component crystalline molecular solid, may display monotropism or enantiotropism or both (Figure 8). In a monotropic system (A and C, B and C), only one polymorph is stable below the melting point and a phase change from the metastable form (A and B) to the stable form (C) is irreversible. For enantiotropic systems (A and B), a reversible phase transition is observed at a definite transition temperature ($T_{t,A\text{-}B}$), where the free energy curves intersect before the melting point (T_m). At temperatures and pressures below $T_{t,A\text{-}B}$ form A will be stable (lower free energy and solubility), whilst in the temperature and pressure range between $T_{t,A\text{-}B}$ and $T_{m,B}$ form B is more stable. Polymorphs in an enantiotropic system are referred to as enantiotropes and those in a monotropic system are monotropes (Bernstein 2002; Grant 1999).

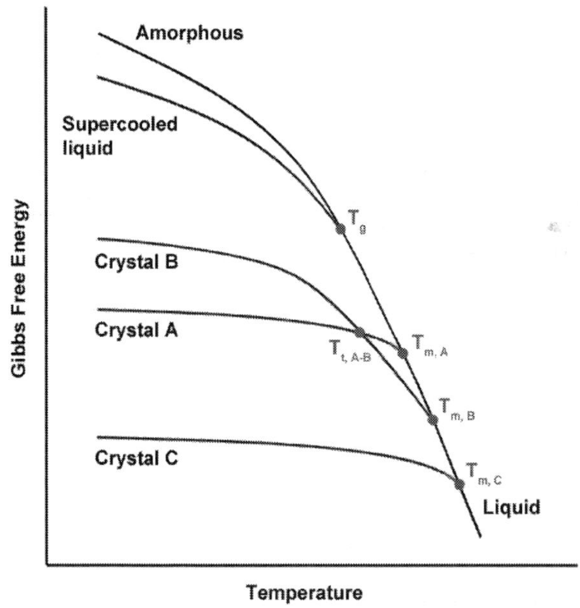

Fig. 8. Gibbs free energy curves for a hypothetical single-component system that exhibits crystalline and amorphous phase transitions. Monotropic systems (A and C, B and C), enantiotropic system (A and B) with a transition temperature T_t, and an amorphous and super-cooled liquid with a glass transition temperature T_g. Melting points, T_m, for the crystalline phases are shown at the intersection of curves for the crystalline and liquid states. (Adapted, with permission of the authors, from Shalaev & Zografi, 2002.)

3.1.1.1 Effects of pharmaceutical processing

Unintentional solid-state conversion of polymorphs sometimes occurs upon exposure to the energetics of pharmaceutical processing and a variety of phase conversions are possible when APIs are exposed to milling, wet granulation, oven drying, and compaction (Brittain & Fiese, 1999):

Grinding or milling is often the final step in the production of bulk APIs and it is performed to reduce particle size and improve particle size homogeneity. Milling can impart a significant amount of energy on a solid and could potentially lead to a full or partial polymorphic conversion or generation of an amorphous substance (or at least a degree of amorphous content). Amorphous forms, being metastable, can in turn reconvert to a crystalline state which may differ from the original product.

Wet granulation, used to improve powder flow and blend homogeneity, is a step that often precedes the production of tablets. During this process, the API is exposed to a solvent (water or an organic solvent with low toxicity, like ethanol) and is once more prone to undergoing solvent-mediated transformations. Exposure to humidity can create similar conditions which could lead to a hydrate being formed.

Drying of APIs is typically achieved with heat and moving air. The possible ramifications of temperature change on polymorph stability have been discussed in the previous section. Drying conditions need to be carefully controlled to avoid possible transformations.

Although not generally a common occurrence, compaction (during tablet manufacturing) can potentially cause metastable polymorphs to convert to the stable form. This can be attributed to the energy applied to bring about compaction. The extent of transformation is dependent on the zone of the tablet, the pressure applied, the compression temperature and the particle size of the API powder.

3.1.1.2 High-pressure polymorphic transitions

Solid-state pressure-induced structural changes in molecular crystals can cause polymorphic transitions (Boldyreva, 2003). However, unlike with high-pressure solvent recrystallization (please refer to section 2.1.5) which is experimentally and mechanistically similar, the conversion tends to be only partial and the data are poor. It is thought that even though the application of high pressure to larger organic molecules may thermodynamically favor the adoption of a new polymorphic form, the solid-state has a substantial kinetic barrier to overcome before the molecules are mobile enough to rearrange (Fabbiani *et al.*, 2004).

3.1.2 Desolvation of hydrates and solvates

With the exception of hydrates that are often used in pharmaceutical dosage forms, most solvates are the penultimate solid form in the production of APIs (Byrn *et al.*, 1999). Solvates may be desolvated by removing the recrystallization solvent and exposing the crystals to air at ambient temperature, leading to a decrease in vapor pressure of the solvent. If a solvate is stable under these conditions, it may be dried under vacuum or in an oven at mild temperatures.

Desolvation of a solvate generally results in one of the following forms (Byrn *et al.*, 1999):

- An unsolvated polymorph with a crystal structure different to that of the solvate.
- An unsolvated polymorph with a crystal structure that is the same as that of the solvate.
- An amorphous material that may or may not recrystallize.

Sometimes, a polymorphic transition may mimic the first of these scenarios. Nevirapine's metastable Form IV is isostructural to a series of solvates prepared from primary alcohols (Stieger *et al.*, 2010b) except that its structure contains no solvent molecules. When removed from its recrystallization medium, Form IV will spontaneously and rapidly convert to the stable Form I in exactly the same way as the aforementioned solvates (Stieger *et al.*, 2009).

3.1.3 Crystallization from amorphous phases

Amorphous pharmaceutical solids may be prepared by common pharmaceutical processes including melt quenching, freeze- and spray-drying, milling, wet granulation and desolvation of solvates (Yu, 2001). Glasses are most often produced from a melt of an API. If crystallization does not occur on cooling the melt below the melting point (T_m) of the crystalline phase, a supercooled liquid is obtained. Further cooling induces a change to a glassy state at the glass transition temperature (T_g), accompanied by a dramatic decrease in molecular mobility and heat capacity. In contrast to melting point, the glass transition temperature of a compound can fluctuate with operating conditions (including the rate of cooling/heating) and as a function of the history of the sample. This indicates that the glass transition is a thermal event affected by kinetic factors (Petit & Coquerel, 2006).

An amorphous or glass form is often an intermediate in the production of crystalline APIs and such processes can be useful for overcoming specific kinds of activation energies. Thermodynamically, amorphous solids are out-of-equilibrium states that contain an excess of stored energy, with reference to the crystalline phases (Figure 9), making them unstable by definition. The excess energy can be released either completely through crystallization or partially by means of irreversible relaxation processes (Petit & Coquerel, 2006).

Crystallization from amorphous solids, as for any crystallization, involves successive nucleation and growth steps. The crystallization rate may be affected by numerous, possibly interdependent, parameters notably including temperature and plasticizers (of which water is one). When the temperature of a supercooled liquid decreases from the melting point to the glass transition temperature, the nucleation rate increases exponentially whereas molecular mobility required for growth decreases exponentially. A maximum crystallization rate therefore occurs between preferred temperatures of nucleation and growth. Cooling rate also affects nucleation, with slow cooling allowing the maintenance of a steady-state nucleation rate. Rapid cooling, on the other hand, prevents full development of nucleation and thereby facilitates glass formation (Petit & Coquerel, 2006; Yu, 2001).

Amorphous solids offer certain pharmaceutical advantages over their crystalline counterparts. They are more soluble, have higher dissolution rates, and in some cases they may even have better compression characteristics than corresponding crystals. Therefore, we may not necessarily want amorphous APIs to recrystallize. Some pharmaceuticals have a natural tendency to exist as amorphous solids, while others require deliberate prevention of recrystallization to enter, and remain in, the amorphous state. Contemporary research into the stabilization of amorphous pharmaceuticals focuses on three key areas: (1) the use of additives for the stabilization of labile substances (e.g. proteins and peptides) during processing and storage; (2) prevention of crystallization of excipients that must remain amorphous to perform their intended functions; and (3) determining the appropriate storage conditions for optimal stability of amorphous materials (Yu, 2001).

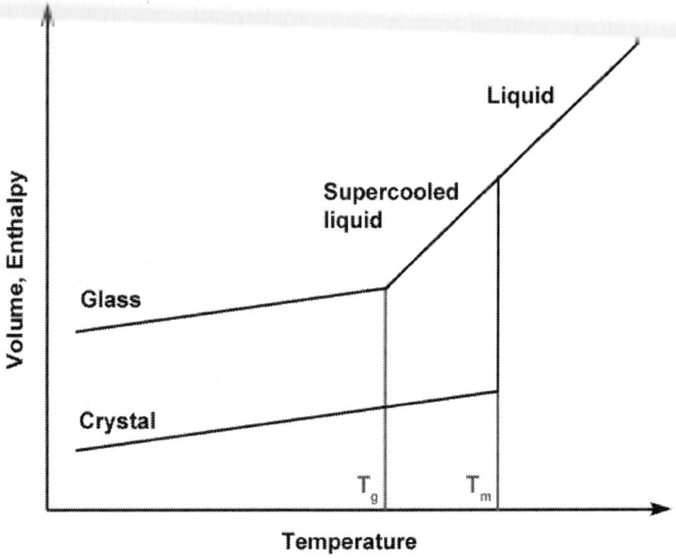

Fig. 9. Schematic representation of enthalpy or volume variations as a function of temperature for condensed materials. (Adapted, with permission of John Wiley and Sons, from Petit & Coquerel, 2006.)

3.2 Recrystallization through Physical Vapor Deposition

Physical vapor deposition (PVD) is an atomistic deposition process in which material is vaporized from a solid or liquid source in the form of atoms or molecules and transported in the form of vapor through a low pressure environment to the substrate where it condenses (Mattox, 1998). During the PVD process, molecules unpack from the original crystal lattice and then recrystallize in the new lattice (Byrn *et al.*, 1999).

The research into pharmaceutical applications of PVD has, to date, largely been limited to the production of amorphous phases - particularly ones with enhanced stability. Although we have long known that sublimation may be used for the production of polymorphs, comparatively few papers have been published on recrystallization *via* PVD. This is surprising when one considers that approximately two-thirds of organic compounds sublime (Guillory, 1999) and most commercially available APIs are crystalline.

Studies on the sublimation-based recrystallization of the following APIs and organic compounds have been published: anthranilic acid (Carter & Ward, 1994); 9,10-anthraquinone-2-carboxilic acid (Tsai *et al.*, 1993); 1,1-bis(4-hydroxyphenyl)cyclohexane

(Sarma *et al.*, 2006); caffeine (Griesser *et al.*, 1999; Carlton, 2011); carbamazepine (Griesser *et al.*, 1999; Zeitler *et al.*, 2007); 1,3-dimethyluracil (Sakiyama & Imamura, 1989); ibuprofen (Perlovich *et al.*, 2004;); malonamide (Sakiyama & Imamura, 1989); nevirapine (Figure 10) (Stieger & Liebenberg, 2009); stanozolol (Karpinska *et al.*, 2011); and theophylline (Fokkens *et al.*, 1983; Griesser *et al.*, 1999).

Fig. 10. Micro-spherical aggregates of nano-crystalline nevirapine prepared by PVD (A), the individual crystallites of which could be obtained through light grinding (B).

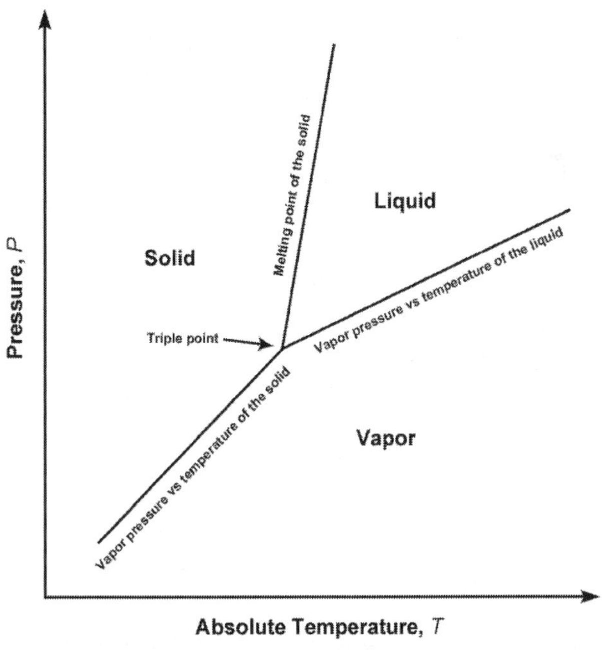

Fig. 11. Pressure-temperature diagram of a one-component crystalline solid for which only one solid phase exists. (Adapted, with permission of Dover Publications, from Ricci, 1966.)

Figure 11 demonstrates how transformations mediated by the vapor phase are highly dependent on vapor pressure and therefore also on temperature (Byrn *et al.*, 1999). Vapor of an API can be generated by its liquid phase once melted (evaporation) or by its solid phase (sublimation). The vast majority of pharmaceutical actives are organic compounds and, as such, they are sensitive to heat degradation. It is well-known that many APIs degrade when melted. It follows that pharmaceutical PVD processes should ideally operate at low pressure and temperature to obtain vapor through sublimation, using compounds with sufficient thermal stability.

Figure 12 illustrates how it is possible to obtain different enantiotropic polymorphs through vapor deposition. At a pressure sufficiently low for sublimation to take place the solid API becomes a vapor, at which point it is no longer Solid A or B because polymorphism is a solid-state phenomenon. The temperature at which the vapor phase recrystallizes determines which polymorph is obtained (provided the pressure is sufficiently low). It can generally be assumed that unstable/metastable polymorphs (in this example, Solid A) will form preferentially at lower temperatures and stable polymorphs (Solid B) can be expected at higher temperatures (Guillory, 1999). This is why the temperature, and distance from the solid, of the collection surface is so important in PVD systems.

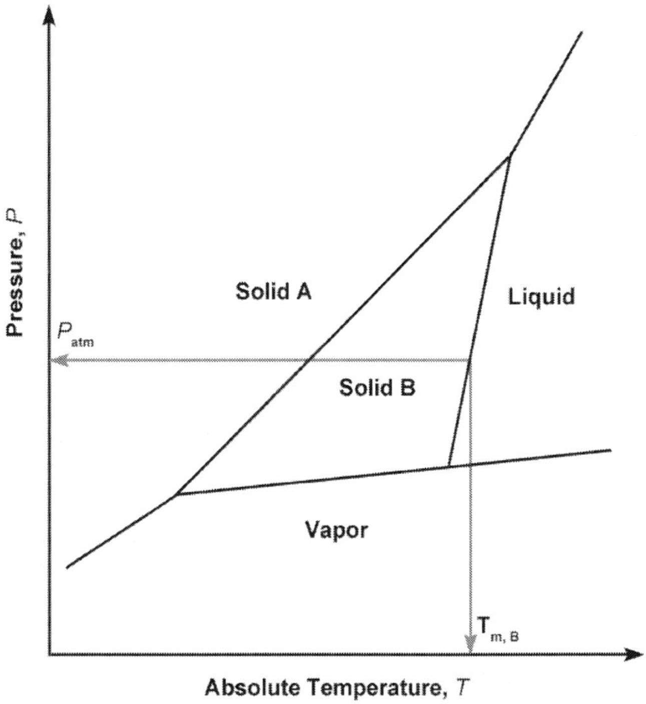

Fig. 12. Pressure-temperature diagram of a one-component crystalline solid with enantiotropic behavior. (Adapted, with permission of John Wiley and Sons, from Lohani & Grant, 2006.)

It has, incidentally, been found that recrystallization from the vapor phase can also be affected or controlled by the nature of the substrate (Carter & Ward, 1994) as has been shown for solvent recrystallization (Rodríguez-Spong *et al.*, 2004).

4. Conclusion

Polymorphs of active pharmaceutical ingredients can be obtained though recrystallization processes based on solvents, solid-state transitions and vapor deposition. When the possible number of experimental variations on each of these is considered, it is easy to become overwhelmed by the task of comprehensive polymorph screening. Indeed, it is no wonder that some polymorphs go undetected for decades or longer. Various semi-automated and high-throughput methods have been developed to assist in this daunting task. The efforts of researchers laboring towards automation are, however "subverted" by colleagues who continually find novel ways of accessing hidden polymorphs! There is, to date, no single method that can identify all possible solid forms of an API.

5. Acknowledgement

We thank North-West University (Potchefstroom Campus) and the National Research Foundation of South Africa for providing research support.

6. References

Banga, S., Chawla, G. & Bansal, A.K. 2004. New Trends in the Crystallisation of Active Pharmaceutical Ingredients. *Business Briefing: Pharmagenerics*, p. 1-5.

Beckman, W. 2000. Seeding the Desired Polymorph: Background, Possibilities, Limitations, and Case Studies. *Organic Process Research & Development*, 4:372-383.

Bernstein, J. 2002. *Polymorphism in Molecular Crystals*. Oxford:Clarendon Press. 410 p.

Blagden, N., De Matas, M., Gavan, P.T. & York, P. 2007. Crystal Engineering of Active Pharmaceutical Ingredients to Improve Solubility and Dissolution Rates. *Advanced Drug Delivery Reviews*, 59(7):617-630.

Boldyreva, E.V. 2003. High-Pressure-Induced Structural Changes in Molecular Crystals Preserving the Space Group Symmetry: Anisotropic Distortion/Isosymmetric Polymorphism. *Crystal Engineering*, 6(4):235-254.

Boistelle, R. & Astier, J.P. 1988. Crystallization Mechanisms in Solution. *Journal of Crystal Growth*, 90(1-3):14-30.

Braatz, R.D. 2002. Advanced Control of Crystallization Processes. *Annual Reviews in Control*, 26(1):87-99.

Brittain, H.G. & Fiese, E.F. 1999. *Effects of Pharmaceutical Processing on Drug Polymorphs and Solvates*. (*In* Brittain, H.G. *Ed*. Polymorphism in Pharmaceutical Solids. Volume 95: Drugs and the Pharmaceutical Sciences. New York: Marcel Dekker.) p. 331-361.

Brittain, H.G. & Grant, D.J.W. 1999. *Effects of Polymorphism and Solid-State Solvation on Solubility and Dissolution Rate*. (*In* Brittain, H.G. *Ed*. Polymorphism in Pharmaceutical Solids. Volume 95: Drugs and the Pharmaceutical Sciences. New York: Marcel Dekker.) p. 279-330.

Byrn, S.R., Pfeiffer, R.R. & Stowell, J.G. 1999. *Solid-State Chemistry of Drugs*. 2nd edition. West Lafayette: SSCI. 574 p.

Capes, J.S. & Cameron, R.E. 2007. Effect of Polymer Addition on the Contact Line Crystallization of Paracetamol. *CrystEngComm*, 9:84-90.

Carlton, R.A. 2011. *Pharmaceutical Microscopy*. New York: Springer. 322 p.

Carter, P.W. & Ward, M.D. 1994. Directing Polymorph Selectivity During Nucleation of Anthranilic Acid on Molecular Substrates. *Journal of the American Chemical Society*, 116(2):769-770.

Carstensen, J.T., Ertell, C. & Geoffrey, J. 1993. Physico-Chemical Properties of Particulate Matter. *Drug Development and Industrial Pharmacy*, 19(1-2):195-219.

Doherty, C. & York, P. 1988. Frusemide Crystal Forms: Solid State and Physicochemical Analyses. *International Journal of Pharmaceutics*, 47(1-3):141-155.

Douillet, J., Stevenson, N., Lee, M., Mallet, F., Ward, R., Aspin, P., Dennehy, D.R. & Camus, L. 2011. Development of a Solvate as an Active Pharmaceutical Ingredient: Developability, Crystallisation and Isolation Challenges. *Journal of Crystal Growth*, article in press.

Fabbiani, F.P.A., Allan, D.R., David, W.I.F., Moggach, S.A., Parsons, S. & Pulham, C.R. 2004. High-Pressure Recrystallization - A Route to New Polymorphs and Solvates. *CrystEngComm*, 6(82):504-511.

Florence, A.T. & Attwood, D. 2006. *Physicochemical Principles of Pharmacy*. 4th edition. London: Pharmaceutical Press. 492 p.

Fokkens, J.G., Van Amelsfoort, J.G.M., De Blaey, C.J., De Kruif, C.G. & Wilting, J. 1983. A Thermodynamic Study of the Solubility of Theophylline and its Hydrate. *International Journal of Pharmaceutics*, 14(1):79-93.

Fujiwara, M., Nagy, Z.K., Chew, J.W. & Braatz, R.D. 2005. First-Principles and Direct Design Approaches for the Control of Pharmaceutical Crystallization. *Journal of Process Control*, 15(5):493-504.

Garetz, B.A., Aber, J.E., Goddard, N.L., Young, R.G. & Myerson, A.S. 1996. Nonphotochemical, Polarization-Dependent, Laser-Induced Nucleation in Supersaturated Aqueous Urea Solutions. *Physical Review Letters*, 77:3475-3476.

Grant, D.J.W. 1999. *Theory and Origin of Polymorphism*. (*In* Brittain, H.G. *Ed*. Polymorphism in Pharmaceutical Solids. Volume 95: Drugs and the Pharmaceutical Sciences. New York: Marcel Dekker.) p. 1-33.

Griesser, U.J. 2006. *The Importance of Solvates*. (*In* Hilfiker, R. *Ed*. Polymorphism in the Pharmaceutical Industry. Germany: Wiley-VCH.) p. 211-233.

Griesser, U.J., Szelagiewicz, M., Hofmeier, U.C., Pitt, C. & Cianferani, S. 1999. Vapor Pressure and Heat of Sublimation of Crystal Polymorphs. *Journal of Thermal Analysis and Calorimetry*, 57(1):45-60.

Guillory, J.K. 1999. *Generation of Polymorphs, Hydrates, Solvates and Amorphous Solids*. (*In* Brittain, H.G. *Ed*. Polymorphism in Pharmaceutical Solids. Volume 95: Drugs and the Pharmaceutical Sciences. New York: Marcel Dekker.) p. 183-225.

Haleblian, J.K. 1975. Characterization of Habits and Crystalline Modifications of Solids and their Pharmaceutical Applications. *Journal of Pharmaceutical Sciences*, 64:1269-1288.

Karpinska, J., Erxleben, A. & McArdle, P. 2011. 17β-Hydroxy-17α-methylandrostanol[3,2-c]pyrazole, Stanozolol: The Crystal Structures of Polymorphs 1 and 2 and 10 Solvates. Crystal Growth & Design, 11(7):2829-2838.

Kim, S., Wei, C. & Kiang, S. 2003. Crystallization Process Development of an Active Pharmaceutical Ingredient and Particle Engineering *via* the Use of Ultrasonics and Temperature Cycling. *Organic Process Research Development*, 7(6):997-1001.

Karpinski, P.H. & Wey, J.S. 2001. *Precipitation Processes.* (*In* Myerson, A.S. *Ed.* Handbook of Industrial Crystallization. 2nd edition. Woburn: Butterworth-Heinemann.) p. 141-159.

Kondepuddi, D.K., Kaufmann, R.J. & Singh, N. 1990. Chiral Symmetry Breaking in Sodium Chlorate Crystallization. *Science*, 250(4983):975-976.

Kondepuddi, D.K. & Sabanayagam, C. 1994. Secondary Nucleation that Leads to Chiral Symmetry Breaking in Stirred Crystallization. *Chemical Physics Letters*, 217(4):364-368.

Kuzmenko, I., Rapaport, H., Kjaer, K., Als-Nielsen, J., Weissbuch, I., Lahav, M. & Leiserowitz, L. 2001. Design and Characterization of Crystalline Thin Film Architectures at the Air-Liquid Interface: Simplicity to Complexity. *Chemical Reviews*, 101(6):1659-1696.

Lipinski, C.A., Lombardo, F., Dominy, B.W. & Feeney, P.J. 1997. Experimental and Computational Approaches to Estimate Solubility and Permeability in Drug Discovery and Development Settings. *Advanced Drug Delivery Reviews*, 46:3-26.

Lohani, S. & Grant, J.W.G. 2006. *Thermodynamics of Polymorphs.* (*In* Hilfiker, R. *Ed.* Polymorphism in the Pharmaceutical Industry. Germany: Wiley-VCH.) p. 21-42.

Manek, R.V. & Kolling, W.M. 2004. Influence of Moisture on the Crystal Forms of Niclosamide Obtained from Acetone and Ethyl Acetate. *AAPS PharmSciTech*, 5(1):101-108.

Mattox, D.M. 2010. *Handbook of Physical Vapor Deposition (PVD) Processing.* 2nd edition. Oxford: William Andrew. 792 p.

McBride, J.M. & Carter, R.L. 1991. Spontaneous Resolution by Stirred Crystallization. *Angewandte Chemie*, 30(3):293-295.

Mullin, J.W. 2001. *Crystallization.* 4th edition. Oxford: Butterworth-Heinemann. p. 322.

Ostwald, F.W. 1897. Studien uber die Bildung und Umwandlung fester Korper. *Zeitschrift für Physikalische Chemie*, 22:289.

Nonoyama, N., Hanaki, K. & Yabuki, Y. 2006. Constant Supersaturation Control of Antisolvent-Addition Batch Crystallization. *Organic Process Research & Development*, 10(4):727-732.

Perlovich, G.L., Kurkov, S.V., Hansen, L.K. & Bauer-Brandl, A. 2004. Thermodynamics of Sublimation, Crystal Lattice Energies, and Crystal Structures of Racemates and Enantiomers: (+)- and (-)-Ibuprofen. *Journal of Pharmaceutical Sciences*, 93(3):654-666.

Petit, S. & Coquerel, G. 2006. *The Amorphous State.* (*In* Hilfiker, R. *Ed.* Polymorphism in the Pharmaceutical Industry. Germany: Wiley-VCH.) p. 259-285.

Ricci, J.E. 1966. *The Phase Rule and Heterogeneous Equilibrium.* New York: Dover Publications. 505 p.

Rodríguez-Homedo, N., Kelly, R.C., Sinclair, B.D. & Miller, J.M. 2006. *Crystallization: General Principles and Significance on Product Development.* (*In* Swarbrick, J. *Ed.* Encyclopedia of Pharmaceutical Technology. 3rd edition. New York: Informa Healthcare.) p. 834-857.

Rodríguez-Spong, B., Price, C.P., Jayasankar, A., Matzger, A.J. & Rodríguez-Homedo, N. 2004. General Principles of Pharmaceutical Solid Polymorphism: A Supramolecular Perspective. *Advanced Drug Delivery Reviews*, 56:241-274.

Rouhi, A.M. 2003. The Right Stuff. *Chemical & Engineering News*, 81(8):32-35.

Sakiyama, M. & Imamura, A. 1989. Thermoanalytical Characterization of 1,3-Dimethyluracil and Malonamide Crystals. *Thermochimica Acta*, 142(2):365-370.

Sarma, B., Roy, S. & Nangia, A. 2006. Polymorphs of 1,1-Bis(4-hydroxyphenyl)cyclohexane and Multiple Z′ Crystal Structures by Melt and Sublimation Crystallization. *Chemical Communications*, 47:4918-4920.

Shalaev, E. & Zografi, G. 2002. *The Concept of 'Structure' in Amorphous Solids from the Perspective of the Pharmaceutical Sciences*. (*In* Levine, H. *Ed.* Amorphous Food and Pharmaceutical Systems. Cambridge: The Royal Society of Chemistry.) p. 11-30.

Shekunov, B.Y. & York, P. 2000. Crystallization Processes in Pharmaceutical Technology and Drug Delivery Design. *Journal of Crystal Growth*, 211(1-4):122-136.

Shinde, K.R., Patil, S.D. & Dhake, A.S. 2011. Recrystallization. *Indian Streams Research Journal*, 1(4).

Skonieczny, S. 2009. Crystallization. *University of Toronto* [illegible] *CHM210*, p. 1-18.

Stieger, N., Caira, M.R., Liebenberg, W., Tiedt, L.R., Wessels, J.C. & De Villiers, M.M. 2010a. Influence of the Composition of Water/Ethanol Mixtures on the Solubility and Recrystallization of Nevirapine. *Crystal Growth & Design*, 10(9):3859-3868.

Stieger, N. & Liebenberg, W. 2009. Method for Increasing the Solubility of a Transcriptase Inhibitor Composition. *Patent application PCT/IB2010/055077* (priority date 2009-11-10).

Stieger, N., Liebenberg, W. & Caira, M.R. 2009. Method of Producing a Polymorph Form. *Patent application PCT/IB2010/055808* (priority date 2009-12-17).

Stieger, N., Liebenberg, W., Wessels, J.C., Samsodien, H. & Caira, M.R. 2010b. Channel Inclusion of Primary Alcohols in Isostructural Solvates of the Antiretroviral Nevirapine: an X-Ray an Thermal Analysis Study. *Structural Chemistry*, 21(4):771-777.

Stoica, C., Verwer, P., Meekes, H., Van Hoof, P.J.C.M., Kaspersen, F.M. & Vlieg, E. 2004. Understanding the Effect of a Solvent on the Crystal Habit. *Crystal Growth & Design*, 4(4):765-768.

Threlfall, T. 2000. Crystallisation of Polymorphs: Thermodynamic Insight into the Role of Solvent. *Organic Process Research & Development*, 4(5):384-390.

Tiwary, A.K. 2006. *Crystal Habit Changes and Dosage Form Performance*. (*In* Swarbrick, J. *Ed.* Encyclopedia of Pharmaceutical Technology. 3rd edition. New York: Informa Healthcare.) p. 822.

Togkalidou, T., Tung, H., Sun, Y., Andrews, A. & Braatz, R.D. 2002. Solution Concentration Prediction for Pharmaceutical Crystallization Processes Using Robust Chemometrics and ATR FTIR Spectroscopy. *Organic Process Research Development*, 6(3):317-322.

Tros de Ilarduya, M.C., Martin, C., Goñi, M.M. & Martinez-Uhárriz, M.C. 1997. Dissolution Rates of Polymorphs and Two New Pseudopolymorphs of Sulindac. *Drug Development and Industrial Pharmacy*, 23:1095-1098.

Tsai, S., Kuo, S. & Lin, S. 1993. Physicochemical Characterization of 9,10-anthraquinone-2-carboxilic acid. *Journal of Pharmaceutical Sciences*, 82(12):1250-1254.

Tung, H., Paul, E.L., Midler, M. & McCauly, J.A. 2009. *Crystallization of Organic Compounds: An Industrial Perspective*. New Jersey: John Wiley & Sons. 289 p.

Yu, L. 2001. Amorphous Pharmaceutical Solids: Preparation, Characterization and Stabilization. *Advanced Drug Delivery Reviews*, 48(1):27-42.

Zaccaro, J., Matic, J., Myerson, A.S. & Garetz, B.A. 2001. Nonphotochemical, Laser-Induced Nucleation of Supersaturated Aqueous Glycine Produces Unexpected γ-Polymorph. *Crystal Growth & Design*, 1(1):5-8.

Zeitler, J.A., Taday, P.F., Gordon, K.C., Pepper, M. & Rades, T. 2007. New Insights into the Solid-State Transition Mechanism in Carbamazepine Polymorphs by Time-Resolved Terahertz Spectroscopy. *ChemPhysChem*, 8(13):1924-1927.

Zhou, G.X., Fujiwara, M., Woo, X.Y., Rusli, E., Tung, H., Starbuck, C., Davidson, O., Ge, Z. & Braatz, R.D. 2006. Direct Design of Pharmaceutical Antisolvent Crystallization through Concentration Control. *Crystal Growth & Design*, 6(4):892-898.

Section 2

Applications, Techniques and Mineral Formation

8

Preparation of Selected Ceramic Compounds by Controlled Crystallization Under Hydrothermal Conditions

Juan Carlos Rendón-Angeles[1], Zully Matamoros-Veloza[2] and Kazumichi Yanagisawa[3]

[1]*Research Institute for Advanced Studies of the NPI Saltillo Campus*
[2]*Technological Institute of Saltillo*
[3]*Research Laboratory for Hydrothermal Chemistry, Kochi University*
[1,2]*México*
[3]*Japan*

1. Introduction

The main aim of this chapter is to review some particular aspects related with the hydrothermal crystallization process for preparing some selected oxide and non-oxide powders, namely functional compound with electric, piezoelectric, ionic conducting and catalytic properties. The literature on the hydrothermal crystallization of oxide and non-oxide powders is vast; the references cited here are those more appropriated to illustrate this review. A particular attempt is made to broaden the traditional concepts of processing perovskite powders with controlled chemical composition and particle morphology, those aspects will be discussed based on the chemical reactivity of the precursor reactants (gels). Furthermore, an additional approach that takes into account the solubility of the solid species (mineral reagents) that are employed as a precursor in the hydrothermal systems on controlling the crystallization process of oxide particles was further investigated for preparing titanates oxide ceramic. The specific reaction pathways and kinetic aspects are discussed and illustrated by experimental setups for the solution of selected problems in hydrothermal crystallization. Also the chapter includes recent work on the formation of inorganic salts of Sr or Ba, under ordinary hydrothermal treatments via the anionic replacement in sulphate minerals, because these particular reactions have promoted peculiar microstructure on the crystallized material that preserves the bulk geometrical features of particular mineral specie. Therefore, this method could be attractive for preparing net-shaped materials with controlled porosity, with optimized functional properties, because can be used as gas sensor, substrates for porous catalytic materials, filters, among other potential applications.

2. Hydrothermal crystallization as technique for oxide and non-oxide synthesis

2.1 Brief history of the hydrothermal technique development

The studies recorded at the scientific annals indicate that the pioneering research on hydrothermal systems was initiated in the middle of the 19th century (Schafthaul, 1845, as cited in Suchanek et. al., 2004). The term of hydrothermal is purely of geological origin; this was firstly used by the geologist Sir Roderick Murchinson (1792–1871), whom analysed the reactivity of water at elevated pressure and temperature, to explain the mineral formation of several rocks and minerals (Byrappa & Yoshimura, 2001). Continuous progress in material synthesis was initially accelerated by notorious developments in hydrothermal pressure vessels apparatus. Hence, at the first decade of the 20th century, geologist achieved a marked technological expansion on the hydrothermal research field, namely efforts were directed towards for designing the first metal vessels and carried out preliminary experiments at laboratory scale related to material synthesis area, mainly in the field of inorganic single crystal growth (Riman et. al., 2002; Suchanek et. al., 2004). However, the development of chemical reactions via hydrothermal processing was further limited because of the severe treatment (supercritical) conditions, which were normally required for single crystal growth. This resulted on discourage extensive research and commercialization of various materials. Hydrothermal epitaxial growth is one example, it was popular during the 1970s, but it did not reach commercial success due to the high temperatures (> 500 °C) and pressures (> 100 MPa), which were involved to achieve the epitaxial crystal growth process. In the middle of the 1980s, the commercial interest in the hydrothermal technology was revived, because the steadily increasing of a large group of materials, manly ceramic powder, had emerged that can be produced under more environmentally friendly conditions (T < 350 °C, P < 100 MPa). At present, the major developments on the hydrothermal synthesis technology including in particular the hydrothermal crystallization has been accomplished in several countries around the world like: China, Japan, USA, UK, Germany and some others with minor contributions (Byrappa & Yoshimura, 2001).

2.2 Definition of hydrothermal synthesis

The term *"hydrothermal"* is difficult to define, based in the etymologic root of the Greek word, *"hydrous"* means water and *"thermal"* means heat. One of the accepted statements for hydrothermal defines it as any heterogeneous chemical reaction that occurs in the presence of a solvent media at above the room temperature (> 25 °C) and pressure levels greater than 0.1 MPa in a closed system, at these conditions does not matter whether the solvent is aqueous or non-aqueous. Hitherto, there is still a bit of confusion regarding the correct use of the term hydrothermal, because in the case of the chemistry field, the chemists prefer to use a different term, namely *solvothermal*, which means any chemical reaction conducted with a non-aqueous solvent or solvent in supercritical conditions. Additionally, another similar related terms widely used amongst the physicist, chemist and material scientist communities are: *glycothermal, alcothermal, ammonothermal,* and so on. However, some researchers also use hydrothermal for describing processes conducted at ambient conditions. The *crystallization* process of solid phases under hydrothermal conditions is usually conducted at autogeneous pressure, achieving a particular saturated vapour pressure of the solution at the specified temperature and composition of the hydrothermal solution. In this concern, in terms of industrial commercial processes mild operating conditions are preferred, for example treatment temperatures below than 350 °C and pressures > 50 MPa (Byrappa, 2005). The limit that indicates the transition from mild to severe conditions during a hydrothermal treatment is normally determined by the strength of the inner vessel materials, which at severe treatment conditions undergoes into a corrosion process. The continuous research in this field has led the way to a better understanding for controlling chemical reactions in various hydrothermal medias, which has remarkably reduced the processing parameters such as: reaction time, temperature, and pressure for hydrothermal crystallization of several oxide and non-oxide materials (T<200 °C, P<1.5 MPa). At present, the recent scientific and technological achievements have made hydrothermal synthesis more economical, because powder preparation can be carried out by a single step cost-effective process, in advanced pressure reactor technology coupled with processing methodologies proposed for a wide number of inorganic compounds (Suchanek et al., 2004).

2.3 Hydrothermal processes for synthesis of inorganic compounds

2.3.1 Hydrothermal single crystal growth

Regardless of the numerous investigations on hydrothermal single crystal growth, quartz crystal is one of the materials extensively investigated up to now. Nowadays, the electronic industry requires the use of pure large single crystals of synthetic origin, because natural quartz crystals are generally irregular in shape, and is difficult to obtain large-scale single crystals wafers by automatic cutting. The crystal growth of oxide species, namely α-quartz; is conducted by the conventional hydrothermal temperature gradient method, thus, the autoclaves frequently used consist of two chambers were different process occurs. One of the important parameter to consider for the single crystal growth of α–quartz is the selection of the proper nutrient material; among the most employed are small particle size α–quartz, silica glass, high quality silica sand, or silica gel (Byrappa 2005). The nutrient reactant is placed at the liner chamber (vessel bottom, see Figure 1) made up of iron, silver or titanium that is less employed, with a suitable baffle and a frame (at the top of the vessel) holding the seed of the material that is been grown. The mineralizer solution is other factor to select coupled with a definite molarity; this is poured into the liner to make the required filling volume and achieves the dissolution of the nutrient. Under these conditions, matter transport proceeds from the nutrient chamber and the crystallization and growth of the crystals is achieved due to the temperature difference at the top-seeded part of the autoclave. The particular optimum conditions determined at Bell Laboratories for growing quartz single crystals are dissolution temperature of 425 °C at a pressure in the range of 100–175 MPa, the crystallization and simultaneous growth proceeds at 375 °C whilst the temperature gradient from the nutrient to the growing chamber was of 50 °C. The mineralizer employed was an alkaline solution of NaOH with concentration varying between 0.5 up to 1 M, and the volume of poured mineralizer solution varied in the range of 78–85 % of the total volume of the vessel (Laudies, 1970; as cited in Byrappa, 2005).

The crystallization and growth of other single crystals rather than SiO_2 was firstly investigated for oxide species, namely TiO_2, ZrO_2, HfO_2 and some related perovskite oxides $PbTiO_3$ and $PbZrO_3$. Experimental results evidenced that the mineralizer employed for dissolving and transport the nutrient reactant markedly affected the crystallization of the oxides and the crystal growth. Thus, the single oxide species were found to dissolve and recrystallize faster in fluoride solutions (NaF, KF and NH_4F) in comparison with alkalis, KOH and K_2CO_3, because of the high chemical reactivity of these nutrient oxides in fluoride solutions. However, the complete dissolution and transport of the nutrient was preferentially achieved in NH_4F solutions rather than NaF or KF mineralizers, even at a low temperature of 470 °C for a reaction interval up to 6 days (Kuznetzov, 1968), and the TiO_2, ZrO_2, HfO_2 single crystal growth was achieved at a temperature of 520 °C with a positive temperature gradient

(30 °C). This is a popular method that promotes the crystal growth and has been widely used on the preparation of berlinite ($AlPO_4$) crystals (Byrappa & Yoshimura 2001).

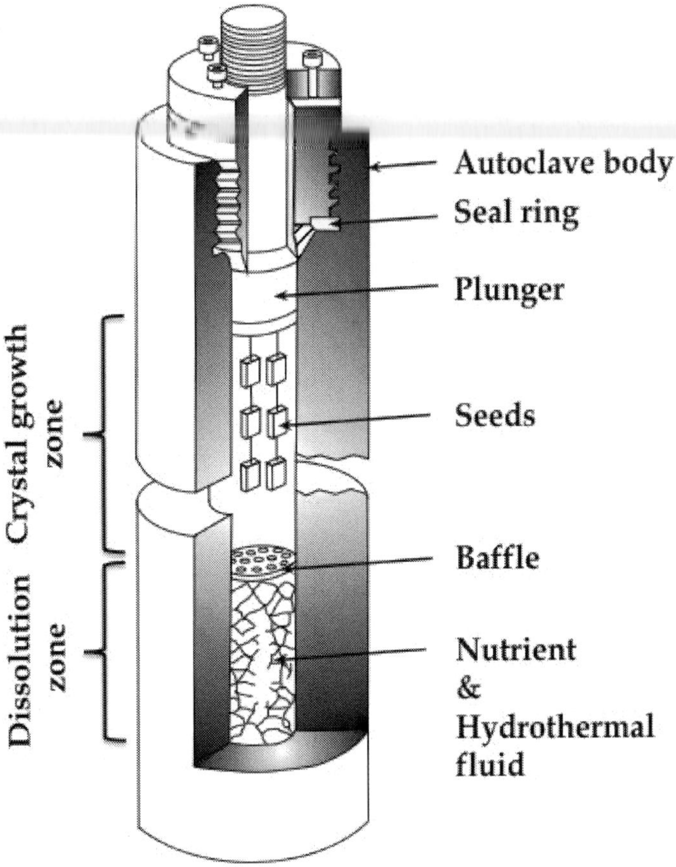

Fig. 1. Scheme of the conventional autoclave employed for single crystal growth under hydrothermal conditions (Schubert, 2000; as cited in Byrappa & Yoshimura, 2001).

A particular emphasize on the hydrothermal crystal growth research has been focused for the crystallization of materials that melts incongruently because such materials cannot be grown with compositional and phase uniformity. The relevant examples of compounds that melts incongruently are those belonging to lead family oxides titanate ($PbTiO_3$), zirconate ($PbZrO_3$), single crystals with millimetre size of both compounds were found to grown satisfactorily in NH_4F hydrothermal solution by seedless grown crystallizing conditions at temperatures between 500-600 °C for several days (Kuznetzov, 1968). However, the hydrothermal crystal grown of some lead related solid solutions, $PbZr_{1-x}Ti_xO_3$ (PZT), $PbMg_{1/3}Nb_{2/3}O_3$ (PMN) and $PbSc_{0.5}Nb_{0.5}O_3$ (PSN), is limited due to the low chemical stability of the perovskite structure even in KF solution, and results on the grown of small single crystals with pyrochlore structure $Pb_{1.83}Mg_{0.29}Nb_{1.71}O_{6.39}$ and $Pb_{1.83}Sc_{0.29}Nb_{1.71}O_{6.39}$, as shown in Figure 2 (Yanagisawa et. al., 1999, 2000). This incongruence has not been determined thermodynamically and experimentally for the PZT, because this compound has been used to grown thin and thick single crystals on different seeds ($SrTiO_3$) with excellent compositional control (Oledzka et. al., 2003). In general, the crystallization and crystal growth of a specific compound can be carried out using different mineralizer or solvents, e.g., water, soluble salts, acid solutions, non aqueous solutions; but some physicochemical factors must be taking into account to fulfil suitable conditions that facilitate both processes.

Fig. 2. Pyrochlore single crystals, of (a) PMN, (b) PSNT and (c) PSN; grown at the top of platinum capsule at 600 °C with a gradient temperature gradient of 40 °C in KF solutions 5, 4.2 and 6.2 M for reaction intervals of 3 (a) and 5 days (c,d), respectively (Yanagisawa et. al., 1999).

2.3.2 Advanced hydrothermal processing methods

Remarkable achievements for developing the *Hydrothermal Microwave Assisted Synthesis* process have been conducted at Pennsylvania State University (Komarneni et. al., 1992); this method enhances solid crystallization kinetics 1-4 order of magnitude faster than that occurring on the conventional hydrothermal processing for a wide solutions. In addition, other advantages of the hydrothermal microwave assisted technique are very high heating rates and the synthesis of novel phases. A great variety of ceramic powders with particular morphologies and controlled particle size, have been produced, e.g. TiO_2, ZrO_2, Fe_2O_3, $BaTiO_3$, $Ca_{10}(PO_4)_6(OH)_2$, etc (Komarneni et. al., 1992; Roy, 1994; Komarneni et. al., 2002).

Another variation implemented to the hydrothermal conventional process resulted on the development of a different hybrid method denominated *Hydrothermal-Electrochemical Synthesis*. This technique was tailored to deposit polycrystalline oxide films on reactive metal substrates. This technique becomes very important when the crystallization of oxide products from supersaturated hydrothermal solutions is hinder in absence of an applied electrical potential. Nowadays, highly crystallized ceramic thin films, such as a $BaTiO_3$, $SrTiO_3$, $LiNiO_2$, $PbTiO_3$, $CaWO_4$ and $BaMoO_4$ can be deposited on metal substrates from aqueous solutions at relatively low temperature 50–200 °C for several hours under a continuous applied voltage charge (Yoshimura & Suchanek, 1997). Additionally, semiconductor thin layers of GaAs, CdTe, CdSe and CdS have been successfully prepared by electrochemical atomic layer epitaxy growth, this mechanism is analogous to molecular epitaxy, however, the crystallization and growth of the layer is enhanced from a saturated hydrothermal media instead of a vapour phase that transport the growing species (Colletti et. al., 1998; as cited in Suchanek et. al., 2004).

Among other tailored techniques, the high energy milling technique has been challenged to hybridize the conventional hydrothermal processing, resulting in a coupled process that involves the classical powder mechanochemical and hydrothermal syntheses. The *Mechanochemical-Hydrothermal* route utilizes the solvency of an aqueous solution, which facilities the crystallization of solid species due to the pressure environment generated inside the mechano-chemical autoclave; it mainly accelerates the rate-determining steps of those factors, for instance: interfacial reaction, solute dissolution or dehydroxylation; that limit hydrothermal chemical reactions at low temperature. These reactions enhance the crystallization to occur locally at the particle surface because of the perturbation of superficial bonded species in the solid coupled with high temperature gradients (400–700 °C) and pressure localized zones that are generated during the mechanical activation of slurries. This is mainly promoted by the friction and adiabatic heating of gas bubbles while maintaining the average temperature close to room temperature (Kosova et. al., 1997; as cited in Suchanek et. al., 2004). Hitherto, this technique has been used for preparing bioceramic materials such as hydroxyapatite (Suchanek et. al., 2002; Chen et. al., 2004), another tailored material like $PbTiO_3$ was also crystallized by this technique.

Since Rustum Roy reported that the use of ultrasonic devices are feasible for improving low temperature inorganic syntheses because reaction kinetics is two orders of magnitude faster than that for the conventional wet chemistry synthesis methods (Roy, 1994). Some attempts to adapt ultrasonic devices emitting acoustic signals of 20 kHz up to 10 MHz have been conducted, because the acoustic signal produce very sharp temperature gradients with localized peak temperature zones, speculated as high as 5000 K, as well as, localized pressure zones of up to 100's MPa. The sonochemical environment also alters the molecular chemistry (chemical bond scission, generate excited states and accelerate electron transfer steps in chemical reactions), and enhances mass transport and crystallization kinetics due to the high convection of the fluid. (Peters, 1996, as cited in Riman et. al., 2002). *Hydrothermal-sonochemical synthesis* method have been used for preparing various ceramic powders

($Ca_{10}(PO_4)_6(OH)_2$, $AlPO_4$, InSb, CdS) and thin films ($Li_2B_4O_7$, $Ba_2TiSi_2O_8$).while the average temperature of the reactors is maintained close to room temperature.

The advanced hydrothermal techniques discussed in this section, even though wet chemical synthesis is offer in conjunction with so many significant advantages over the conventional method, might be able to have wide application in the industry. Compared to solid-state materials processing, these technologies might be more facile for scaling up. This situation, however, still cannot guarantee successful application of all these technologies to industries. Hence, much effort is needed to obtain a more comprehensive understanding of several physicochemical phenomena related to each of the new technologies hybridizing the conventional hydrothermal synthesis, in order to establish a relationship between science and technology that could lead to optimize these methods for their employment at a industrial scale (Shi & Hwang, 2003).

2.4 Conventional hydrothermal crystallization process for advanced ceramic materials

2.4.1 Factors that affect optimizing hydrothermal crystallization experiments

In general, a majority of the hydrothermal synthesis research work that has been done in the past was based on Edisonian trial and error design for process development, but this is not the best experimental approach for discerning between processes that are controlled by either thermodynamics or kinetics. In contrast, much effort has been paid to use thermodynamic modelling for processing design; this approach is based on fundamental physicochemical principles instead of the Edisonian methods (Riman et. al., 2002). Therefore, a great number of hydrothermal fundamental works for some particular solid-aqueous systems have provided sufficient experimental hydrothermal physical chemistry data. An important point derived from these works is related with the behaviour of the solvent under hydrothermal conditions, because it has a relationship with aspects like structure at critical, supercritical and sub-critical conditions, solution dielectric constant, pH variation, viscosity, expansion coefficient, density, etc., all these parameters depend markedly on thermodynamic variables such as pressure and temperature. At present, hydrothermal crystallization process is the only one where a fundamental understanding of kinetics is lacking due to the absence of physicochemical data of the intermediate phases forming in specific aqueous solutions. Although, fundamental research works related to synthesis of specific compounds demonstrated the importance of crystallization kinetics, however, a better understanding of crystallization kinetics still in an early stage of development. In this case, due to the absence of predictive methodology models, ones must estimated on terms of chemical equilibrium of the reaction the effect of temperature, pressure, precursor, and time to achieve solid crystallization and improve the reaction kinetics. Insight into this would enable us to understand how to control the formation of ionic species in the solution, the crystallization of solid phases and the rate of their growth.

In the last decade, much effort has directed towards for developing thermochemical models based in fundamental knowledge of thermodynamic and the Hegelson-Kirkham-Flowers-Tanger equation of state, which allows to represent standard-state properties across substantial temperature and pressure ranges in order to estimate chemical reactions path ways under hydrothermal conditions (Lencka & Riman, 2002). Modelling can be successfully applied to very complex aqueous electrolytes over specific ranges of temperature and reactant and solvent concentration and non-aqueous systems can also be model as well. Practical computer software for conducting thermochemical modelling studies was recently developed by OLI System Inc., USA. This software can be used to conduct studies for chemical reactions for hydrothermal systems within the temperature range of -50 to 300 °C, at pressures ranging from 0 to 150 MPa and concentration of 0 up to 30 m in molar ionic strength; for the non-aqueous systems the temperature range covered is from 0 to 1200 °C and pressure from 0 to 150 MPa with species concentration from 0 to 1.0 mole fraction. Major predictions using the thermodynamic model have been done to determine the optimum hydrothermal conditions for achieving the crystallization of a wide variety of oxides such as, $BaTiO_3$, $PbTiO_3$, $CaTiO_3$, $SrZrO_3$ and $(BaSr)TiO_3$ (Lencka & Riman, 1993, 1995; Gersten et. al., 2004).

Currently, the research efforts on the thermochemical-modelling topic are being focused to establish an overall rational engineering-based methodology that will speed up process development. The proposed methodology for conducting this study involves the following four steps:

1. Compute thermodynamic equilibrium as a function of chemical processing variables.
2. Generate equilibrium diagrams to draw the process variable space for the phases of interest.
3. Design hydrothermal experiments to test and validate the computed diagrams.
4. Utilize the processing variables to explore opportunities for controlling reactions and crystallization kinetics.

Recently, population balance modelling has received much attention from both academic and industrial areas because of its applicability to a wide variety of particular processes. In general, a population balance model can be proposed by the collective phenomenology contained in entities displacement through their state space and the birth-and-death processes that terminate existing entities and produce new entities. The phenomenology concerns the behaviour of any single entity in conjunction with other entities, which is for the population balance modelling a reasonable description of the system (Ramkrishna & Mahoney, 2002). Regarding the hydrothermal crystallization process, a rigorous kinetic model for the solution precipitation and hydrothermal synthesis of $BaTiO_3$ particles based on mass and population balances has been recently proposed (Testino et. al., 2005). The population balances considered three elementary chemical kinetic processes: primary nucleation, secondary nucleation, and diffusion-controlled particle growth. Secondary nucleation accounts for the acceleration of the formation kinetics after an initial slow crystallization of $BaTiO_3$ particles. The time evolution of yield and particle size is represented by means of discretized mass and population balance equations. Hence, the algorithm is capable for calculating a generic number of conditions involved in the chemical reaction, which can optimize the crystallization and control the growth of $BaTiO_3$ particles. Another different approach based in a bi-variate population balance equations have prove to give more accurate results for the modelling of the barium titanate hydrothermal crystallization. The results obtained from the proposed population balance mathematical model clearly showed that such an approach can overcome the limitations of previous modelling work, and provide a useful tool for more detailed kinetic parameters estimation.

Moreover the model shows that secondary nucleation is indeed very important but particle aggregation cannot be neglected, these variables were not considered in the previous model (Marchizo, 2009).

Despite of the remarkable achievement to develop new strategies based on fundamental principles of thermodynamics and chemistry (reaction equilibrium and kinetics), caution should bear in mind for applying the proposed models to diverse hydrothermal systems and experimental situations. Hence, the estimation of kinetic parameters using population balance derived models could not be correct, these parameters can exhibit variations due to differences on temperature and concentration in the fluid media, which would lead to erroneous conclusions, because the reaction mechanism at certain conditions might change. Likewise, the population balance derived models assume that the system is well mixed and the crystallization rate is uniquely controlled by the chemical kinetics. These assumptions are generally valid for some small-scale laboratory reactors, but they fail in larger industrial plants where mixing is not well controlled. A mixing-limited precipitation rate is one of the problems commonly encountered in the scale-up of solid precipitation processes.

2.4.2 Hydrothermal crystallization process of selected oxide compounds

In the past, the term *hydrothermal crystallization* was used to referred it as nonconventional chemical process, this process involves heating an aqueous suspension of insoluble salts in an autoclave at temperatures and pressures greater than 100 °C and 0.1 MPa, respectively; resulting in the crystallization of the desired well-crystallized phases. This process, however, can be analogous to the term *hydrothermal synthesis* broadly used for physicists and chemists involved in this type of research, because both terms are related with the *genesis* of a specific compound. The crystallization process can mainly occur under very precise conditions of reaction, i.e. temperature, pressure, pH, concentration mineralizer or solvent solution. Thus, the hydrothermal crystallization method using inexpensive and chemicals easy to handle, can produce single or multicomponent oxides. The advantages of hydrothermal crystallization are the reduced energy costs due to the mild temperatures sufficient to achieve chemical reactions, less pollution, simplicity in the process equipment, and the fast rate of solid precipitation reactions (Vivekanandan & Kutty, 1989; Yoshimura & Suchanek, 1997; Riman et. al., 2002).

There are thousands of research works in the literature that include a vast experimental data related with the hydrothermal crystallization of metal oxides or inorganic compounds. The most popular among the metal oxides are those of perovskite oxides group, because of their wide application at the electronic industry. Hence, fundamental principles of crystallization of these materials are discussed in detail and compared with the present research experimental of the present authors regarding the crystallization of perovskite particles of $SrTiO_3$, $BaTiO_3$ and $La_{1-X}M_XCr_{1-Y}N_YO_3$.

2.4.3 Hydrothermal crystallization of perovskite derived oxides from gel precursors

In this section, we focus on the hydrothermal crystallization of perovskite-type structure oxide materials, because this particular group of oxides offers stupendous functional properties for a wide number of applications, in particular electronic applications. Perovskite compounds have the general formula ABO_3, and Perovskite is cubic structure in nature which is shown in Figure 3, where the large A cations (Ba^{2+}, Sr^{2+}, Ca^{2+}, Pb^{2+}, La^{3+}, Bi^{3+} and K^{+}), but low in electric charge, are surrounded by 12 oxygens, whilst B ions relatively small in size (Ti^{4+}, Zr^{4+}, Sn^{4+}, W^{6+}, Nb^{5+}, Mn^{3+}, Mg^{2+}) are coordinated by 6 oxygens. The crystallization of the most representative compounds in this group, namely $BaTiO_3$ (BT), $SrTiO_3$ (ST) and solid solutions of $Ba_{1-X}Sr_XTiO_3$ (BST) or $PbZr_{1-X}Ti_XO_3$ (PZT) has been extensively studied under hydrothermal conditions.

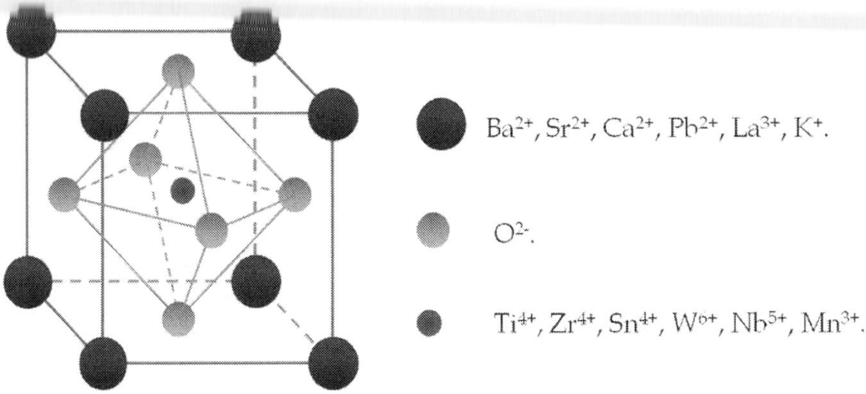

Fig. 3. Schematic representation of the unit lattice cell of the Perovskite cubic structure.

2.4.3.1 Hydrothermal crystallization of perovskite barium and strontium titanate oxides

$BaTiO_3$ (BT) and $SrTiO_3$ (ST) are the metal earth alkaline perovskite-like structures most prepared species under hydrothermal conditions, these compounds has similarity on the crystallization process as that proposed for PT. In terms of the chemical reaction equilibrium that can be produced on the hydrothermal system, and in accordance with the chemical reagent reactants that have been used the chemical reactions that are related with the crystallization of BT and ST powders are as follows:

Where M on chemical equations 3 and 4 is related to earth alkaline metals of the group IIA of the periodic table of elements, *viz.* Ca, Sr and Ba. The chemical reactions from 1 to 9 represent all the reaction equilibrium that are able to proceed in the system M-Ti-Na-H_2O and these reactions serve to produce the crystallization process, because they stem from the principal chemical reactions 1-12 that have widely investigated on the crystallization of BT under hydrothermal alkaline conditions and can operate on the case of the ST compound as well.

$$H_2O = H^+ + OH^- \quad (1)$$

$$H_2O(g) = H_2O \quad (2)$$

$$MTiO_3(s) + H_2O = M^{2+} + 2OH^- + TiO_2(s) \quad (3)$$

$$MOH^+ = M^{2+} + OH^- \quad (4)$$

$$Ti^{4+} + H_2O = TiOH_3^+ + H^+ \quad (5)$$

$$TiOH^{3+} + H_2O = Ti(OH)_2^{2+} + H^+ \quad (6)$$

$$Ti(OH)_2^{2+} + H_2O = Ti(OH)_3^+ + H^+ \quad (7)$$

$$Ti(OH)_3^+ + H_2O = Ti(OH)_4(aq) + H^+ \quad (8)$$

$$Ti(OH)_4(aq) = TiO_2(s) + 2H_2O \quad (9)$$

$$BaCl_2(aq) + TiCl_4(aq) + 6NaOH(aq) = BaTiO_3(s) + 6NaCl(aq) + 3H_2O(l) \quad (10)$$

$$Ba(OH)_2(aq) + TiO_2{\bullet}xH_2O(s) + NaOH(aq) = BaTiO_3(s) + NaOH(aq) + xH_2O(l) \quad (11)$$

$$Ba(OH)_2(aq) + Ti(OH)_4{\bullet}xH_2O(aq) + NaOH(aq) = BaTiO_3(s) + NaOH(aq) + 4xH_2O(l) \quad (12)$$

The chemical synthesis of BT particles has been studied over a broad range of experimental hydrothermal conditions; this compound is preferentially formed in highly concentrated alkaline solutions (pH > 10), as indicated by thermodynamic calculations in BT phase stability diagram proposed elsewhere (Lencka et. al., 1993; Riman et. al., 2002). The normal intervals of reaction time and temperature are 2–96 h and 120–250 °C where the crystallization is normally achieved, thus, the preparation of this type of powder is carried out in typical stainless steel 304 Teflon-lined autoclaves. Additional parametric variable studies recently conducted accounts for other variables that also have an effect on the formation of the BT pure phase like the molar ratio of precursor alkaline (KOH or NaOH) solvent (Vivekanandan & Kutty, 1989; Lee et. al, 2003) and the reactants ($Ba(OH)_2$ or titania source) (Moon et. al., 2003, Qi et. al., 2004), the Ba/Ti molar ratio of the raw materials (Wada et. al., 1995). Another specific studies recently conducted lead to remarkable analysis of the kinetic process related to the crystallization of BT particles and their growth (Walton et. al., 2001), and the control of the particle size and shape of the BT particles have been carried out by stirring the autoclave (Kubo et. al., 2009) and microwave assisted (Moreira et. al., 2008) hydrothermal methods.

Hitherto, in accordance with the analyses conducted based on experimental observations, indicate that the crystallization mechanism of BT particles is dissolution-precipitation in nature, and this mechanism is mainly controlled by the dissolving rate of the reactant solid specie, namely titania (oxide or amorphous gel). Thus, the presence of a high concentration of OH^- ions in the hydrothermal system is required to produce the hydrolysis of aqueous species, and these ions also seemed to act as catalysts to accelerate the transition from Ba-OH bonds, usually resulting on the crystallization of BT via precipitation that proceeds via the chemical reaction (13), resulting from the previous reaction (10–12) that is carried out during the earlier and intermediate stages of the hydrothermal treatment and are linked with the dissolution process.

$$BaOH^+(aq) + Ti(OH)_4^0(aq) = BaTiO_3(s) + 2H_2O + H^+(aq) \quad (13)$$

Once the solution is supersaturated due to dissolution of the precursor, precipitation of BT from the homogeneous solutions spontaneously occurs, yielding an abundant number of nuclei. Therefore, the dissolution of either titania hydrous amorphous or crystalline powders is the rate limiting step that controls the early stage of BT nucleation and growth during the hydrothermal crystallization process, for anhydrous TiO_2 precursor, Ti-O bonds must be broken via hydrolytic attack to for hydroxyl-titanium complexes ($(Ti(OH)_X^{4-X})$) capable of dissolution and reaction with barium ions or complexes (Ba^{2+} or $BaOH^+$) in solution to precipitate BT. Subsequently, the nuclei grow rapidly, resulting in an accelerated crystallization kinetics rate at the initial step of the process. In the solvent, the solute concentration decreases to below the supersaturation point as a result of this event, but remains sufficiently high for the particles to growth, avoiding a secondary nucleation. The particle coarsening proceeds during intermediate and final steps of the crystallization process, and mainly depends on the dissociation of the remaining reactants, or in some cases the separation of terminal organic groups that are liked to the Ti or Ba, e.g. acetiylacetone or acetate, these crystallization barriers serves to slow, if not halt, the kinetic rate in the final stage of the hydrothermal treatment (Eckert et. al., 1996). The crystallization studies indicates that the activation energy fro barium titanate crystallization under hydrothermal conditions varies on the range of 21-105 $kJmol^{-1}$, and the value of activation energy specifically depends on the type of compound used as titania source (Eckert et. al.; 1996; Walton et. al., 2001), but these values of activation are within the range of activation energy values for chemical reaction that proceed on the liquid stage.

On the other hand, major attempts recently conducted to investigate the crystallization of $SrTiO_3$ particles under hydrothermal conditions, have been designed by considering the fundamental principles derived from the synthesis of its related compounds, *viz.* BT or PT perovskite-like structure. A broad type of strontium titanate particle shapes, e.g. nanotube, dendrite, cuboidal and spherical; had been produced through optimizing the hydrothermal crystallization conditions by controlling parameter such as: the pH of the hydrothermal alkaline media in the range between 10–12 (Wendelbo et. al, 2006), reaction temperature and time. Furthermore, different chemical reagents including inorganic and organometallic had been employed as source of Sr^{2+} ($Sr(NO_3)_2$, $SrCl_2$, $Sr(OH)_2 \bullet 6H_2O$) and Ti^{4+} ($TiCl_4$, $TiO_2 \bullet H_2O$, $Ti(SO_4)_2$, $Ti(OC_4H_9)_4$, $Ti(OPr^i)_4$) for producing ST particles (Moon et. al., 1999; Zhang et. al., 2001; Wang et. al., 2009). Another factor that has been studied is related with the use of organic dispersants, namely polyvinyl alcohol, for controlling the crystallization of fine regular shaped ST particles. The dispersant that have large polymeric changes can operate as micelles that attract the hydrolysed ionic species that are formed during the hydrothermal treatment, this lead to and homogeneous nucleation and limited particle growth, resulting in the preparation of nanometer sized ST regular particles (Wei et. al., 2008). Regarding the mechanism correlated with the crystallization of ST particles, one model derived from further experimental data states that the dissolution-crystallization coupled with a second aggregative growth-recrystallization mechanisms are related with the bulk crystallization stage of ST particles under alkaline hydrothermal conditions, but is similar to that proposed for the formation of BT particles (Fig. 4). The second mechanism is achieved when heterogeneous nucleation promotes the formation of ST particles, the surface of the TiO_2 particles acts as the precipitation site for ST nuclei, and when the size of TiO_2 particles is reduced by the progressive dissolution, the aggregation growth of the ST particles proceeds in the reaction media. This phenomenon produces a marked particle agglomeration and also the coarsening of the particles can be promoted by the Ostwald recrystallization mechanism (Zhang et. al., 2004).

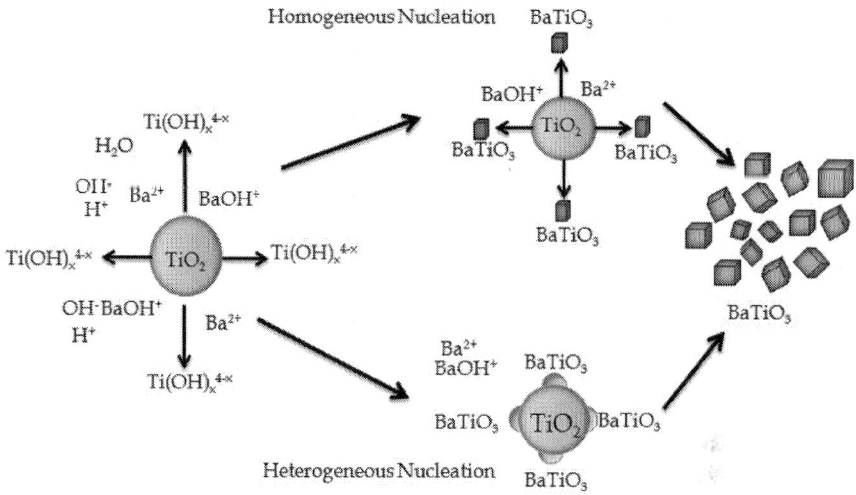

Fig. 4. Schematic sketch of the dissolution/precipitation mechanism that conduces to the hydrothermal crystallization of $BaTiO_3$ (Ecker et. al. 1996, Zhang et. al., 2004).

2.4.3.2 Single step hydrothermal crystallization of perovskite strontium and barium oxides using mineral precursors

In accordance to the former literature, in the last decade, the interest for using pure mineral species to produce pure synthetic inorganic compounds has been rising. In general, the synthesis of $SrTiO_3$ or $BaTiO_3$ has been broadly conducted by employing $Ti(OH)_4$ gel (Ti-gel), while several strontium soluble salts had been used as a source of Sr^{2+} ions. However, the challenge for employing low-grade chemical reagent precursors for preparing ABO_3 particles has not been considered yet. The approach for employing a pure mineral ore, like celestite ($SrSO_4$) or barite ($BaSO_4$) with a low grade of impurities; to produce strontium or barium compounds under hydrothermal conditions, was considered based on the analysis of the proposed chemical methods for producing functional ceramic materials and the former information related to the ionic substitution on mineral species, which are analogous to the mechanistic principles of hydrothermal crystallization discussed in this section. The employment of a low cost precursor may provide an additional advantage in order to propose an economical effective processing method.

Recently, the present authors have conducted an exhaustive efforts to develop a simple single step reaction method for the preparation of $SrTiO_3$, which involves the employment of a mineral $SrSO_4$ crystal plate (0.2 ± 0.0010 g, 6 ± 1 mm side and 2 ± 0.5 mm thick) with $Ti(OH)_4$•$4.5H_2O$ gel (1 g, stoichiometric ratio Sr/Ti=1) under hydrothermal conditions, at various temperatures (150–250 °C) for different reaction intervals (0.08–96 h) in KOH solutions with different concentrations (5–10 M). The hydrothermal treatments were carried out in stainless steel micro-autoclaves lined with Teflon with a filling volume ratio of 50% of the total inner volume (30 ml). This process involves a complex solute dissolution stage because $SrSO_4$ is chemically stable even at acid, neutral and mild basic conditions. However, the use of a mineral single crystal favours to analyse systematically the effect of the dissolution of this reactant coupled with the $Ti(OH)_4$ dehydration and serve to control the synthesis and crystallization kinetics of ST particles. Thus, the complete crystallization of ST particles in a single step of reaction occurred with the complete dissolution of the $SrSO_4$ crystal obtained at 250 °C for 96 h in a 5 M KOH solution, resulting in the formation of $SrTiO_3$ particles with two different shapes (peanut-like and cubic) as is shown in Figure 4. The parameter that has a marked effect on control the particle size and morphology is the temperature rather than the interval of reaction. ST particles having a bimodal size distribution (0.5–1 μm and 0.2–0.5 μm) were prepared at different temperatures for 96 h in a 5 M KOH solution from $SrSO_4$ crystals. The ST powders produced at mild temperatures (200 °C) were constituted by a large amount of agglomerated peanut-like shape irregular particles (Fig. 5a), together with a small amount of bulky crystals with a regular pseudo-cubic shape (1–4 μm). In contrast, at high temperature (250 °C), the pseudo-cubic (3–6 μm) ST particles exhibited a coarsening, while the amount of peanut-like shape (length = 1–5 μm and width = 0.5–1 μm) particles was significantly reduced (Fig. 5b). Hence, the coarsening of pseudo-cubic shaped particles is attributed to the Ostwald ripening particle growth mechanism (Peterson & Slamovich, 1999). Indeed, the growth of the aggregated particles is due to a recrystallization mechanism, which involves the dissolution of the primary small ST peanut-like particles produced at lower temperatures (< 200 °C). This process is promoted due to lower chemical stability that exhibits the strontium titanate oxide in highly concentrated (> 5 M) KOH solvent solutions at 250 °C. Another point that deserves to be emphasized is that related with the structural features of either peanut-like or cubic-like shaped particles. EDX spectra obtained on both particles did not show a marked difference on the molar Sr/Ti ratio, because the particles contain similar Sr and Ti amounts as it is suggested by the corresponding EDX spectrum in Fig. 5c (Rangel-Hernandez et. al., 2009).

In contrast, the increase on the surface area of the precursor $SrSO_4$ produces the control of morphology and size homogeneity for the ST particles. In general, the use of $SrSO_4$ powders with a narrow particle size distribution in the range from 25-38 μm, favoured the crystallization of very fine cubic ST particles, even at very short reaction intervals (0.08 h, Fig. 6a). However, ST powders with a bimodal size distribution, consisting of a large amount of pseudo cubic particles (0.75 μm) and a small quantity of large cubic particles (1.5 μm) were produced during intermediate reaction intervals between 3-12 h. In contrast, at the longest reaction interval of 24 h, the preferential formation of very fine pseudo-spherical ST particles (average size of 0.4 μm) was observed (Fig. 6b). The formation of the cubic particles is due to the crystallographic habit growth, which proceeded when the alkaline solvent reaches a supersaturation steady-state with the ionic species $Sr(OH)^+$ and $Ti(OH)_4^0$. Hence, a massive homogeneous nuclei formation and fast growth of the cubic particles proceeds in the hydrothermal system during intermediate reaction intervals. However, the ST cubic particles underwent into a preferential dissolution by increasing the reaction interval at long reaction period (24 h). The ST particle dissolution proceeds at high-energy faceted edges on the fine cubic particles, because of the relative low chemical stability of the ST perovskite structure in the concentrated alkaline media. This fact leads to further dissolution of the cubic particles and the recrystallization of the pseudo-spherical shaped particles from the solution.

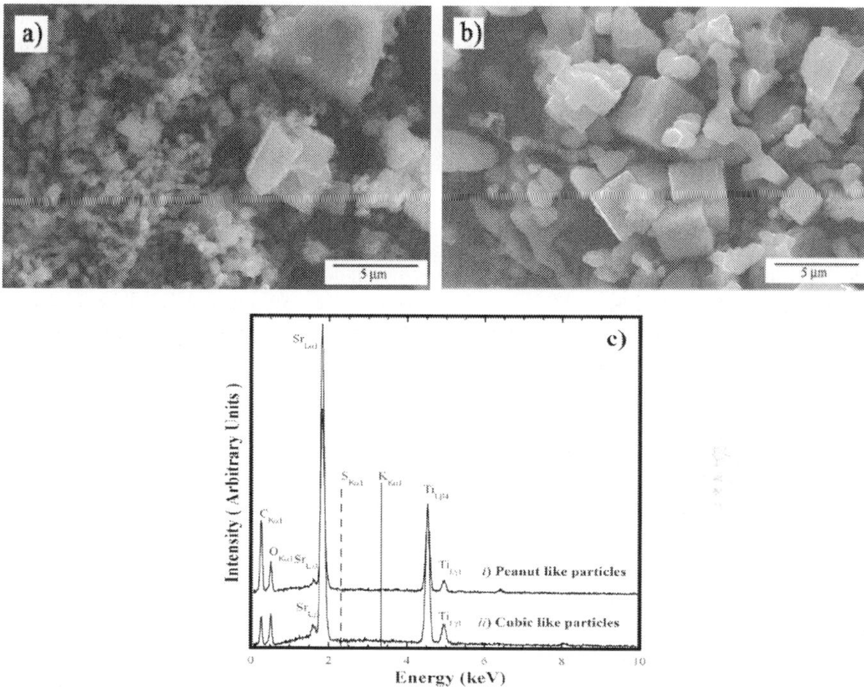

Fig. 5. Morphologies of ST particles obtained after hydrothermal treatments of $SrSO_4$ crystal plates, carried out at temperatures of (a) 200 °C, (b) 250 °C; and (c) EDX spectra of ST (*i*) peanut-like and (*ii*) cubic particles shown in Fig. 3b (Rangel-Hernandez et. al., 2009).

Fig. 6. $SrTiO_3$ particles hydrothermally produced using $SrSO_4$ powders at 250 °C in 5 M KOH solution for intervals of (a) 0.08 and (b) 24 h (Rangel-Hernandez et. al., 2009).

The bulk crystallization process of ST particles is dissolution-precipitation in nature, and the reaction path involving this mechanism that produces nuclei formation and crystal growth, is represented physically in the Figures 7a and 7b. These Figures gave the set of the original

$SrSO_4$ crystal plate embedded in the $Ti(OH)_4 \bullet 4.5H_2O$ gel prior the hydrothermal treatment and after 6 h of reaction at 200 °C. In terms of fundamental chemistry the reaction path way occurs trough the chemical equations 14 and 15:

$$SrSO_4(crystal) + Ti(OH)_4 \bullet 4.5H_2O(gel) + 2KOH = SrOH^+(aq) + K_2SO_4(aq) + Ti(OH)_4^0(aq) + 4.5H_2O + OH^-(aq) \quad (14)$$

$$SrOH^+(aq) + Ti(OH)_4^0(aq) + OH^-(aq) = SrTiO_3(s) + 3H_2O \quad (15)$$

Macroscopic aspects related with the single-step synthesis process were revealed during the early stages of the reaction at 200 °C for 6 h in a 5 M KOH solution (Fig. 7b). In general, the reaction gradually proceeds by the $SrSO_4$ crystal dissolution, the consumption of $SrSO_4$ occurs locally on those crystal surfaces exposed to the alkaline fluid and also on those crystal surfaces in contact with the Ti-gel. This phenomenon agrees with those results previously reported and provided evidences of the solubility of $SrSO_4$ compound in alkaline hydrothermal solutions (5 and 10 M NaOH) (Rendón-Angeles et. al., 2006). Hence, the continuous dissolution of the $SrSO_4$ crystal must yield the formation of $SrOH^+$ species (Eq. 14), because of the saturated alkaline conditions of the solvent. Indeed, the solubility of the Ti-gel is low in high concentrated KOH (> 0.1 M) solutions (Wang et. al., 2009, Rangel-Hernandez et. al., 2009). However, the consumption of the $SrSO_4$ crystal dissolution (α) was markedly reduced due to the presence of the Ti-gel (Fig. 7c). This fact let us to conclude that the reactivity of Ti-gel is the limiting rate variable for crystallization, because reduces the dissolution rate of the $SrSO_4$ phase. Thus, the equation (4) is proposed for the crystallization of ST and is similar to that discussed at the early part of the present section for PT and BT compounds. The nucleation and growth processes of the ST particles occur when the alkaline (KOH) solution reaches a supersaturation steady-state of the species $Sr(OH)^+$ and $Ti(OH)_4^0$. The ST particles precipitation locally occurred at Ti-gel surface and between the spaces with the remaining $SrSO_4$ crystal, because the mass transport was limited during the treatments, which were conducted under neither agitation nor thermal gradients. Additionally, kinetic data obtained from $SrSO_4$ consumption curves depicted that the activation energy required for the synthesis of $SrTiO_3$ powders from the complete consumption of an $SrSO_4$ crystal plate under hydrothermal conditions, is 27.9 kJ mol^{-1}. One particular fact that must be emphasised is related with the incorporation of the major impurities on the ST particles during the synthesis process, namely Ba (5.7 wt.%) and CO_3^{-2} (0.6 wt.%), contained in the $SrSO_4$ crystals, does not proceed during the crystallization event because neither the presence of barium nor carbonate compounds were found on the compositional analysis conducted by EDX and DRX techniques in the ST particles.

On the other hand, the crystallization of perovskite related compounds like ST or BT via the transformation of sulphate alkaline earth metal mineral species is hinder for the barite mineral. The crystalline transformation of $BaSO_4$ to $BaTiO_3$ was studied using the single step reaction route aforementioned; and the high chemical stability (low solubility) of the orthorhombic structure limited the dissolution of the barite crystals under hydrothermal conditions even in highly concentrated KOH solutions (>5 M). The reactivity of the orthorhombic structure with the alkaline solvent can be influenced by the size of the alkaline earth metal ion incorporated in the structure, these results are in agreement with those early reported on the transformation of Ca, Sr and Ba-chlorapatite crystals into their respective hydroxyapatite species (Rendón-Angeles et. al., 2000a). This factor reduces the rate of the BT particle crystallization process, but the reaction was conducted by increasing the concentration of the KOH solution up to 10 M, and the complete dissolution of the $BaSO_4$ crystal occurred for large reaction intervals of 144 h above 200 °C. Optimum steady-state supersaturation of the solvent media with Ba^{2+} and Ti^{4+} ions that achieve the conditions for homogeneous nucleation of BT particles were found to proceed at temperatures below 200 °C (Figure 8). Fine BT particles with pseudo-cubic, star-like and dendrite shapes were preferentially formed in 10 KOH. At sever treatment conditions of temperature 250 °C for 144 h, the $BaSO_4$ crystals were completely dissolved, and the formation of reaction by-products Ba_2TiO_4 (needle crystals in Figure 8c) and a few amount of TiO_2 simultaneously occurred with the excessive growth of polyhedral aggregated BT crystals. This particular behaviour is attributed to the differences in the chemical reactivity of the alkaline solution resulting in different dissolving rates for solid species, this can shift the reaction equilibrium coupled with the saturation conditions. Therefore, other crystalline phases are able to precipitate because thermodynamically are more stable than BT.

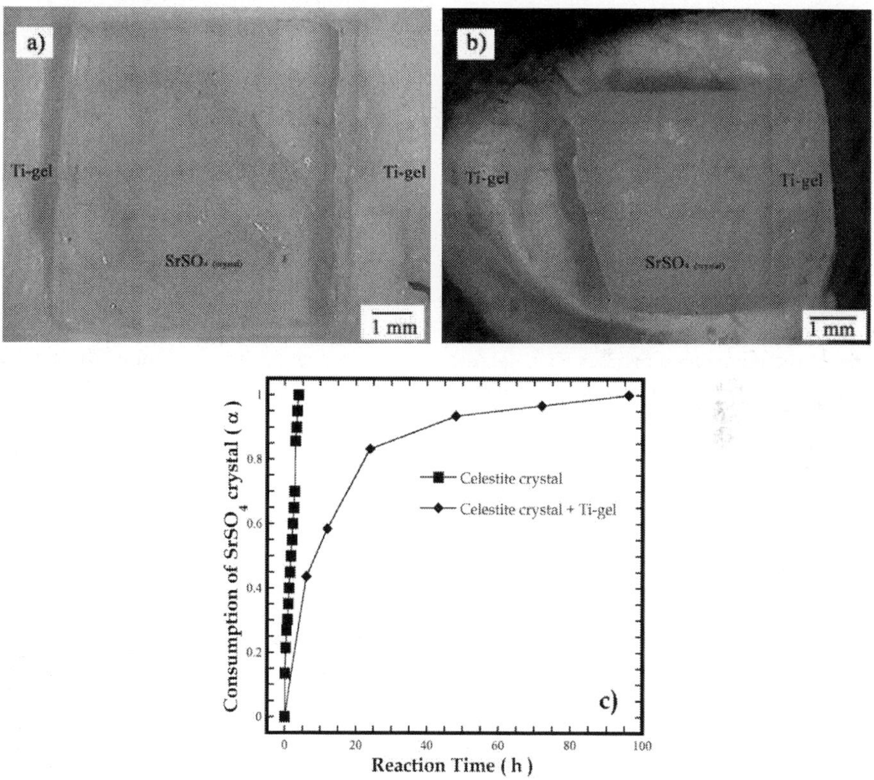

Fig. 7. Aspects of (a) original $SrSO_4$ crystal plate embedded in Ti-gel, and (b) the partially reacted $SrSO_4$ crystal and Ti-gel at 200 °C for 6 h in a 5 M KOH solution. (c) Consumption curves (α) of $SrSO_4$ crystal vs. the reaction interval obtained at 250 °C in a 5 M KOH solution, (■) without Ti-gel and (◆) with Ti-gel additions (Rangel-Hernandez et. al., 2009).

2.4.3.3 Elimination of mineral impurities during the synthesis of strontium titanate particles under hydrothermal conditions

Regarding the presence of high content of impurities during the crystallization of ST particles from mineral celestite ores. The experimental research work conducted recently by the authors of the present chapter; clearly contribute to demonstrate that refining of the major metal ion contained in the mineral ore proceeds during the hydrothermal crystallization of ST powders. Thus, the chemical stability of the barite-celestite mineral specie was investigated to elucidate the feasibility to conduct simultaneously the crystallization of ST particles and the release of major impurities such as Ba, and the preparation of BT powders as well. In particular, the barite-celestite mineral consists in a solid solution with a general chemical formula of $Sr_{0.70}Ba_{0.30}SO_4$, which was determined by wet chemistry using ICP, the total content of major constituents $SrSO_4$ and $BaSO_4$ was 64.7 wt.% and 35.3 wt.%, respectively. Additional microprobe wavelength X-ray diffraction analyses conducted by scanning electron microscope observations indicated that the microstructure of the mineral consists of two major solid solutions, one rich in strontium $Sr_{0.95}Ba_{0.05}SO_4$ and a less amount of an intermediate $Sr_{0.75}Ba_{0.25}SO_4$. The hydrothermal treatments of barite-celestite crystal plate samples were conducted in accordance with the experimental procedure explained before for high pure celestite crystals.

Fig. 8. $BaTiO_3$ particles crystallized from barite mineral plates in a 10 M KOH solution for 24 h at temperatures of (a) 150, (b) 200 and (c) 250 °C.

The synthesis of ST particles was preferentially obtained using the barite-celestite crystals in a feedstock alkaline 5 M KOH solution; the transformation of the mineral specie into the perovskite oxide is strongly affected by the temperature and concentration of the solvent media rather than the reaction interval. Marked morphological and particle size differences were derived due to the increase of reaction temperature, a mixture of fine particles resembling cubic and start-like shapes were formed at mild temperatures (150 and 200 °C) for 12 h, increasing the temperature at 250 °C bulky aggregated cubic ST particles (20 μm size) were crystallized (Figure 9). During the hydrothermal crystallization of the ST compound the incorporation of the Ba^{2+} ions was avoided to occur in the structure of ST particles, this is supported by the EDX spectra of the cubic and star-like ST particles where no traces of Ba are visible (Fig. 9d).

Fig. 9. $SrTiO_3$ particles crystallized from barite-celestite mineral plates in a 5 M KOH solution for 12 h at different temperatures of (a) 150, (b) 200 and (c) 250 °C. (d) EDX spectra of the ST particles shown in (b).

Two main factors are associated with; one is related to the marked compositional gradient differences of the metal ions Sr^{2+} and Ba^{2+} produced in the solvent media during mineral dissolution, coupled with the stability of Ba^{2+} to form complexes ions in alkaline solutions ($BaOH^{+}$, Eq. 4). Indeed, wet chemical analyses of the remaining solutions after the treatments gave evidences that support the above inference, because the Ba^{2+} ions content gradually increased as far as the complete mineral dissolution was concluded for 96 h at 250 °C. The content of Ba determined by in the solution was 20.0 ± 0.8 wt%, which is nearly the same, measured in the original barite-celestite crystals (20.8 wt%). The ST particles crystallization proceeds by the same model established previously on the case of the transformation of high pure celestite crystal to ST particles, the control of morphology and particle size for the ST powders can be optimized by limiting the dissolution-recrystallization mechanism that operates at sever treatment conditions of temperature (T> 200 °C) and long periods (t> 72 h). Therefore, these results probe that the hydrothermal technique combined with the use of mineral (pure and contaminated) as precursor for the synthesis of inorganic materials can be an attractive technique to be explored at an industrial scale.

2.4.3.4 Hydrothermal crystallization of perovskite lanthanum chromite oxides

Lanthanum chromite ($LaCrO_3$, LC) powders substituted with alkaline metals (Ca or Sr) have been widely accepted as the candidate for interconnection materials in Solid Oxide Fuel Cells (SOFCs) and also particular solid solutions can be used as oxygen sensor at high temperature (Chakraborty et. al., 2000) The partial substitution of lanthanum ions with alkaline metal ions, Ca or Sr in the A site of La, increases the chemical stability and electric conductivity, whilst Al or Ni in the B site reduces the thermal expansion coefficient at high temperature, when compared with the properties of pure lanthanum chromite (Ianculescu, et. al., 2001). Hitherto, various chemical routes have been used to process lanthanum chromite powders (Bliger. et. al., 1997). However, these chemical processes involve heat treatments at temperatures beyond 700 °C, in order to obtain the crystalline phase pure phase and their solid solutions with Ca or Sr in the A site and Ni or Al in the B in the sites of the perovskite-like structure ABO_3. Hence, the present authors recently determined through exhaustive research work, the aspects related to the crystallization of precursor complex gel of the $La_{1-X}M_XCr_{1-Y}N_YOH_{6-x}$ compound into their respective perovskite solid solutions $La_{1-X}M_XCr_{1-Y}N_YO_3$ (where M = Ca or Sr and N = Al or Ni) with orthorhombic crystalline structure under hydrothermal conditions, at a temperature range between 350–500 °C for short reaction intervals (0.5–2 h).

The precursor complex gel was prepared by the alkaline coprecipitation method widely used to prepare this type of materials; the details of the chemical preparative method are given elsewhere (Inagaki et. al., 1990). Precursor lanthanum chromite complex gels were prepared by employing reagent grade chemicals of: $LaCl_3 \bullet 7H_2O$ (99.998%), $Cr(NO_3)_3 \bullet 9H_2O$ (99.9%), $CaCl_2 \bullet 2H_2O$ (99%), $SrCl_2 \bullet 2H_2O$ (99%), $Ni(NO_3)_2 \bullet 6H_2O$, $Al(NO_3)_3 \bullet 9H_2O$ and NaOH (99.998%) (Wako Pure Chemical Industries, Ltd., Japan). Aqueous solutions with a concentration 0.05 M of $LaCl_3$, $Cr(NO_3)_3$ and $CaCl_2$ were prepared with deionized water, and a solution of 0.5 M of NaOH was employed as coprecipitation media. In a typical procedure, a volume of 475 ml of the precipitating solution (NaOH) was poured in a biker, and chromium or the mixtures of Cr+Al, Cr+Ni; solution (500 ml) was then mixed, which results in the formation of an opaque whitish green precipitate ($Cr(OH)_3$, or $Al(OH)_3$, $Ni(OH)_2$), which was subsequently dissolved by vigorous stirring. Finally, the coprecipitation of the complex gel was carried out by the addition of the same volume (500 ml) of the solution containing the other elements, La or the mixture of La+Ca, La+Sr. The solutions were mixed in different volumetric ratios, La:M:Cr:N, 1:0:1:0, 0.9:0.1:1:0 and 0.8:0.2:1:0, 1:0:0.95:0.05; 1:0:0.9:0.1; 0.8:0.2:0.95:0.05, 0.8:0.2:0.9:0.1; which matches the compositional stoichiometric of the solid solutions, $LaCrO_3$, $La_{0.9}Ca_{0.1}CrO_3$, $La_{0.8}Ca_{0.2}CrO_3$, $La_{0.8}Sr_{0.1}CrO_3$, $La_{0.8}Sr_{0.2}CrO_3$, $LaCr_{0.95}Al_{0.05}O_3$, $LaCr_{0.9}Al_{0.1}O_3$, $LaCa_{0.2}Cr_{0.95}Al_{0.05}O_3$, $LaCa_{0.2}Cr_{0.9}Al_{0.1}O_3$, $LaSr_{0.2}Cr_{0.95}Al_{0.05}O_3$, $LaSr_{0.2}Cr_{0.9}Al_{0.1}O_3$, $LaSr_{0.2}Cr_{0.95}Ni_{0.05}O_3$, $LaSr_{0.2}Cr_{0.9}Ni_{0.1}O_3$. The coprecipitated gel was centrifuged and a volume of 20 ml was then poured into a hydrothermal Hastelloy C-lined microautoclave (40 ml capacity, Figure 10). The vessel was heated at a constant rate of 20 °C/min up at various temperatures (350–500

°C) for a reaction interval between 0.5 and 2 h. After the treatments, the precipitates were well washed with distilled water, decanted and then dried in an oven at 100 °C overnight (Rivas-Vázquez et. al., 2004, 2006; Rendón-Angeles et. al., 2009).

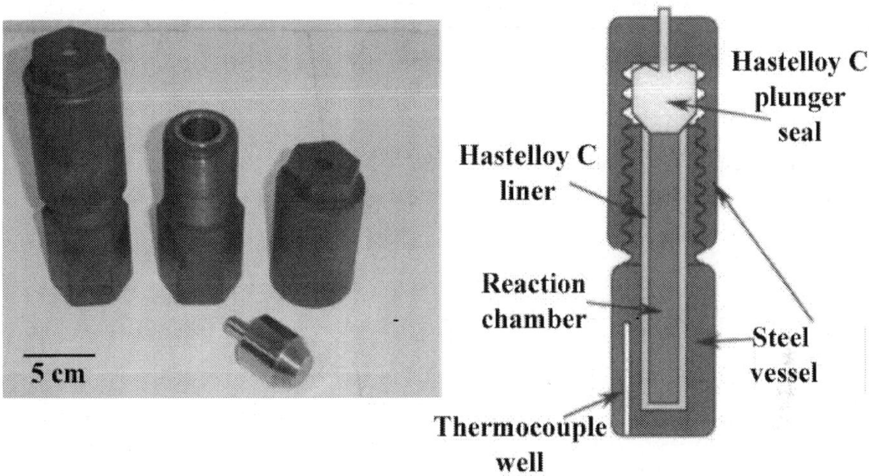

Fig. 10. Typical Hastelloy C type lined microautoclave and scheme of the cross section of the vessel, which was employed for conducting the synthesis of perovskite lanthanum chromite powders under hydrothermal conditions.

The minimum temperature for conducting the crystallization of the pure LC powders was 375 °C for a reaction interval of 1 h. This parameter, however, is mainly affected by the incorporation of the metal ions in both A and B sites of the perovskite structure ABO_3, and also with the amount of metal dopant ion inserted. Thus, the crystallization of powders corresponding to the solid solutions $La_{1-X}M_XCr_{1-Y}N_YO_3$ (where M = Ca or Sr and N = Al or Ni) was observed to proceeds at temperatures above 400 °C, when only a 10 mol% of Ca^{2+} or Sr^{2+} were partially substituting La^{3+} site. Thus, in the particular case of the $La_{0.9}Ca_{0.1}CrO_3$ and $La_{0.8}Ca_{0.2}CrO_3$ solid solutions, the structural analyses in Figure 11a depicted that at a constant reaction interval 1 h, the powders of $La_{0.9}Ca_{0.1}CrO_3$ were produced at a temperature of 400 °C without contaminant formation, in contrast with the $La_{0.8}Ca_{0.2}CrO_3$ powders that were produced together with a marked amount of secondary crystalline phases of $La(OH)_3$, CrOOH, and $CaCrO_4$, these phases were eliminated by increasing the reaction temperature up to 425 °C resulting in the crystallization of the pure phase $La_{0.8}Ca_{0.2}CrO_3$. This behaviour was also determined for the crystallization of the $La_{0.9}Ca_{0.1}CrO_3$ and $La_{0.8}Ca_{0.2}CrO_3$ powders. The minimum temperature that achieves the perovskite powders formation was found to increase at 450 °C, when Al^{3+} partially substituted Cr^{3+} sites. This situation is more critical when two different metal dopants were simultaneously incorporated at the same time to form a complex oxide such as; $LaCa_{0.2}Cr_{0.95}Al_{0.05}O_3$ and $LaCa_{0.2}Cr_{0.9}Al_{0.1}O_3$, the powders corresponding to these LC solid solutions were obtained at minimum temperature of 475 °C, which is a significant increase (100 °C) when compared with the low temperature needed to form the pure $LaCrO_3$. The significant variation on the processing parameters experimentally determined, namely the temperature, can be associated with the thermodynamic fundamental principles, e.g. the Gibbs free energy. The calculus of the ΔG values for the formation of Ca and Sr doped LC solid solutions over a wide range of temperature (Fig. 11b), are in a good agreement with the trend of the experimental results aforementioned. Because, in terms of energy, the crystallization reaction that required less energy consumption is that leads to the formation of pure $LaCrO_3$, compound that exhibit the lowest ΔG values in comparison with those of $La_{0.9}M_{0.1}CrO_3$ and $La_{0.8}M_{0.2}CrO_3$ solid solutions (M= Ca^{2+} or Sr^{2+}).

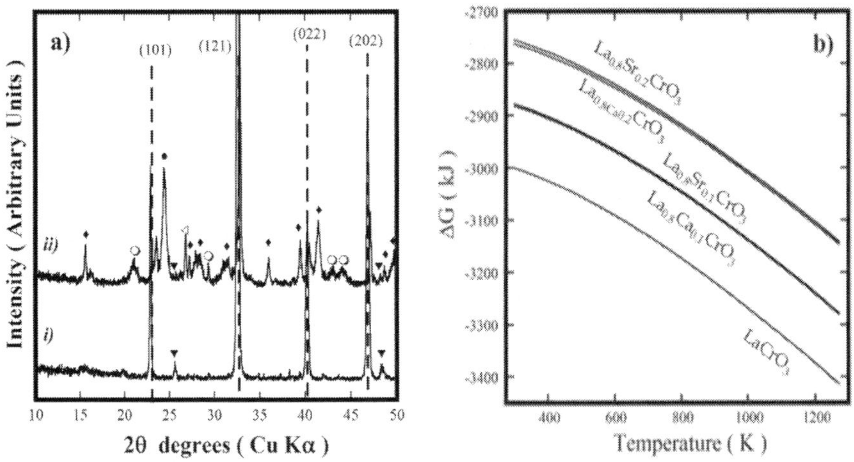

Fig. 11. (a) X-ray diffraction patterns of (*i*) $La_{0.9}Ca_{0.1}CrO_3$ and (*ii*) $La_{0.8}Ca_{0.2}CrO_3$ powders produced at 400 °C for 1 h. $LaCrO_3$ (JCPDS 33-701) compound doted line; (◆) $La(OH)_3$, (◁) CrOOH, (●) $CaCrO_4$, (▼) low intensity peaks of $LaCrO_3$. (b) Variation of the ΔG formation for pure $LaCrO_3$ and $La_{0.9}M_{0.1}CrO_3$ and $La_{0.8}M_{0.2}CrO_3$ solid solutions (M= Ca^{2+} or Sr^{2+}).

Morphological aspects of powders corresponding to the solid solutions of $La_{0.9}M_{0.1}CrO_3$ and $La_{0.8}M_{0.2}CrO_3$ (M= Ca^{2+} or Sr^{2+}) solid solutions obtained at 400 and 425 °C, respectively; are shown in Figure 12. Particles with submicron size and irregular morphology, which resembles a peanut-like shape, were preferentially formed on all cases investigated, it is also indicated that the particle morphology of these compounds is irrespective of the gel crystallization temperature. Furthermore, the particles showed a marked agglomeration due to its particular morphology and particle size aspect. One point that deserves emphasis is that related with the marked tendency for particle bounding that underwent the particles. These are reliable evidences that indicate the particles of the formed LC solid solutions were partially dissolved in the fluid at an intermediate reaction stage (>1 h) of the hydrothermal treatment, once the secondary crystalline phases were completely exhausted.

This phenomenon seems to occur markedly on the $La_{0.8}Sr_{0.2}CrO_3$ particles (Fig. 12d) hydrothermally synthesized 1 h at 425 °C, because these particles exhibit the smallest particle size (average 250 nm) in comparison with the other three powders that have an average particle size of 350 nm (Figs. 12a-12c). This fact can be explained based on the chemical stability of the crystalline phase, which must be low and therefore the dissolution proceeds rapidly. Indeed, this inference is supported by the fact that the amount of regular $La_{0.8}Sr_{0.2}CrO_3$ particles is further decreased in this particular powder. Although, the particles exhibited a non-peculiar morphology when compared with those used for preparing dense advanced ceramic materials, it was experimentally probed that these fine powders have better sinterability features even in oxidizing atmospheres (Rivas-Vázquez et. al., 2004, 2006; Rendón-Angeles et. al., 2009).

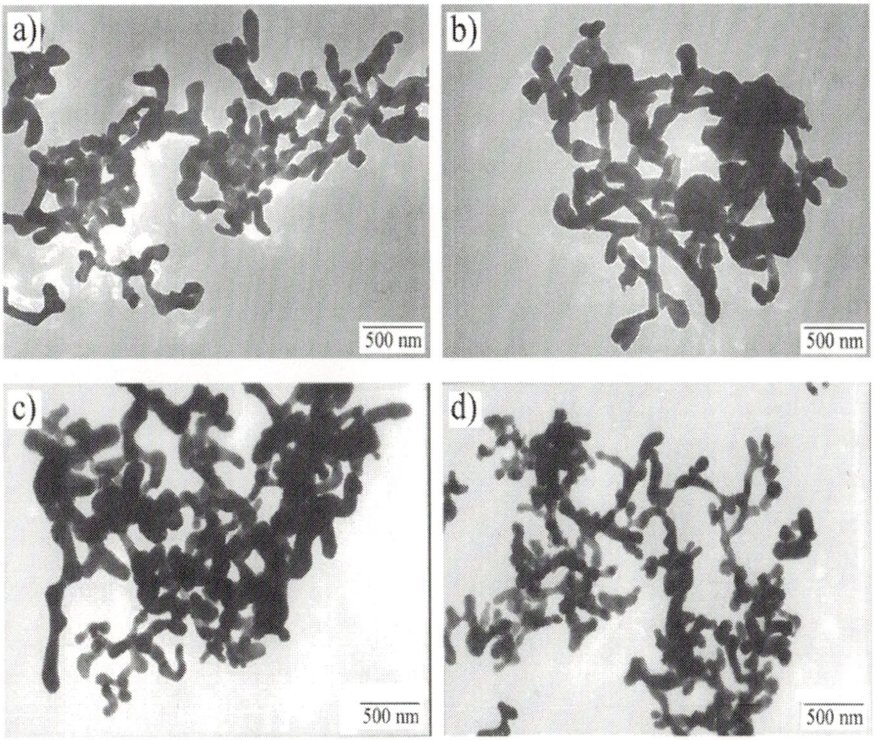

Fig. 12. Transmission electron micrographs of hydrothermally produced powders for 1 h at 400 °C (a) $La_{0.9}Ca_{0.1}CrO_3$ and (c) $La_{0.9}Sr_{0.1}CrO_3$; and 425 °C (b) $La_{0.8}M_{0.2}CrO_3$ and (d) $La_{0.8}M_{0.2}CrO_3$.

The details of the reaction pathway that are linked to the crystallization of the final tailored LC composition were determined during the preparation of $La_{0.9}Ca_{0.1}CrO_3$ solid solution. Preliminary observations conducted even during the heating stage of treatment confirmed that the hydrothermal crystallization involves a preliminary reaction stage, which is related with the dehydration process of the complex gel and proceeds at temperatures above 300 °C, resulting on the autogeneous formation of hydrothermal solvent inside the reaction vessel. The pH of the remaining solutions after the hydrothermal treatment varied in the range between 7.2-7.8, which confirms that the crystallization of the lanthanum chromite powders was conducted in low alkaline conditions. Once the treatment temperature is reached (400 °C), as a result of the complex gel dehydration process, the crystallization of intermediate secondary phases proceeded during the earlier stages of the reaction (10–30 min). The major stable secondary phases that were produced during this reaction interval range were $La(OH)_3$, $CrOOH$, which correspond to the needle and platelets, respectively (Figs. 13a–13b); these observations are in a good consistence with the X-ray diffraction patterns of the powders produced at 400 °C for different reaction intervals (Fig. 13d), these results also showed the presence of a slight amount of $CaCrO_4$. The dissolution of the reaction by-products is fast in such a diluted alkaline hydrothermal media. Hence, the nucleation of the oxide particles proceeds preferentially at the surface of the remaining complex gel during the dehydration process, because the crystallization by the process employed in this study proceeded at higher temperatures (300–450 °C) than those determined for $LaCrO_3$ and $La_{0.5}Sr_{0.5}MnO_3$ particles under concentrated alkaline hydrothermal conditions (8 M KOH), 240 and 260 °C, respectively (Zheng et. al., 1999; Spooren et. al., 2003). The complete crystallization of $La_{0.9}Ca_{0.1}CrO_3$ particles was even promoted for a shorter interval as 1 h, resulting in the formation of irregular particles resembling peanut-like shapes (Fig. 13c).

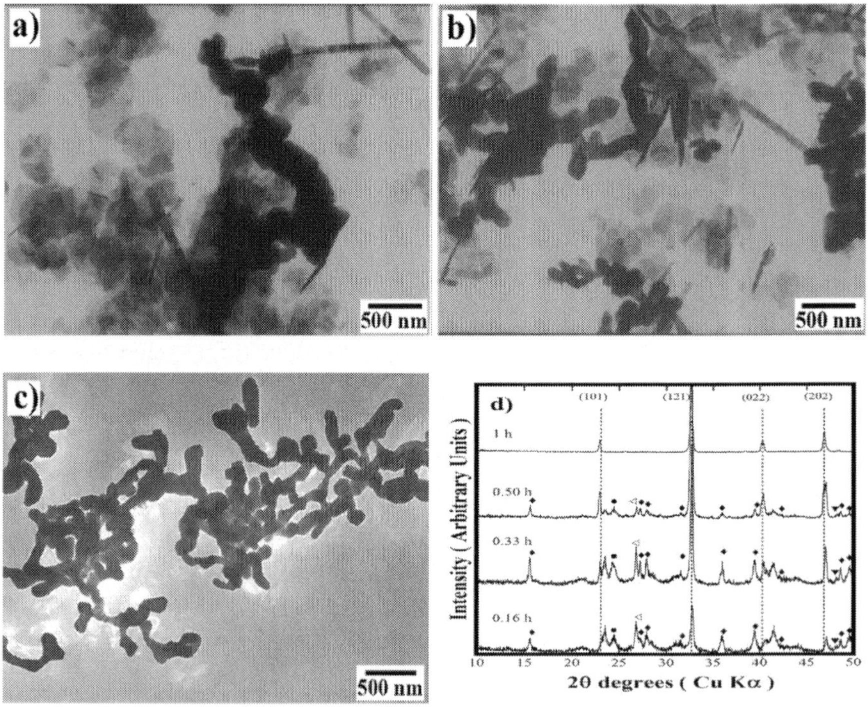

Fig. 13. TEM micrographs of the reaction products obtained under hydrothermal conditions at 400 °C for different intervals (a) 0.33, (b) 0.5 and (c) 1 h. (d) X-ray diffraction patterns of complex gel corresponding to the above micrographs. $LaCrO_3$ (JCPDS 33-701) compound doted line; (◆) $La(OH)_3$, (◁) $CrOOH$, (●) $CaCrO_4$, (▼) low intensity peaks of $LaCrO_3$.

Regarding the peculiar morphology of the $La_{0.9}M_{0.1}CrO_3$ and $La_{0.8}M_{0.2}CrO_3$ solid solutions ($M= Ca^{2+}$ or Sr^{2+}). Two main factors promoted the formation of the peanut-like shaped particles (Figure 14); the first is related with the nucleation and growth process which proceeded very fast for reaction intervals of 0.5 up to 2 h at interval of temperatures (400–450 ∘C) and limited the growth of $La_{1-X}M_XCrO_3$ particles. The second one is associated with the concentration of the solvent media; the alkaline fluid autogenously formed during the hydrothermal treatment is not capable of achieving the actuated conditions that lead the crystallization of particles with cubic habit (Zheng et. al., 1999; Spooren et. al., 2003). However, the $La_{1-X}Sr_XCrO_3$ particles were partially dissolved in the solvent media at high temperature (400–450 °C), resulting in a marked particle joining by developing necks on the surfaces of primary synthesized particles in contact, as shown in Figure 14 for $La_{0.8}Sr_{0.2}CrO_3$ powders.

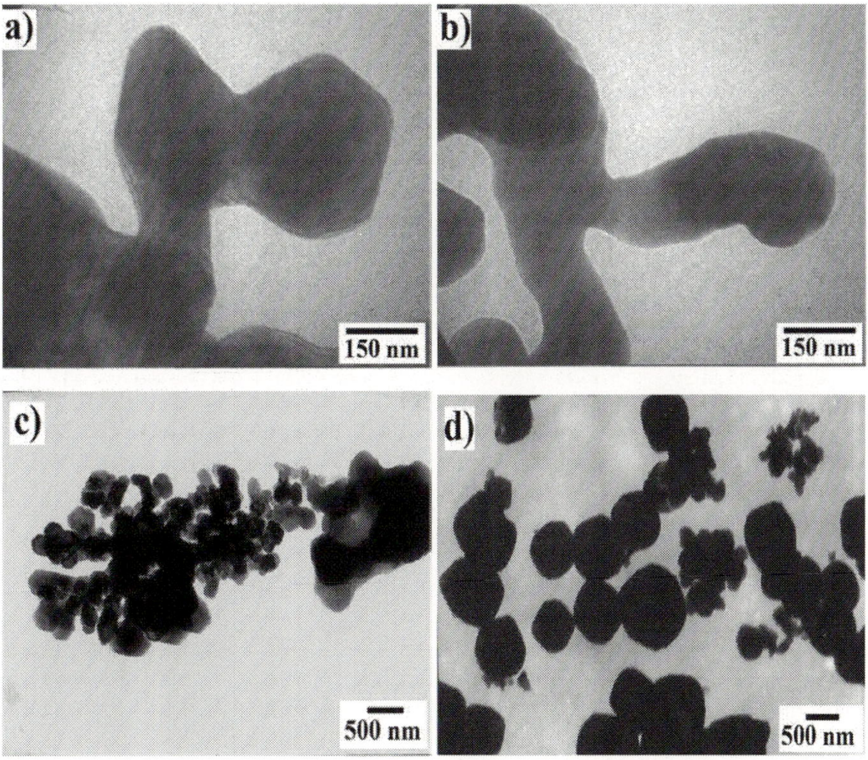

Fig. 14. Transmission electron micrographs of $La_{0.8}Sr_{0.2}CrO_3$ powders obtained at 450 °C for reaction intervals of (a) 1 and (b) 2 h. $La_{0.8}Sr_{0.2}Cr_{0.9}Al_{0.9}O_3$ powders prepared at 475 °C for 1 h in NaOH solutions of (c) 0.1 and (d) 5 M using as a precursor dried complex gel.

The local recrystallization of the solute is mainly promoted inside the reaction vessel, due to the fact that no fluid convection occurred during the hydrothermal treatment, because the treatments were conducted under static conditions at constant temperature, therefore, mass transfer due to convection was further limited (Bayrappa & Yoshimura, 2001). Hence, the crystallization mechanism is similar to that described previously for other perovskite species, but it differs on the rate of kinetics that lead to produce particles particular aspect. In contrast, the particle bonding was limited by modifying the treatment conditions, using a dried complex gel $La_{0.8}Sr_{0.2}Cr_{0.9}Al_{0.1}O_3$ and different aqueous solvent media (water, KOH, NaOH and KF), resulting in the optimum control of dissolution and crystallization of fine particles with regular. A marked particle growth, however, was determined as a result of increasing the concentration of the alkaline solution from 0.1 to 5 M (NaOH or KOH). The formation of submicron particles monodispersed (0.5 - 0.75 μm) was found to proceed on these solvents (Figs. 14c, 14d). Furthermore, the employment of the alkaline solvents NaOH and KF leads to the crystallization of pseudo-cubic shaped and hexagonal plate particles, respectively; the differences on the particle morphology control are due to the chemical reactivity of the different solutions with the dried complex gel.

2.5 Replacement reactions on minerals species under hydrothermal conditions as a new approach for preparing inorganic materials

Another different approach similar to the mineral transformation into perovskite oxide powders was investigated to establish an alternative processing route for mineral ores via controlled dissolution-precipitation. The precursor mineral considered were the alkaline earth sulphates of $SrSO_4$ and $BaSO_4$, for preparing high pure strontium inorganic compounds, namely $SrCO_3$, $BaCO_3$, $SrCrO_4$, SrF_2, $Sr(OH)_2$. This chemical preparative method involves the dissolution–recrystallization mechanism, similar to the process that achieves the ionic replacement reaction in mineral ores, and promotes the conversion of natural ores at the earth's crust into more chemically stable mineral species or inorganic compounds (Putnis, 2002, 2009). The development of more efficient and environmental friendly chemical routes has recently been under concern of some researchers. The new chemical techniques or the optimized conventional routes might lead to reduce the pollution grade of contaminants, which is produced during mineral processing stages. The related alkaline earth sulphate minerals, such as $SrSO_4$ and $BaSO_4$, have been exploited since several decades, because these are the main source of the alkaline earth metals elements, Sr and Ba, and are used as a main source for preparing inorganic compounds of Sr and Ba. During the last decade, some attention has been paid to the use of celestite ore for producing functional ceramic compounds with magnetic properties, e.g. strontium hexaferrite ($SrFe_{12}O_{19}$); two different methods were proposed, coprecipitation in aqueous solutions via mineral powder leaching and powder mechanochemical activation techniques (Hessien et. al., 2009; Tiwary, et. al. 2008). Hence, since a decade, the present authors have devoted efforts in order to investigate, from a different approach of preparative chemistry; the nature of the possible replacement reactions that can be achieved under hydrothermal conditions, as was found to proceed even in synthetic inorganic compounds (Rendón-Angeles et. al., 2000b). This could derive in an optimum method for preparing high grade Sr and Ba compounds. The most relevant aspects found regarding this topic are given in this section.

2.5.1 Aspects of the compositional and structural transformation on sulphate minerals under alkaline hydrothermal conditions

Hitherto, the ion exchange reaction of SO_4^{2-} ions with CO_3^{2-} ions has been one of the subjects of research work, which involves both mineral celestite ($SrSO_4$) and barite ($BaSO_4$) species.

This particular transformation involves the conversion of MSO_4 into MCO_3, M= Sr or Ba, because these carbonated compound are widely used as a precursors for the preparation of strontium and barium inorganic compounds (Suaréz-Orduña et. al., 2004a; Rendón-Angeles et. al., 2008). Preliminary evidences of the ion exchange replacement process were investigated on the mineral celestite specie, thermodynamic and kinetic details were reported elsewhere (Yoshino et. al., 1985). The exchange of SO_4^{2-} ions with CO_3^{2-} ions was investigated on large mineral $SrSO_4$ bulky crystal plates, which were leached at low temperature (55 °C). The ion replacement was achieved by two reaction mechanisms, at initial and intermediate stages, the superficial reaction and the diffusion of SO_4^{2-} ions produced a dense $SrCO_3$, the complete ion replacement process proceeded by a second mechanism by a solid-state ion exchange mechanism even under hydrothermal conditions (Yoshino et. al., 1985).

Recently, a better approach conducted to elucidate the source of the conversion $M^{2+}SO_4$ to $M^{2+}CO_3$, M= Sr or Ba; was proposed based on the crystalline structural differences associated to the physical bulk molar volume change, which also must occur due to the replacement of a large anion SO_4^{2-} by a smaller one CO_3^{2-}. Thus, the study was conducted in large single crystals of celestite $SrSO_4$ and barite $BaCO_3$, which were treated under alkaline hydrothermal conditions using high concentrated carbonated solutions of Na_2CO_3 and K_2CO_3. The mineral celestite single crystals ($SrSO_4$, square plates 10 mm wide and 3 mm thick) were topotaxially converted to strontianite ($SrCO_3$) under alkaline hydrothermal conditions. The reaction was completed in a short reaction time (such as 24 h) at a temperature of 250 °C. Increasing the treatment temperature and the molar ratio CO_3^{2-}/SO_4^{2-} accelerated the exchange of SO_4^{2-} with CO_3^{2-} ions. Under these conditions the topotaxial hydrothermal conversion to strontianite is carried out with the formation of intermediate CO_3^{2-}-rich solid solutions in the system $SrCO_3$–$SrSO_4$, this is shown in the Figure 15; and these solid solutions were formed by a controlled crystallization process achieved by the cluster dissolution–recrystallization. An anisotropic dissolution gave a characteristic texture in the converted crystals, and the difference on the reactivity of the celestite crystals in Na_2CO_3 and K_2CO_3 solutions resulted in a different texture inside the pseudomorphical converted strontianite crystals (Rendón-Angeles et. al., 2000b).

Despite the aspect of the converted crystals remained without any change after the conversion process (Fig. 15c), the formation of small holes randomly distributed were produced on outside layer produced as a result of the transformed into the MCO_3 phase Figure 16. However, the morphology of these holes underwent a change due to a simultaneous dissolution of the solid product, the recrystallization process of this phase also occurred at the same place where the previous dissolution occurred. One point that must be emphasized is that the two phases are separated by a sharp boundary in texture, as well as composition. From this observation it is clear that the replacement reaction begins from the surface of the mineral that was in contact with the hydrothermal media (Fig. 16a). In addition, the reaction proceeded by the incorporation of the solvent through the inner porosity. The holes form a zig-zag network inside the crystals which allows incorporating fresh solvent media at the reaction front (Suaréz-Orduña et. al., 2004a; Rendón-Angeles et. al., 2008). This reaction is similar to the ionic replacement process that was found for the conversion of chlorapatite and hydroxyapatite single crystals into fluorapatite single crystals under hydrothermal conditions (Rendón-Angeles et. al., 2000b). In terms of the macroscopic aspects and crystalline structural differences associated with the replacement of a large ion (SO_4^{2-} = 4.32 Å) by a smaller (CO_3^{2-} = 1.55 Å). However, the completely converted MCO_3 crystals are constituted for very tiny crystals randomly oriented which resemble a polycrystalline arrangement on the converted $M^{2+}CO_3$. Thus in this particular case, the conversion process proceeds with the formation of a converted layer that has a peculiar texture (holes, Fig. 16b) and a moving reaction front. Hence, the conversion of the $M^{2+}SO_4$ mineral species into their related carbonated inorganic compounds; is associated to a pseudomorphic replacement process rather than the ion-exchange process. In addition, at the hydrothermal conditions in which the mineral conversion can be achieved, the pseudomorphic replacement process is mainly achieved by a mechanism of coupled bulk dissolution and precipitation. It is well known that this mechanism promotes the formation of a great wide type of inorganic compounds (Putnis, 2002).

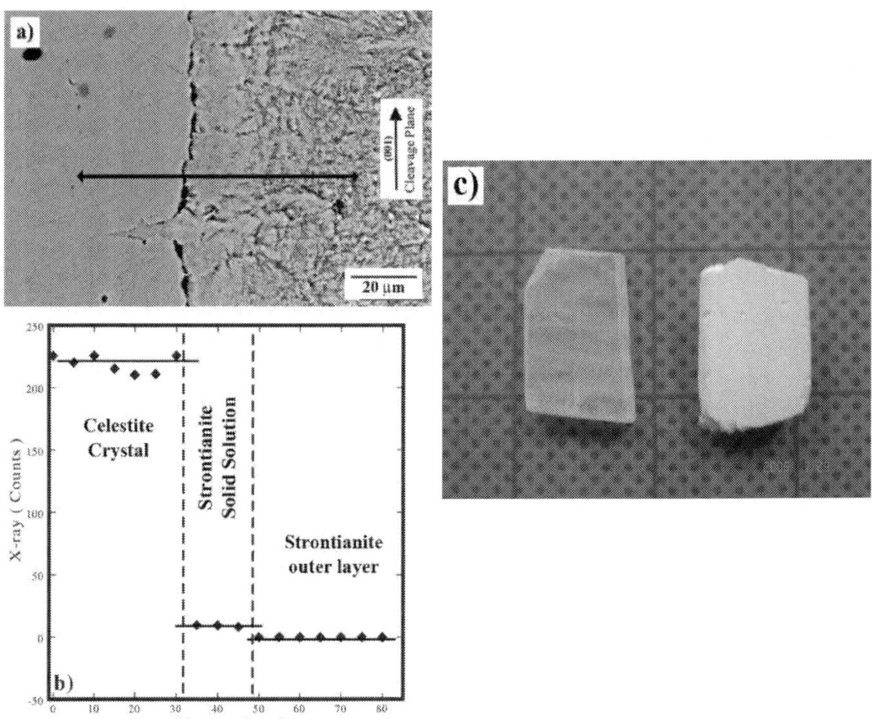

Fig. 15. (a) SEM image of the reaction interface and (b) sulfur concentration in a partially converted celestite crystal obtained by hydrothermal treatment in a Na_2CO_3 solution with a molar ratio CO_3^{2-}/SO_4^{2-} = 10, at 250 °C for 1 h. and (c) 96 h, aspects of the original $SrSO_4$ (left) and converted $SrCO_3$ (right). Grid size= 10 mm

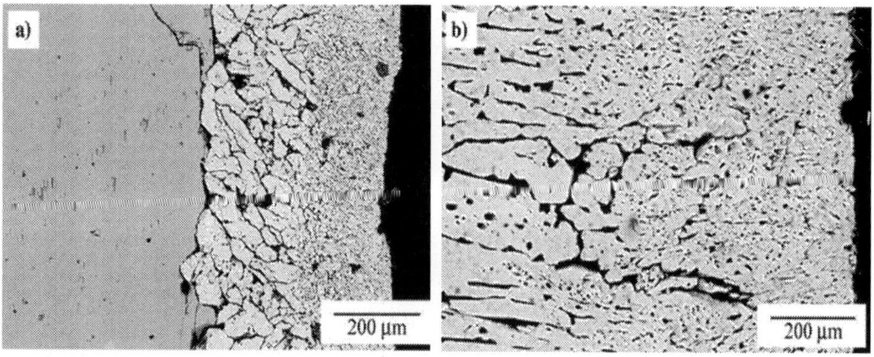

Fig. 16. SEM micrographs of barite crystals converted to barium carbonate after hydrothermal treatment at 250 °C in a Na_2CO_3 solution with a molar ratio CO_3^{2-}/SO_4^{2-} of 10 for different reaction intervals (a) 24 h (b) 192 h (Rendón-Angeles et. al., 2008).

2.5.2 Aspects associated with the formation and control of the residual porosity during the replacement reaction of mineral sulphate species

In terms of the porosity resulted as consequence of the crystalline transformation, this factor has also been the subject of controversy, because the causes that promote its formation are not clarified yet. However, we have recently found that the formation of the residual porosity depends strongly of two principal factors: i) the differences on the molar volume associated with the crystalline structural differences and ii) the chemical stability (solubility) of the new converted crystalline phase in the hydrothermal media. These inferences were established on partially and completely converted $SrCO_3$ specimens obtained by hydrothermal treatments at 250 °C for a interval of 24 h with a molar ratio $CO_3^{2-}/SO_4^{2-} = 10$. The inner volume on the converted crystals was determined by helium picnometry measurements and those results are giving in Figure 17. It is clear that the volume measurements conducted on the completed converted $SrCO_3$ plates at 250 °C for 96 h, revealed that the residual inner porosity value obtained on the specimens treated in Na_2CO_3 solutions is nearly similar to the theoretical value (dotted line in Fig. 17), this value (15.62 %) associated with the reduction of the molar volume, was calculated by considering the unit cell volume values of the parent (celestite, 307.06 Å^3) and the product (strontianite, 259.07 Å^3). This crystalline structural variation is related with the formation of the residual inner porosity, because macroscopically the crystal plate remains without any change regarding its shape and dimension, therefore, the bulk molar volume reduction does not proceed on the crystal plate and this must be compensated by which is like to proceed the residual porosity. Moreover, the control of the porosity depends on the chemical stability of the phase crystallized with the ion exchange media, once the replacement reaction was completed (Suaréz-Orduña et. al., 2004b). This inference is supported by the fact that when the conversion was carried out in K_2CO_3 solutions, the residual inner porosity on the completed converted $SrCO_3$ plate was markedly increased due to a further dissolution of the converted $SrCO_3$ crystal, in comparison with the porosity obtained the $SrCO_3$ plates transformed with highly concentrated Na_2CO_3 solutions at $CO_3^{2-}/SO_4^{2-} > 5$.

On the other hand, the remaining porosity is limited by the differences on the crystalline structure between the parent and the product. This is likely to occur when a large one replaces a small anion and a change on the structural structure also proceeds. One example was found in the case of the conversion of celestite to $SrCrO_4$, this reaction proceeds at low temperatures (200 °C) in relatively alkaline hydrothermal conditions (K_2CrO_4 solution), with the formation of a peculiar phase on the surface of the partially converted crystals. It is well known that dissolution of mineral species in hydrothermal fluids normally proceeds anisotropically, producing a peculiar texture with holes inside the recrystallized mineral specie (Suaréz-Orduña et. al., 2004a; Rendón-Angeles et. al., 2008). The holes might not be inherited from etch pits produced during the dissolution process, because they did not penetrate the crystals. The factor that has a marked influence in limiting the formation of a residual porosity is related with the replacement of SO_4^{2-} ions by CrO_4^{2-} ions in the $SrSO_4$ crystals. This fact is suggested from the structural change of the orthorhombic to monoclinic structure, which occurs during the conversion. In terms of the global unit cell volume, an expansion process is likely to proceed, therefore, in accordance with the differences on the unit cell volume between that for celestite (312.37 Å^3) and that of $SrCrO_4$ (354.11 Å^3), a volume increase of 41.74 Å^3 is attained in the converted new layer of $SrCrO_4$ as is seen in Figure 18. Hence, the global volumetric unit cell expansion must be compensated by the formation of a continuous solid phase and the formation of some microcracks in this new phase (Rendón-Angeles et. al., 2000b). The formation of a solid phase covering the partially reacted $SrSO_4$ might reduce the transfer of fresh ion exchange media, because under hydrothermal conditions the absence of a texture (small porosity) in the converted phase avoids the penetration of the hydrothermal fluid, coming to an abrupt halt of the replacement reaction

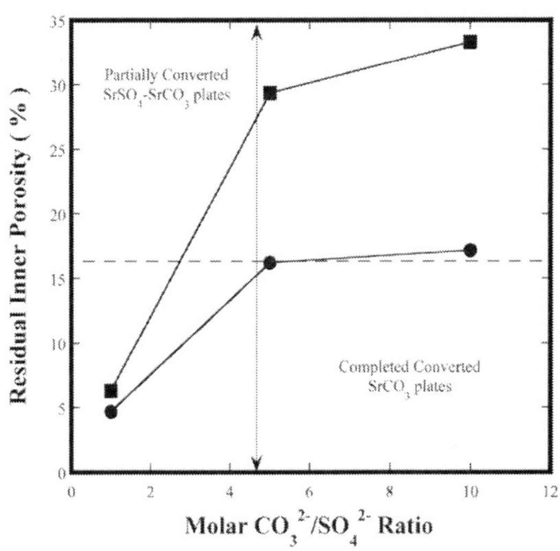

Fig. 17. Variation of the residual porosity on partial and completely converted $SrCO_3$ crystal plates, under hydrothermal conditions at 250 °C for 96 h in (●) Na_2CO_3 and (■) K_2CO_3 solutions with different molar CO_3^{2-}/SO_4^{2-} ratios. Dotted line= theoretical value calculated from the variation of the cell unit lattice associated with the structural conversion.

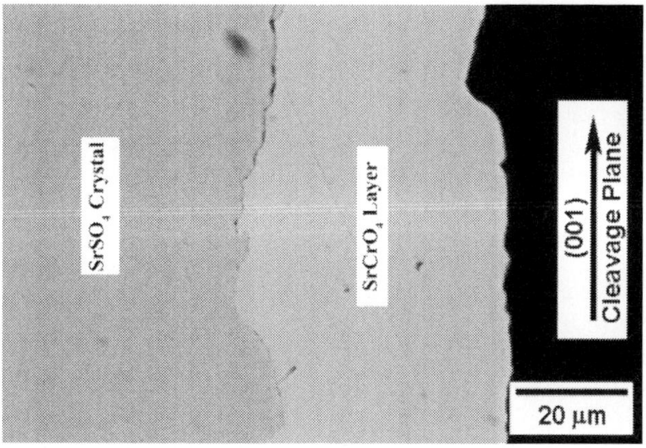

Fig. 18. Partially reacted $SrSO_4$ crystal at 200 °C for 96 h in a hydrothermal media with a molar ratio CrO_4^{2-}/SO_4^{2-}=1.

3. Conclusions

The hydrothermal crystallization of advanced materials is an important branch of chemical science and technology, this process has advantages over conventional technologies, namely the purity of products, quality, and performance, and can be consider as a green chemistry process because is environmentally friendly, because reactions consume lesser energy and these can be carried out under controlled parameters in a closed system. Although, the study of a wide number of oxides have provided physico-chemical experimental information related to solid-aqueous interactions, that affect the dissolution-precipitation mechanism main driving force for achieving bulk solid crystallization, coupled with thermodynamic modelling analysis; have contributed to establish the optimum hydrothermal processing conditions that favours the crystallization proceeds at high yield rates. However, still crystallization research work to carry out on those solid-aqueous systems associated to a specific compound that departs from the thermodynamic modelling limits. In addition, the employment of mineral ore as reactant precursors emerges as an interesting route to explore in order to produce other advanced ceramic compounds via the hydrothermal crystallization. Among the new research routes explored for hydrothermal processing, the ionic replacement is able to become common knowledge in the near future as alternative processing route, because this reaction promotes a peculiar microstructure in the transformed material that preserves the bulk original geometrical aspects during a single step hydrothermal crystallization reaction. This processing route can lead to prepare pore net-shaped materials with controlled porosity, which depending on their functional properties of the compound, can be used as gas sensor, substrates for porous catalytic materials, filters, and other applications. The conversion process focused toward a particular preparation of a functional inorganic compound from the mineral transformation requires a proper appreciation of the physico-chemical aspects involved for the solid solution-aqueous solution system. Likewise, to evaluate the likely porosity development in any mineral replacement reaction requires knowledge of the coexisting solid and fluid phases involved and their relative solubility.

4. Acknowledgment

One of the authors JCRA wish to acknowledge CONACYT and COECYT for the financial support trough research grants (Project 34830-U) and (FOMIX-COAH 2003-C02-02), respectively. JCRA and ZMV are indebt to the SNI for the financial support. Many thanks are particularly offered to former Ph D. students that collaborate with the experimental work, Roberto Suarez-Orduña, Laura Patricia Vazquez-Rivas and Yadira Marlen Rangel-Hernandez, also to technicians MSc. Martha Rivas Aguilar and Ing. Felipe de Jesus Márquez-Torres, whom helped in the preparation of samples and its observation by SEM.

5. References

Bilger, S.; Blab, G. & Förthmann, R. (1997). Sol-gel Synthesis of Lanthanum Chromite Powder, *Journal of the European Ceramic Society*, Vol. 17, No. 8, pp. 1027-1031, ISSN 0955-2219

Byrappa, K. & Yoshimura, M. (2001). *Handbook of Hydrothermal Technology*, William Andrew Publishing/Noyes, ISBN 0-8155-1445-X, New York, USA

Byrappa, K. (2005). *Hydrothermal Processing*, Kirk-Othmer Encyclopedia of Chemical Technology, pp. Copyright John Wiley & Sons, Inc., ISBN 9780471238966

Chakraborty, A.; Basu, R. N. & Maiti, H. S. (2000). Low Temperature Sintering of La(Ca)CrO_3 Prepared by an Autoignition Process, *Materials Letters*, Vol. 45, No. 3-4, (September 2000), pp. 162-166, ISSN 0167-577X

Chen, C. W.; Riman, R. E; TenHuisenb, K. S. & Brown. (2004). Mechanochemical–Hydrothermal Synthesis of Hydroxyapatite from Nonionic Surfactant Emulsion Precursors, *Journal of Crystal Growth*, Vol. 270, pp. 615–623, ISSN 0022-0248

Eckert J. O.; Hung-Houston, C. C.; Gersten, B. L.; Lencka, M. M. & Riman, R. E. (1996). Kinetics and Mechanism of Hydrothermal Synthesis of Barium Titanate, *Journal of the American Ceramic Society* Vol. 79, No. 11, (November 1996), pp. 2929-2939, ISSN 0002-7820

Gersten, B. l.; Lencka, M. M. & Riman R. E. (2004). Low Temperature Hydrothermal Synthesis of Phase-Pure (Br,Sr)TiO_3 Perovskite Using EDTA, *Journal of the American Ceramic Society*, Vol. 87, No. 11, (January 2004), pp. 2025-2032, ISSN 0002-7820

Hessien, M. M.; Hassan, M. M. & El-Barawy, K. (2009). Synthesis and Magnetic Properties of Strontium Hexaferrite from Celestite Ore, *Journal of Alloys and Compounds*, Vol. 476, No. 1-2, 12 (May 2009), pp. 373-378, ISSN 0925-8388

Ianculescu, A.; Braileanu, A.; Pasuk, I. & Zaharescu, M. (2001). Phase Formation Study of Alkaline Earth-doped Lanthanum Chromites, *J. Therm. Anal. Calorim.*, Vol. 66, No. 2, (November 2001), pp. 501-507, ISSN: 1388-6150

Inagaki, M.; Yamamoto, O. & Hirohara, M., (1990). Synthesis of $LaCrO_3$ from Complex Precipitation and its Electrical Conductivity, *Journal of the Ceramic Society of Japan*, Vol. 98, pp. 675–678, ISSN 1882-0743

Kashkurov, K. F., Nikitichev, P. I., Osipov, V. V., Sizova, L. D., & Simonov, A. V. (1968). Growth of Large Corundum Crystals by the Hydrothermal Method, *Soviet Physics Crystallografia*, Vol. 12, pp. 837– 839, ISSN 0038-5638

Kobe K. A. & Deiglmeier N. J. (1943). Strontium Carbonate, Conversion from Strontium Sulfate by Metathesis with Alkali Carbonate Solution, *Industrial and Engineering Chemistry*, Vol. 35, No. 3, pp. 323-325, ISSN 1226-086X

Komarneni S., Roy R. & Li, Q. H. (1992). Microwave-Hydrothermal Synthesis of Ceramic Powders, *Materials Research Bulletin*, Vol. 27, pp. 1393-1405, ISSN 0025-5408

Komarneni, S. ; Komarneni Y. S.; Newalkar, B.; Stour, S. (2002). Microwave-Hydrothermal Synthesis of Al-Sustituted Tobermorite from Zeolites. *Materials Research Bulletin*, Vol. 3, (March 2002), pp. 1025-1032, ISSN 0025-5408

Kubo, T.; Hogiri, M.; Kagata, H. & Nakahira, A. (2009). Synthesis of Nano-Sized $BaTiO_3$ Powders by The Rotatory-Hydrothermal Process, *Journal of the American Ceramic Society*, Vol. 92, No. S1, (January 2009), pp. S172-S176, ISSN 0002-7820

Lee, S. K.; Choi, G.J.; Hwang, U. Y.; Koo, K. K. & Park, T. J. (2003). Effect of Molar Ratio of KOH to Ti-Isopropoxide on the Formation of $BaTiO_3$ Powders by Hydrothermal Method, *Materials Letters*, Vol. 57, No. 15, (April 2003), pp. 2201-2207, ISSN 0167-577X

Lencka, M. M. & Riman, R.E . (1993). Thermodynamic Modeling of Hydrothermal Synthesis of Ceramic Powders, *Chemistry of Materials*, Vol. 5, (October 1993), pp. 61-70, *ISSN* 0897-4756

Lencka, M.M. & Riman, R. E. (1995). Thermodynamics of the Hydrothermal Synthesis of Calcium Titanate with Reference to Other Alkaline-Earth Titanates, *Chemistry of Materials*, Vol. 7, (September 1995), pp. 18-25, *ISSN* 0897-4756

Lencka, M.M. & Riman, R.E. (2002). *Intelligent Systems of Smart Ceramics*, Encyclopedia of Smart Materials, Vol. 1, John Wiley & Sons, Inc., ISBN 9780-471216278, N.J, USA

Marchisio, D. L. (2009). On the Use of Bi-Variate Population Balance Equations for Modelling Barium Titanate Nanoparticle Precipitation, *Chemical Engineering Science*, Vol. 64, (February 2009), pp. 697-708, ISSN 0009-2509

Moreira, M. L.; Mambrini, G. P.; Volanti, D. P.; Leite, E. R.; Orlandi, M. O.; Pizani, V. R. Mastelaro, P.S.; Paiva-Santos, C. O.; Longo, E. & Varela, J. A. (2008). Hydrothermal Microwave: A New Route to Obtain Photoluminescent Crystalline $BaTiO_3$ Nanoparticles, *Chemistry of Materials*, Vol. 20, pp. 5381–5387, *ISSN* 0897-4756

Moon, J.; Kerchener, J. A.; Krarup, H. & Adair, H. (1999). Hydrothermal Synthesis of Ferroelectric Perovskites from Chemically Modified Titanium Isopropoxide and Acetate Salts, *Journal of Materials Research*, Vol. 14, No. 2, (February 1999), pp. 425-434, ISSN 0884-2914

Oledzka, M.; Lencka, M. M.; Pincelup, M. P.; Mikulka-Bulen, K.; McCandlish, L. & Riman, R. (2003). Influence of Precursor on Microstructure and Phase Composition of Epitaxial Hydrothermal $PbZr_{0.7}Ti_{0.3}O_3$ Films, *Chemistry of Materials*, Vol. 15, (March 2003), pp. 1090-1098, ISSN 0897-4756

Peterson, C. R. & Slamovich, E. B. (1999). Effect of Processing Parameters on the Morphology of Hydrothermally Derived $PbTiO_3$ Powders, *Journal of the American Ceramic Society*, Vol. 82, No. 7, (July1999), pp. 1702-1710, ISSN 0002-7820

Putnis, A. (2002). Mineral Replacement Reactions: From Macroscopic Observations to Microscopic Mechanisms, *Mineral Magazine*, Vol. 66, No.5, (October 2002), pp. 689-708, ISSN 0026-461X

Putnis A. (2009). Replacement Reactions, *Review Mineral Geochemistry*, Vol. 70, pp. 87-124, ISSN 1529-6466

Qi, L.; Lee, B. I.; Badheka, P.; Yoon, D. H.; Samuels, W. D. & Exarhos, G. J., (2004). Short-range Dissolution–Precipitation Crystallization of Hydrothermal Barium Titanate, *Journal of the European Ceramic Society*, Vol. 24, No. 13, (October 2004), pp. 3553–3557, ISSN 0955-2219

Ramkrishna, D. & Mahoney, A. W. (2002) Population Balance Modeling Promise for the Future, *Chemical Engineering Science*, Vol. 57, (February 2002), pp. 595 - 606, ISSN 0009-2509

Rangel-Hernandez, Y. M.; Rendón-Angeles, J. C.; Matamoros-Veloza, Z.; Pech-Canul, M. I.; Diaz-de la Torre, S. & Yanagisawa, K. (2009). One-step Synthesis of Fine $SrTiO_3$ Particles Using $SrSO_4$ Ore Under Alkaline Hydrothermal Conditions, *Chemical Engineering Journal*, Vol. 155, No. 2, (January 2009), pp. 483 492, ISSN 1385-8947

Riman R.E .; Suchanek, W. L. & Lencka, M. M. (2002). Hydrothermal Crystallization of Ceramics, *Annales de Chimie Science des Materiaux*, Vol. 27, No. 6, (November 2002), pp. 15-36, ISSN 0151-9107

Rivas-Vázquez, L. P.; Rendón-Angeles, J. C.; Rodríguez-Galicia, J. L.; Zhu, K. & Yanagisawa, K. (2004). Hydrothermal Synthesis and Sintering of Lanthanum Chromite Powders Doped With Calcium, *Solid State Ionics*, Vol. 172, No. 1-4, (August 2004), pp. 597-600, ISSN 0167-2738

Rivas-Vázquez, L. P.; Rendón-Angeles, J. C.; Rodríguez-Galicia, J. L.; Gutiérrez-Chavarria, C. A.; Zhu, K. & Yanagisawa, K. (2006). Preparation of Calcium Doped $LaCrO_3$ Fine Powders by Hydrothermal Method and its Sintering, *Journal of the European Ceramic Society*, Vol. 26, No. 1-2, (October 2006), pp. 81-88, ISSN 0955-2219

Rendón-Angeles, J. C.; Yanagisawa, K.; Ishizawa, N. and Oishi, S. (2000a). Effect of Metal Ions of Chlorapatites on the Topotaxial Replacement by Hydroxyapatite under Hydrothermal Conditions, *Journal of Solid State Chemistry*, Vol. 154, No. 2, (November 2000), pp. 569-578, ISSN 0022-4596

Rendón-Angeles, J. C.; Yanagisawa, K.; Ishizawa, N. and Oishi, S. (2000b). Topotaxial Conversion of Chlorapatite and Hydroxyapatite to Fluorapatite by Hydrothermal Ion Exchange, *Chemistry of Materials* Vol. 12, No. 8, pp. 2143-2150, (June 2000), *ISSN* 0897-4756

Rendón-Angeles, J. C.; Pech-Canul, M. I.; López-Cuevas, J.; Matamoros-Veloza, Z. & Yanagisawa., K. (2006). Differences on the Conversion of Celestite in Solutions Bearing Monovalent Ions Under Hydrothermal Conditions, *Journal of Solid State Chemistry*, Vol. 179, No. 12, (December 2006), pp. 3645-3652, ISSN 0022-4596

Rendón-Angeles, J. C.; Matamoros-Veloza, Z.; López-Cuevas J.; Pech-Canul, M. I & Yanagisawa, K. (2008). Stability and Direct Conversion of Mineral Barite Crystals in Carbonated Hydrothermal Fluids, *Journal of Material Science*, Vol. 43, pp. 2189, ISSN 0022-2461

Rendón-Angeles, J. C., Yanagisawa, K.; Matamoros-Veloza, Z.; Pech-Canul, M. I.; Mendez-Nonell, J. & Diaz-de la Torre, S. (2010). Hydrothermal Synthesis of Perovskite Strontium Doped Lanthanum Chromite Fine Powders and its Sintering, *Journal of Alloys and Compounds*, Vol. 504, No. 1, (August 2010), pp. 251-256, ISSN 0925-8388

Roy, R. (1994). Accelerating the Kinetics of Low-Temperature Inorganic Syntheses, *Journal of Solid State Chemistry*, Vol. 111, (July 1994), pp. 11-17, ISSN 0022-4596

Schubert, U. and Hüsing, N., 2000. Synthesis of Inorganic Materials

Shi, S. &. Hwang, J. Y. (2003). Microwave-Assisted Wet Chemical Synthesis: Advantages, Significance, and Steps to Industrialization, *Journal of Mineral & Materials & Characterization & Engineering*, Vol. 2, No.2, pp. 101-110, ISSN: 1539-2511

Suárez-Orduña, R; Rendón-Angeles, J. C.; López-Cuevas, J; & Yanagisawa, K. (2004a). The Conversion of Mineral Celestite to Strontianite Under Alkaline Hydrothermal

Conditions, *Journal of Physics: Condensed Matter,* Vol. 16, pp. S1331-S1344, ISSN 0953-8984

Suárez-Orduña, R.; Rendón-Angeles, J. C.; Matamoros-Veloza, Z. & Yanagisawa, K. (2004b). Exchange of SO_4^{2-} Ions With F^- Ions In Mineral Celestite Under Hydrothermal Conditions, *Solid State Ionics,* Vol. 172, pp. 393-393, ISSN 0167-2738

Suchanek, W. L.; Shuka, P.; Byrappa, K.; Riman, R. E.; TenHuisen, K. S. & Janas, V. F. (2002). Mechanochemical-hydrothermal synthesis of Carbonated Apatite Powders at Room Temperature, *Biomaterials,* Vol. 23, (April 2001), pp. 699–710, ISSN 0142-9612

Suchanek, W. L.; Lencka M. M; & Riman, R. E. (2004). Hydrothermal Synthesis of Ceramics Materials, *Aqueous System at Elevated Temperature and Pressure Physical Chemistry in Water, Steam and Hydrothermal Solutions,* Chapter 18, ISBN 0-12-544461-3, Elsevier Ltd., New Jersey, USA

Shi, S. &. Hwang, J. Y. (2003). Microwave-Assisted Wet Chemical Synthesis: Advantages, Significance, and Steps to Industrialization, *Journal of Mineral & Materials & Characterization & Engineering,* Vol. 2, No.2, pp. 101-110, ISSN 1539-2511

Spooren, J.; Rumplecker, A.; Millange, F. & Walton, R. I. (2003). Subcritical Hydrothermal Synthesis of Perovskite Manganites: A Direct and Rapid Route to Complex Transition-Metal Oxides, *Chemistry of Materials* Vol. 15, (March 2003), pp. 1401-1403, ISNN 0897-4756

Testino, A.; Buscaglia, V.; Buscaglia, M. T.; Viviani, M. & Nanni, P. (2005). Kinetic Modeling of Aqueous and Hydrothermal Synthesis of Barium Titanate ($BaTiO_3$), *Chemistry of Materials,* Vol. 17, (September 2005), pp. 5346-5356, *ISSN* 0897-4756

Tiwary, R. K.; Narayan, S. P. & Pandey, O.P. (2008). Preparation of Strontium Hexaferrite Magnets from Celestite and Blue Dust by Mechanochemical Route, *Journal of Mining and Metallurgy, Section B,* Vol. 44, (February 2008), pp. 91 - 100, ISSN 1450-5339

Vivekanandan, R. & Kutty, T. R. N. (1989). Characterization of Barium Titanate Fine Powders Formed From Hydrothermal Crystallization, *Powder Technology,* Vol. 57, No. 3, (March 1989), pp. 181 - 192, ISSN 1570-1468

Wada, S. Suzuki T. & Noma T. (1995). Preparation of Barium Titanate Particles by Hydrothermal Method and Their Characterization, *Journal of the Ceramic Society of Japan,* Vol. 103, No. 12, (December 1995) pp. 1220-1227, ISSN 1882-0743

Walton, R. I.; Millange, F. Dmith, R. I., Hansen, T. C. & O'hare, D. (2001). Real Time Observation of the Hydrothermal Crystallization of Barium Titanate Using In Situ Neutron Powder Difracction. *Journal of the American Chemical Society,* Vol. 123, pp. 12547-12555, ISSN 0002-7863

Wang, Y.; Xu, G.; Yang, L.; Ren, Z.; Wei, X.; Weng, W.; Du, P.; Shen, G. & Han, G. (2009). Formation of Single-Crystal $SrTiO_3$ Dendritic Nanostructures via a Simple Hydrothermal Method, *Journal of Crystal Growth,* Vol. 311, No. 8, (April 2009), pp. 2519-2523, ISSN 0022-0248

Wei, X.; Xu, G.; Ren, Z.; Xu, C.; Sheng, G. & Han, G. (2008). PVA-Assited Hydrothermal Sinthesis of $SrTiO_3$ Nanoparticles with Enhanced Photocatalytic Activity for Degradation of RhB, *Journal of the American Ceramic Society,* Vol. 91, No. 11, (November 2008), pp. 11-17, ISSN 0002-7820

Wendelbo, R.; Akporiaye, D. E.; Karlsson, A.; Plassen, M. & Olafsen, A. (2006). Combinatorial Hydrothermal Synthesis and Characterization of Perovskites, *Journal of the European Ceramic Society,* Vol. 26, No. 6, pp. 849-859, ISSN 0955-2219

Yanagisawa, K.; Rendón-Angeles, J. C.; Kanai H. & Yamashita, Y. (1999). Stability and Single Crystal Growth of Dielectric Materials Containing Lead Under Hydrothermal Conditions, *Journal of the European Ceramic Society*, Vol. 19, No. 6-7, (June 1999), pp. 1033-1036, ISSN 0955-2219

Yoshimura, M. & Suchanek, W. (1997). In Situ Fabrication of Morphology-Controlled Advanced Ceramic Materials by Soft Solution Processing, *Solid State Ionics*, Vol. 98, No. 3-4, (June 1997), pp. 197-208, ISSN 0167-2738

Yoshino, K.; Nishino, T.; Yoshino, M. & Yoshimura, S. (1985). Exchange Reaction between Alkaline Earth Carbonate and Sulfate under Hydrothermal Condition, *Yogyo-Kyokai-Shi*, Vol. 93, No. 6, pp. 334-337, ISSN 0372-7718

Zhang, S.; Han, Y.; Chen, B. & Song, X. (2001). The Influence of $TiO_2 \cdot H_2O$ Gel on Hydrothermal Synthesis of $SrTiO_3$ Powders, *Materials Letters*, Vol. 51, No. 4, (November 2001), pp. 368-370, ISSN 0167-577X

Zhang, S.; Liu, J.; Han, Y.; Chen, B. & Li, X. (2004). Formation Mechanisms of $SrTiO_3$ Nanoparticles Under Hydrothermal Synthesis, *Materials Science and Engineering B*, Vol. 110, No. 1, (June 2004), pp. 11-17, ISSN 0921-5107

Zheng, W.; Pang, W.; Meng, G. & Peng, D. (1999). Hydrothermal Synthesis and Characterization of $LaCrO_3$, *Journal of Materials Chemistry*, Vol. 9, pp. 2833–2836, ISNN 0959-9428

Fe-Based Nanocomposite Formed by Thermal Treatment of Rapid-Quenched $Fe_{81}B_{13}Si_4C_2$ Alloy

Dragica M. Minić[1], Vladimir A. Blagojević[1] and Dušan M. Minić[2]
[1]*University of Belgrade, Faculty for Physical Chemistry,*
[2]*Military Technical Institute, Belgrade,*
Serbia

1. Introduction

Amorphous alloys have been a focus of considerable scientific interest, both from fundamental and practical point of view, ever since the first of its kind ($Al_{75}Si_{25}$) was produced by Klement, Willens and Duwez in 1960 (Klement et al., 1960). It has been shown that the amorphous alloys have features that are different from those of crystalline alloys in both alloy compositions and atomic configurations. This enabled the exhibition of various characteristics which were not obtained for conventional crystalline alloys. Their soft ferromagnetic properties (saturation magnetization, high permeability, low coercivity and loss), high corrosion resistance and good mechanical properties make them suitable for use in a variety of applications, such as power devices, information handling technology, magnetic sensors, anti-theft security systems and construction materials (Minić et al., 2007). Since amorphous alloys are meta-stable, elevated temperature or prolonged performance can induce a transformation into a crystalline state, which could lead to a loss of their advantageous physical properties limiting them to single-use applications. Commercial soft magnetic nanocrystalline materials have recently been successfully obtained by crystallization of amorphous precursors. Materials like this are characterized by a microstructure of nanocrystals embedded into an amorphous matrix, exhibiting superior soft magnetic and mechanical properties to both amorphous and crystalline magnetic alloys (Blagojević et al., 2011; Minić et al., 2011a). This dependence of functional properties on microstructure can be used to produce functional materials with tailored properties (Maričić et al., 2012).

The production of early amorphous alloys required very high cooling rates (as much as 10^6 K/s) to avoid crystallization. This limited the form in which they could be produced, as one dimension had to remain small enough to allow sufficiently rapid heat extraction, in order to achieve the necessary high cooling rates. The result was that the thickness of amorphous metal specimens was limited to less than 100μm. In 1976, Liebermann and Graham developed a new method of manufacturing thin ribbons of amorphous metals on a supercooled rapidly rotating disc – the melt-spinning method (Liebermann & Graham, 1976). For the next thirty years, with the production of new materials, the required cooling rate diminished, until, in 1990s, materials were developed, whose production required cooling rates as low as 1 K/s, allowing these alloys to be cast into metallic moulds to produce specimens up to 100mm thick (Ponnambalam et al., 2004). The alloys that require cooling rates below 10^3 K/s are known as bulk amorphous alloys, as the production process allows for higher specimen thickness (above 1mm), classifying them as bulk materials.

2. Experimental procedures

The ribbon shaped samples of $Fe_{81}B_{13}Si_4C_2$ amorphous alloy were obtained using the standard procedure of rapid quenching of the melt on a rotating disc (melt spinning method). The obtained ribbon was 2 cm wide and 35 μm thick. During the preparation process of the amorphous alloy ribbon, one of the sides was in direct contact with the cooled rotating disc, while the other was in inert atmosphere. As a result, the two sides of the ribbon show an easily observable difference in reflectivity, surface morphology and structure, as can be seen in X-ray diffraction (XRD) spectra, Fig.1, as well as SEM images of the sample in the as-prepared alloy (Fig.2). The side that was in the contact with the cooled spinning disc is usually labeled as fishy or matte side and the other side, free of contact, as shiny side.

DSC was obtained using SHIMADZU DSC-50 analyzer. In this case, samples weighting several milligrams were heated in the DSC cell from the room temperature to 650°C in a stream of nitrogen with nitrogen flowing at a rate of 20 mL min^{-1} at the heating rates of 5, 10, 20 and 30°Cmin^{-1}.

Mössbauer spectra were taken in the standard transmission geometry using a $^{57}Co(Rh)$ source at room temperature. The calibration was done against α-Fe foil data. "CONFIT" program package was used for the spectra fitting and decomposition (T. Žák, 1999).

The X-ray diffraction (XRD) patterns were recorded on an X'Pert PROMPD diffractometer from PANalytical with CoKα radiation operated at 40kVand 30mA. For routine characterization, diffraction data was collected in the range of 2θ Bragg angles (15–135°, step 0.0081). For a quantitative analysis and determination of crystallite size from XRD spectra, TOPAS V3 general profile and structure analysis software for powder diffraction data was used (BrukerAXS, general profile and structure analysis software for powder diffraction data, Karlsruhe, 2005). Dislocation density was obtained from the Rietveld analysis, while microstrain was calculated using Williamson-Hall method (Williamson, Hall 1953), using the XRD data. Lattice parameters obtained through XRD spectra were used to calculate the unit cell volumes, which were then compared to the standard values in JCPDS database.

Vickers microhardness tests were performed using MHT-10 (Anton Paar, Austria) microhardness tester, with loads of 0.4N and loading time of 10s (Minić, et al. 2011c). Up to seven measurements were performed on each individual sample, using the average value of microhardness for each sample. The measurements were performed on the cross-section of the ribbons, rather than on any of the sides. Error was calculated as standard deviation for each series of measurements.

2.1 Microstructure of as-prepared $Fe_{81}B_{13}Si_4C_2$ amorphous alloy

XRD spectra of the as-prepared alloy ribbon, Fig.1, showed that the as-prepared alloy already had a degree of crystallinity caused by presence of α-Fe phase (JCPDS-PDF 06-0696). The degree of crystallinity is much higher on the shiny side, with peak intensity in XRD spectrum being about 8 times higher on the shiny side.

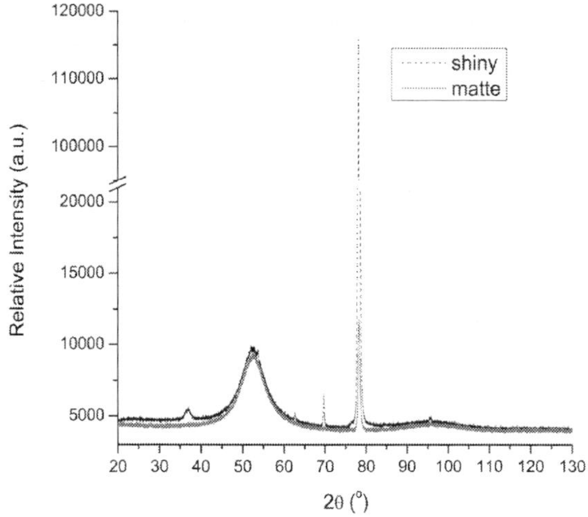

Fig. 1. X-ray diffraction spectra of as-prepared alloy

The nanocrystalline phase can be identified as α-Fe on matte side and a mixture of α-Fe and Fe_3Si on shiny side, with α-Fe being the major component. In addition to sharp crystalline peaks (37, 69 and 78°) in the XRD spectra of the as-prepared alloy, a broad spread halo around 53°, corresponding to domains of short-range ordering in the sample, was also observed. The position of the spread halo corresponds to approximate position of Fe_3Si peak and, using the Scherrer equation, we estimated the size of these domains to be 1-1.5 nm. The entire structure could best be described as a combination of nanocrystals and short-range ordered domains embedded in an amorphous matrix. Recent theoretical studies of iron-based binary systems predict existence of short-range ordering in iron-based amorphous alloys (Lass et al., 2010). Therefore, appearance of domains of short-range ordering in $Fe_{81}B_{13}Si_4C_2$ amorphous alloy could be expected.

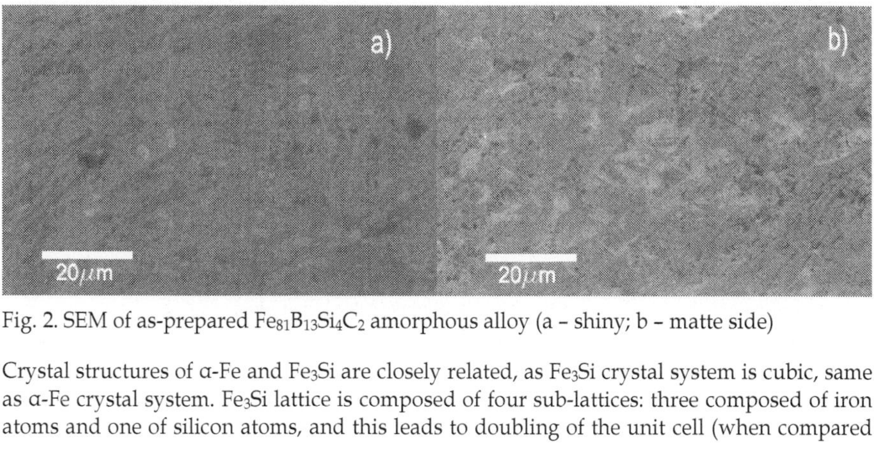

Fig. 2. SEM of as-prepared $Fe_{81}B_{13}Si_4C_2$ amorphous alloy (a – shiny; b – matte side)

Crystal structures of α-Fe and Fe_3Si are closely related, as Fe_3Si crystal system is cubic, same as α-Fe crystal system. Fe_3Si lattice is composed of four sub-lattices: three composed of iron atoms and one of silicon atoms, and this leads to doubling of the unit cell (when compared to α-Fe). Fe_3Si lattice is, also, slightly distorted, so the value of its lattice parameter is slightly higher than twice the value of lattice parameter of α-Fe.

2.2 Thermal stability of alloy

The thermal stability of the alloy was investigated using differential scanning calorimetry (DSC) in a nitrogen atmosphere (Minić, et al., 2009a). Typical DSC scan obtained during heating and cooling cycle is presented in Fig. 3. DSC scan involves a series of endothermic and exothermic peaks indicating a stepwise process of structural stabilization of the alloy in the temperature range 170-560°C. A broad exothermic peak, indicated as (T_{sr}), in temperature range 170-400°C, coresponding to structural relaxation, is followed by endothermic hump (temperature of glass transition T_g) and a short supercooled liquid region before the sharp exothermic crystallization peak (T_k) in temperature range 500-540°C. The enthalpy of crystallization is 83.5 J/g as determined from area of corresponding peak obtained at heating rate of 20 °Cmin^{-1}.

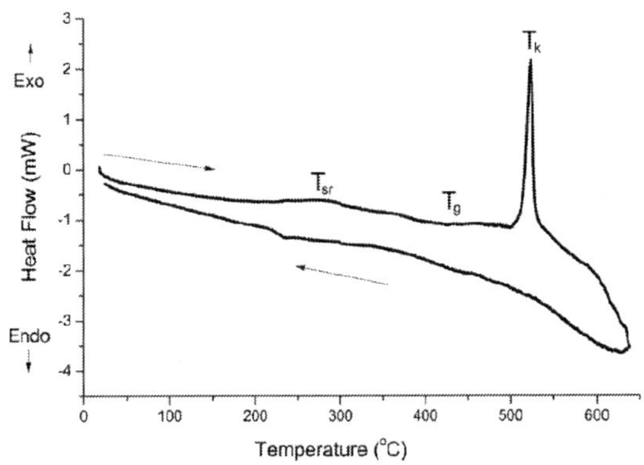

Fig. 3. DSC scan of heating and cooling cycle in nitrogen atmosphere; heating rate 10°C/min.

2.3 Magnetic properties

2.3.1 Thermo-magnetic behavior

Thermally induced processes were also studied using the thermo-magnetic scan, where the sample is heated, annealed and then cooled in vacuum furnace at low magnetic field of 4 kA m^{-1} while its magnetic moment is monitored (Minić, et al., 2011b). Both heating and cooling rate were 4°C min^{-1}, dwell time at maximum temperature of 800°C was 30 minutes. The shape of the thermo-magnetic curve, Fig. 4, reflects changes in the magnetic moment of the sample, caused by phase or structural transitions.

Most pronounced change represents the Curie point (T_c = 420°C), where the magnetization of the respective phase drops to almost zero, because the thermal motion overcomes magnetic interaction. Annealing at the temperature near 200°C is sometimes called stress-relieving (or structural relaxation). Temperatures marked by arrows in Fig. 4 have been identified as points of interest for further study of structural transformations.

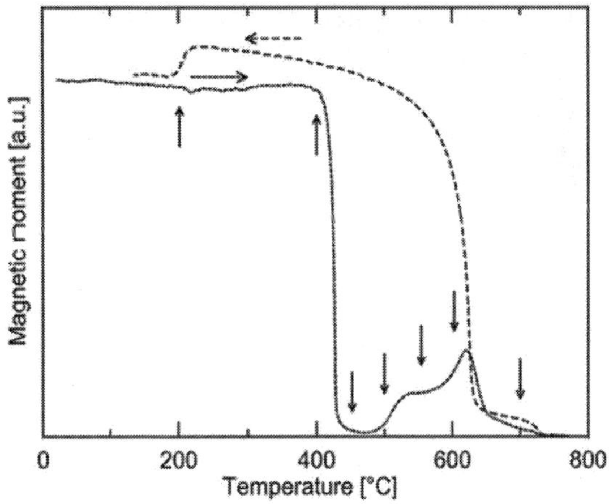

Fig. 4. Thermomagnetic scans for increasing (dotted line) and decreasing (dashed line) temperature.

2.3.2 Magnetic susceptibility

Measurements of relative magnetic susceptibility were performed using a modified Maxwell method, based on the action of an inhomogeneous field on the magnetic sample. The magnetic force measurements were performed with a sensitivity of 10^{-6}N in an argon atmosphere (Minić, et al. 2009b).

The temperature dependence of the relative magnetic susceptibility of the as-prepared $Fe_{81}B_{13}Si_4C_2$ amorphous alloy during three thermal treatments to different temperatures is presented in Fig. 5. During the first and second treatment, the decrease in the magnetic susceptibility in the temperature region from 320°C to 380°C is the result of proximity to the Curie temperature of the amorphous alloy. Before the start of the second treatment it was observed that magnetic susceptibility increased slightly. This was caused by the structural relaxation of an amorphous structure during the first treatment. During this process, internal strains and the free volume are reduced in the starting material. These changes are accompanied by subtle inter-atomic movements, causing the changes in the electron structure and leading to an increase in the number of electrons with unpaired spin in the direction of the outer magnetic field (Minić et al., 2009b). This also leads to a decrease in the number of electrons spinning in the reverse direction and causes an increase in the magnetic susceptibility upon cooling. At the same time, strains and decrease in the free volume enable greater mobility of the walls of the magnetic domains and this behavior further contributes to the increase in the magnetic susceptibility.

During the second treatment, the alloy loses its ferromagnetic properties in the temperature region from 400°C to 470°C. With further heating, the magnetic susceptibility starts to rise, and the alloy regains its ferromagnetic properties once the crystallization process starts at about 490°C. After the second heating to 490°C, the magnetic susceptibility decreases by 23 % when compared to the value of as-prepared alloy and to the value of the relaxed state of the lattice after the first heating cycle. During the third treatment, above the crystallization temperature, the alloy maintains its ferromagnetic features in the whole temperature region, whereas the maximum change in the magnetic susceptibility occurs at about 190°C as a consequence of further phase transformation of the crystallized alloy.

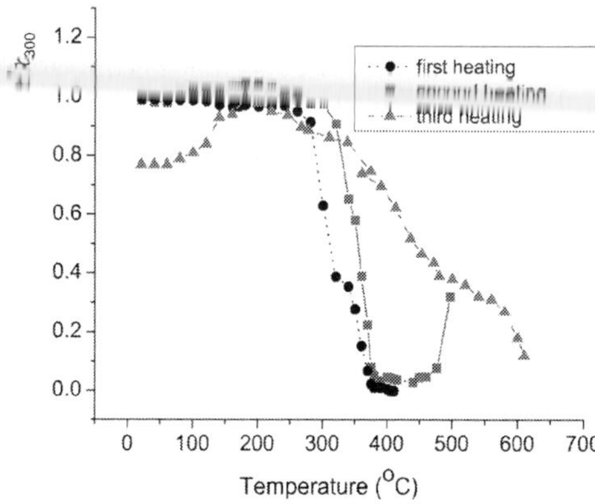

Fig. 5. Temperature dependence of relative magnetic susceptibility of as-prepared $Fe_{81}B_{13}Si_4C_2$ amorphous alloy during three thermal treatments up to different temperatures: a) 420°C; b) 500°C; c) 630°C.

2.3.3 Mössbauer spectra

In the Fig. 6, spectra illustrate the ability of Mössbauer effect to distinguish between individual iron-containing phases of different structure. Broad-line components (Fig. 6a) are

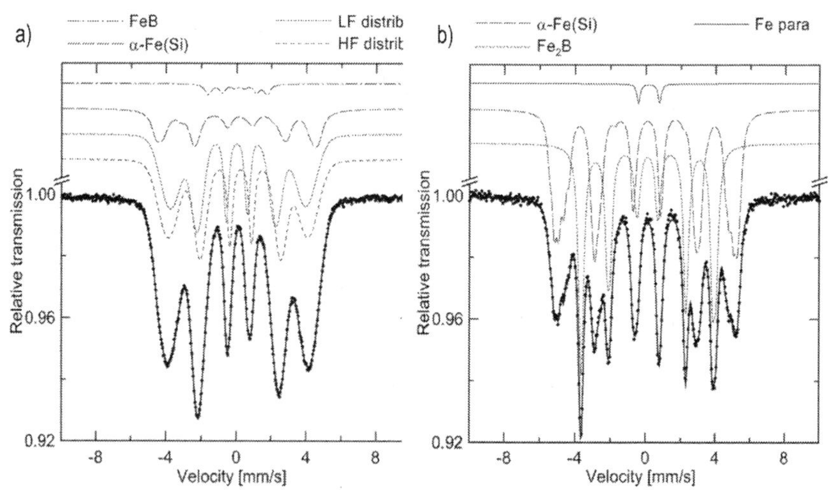

Fig. 6. Mössbauer spectra of the as-prepared material (left) and of material after final annealing at 700°C (right), including components of iron containing phases.

typical for the amorphous volume of the as-prepared sample, while the sharp lines (Fig. 6b) characterize the crystalline structure with well defined position of atoms, which results from the thermally induced crystallization process during annealing of the sample (Minić, et al. 2011b). The computer processing of Mössbauer spectra yielded intensities, *I*, of components, their hyperfine inductions, B_{hf}, isomer shifts, δ, and quadrupole splitting, σ (T. Žák & Y. Jirásková, 2006). The contents of the iron-containing phases are determined as proportional to the relative areas of the corresponding spectral components. The phase percentages correspond to the distribution of Mössbauer iron atoms among phases (Table 1).

Annealing temperature	Amorphous at%	α-Fe(Si) at%	Fe_2B at%	Fe_3B at%	FeB at%	α-Fe at%	Fe para at%
as-prepared alloy	0.95	0.03	–	–	0.02	–	–
200°C/30 min.	0.94	0.02	–	–	0.02	0.02	–
450°C/30 min.	0.83	0.14	–	0.03	–	–	–
500°C/30 min.	–	0.42	0.42	0.15	–	–	0.01
550 °C/30 min.	–	0.52	0.47	–	–	–	0.01
600°C/30 min.	–	0.54	0.45	–	–	–	0.01
700°C/30 min.	–	0.55	0.44	–	–	–	0.01

Table 1. Mössbauer tentative phase analysis (distribution of Mössbauer iron atoms among phases) (Minić, et al. 2011e)

In the as-prepared alloy, the amorphous structure, having a high-field and a low-field component is accompanied by a small amount of α-Fe(Si) solid solution and a FeB phase. Mössbauer phase analysis at higher temperatures reveals α-Fe(Si) solid solution and Fe_2B phase to be the most important final crystallization products, although metastable phase Fe_3B, was detected initially at 450°C and in higher percentage at 500°C. Amount of iron atoms in paramagnetic positions is almost below the sensitivity threshold. Content of silicon in the α-Fe(Si) solid solution seems to be about 9 at.%, which is close to 7 at.% published in (Saegusa & Morrish, 1982).

2.4 Electrical properties

2.4.1 Electrical resistance

The electrical resistance of the ribbon was measured using the four-point method within a temperature interval of 20–630°C in an argon atmosphere (Minić, et al., 2011d). Fig. 7a shows the temperature dependence of the electrical resistivity of the alloy in the temperature range of 25-630°C. The dependence clearly shows each structural stabilization step which causes the change in the ordering of the investigated material. These changes are more obvious in the derivative curve (Fig. 7b).

The slow increase of electrical resistivity was caused by the structural relaxation process in the temperature range of 200-380°C. This process is followed by an increase of electrical resistivity in the vicinity of Curie temperature T_c at 420°C, corresponding to the first maximum of the differential curve. At this point the effect that scattering of conductive electrons had on the magnons disappeared (I. Balberg & J. S. Helman, 1978; G. Bohnke et al., 1983) and the amorphous alloy loses its ferromagnetic properties (D. M. Minić et al. 2010). This is in excellent agreement with the results of the thermo-magnetic measurements (Fig.

4). The beginning of crystallization at about 520°C caused a sharp decrease of electrical resistivity. The appearance of two clearly separated maxima, T_{k1} and T_{k2} (490 and 510°C respectively) on the differential curve of electrical resistivity (Fig. 7b), suggests that crystallization of the amorphous alloy is a complex process, occurring in two steps, which appear as a single overlapping peak in the DSC scans, Fig. 3.

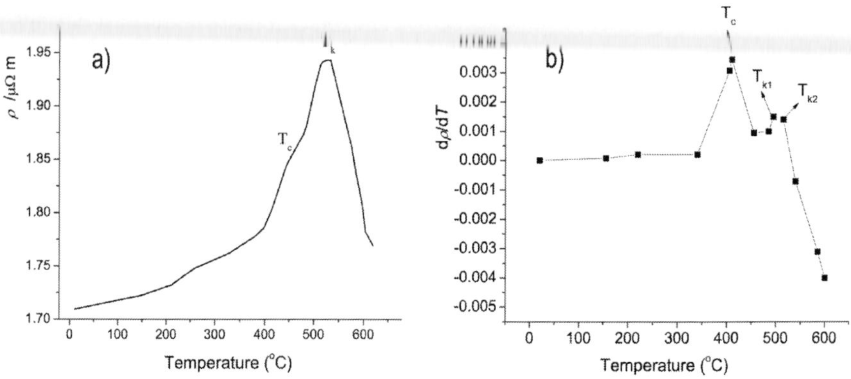

Fig. 7. Temperature dependence of the electrical resistivity of amorphous alloy.

The electrical resistivity of the crystalline alloy is lower than that of the amorphous alloy of the same composition, as a result of the increase in electron free path. The linear change of electrical resistivity with increasing temperature during the second treatment shows that crystallization was completed during the first heating cycle (Fig. 8).

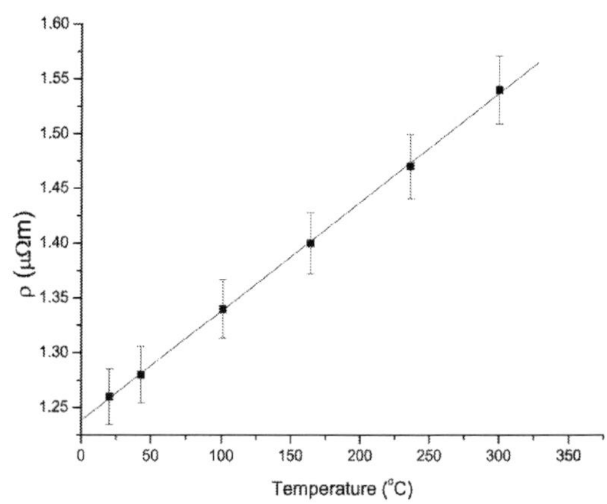

Fig. 8. Temperature dependence of the electrical resistivity of second thermal treatment

2.4.2 Thermo-electromotor force

The structural relaxation processes, as well as the crystallization, in the temperature interval of 25-680°C, were also investigated by measuring the thermo-electromotor force (TEMF) of a thermocouple made by coupling a copper conductor to the amorphous alloy (D.M. Minić, et al. 2009b). The alloy sample was mechanically attached to a copper conductor, forming the Cu - $Fe_{81}B_{13}Si_4C_2$ thermocouple, which was placed into a specially designed furnace, while the other end of the sample was submerged into a mixture of water and ice. The TEMF produced by the thermocouple during the heating process was measured by a voltmeter with the sensitivity of 10^{-5}V.

The temperature dependence of a thermo-electromotor force (Fig. 9) shows three linear regions corresponding to the structural transformations of the alloy. Different slopes of these linear dependences correspond to structural changes involving a structural relaxation, the loss of ferromagnetic properties, and the crystallization, respectively. The temperature coefficient of TEMF is a function of the electron state density at the Fermi level:

$$\alpha = \frac{k}{2e}\left(\frac{N_{1E_F}}{N_{2E_F}} - \frac{N_{2E_F}}{N_{1E_F}}\right) \tag{1}$$

where k is Boltzmann's constant, e is electron charge, $N_{1(Ef)}$ is the electron state density in copper and $N_{2(Ef)}$ is the electron state density in the alloy.

The electron density of states in copper remained unchanged during heating to 680°C, meaning that the change in the temperature coefficient during the heating of the thermocouple was caused only by the change of the electron density of states at the Fermi level of the alloy. Based on the slope of the first linear segment in Fig. 9, temperature coefficient a_1= 9.4 μV/K, and the relative change in the electron density of states of the alloy caused by the structural relaxation process was determined to be $\frac{\Delta N_1}{N} = 3.53\%$. The temperature coefficient for the second linear segment is a_2= 8.36 μV/K, and $\frac{\Delta N_2}{N} = 5.33\%$

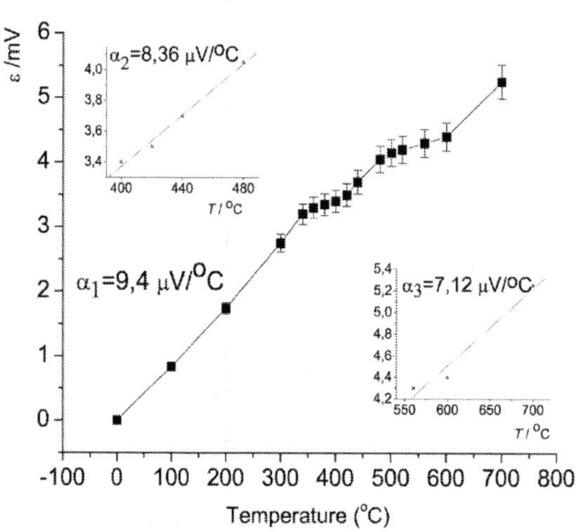

Fig. 9. Temperature dependence of thermo-electromotor force.

was determined to be 5.33% and for the third linear segment a_3= 7.12 μV/K and $\frac{\Delta N_3}{N} = 7.81\%$ was 7.81%. The overall change in the electron density of states at the Fermi level caused by the structural transformations during heating the alloy in temperature range 25-680°C is the sum of the three ΔN/N values and equals 16.67%.

The increase in the electron density of states at the Fermi level and the above mentioned increase in free path of the electron combine to diminish resistivity of the crystalline alloy (Fig. 8).

2.5 Structural transformations induced by thermal treatment

X-ray diffraction (XRD) spectra of the alloy ribbon samples (Fig. 10) show that thermal treatment (200-700°C) caused a series of structural transformations of the amorphous alloy leading to formation of more than one crystalline phase: the stable α-Fe, Fe_3Si and Fe_2B as well as metastable Fe_3B. Crystallization initially leads to formation of a nanocomposite structure of nanocrystals dispersed in the amorphous matrix. After thermal treatment at 700°C, the alloy ribbon sample was fully crystallized and composed of interdispersed nanocrystals of three crystalline phases: α-Fe, Fe_3Si and Fe_2B. The analysis of crystallite orientation on the two sides of the ribbon showed that Fe_3Si and α-Fe crystallites exhibit a degree of preferential orientation after thermal treatment. In the as-prepared alloy, Fe_3Si and α-Fe crystallites on the shiny side are oriented preferentially in [100] direction, while those on the matte side are not. This is probably the reason for higher reflectivity of the shiny side of the ribbon and the consistently high intensity of peak around 78° in the XRD spectra of the shiny side. After the treatment at 500°C, the degree of preferential orientation increases on the shiny side and forms on the matte side, suggesting asymmetric growth of the crystallites. After treatment at 700°C, amplitudes of the degree of preferential orientation on both sides decreased, suggesting that the crystallites were growing in more symmetrical manner. All of these changes are much more pronounced on the shiny side than on the matte side.

Rietveld analysis (Minić, et al. 2011c, Fig. 11) of XRD spectra yielded phase composition of the crystalline portion of the samples, average crystal sizes, dislocation density and microstrain for individual phases (Table 2). R^2 was greater than 0.98 for all XRD measurements. Individual phase contributions of α-Fe and Fe_3Si could not be completely separated, because of the overlap of their peaks. The phase composition data shows that shiny side has higher percentage of metastable Fe_3B phase, but lower percentage of Fe_2B phase. After heating at 700°C, the final phase composition on the matte side shows more crystalline Fe_2B than on the shiny side. The phase content of Fe_3B declines in two steps, with increase in heating temperature, to disappear completely after thermal treatment at 650°C. The first sharp decrease in percentage of Fe_3B (550°C) coincides with an increase in percentage of combined α-Fe and Fe_3Si. After thermal treatment at 650°C, Fe_3B phase disappears completely, and this coincides with increase in phase content of Fe_2B. Phase content of Fe_2B phase shows an increase after treatment at temperatures above 550°C.

The evolution of average crystal sizes for the respective phases shows that while there is some difference between shiny and matte side of the ribbon at the onset of crystallization, those differences become almost negligible after heating at 700°C. On closer examination, Fe_3Si and α-Fe have lower crystal sizes on the matte side, at heating temperatures of 500°C and below, and then grow faster than they do on the shiny side. As a consequence, crystal sizes of Fe_3Si and α-Fe, respectively, at the end of crystallization are almost the same on both sides of the ribbon. The evolution of average crystal size of Fe_2B shows the same trend of smaller initial crystal size and then faster crystal growth on the matte side, with increase in temperature. Metastable Fe_3B phase shows the same trend with regards to growth, except, it has a higher average crystal size on the matte side, after the samples are treated at 500°C.

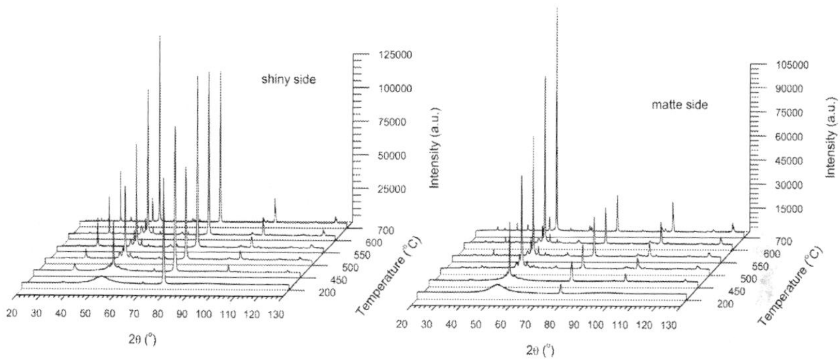

Fig. 10. XRD spectra of alloy samples after thermal treatment (left - shiny side; right - matte side)

In order to further comprehend the changes in phase composition, unit cell volumes for individual phases were determined using XRD data, and compared with standard values in JCPDS database (Minić, et al. 2011c). This way, we estimated the lattice distortion caused by the presence of boron in α-Fe and Fe_3Si lattices. The change in unit cell volume was negative for α-Fe and positive for Fe_3Si phase. The diagram (Fig 12) shows that the distortion of the unit cell of α-Fe was the greatest before the crystallization started; it decreased during the crystallization and remained relatively stable after treatment at 500°C and higher temperatures.

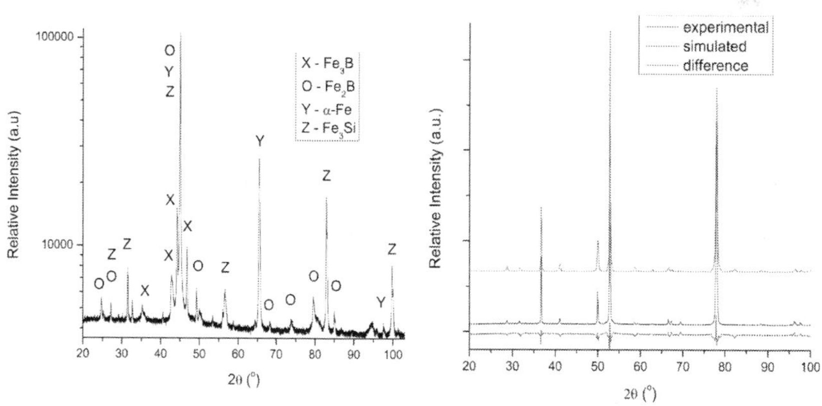

Fig. 11. XRD spectrum of alloy sample treated at 600°C with individual peak assignments (left); Rietveld analysis of XRD spectrum of alloy sample treated at 700°C (right)

The change in Fe_3B percentage after treatment at 550°C was accompanied by a decreased distortion of the unit cell of both α-Fe and Fe_3Si and increase in their combined phase content. This indicates that iron from Fe_3B was transformed into these two phases, with boron being incorporated back into the α-Fe/Fe_3Si matrix. As Fe_3Si showed greater change, Fe_3B probably transformed more to Fe_3Si than it did to α-Fe.

Temperature °C	Matte				Shiny			
	α-Fe	Fe_3Si	Fe_2B	Fe_3B	α-Fe	Fe_3Si	Fe_2B	Fe_3B
Phase composition (% mass of crystalline phase)								
500	84± 4		8± 2	8± 4	83± 5		6± 2	11± 2
550	88± 3		8± 2	4± 1	88± 3		8± 2	4± 1
600	86± 3		10± 2	4± 1	89± 3		8± 2	3± 1
700	87± 2		13± 2	-	90± 2		10± 2	
Average crystal size (nm)								
25	26.2				36.9			
200	24.6				36.5			
450	19.7				32.9			
500	26.7	22.3	49.3	16	33.1	35.1	37.3	15.8
550	32.1	35.4	90.6	46.4	36.4	45.6	82.7	44.8
600	38.6	51.6	97.7	56.6	39.0	52.7	111.1	67.7
700	44.6	71.9	107.5	-	44.9	67.5	133.5	
Dislocation density (10^{15} m^{-2})								
25	4.37				2.2			
200	4.96				2.25			
450	7.73				2.77			
500	4.21	6.03	1.23	11.72	2.74	2.43	2.16	12.02
550	2.91	2.39	0.37	1.39	2.26	1.44	0.44	1.50
600	2.01	1.13	0.31	0.94	1.97	1.08	0.24	0.66
700	1.51	0.58	0.26	-	1.49	0.66	0.17	
Microstrain (%)								
25	2.19				2.47			
200	1.85				2.63			
500	2.53				3.56	4.73		
550	2.00	4.27			2.26	3.76		
600	1.94	4.03			1.70	5.69		
700	1.26	3.47			1.37	2.66		

Table 2. Phase composition and dislocation density in the samples, obtained by analysis of XRD data

Reduced lattice distortion in these phases is probably due to the fact that mass percentage of boron in Fe_3B (6.06%) is lower than the mass percentage of boron in the as-prepared alloy (13%). Therefore, transformation of Fe_3B would lead to dilution of boron in α-Fe/Fe_3Si matrix, leading to stabilization of its crystal structure. The change in unit cell volumes of Fe_3Si and α-Fe showed increase at higher heating temperatures and it is possible that the distortion of the lattice in α-Fe and Fe_3Si is caused by the increasing occurrence of crystal/crystal interfaces between α-Fe and Fe_3Si on one side and Fe_2B on the other.

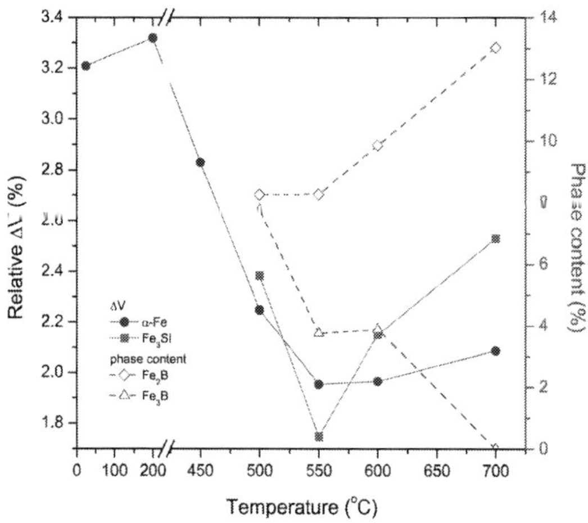

Fig. 12. Change in unit cell volumes of α-Fe and Fe_3Si against change in phase contents of Fe_2B and Fe_3B

2.6 Mechanical properties

Microhardness was measured on the cross-section of the alloy ribbon samples, rather than on any of the sides, giving a good measure of the average properties of the alloy samples. This was done due to extreme brittleness of the ribbon samples after crystallization. In order to clearly show the influence of the change in microstructure, induced by thermal treatment, on microhardness, we presented the microhardness data in combination with DSC scan (Fig. 12). In terms of microstructure, the as-prepared alloy contains a small percentage (less than 5%) of crystalline α-Fe phase in form of nanocrystals dispersed in the amorphous matrix (Fig. 1) and small domains (1-1.5 nm in size) of short-range crystalline ordering. This structure can be best described as a nanocomposite of nanocrystals and nanoclusters embedded in the amorphous matrix. The nanocomposite structure combines with chemical composition, involving significant percentage of boron, silicon and carbon, all of which are known to increase hardness in iron alloys, to produce high hardness of the as-prepared alloy sample (909HV).

The change of microhardness with respect to heating temperature showed three distinct temperature regions with completely different behavior. Before the onset of crystallization around 450°C, microhardness exhibited slight growth, from 909 to 931HV, which corresponds to lattice relaxation, as shown by the broad exothermic peaks in 200-400°C region in the DSC (Fig. 13a). In the second region, after thermal treatment at 450°C, it increased to 951HV and then to 1250HV, after thermal treatment at 500°C. This corresponds to crystallization involving formation of disordered Fe_3Si crystal phase (around 450°C), stable crystalline phases Fe_3Si, Fe_2B, α-Fe and metastable Fe_3B (exothermic peak around 512°C). These phases form a nanocomposite of nanocrystals dispersed in the amorphous matrix, which is the main cause of the increased microhardness. In the last temperature region, after the sample was treated at 700°C, microhardness decreased to 908HV, as a consequence of further crystal growth, leading to the change in the nature of the interfaces.

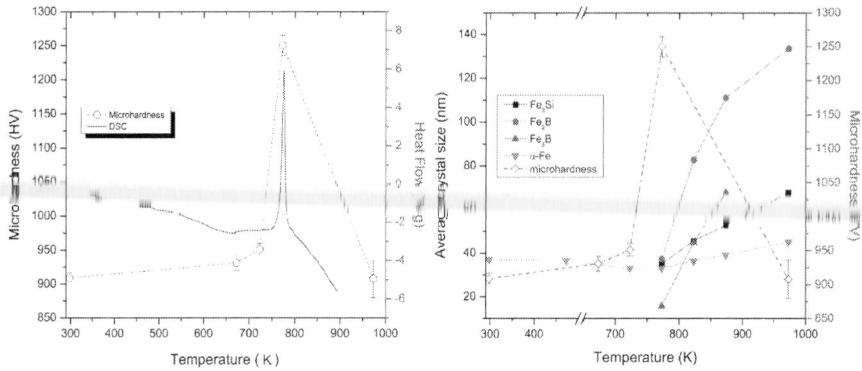

Fig. 13. Microhardness shown with DSC scan (a) and average crystal size on matte side (b).

These changes in microhardness can also be correlated with change in microstructure, induced by thermal treatment, through evolution of average crystal size of the crystalline phases in the alloy (Fig. 13b) (Minić, et al. 2011c). It can be seen that average crystal size has a significant influence on the change in microhardness. When the formed nanocrystals are relatively small (below 50 nm), microhardness remains high, while appearance of larger nanocrystals (over 100 nm) leads to a sharp decrease in microhardness. After thermal treatment at temperatures from 200 to 700°C, the alloy structure gradually transformed from a relatively homogeneous to a granulated structure with larger crystalline domains. This is consistent with creation, at the beginning of the crystallization, of a nanocomposite of small nanocrystals dispersed in amorphous matrix, which significantly increases microhardness, due to the fact that the dominant type of interface in the alloy is crystal/amorphous, as opposed to crystal/crystal in a completely crystalline alloy. As the average crystal size increased, crystal/crystal interfaces became dominant and microhardness decreased, due to increased interfacial energy and more successful propagation of shear bands and cracks along these interfaces. During the course of the observed structural transformations, the overall composition of the alloy did not change drastically, while microhardness showed significant fluctuation. This means that the changes of microhardness in our alloy were caused primarily by changes in its microstructure, rather than changes in its composition, with the main factor being the change in the average crystal size and creation of a granulated structure, as opposed to nanocrystals embedded in amorphous matrix (Table 2). The granulated structure, in addition to containing larger crystals and more crystal/crystal interfaces, is also much more porous than the original nanocomposite crystal/amorphous structure of the as-prepared alloy.

2.7 Kinetics of crystallization

All kinetic studies assume that the isothermal rate of conversion $d\alpha/dt$ is a linear function of the temperature-dependent reaction rate constant, $k(T)$, and a temperature-independent function of the conversion, $f(\alpha)$ (Vyazovkin, 2000)

$$\frac{d\alpha}{dt} = k(T) f(\alpha), \tag{2}$$

where a is the fractional extent of reaction (conversion degree), t is time and the function $f(\alpha)$ depends on the particular crystallization mechanism.

The temperature dependence of the rate conversion is introduced by replacing $k(T)$ with the Arrhenius equation, which gives the relation

$$\frac{d\alpha}{dt} = A\exp\left(-\frac{E}{RT}\right)f(\alpha) \tag{3}$$

where A (pre-exponential factor) and E (activation energy) are the Arrhenius parameters and R is the gas constant.

Kinetic description of solid state transformations usually includes a kinetic triplet, involving Arrhenius parameters (activation energy, E and pre-exponential factor, A) as well as an algebraic expression of the conversion function, $f(a)$ (presented in Table 3), which describes the dependence of the reaction rate on the conversion degree, a.

In solid state reactions, the constant value of activation energy can be expected only for a single-step reaction, therefore E in equation (3) becomes an apparent quantity (E_a), based on a quasi-single-step reaction. In non-isothermal measurements at constant heating rate, β, the equation (3) transforms to:

$$\beta\frac{d\alpha}{dT} = A\exp\left(-\frac{E_a}{RT}\right)f(\alpha) \tag{4}$$

where $d\alpha/dt \equiv \beta\,(d\alpha/dT)$.

The integral form of the reaction model can be obtained by integration of equation (4)

$$g(\alpha) = \int_0^{\alpha}\frac{d\alpha}{f(\alpha)} = \frac{AE_a}{R\beta}p(x) \tag{5}$$

where $p(x)$ is the temperature integral for $x=E_a/RT$ which does not have analytical solution.

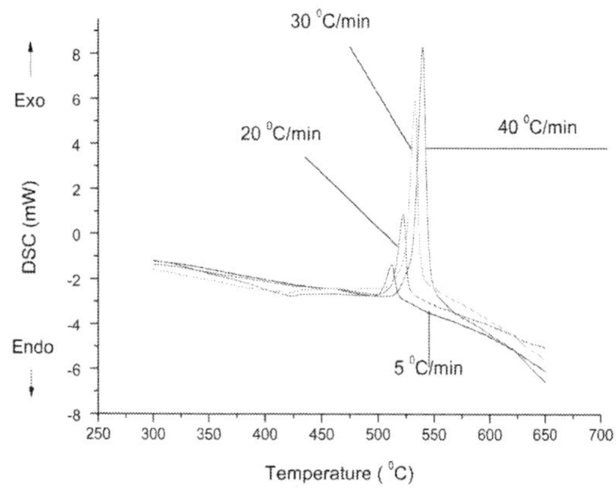

Fig. 14. DSC curves at different heating rates

The different algebraic expressions of conversion functions for solid state transformations are given in Table 3.

Fig. 14 shows the continuous DSC curves of $Fe_{81}B_{13}Si_4C_2$ ribbon in temperature range 300-650°C taken at four different heating rates. All curves contain a single well formed exothermic peak representing crystallization in the temperature range 500-560°C.

Mechanism label	$f(\alpha)$	$g(\alpha)$
Power law P4	$4\alpha^{3/4}$	$\alpha^{1/4}$
Power law P3	$3\alpha^{2/3}$	$\alpha^{1/3}$
Power law P2	$2\alpha^{1/2}$	$\alpha^{1/2}$
Power law P3/2	$3/2\alpha^{1/3}$	$\alpha^{2/3}$
Avrami-Erofeev A3/2	$3/2(1-\alpha)[-\ln(1-\alpha)]^{1/3}$	$[-\ln(1-\alpha)]^{2/3}$
Avrami-Erofeev A2	$2(1-\alpha)[-\ln(1-\alpha)]^{1/2}$	$[-\ln(1-\alpha)]^{1/2}$
Avrami-Erofeev A3	$3(1-\alpha)[-\ln(1-\alpha)]^{3/2}$	$[-\ln(1-\alpha)]^{1/3}$
Avrami-Erofeev A4	$4(1-\alpha)[-\ln(1-\alpha)]^{3/4}$	$[-\ln(1-\alpha)]^{1/4}$
Prout-Tompkins B1	$\alpha(1-\alpha)$	$\ln[\alpha(1-\alpha)^{-1}]$
Šesták-Berggren	$\alpha^N(1-\alpha)^M$	-
One dimensional phase boundary R1	1	α
Contracting cylinder R2	$2(1-\alpha)^{1/2}$	$1-(1-\alpha)^{1/2}$
Contracting sphere R3	$3(1-\alpha)^{2/3}$	$1-(1-\alpha)^{1/3}$
D1 One-dimensional diffusion	$1/2\alpha$	α^2
D2 Two-dimensional diffusion	$[-\ln(1-\alpha)]^{-1}$	$(1-\alpha)\ln(1-\alpha)+\alpha$
D3 three-dimensional diffusion	$3/2(1-\alpha)^{2/3}[1-(1-\alpha)^{1/3}]^{-1}$	$[1-(1-\alpha)^{1/3}]^2$
D4 Ginstling-Brounshtein	$3/2[(1-\alpha)^{-1/3}-1]^{-1}$	$(1-2\alpha/3)-(1-\alpha)^{2/3}$
F1 First-order	$1-\alpha$	$-\ln(1-\alpha)$
F2 Second-order	$(1-\alpha)^2$	$(1-\alpha)^{-1}$
F3 Third-order	$1/2(1-\alpha)^3$	$(1-\alpha)^{-2}$

Table 3. Algebraic expressions of conversion functions for solid state transformations

All values of initial (T_i), maximal (T_p) and final temperature (T_f) for both exo-peaks are shifted to higher values with increasing heating rate indicating the presence of kinetic effects (Table 4). The shape of DSC curves depends on the heating rate, because, as the values of shape factor S show, asymmetry of peaks increases with the decrease of heating rates indicating that the heating rate has a strong influence on the crystallization process. The values of shape factor S is obtained as the ratio of half-widths (left and right) for individual crystallization peaks for particular heating rate. The fractional extent of the sample transformed into crystalline phase, α, has been obtained from the DSC curve as a function of temperature (T) (Minić & Adnađević, 2008). At any temperature T, α is defined as $\alpha = S_T/S$, where S is the total area of the exotherm, between the temperature T_i, of the onset of crystallization and the temperature T_f, of the end of crystallization, and S_T is the area between the initial temperature T_i and a generic temperature, T, ranging between T_i and T_f (Friedman, 1964). The plots of α versus T at different heating rates for the considered crystallization process are shown in Fig. 15.

The sigmoid shape of fractional conversion curves in Fig. 15 indicates, for all heating rates, a slow initial period corresponding to nucleation with increasing rate caused by the increase of specific surface area of nuclei. The saturation part that follows is the consequence of nuclei merging together causing a decrease in their surface area.

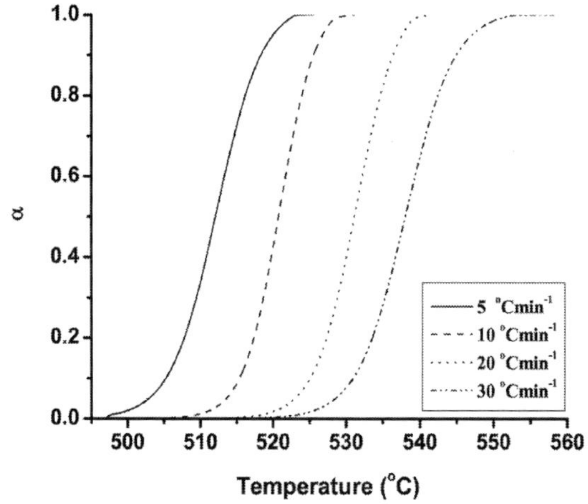

Fig. 15. Fractional conversion (α) as a function of temperature (*T*) for the crystallization alloy at different heating rates (5, 10, 20 and 30°Cmin^{-1}).

2.7.1 The overall apparent activation energy of crystallization

The overall activation energy of crystallization of an amorphous alloy under linear heating condition can be determined by the Kissinger as well as by the Ozawa peak method relating to the dependence of exothermic peak temperature Tp on heating rate β.

Kissinger (Kissinger, 1957) proposed that the activation energy can be determined according the equation

$$\ln\left(\frac{\beta}{T_p^2}\right) = \ln\left(\frac{AR}{E_a}\right) - \frac{E_a}{RT} \tag{6}$$

For the determination the activation energy in non-isothermal conditions Ozawa (Ozawa, 1970) proposed the equation:

$$\ln\beta = \ln\frac{AE_a}{R}C - 1.0516\frac{E_a}{RT_p} \tag{7}$$

The values of peak temperatures together with values of kinetic parameters (E_a and $\ln A$) calculated by both methods and the symmetry factors (SF) are given in the Table 4.

β °Cmin⁻¹	T_i °C	T_p °C	T_f °C	S	Ozawa		Kissinger	
					E_a kJmol^{-1}	A × 10^{22} min^{-1}	E_a kJmol^{-1}	A × 10^{21} min^{-1}
5	492	512	542	0.59	338.0 +1.8	1.10 ± 2.3	351.2±1.8	3.06 ±2.3
10	501	520	546	0.59				
20	509	531	554	0.70				
30	513	[illegible]	[illegible]	0.75				

Table 4. Values of T_i, T_p, T_f and S for crystallization of the amorphous $Fe_{81}B_{13}Si_4C_2$ alloy upon continuous heating at different heating rates.

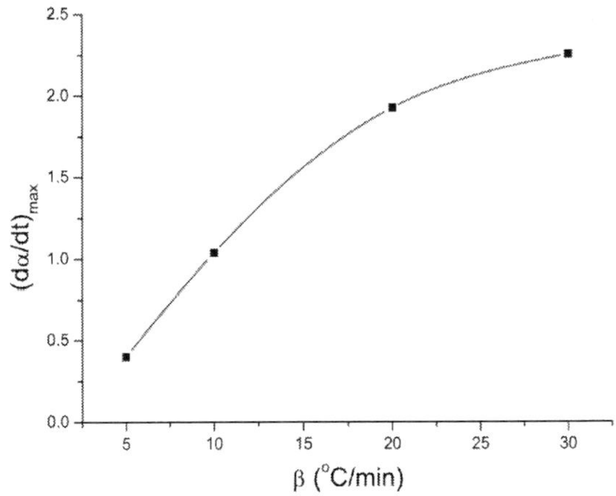

Fig. 16. The dependence of maximum rate of crystallization on heating rate.

The dependence of maximum of rate on rate of heating (Fig. 16) shows the growth of maximum of rate of crystallization with respect to the rate of heating. It indicates that the rate of crystallization reaches a saturation point at high heating rates.

The activation energy of crystallization process involving formation of nuclei and their growth, according to some opinions, has no physical meaning, just empirical significance and only establishes the dependence of the rate of conversion on temperature. This energy can be spent, not just for overcoming the activation barrier but also for its downturn due to cooperative displacement of groups of atoms. Finally, the crystallization of amorphous alloys is a very complicated process accompanied by nucleation and growth of various crystal phases under continuously varied conditions in the conversion zone. With the multitude of possible ways of conversions, only those mechanisms and activated complexes of the crystallization process will be realized that have the highest probability at a given temperature. Any change in crystallization conditions, such as heating rate, can result in change of the mechanism and main activation complex of the crystallization process. Thus, high values of activation energy of crystallization of amorphous alloys, first of all, indicate that a large number of atoms participate in an elementary act of structure reorganization, as well as high complexity of the transformation processes.

2.7.2 The dependence of apparent activation energy on range of conversion

On the basis of dynamic DSC measurements for different heating rates, isoconversional method of Kissinger-Akahira-Sunose method, also known as the „model-free method", enables the determination values of E_a over a wide range of α without the determination of the conversion function (Kissinger, 1957; Akahira, Sunose, 1971). This model involves measuring the temperatures T_a corresponding to fixed values of the crystallized volume fraction, a, for different heating rates, β, and application of the relation:

$$\ln\left(\frac{\beta_i}{T^2_{\alpha,i}}\right) = \ln\left(\frac{AR}{E_{a,\alpha}}F(\alpha)\right) - \frac{E_{a,\alpha}}{RT_{\alpha.i}} \quad (8)$$

where a subscript α designates values related to a given conversion degree, and i is a number of the non-isothermal experiment conducted at the heating rate β_i.

The left-hand side of Eq. (8) is linear with respect to the inverse temperature, $1/T_a$, and enables the activation energy to be evaluated using a linear regression method. In case of a single step process, a constant value of $E_{a(a)}$ is obtained. On the other hand, the dependence of $E_{a(a)}$ on α indicates complex process involving more than one step with different activation energies. It can be observed that the apparent activation energy for the considered crystallization process (Fig. 17) is practically constant in the $0.05 \le \alpha \le 0.7$ range indicating the existence of a single-step reaction. (Vyazovkin, 2000; Opfermann & Flammersheim, 2003). The average value of apparent activation energy was determined as $E_a = 356.5 \pm 5.5$ $kJmol^{-1}$.

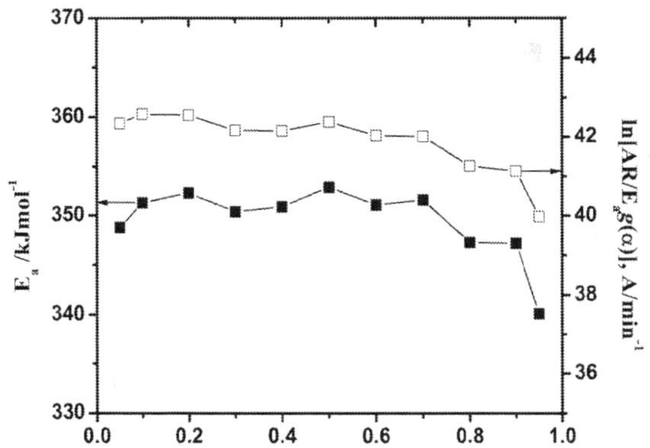

Fig. 17. Apparent activation energies (E_a) and the intercepts as function of the crystallized fraction α.

2.7.3 Preliminary determination of kinetic model

For the preliminary determination of kinetic model of the crystallization process, Dollimore's method was used (D. Dollimore, 1991, 1992; Lee, 1998).

β (°C/min)	α_{max}	$\frac{\Delta LoT}{\Delta HiT}$	Half-width (°C)	T_i	T_f
5	0.51	1.0	9.5	sharp	sharp
10	0.53	1.0	9.5	sharp	sharp
20	0.53	0.9	10.0	sharp	sharp
30	0.55	0.8	11.0	sharp	sharp

Table 5. Parameters describing the asymmetric DSC peaks of crystallization α-Fe phase in amorphous $Fe_{81}B_{13}Si_4C_2$ alloy

This model is based on the "sharpness" of initial and final temperature of differential rate curves, Fig. 18, as well as on its asymmetry. The "sharpness" of the initial and final temperatures is influenced by kinetic factors, and especially by the mechanism of the process. Certain kinetic models lead to an asymptotic or diffuse departure from the base line in the differential form of the thermal curve, while others produce a very sharp approach to the final plateau. The investigation of these parameters that describe geometry and asymmetry of the differential rate curves can be an indicator of the probable kinetic mechanism. According to these parameters, different types of kinetic mechanisms have been listed (Lee, 1998). So, when the crystallization process is not complex the qualitative approach to its kinetics may be obtained using parameters such as α_{max} or $(d\alpha/dt)_{max}$, the shape of the initial and final temperatures as well as peak temperature (T_p) or half-width from differential rate curves.

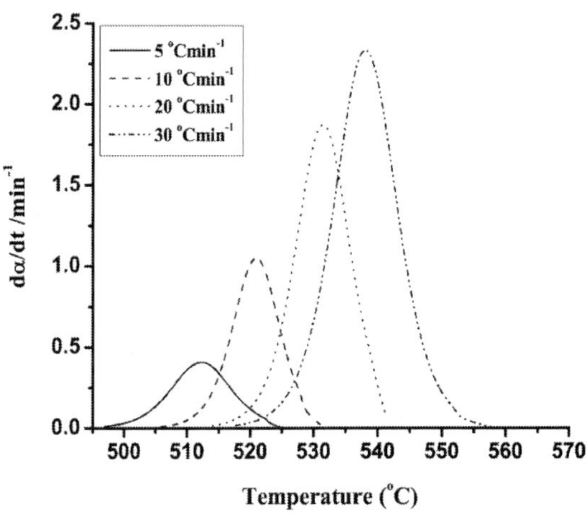

Fig. 18. Differential rate curves ($d\alpha/dt$ vs. T) for different heating rates (β = 5, 10, 20 and 30°Cmin^{-1}).

In this case, Dollimore's procedure was applied to the conversion and differential rate curves (Fig.18) whose slight asymmetry is observed between T_i and T_f. The other parameters such as the conversion at the rate of maximum crystallization, α_{max}, peak temperature, T_p, at $(d\alpha/dt)_{max}$, and the ratio ΔLoT/ΔHiT (shape factor), which is the ratio between the low and high temperature points at half-width of the differential rate curve peak are presented in

Table 5. It is clearly seen that the position of the broadening exotherm, which is connected with the crystallization, was shifted toward higher temperature with the increase of the heating rate as well as asymmetry of peaks. This suggests that the crystallization process should not be characterized by a definite critical temperature independent of the heating rate. The determined values of α_{max} for different heating rates were in the range from 0.51 to 0.55. These results indicate that the non-isothermal crystallization mechanism of α-Fe in amorphous $Fe_{81}B_{13}Si_4C_2$ alloy can not be fully described within the JMA (Johnson-Mehl-Avrami) models (A2, A3 and A4, Group A).

2.7.4 Determination of the kinetic model

General equation, enabling the analysis of conversion kinetics involving nucleation and growth in solid phase was proposed by Avrami (Avrami, 1939):

$$\alpha(t) = 1 - \exp[-(kt)^n], \tag{9}$$

where $\alpha(t)$ is conversion degree, n is kinetic exponent, $k{=}k_o\exp(-E_a/RT)$.

Differentiation of this equation with respect to time gives the rate equation, known as the JMA equation:

$$\left(\frac{d\alpha}{dt}\right) = kn(1-\alpha)[-\ln(1-\alpha)]^{1-1/n}. \tag{10}$$

The JMA equation is based on assumptions of isothermal crystallization, homogenous or heterogeneous nucleation at randomly dispersed particles of the second phase. The growth rate of new phase is independent of time and controlled by temperature and low anisotropy of growing crystals. However if the entire nucleation process takes place during the early stage of transformation and becomes negligible afterwards, JMA equation can also be applied to non-isothermal conditions (Henderson, 1979).

The validity of listed assumptions is not given a priori and simple and reliable testing methods were developed (Málek, 1992, 1995; Gotor et al, 2000; Criado et al, 2003). Once the apparent activation energy has been determined, it is possible to find the kinetic model which best describes the measured set of thermoanalytical data. It can be shown that, for this purpose, it is useful to define two special functions y(α) and z(α), which can easily be obtained by simple transformation of the experimental data. The conversions, in which the y(α) and z(α) functions exhibit the maximum values are designated as α_y^* and α_z^*, respectively. Under non-isothermal conditions these functions can be expressed as follows (Málek, 1992, 1995, 2000):

$$y(\alpha) = \left(\frac{d\alpha}{dt}\right)\exp\left(\frac{E_a}{RT}\right) = Af(\alpha) \tag{11}$$

$$z(\alpha) \approx \left(\frac{d\alpha}{dt}\right)T^2 \tag{12}$$

The maximum of the y(α) function for the JMA model depends on the value of the kinetic exponent:

$$\alpha_y^* = 0 \text{ for } n \le 1$$

$$\alpha_y^* = 1\text{-exp}\,(n^{-1}\text{-}1) \text{ for } n > 1$$

The value of α_y^* is always lower than the maximum of value for α_z^*. For JMA model, α_z^* =0.632. This value is characteristic "fingerprint" of the JMA model and it can be used as a simple test of the applicability of this model.

The obtained normalized functions y(α) and z(α), Fig 19, are independent on the heating rate (β) of the system, and the both functions exhibits the well-defined maxima which were located at exactly defined value of α (α_y^* for the y(α) function and α_z^* for z(α) function, respectively), Table 6.

β °C min^{-1}	α_y^*	α_z^*
5	0.41 ± 0.01	0.53 ± 0.01
10	0.42 ± 0.01	0.51 ± 0.01
20	0.42 ± 0.01	0.55 ± 0.03
30	0.41 ± 0.01	0.52 ± 0.01

Table 6. The maximum of α_y and α_z for the different heating rates

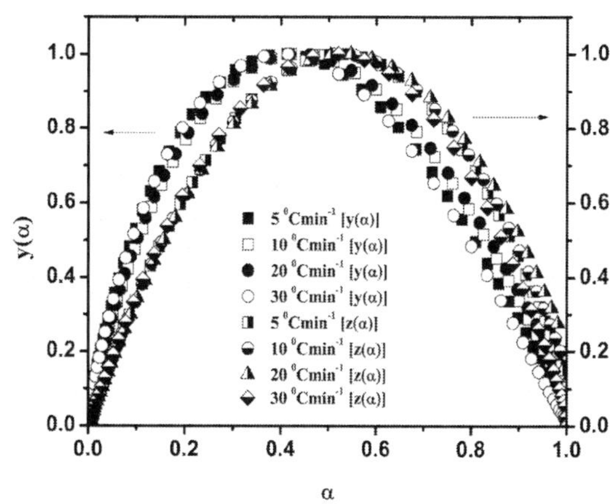

Fig. 19. Normalized *y(a)* and *z(a)* functions at the different heating rates.

From Table 6, it can be seen that the values of α_y^* fall into the range $\alpha_y^* \in (0, \alpha_z)$ ($0.41 \le \alpha_y^* \le 0.42$) and the values of α_z^* are less than 0.632 ($0.51 \le \alpha_z^* \le 0.55$). From the obtained results, it follows that the conditions of validity of the JMA model are not fulfilled for crystallization of *a*-Fe. The displacement of α_z^* in lower value range indicates complexity of the process and can be caused by the influence of surface nucleation or the effect of released crystallization heat on temperature distribution within the sample. However, the relatively high value of α_y^* indicates an increasing effect of the crystallized phase to overall crystallization kinetics, where the crystallized phase further increases the rate of the crystallization. Such autocatalytic behavior can be well described using an empirical two parameter Šesták-Berggren's kinetic model (Šesták, Berggren, 1971; Málek, et al. 1989). This model is based on the equation:

$$f(\alpha) = \alpha^M (1-\alpha)^N, \tag{13}$$

where M and N represents the kinetic exponents.

In this case the expression for reaction rate of the investigated crystallization process can be given as:

$$\frac{d\alpha}{dt} = A\exp\left(-\frac{E_a}{RT}\right)\alpha^M (1-\alpha)^N \tag{14}$$

For this model, the ratio of the kinetic exponents $p = M / N$ can be calculated from the maximum of the y (α) function (Málek, 2000):

$$p = \frac{M}{N} = \frac{\alpha_y^*}{\left(1-\alpha_y^*\right)}. \tag{15}$$

Introducing this equation in equation this (13) gives:

$$\ln\left[\left(\frac{d\alpha}{dt}\right)\exp\left(\frac{E_a}{RT}\right)\right] = \ln A + N\ln\left[\alpha^p (1-\alpha)\right]. \tag{16}$$

This equation describes the processes of nucleation and growth in non-crystalline solids very well. The parameters M and N define relative contributions of acceleratory and decay regions of the kinetic process. From the linear dependence $\ln\left[(d\alpha / dt)\exp(E_a / RT)\right]$ =f($\ln\left[\alpha^p(1-\alpha)\right]$), it is possible to obtain the kinetic exponent N and the pre-exponential factor, ln A. The value of kinetic exponent M can be obtained directly from eq. (15).

β °C/min	M	N	ln*A* (*10²²)
5	0.75 ± 0.03	1.08 ± 0.10	9.0±0.2
10	0.66 ± 0.05	0.92 ± 0.05	10.7±0.3
20	0.64 ± 0.05	0.89 ± 0.07	9.4±0.2
30	0.81 ± 0.10	1.17 ± 0.04	10.6±0.2
Average	0.72 ± 0.06	1.02 ± 0.07	9.9±0.2

Table 7. Kinetic exponents M and N at different heating rates (Adnađević et al., 2010)

Table 7 lists the values of kinetic exponents M and N, as well as the values of ln$A$ obtained by the procedure described above, for the considered crystallization process at different heating rates.

The obtained values of kinetic exponents M and N vary a little with respect to heating rate β. The values of M vary in the range of $0.64 \leq M \leq 0.81$ with average value of $M_{av} = 0.72$. The values of N vary in the range of $0.89 \leq N \leq 1.17$ with average value of $N_{av} = 1.02$. The values of the pre-exponential factor (ln$A$) are independent on the heating rate (β), within the limits of the experimental error. It was shown that this two parameter autocatalytic model has physical meaning only for M<1(Gotor, et al. 2000).

In order to check the established kinetic model we applied the "Master plot" method (Criado, et. al 2003; Gotor, 2000). Using the value at α = 0.5 as a reference point, the following differential master equation is easily derived from eq. (4):

$$\frac{f(\alpha)}{f(0.5)} = \frac{d\alpha / dt}{(d\alpha / dt)_{0.5}} \frac{\exp(E_a / RT)}{\exp(E_a / RT_{0.5})} \tag{17}$$

where $(d\alpha/dt)_{0.5}$, $T_{0.5}$ and f (0.5) are the reaction rate, the temperature reaction and the differential conversion function, respectively at $a = 0.5$.

The left side of Eq. (17) is a reduced theoretical curve which is characteristic of each kinetic function. The right side of the equation is associated with the reduced rate and can be obtained from experimental data if the apparent activation energy is known and remains constant throughout the reaction. Comparison of both sides of Eq. (17) tells us which kinetic model describes an experimental reaction process. It can be seen in Fig. 20, using the average value for the apparent activation energy determined from Kissinger-Akahira-Sunose isoconversional method, the suggested kinetic model works very well in the entire conversion range. Therefore, the Šesták-Berggren autocatalytic model represents the best reaction model for describing the crystallization process of α-Fe in the amorphous $Fe_{81}B_{13}Si_4C_2$ alloy. The higher value of N exponent designates that the formed crystallized phase has the decisive influence on the kinetics of transformation and the rate of growth. In the propagation process, on account of overlapping of nuclei during growth, it acts to retard crystallization rate. Bearing this in mind, we can suppose that in the amorphous alloy, the α-Fe embryos

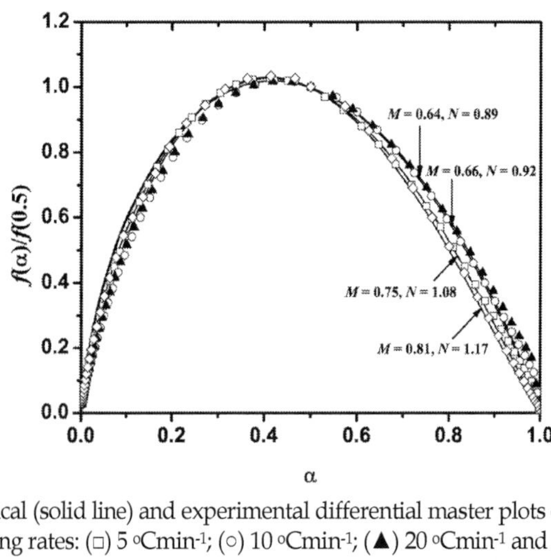

Fig. 20. The theoretical (solid line) and experimental differential master plots of $f(\alpha)/f(0.5)$ versus α for different heating rates: (□) 5 °Cmin⁻¹; (○) 10 °Cmin⁻¹; (▲) 20 °Cmin⁻¹ and (◊) 30 °Cmin⁻¹.

already exist at $T \leq 500°C$ and at $T \geq 500°C$ these embryos are momentarily transformed into nuclei. The established acceleration of crystallization process is a consequence of significant increase of strain in alloy, which arises on account of formation of α-Fe.

2.7.5 Morphology of crystal growth

For non-isothermal crystallization, where the volume fraction of crystalline phase α precipitated in glass heated at a uniform heating rate β is related with the activation energy E_a, Matusita et al. proposed the following relation (Matusita & S. Sakka, 1979, 1980; Matusita, et al. 1984):

$$\ln[-\ln(1-\alpha)] = -n\ln\beta - \frac{1.052mE_a}{RT} + const \qquad (18)$$

where m and n are constants with values between one and four depending on the morphology and kinetics of the growth nuclei, Table 8.

Mechanism	n	m
Bulk nucleation		
Three dimensional growth	4	3
Two dimensional growth	3	2
One dimensional growth	2	1
Surface nucleation	1	1

Table 8. Values of constants n and m for different crystallization mechanisms

The values of n obtained from the slopes of linear plots ln[-ln (1-α)] versus -lnβ at different temperatures for considered crystallization process are given in Table 9. For all considered temperatures, the value of n is ≈ 4.0, within the limits of experimental errors. It follows, then, that the kinetics of crystallization process is independent from the temperature.

The crystallization exponent n is connected with the number of growth dimensions (m) and the number of nuclei forming stages (s) (Matusita, Sakka, 1979) by the following equation

$$n = m + s \qquad (19)$$

where m is the number of growth dimensions as defined in Table 8, s is the number of the nuclei forming stages (s = 0 - at instantaneously nucleation; s = 1 - at constant nucleation rate and s > 1 at self-acceleratory nucleation rate).

In order to describe the crystallization process in detail, the value of parameter m should be determined from the plot of ln [-ln (1-α)], because a function of reciprocal temperature is linear with a slope of 1.051× $(m + 1)E_a/R$, using the value of activation energy determined above.

Temperature °C	n	β °C/min	m	s
518	3.92	5	2.84	1.16
520	4.08	10	3.08	0.92
522	4.07	20	3.22	0.78
		30	2.84	1.16

Table 9. The values of n at three temperatures and values of m and s for four different heating rates

The values of parameters *m* and *s* obtained at the different heating rates for the investigated crystallization process of α-Fe in $Fe_{81}B_{13}Si_4C_2$ amorphous alloy are given in Table 9. Based on the obtained values of parameters *m* and *s*, at the different heating rates, (Table 9), we asserted with high degree of reliability, that the nucleation process of α-Fe occurs within amorphous alloy with a constant rate, in three effective directions (three-dimensional growth) proceeding with constant nucleation rate.

3. Conclusion

Metallic amorphous alloy are a class of materials which has seen dramatic developments in recent times with design of materials stable enough to allow bulk production. While the functional properties of amorphous alloys allow for many possible fields of application, they can also be used as precursors in preparation of nanocomposite materials composed of nanocrystals in the amorphous matrix. These nanocomposites often exhibit superior mechanical, electrical and magnetic properties to both purely amorphous and purely crystalline materials. Thermal treatment of amorphous alloys can allow for controlled crystallization, leading to formation of nanocomposite materials with targeted properties.

$Fe_{81}B_{13}Si_4C_2$ alloy undergoes multi-step crystallization process as a result of thermal treatment. The changes in microstructure cause changes in electrical, magnetic and mechanical properties of the alloy as the alloy structure changes from predominantly amorphous to crystal/amorphous nanocomposite to nanocrystalline composite of α-Fe and Fe_2B phases.

4. References

Adnađević, B.; Janković B. & Minić D. M.; (2010), Kinetics of the apparent isothermal and non-isothermal crystallization of the α-Fe phase within the amorphous $Fe_{81}B_{13}Si_4C_2$ alloy, Journal of Physical Chemistry of Solids, vol. 71, pp. 927-934

Akahira, T. & Sunose, T. (1971) Trans. Joint Convention of Four Electrical Institutes, paper no. 246 (1969), Research Report, Chiba Institute of Technology, vol. 16, pp. 22-31

Avrami, M. (1939), Kinetics of Phase Change. I General Theory, Journal of Chemical Physics, vol. 7, No. 12, pp. 1103-1113

Blagojević, V. A.; Minić, D. M.; Žak, T.; Minić, D. M. (2011), Influence of thermal treatment on structure and microhardness of $Fe_{75}Ni_2Si_8B_{13}C_2$ amorphous alloy, Intermetallics vol. 19, pp. 1780-1785

Balberg, I. & Helman, J. S. (1978), Critical behavior of the resistivity in magnetic systems. II. Below Tc and in the presence of a magnetic field, Physical Review B, vol. 18, No. 1, pp. 303-318

Böhnke, G.; Kaul, S. N.; Kettler, W. & Rosenberg, M. (1983), Critical behaviour of the electrical resistivity in amorphous ferromagnetic alloys, Solid State Communications, vol. 48, No. 9, pp. 743-746

Criado, J. M.; Pérez-Maqueda, L. A.; Gotor, F. J.; Málek, J. & Koga, N. (2003), A unified theory for the kinetic analysis of solid state reactions under any thermal pathway, Journal of Thermal Analysis and Calorimetry, vol. 72, No. 3, pp. 901-906

Dollimore, D.; Evans, T.A.; Lee, Y.F. & Wilburn, F.W. (1991), Calculation of activation energy and pre-exponential factors from rising temperature data and the generation of TG and DTG curves from A and E values, Thermochimica Acta, vol. 188, No. 1, pp. 77-85

Dollimore, D.; Evans, T.A.; Lee, Y.F.; Pee, G.P. & Wilburn, F.W. (1992), The significance of the onset and final temperatures in the kinetic analysis of TG curves, Thermochimica Acta, vol. 196, No. 2, pp. 255-265

Friedman, H. L. (1964), Kinetics of thermal degradation of char-forming plastics from thermogravimetry. Application to a phenolic plastic, Journal of Polymer Science Part C: Polymer Symposia, vol. 6, No. 1, pp. 183-195

Gotor, F. J.; Criado, J. M.; Málek, J. & Koga, N. (2000), Kinetic Analysis of Solid-State Reactions: The Universality of Master Plots for Analyzing Isothermal and Nonisothermal Experiments, Journal of Physical Chemistry A, vol. 104, No. 46, pp. 10777-10782

Henderson, D. W. (1979), Experimental analysis of non-isothermal transformations involving nucleation and growth, Journal of Thermal Analysis and Calorimetry, vol. 15, No. 2, pp. 325-331

Kissinger, H. E. (1957), Reaction Kinetics in Differential Thermal Analysis, Analytical Chemistry, vol. 29, No. 11, pp. 1702-1706

Klement, W.; Willens, R. H & Duwez, P.O.L. (1960), Non-crystalline Structure in Solidified Gold-Silicon Alloys, Nature, vol. 187, No. 4740, pp. 869–870

Lass, E. A.; Zhu, A.; Shiflet, G. J. & Poon, S. J. (2010), A short-range ordering description of amorphous metal alloys using the central atoms model, Acta Materialia, vol. 58, No. 16 pp. 5460-5470

Lee, Y.F. & Dollimore, D. (1998), The identification of the reaction mechanism in rising temperature kinetic studies based on the shape of the DTG curve, Thermochimica Acta, vol. 323, No. 1-2, pp. 75-81

Libermann H. & Graham C. (1976), Production Of Amorphous Alloy Ribbons And Effects Of Apparatus Parameters On Ribbon Dimensions, IEEE Transactions on Magnetics, vol. 12, No. 6, pp. 921

Málek, J.; Criado, J. M.; Šesták, J. & Militky, J. (1989), The boundary conditions for kinetic models, Thermochimica Acta, vol. 153, pp. 429-432

Málek, J. (1992), The kinetic analysis of non-isothermal data, Thermochimica Acta, vol. 200, pp. 257-269

Málek, J. (1995), The applicability of Johnson-Mehl-Avrami model in the thermal analysis of the crystallization kinetics of glasses, Thermochimica Acta, vol. 267, pp. 61-73

Málek, J. (2000), Kinetic analysis of crystallization processes in amorphous materials, Thermochimica Acta, vol. 355, No. 1-2, pp. 239-253

Maričić, A.; Minić, D. M.; Blagojević, V. A.; Kalezić-Glišović, A.; Minić, D. M. (2012), *Effects of structural relaxation on functional properties of amorphous alloy* $Fe_{73.5}Cu_1Nb_3Si_{15.5}B_7$, Intermetallics vol. 21, pp. 45-49

Matusita, K. & Sakka, S. (1979), Kinetic study of the crystallisation of glass by differential scanning calorimetry, Physics and Chemistry of Glasses, vol. 20, pp. 81-85

Matusita, K. & Sakka, S. (1980), Kinetic study of crystallization of glass by differential thermal analysis—criterion on application of Kissinger plot, Journal of Non-Crystalline Solids, vol. 38-39, No. 2, pp. 741-746

Matusita, K.; Konatsu, T. & Yokota, R. (1984), Kinetics of non-isothermal crystallization process and activation energy for crystal growth in amorphous materials, Journal of Materials Science, vol. 19, No. 1, pp. 291-296

Minić, D. M.; Maričić, A., Dimitrijevic, R. Z.; Ristić, M. M. (2007), Journal of Alloys and Compdounds, vol. 430, pp. 241-245

Minić, D. M.; & Adnađević, B. (2008), Mechanism and kinetics of crystallization of α-Fe in amorphous $Fe_{81}B_{13}Si_4C_2$ alloy, Thermochimica Acta, vol. 474, No. 1-2, pp. 41-46

Minić, D. M.; Maričić, A. & Adnadjević, B. (2009), Crystallization of α-Fe phase in amorphous $Fe_{81}B_{13}Si_4C_2$ alloy, Journal of Alloys and Compounds, vol. 473, pp. 363-367

Minić, D. M.; Minić, D. G. & Maričić, A. (2009), Stability and crystallization of $Fe_{81}B_{13}Si_4C_2$ amorphous alloy, Journal of Non-Crystalline Solids, vol. 355, No. 50-51, pp. 2503-2507

Minić, D. M.; Plazinić, M.; Živanić, J. & Maričić, A. (2010), Influence of Structural Transformations on Electric and Magnetic Properties of $Fe_{81}B_{13}Si_4C_2$ Amorphous Alloy, Science of Sintering, vol. 42, pp. 61-68

Minić, D. M.; Blagojević, V.; Minić, D. G.; Gavrilović, A.; Rafailović, L. (2011), Influence of thermally induced structural transformations on hardness in $Fe_{89.8}Ni_{1.5}Si_{5.2}B_3C_{0.5}$ amorphous alloy, Journal of Alloys and Compounds vol. 509, pp. 8350-8355

Minić, D. M.; Minić, D. M.; Žák, T.; Roupcova, P. & David, B. (2011), Structural transformations of $Fe_{81}B_{13}Si_4C_2$ amorphous alloy induced by heating, Journal of Magnetism and Magnetic Materials, vol. 323, No. 5, pp. 400-404

Minić, D. M.; Blagojević, V. A.; Minić, D. M.; Gavrilović, A. & Žak, T. (2011), Influence of microstructure on microhardness of $Fe_{81}Si_4B_{13}C_2$ amorphous alloy after thermal treatment, Metallurgical and Materials Transactions A, vol 42, pp. 4106-4112

Minić D. M; Blagojević V. A; Minić D. M.; Žák T(2011), Influence of microstructural inhomogeneouity of individual sides of $Fe_{81}Si_4B_{13}C_2$ amorphous alloy ribbon on thermally induced structural transformations, Materials Chemistry and Physics, vol. 130, pp. 980-985

Opfermann, J. & Flammersheim, H. J. (2003), Some comments to the paper of J.D. Sewry and M.E. Brown: "Model-free" kinetic analysis, Thermochimica Acta, vol. 397, No. 1-2, pp. 1-3

Ozawa, T. (1970), Kinetic analysis of derivative curves in thermal analysis, Journal of Thermal Analysis and Calorimetry, vol. 2, No. 3, pp. 301-324

Ponnambalam, V.; Poon, S. J. & Shiflet, G. J. (2004), Fe-based bulk metallic glasses with diameter thickness larger than one centimeter, Journal of Materials Research vol. 19, No. 5, pp. 1320-1323

Ribić-Zelenović, L.; Rafailović, L.; Spasojević, M. & Maričić, A. (2008), Correlation between electron state density change and the electrical resistivity and magnetic permeability changes in the nanostructured powder of the NiMo alloy, Physica B: Condensed Matter, vol. 403, No. 12, pp. 2148-2154

Saegusa, N. & Morrish, A. H. (1982), Atomic rearrangements and crystalline transformations in amorphous Fe81B13.5Si3.5C2, Physical Review B, vol. 26, No. 1, pp. 305-314

Šesták, J. & Berggren, G. (1971), Study of the kinetics of the mechanism of solid-state reactions at increasing temperatures, Thermochimica Acta, vol. 3, No. 1, pp. 1-12

Vyazovkin, S. (2000), Computational aspects of kinetic analysis.: Part C. The ICTAC Kinetics Project — the light at the end of the tunnel?, Thermochimica Acta, vol. 355, No. 1-2, pp. 155-163

Williamson, G. K. & Hall, W. (1953), X-ray line broadening from filed aluminium and wolfram, Acta Metallurgica, vol. 1, No. 1, pp. 22-31

Žák, T. (1999) In: Mössbauer spectroscopy in Material Science, Eds. M. Miglierini and D. Petridis, NATO Science Series, Kluwer, Dordrecht, 1999, pp. 385

Žák, T. & Jirásková, Y. (2006), CONFIT: Mössbauer spectra fitting program, Surface Interface Analysis, vol. 38, No. 4, pp. 710-714

10

Crystallization of Iron-Containing Oxide-Sulphide Melts

Evgeniy Selivanov and Roza Gulyaeva
Institute of Metallurgy of the Ural Branch of the Russian Academy of Sciences
Russia

1. Introduction

The processing of the sulphide raw materials (ores, concentrates and mattes) of non-ferrous metallurgy is related to the formation of a large amount of iron containing slags. The initial product of the oxidation of sulphides in real commercial plants is an oxide–sulphide melt, in which decomposition under the action of fluxes is accompanied by matte and slag formation (Selivanov et al., 2009a). The fraction of oxygen in a sulphide melt and the fraction of sulphur in an oxide melt are each controlled by the contents of silicon dioxide and iron oxides in a slag and the contents of non-ferrous metals in a matte. According to modern concepts, the heterogeneity of slags is caused by mechanical matte, magnetite and spinel inclusions, where the spinel inclusions form during oxidation processes (Selivanov et al., 2000; Spira & Themelis, 1969; Tokeda et al., 1983; Vanyukov & Zaitsev, 1969, 1973). The cooling (i.e., the crystallization) of a slag leads to the formation of new oxide and sulphide phases within it. Information on the available forms of the useful components is important for the reduction of metal loss through a slag and for the selection of their re-extraction methods.

A number of works are devoted to the study of the kinds of copper existing in slags. Major results are generalized in the monographs of (Ruddle, 1953; Vanyukov et al., 1988; Vanyukov & Zaitsev, 1969; 1973). Phase equilibria in the systems relevant to copper pyrometallurgy have been discussed mostly for molten states (Elliott, 1976; Kopylov, 2001; Yazawa, 1974). It is considered that the loss of non-ferrous metals through slags is caused by their oxide, sulphide and metal solubility. It was discovered that a part of copper is presented in the crystallized slag by matte mechanical inclusions (Vanyukov & Zaitsev, 1969; 1973). Data on the copper sulphide solubility in a slag was reported by (Mohapatra, 1994; Nagamori, 1974; Vanyukov et al., 1988; Vaysburd, 1996). There is no valid confirmation of the presence of individual copper oxide inclusions or copper silicates and ferrites in a slag. Information on the existence of other metals (Zn, Pb, As, etc.) in a slag needs to be specified more exactly in each separate case. The bulk of the zinc is transferred into the slag during the smelting of sulphide copper-zinc concentrates in the Vanyukov furnace for a rich matte and crude metal (Vanyukov et al., 1988). It is assumed herein that zinc is present in a slag in the form of an oxide. Some questions concerning the constituent phases of crystallization during the rapid cooling of a non-ferrous metallurgy slag are partially disclosed by (Cardona et al., 2011). However, no task-oriented studies devoted to the estimation of the cooling rate's effect on the formation of phases and the presence of different kinds of non-ferrous metals in a non-ferrous metallurgy slag have been found.

The goal of this work is to study phase composition and the kinds of metals present in the slag samples of copper-zinc concentrates of pyrometallurgical processing and the nickel oxide ores of smelting. The main task of the study lies in the estimation of the cooling rate and iron's oxidation level's influence on the phase composition, structure, thermal properties and forms of non ferrous metals extant in the crystallized oxide sulphide systems FeO_x–SiO_2–FeS-Cu_2O-ZnO and SiO_2-FeO_x-MgO-CaO-NiO-FeS.

2. Methods of investigation

The chemical analysis accuracy resolution is 0.1% for the elements' content in the slag samples over 1% (Fe, S, Zn, SiO, CaO, Al_2O_3). It is equal to 0.02% when the elements concentration in the slag is less than 1% (Cu, Sb, Pb, As). The phase composition of the samples has been determined by using an X-ray diffractometer (Cu-K_α - radiation). The temperatures and heats of the phase transformations are determined by means of differential-scanning calorimetry with a Netzsch STA 449 C Jupiter thermo-analyser with a heating rate of 20 °C/min in an argon flow. The determination of phase element composition is performed with a JSM-59000LV raster electronic microscope (ESM) and an Oxford INCA Energy 200 dispersion X-ray spectrometer (EDX). The results of the X-ray spectrum microanalysis have (EPMA) a relative error of 2% where the content of the elements is greater than 10%. The relative error is close to 5% at concentrations of elements are from 1% to 10%. This relative error is 10% than concentrations of elements are less 1%. The microstructure of the samples is studied by an Olympus optical microscope using the Simagic application program.

The analysis of the gases evolving in the heating of materials was carried out by a QMS 403C Aëolos mass - a spectrometer connected with the thermo analyser. To perform the thermodynamic simulation (TDS) of the equilibrium phases during the cooling of working bodies whose compositions corresponded to the initial slag samples, we used the HSC 5.1 Chemistry (Outokumpu) software package based on the minimization of the Gibbs energy and variational thermodynamics principles (HSC Chemistry, 2002; Moiseev & Vyatkin, 1999).

The initial slags were put in Al_2O_3 crucibles and melted (1300 °C) in a resistance furnace with an electrographite heater for the investigation of the cooling rate's influence on the crystallization of melts. The direct cooling of the slag was carried out in a furnace and it provided for a decrease of the temperature rate up to crystallization (solidus) at about 0.3 °C/s; in the air after removing the crucible from the furnace - 1.7 °C/s; by means of the pouring of the melt from a crucible into a water basin - 900 °C/s. With the water granulation of the slags, we fabricated particles with an average size of 1.5–2.0 mm. The calculation of the cooling time of these particles was carried out using the expression (Naboichenko et al., 1997):

$$\tau_{cool} = d_d \, (c_p \, \rho_{sl} \, / \, 6 \, \alpha) \, \ln[(T_m - T_s)/(T_d - T_s)], \quad (1)$$

where τ_{cool} is the drop solidification time; d_d is the drop size; c_p is the heat capacity; ρ_{sl} is the melt density; α is the heat-transfer coefficient of the melt-water system; T_m, T_d and T_s are the temperatures (K) of the slag melt, drop and vapour, respectively. The granules obtained from the slag were subjected to isothermal annealing in an electric resistance furnace (for 5 and 60 minutes) at temperatures of 750 °C and 1000 °C in an inert atmosphere.

The overall strategy of this investigation is presented schematically (Fig. 1).

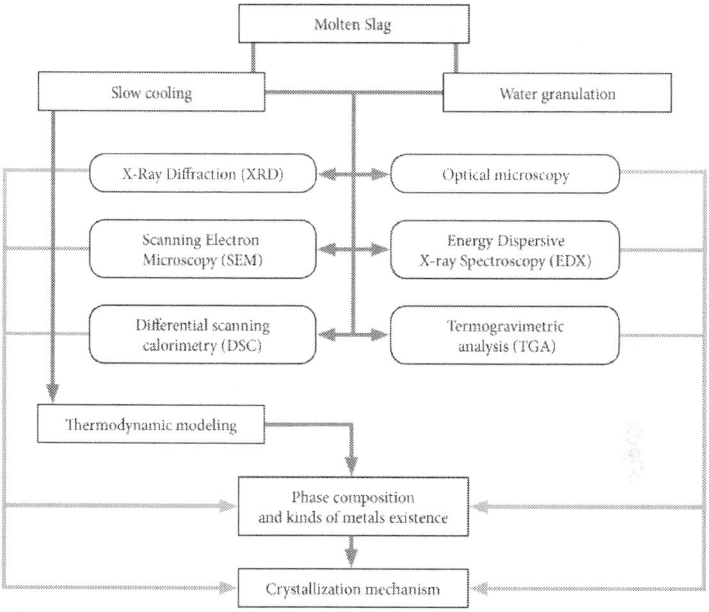

Fig. 1. Overall strategy of this investigation

3. Effect of the cooling rate on the structure of slag from the melt of copper-zinc concentrates in a Vanyukov furnace

The autogenous smelting technology of sulphide copper–zinc concentrates in a Vanyukov furnace was developed in "Sredneuralsky Copper Smelter Plant" JSC (Russia, Ural) (Vanyukov & Zaitsev, 1969, 1973; Vanyukov et al., 1988). Concentrates (14 - 16% Cu) are melted for the mattes contents with 45 – 55% copper.

The degree of copper concentration, defined as the ratio of metal content in the matte to its content in the charge, is within the range 3.0 - 4.0. The relatively low quality of the incoming concentrate and the desire to increase the copper content in the matte predetermine the high flow of the oxygen-air mixture and the large amount of slag which is produced. The slag contains iron oxide (III) in the form of magnetite, which largely determines the matte-slag emulsion delamination.

A large number of studies (Jalkann, 1991; Rüffler & Dávalos, 1998; Selivanov et al., 2000, 2004; Vanyukov & Zaitsev, 1969, 1973) have been devoted to the evaluation of slag structure and the metal forms of these of non-ferrous metals are presented in the literature. However, a common law for such complex systems as metallurgical slags does not allow us to extrapolate the known data on the studied samples because new objects require additional study.

The object of the research is the slag from the melting of copper-zinc concentrates in a Vanyukov furnace which contains, %: 40.5 Fe, 2.4 S, 0.8 Cu, 3.9 Zn, 32.1 SiO_2, 2.8 CaO, 0.8 MgO, 2.6 Al_2O_3, 0.1 Sb, 0.5 Pb, 0.1 As (Selivanov et al., 2009b, 2010).

A slag sample is taken from the furnace slag siphon at its overflow into the drain trough. The slag was in contact with a matte containing, %: 44.9 Cu, 23.8 Fe, 2.3 Zn, 22.8 S, 0.1 Sb, 2.0 Pb, until its discharge.

Reflexes which correspond to (Fe_2SiO_4) fayalite, (Fe_3O_4) magnetite and zinc sulphide (sphalerite) are identified in the initial slag through X-ray analysis (Fig. 2). The melting of the slag followed by its cooling reduces the intensity of the X-ray reflexes of the identified phases. Amorphization (glass formation) is reached throughout the mass of the sample when the cooling rate of the slag is equal to 900 °C/s.

Thermograms (Selivanov et al., 2009) of the samples (Fig. 3) allow us to estimate the melting and crystallization temperatures of the samples. Two endothermic effects are observed with the heating of the initial slag, which is begun at 972 °C and 1067 °C.

The first of these characterizes the melting of the eutectic and the second of the entire mass of the slag. The temperature of initial crystallization is equal to 1021 °C. According to the mass spectrometry data an evolution of a certain amount of SO_2 occurred under the sample's heating (from 300 °C - 400 °C). This evolution is caused by interaction of sulphides with iron oxides of higher valence. Slag mass loss does not exceed 1.0% with heating up to 1200 °C. The view of the sample thermogram crystallized at 0.3 °C/s is essentially identical to the results obtained for the original slag.

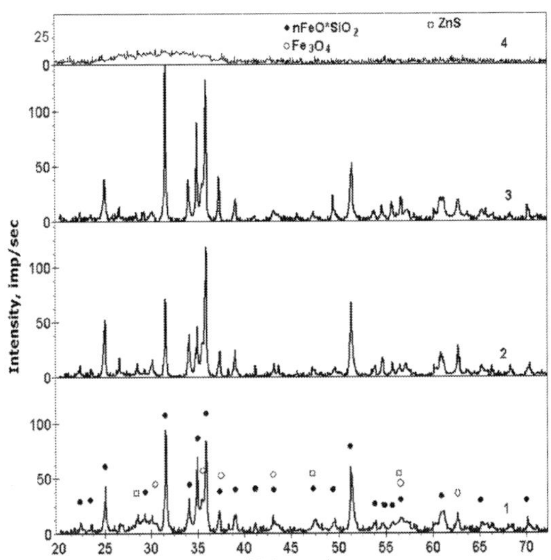

Fig. 2. Diffractograms of the initial (1) slag of melting of copper-zinc concentrates and samples obtained after their melting and cooling rates: 0.3 (2), 1.7 (3), 900 °C/s (4).

We may note the proximity of the starting temperatures of the thermal effects associated with melting (1062 °C) and melt crystallization (1045 °C). The appearance of the effect on the DSC curve is characterized for a sample cooled at a rate of 900 °C/s (Fig. 3). The effect starts from 507 (T_{ons}) and its middle is at 533 °C (Tg), which is connected with a second-order phase transition and the resulting process of slag devitrification (Mazurin, 1986) for the sample cooled at 900 °C/s (Fig. 3). Two exothermal heating effects are revealed on further heating with the onset/maximum at 541/577 °C and 628/644 °C. Apparently, the "cold" crystallization of the slag - the ordering of its structure - takes place at these temperatures and the presence of a doublet of peaks is caused by its two phases.

Devitrification observed on heating the sample containing glass and further "cold" crystallization is connected with the formation of magnetite (exothermal effect) and the isolation of crystals of the iron-silicate phase with a slightly lower (in comparison with glass) quantity of iron oxide:

$$Fe_{1+x}SiO_{3+x+y} = Fe_{1+x-3y/4}SiO_{3+x} + y/4Fe_3O_4. \quad (2)$$

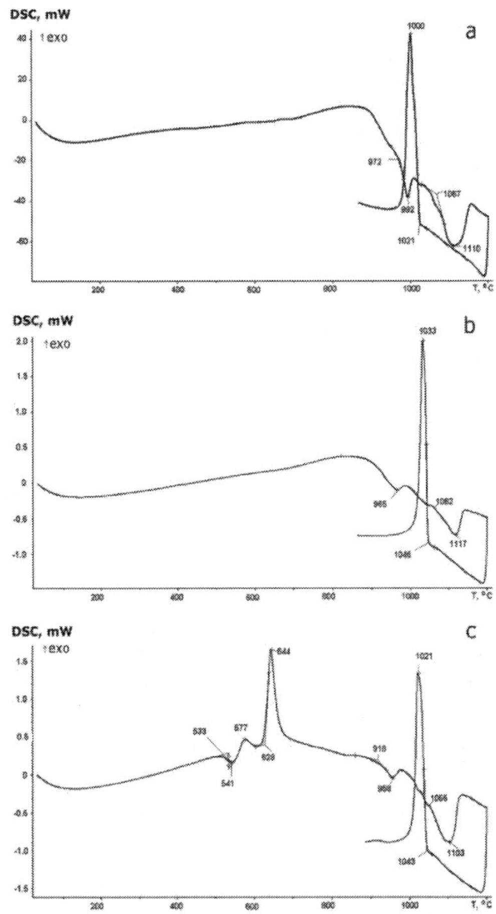

Fig. 3. Thermograms (20 °C/min, argon): of initial slag (a) from the melt of copper-zinc concentrates and the samples obtained after its melting and cooling at the rates of : 0.3 (b) и 900 °C/s (c)

The endothermic effects which started at 918 and 1055 °C point to the melting of the phase components of the slag. The temperature of the initiation of crystallization melting is 1043 °C which agrees with the temperature determined for the sample cooled at a rate of 0.3 °C/s. The temperature values and the change of heat capacity at devitrification (Δc_p) calculated from the experimental data and from the heat values of the 'cold' ($L_{c.cr.}$) and high ($L_{h.cr.}$) temperature crystallization of the hardened sample of slag are given in table 1. According to the data obtained, the heating of high-iron vitreous slag completely transforms it from an amorphous state to a crystalline state. The heat of slag melting is 167 J/g (Selivanov et al., 2009b).

Devitrification			$L_{c.cr.}$, J/g		$L_{m.}$, J/g		$L_{h.cr.}$, J/g
T_{ons}, °C	T_g, °C	Δc_p, J/(g·K)	1 peak	2 peak	1 peak	2 peak	
507	533	0.756	15	98	13	152	164

Table 1. The values of heat effect enthalpies of slags from a melt of copper-zinc concentrates at samples cooled at a rate of 900 °C/s.

The microstructure (Fig. 4) of the initial slag is represented by iron-silicates, magnetite and matte particles. Magnetite has been formed as fine-dispersed branchy dendrites. The isolated coarse matte particles are mechanically carried out together with the slag, which reach a size of up to 150 μm. The silicate constituent of the slag has small sulphide patches, which reach a size of 1.0-2.0 μm; they are concentrated along the boundaries of large iron-silicate aggregates. According to the X-ray spectral microanalysis data, the iron-silicate phase (Table 2) is heterogeneous, both in the main elements (silicon and iron) and the impurities dissolved in it. The calculated composition of the iron-silicates ranges from $Fe_2Si_3O_8$ to $Fe_3Si_2O_7$. With the elevation of the Fe/Si proportion in the iron-silicate phases, the content of calcium, sulphur, lead and zinc oxides in them decreases:

Fe	Ca	S	Pb	Zn	
$Fe_3Si_2O_7$	43.4	0.5	0.1	-	3.2
$FeSiO_3$	34.5	1.7	0.8	0.3	3.6
$Fe_2Si_3O_8$	22.1	7.9	1.6	0.6	4.4

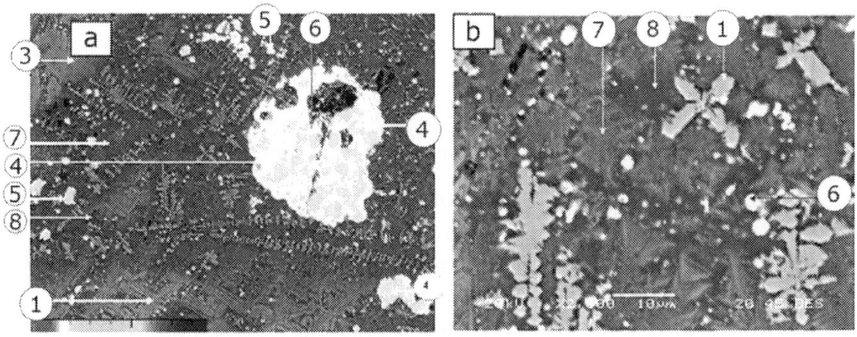

Fig. 4. The microstructure of the initial slag taken from smelting of copper-zinc concentrates and the point of local phase probing

The magnetic crystals (60.5 – 61.9% Fe) which are in the plane of the section also contain impurity elements in %: 1.0 Al; 2.5 - 3.0 Si; 0.3 - 0.4 Ti; 0.1 Cr; 0.1 Mg; 2.2 Zn and up to 0.2 Cu. The sulphide constituents of the slag are represented by the matte particles (48% Cu) with inclusions of zinc and lead sulphides. Solid solutions on the base of a ZnS-FeS system have a composition within the limits of $Zn_{0.24}Fe_{0.76}S$ to $Zn_{0.45}Fe_{0.55}S$ and, apart from the primary elements, contain 4.7-6.1% Cu and up to 0.5% Pb.

The iron-silicate phases which have different compositions in which magnetite and sulphides are found (Fig. 5) also constitute the base of the sample cooled at the rate of 0.3 °C/s. In comparison with the initial slag, the enlargement of iron, magnetite and sulphide silicate crystals has been marked. The main area of the matte is occupied by the conglomerates of coarse crystals of the iron-silicate phase, having a composition close to $Fe_3Si_2O_7$, and the spaces among them are filled up by $Fe_2Si_3O_8$ with small dendrites of $FeSiO_3$. As well as for the initial slag, the content of the impurity elements correlates with the iron content in the silicate:

Fe	Ca	S	Pb	Zn	Mg	
$Fe_3Si_2O_7$...	44.8	0.3	-	-	2.8	1.1
$FeSiO_3$	36.9	1.7	0.3	-	3.2	0.2
$Fe_2Si_3O_8$	23.5	4.3	1.0	0.4	4.0	-

№	Content, mas.%											Composition
	Mg	Al	Si	S	Ca	Ti	Fe	Cu	Zn	Pb	O	
1	0.1	1.0	2.5-3.0	0.1	0.3	0.3-0.4	60.5-61.9	0.2	2.2	-	31.5-31.7	Fe_3O_4
2	0.9-1.0	0.3-0.4	14.0-14.7	0.1	0.4-0.5	-	42.8-44.0	-	3.1-3.3	-	36.8-37.3	$Fe_3Si_2O_7$
3	-	-	0.2	27.1		-	20.6	48.8	3.0	-	-	$Cu_{2.1}FeS_{2.3}$ (matte)
4	-	-	0.2-0.3	32.4-32.5	0.1	-	38.2-38.6	25.1-26.0	2.7-3.2	0.5	-	$CuFe_{1.6}S_{2.4}$
5	-	-	0.5 1.4	33.2 32.6	0.2 0.5	-	43.0-30.8	6.1-4.7	16.9-29.8	0.5	-	$Zn_{0.24}Fe_{0.76}S$ $Zn_{0.45}Fe_{0.55}S$
6	-	-	0.2-0.5	20.0-23.3	-	-	12.7-19.5	19.1-23.8	0-4.8	33.3-41.9	-	(Pb,Cu,Fe)S
7	0.4	1.2	17.5	0.7-0.9	1.4-1.9	-	33.9-35.0	0.2	3.3-3.9	0.2-0.3	39.1-39.8	$FeSiO_3$
8	0.1	2.0-2.5	19.7	1.5-1.7	7.1-8.7	0.3	21.9-22.2	0.2	4.4	0.5-0.6	40.7-41.0	$Fe_2CaSi_3O_8$

Table 2. EPMA data on the phase composition of the initial slag from the smelting of copper-zinc concentrates (according to Fig. 4)

Matte particles with sizes from 1 to 15-30 μm are found mainly between iron-silicate blocks of a $Fe_3Si_2O_7$ composition. Matte decomposition into sulphides (bornite, sphalerite, galenite) occurs at cooling and their compositions - according to the analysis data - fluctuate widely (Table 3). More easily melted is lead containing a sulphide alloy from the margin along the surfaces of the matte particles. High-ferrous sphalerite ($Zn_{0.4}Fe_{0.6}S$) has been revealed both as an independent phase of around 2-10 μm in size and the inside of matte particles.

Magnetite (60.0-60.4% Fe) takes the form of both geometrical crystals and the form of dendrites arranged between $Fe_3Si_2O_7$ blocks and in direct contact with $Fe_2Si_3O_8$, both as in the initial slag and in the magnetite apart from the iron, which have revealed zinc, silicon and aluminium impurities as well as titanium (0.6%) and chromium (0.1%).

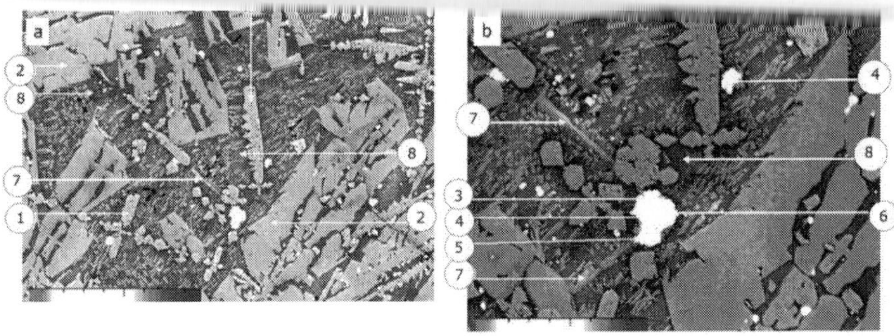

Fig. 5. Microstructure of the slag of copper-zinc concentrates' melting cooled from the melt at the rate of 0.3 °C/sec and the points of local phases' probing: an increase of x200 (a) and x500 (b)

The structure of a slag sample cooled at a rate of 900 °C/s is represented (Fig. 6) by glass and sulphide inclusions (up to 20 μm) with round forms. According to EPMA, the glass contains about 30% SiO_2 and 50% FeO_{1+x} (Table 4). The sulphide phase (inclusion of more than 15 μm in size) is inhomogeneous in its composition - its central part closely corresponds to Cu_5FeS_4. Lead and zinc sulphides are revealed inside matte particles.

№	Content, mas.%									Composition
	Al	Si	S	Ca	Fe	Cu	Zn	Pb	O	
1	2.5-2.6	1.7-2.0	-	0.2	60.0-60.4	-	2.8	-	31.4-31.6	Fe_3O_4
2	-	13.9-14.3	-	0.3	44.3-45.2	-	2.8-2.9	-	36.7-37.0	$Fe_3Si_2O_7$
3	-	0.3	26.7	-	18.9	53.0	0.6	0.8	-	$Cu_5Fe_2S_{4.5}$ (matte)
4	-	0.3-0.5	32.5	0.1	37.8	27.6	1.4	0.6	-	$Cu_2Fe_3S_{4.5}$ (matte)
5	-	0.6	27.8	0.1	35.7	6.1	26.7	3.0	-	$Zn_{0.4}Fe_{0.6}S$
6	-	0.3	18.2	-	13.0	17.0	1.0	50.6	-	$PbS-Cu_2S-FeS$
7	1.1-2.3	16.0-18.1	0.2-0.4	1.1-1.4	34.7-39.0	-	3.0-3.4	-	38.1-39.5	$FeSiO_3$
8	4.0-4.3	20.3-20.6	0.9-1.1	4.2-4.4	23.4-23.6	-	3.8-4.1	0.4	41.4-41.6	$Fe_2Si_3O_8$

Table 3. EPMA data on the composition of slag sample phases after their melting and cooling at a rate of 0.3 °C/s (according to Fig. 5)

The phase close to fayalite (Fe_2SiO_4) has not been revealed in any of study samples. All of the complexes of iron-silicate phases that were formed during slag cooling correspond to the atomic relations of Fe/Si within the limits of 0.7-1.5. Slag cooling at the high (900 °C/s) rate results in the formation of glass with the proportion of Fe/Si equal to about 1.4, without isolating magnetite in a self-dependent phase.

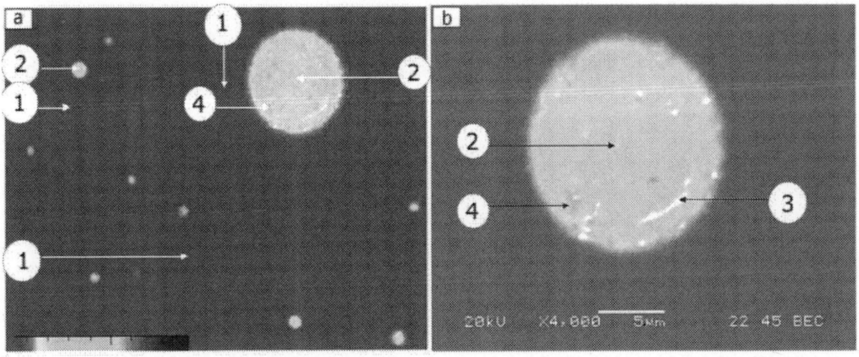

Fig. 6. Microstructure of the slag of copper-zinc concentrates' melting cooled from the melt at the rate of 900 °C/s and EPMA points

№	Content, mas.%										Composition
	Al	Si	S	Fe	Cu	Zn	Pb	O	Mg	Ca	
1	1.8-2.1	13.9-14.0	0.8	38.4-38.9	0.2-0.6	3.7-3.9	0.3-0.4	37.5-37.6	0.4	1.85	$Fe_{1.4}SiO_{3.4}$ (glass)
2	-	0.3-2.0	21.9-23.7	12.5-14.6	59.6-61.4	0.3-0.6	0.8-1.7	-	-	-	$Cu_5FeS_{3.3}$
3	-	0.4	23.3	11.6	44.7	16.3	1.7	-	-	-	Cu_5FeS_4-(Zn,Fe)S
4	-	0.3	21.1	9.7	57.9	0.3	8.0	-	-	-	Cu_5FeS_4-PbS

Table 4. EPMA data on the phase composition of the slag samples cooled from the melt at the rate of 900 °C/s (according to Fig. 6)

Proceeding from the fact that oxide phases with a high content of iron contain a lower quantity of CaO, one can draw a conclusion about the influence of a lime flux on phase formation (Selivanov et al., 2009a). For those slags with a Fe/SiO_2 ratio higher than 1, the increase of calcium oxide will not cause its solution in $Fe_3Si_2O_7$ but rather will favour the decomposition of this compound, which proceeds - in the limiting case - with the formation of calcium silicate and iron oxides. If the Fe/SiO_2 ratio in the slag is less than 1, then the CaO will dissolve in iron-silicate phases ($FeSiO_3$ and $Fe_2Si_3O_8$), reducing their melting temperature. One should bear in mind that these points are applied to those oxide melts which do not contain iron oxides of the highest valency. As has been shown in the works of (Okunev & Galimov, 1983; Tokeda et al., 1983), in oxide melts with a high degree of iron oxidation, CaO and Fe_2O_3 interaction establishes the formation of calcium ferrites.

Slow slag precipitation leads to the concentration of the matte particles among large grains of $Fe_3Si_2O_7$. On cooling, sulphide phases - the bulk of which is bornite - and crystallize from the matte. This is besides the fact that small crystals of sulphides form in the course of slag cooling, which can be explained by the peculiarities of the segregation of the oxide-sulphide system and by the change of the sulphides' solubility in iron-silicate melts.

The zinc on slag crystallization is distributed between oxide and sulphide phases. A (Zn,Fe)S independent phase containing 17-38% zinc has been revealed only at low rates of slag cooling. Lead in the slag takes both oxide and sulphide forms. The lead content in the iron-silicate phase increases as the content of silicon dioxide grows in it. Lead forms sulphide phases with [illegible]% Pb which precipitate out of the sulphide melt (matte) on cooling (Selivanov et al., 2009b).

Thus, the rate of cooling of the melted slag influences the size and the number of forming phases, which defines the copper, zinc and lead distribution between oxide and sulphide forms. Changing the content of the calcium oxide in the slag and the rate of cooling, one can provide the preparation of the material for the subsequent redistribution of the precious metal's re-extraction. For example, in order to finish the slag by the floatation method (Dovshenko et al., 1997; Korukin et al., 2002; Sarrafi, 2004) it is necessary to form rather large sulphide particles, which is achieved by the familiar processes of reducing the cooling melts rate. The isolated concentrates besides copper will contain other non-ferrous metals, which designates their accumulation and concentration in the semi-products of closed-circuit processing schemes. The calcium oxide content and the rate of slag cooling also influence the composition of iron-silicate forming phases, the properties of which determine the expenditure of energy on slag grinding. The information about the structure of the high-ferriferous slag cooled at the high rate allows us to define characteristics of the glassy condition, including the devitrification temperature, heat and temperature, meaning the solid and liquid phases' crystallization of the quenched slag.

Oxide materials are important for pyrometallurgical processes for the production of non-ferrous metals, which are characterized by composition complexity and - apart from iron, silicon, calcium and aluminium oxides - contain the impurity elements Cu, S, Zn, Pb etc. Depending upon the composition and cooling rate from the melted state, the solid samples of oxide systems can be singled out both in crystalline and amorphous states. It is known that in oxide systems containing more than 50% SiO_2 and up to 20% of Fe_2O_3, a glassy state is formed even at a low cooling rate (Karamanov & Pelino, 2001). It is shown (Selivanov et al., 2009b, 2010) that all of the complex of iron-silicate phases generated during sample slow cooling corresponds to a molar Fe/Si ratio within the limits of 0.9-1.6. Sample cooling with a high (900 °C/s) rate results in the increase of this ratio by up to 3.4. It should be noted that a decrease of the cooling rate results in the increase of the portion and size of magnetite crystals and sulphide particles.

In connection with the study, the conditions of crystals' formation from amorphous high iron oxides based on FeO_x-SiO_2-MeS (Me - Cu, Zn) systems and the determination of the forming phases' composition are of great interest.

The sample of slag from the melt of copper-zinc concentrates in the Vanyukov furnace was melted in the furnace at a temperature of 1250 °C and cooled down using the water granulation method. The cooling rate of the oxide melt calculated from equation 1 is 900 °C/s. The granules obtained are isothermally annealed in the resistance electric furnace (during 5 and 60 min) at a temperature of 750 °C (Gulyaeva et al., 2011).

X-ray analysis results (Selivanov et al., 2009b) have shown that with the cooling of the oxide melt at a rate of about 900 °C/s, an amorphous product is formed (Fig. 2). The reflexes corresponding to the crystal phase are virtually absent at the diffractogram; however, the diffusion scattering characterizing the availability of a short-range order in the formed glass was observed.

The thermogram received by means of a quenched sample heating is shown in Fig. 3. Based on the data received, the heating of high iron vitrified oxides up to 1230 °C and their cooling at a rate of 20 °C/min results in their transition from an amorphous to a crystalline state. The stability factor of the vitreous state is shown in the following equation (Biswas et al., 2010):

$$\Delta T = T_g - T_{c.cr.}, \tag{3}$$

where T_g and $T_{c.cr.}$ - devitrification temperatures (533 °C) and the onset of "cold" crystallization (555 °C). For the sample under study, the ΔT value amounts to 22 degrees, which points to the non-stability of the amorphous state.

The annealing of sampled granular particles of slag at a temperature of 750 °C does not result in the external change of their forms. The data (Gulyaeva et al., 2011) of the X-ray analysis of amorphous oxide samples after annealing for durations of 5 and 60 min showed that the material heating results in the formation of a crystalline state. In diffractograms (Fig. 7), reflexes of annealed samples typical for fayalite, quartz and magnetite (very weak) were detected. Fayalite reflection (d = 2.50 Å) intensity increases from 74 (Fig. 7 a) to 86 pulses/s (Fig. 7 b) with the growth of the annealing duration, whereas magnetite reflection (d = 1.48 Å) intensity does not change and is equal to 11 pulses/s.

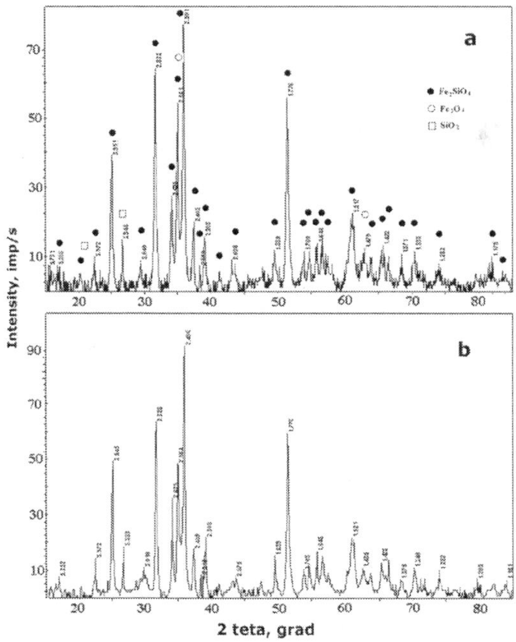

Fig. 7. X-ray spectra of granular slag samples from the melts of copper-zinc concentrates annealed at 750 °C within 5 (a) and 60 min (b)

The microstructure analysis of the samples (Fig. 8) received after annealing showed that initially (τ = 5 min) single faceted crystals of magnetite are presumably formed. Over the entire area of the metallographic section, a thin grid of crystal is formed with a cell size of 0.3 - 0.5 µm. It should be noted that an increase of the annealing duration (τ = 60 min) of an amorphous sample results in formation of certain particles of larger dendrites with the axes of the second- and third-order in the area close to the surface. Closer to the particles' border, the dendrites' axes' length decreases, the phases having - in the metallographic section plane - a structure in the form of triangles and stars sized of 5-10 µm form around them. Proceeding from the thermal analysis data (Sycheva & Polyakova, 2004), one can evaluate the viscosity crystallization criterion, which is equal to:

$$\alpha = T_g/T_m, \tag{4}$$

where T_m - temperature of the dissolution of the crystalline phases in the melt (1103 °C), defining together with the forces of surface tension the capacity of glasses to volume crystallization. The value of α calculated by this equation reached 0.6, which shows the inclination of the studied oxide glass-to-surface crystallization. The mechanism of the initial formation of crystalline phases is apparently explained by the annealing temperature (750 °C) which is higher than that of cold crystallization. During the heating of a quenched sample from an oversaturated silicate phase forming the glass diffusion separation of magnetite and the formation of iron-silicate phase crystals, there occurs a little less (compared to glass) iron content through equation 2.

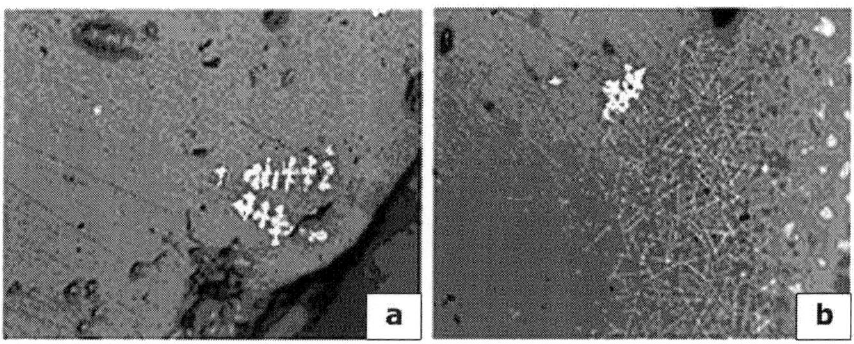

Fig. 8. Microstructure of the slag sample (x500) annealed at 750 °C during 5 (a) and 60 min (b)

The analysis of the sample annealed within 60 min by the X-ray spectral microanalysis of the characteristic radiation of the elements showed that the basis of the dendrite phase is formed by iron oxides while the basis of glass is formed by iron-silicates. The composition of the dendrite phases depending on crystal forms is variable - one of the phases in the form of triangles and stars is richer in iron and close to magnetite in its composition. They contain fewer impurity elements of silicon (0.9%) and calcium (0.2%) (Fig. 9, Table 5). Copper (0.2-0.7%) and zinc (2.7-3.7%) concentrations in magnetite are virtually unconnected with the geometric form of crystals. The silicate phase is close in composition to $Fe_{1.36}SiO_{3.36}$: the content of copper in it reaches 0.8% and that of zinc reaches 4.1%. The presence of sulphur determines the formation of fine bornite phases of non-spherical forms (1-6 µm) in the sample and larger chalcosine phases with a size of 15 - 25 µm.

Thus, the granulation of iron-silicate oxide melts containing more than 30% SiO_2 results in the formation of glass. The annealing of the granuled slag of the autogenous smelting of copper-zinc concentrates at 750 °C results in the initial separation of magnetite in the form of geometric crystals from an oversaturated iron-silicate matrix and - furthermore - in the form of a dendrite structure. The diffusion mechanism of magnetite crystals' growth has a superficial character. The composition of the crystallizing phases and the dissolubility of nonferrous metals in them have been established.

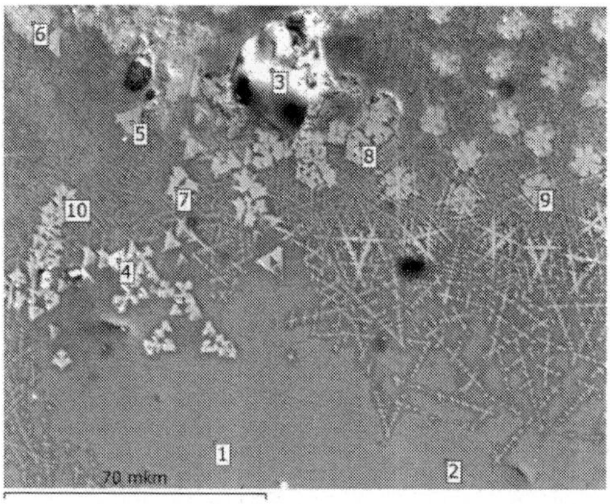

Fig. 9. Microstructure of the sample after annealing at 750 °C within 60 min and EPMA points

№	Content, mass, %					
	Si	Fe	S	Cu	Zn	Al
1	14.9	40.4	0.5	0.8	3.8	2.7
2	14.6	41.1	0.4	0.5	4.1	2.8
3	0.2	3.5	21.5	76.5	-	0.5
4	5.7	26.2	7.9	20.9	2.7	1.7
5	0.9	69.6	-	0.4	2.7	1.9
6	4.1	61.9	0.3	0.6	3.7	2.0
7	4.9	60.7	0.1	0.2	2.6	2.8
8	2.3	64.5	0.1	0.7	3.8	2.2
9	3.1	63.3	-	0.2	3.7	2.7
10	2.4	64.7	-	0.4	3.4	2.7

Table 5. EPMA data on the phase composition of a quenched slag sample annealed at 750 °C during 60 min (Fig. 9)

4. Effect of the cooling rate on the phase composition and structure of copper matter converting slags

As was mentioned (Selivanov et al., 2009a, 2009b), the distinguishing feature of the production process in the "Sredneuralsky Copper Smelter Plant" lies in the fact that mattes containing 45-55% copper recovered upon the smelting of copper-zinc concentrates in the

Vanyukov furnace are converter. Apart from copper and precious metals, the mattes concentrate zinc, lead, arsenic and antimony. During the conversion, a part of these metals passes into a gas phase and dust, and the other part of the metals is redistributed between white matte and slag and then between copper and slag. Thus, the precipitating converter slags have a high content of precious metals, the re-extraction of which increases the coefficient of an integrated approach to the raw materials' use. The choice of the slag processing method is determined based upon the forms of the metals.

A thermodynamic simulation (TDS) was performed for the working body of the following composition,%: 21.0 FeO; 16.0 Fe_2O_3; 20.3 SiO_2; 5.1 ZnO; 11.0 CuO; 2.6 Al_2O_3; 2.0 CaO; 1.2 MgO; 2.0 Sb_2O_3; 1.2 Pb, 3.1 S. The degree of iron oxidation (α) in the working body determined Fe^{3+}/Fe^{2+} the ratio amounts as 0.4. During the thermodynamic simulation, we used the compounds inherent in the FeO_x-SiO_2-FeS-Cu_2O-ZnO system. The calculations were carried out for the 100 kg working body at a gas phase (nitrogen) volume of 2.24 m^3 over the slag and at a pressure in the system equal to 0.1 MPa.

According to the TDS results, the main components of the equilibrium system which was cooled under equilibrium conditions were: Fe_3O_4, Fe_2SiO_4, SiO_2, ZnS, Cu_2S, Cu_5FeS_4, $CuFeS_2$, as well as metallic copper and Cu_2Sb; the possibility of the solidification of the last two components in such systems was noted earlier (Selivanov et al., 2000, 2004). A change in the degree of iron oxidation in the working body does not introduce qualitative changes in the phase composition of the slag but it does affect the interfacial distribution of non-ferrous metals. Copper occurs predominantly in the form of a metal or a sulphide (Fig. 10), and the temperature changes influence their mass ratios. As such, a larger amount of Cu_2S forms within the temperature range of 700-1100 °C, whereas the production of metallic copper in this temperature range decreases substantially. The reduction of the temperature favours the formation of complex copper sulphides, mainly bornite.

Zinc at high temperatures is represented by the oxide compounds ZnO, $ZnFe_2O_4$, $ZnSiO_3$ and $ZnAl_2O_4$, the fractions of which decrease as the slag cools down. At low temperatures, the probability of the solidification of zinc sulphide - ZnS - is high.

The samples for investigation have been prepared from a converter slag containing %: 40.7 Fe, 3.0 S, 8.5 Cu, 4.0 Zn, 19.5 SiO_2, 1.9 CaO, 0.7 MgO, 2.3 Al_2O_3, 0.2 Sb, 0.9 Pb, 0.1 As. According to the data of the X-ray diffraction phase analysis, the base phases in the initial converter slag are a fayalite of Fe_2SiO_4 and a magnetite of Fe_3O_4. Copper is mainly detected as bornite and copper sulphide. Zinc is revealed in the form of a sulphide with a sphalerite structure. The melting of the samples followed by rapid cooling decreases the diffraction reflex intensities and preserves the non-equilibrium high-temperature phases. With the cooling of the sample at the rate of 900 °C/s, a significant amount of amorphous phase forms. No reflections of the ZnS phase have been detected at a high cooling rate.

When the initial converter slag sample is subjected to differential thermal analysis during heating in an argon atmosphere (Selivanov et al., 2009a), the following melting-induced endothermic heat effect is observed: it begins at 1079 °C and has a maximum value at 1122 °C (Fig. 11). During its cooling, we established the solidification temperature of the slag melt, which is 1108/1078 °C. According to the mass-spectrometry data, a certain amount of SO_2 precipitates in the heating of the sample (beginning from 300–400 °C), which results from the interaction of sulphides with the oxides of iron of the highest valence. The slag weight loss upon heating up to 1300 °C is 1.2%.

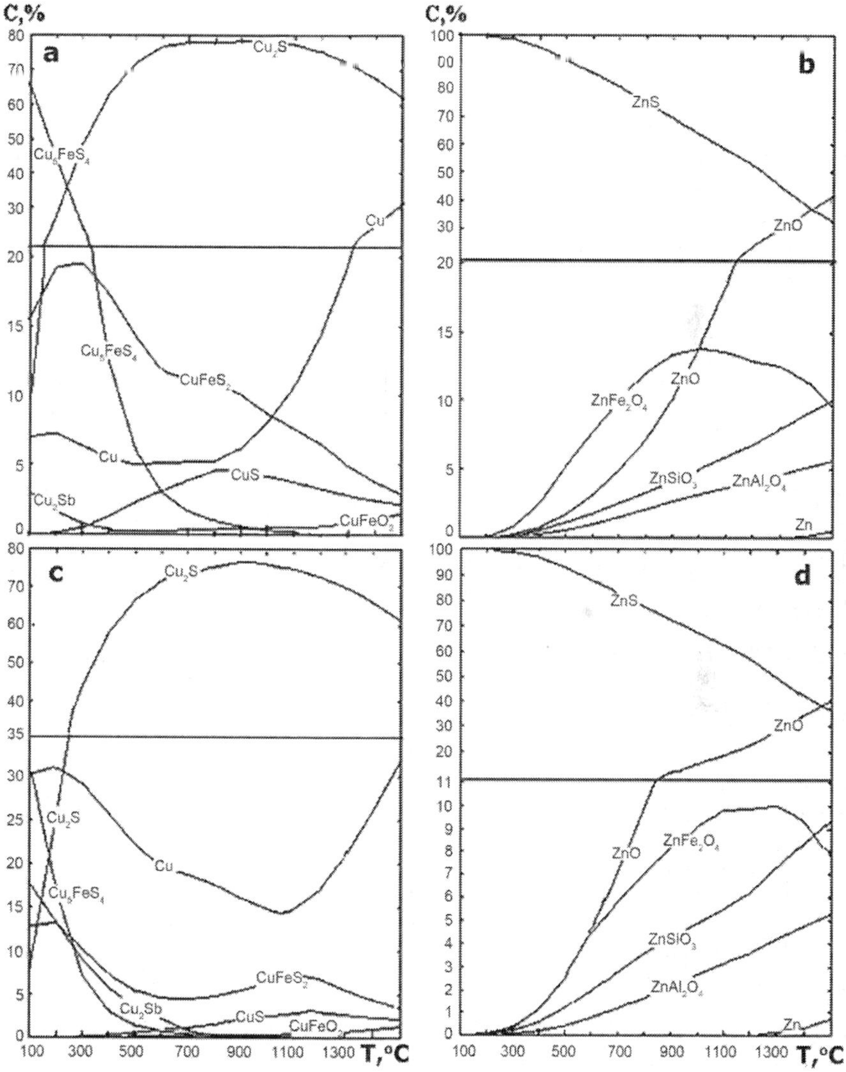

Fig. 10. Distributions (relative %) of copper (a, b) and zinc (c, d) in the components of a condensed phase, depending upon the temperature at the oxidizing degree of iron in the converter slag: 0.4 (a, b) and 0.1 (c, d)

A comparison of the thermograms of the initial slag sample and a sample cooled at a rate of 0.3 °C/s indicates that they are identical. However, a sample cooled at a rate of 900 °C/s is characterized by the appearance of an effect at 533 °C in the DSC curve, which is caused by a second-order phase transformation during devitrification. When this sample is heated further, we detect an exothermic heat effect with an onset/maximum at 608/635 °C. This effect is interpreted as "cold" slag crystallization (ordering its structure). The endothermic effects at 918/963 °C and 1064/1127 °C point to the melting of the sulphide and oxide components of the sample. On the DSC curves of the samples cooled at the rates of 0.3 and 900 °C/s, the solidification effects were equal to 1055/1049 °C and 1085/1072 °C, correspondingly.

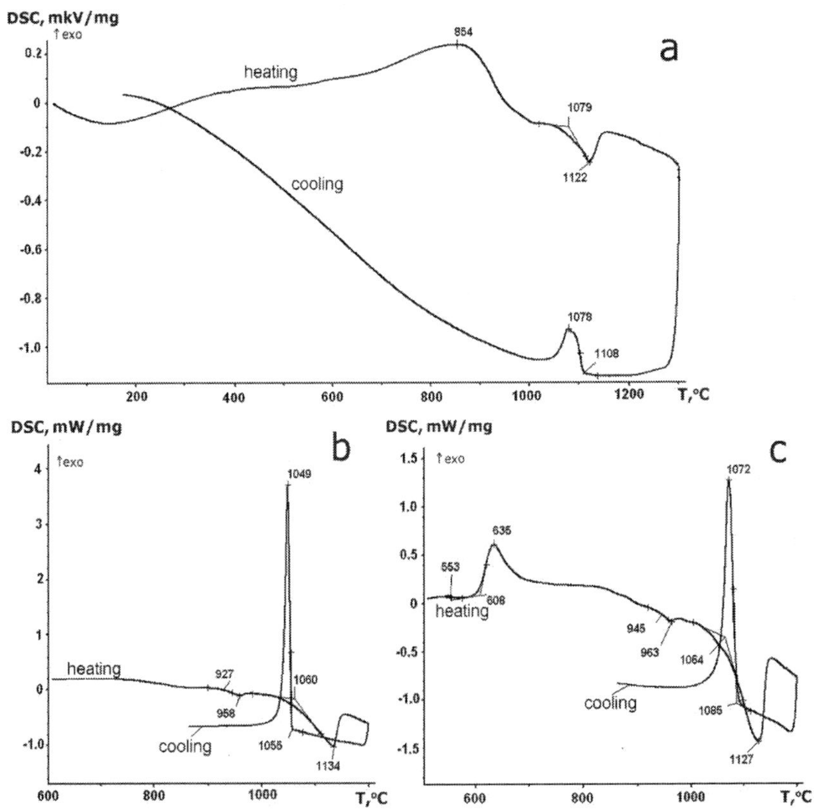

Fig. 11. Sample thermograms of (a) the initial converter slag and the slags that form upon the cooling of the melt at the rates of (b) 0.3 and (c) 900 °C/s

In essence, the temperatures and enthalpies of the thermal effects of the heating of the granulated slags from melts of copper-zinc concentrates and the converting of the matte have similar values (Table 6).

Devitrification			$T_{c.cr.}$, °C	$L_{c.cr.}$, J/g	L_m, J/g		$L_{h.cr.}$, J/g
T_o, °C	T_g, °C	Δc, J/g K			1 peak	2 peak	
533	553	0.150	608	66	4	164	157

Table 6. Temperatures and enthalpies of the thermal effects at the heating of the granulated and converter slags

The microstructure of the initial converter slag is represented by iron-silicates and matte particles (Fig. 12). The slag contains a large number of 100 μm magnetite crystals of a regular shape and spherical matte particles smaller than 300 μm. The matte particles have an eutectic structure (copper sulphides, bornite, metallic copper). The silicate constituent of the slag has a small amount of metallic and sulphide copper. These inclusions have sizes equal to 0.1 - 2.0 μm and are concentrated along the boundaries of large iron-silicate aggregates.

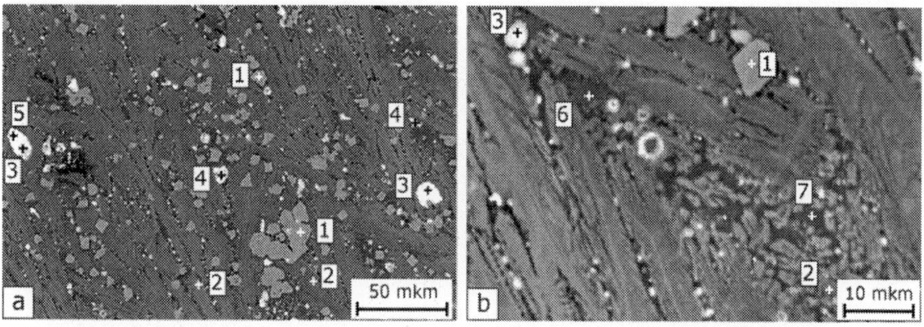

Fig. 12. Microstructure of the sample of industrial converter slag and EPMA points

According to the electron-probe microanalysis of the commercial sample, the iron-silicate phase is heterogeneous in terms of both major (iron, silicon) and dissolved impurities (Table 7). The calculated composition of the iron-silicates ranges from $Fe_{0.94}SiO_{2.94}$ to $Fe_{1.59}SiO_{3.59}$. The iron-silicates contain 0.5-2.0 Al_2O_3, 0.5-1.0 MgO, 0.2-0.5 K_2O, 0.2-1.1 CaO, 5.3-6.6 Zn, and 0.1-0.6% S. Moreover, we detected a silicate phase that corresponds to the SiO_2-FeO-CaO-Al_2O_3-Zn(Pb)O system and has a low iron content with a high calcium content. The magnetite crystals (61.4-63.3% Fe) located in the plane of the section contain the following impurity elements: 1.2-1.3 Al, 0.4-1.2 Si, 0.3-0.2 Ti, 0.1-2.5 Cr and 2.7-3.2% Zn. The sulphide phases in the slag are represented by bornite- and sphalerite-based solid solutions. The copper content in the bornite solid solution is lower than the stoichiometric copper content. The ZnS-based phase (sphalerite) contains 37.9-53.3 Zn, 0.9-7.3 Cu, 11.5-21.6% Fe, and a near-stoichiometric sulphur content. The regions of the PbS-Cu_2S-FeS solid solution are located along the periphery of the sulphide phases.

The base of the slag sample cooled at the rate of 0.3 °C/s also consists of iron-silicate phases, magnetite and sulphides (Fig. 13). The re-melting and slow cooling of the slag result in a significant coarsening of the formed crystals. The area of its polished section is a mainly occupied silicate phase with a composition close to $Fe_{1.59}SiO_{3.59}$; between them there are small $Fe_{1.12}SiO_{3.12}$ dendrites and calcium- and silicon-rich phases (Table 8). Apart from iron, the magnetite also contains zinc, titanium, silicon and aluminium impurities as in the initial converter slag.

№ point	Content, mas.%													Phases
	Mg	Al	Si	S	K	Ca	Ti	Cr	Fe	Cu	Zn	Pb	O	
1	-	1.2-1.3	0.4-1.2	-	-	0.1	0.2-0.3	0.1-2.5	61.4-63.3	-	2.7-3.3	-	30.4-30.8	Fe_3O_4
2	0.3-[illegible]	0.3-[illegible]	13.8-[illegible]	0.1-[illegible]	0.1-[illegible]	0.1-0.8	-	-	38.1-[illegible]	[illegible]	4.3-5.3	-	36.5-38.3	Iron-silicate
3	-	-	0.3	26.2-27.6	-	0.1	-	-	17.0-20.8	50.1-55.1	0.3-0.4	1.0-1.2	-	Cu_5FeS_4 solid solution
4	-	до 0.5	0.4-3.4	30.0-32.7	0.1	0,2	-	-	11.5-21.6	0.9-7.3	37.9-53.3	-	-	(Zn,Fe,Cu)S
5	-	-	0.2	16.6	-	-	-	-	9.4	23.4	0.6	49.8	-	(Pb,Cu,Fe)S
6	-	3.0	17.7	0.4	0.7	0.9	0.4	-	33.2		3.5	0.6	39.4	$Fe_{0.94}SiO_{3.47}$
7	0.3	4.6	21.9	1.6	0.2	6.3	0.4	-	14.6	0.5	4.4	3.1	42.1	Iron-silicate

Table 7. EPMA data on the phase composition of the initial converter slag (according to Fig.12)

The coarse sulphide particles of a size of 15–30 μm consist of bornite - the composition of which varies from $Cu_{7.2}FeS_{6.4}$ to $Cu_{3.2}FeS_{3.3}$ - and a PbS-Cu_2S-FeS alloy (Fig. 13). The bornite is located in the centres of the particles, while the lead-containing sulphide alloy with a lower melting point forms on the fringes on its surface.

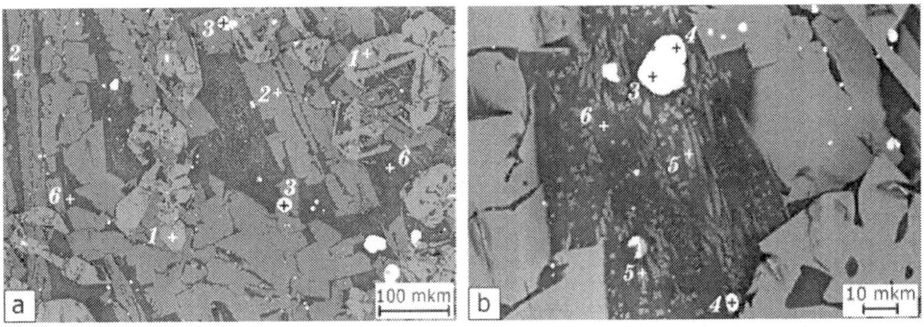

Fig. 13. The sample microstructure of the converter slag cooled at a rate of 0.3 °C/s and EPMA points

The structure of the slag sample cooled at a rate of 900 °C/s is represented by glass, magnetite, iron-silicate phase crystals and sulphide inclusions of a spherical form up to 10 μm (Fig. 14). According to the EPMA data (Table 9), the glass has about 36% SiO_2 and 51% FeO_{1+x}. Acicular crystals 5-15 μm long and about 1 μm thick are clearly visible against the background of the glass; their composition is close to that of iron-silicate of $Fe_{3.4}SiO_{4.4}$. Magnetite (60.4–61.6% Fe) is present in the form of dendrites. The sulphide phase (its coarsest particle is 6.4 μm in size) is inhomogeneous and its central portion corresponds to the formula of $Cu_{5.4}FeS_{3.4}$. The distribution of nonferrous metals in the sulphide particle is also non-uniform: the centre contains 6.2% Zn, 1.6% Pb, and 1.4% As, and the periphery contains 1.3% Zn, 2.0% Pb and 4.7% As.

№ point	Content, mas.%									Phases
	Al	Si	S	Ca	Fe	Cu	Zn	Pb	O	
1	2.4-2.5	0.4	-	0.1	63.3	-	2.4-2.5	-	30.8	Fe_3O_4
2	-	15.0	-	0.3	42.1-42.9	-	3.4	-	37.2-37.3	Iron-silicate
3	-	0.2-0,5	21.5-28.5	0.1	8.2-14.9	55.4-69.2	0.3	0.5-0.6	-	Cu_5FeS_4 solid solution
4	-	0.3-0.8	18.8-18.9	0.1-0.2	9.3-13.1	23.4-25.3	0.5-0.8	40.6-47.8	-	(Pb,Cu,Fe)S
5	2.4-2.7	16.4-16.9	0.4	1.9-2.1	32.0-33.4	< 0.2	5.3-5.5	0.9-1.0	38.1-38.5	Iron-silicate
6	5.8-7.2	19.9-22.5	0.6-1.1	5.7-6.9	13.0-15.3	-	3.9-6,2	2.3-2.7	40.7-42.5	Iron-silicate

Table 8. EPMA data on the phase composition of the converter slag cooled at a rate of 0.3 °C/s

This data indicates that an increase in the cooling rate leads to vitrification. However, even at a cooling rate of 900 °C/s an iron-silicate phase and magnetite solidify and sulphides precipitate. The compositions of the iron-silicate crystalline phases vary over wide limits: as the cooling rate increases, high-iron modifications form and the fraction of magnetite

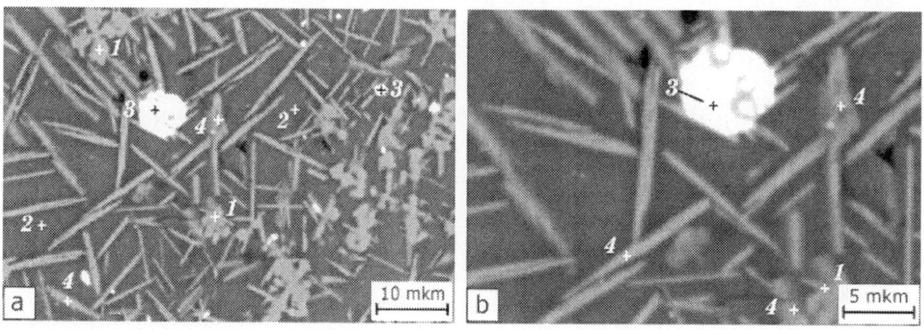

Fig. 14. Microstructure of the converter slag cooled at a rate of 900 °C/s and EPMA points

№ point	Content, mas.%										Phases
	Al	Si	S	Ti	Fe	Cu	Zn	Pb	O	As	
1	2.7-3.1	1.4	-	0.2	60.4-61.1	-	2.4-2.5	0.1	31.4-31.5	-	Fe_3O_4
2	2.3-2.6	15.5-17.3	0.8-0.9	0.1	34.2-41.0	0.2	4.7-5.4	0.9	28.0-38.1	0.3	Glass
3	-	0.3-0.9	17.2-20.3	-	9.8-10.8	62.6-64.8	0.4-6.2	0.9-2.0	-	0.6-4.7	Cu_5FeS_4
4	1.5-1.9	7.4-7.8	0.2-0.3	0.1	51.1-51.3	0.2-0.03	3.2-3.4	0.2-0.3	34.0-34.1	<0.1	Iron-silicate

Table 9. EPMA data on the phase composition of the converter slag cooled at rate of 900 °C/s (according to Fig. 14)

crystals decreases. A decrease in the cooling rate of the slag is accompanied by magnetite formation (endothermic effect) and the precipitation of iron-silicate crystals with a lower (compared to glass) iron content.

According to the EPMA data, none of the samples contains a phase close to stoichiometric fayalite (Fe_2SiO_4). The whole set of the iron-silicate phases that form during the slow cooling of the converter slag corresponds to an Fe/Si atomic ratio of 0.9–1.6. The cooling of the slag at the high (900 °C/s) rate is a result of the increase in this ratio to 3.4. A decrease in the cooling rate of the molten slag favours an increase in the fraction and sizes of magnetite crystals and sulphide particles. The distribution of non-ferrous metals between phases changes according to the fraction of the sulphides.

Copper is concentrated in the bornite-based solid solution, which forms in all the samples. The bornite content increases as the cooling rate decreases, which can be explained by the specific features of the solidification and separation of oxide–sulphide systems that are related to changes in the sulphides' solubilities. The results obtained agree with the TDS data with regard to the predominant formation of copper sulphides during the cooling of the slag. We failed to detect metallic copper in real slags, irrespective of the cooling rate (see Fig. 10). During slag solidification, zinc is distributed between oxide and sulphide phases. An individual (Zn,Fe)S phase containing 38–53% zinc was revealed only at the low cooling rate. Lead in the slag is present in both oxide and sulphide forms. Its content in the iron-silicate phase correlates (increases) with the silicon dioxide content. Lead forms the regions of a sulphide phase (40–48% Pb) of 1–2 μm in size at the low cooling rate. The formation of the oxide compounds of zinc and lead supports the absence of an equilibrium state in all of the investigated slag samples. Sulphide phases have a high arsenic content. The arsenic content in a bornite-based solid solution reaches 4.7% in a slag sample cooled at a rate of 900 °C/s and the arsenic content in iron-silicates is 0.1%.

Thus, when changing the cooling rate of the slag, we can affect the forms of copper, zinc, arsenic and lead in it in order to prepare the slag for the additional recovery of precious metals (Selivanov, 2009a). Moreover, the slag cooling conditions affect the composition of the iron-silicate forming phases, and its properties control the energy consumed for grinding as well as the possibility of using magnetic separation methods for the precipitation of iron oxides, and so on. The converter slag contains both mechanically-introduced coarse matte particles and fine sulphides, which precipitate during the solidification of an oxide–sulphide melt. The cooling rate of the molten slag controls both the phase composition and the particle size of the oxide and sulphide forming phases. The oxide component transforms into a glassy state at a high (900 °C/s) cooling rate of the slag. The devitrification and cold crystallization of the glass falls within the temperature range of 533 - 635 °C.

The copper in the slag is mainly represented by the bornite-based solid solution, the content of which increases as the cooling rate decreases. Zinc and lead are distributed between the oxide and sulphide components. The individual sulphides of these metals are only revealed with the low cooling rates of the slag.

5. Forms of metals finding in the slag of combined melting – The converting of copper concentrates

The autogenous processes of the converting of copper-containing raw materials, including the application of both «Noranda» and combined melting-converting (CMC) units have been widely used in non-ferrous metallurgy. However, the copper content in the slags which are formed during the smelting of concentrates in these units is rather large. In order to decrease the loss of metals and to choose the methods of the processing of slags, it is of the prime importance to reveal the forms of the existence of the precious components in them. The molten slags are - in their compositions - close to the FeO_x-SiO_2 system and, at the temperatures corresponding to pyrometallurgical processes, agree with homogeneic melts (Vanyukov & Zaitsev, 1969, 1973). During slag cooling, a number of micro-processes connected with compound crystallization, liquation phenomena and the change of the detection of forms of non-ferrous metals occurs (Kukoev et al., 1979). The last of these, in turn, determine the choice of methods for the re-extraction of precious components from the slag.

As a starting sample – the slag of a pilot unit of CMC cooled at a rate of about 0.5 °C/min (Selivanov et al., 2004). According to the data of the chemical analysis, the slag contained, %: 1.2 Cu, 55.9 Fe_{total}, 4.9 Fe^{3+}, 53.3 Fe^{2+}, 0.4 Fe_{met}, 3.1 S, 16.0 SiO_2, 3.6 Zn, 0.1 Pb, 0.1 As, 0.1 Sb, 0.5 CaO and 0.5 Al_2O_3. The studied slag sample in its chemical composition is close to the slag of copper matte converting. The relatively high sulphur content in the slag allows it to be referred to the oxide-sulphide melts class, the crystallization of which must be accompanied by a number of complex interactions changing the form of the metals' detection (Kukoev et al., 1979; Selivanov et al., 2000).

The thermodynamic modelling of the processes (Moiseev & Vyatkin, 1999; Selivanov et al., 2004) occurring during slag cooling was carried out for the working body and is in substantial agreement with the slag composition taken for the investigation. The thermodynamic functions of elements and compounds in the condensed (Cu, Cu_2O, CuO, $CuFe_2O_4$, $Cu_2Fe_2O_4$, Cu_2S, $CuFeS_2$, Cu_5FeS_4, Cu_3As, Cu_2Sb, Zn, ZnO, ZnS, $ZnSiO_3$, Zn_2SiO_4, $ZnAl_2O_4$, $ZnFe_2O_4$, Fe_{met}, FeO, Fe_3O_4 , Fe_2O_3, FeS, FeS_2, $FeSiO_3$, Fe_2SiO_4, $FeAl_2O_4$, Fe_2ZnO_4, Pb, PbO, Pb_2O_3, PbS, $PbSiO_3$, As, As_2O_3, As_2O_5, As_2S, As_2S_3, Sb, Sb_2O_3, Sb_2O_5, Sb_2S, Al_2O_3, Al_2SiO_5) and gaseous (S_2, SO_2, SO_3, Zn, ZnO, N_2, O_2, Pb, PbS, As_2O_3, As_2O_5, Sb_2O_5, Sb_2O_3, etc.) states have been used for the calculations. The modelling was carried out during the changing of the temperature from 1520 to 25 °C with steps of 50 degrees.

According to the TDS data, slag cooling leads to the changing of the parts of phases and the forms of the metals which exist (Fig. 15). Accordingly, the working body temperature decrease increases the crystallization probability of $FeSiO_3$ and Fe_2SiO_4 iron-silicate compounds and favours magnetite formation, which can be explained by the disproportionate amount of iron oxide (II) and by the interactions between non-ferrous metals oxides and iron oxide:

$$4FeO = Fe_3O_4 + Fe, \tag{5}$$

$$FeO + MeO = Me + Fe_3O_4. \tag{6}$$

For the non-ferrous metals in the slag, one would expect the changing of their forms of existence at the expense of reactions between sulphides and oxides (Belyaev et al., 2001; Spira & Themelis, 1969). If, at a high temperature, the copper in the slag is preferably in the form of sulphide, then cooling can lead to its transition into its metallic state:

$$3Cu_2S + FeS = Cu_5FeS_4 + Cu, \tag{7}$$

$$Cu_2S + 4FeO \rightarrow FeS + 2Cu + Fe_3O_4. \tag{8}$$

The working body (slag) cooling favours a ZnO → ZnS transformation according to the reaction:

$$ZnO + FeS \rightarrow ZnS + FeO. \quad (9)$$

This means that copper, antimony and arsenic are the most electropositive metals for the problem at hand and one should also expect the formation of Cu-Sb-As alloys as well as copper.

Iron-silicates and oxides, as well as bornite, were found in the slag by way of X-ray diffraction analysis (Fig. 16) and they were identified on the basis of the data from (PC-PDF, 2003) for Fe_2SiO_4, FeO, Cu_5FeS_4 and Fe_7SiO_{10}. Some discrepancies in the X-ray reflexes between their meanings for pure compounds is suggestive of the formation of solid solutions which distort the minerals' lattice.

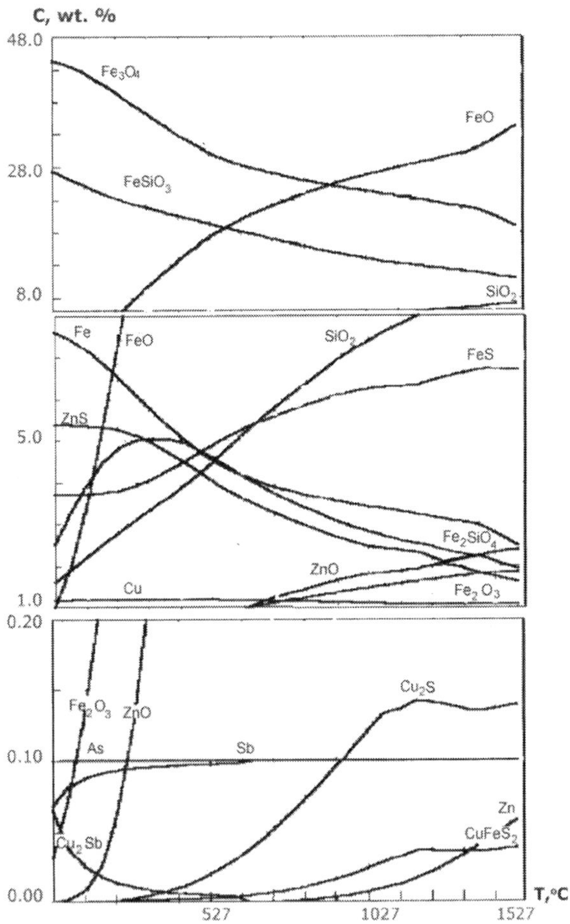

Fig. 15. Changing of the forms of metals present in the CMC slag according to the TDM data

The main structural components of a slag are fayalite and an iron-silicate phase with a high FeO content (Fig. 17). Fayalite, which is close to Fe_nSiO_{2+n} in composition, is presented by a solid solution of iron, zinc and calcium silicates. The iron-silicate phase (83% FeO) conforms to the $nFeO \cdot mSiO_2$ formula. The micro-hardness of fayalite varies within the limits of 5300 to 8400 MPa but that of the micro-hardness of the iron-silicate phase within the limits of 4100 to 6100 MPa. Apart from the main phases in the slag, we have revealed iron oxides with a micro-hardness of 5100 to 8400 MPa, which exceeds the values characteristic of pure wustite (~ 4300 MPa), iron sulphides which deposit in the form of FeS troilite (2440-2800 MPa) and FeS-FeO eutectic (3400 to 4900 MPa) of different dispersivity, zinc ferriferous sulphides (christophite) and, in lesser amounts, bornite and a solid solution on its base.

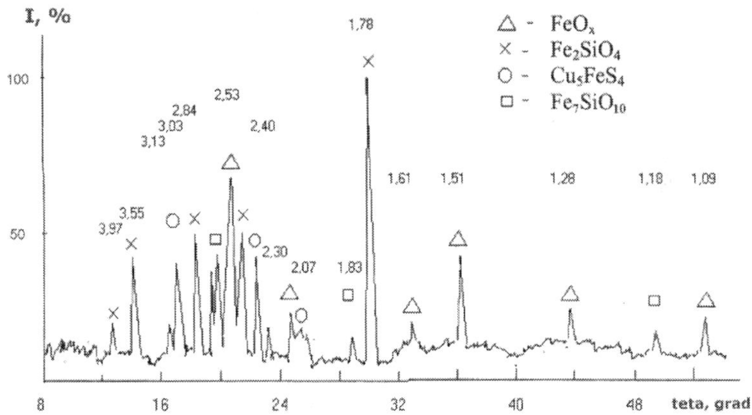

Fig. 16. X-ray diffractogram of the CMC slag

The crystallization of slag containing 16% silica, according to the diagram of the FeO-SiO_2 condition, proceeds from a temperature of 1250 °C and is accompanied by the rejection of iron-silicate phase crystals and excess wustite (Selivanov et al., 2004). X-ray spectral microanalyses of the slag (Fig. 18) enable us to establish the composition of the phase components (Table 10). As follows from the data obtained, the oxide and iron-silicate phases do not contain copper. A small quantity of non-ferrous metals (Zn, As - 0.2%) was dissolved in the FeO_x phase. Unlike FeO_x, the silicate phases (the first - with 13.4% and the second - with 5% of Si) have a zinc content of up to 3 - 4%. The silicate phases conform to compositions of Fe_2SiO_4 (fayalite) and Fe_7SiO_{10} (a solid solution with a high content of FeO).

Non-ferrous metals are preferably concentrated in metallic (a size up to 40 μm) and sulphide (up to 100 μm) inclusions. The metallic phase is represented by a Cu-Sb alloy with Sn (3-5%), As (5 - 10%) and Ni (1 - 7%) dissolved in it. The sulphide phases are formed with the participation of Cu, Zn and small quantities of Pb (0.2-0.8%) and As (0.2%). A conglomerate of sulphides and wustite occurs between the crystals of fayalite, and this conglomerate is revealed in the part of the section which is sized 200 μm (Fig. 18). There is a Cu-Sb particle in the centre of the conglomerate. The composition of this particle changes from the surface to the centre. The surface of the particle contains 50% copper and about 30% of Sb, whereas the internal part is more than 50% of the antimony. The heterogeneity of the particle is seen in the micro-structure obtained by the absorbed electrons during X-ray spectral microanalysis. The coefficients of distribution during liquation (K_l - the proportion of elements contents in copper $[C]_{Cu}$ and antimony $[C]_{Sb}$ parts) have the following meanings:

Element	Cu	Fe	Sb	Zn	Pb	Sn	Ni	As
$K_l=[C]_{Cu}/[C]_{Sb}$	10	0.2	0.5	0.5-1.0	0.5	1.2-2.5	0.2-0.4	0.3-0.5

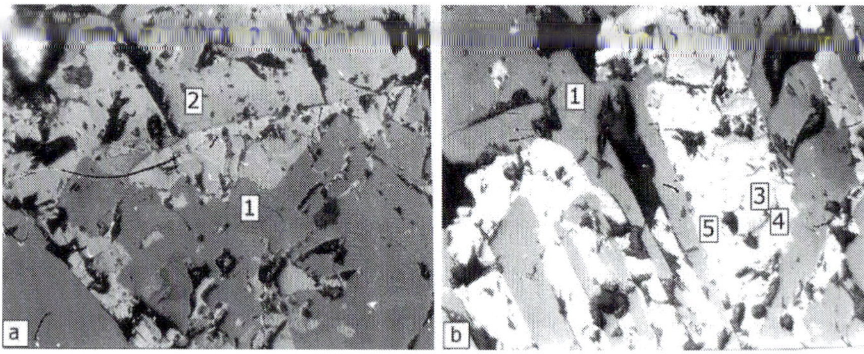

Fig. 17. Microstructure of the CMC slag : a) – x100; б) – x 200: 1 – fayalite, 2 – iron-silicate phase, 3 - wustite, 4 - christophite, 5 – eutectic of FeS-FeO

Scanning the area revealed that the sulphide-metallic part in the section space (Fig. 18) consists of a conglomerate of copper, zinc and iron sulphides as well as an intermetallic phase, closely conforming to the Cu_6Sb composition. The metallic phases stand out in close proximity or else together with wustite and sulphides. The phases are basically solid solutions and contain a significant proportion of impurities. Solid solutions of iron and zinc sulphides are enriched with copper (0.3-3.1%) in contrast with oxide-silicate phases.

As is known, zinc, iron sulphides and FeO are isolated during cooling from FeO-ZnS melts, having unlimited solubility in the liquid state (Kopilov et al., 2002; Toguzov et al., 1982). Apart from this, the double FeO-ZnS eutectic and below at a temperature 920 °C, a threefold FeO-FeS-ZnS eutectic crystallizes from the high-sulphurous residual melt. The last eutectic has the theoretical composition,%: 61.0 FeS, 36.0 FeO, 2.5 ZnS. During slow cooling, the formation of both the threefold and the double eutectic FeS-FeO and FeO–ZnS is possible. In the investigated sample, the FeO-ZnS eutectic is represented by christophite and wustite, which were evolved in turn, and the FeS-FeO eutectic by the primary troilite and wustite of a different dispersion (Fig. 17). Close contact with christophite and wustite in the field of the section confirms the progress of the reaction (9).

Complete (solidus) sulphide crystallization initiates at a temperature below 850 °C, with the formation of another threefold eutectic (30% Cu_2S, 45% FeS and 25% FeO) consisting of FeO, FeS and a solid solution of bornite. In cooling, the solid solution of bornite dissociates - partially or fully - forming a lattice structure of a chalcopyrite decomposition in bornite. The conglomerate illustrated in Fig. 18 is demonstrated by a bornite solid solution, a ferrous sulphide of zinc and copper, troilite, wustite and christophite, which provides evidence of both threefold eutectics. The phase components of the threefold eutectics are more dispersed than those of the double ones.

From the phase diagram of $FeO-Cu_2S$ it follows that copper sulphide is soluble in those oxide melts containing FeO, but if the temperature is below 1100 °C then the interaction of the components with the formation of a solid solution of bornite, metallic copper and iron oxide occurs, which is due to the course of the reaction:

$$4Cu_2S + 4FeO = Cu_5FeS_4 + Fe_3O_4 + 3Cu. \tag{10}$$

In Fig. 18 Sulphide and metallic phases are represented by metallic copper as Cu_6Sb, with two kinds of sulphides containing iron, bornite and wustite as well.

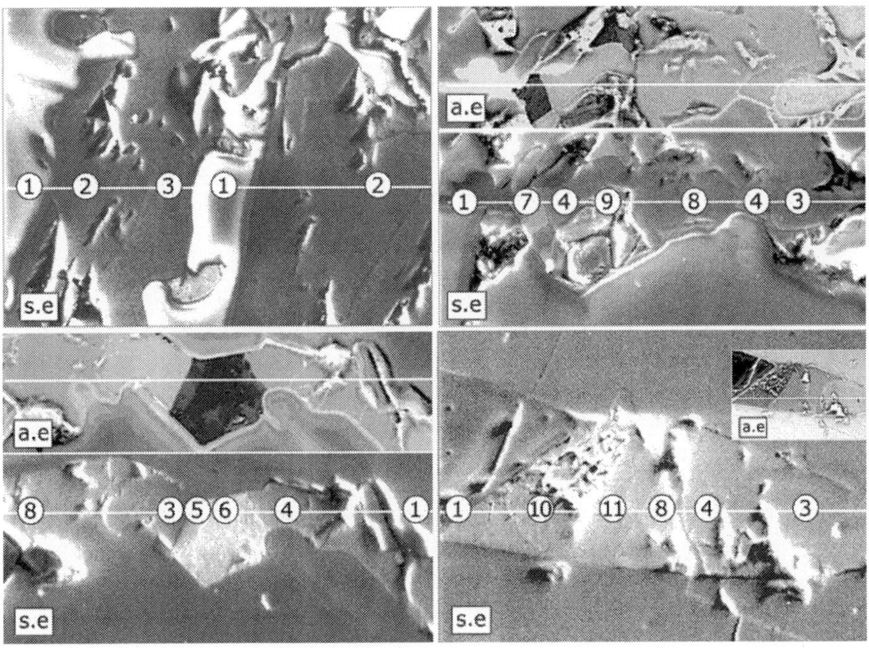

Fig. 18. Structure of the CMC slag in the absorbed (a.e.) and secondary (s.e.) electrons in the sections of 200 μm: 1 - Fe_2SiO_4; 2 - Fe_7SiO_{10}; 3 - FeO_x; 4 - FeS; 5 - Cu-Sb alloy (the particle surface), 6 - Sb-Cu alloy (the particle centre), 7 - Cu_6Sb; 8 - (Fe, Zn)S; 9 - Cu_5FeS_4, 10 - a solid solution on the base of the bornite, 11 - a solid solution of copper and zinc sulphides

Proceeding from the composition (according to the EPMA data) of clearly marked crystals, the phases containing copper correspond to $Cu_{4.4}FeS_3$, 2 and Cu_5FeS_4, the phases containing zinc to $Zn_{0.4}Fe_{0.6}S$ and the iron sulphide phase to FeS. The coefficient of the elements' distribution between the sulphide phases has the following meanings ($K_{b/t}$ and $K_{c/t}$ - the coefficients of the distribution are determined as the ratio of the metals' content in the bornite/troilite and christophite/troilite phases):

Element	Cu	Zn	Pb	Ni	As
$K_{b/t}$	104 - 149	2.5 – 5.0	1.8	2 - 5	1 - 2
$K_{c/t}$	1 - 3	1.5 - 2.5	0.6 - 0.9	1 - 3	1 – 3

Thus, the slag melts of the integrated melting-converting of copper concentrates refer to oxide-sulphide systems' crystallization, which is accompanied by a phase transformation characteristic of both oxide and sulphide systems. Besides this, the phase changes fraught with the interaction of oxides and sulphides proceed in oxisulphide systems, resulting in the extraction of metals and oxide-sulphide eutectics.

The oxide component of slowly cooled slag is represented by the solid solutions of iron oxides and iron silicates. The number of non-ferrous metals dissolved in them (apart from zinc) is less than a tenth part of one percent. The cooling and crystallization of slag leads to the concentrating of non-ferrous metals in metallic and sulphide phases. The metallic phase is represented by Cu-Sb alloy with Fe, Ni, Sn, As and other elements dissolved in it. The metallic phase serves as the collector of non-ferrous metals and contains up to 50% antimony. In separating the metal from the solid slag by the methods concentration (flotation, gravitation, etc.) one should take into account the formation of copper alloys but not copper, as has been assumed before.

№	Content, mas, %											Phases
point	Fe	Cu	Ni	Zn	S	Sb	Pb	Sn	Si	As	Ca	
1	50.9	0	0	3.5	0	0	0	0	13.4	0	0.9	Fe_2SiO_4
2	64.7	0	0	3.3	0	0	0	0	5.0	0	0.1	Fe_7SiO_{10}
3	76.4	0	0	0.2	0	0	01	0	0.2	0.1	0.1	FeO_x
4	62.5	0.4	0	0.1	36.4	0	0.3	0	0.1	0	0	FeS
5	3.7	50.9	0.9	0.2	0.2	29.2	0.1	5.5	0.1	5.7	0.1	Cu-Sb (surface)
6	21.7	4.8	7.2	0.4	0.1	50.2	0.3	2.7	0.1	10.7	0.1	Sb-Cu (centre)
7	0.9	60.5	3.3	0.4	0.1	23.8	0.1	6.7	0.3	3.0	0.1	Cu_6Sb
8	37.3	0.3	0.1	27.2	34.2	0	0.3	0	0	0	0	(Zn,Fe)S
9	12.6	62.6	0.1	0.3	23.3	0.1	0.6	0	0	0.2	0.1	Cu_5FeS_4
10	21.5	56.4	0.1	0.4	20.8	0	0.8	0	0.1	0.1	0.1	Cu_5FeS_4 solid solution
11	27.8	11.0	0.1	27.4	31.2	0	0.3	0	0	0.1	0.1	Solid solution of copper and zinc sulphides

Table 10. The composition of phases identified in the CMC slag (Fig. 18)

The sulphide phases are represented mainly by solid solutions containing copper, iron, zinc and small amounts of lead and arsenic as well. The main phases are bornite, christophite, troilite and the products of the eutectics' decomposition. The size of the metallic and sulphide phases in the slag amounts to 10 - 100 μm, which will allow their breakdown to be carried out by way of grinding with the standard equipment.

6. Peculiarities of the crystallization of high-magnesian iron-silicate slags

The processing of oxide nickel ores with the smelting technique for mattes is connected with the formation of a great amount of high-magnesian silicate slags. The investigations presented here are devoted to the study of forms for finding metals and structures of granulated slags of nickel production which are in substantial agreement with the SiO_2–

FeO_X-MgO–CaO system, with the phase transformations' peculiarities occurring during their heating as well as during their crystallization during the course of annealing.

The commercial slag sample of the shaft smelting of oxide nickel ores from "Ufaleynickel" JSC (Russia, Ural), obtained by way of the granulation of an oxide melt in the water pond and containing, %: 13.0 MgO, 42.1 SiO_2, 20.8 Fe, 7.2 CaO, 5.8 Al_2O_3, 0.2 Ni, 0.3 S, was taken for investigation (Sergeeva et al., 2011). The sample slag annealing was carried out in a furnace by heating it up to 1000 °C and upon the subsequent maintenance of this temperature for 60 minutes.

The results of the X-ray-phase analysis showed that the initial granulated sample of slag almost completely consisted of glass (Fig. 19). There are no reflexes conforming to crystal phases in the diffractogram. Responses conforming to $Ca(Mg,Fe)Si_2O_6$ diopside and $(Fe,Mg)_2SiO_4$ fayalite, with the structure of olivine, were revealed in the slag diffractogram after the annealing.

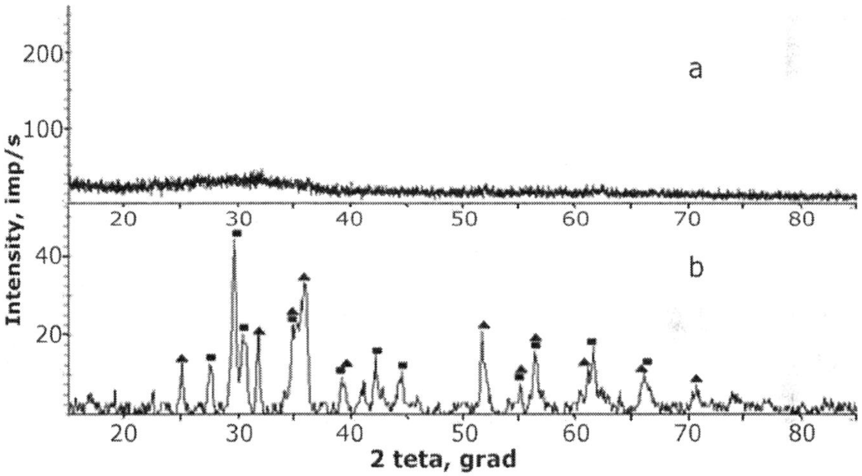

Fig. 19. Slag diffractograms: a – initial, granulated; b – after annealing: ■ – $Ca(Mg,Fe)Si_2O_6$; ▲ – $(Fe,Mg)_2SiO_4$

During the heating (20 °C/min) of the granulated slag in the argon flow on the curve of the heat flow, we revealed the effect at 646 °C due to the phase transition of the second sort's conformity to the process of devitrification (Fig. 20). The heat capacity change of devitrification was equal to 0.513 J/(g·K). A complex exothermal effect beginning at 744 °C and with the maximum meaning 766 °C and 786 °C was due to the "cold" phases' crystallization - probably to the formation of diopside and olivine crystals. These two peaks show that the crystallization temperatures of these phases are shifted with respect to each other. The endothermic effect with a beginning/maximum of 1102/1171 °C is connected with the slag melting. During the cooling of the melted sample, the temperature of the crystallization is determined - it is 1159 °C. The practically complete amorphous state of the phases in the initial granulated slag confirms the relationship of the heat meanings of "cold" crystallization (153 J/g), melting (152 J/g) and melt crystallization at cooling (154 J/g). According to the data obtained, the heating of high-magnesian glassy slag up to 1250 °C and its cooling at a rate of 20 °C/min lead to its transition from an amorphous state to a crystal state. The value of the stability factor of the glassy state of ΔT for the investigated slag sample is equal to 98 °C, which points to the stability of the amorphous state.

The microstructure of the initial granulated slag is represented mainly by glass (Fig. 21). There are multiphase areas on the surface of the section and globular particles with a size of 5 - 15 µm are concentrated in them. The representation of the section in the characteristic emission of elements is evidence of the fact that the main slag component consists of magnesium, silicon, iron, aluminium and calcium oxides. The other part is represented by small insular edged parts (10 - 20 µm) with a high iron, silicon and magnesium oxide content. The globular phases (less than 1% of the section area) consist of solid solutions of iron and nickel sulphides and their compounds. The results of the determination of the elements' composition of the parts of the granulated slag showed (Table 11) that the main part of it is represented by glass, containing up to 23.0% Fe, 20.0% Si, 6.3% Ca, 6.8% Mg and 2.7% Al. Another silicate phase with a high magnesium (14.0 - 23.0%), silicon (16.6 - 18.3%) and iron (18 - 30%) oxides content is in close agreement with the ferrous olivine.

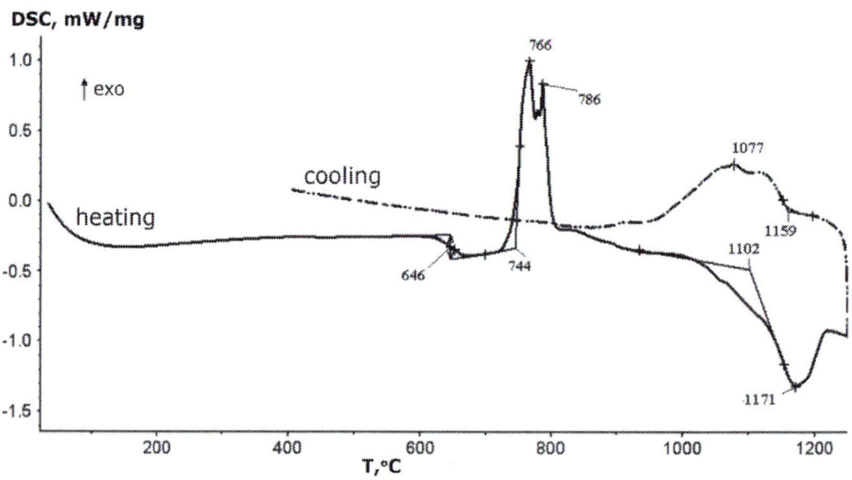

Fig. 20. DCK curves of granulated high-magnesian iron-silicate slag under heating and cooling at a rate of 20 °C/min

The nickel content in the glass is negligible (less than 0.1%), whereas its concentration in olivines reaches 0.2 - 1.2%. Sulphide metal shorts are either a mechanical suspension of matte or else are those solid solution phases which were isolated as the result of unbalanced slag cooling. The detailed analysis of the impurities' composition showed their heterogeneity. The grains consisting of the sulphide phases (27.8 - 29.1% S) concentrated in both nickel (up to 30.9%) and iron (up to 54.0%) are revealed in the impurities. The sulphide grain boundaries are metalized and the sulphur content in them is within the limits of 7.1 - 19.1%, Fe - 31.0 - 43.9%, Ni - 43.2 - 45.0% and Co - 1.6 - 2.6%.

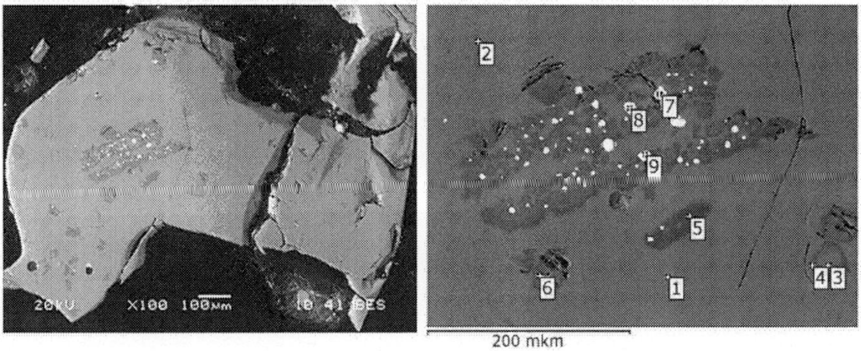

Fig. 21. Microstructure of the particle of high-magnesian iron-silicate granulated slag and EPMA points

The annealing at 1000 °C leads to the change of samples microstructure; the starting processes of solid crystallization can be seen in them, in contrast with the ferrous magnesian-free slags (Gulyaeva et al., 2011) where these phenomena proceed on the surface. The phases of the needle forms (Fig. 22) form in the heterogeneous part of the section - they have an average composition of $(Mg_{0.79}Fe_{0.71}Al_{0.09}Ca_{0.16})Si_{1.10}O_4$. Thus, the cut of olivine

№	Composition in the points of zonding	Content, %							
		Ni	Mg	Si	Al	Fe	S	Co	Ca
Before annealing (Fig. 21)									
1	- $Fe(Mg_{0.67}Ca_{0,37})(Si_{1.74}Al_{0.24})O_6$	0	6.7	20.3	2.7	23.0	0.4	0.2	6.3
2	$Fe(Mg_{0.67}Ca_{0,37})(Si_{1.74}Al_{0.23})O_6$	0.1	6.8	20.1	2.6	22.3	0.4	0.2	6.1
3	$(Mg_{1.03}Fe_{0,97})Si_{1.05}O_4$	1.2	19.9	17.8	0.1	30.4	0	0.5	0.3
4	$(Mg_{1.40}Fe_{0,70})Si_{1.09}O_4$	0.4	14.0	16.6	0	22.9	0	0.2	0.5
5	$(Mg_{1.54}Fe_{0,53})Si_{1.07}O_4$	0.2	22.7	18.3	0	18.0	0	0.2	0.7
6	$(Mg_{1.20}Fe_{0,84})Si_{1.03}O_4$	0.5	17.1	17.1	0	26.3	0	0.4	0.4
7	Matte	16.8	0.2	0.4	0.1	51.0	26.8	1.4	0.1
8	Matte	21.3	0.2	0.4	0.1	47.0	26.2	1.5	0.2
9	Matte	33.0	0.2	0.4	0.1	38.2	24.7	1.3	0.2
After annealing (Fig. 22)									
10	- $(Mg_{1.53}Fe_{0.52})Si_{1.05}O_4$	0.2	23.0	18.1	0.1	18.0	0	0.1	0.3
11	$(Mg_{1.46}Fe_{0.55})Si_{1.03}O_4$	0.2	22.1	17.8	0.1	18.9	0	0.2	0.6
12	$(Mg_{1.37}Fe_{0.67})Si_{1.05}O_4$	1.6	18.9	17.5	0.1	23.0	0	0.2	0.5
13	$(Mg_{1.15}Ca_{0.11}Fe_{0.64})Si_{1.07}O_4$	0.1	14.6	18.2	1.1	22.4	0.2	0.2	3.0
14	$Fe_{0.97}(Mg_{0.74}Ca_{0.49})(Si_{1.80}Al_{0.24})O_6$	0.1	7.2	20.2	2.7	21.3	0.4	0.2	7.8
15	Matte	23.3	0.2	0.4	0.1	40.7	31.1	1.2	0.2
16	$Fe_{0.75}Ni_{0.42}S$ $(Mg_{1.42}Fe_{0.59})Si_{1.04}O_4$	0.3	21.0	17.7	0	20.2	0	0.1	0.8
17-20	$Fe_{0.94}(Mg_{0.74}Ca_{0.37})(Si_{1.75}Al_{0.24})O_6$	0.1	7.5	20.4	2.7	21.8	0.3	0	6.2
21-23	$Fe_{1.09}(Mg_{0.70}Ca_{0.39})(Si_{1.73}Al_{0.22})O_6$	0	6.7	19.6	2.5	24.5	0.5	0.1	6.3

Table 11. Elemental phase composition of the samples of granulated and annealed high-magnesian slags at EPMA points

crystals is shown clearly and their sizes increase to 60 μm. The homogeneous glass breaks down into fine-dispersed (less than 0.5 μm) phases in which the $Fe_{1.09}(Mg_{0.70}Ca_{0.39})(Si_{1.73}Al_{0.22})O_6$ composition differs slightly from the matrix, with a slight deviation in the iron (Table 11). The annealing also leads to changes in the microstructure of sulphide impurities, increasing their inhomogeneity. Fine-dispersed metallic phases containing 52.9 - 57.6% Ni, 36 - 38.9% Fe, 2.6 - 2.9% Co and 1.2 - 4.6% S also precipitate from an unbalanced matte and a monosulphide solid solution close in the composition to $Fe_{1.00}Ni_{0.06}S$ forms.

Thus, the processes at the beginning of glass crystallization and the extraction of diopside and olivin are registered as a result of the annealing of the granulated slag of the shaft melting of oxide nickel ores. The main part of the nickel and cobalt is in a granulated slag as globules of a matte, while the sample annealing leads to the precipitation of fine-dispersed metallic phases enriched in nickel. Thermic effects conforming to devitrification processes, "cold" crystallization, melting and melting crystallization are determined and so they can be taken as basic effects for metallurgical calculation.

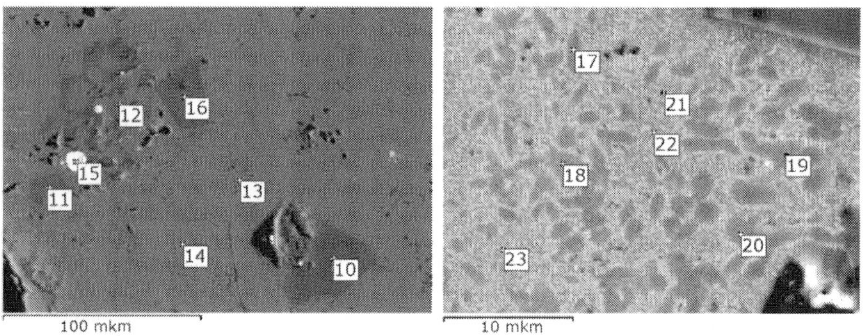

Fig. 22. Microstructure of particles of high-magnesian granulated slag after annealing at 1000 °C and EPMA points

7. Conclusion

The effect of melts' cooling rates on the phases' structure, composition and size, as well as the inter-phase distribution of impurity elements, have been estimated using exemplary slag samples taken from copper-zinc concentrates smelted in a Vanyukov furnace ("Sredneuralsky Copper Smelter Plant" JSC), combined smelting-converting ("Svyatogor" JSC), copper matte converting ("Sredneuralsky Copper Smelter Plant" JSC) and oxide ores smelting for matte ("Ufaleynickel" JSC). The cooling rate of iron oxide-sulphide melts affects the crystal structure, particle size and the number of forming phase. The cooling of the melt by using water granulation leads to the formation of glassy phases, the number of which increases with the content of SiO_2. The iron-silicate phases of variable composition, magnetite and sulphide components constitute the basis of the samples under study. Sulphides are represented by matte mechanical losses (particle size 50 μm) and fine inclusions (2-10 μm) formed from the melt during crystallization.

The amorphous constituent (glass) was observed in all the rapidly quenched samples. The devitrification temperature and the "cold" crystallization of the glassy phases of slags from the melting of copper-zinc concentrates and the converting of a copper matte are within the range of 530-550 °C and 630 °C, respectively. The devitrification and cold crystallization of the granulated and high manganese slag processes from nickel production occur around temperatures of 650 to 740 °C.

The modes of the cooling of the melted slag determine the distribution of non-ferrous metals (copper, zinc, lead, nickel) between oxide and sulphide forms. The possibility of christophite Zn(Fe)S formation under rapid and slow slag cooling has been demonstrated for the first time. It was shown that the content of impurity elements in the phases arising depends upon the cooling rate and upon the Fe/SiO_2 ratio in these phases. The metal, copper-based, component accumulating antimony and arsenic was revealed in the slag sample taken from the copper matte conversion.

The forms of non-ferrous metals in the crystallized slag are able to be regulated by changing the Fe^{+3}/Fe^{+2} and CaO/SiO_2 ratios, the amounts of iron, CaO and MgO oxides. The data obtained is useful for the justification of processes for the re-extracting of precious metals.

8. References

Biswas, K., Sontakke, A.D., Majumder, M. & Annapurna, K. (2010). Nonisothermal crystallization kinetics and microstructure evolution of calcium lanthanum metaborate glass. *J. Therm. Anal. Calorim*, Vol.101, pp. 143–151, ISSN 1388-6150.

Cardona, N., Coursol, P., Vargas, J. & Parra, R. (2011). The Physical Chemistry of Copper Smelting Slags and Copper Losses at the Paipote Smelter. Part 2 - Characterization of industrial slags, *Canadian Metallurgical Quarterly*, Vol.50, No.4, pp. 330-340, ISSN 0008-4433.

Dovchenko, V.A., Paretskii, V.M., Chakhotin, V.S. et al. (1997). Flotation Depletion of the Self-Decomposing Slags of the Slags of Melting of Copper Concentrates for Producing White Matte, *Tsvetn. Met.*, No.1, pp. 27–31, ISSN 0372-2929.

Elliott, J. (1976). Phase relationships in the pyrometallurgy of copper. *Metallurgical Transactions*, Vol.7B, pp. 17-33, ISSN: 1073-5623.

Gulyaeva, R.I., Selivanov, E.N. & Selmenskikh, N.I. (2011). Crystallization of oxide high iron melts. EPJ Web of Conferences, Vol.15, 01010. Published online 18.05.2011, available from http://www.epj-conferences.org.

Karamanov, A. & Pelino, M. (2001). Vitrification of copper flotation waste. *J. Non-Crystalline Solids*, Vol.281, No.1-2, pp. 333-339, ISSN 0022-3093.

Kopylov, N.I., Lata, V.A. & Toguzov, M.Z. (2001). *Interactions and Phase States in Molts Sulfides Systems*, ISBN 5-628-02752-9, Gilim, Almaty, in Russian.

Koryukin, E.B., Litovskikh, S.N. & Kireeva, O.V. (2002). Flotation-Magnetic Scheme of Processing of Converter Slags. *Tsvetn. Met.*, No.8, pp. 18–20, ISSN 0372-2929.

Mazurin, O.V. (1986). *Vitrification*, Nauka, Moscow, Russia.

Mohapatra, B.K., Nayak, B.D. & Rao, G.V. (1994). Microstructure of and metal distribution in Indian copper converter slags - study by scanning-electron microscopy. *Mineral Processing and Extractive Metallurgy*, Vol.103, pp. C217-C220, ISSN 0371-9553.

Moiseev, G.K. & Vyatkin, G.P. (1999). *Thermodynamic Simulation in Inorganic Systems*, YuUrGU , Chelyabinsk, Russia.

Naboichenko, S.S., Nichiporenko, O.S., Murashev, I.B. et al. (1997). *Nonferrous Metal Powders*, Metallurgiya, Moscow, Russia.

Nagamori, M. (1974). Metal loss to slag: Part I. Sulfidic and oxidic dissolution of copper in fayalite slag from low grade matte, *Metallurgical Transactions*, Vol.5, pp. 531-538. ISSN 1073-5623.

Okunev, A.I. & Galimov, M.D. (1983). *Oxidation of iron and sulfur in the oxide-sulfide systems,* Nauka, Moscow, Russia.

Roine, A. (2002). *Outokumpu HSC Chemistry for Windows,* (Outokumpu Research OY), Pori, www.outotec.com.

Ruddle, R.W. (1953). *The Physical Chemistry of Copper Smelting,* Institution of mining and metallurgy, London.

Rüffler, R. & Dávalos, J. (1998). Phase Analytical Studies of Industrial Copper Smelting slags. Part II Silicate slags. *Hyperfine Interact.*, Vol.111, pp. 299–307, ISSN 0304-3843

Sarrafi, A., Rahmati, B., Hassani, H.R. & Shirazi, H.A. (2004). Recovery of Copper from Reverbatory Furnace Slag by Flotation. *Mineral Engineering,* Vol.17, No.3, pp. 457–459, ISSN 0892-6875.

Selivanov, E.N., Okunev, A.I. & Moiseev, G.K. (2000). Phase Transformations during Cooling of the Slags of Melting of Copper Concentrates for Producing High-Grade Matte. *Rasplavy,* No.2, pp. 18–24, ISSN 0235-0106.

Selivanov, E.N., Belyaev, V.V., Selmenskikh, N.I. & Pankratov, A.A. (2004). Forms of Metals in the Slag of Combined Melting–Converting of Copper Concentrates. *Rasplavy,* No.1, pp. 33–41, ISSN 0235-0106.

Selivanov, E.N., Gulyaeva, R.I., Udoeva, L.Yu. et al. (2009a). Effect of the Cooling Rate on the Phase Composition and Structure of Copper Matte Converting Slags. *Russian Metallurgy (Metally),* No.4, pp. 281–288, ISSN 0036-0295.

Selivanov, E.N., Gulyaeva, R.I., Zelytin D.I. et al. (2009b). Effect of speed of cooling on structure of slag from of copper-zinc concentrates in Vanukov furnace. *Tsvetnye Metally,* No.12, pp. 27-31, ISSN 0372-2929.

Selivanov, E.N., Gulyaeva, R.I. & Selmenskich, N.I. (2010). Phase Formation in FeO_x-SiO_2-Cu_2O-ZnO-FeS System during Melts Crystallization. *Defect and Diffusion Forum,* Vols. 297-301, pp. 602-607, ISSN 1662-9507.

Sergeeva, S.V., Gulyaeva, R.I. & Selivanov, E.N. (2011). Features of crystallization of high-magnesian silicate slag. *13th Russian Conference on Composition and Properties of Melts of Metals and Slags.* Vol.3, pp. 99-102, Yekaterinburg, Russia, September 12-16, 2011.

Spira, P. & Themelis, N.J. (1969). The Solubility of Copper in Slags. *Journal of Metals*, Vol.29, No.4, pp. 35-42, ISSN 0148-6608.

Sycheva, G.A. & Polyakova, I.G. (2004). Nucleation of crystals in glasses based on blast furnace slag, *11th Russian Conference on Composition and Properties of Melts of Metals and Slags.* Vol.3, pp. 186-190, Yekaterinburg, Russia, September 14-16, 2004.

Tokeda, Y., Ishiwata, S. & Yazawa, A. (1983). Distribution Equilibria of Minor Elements between Liquid Copper and Calcium Ferrite Slag. *Trans. Japan Inst. Metals*. Vol.24. №10. pp. 518–528, ISSN 0021-4434.

Vanyukov, A.V., Bystrov, V.P., Vaskevich, A.D. et al. (1988). *Melting in the Liquid Bath,* Metallurgiya, ISBN 5-229-00062-7, Moscow, Russia.

Vanyukov, A.V. &. Zaitsev, V.Ya. (1969). *Slags and Mattes of Nonferrous Metallurgy,* Metallurgiya, Moscow, Russia.

Vanyukov, A.V. &. Zaitsev, V.Ya. (1973). *Theory of Pyrometallurgical Processes,* Metallurgiya, Moscow, Russia.

Vaysburd, S.E. (1996). *The Physical Chemistry Properties and Structural Features of Sulfide Melts.* Metallurgiya, ISBN 5-229-00903-9, Moscow, Russia.

Yazawa, A. (1974). Thermodynamic considerations of copper smelting, *Canadian Metallurgical Quarterly*, Vol.13, No.3, pp. 443-455, ISSN 0008-4433.

11

Real-Time Analysis to Evaluate Crystallization Processes

João F.Cajaiba da Silva, Andréia P. M. da Silva and Rodrigo C. de Sena
Instituto de Química – Universidade Federal do Rio de Janeiro
Brazil

1. Introduction

Crystallization is one of the most important unit operations employed by the pharmaceutical, microelectronics, food and fine chemicals industries for the production of solid products with high added value. This operation can be used as a method to perform separation and purification of crystalline compounds such as proteins, polymers, pharmaceuticals, inorganic salts, etc. (Févotte & Klein, 1995; Feng & Berglund, 2002; Mersmann, 2001; Lewiner et al., 2001; Mullin, 2001; Liotta & Sabesan, 2004; Joung et al., 2005; Pelberg et al., 2005; Derdour et al., 2011).

The experimental conditions used during a crystallization process may alter the physical properties of the final product such as its chemical purity, crystal size distribution and morphology. These properties can have impacts on the subsequent purification operations such as filtration, washing and drying, and can also alter the bioavailability of pharmaceuticals (Sabesan & Liotta, 2004). Polymorphism is another issue related to the crystallization process that has profound importance in the pharmaceutical industry because it can alter the kinetics of a crystal's solubilization and in some cases polymorphs present problems of toxicity (Fujiwara et al., 2005).

The challenges involved in controlling crystallization are significant, since the kinetic parameters of the process are strongly affected by several factors such as the presence of impurities (Gunawan et al., 2002; Ma et al., 1999; Rauls et al., 2000; Poddar, 2002), breaking of crystals (Kougoulos et al., 2005; Gahn & Mersmann, 1995) and clustering (Yu et al., 2005; Paulaime et al., 2003) among other effects that are difficult to characterize.

The identification and control of factors that affect the final quality of crystals are essential to ensure uniformity among different batches and to improve product quality. The development of more accurate and sensitive sensors for real-time analysis of crystallization must allow significant advances in monitoring, control and optimization of crystallization processes (Liotta & Sabesan, 2004).

The accuracy of off line methods for evaluating crystallization processes is strongly dependent on sampling. When the collected samples do not represent the whole, the errors introduced may cause a misinterpretation of the crystallization process. By using real-time analysis these types of errors can be greatly reduced. The possibility of performing in situ analysis of crystal size distribution, crystal shape, crystal habit, agglomeration and breakage can indicate what changes should be made to the crystallization process parameters such as cooling and stirring rates and time for seeding. The optimization of these parameters allows crystals of the desired characteristics to be obtained (Yu et al., 2004).

This section is intended to give a brief overview of crystallization processes and the methods dedicated to monitor them in real-time. Additionally, a comparison between four in line methods to determine the onset of adipic acid crystallization was performed.

1.1 Solubility and supersaturation

The determination of the solubility of a solid in a specific solvent is a key step in the study of crystallization processes. The solubility curve is used as a benchmark to assess the degree of supersaturation and the metastable zone limits. The solubility or condition of saturation is determined experimentally by heating a suspension and observing the temperature at which the solid phase is completely dissolved. The cooling of a saturated solution results in a system that is not in thermodynamic equilibrium, a supersaturated solution (Mullin, 2001; Giulietti et al., 2001).

The supersaturation is the driving force for crystallization processes and can be defined as the difference between the chemical potential of a solute in a supersaturated solution and the chemical potential of the saturated solution. Supersaturation can be created by cooling, by adding an anti-solvent, by performing a chemical reaction that generates a product of lower solubility, by solvent evaporation, etc. Among the methods used to create supersaturation, cooling is the most used. The usage of this method is restricted to substances whose solubility changes significantly during a temperature variation. The expected properties of the solid material, as well as economic aspects, form the basis for making a decision about which method should be used to create supersaturation (Mullin, 2001; Giulietti et al., 2001).

The difference between the concentration of a compound in a supersaturated solution, c, and its concentration in a saturated solution, c*, is known as absolute supersaturation and is expressed by equation 1.

$$\Delta c = c - c^* \qquad (1)$$

When Δc is greater than 0, the system is supersaturated. Another important parameter is the supersaturation ratio, S, which is defined by equation 2. In this case the system is said to be supersaturated when S is greater than 1.

$$S = \frac{c}{c^*} \qquad (2)$$

1.2 Metastable zone width

Supersaturated solutions exhibit a metastable region, where despite the instability of the system, there is no separation of a solid phase. The determination of this region is, in general, the first phase in the design of a batch cooling crystallization process. The metastable zone width (MZW) is a property that depends on several characteristics of the system (cooling rate, solute concentration, stirring rate, thermal history of the solution, presence of impurities, etc.) (Liotta & Sabesan, 2004).

The metastable zone width is defined as the difference between saturation temperature and the temperature which is detected in the formation of the first crystals. This temperature difference is known as the maximum undercooling, ΔT_{max}. Figure 1 illustrates schematically the solubility curve and metastable zone boundary for a hypothetical case of a system cooled from an under saturated condition until the condition of supersaturation.

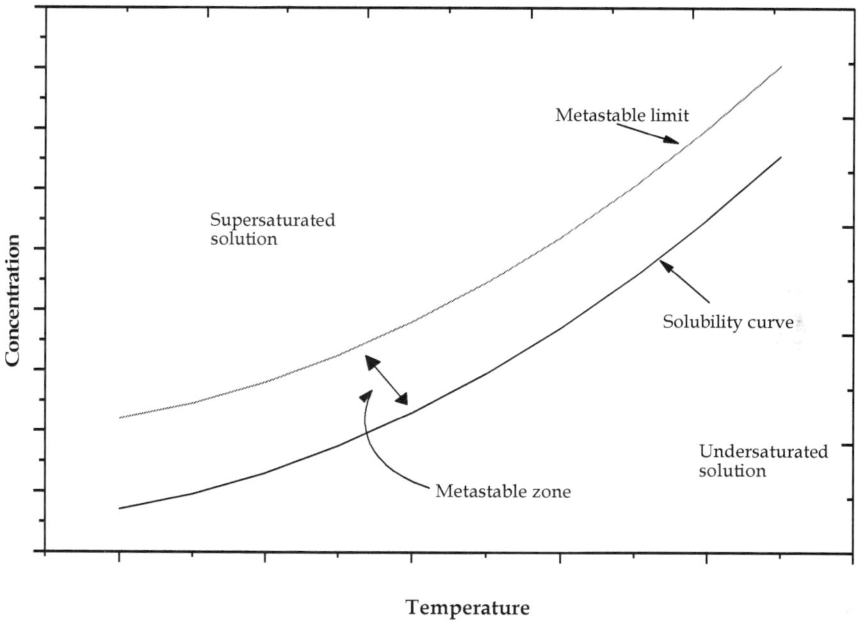

Fig. 1. Solubility curve and metastable zone.

In batch cooling crystallizations for example, maintaining the solution concentration profile within the metastable zone and close to the solubility curve promotes crystal growth and helps to avoid secondary nucleation (Beckmann, 2000). In the case of industrial crystallizers as a general rule, the level of supersaturation is maintained at about half the metastable zone (Marciniak, 2002).

1.3 Nucleation and crystal growth

The supersaturation is the main requirement for crystallization to occur and its creation does not imply the immediate separation of the phases. In a supersaturated solution, part of the dissolved solute tends to reorganize again to form the solid phase. However, the formation of the solid phase (positive energy) implies the generation of an interface (energetically unfavourable). Therefore, for the formation of nuclei to occur within the solution, it is necessary that this barrier is overcome (Mullin, 2001; Ulrich & Strega, 2002; Giulietti, 2001).

The nucleation rate depends on the supersaturation. When the supersaturation is extremely high, the nucleus formation is a random process and difficult to reproduce. For this reason, whenever possible, this condition is avoided in industrial applications (Mullin, 2001).

The crystal growth also depends on the supersaturation level. High growth rates lead to products with a degree of purity lower than those generated in conditions of lower supersaturation owing to the inclusion of liquids and other impurities. From the industrial point of view there is a compromise between the desired characteristics of the products and the economic efficiency of the process. This means that is not always possible to carry out crystallizations with a low rate of crystal growth because such a condition significantly increases the residence time of the product inside the crystallizer (Mullin, 2001; Ulrich & Strege, 2002).

1.4 Sensor technologies for monitoring crystallization processes

Continuous monitoring and control of crystallization processes in real-time require the use of sensors able to provide information regarding product quality and critical process variables.

The methods used to assess information of the products obtained from crystallization processes can be divided into four main groups (Yu et al., 2004):

- off line methods: the analysis is performed after sampling;
- on line methods: the sample stream is diverted from the crystallizer for analysis and subsequently returned to the system;
- in line methods: sensors are integrated into the crystallizer and provide real-time information about the process. The sensors are in direct contact with the material and can cause disturbances in the system;
- non-invasive methods: sensors are integrated into the crystallizer and provide real-time information of the process. In this case, the sensors do not come into direct contact with the material.

The real-time analyses described in this chapter can be included in the in line and the non-invasive methods.

1.4.1 Attenuated total reflectance Fourier transform infrared spectroscopy (ATR-FTIR)

ATR-FTIR is a consolidated technique in the monitoring of crystallization processes (Chen et al., 2009; Kadam et al., 2011; Pöllänen et al., 2006; Qu et al., 2009; Sheikhzadeh et al., 2008; Liotta & Sabesan, 2004). The method allows estimating the degree of supersaturation of the system by measuring the concentration of solute in the solution. The only prerequisite for application of the technique is the absorption of infrared radiation in the medium by the solute. The sensor used in this method has an element of internal reflection of high refractive index. The radiation passes through the element of reflection, being reflected when it encounters a material with lower refractive index. The amount of reflected radiation depends on the angle of incidence on the interface and when this angle is greater than a critical angle (depending on the ratio between the two refraction index) the refraction is attainable. However, radiation penetrates only a short distance into material of lower refractive index; this radiation is called the evanescent radiation (evanescent wave). Thus, if a sample is able to absorb infrared radiation, the beam is attenuated at frequencies absorbed by the sample (Man et al., 2010; Dunuwilaa et al., 1994).

1.4.2 Focused beam reflectance measurement (FBRM)

The FBRM measurement principle is based on backward light scattering. A laser beam is coupled to a probe via an optical fibre. This laser beam is deviated from the probe's central axis and focused into a disperse medium with an optical conduit. When this laser beam intersects with a particle, light scattering occurs. A certain fraction of the light is scattered back into the system. This back scattered light is coupled via a beam splitter to a second fibre and conduced to a detector. The rotational velocity of the laser is constant. The time span in which back scattering is detected is therefore directly proportional to the path length of the laser on the particle. It is assumed that the particle velocity is small compared to the laser rotational velocity. The length of the laser path on the particle is therefore proportional to the time span in which scattering is detected. This path length is called a chord length. Depending on the laser position, different chord lengths are measured even for a single particle. Those chord lengths are generally different from any characteristic particle length. In order to calculate the particle size distribution from the chord length distribution, a model is needed. This model has to cover all relevant aspects of the measurement technology (Kail et al., 2008; Barrett & Glennon, 2002). Figure 2 presents a schematic view of the Mettler Toledo FBRM probe.

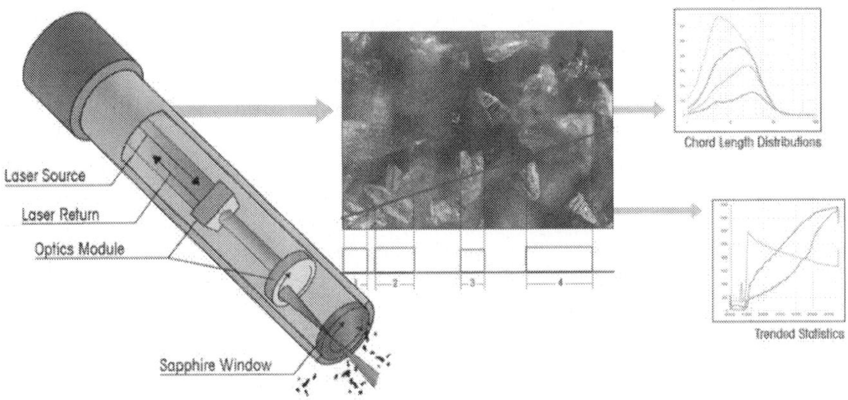

Fig. 2. Measurement principles of the Lasentec FBRM probe.

Just as with the ATR-FTIR technique, the FBRM is a well-established method of monitoring crystallization processes (Sun et al., 2010; Barrett & Glennon, 2002; Czapla et al., 2010; Sheikhzadeh et al., 2008; Nguyen & Kim, 2008).

1.4.3 Calorimetry

The principle of reaction calorimetry is based on the heat flow in processes involving changes in chemical or physical properties. The rate of release or absorption of heat is a parameter dependent on the kinetics and thermodynamics of the process under study (Gesthuisen et al., 2005; Mantel & Meyer, 2008). The combination of heat flow with the mass balance allows us to estimate the conversion of the instantaneous and cumulative process under study. Reaction calorimetry is non-invasive, fast, robust, relatively simple and its principle is based on the measurement of temperature differences (Gesthuisen et al., 2005; Mantel & Meyer, 2008). From the industrial point of view, calorimetry is a technique of great importance because it allows safely scaling up a process from pilot to industrial scale (Gesthuisen et al., 2005). The method finds application in the study of polymerization reactions (Benamor et al., 2002; Elizalde et al., 2005), biotechnological processes (Marison et al., 1985), study of supercritical fluids (Lavanchy et al., 2004; Mantel & Meyer, 2008), optimization of chemical reactions (Barton et al., 2003) and determination of kinetic parameters of chemical reactions (Silva et al., 2003, [illegible] et al., 2006) among other applications. There are few articles where heat flow calorimetry is applied to the study of crystallization processes (Févotte & Klein, 1995)

1.4.4 Image analysis

The use of Complementary Metal Oxide Semiconductor (CMOS) and Charge Coupled Device (CCD) cameras has been widely introduced in analytical chemistry for different reasons such as fast image capturing, stable background and good linearity (Jolling et al., 2007). These sensors are capable of converting the intensity of light that focuses on it in digital storable values as bits. The analytical response that generates an image representing the patterns of the colours Red (R), Green (G) and Blue (B). These patterns are known as RGB 8 bits for each channel, totalling 256 levels. The combination of the three matrices (R, G and B) allows the acquisition of 16 million colours (Gaiao et al., 2006; Safavi et al., 2007). Different methodologies employing this image analysis has been described in literature. A digital camera was used as a sensor for simultaneous determination of Al(III) and Fe(III) in alloys using the chrome azurol S(CAS) as chromogenic reagent (Maleki et al., 2004). An instrumental detection technique for titration based on digital images was proposed (Gaiao et al., 2006). A similar method for the measurement of lithium, calcium and sodium through the radiation emitted by the sample into an air-butane flame was developed (Lyra et al. (2009). Image analysis was also used for a real-time assessment of the coffee roasting process (Hernández et al., 2008). An approach employing a CCD camera as a sensor for recognizing volatile alcohols was described (Shirshov et al., 2007). A method based on external bulk video imaging was proposed for metastable zone identification in food and pharmaceutical crystallization processes, and showed good performance when compared to FBRM and ultra-violet visible spectroscopy (Simon et al., 2009). Additionally, CCD cameras have been used as detectors in clinical analysis and showed high detection sensitivity (Liang et al., 2004; Alexandre et al., 2001). The rapid improvements in digital camera technology provide the opportunity for the development of new methodologies employing digital cameras as an analytical sensor with high sensitivity, robustness, speed and low cost for implementation that reduces the analysis time.

2. Experimental part

In this section batch cooling crystallization of adipic acid will be used as a model to demonstrate the use of in line and non-invasive techniques for monitoring crystallization. For this purpose, four analytical tools with different physical principles were used. The procedures and techniques used in the experiments are described in the following subsections.

2.1 Materials and methods

2.2.1 Solubility

The solubility of adipic acid (99.8%) in ethanol (analytical grade) was determined for twelve different temperatures ranging from 16.0 to 64.0°C. The experiments were performed in a 1.8-L Hastelloy jacketed reactor vessel connected to an RC1e reaction calorimeter. The solutions were prepared by successive additions of adipic acid to a solution containing ethanol at a stirring rate of 300 rpm. For measurements at 44.0°C, the mass of ethanol used was 524.0g. The used mass for other temperatures are presented in Table 1. The solubilization temperature was maintained constant during the whole process.

Temperature (K)	Temperature (°C)	Ethanol mass (g)
289.15	16.0	590.3
293.95	20.8	582.7
298.15	25.0	574.7
303.95	30.8	564.3
306.15	33.0	556.3
308.25	35.1	550.8
312.15	39.0	537.0
317.15	44.0	524.0
323.15	50.0	492.4
330,15	57.0	459.5
334.15	61.0	428.1
337.15	64.0	405.3

Table 1. Experimental conditions for testing the solubility of the adipic acid in ethanol.

The ATR-FTIR measurements were performed by using a Mettler-Toledo ReactIR IC10 spectrometer. The base unit contains the Fourier transform mid-infrared source and the mercuric cadmium telluride (MCT) detector that should be cooled with liquid nitrogen. The sample interface module (SIM) is the interface on the instrument base unit where the K6 (16 mm diameter) conduit connects. It contains the optics that transfer the infrared source light from the base unit to the probe in contact with the chemical materials contained in the vessel and then back to the detector. Measurements are taken optically using a diamond sensing element that uses a multiple reflection ATR crystal and a gold seal between the metal housing and the sensor. Figure 3 illustrates schematically the K6 Mettler Toledo probe tip.

The focused beam reflectance measurements were obtained by using a Mettler-Toledo Lasentec D600L probe consisting of a Hastelloy C-22 tube with the sensor at one end with an optical diameter of 19 mm and a length of ~406 mm. The FBRM laser provides a continuous beam of monochromatic light with a wavelength of 780 nm. The beam is located approximately 3 mm to the focal point.

The ATR-FTIR and FBRM probes were kept immersed in the adipic acid solution 5 cm above the propeller stirrer. Infrared spectra obtained from 4000 to 650 cm^{-1} at 4 wave numbers resolution were collected at 15 s intervals with each spectrum averaged over 30 scans.

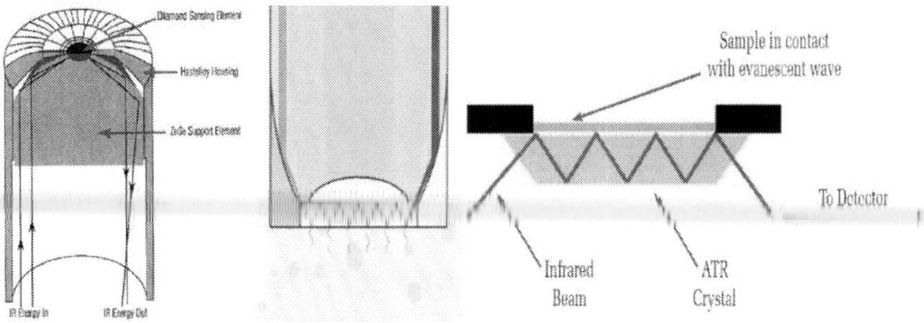

Fig. 3. Schematic draw of the Mettler Toledo K6 probe tip.

2.2.2 Batch cooling crystallization

The batch cooling crystallization experiments were carried out in a double walled glass reactor with a capacity of 2.0L. The solutions were stirred by a propeller stirrer at 300 rpm. The experimental setup is schematically presented in Figure 4.

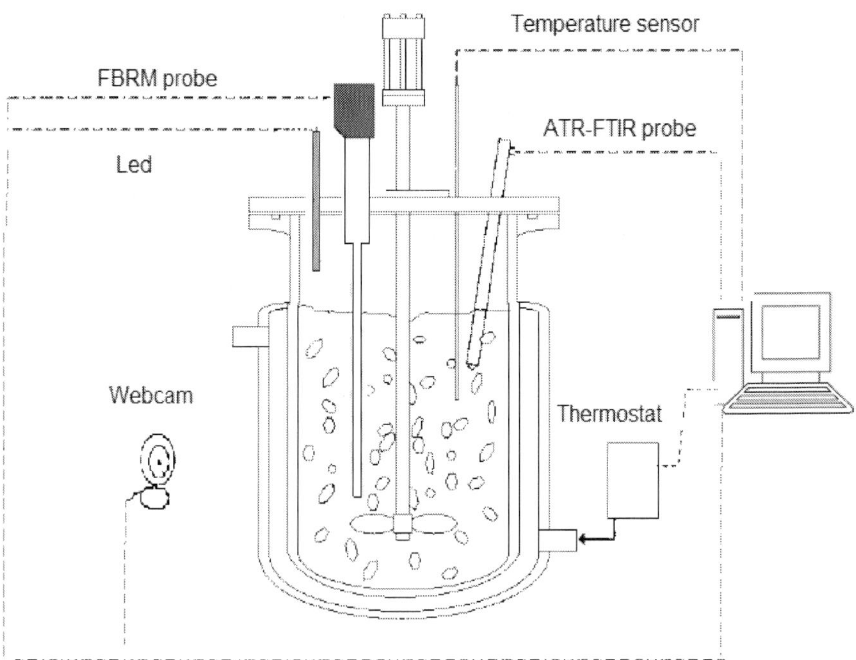

Fig. 4. Experimental setup.

The mass of ethanol used in each run was 600 g and the amount of adipic acid was varied from 91.2 to 279.9 g. The temperature was controlled by the precise RC1e thermostat. The temperature of solution was kept 5°C higher than that saturation condition in order to assure that no crystal was presented in the solution prior to starting the cooling process. After 0.5 h, the solution was cooled at constant rate of 0.2 and 1°C/min, until the onset of crystallization. The experimental conditions of experiments are presented in Table 2.

Concentration (mass %)	Temperature at the start of cooling (°C)	Final temperature (°C)	Mass of adipic acid (g)
13.2	30.0	-2.00	91.20
15.1	35.0	5.00	106.8
17.3	40.0	10.0	125.2
19.7	45.0	15.0	146.8
22.3	50.0	15.0	172.2
25.2	55.0	20.0	202.1
28.4	60.0	25.0	237.6
31.8	65.0	25.0	279.7

Table 2. Experimental conditions for crystallization experiments of adipic acid in ethanol.

In situ methods, FBRM and ATR-FTIR, were used to monitor adipic acid crystallization. Non-invasive measurements were carried out by monitoring the heat released and the image patterns of colours red, green and blue during the crystallization. Images of experiments were acquired by using a low cost PC webcam (Microsoft Life Cam VX-2000). The webcam was placed externally and, in order to avoid interferences by external light and to maintain the CCD noise under controlled conditions, the reactor was enclosed with a black box. The images were captured during whole experiments. A light-emitting diode (LED) was used as a source of illumination. The images acquired were analysed according to their patterns of colour red, green and blue, and for this purpose software was developed which allows the evaluation of alteration in these patterns of colours. The software enables the analysis of the whole image or the user can define a specific region previously selected from an image. The software automatically saves the coordinates of the delimited region for all digital images and calculates the R, G and B values averaging all pixels.

The heat released during the crystallization was monitored using the RC1e. To be able to calculate the heat flow during the crystallization, the total heat transfer coefficient (U) and the heat capacity (Cp) of the solution of adipic acid were measured.

3. Results and discussion

This section presents the results obtained in the study of solubilization of adipic acid in ethanol as well as its crystallization, evaluating the possibility of using calorimetry, infrared (ATR-FTIR) and FBRM as techniques for determining the width of the metastable zone.

3.1 Determination of the solubility of adipic acid

The solubility curve of adipic acid in ethanol, presented in Figure 5, was prepared according to procedure in literature that used ATR-FTIR and heat flow to calculate the solubility curve of adipic acid in acetone (Silva & Silva, 2011).

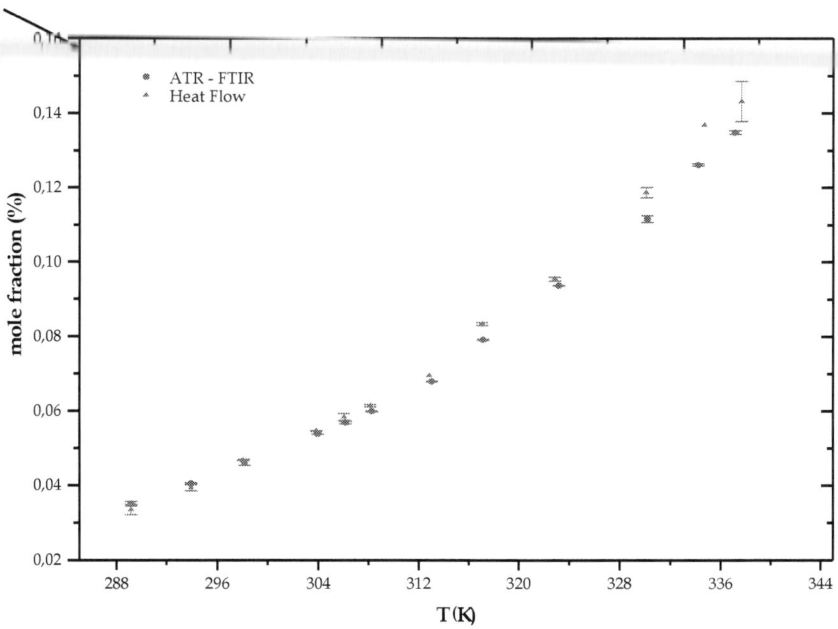

Fig. 5. Solubility curves of adipic acid in ethanol by using ATR-FTIR and heat flow calorimetry.

3.2 Determination of the onset temperature for the crystallization of adipic acid in ethanol by heat flow calorimetry

Since during crystallization there is a decrease in entropy, the second term of the Gibbs free energy (equation 3) becomes positive and therefore, for ΔG to be negative, it is necessary that the change in enthalpy is negative, which shows that the crystallization processes are exothermic.

$$\Delta G = \Delta H - T\Delta S \tag{3}$$

The heat flow for cooling at 1^0C/min is presented in Figure 6.

The analysis of Figure 6 shows that when the solution temperature was approximately 37.2°C, the jacket temperature was reduced by about 10°C, so that the cooling rate of 1°C/min could be maintained, meaning that an exothermic process had started. The heat flow measurement had a fast increase at the same moment indicating that the crystallization of the adipic acid (the exothermic process) had begun. In the initial moments of the crystallization there was an intense release of heat and soon after that the release decreased and became nearly constant. The region where heat release is constant may be associated with the growth of crystals (Riesen, 2005). The heat release continued until the moment when the reactor temperature was maintained at 15.0 °C. With the constant temperature the heat flow remained stable, indicating that the process that was responsible for the release of heat had ceased. As the only process that was taking place inside the reactor was the crystallization of adipic acid, this release is solely related to this process.

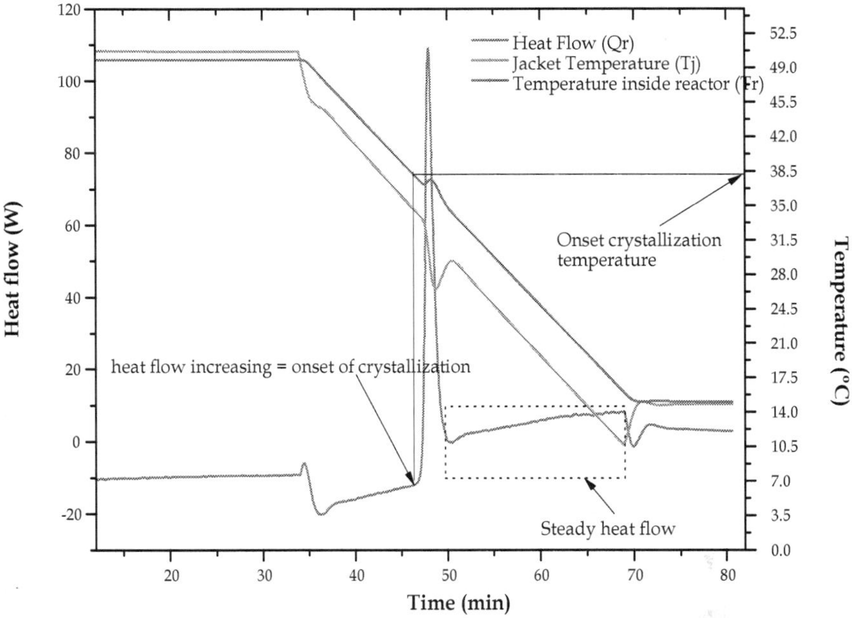

Fig. 6. Heat release curve (Qr), the temperature inside the reactor (Tr) and temperature of the jacket (Tj), obtained from crystallization experiments using an ethanol solution with initial concentration of 22%, cooled at a rate of 1°C/min and with a stirring rate of 300 rpm.

The maximum supersaturation achieved in the solution is a function of the maximum cooling achieved by the system. Thus, it is expected that the higher the maximum cooling achieved by a system, the greater the value of supersaturation in the medium. In general the smaller nucleation rates are obtained at lower cooling rates (Mullin, 2001). In order to check this information, the crystallization of adipic acid was performed at a lower cooling rate as presented in Figure 7.

The rate at which heat is released when employing a cooling rate of 1°C/min is higher than when using a cooling rate of 0.2°C/min.

The comparison between the results presented in Figures 6 and 7 confirms that the maximum heat released is a function of cooling rate.

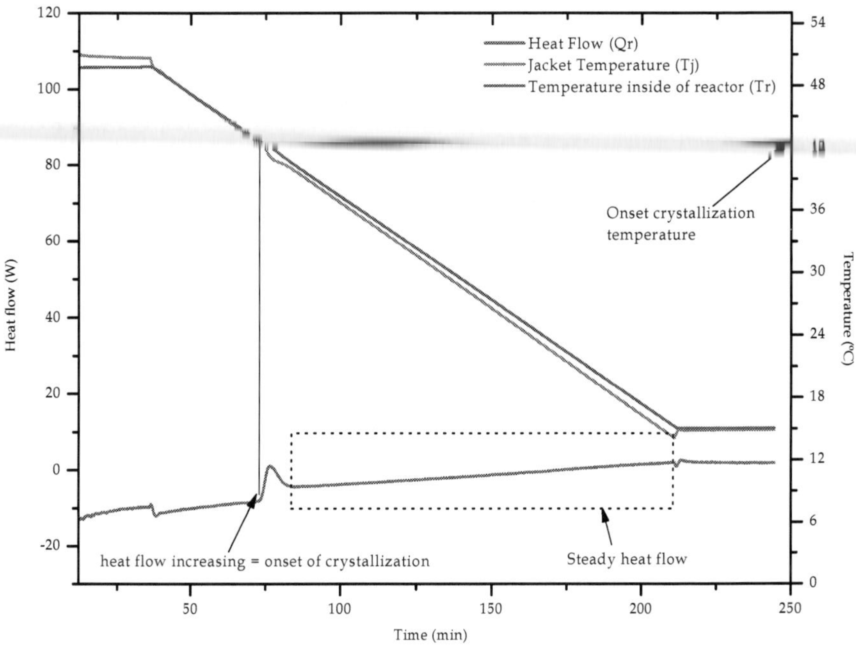

Fig. 7. Heat release curve (Qr), the temperature inside the reactor (Tr) and temperature of the jacket (Tj), obtained from crystallization experiments using an ethanol solution with initial concentration of 22%, cooled at a rate of 0.2°C/min and with a stirring rate of 300 rpm.

3.3 Determination of the onset temperature for the crystallization of adipic acid in ethanol by RGB image analysis

The image video analysis was performed by using software named MasterView RGB that captures images in real-time from a PC webcam by evaluating changes in the component's RGB colour pixel by pixel. The software automatically saves the coordinates of the delimited region for all digital images and calculates the RGB values averaging all pixels (de Sena et al., 2011).

Figure 8 shows the graphic interface of MasterView RGB for an adipic acid solution and Figure 9 presents the crystallization onset by measuring the RGB variation.

The method of image analysis is based on comparison of images. This procedure is comparable to human visual inspection and the method of image analysis has greater sensitivity and is not subject to misinterpretation.

The variation of the red channel during the crystallization of adipic acid is presented in Figure 10.

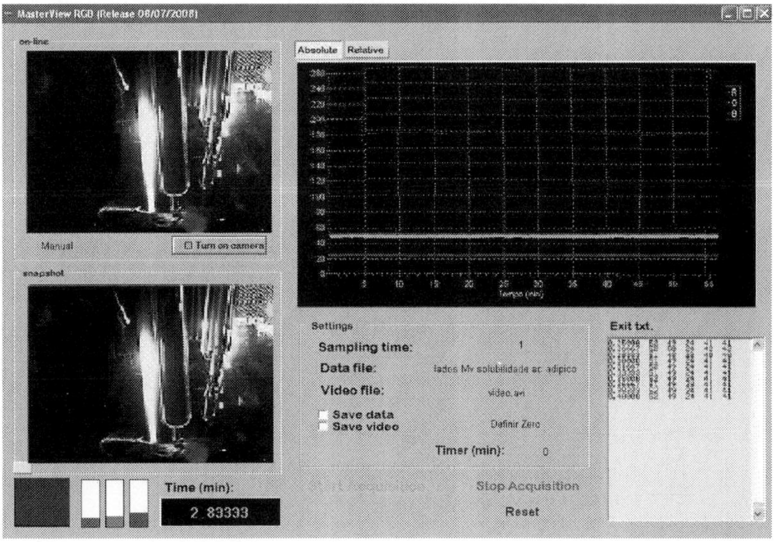

Fig. 8. Monitoring of the onset of the crystallization of adipic acid using the image analysis method (saturated solution).

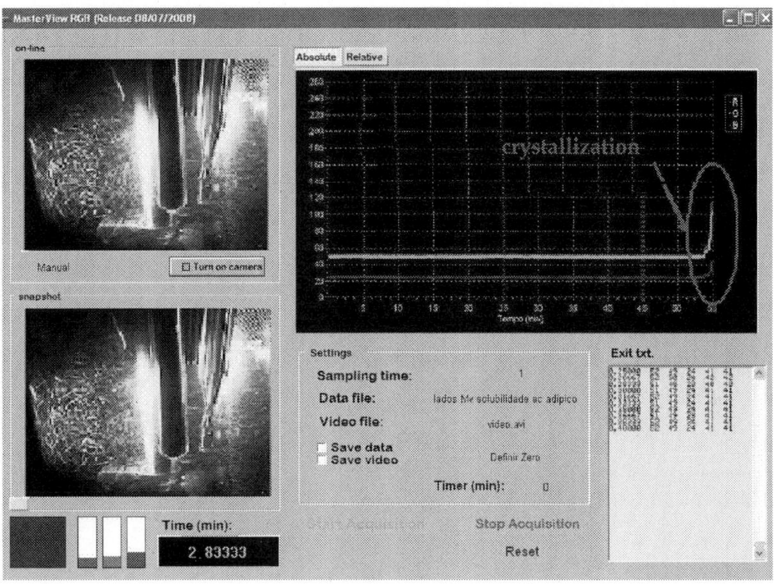

Fig. 9. Monitoring of the onset of the crystallization of adipic acid using the image analysis method (increasing the mass of adipic acid crystals due to cooling).

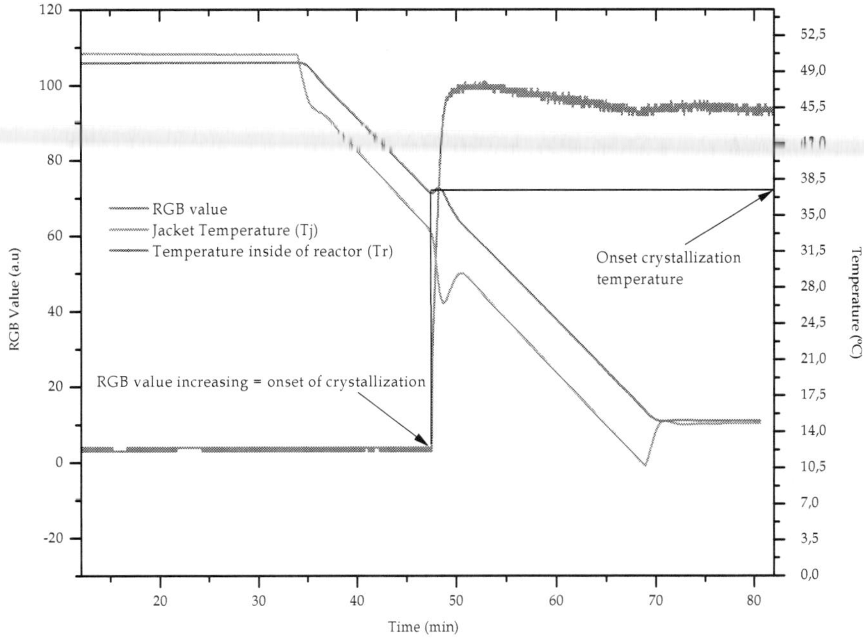

Fig. 10. Image analysis red channel, the temperature inside the reactor (Tr) and temperature of the jacket (Tj), obtained from crystallization experiments using ethanol solution with initial concentration of 22%, cooled at a rate of 1°C/min and with a stirring rate of 300 rpm.

As can be seen in Figure 8, before the crystallization of adipic acid (red signal channel close to zero), the red value increases immediately when the adipic acid crystallization starts. This result is in total accordance with the results obtained by heat flow calorimetry. It was established that the onset temperature of crystallization, determined by the variation in RGB signal, would be that corresponding to the time when the RGB signal began to increase.

3.4 Determination of the onset temperature of crystallization using infrared (ATR-FTIR)

The infrared absorption spectrum of adipic acid is presented in Figure 11.

The carbonyl stretching absorption region of carboxylic acids used in the ATR-FTIR analysis is indicated in Figure 11.

The analysis of crystallization of a solution of adipic acid in ethanol was accompanied by infrared (ATR-FTIR), through the decrease of the signal of the peak area for the absorption of carbonyl (C=O) in the medium. Figure 12 presents the carbonyl absorption variation during a crystallization process that started at 50 °C.

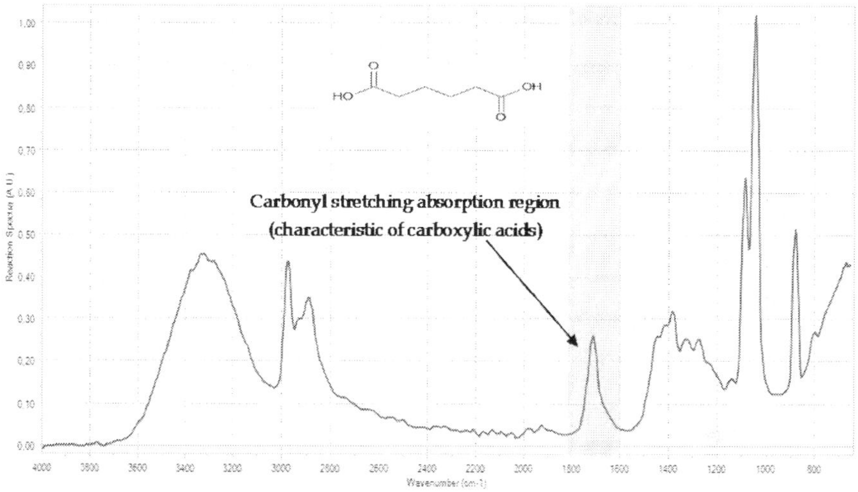

Fig. 11. ATR/FTIR spectra of adipic acid in the infrared region.

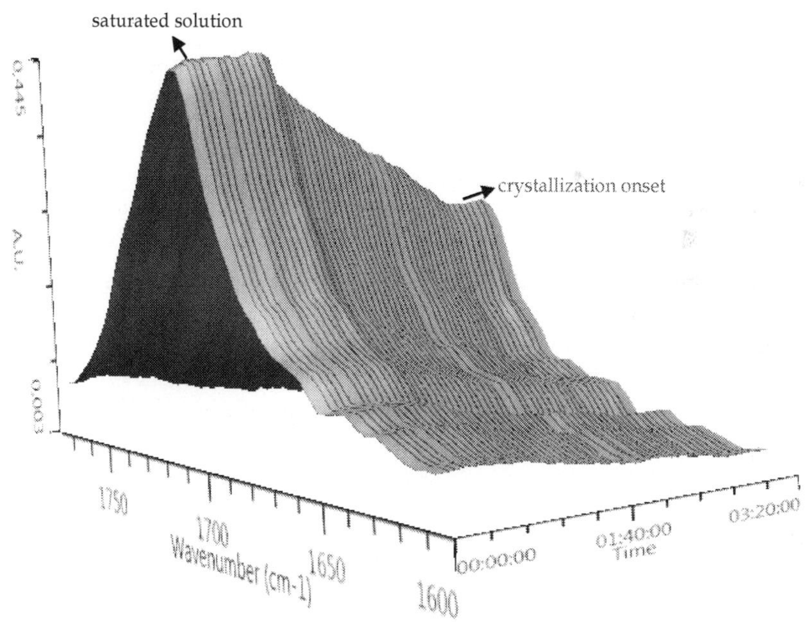

Fig. 12. Adipic acid carbonyl absorption variation measured in the interval of 1750-1600 cm^{-1}.

The carbonyl absorption trend related to the data presented in Figure 12 is presented in Figure 13.

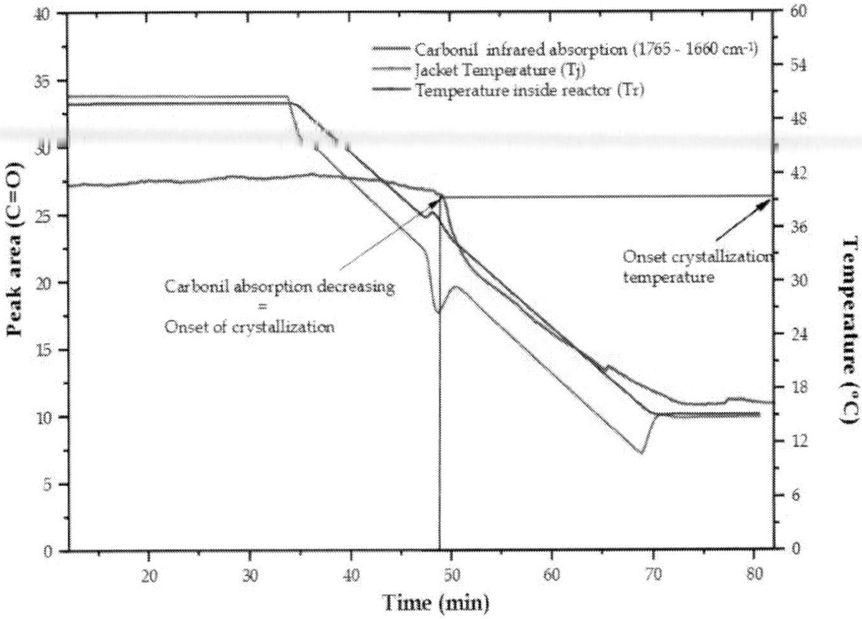

Fig. 13. Absorption curve of the carbonyl C=O (1765-1660 cm 1) in ethanol, the temperature inside the reactor (Tr) and jacket temperature (Tj), obtained from crystallization experiments using an ethanol solution with initial concentration of 22% , cooled at a rate of 1°C/min and stirring rate of 300 rpm.

The carbonyl absorption area presented a decrease at the precise moment the crystallization started. This observation is in total agreement with the fact that the ATR probe detects the concentration of adipic acid soluble in ethanol. A comparison between the infrared absorption for the cooling rates of 1 and 0.2^{0}C/min is presented in Figure 14.

The FTIR analysis may be considered as an indirect measurement of the kinetics of crystallization, since the ATR probe cannot detect the crystallized adipic acid. As expected, the diminution of the carbonyl absorption was more pronounced when the higher cooling rate was used.

3.5 Determination of onset temperature of the crystallization using Focused Beam Reflectance Measurement

The variation in the total chord counts obtained from the FBRM analysis can detect the crystallization onset. This signal remained in the form of a stable baseline, with a few counts per second before the formation of crystals. By the time the crystallization of adipic acid had started the chord counts increased dramatically. This behaviour can be seen in Figure 15.

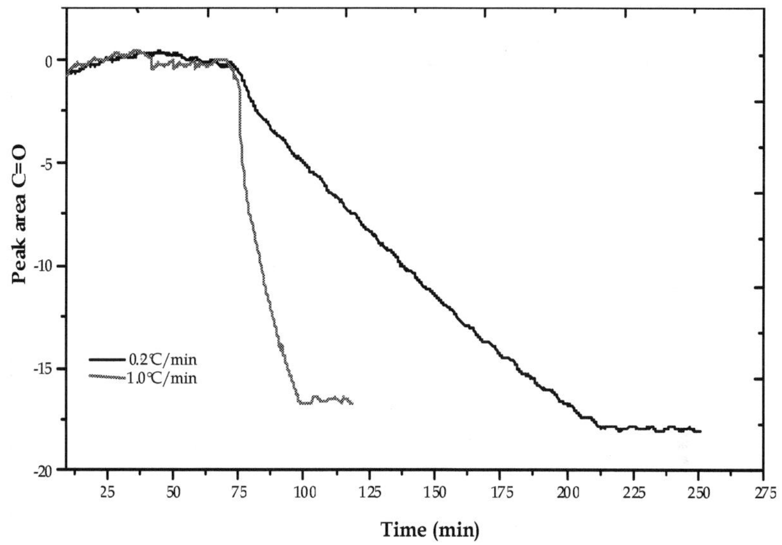

Fig. 14. Infrared absorption curve of the carbonyl C=O (1765-1660 cm^{-1}) of adipic acid during crystallizations performed at cooling rates of 1.0 and 0.2°C/min.

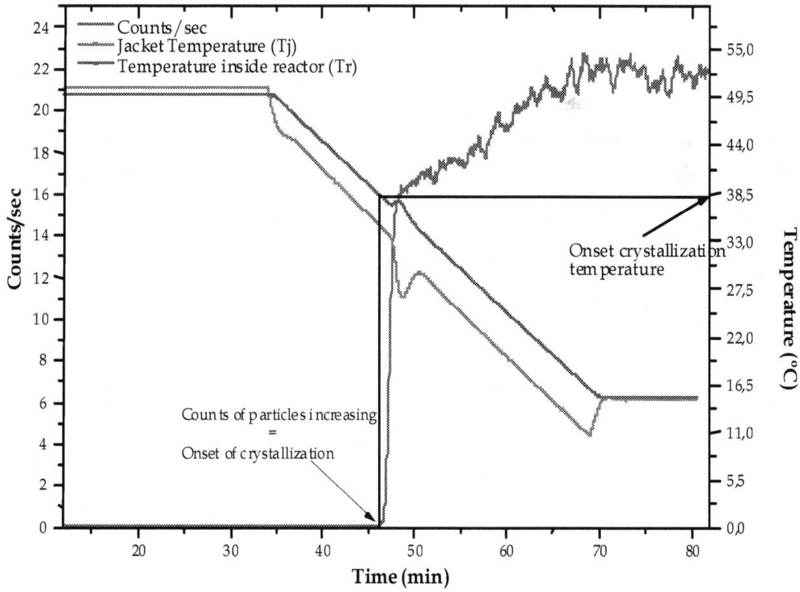

Fig. 15. Total chord counts (FBRM), the temperature inside the reactor (Tr) and temperature of the jacket (Tj), obtained from the crystallization experiments using a solution of ethanol with initial concentration of 22%, cooled at a rate of 1°C/min and stirring rate of 300 rpm.

The comparison between the chord length distribution in the crystallization onset with the values obtained in the end cooling, presented in Figure 16, can furnish information about crystal growth and this variation can also be detected by analysing the particle video microscopy images correspondent to these different moments that are shown in Figure 17.

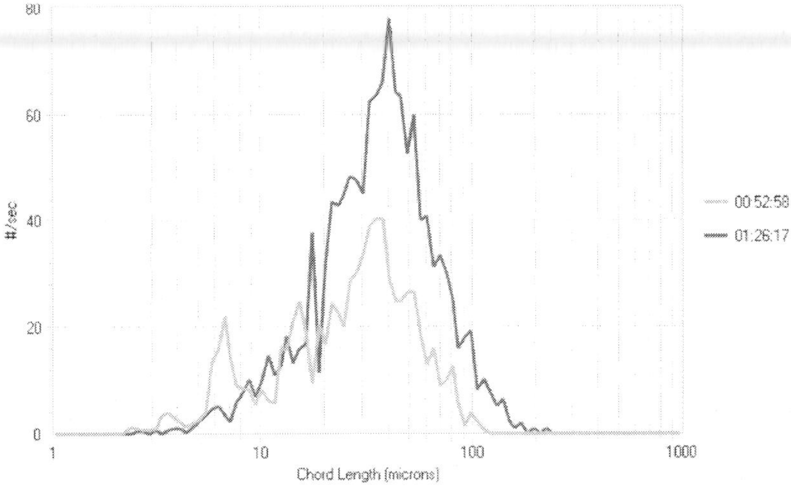

Fig. 16. Chord length distribution obtained in the crystallization onset, green curve and at the end of the experiment.

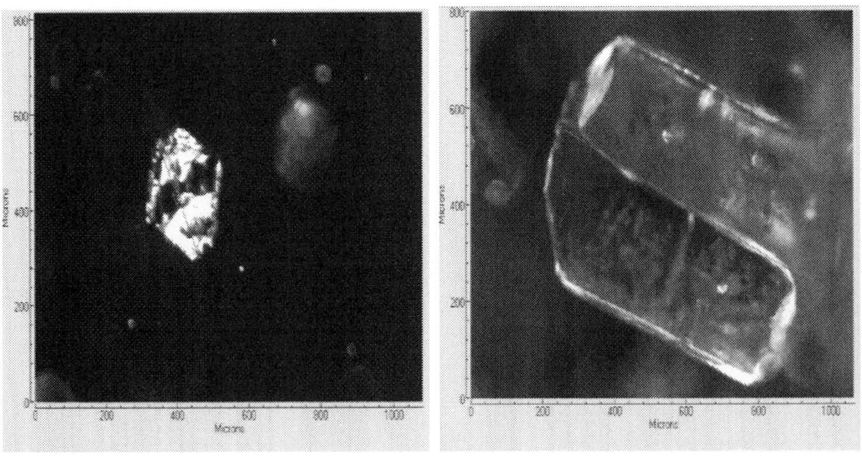

Fig. 17. Real-time particle video microscopy obtained in the crystallization onset (a) and in the end of the experiment (b).

The distributions shown in Figure 16 were displayed on a logarithmic scale on the horizontal axis (length of chords) for easy viewing. The shift to the right side in the chord length distribution indicates crystal growth. The crystal sizes presented in Figure 17 confirm the crystal growth.

The average size of the chord length of adipic acid particles produced in the set of experiments is presented in Table 3.

concentration (% mass)	cooling rate (°C/min)	average size (μm)
13.2	0.2	75.20
13.2	1.0	65.24
15.1	0.2	76.91
15.1	1.0	73.11
17.3	0.2	84.36
17.3	1.0	73.59
19.7	0.2	73.48
19.7	1.0	70.76
22.3	0.2	119.4
22.3	1.0	77.47
25.2	0.2	106.7
25.2	1.0	88.88
28.4	0.2	127.9
28.4	1.0	80.31
31.8	0.2	132.2
31.8	1.0	88.78

Table 3. Average size of the chord length of adipic acid crystals in the different experimental conditions.

The results presented in Table 3 show that the average size of adipic acid crystals increases as the cooling rate decreases. This fact is directly correlated with the metastable zone width, which is directly proportional to the rate of nucleation (Nyvlt et al., 2001). Thus, the largest crystals were observed when the cooling rate was 0.2°C/min, the rate at which the narrowest metastable zone was maintained.

3.6 Determination of metastable zone width of adipic acid in ethanol

The metastable zone width of a system is dependent on the methodology employed for its determination. Depending on the sensitivity of the technique used to detect the onset of crystallization, significant deviations can occur between two different methodologies (Marciniak, 2002). For a satisfactory evaluation of the metastable zone limit, it is necessary that the reactor be cooled at a constant rate. This constant cooling can be seen through the linear behaviour of both the crystallizer temperature and the jacket of the reactor.

The onset crystallization temperatures determined by the different real-time methods are presented in Table 4 and the agreement between these data can be seen in Figure 18.

Adipic acid concentration (%)	T calorimetry (°C)	T infrared (°C)	T FBRM (°C)
13.2	16.45	16.13	16.23
15.1	18.02	17.78	17.90
17.3	24.44	24.34	24.20
[illegible]	[illegible]	[illegible]	[illegible]
22.3	37.34	37.16	37.28
25.2	42.77	42.70	42.72
28.4	50.30	50.24	50.27
31.8	57.22	57.11	57.19

Table 4. Comparison of the calorimetric method, infrared and total chord counts for the determination of the onset temperature of crystallization of adipic acid in ethanol using a cooling rate of 1°C/min.

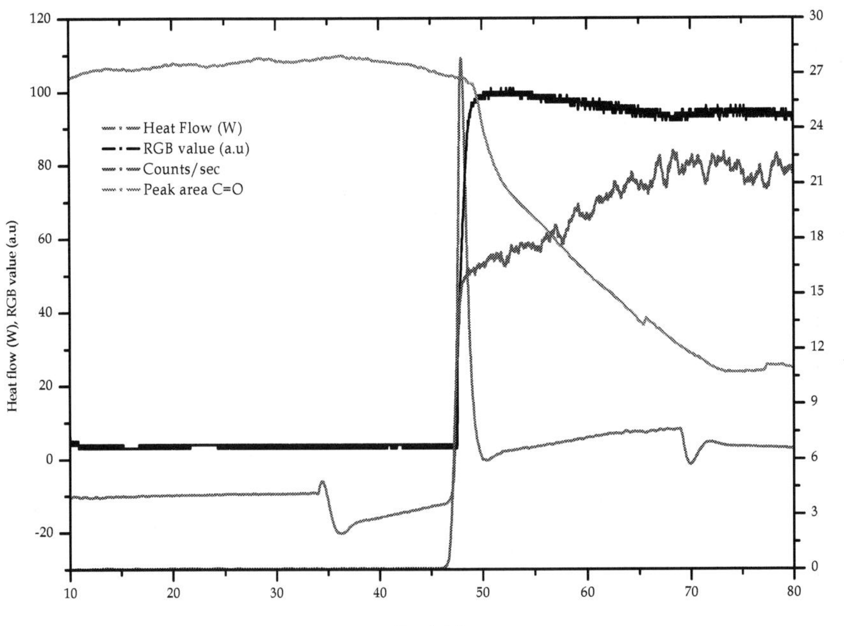

Fig. 18. Curves of the peak area of absorption of carbonyl (C=O), total count (FBRM), image analysis by RGB (**RGB**), calorimetry (heat flow) obtained from the cooling of a solution of 22% adipic acid in ethanol, cooled to 1.0 ° C / min with a stirring rate of 300 rpm.

The results presented in Figure 18, obtained by the different real-time analyses, were similar. In this case, the heat flow measurement presented higher sensibility than the other real-time analyses, but all techniques can be considered equivalent.

3.7 Study of the effect of the cooling rate on the metastable zone width of adipic acid in ethanol

Considering that the calorimetric method presented the highest sensibility, this was the chosen method to present the metastable zone limits for the crystallization of adipic acid by using cooling rates of 1.0 and 0.2^{0}C/min as presented in Figure 19.

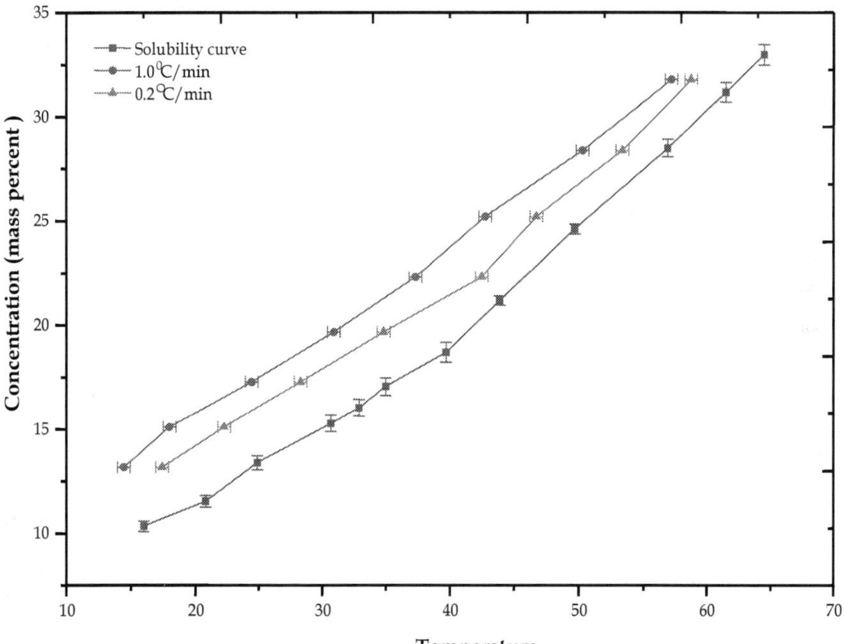

Fig. 19. Solubility curve of adipic acid in ethanol and the metastable zone limits determined at 1.0 and 0.2^{0}C/min.

The metastable zone width, expressed in terms of maximum cooling achieved, Δt, which is the difference between the saturation temperature, T saturation, and the onset temperature of crystallization, is reduced when the cooling rate is diminished. This statement can be confirmed by the data presented in Figure 19.

4. Conclusion

Considering that the experimental conditions employed during crystallization may affect physical properties of the product, such as its chemical purity, crystal size distribution and polymorphism and that these properties are completely related to the metastable zone width, accurate measurement techniques capable of providing data to construct solubility curves, as well as the onset of crystallization, are of vital importance. Off line analytical methods used to determine solubility curves and to detect the onset of crystallization present much higher errors than those that use real-time analysis for the same purpose. The four real-time analyses used in this work, even though working in different ways, presented equivalent performance, indicating that all of them can be used as process analytical tools to evaluate crystallization processes.

5. References

Alexandre, I.; Hamels, S.; Dufour, S.; Collet, J.; Zammatteo, N.; De Longueville, F.; Gala, JL. & Remacle, J. (2001). Colorimetric Silver Detection of DNA Microarrays. *Analytical Biochemistry*, Vol.295, No.1, (August 2001), pp. 1-8, ISSN 0003-2697

Barton, B.; Gouws, S.; Shaefer, MC. & Zeelie, B. (2003). Evaluation and optimization of the reagent addition sequence during the synthesis of Atrazine (6-chloro-N^2-ethyl-N^4-isopropyl-1,3,5-triazine-2,4-diamine) using reaction calorimetry. *Organic Process Research and Development*, Vol.7, No.6, pp. 1071-1076, ISSN 1083-6160.

Barrett, P. & Glennon, B. (2002). Characterizing the Metastable Zone Width and Solubility Curve Using Lasentec FBRM and PVM Original. *Chemical Engineering Research and Design*, Vol.80, No.7, (October 2002), pp. 799-805 ISSN 0263-8762

Beckmann, W. (2000). Seeding the desired polymorph: Background, possibilities, limitations, and case studies. *Organic Process Research and Development*, Vol.4, No.5, (September 2000), pp. 372-383, ISSN 1083-6160

BenAmor, S.; Colombié, D. & McKenna, T (2002) On line calorimetry: Applications to the monitoring of emulsion polymerization without samples or models of the heat-transfer coefficient. *Industrial and Engineering Chemistry Research*, (May 2002), Vol.41, No.17, pp. 4233-4241, ISSN 0888-5885.

Chen, ZP.; Morris, J.; Borissova, A.; Khan, S.; Mahmud, T.; Penchev, R. & Roberts, KJ. (2009). On-line monitoring of batch cooling crystallization of organic compounds using ATR-FTIR spectroscopy coupled with an advanced calibration method. *Chemometrics and Intelligent Laboratory Systems*, Vol.96, No.1, (March 2009), pp. 49-58, ISSN 0169-7439.

Czapla, F.; Kail, N.; Öncül, A.; Lorenz, H.; Briesen, H. & Seidel-Morgenstern, A. (2010). Application of a recent FBRM-probe model to quantify preferential crystallization of dl-threonine. *Chemical Engineering Research and Design*, Vol.88, No.11, (November 2010), pp. 1494-1504, ISSN 0263-8762

Derdour, L.; Pack, S. K.; Skliar, D.; Lai, C. J. & Kiang, S. (2011). Crystallization from solutions containing multiple conformers: A new modeling approach for solubility and supersaturation. *Chemical Engineering Science*, Vol.66, No.1, (January 2011) pp. 88-102, ISSN 0009-2509.

Dunuwila, D. D.; Carroll II, L. B.; Berglund, K.A. (1994). An investigation of the applicability of attenuated total reflection infrared spectroscopy for measurement of solubility and supersaturation of aqueous citric acid solutions. *Journal of Crystal Growth*, Vol.137 Nos.3-4 (April 1994) pp. 561-568 ISSN 0022-0248

Elizalde, O.; Azpeitia, M.; Reis, M. M.; Asua, J. M.; & Leiza, J. R. (2005). Monitoring emulsion polymerization reactors: Calorimetry versus Raman spectroscopy. *Industrial and Engineering Chemistry Research*, Vol.44 No.18 (July 2005) pp. 7200-7207. ISSN 0888-5885

Fakatselis, T.E. (2002). Residence time optimization in continuous crystallizers. *Crystal Growth and Design*, Vol.2, No.5, (September 2002), pp. 375-379, ISSN 1528-7483

Feng, L. & Berglund, K. A. (2002). ATR-FTIR for determining optimal cooling curves for batch crystallization of succinic acid. *Crystal Growth and Design*, Vol.2, No.5, (September 2002), pp. 449-452, ISSN 1528-7483

Févotte, G. & Klein, J. P. (1995). Crystallization calorimetry for the assessment of batch seeding and cooling police. *The Chemical Engineering Journal*, Vol.59, No.2, (October 1995), pp.143-152, ISSN 1385-8947

Fujiwara, M.; Nagy, Z. K.; Chew, J. W. & Braatz, R. D. (2005). First-principles and direct design approaches for the control of pharmaceutical crystallization. *Journal of Process Control*, Vol.15, No.5, (August 2005), pp. 493-504, ISSN 0959-1524

Gahn, C. & Mersmann, A. (1995). The brittleness of substances crystallized in industrial-processes. *Powder Technology*, Vol.85, No.1, (October 1995), pp. 71-81, ISSN 0032-5910

Gaiao, EN.; Martins, VL.; Lyra, WS.; Almeida, LFA.; Silva, ECS. & Araújo, MCU. (2006). Digital image-based titrations. *Analytical Chimica Acta*. Vol.570, No.2, (Juno 2006), pp. 283-290, ISSN 0003-2670.

Gesthuisen, R.; Krämer, S.; Niggemann, G.; Leiza, J. R.; & Asua, J. M. (2005). Determining the best reaction calorimetry technique: theoretical development. *Computers and Chemical Engineering*, Vol.29, No.2,(January 2005) pp. 349-365 ISSN 0098-1354.

Giulietti, M.; Seckler, M. M.; Derenzo, S.; Ré, M. I. & Cekinski, E. (2001). Industrial crystallization and precipitation from solutions: State of the technique. *Brazilian Journal of Chemical Engineering*, Vol.18, No.4, pp. 423-440, ISSN 0104-6632

Gunawan, R.; Ma, D.L.; Fujiwara, M. & Braatz, R.D. (2002). Identification of kinetics parameters in multidimensional crystallization processes. *International Journal of Modern Physics B*, Vol.16, Nos.1-2, (January 2002), pp. 367-374, ISSN 0217-9792

Hernández, JA.; Heyd, B. & Trystram, G. (2008). On-line assessment of brightness and surface kinetics during coffee roasting. *Journal of Food Engineering*. Vol.87, No.3, (August 2008), pp. 314-322, ISSN 0260-8774.

Jolling, K.; Vandeven, M.; Eynden, JVD.; Amellot, M. & Kerkhove, EV. (2007). A highly reliable and budget-friendly Peltier-cooled camera for biological fluorescence imaging microscopy. *Journal of Microscopy*. Vol.228, No.3, (December 2007), pp. 264-271, ISSN 0022-2720.

Joung, O. J.; Kim, Y. H. & Fukui, K. (2005). Determination of metastable zone width in cooling crystallization with a quartz crystal sensor. *Sensors and Actuators B: Chemical*, Vol.105, No. 2, (March 2005), pp. 464-472, ISSN 0924-4005.

Kadam, S. S.; Mesbah, A.; der Windt, E. V. & Kramer, H. J. M. (2011). Rapid online calibration for ATR-FTIR spectroscopy during batch crystallization of ammonium sulphate in a semi-industrial scale crystallizer. *Chemical Engineering Research and Design*, Vol.89, No.7, (July 2011), pp. 995-1005, ISSN0263-8762.

Kail, N.; Briesen, H.; & Marquardt, W. (2008). Analysis of FBRM measurement by means of a 3D optical model. *Powder Technology* Vol.185 No.3 (July 2008) pp. 211-222, ISSN 0032-5910.

Kougoulos, E.; Jones, A. G.; Jennings, K. H.; & Wood-Kaczmar, M. W. (2005). Use of focused beam reflectance measurement (FBRM) and process video imaging (PVI) in a modified mixed suspension mixed removal (MSMPR) cooling crystallizer. *Journal of Crystal* Growth, Vol.273, Nos.3-4, (January 2005), pp. 529-534, ISSN 0022-0248

Lavanchy, F.; Fortini, S. & Meyer, T. (2004). Reaction calorimetry as a new tool for supercritical fluids. *Organic Process Research and Development,* Vol.8, No.3, pp. 504-510, ISSN 1083-6160.

Lewiner, F.; Klein, J.P.; Puel, F. & Fevotte, G. (2001). On-line ATR FTIR measurement of supersaturation during solution crystallization processes. Calibration and applications on three solute/solvent systems. *Chemical Engineering Science,* Vol.56, No.6, (March 2001), pp. 2069-2084, ISSN 0009-2509

Liang, RQ.; Tan, CY. & Ruan, KC. (2004). Colorimetric detection of protein microarrays based on nanogold probe coupled with silver enhancement. *Journal of Immunological Methods,* (February 2004), Vol.285, No.2, pp. 154-163, ISSN 0022-1759.

Liotta, V. & Sabesan, V. (2004). Monitoring and feedback control of supersaturation using ATR-FTIR to produce an active pharmaceutical ingredient of a desired crystal size. *Organic Process Research and Development,* Vol.8, No. 3, (May 2004), pp. 488-494, ISSN 1083-6160

Lyra, WS.; dos Santos, VB.; Dionízio, AGG.; Martins, VL.; Almeida, LF.; Gaião, EN.; Diniz, PHGD.; Silva, EC. & Araújo, MCU. (2009) Digital image-based flame emission spectrometry. *Talanta.* Vol.77, No.5, (May 2011), pp. 1584-1589, ISSN 0039-9140.

Ma, D. L.; Chung, S.H.; Braatz, R. D. (1999). Worst-case performance analysis of optimal batch control trajectories. *Aiche Journal,* Vol.45, No.7, (July 1999), pp. 1469-1486, ISSN 0001-1541

Maleki, N.; Safavi, A. & Sedaghatpour, F. (2004). Single-step calibration, prediction and real samples data acquisition for artificial neural network using a CCD camera. *Talanta.* Vol.64, No.4, (November 2004), pp. 830-835, ISSN 0039-9140.

Marciniak, B. (2002). Density and ultrasonic velocity of undersaturated and supersaturated solutions of fluoranthene in trichloroethylene, and study of their metastable zone width. *Journal of Crystal Growth,* Vol.236, No.1-3, (March 2002), pp. 347-356, ISSN 0022-0248

Marisol, IW.; Bijou, B. & Von Stockar, U. (1985). Use of a novel heat-flux calorimeter for the measurement of heat generated during growth of *K. fragilis* on lactose. *Thermochimica Acta* Vol.85, No.93, (April 1985), pp. 497-500, ISSN 0040-6031.

Man, Y. B. C.; Syahariza, Z. A. & Rohman, A., Fourier Transform Infrared (FTIR) Spectroscopy: Development, Techniques and Application in the Analyses of Fats and Oils, editor Oliver J. Rees, Nova Science Publishers, Inc. (2010).

Mantelis, CA. & Meyer, T. (2008). Supercritical reaction calorimetry: Versatile tool for measuring heat transfer properties and monitoring chemical reactions in supercritical fluids. *Industrial and Engineering Chemistry Research,* (May 2008), Vol.47, No.10, pp. 3372-3379, ISSN 0888-5885.

Mersmann, A. (2001) *Crystallization Technology Handbook.* Marcel Dekker Inc., New York.

Mullin, J. W. (2001). *Crystallization.* (4thed), Butterwoth-Heinemann, ISBN 075064833.3, Oxford, UK.

Nguyen, TNP. & Kim, KJ. (2008). Kinetic study on hemipenta hydrate risedronate monosodium in batch crystallization by cooling mode Original. *International Journal of Pharmaceutics,* (November 2008), Vol.364, No.1, pp. 1-8, ISSN 0378-5173.

Nýlvt, J.; Hostomsky, J. & Giulietti, M. (2001). *Cristalização.* IPT/UFSCar, ISBN85-85173-63-7, São Paulo, Brazil.

Paulaime, A. M.; Seyssiecq, I. & Veesler, S. (2003). The influence of organic additives on the crystallization and agglomeration of gibbsite. *Powder Technology*, Vol.130, Nos.1-3, (February 2003), pp. 345-351, ISSN 0032-5910

Perlberg, A.; Lorenz, H. & Seidel-Morgenstern, A. (2005). Crystal growth kinetics via isothermal seeded batch crystallization: Evaluation of measurement techniques and application to mandelic acid in water. *Industrial and Engineering Chemistry Research*, Vol.44, No. 4, (February 2005), pp. 1012-1020, ISSN 0888-5885

Podder, J. (2002). The study of impurities effect on the growth and nucleation kinetics of potassium dihydrogen phosphate. *Journal of Crystal Growth*, Vol.237-239 Part 1, (April 2002), pp. 70-75, ISSN 0022-0248.

Pöllänen, K.; Häkkinen, A.; Reinikainen, S. P.; Louhi-Kultanen, M.; & Nyström, L. (2006). A Study on Batch Cooling Crystallization of Sulphathiazole: Process Monitoring Using ATR-FTIR and Product Characterization by Automated Image Analysis. *Chemical Engineering Research and Design*, Vol.84, (January 2006), No.1, pp. 47-59, ISSN 0263-8762.

Qu, H.; Alatalo, H.; Hatakka, H.; Kohonen, J.; Louhi-Kultanen, M.; Reinikainen, S. P.; & Kallas, J. (2009). Raman and ATR FTIR spectroscopy in reactive crystallization: Simultaneous monitoring of solute concentration and polymorphic state of the crystals. *Journal of Crystal Growth* Vol.311, (June 2009), No.13, pp.3466-3475, ISSN 0022-0248

Rauls, M.; Bastosch, K.; Kind, M.; Kuch, St.; Lacmann, R. & Mersmann, A. (2000). The influence of impurities on crystallization kinetics - a case study on ammonium sulfate. *Journal of Crystal Growth*, Vol.213, Nos.1-2, (May 2002), pp.116-128, ISSN 0022-0248.

Safavi, A.; Maleki, N.; Rostamzadeh, A. & Maesum, S. (2007). CCD camera full range pH sensor array. *Talanta*. Vol.71, No.1, (January 2007), pp. 498-501, ISSN 0039-9140.

Seiceira, R.; Higa, CM.; Barreto, AG. & da Silva, JFC. (2005). Determination of kinetic parameters and Hammett ρ from the synthesis of triaryl phosphites using reaction calorimetry. *Thermochimica Acta* Vol.428, No.1-2, pp. 101-104, ISSN 0040-6031.

Sheikhzadeh, M.; Trifkovic, M.; & Rohani, M. (2008). Real-time optimal control of an anti-solvent isothermal semi-batch crystallization process. *Chemical Engineering Science*. Vol.63, (February 2008), No.3, pp.829-839, ISSN 0009-2509.

Shirshov, YM.; Khoruzhenko, VY.; Kostyukevych, KV.; Khristosenko, RV.; Samoylova, IA.; Pavlunchenko, AS.; Samoylov, AV. & Ushenin, YV. Analysis of some alcohols molecules based on the change of RGB components of interferentially colored calixarene films. *Sensors Actuators B: Chemical*. Vol.112, pp. 427-436, ISSN 0925-4005.

Silva, J.F.C. & Silva, A.P.M. (2011). Determination of the Adipic Acid Solubility Curve in Acetone by Using ATR-FTIR and Heat Flow Calorimetry. *Organic Process Research and Development*. Vol.15, No.4, (May 2011), pp. 893-897, ISSN 1083-6160.

Simon, LL.; Nagy, ZK. & Hungerbuhler, K. (2009). Endoscopy-based in situ bulk video imaging of batch crystallization processes. *Organic Process Research and Development*, Vol.13, No.6, pp. 1254-1261, ISSN 1083-6160.

Sun, Y.; Song, X.; Wang, J.; Luo, Y. & Yu, J. (2010). Determination of seeded supersolubility of lithium carbonate using FBRM. *Journal of Crystal Growth*, Vol.312, No.2, (January 2010) pp. 294-300, ISSN 0022-0248.

Teychené, S.; Autret, J. M. & Biscans, B. (2004). Crystallization of efflucimibe drug in a solvent mixture: effects of process conditions on polymorphism. *Crystal Growth and Design*, Vol.4, No.5, (September 2004), pp. 971-977, ISSN 1528-7483

Ulrich, J. & Strege, C. (2002). Some aspects of the importance of metastable zone width and nucleation in industrial crystallizers. *Journal of Crystal Growth*, Vol.237-239 Part 3, (April 2002), pp. 2130-2135, ISSN 0022-0248

Yu, Z. Q.; Tan, R. B. H. & Chow, P. S. (2005). Effects of operating conditions on agglomeration and habit of paracetamol crystals in anti-solvent crystallization. *Journal of Crystal Growth*, Vol.279, No.3-4, (June 2005), pp. 477-488, ISSN 0022-0248

Yu, L. X.; Lionberger, R. A.; Raw, A. S.; D'Costa, R.; Wu, H. & Hussain, A. S. (2004). Applications of process analytical technology to crystallization processes. *Advanced Drug Delivery Reviews*, Vol.56, No. 3, (February 2004), pp. 349-369, ISSN 0169-409X

12

Phenacetin Crystallization: Cooling Regimes and Crystal Morphology

Humphrey A. Moynihan and Dawn M. Kelly
Dept. of Chemistry / Analytical and Biological Chemistry Research Facility, University College Cork, Cork
Ireland

1. Introduction

Phenacetin crystallizations were undertaken to gain an insight into how to better control the various parameters of a crystallization process. Good control of the crystallization process allows the design of experiments to control certain characteristics of the final crystal product. These include crystal morphology, phase, particle size and crystal size distribution (CSD). These physical attributes of a crystal population are thought to be achieved through the control of the nucleation and growth of crystals during the crystallization process. One of the first attempts to control crystallization processes was proposed by Griffiths (Griffiths, 1925), who suggested the idea of "controlled cooling" in batch crystallizations by maintaining the supersaturation in the metastable region in order to improve the product CSD. Garside et al. performed experimental studies employing this concept of controlled cooling in crystallization processes, which included simple strategies such as isothermal operation and linear cooling (Garside et al., 1972). Since these early crystallization control approaches, there have been many strategies that indirectly affect the CSD of a crystallization process, often involving a change in temperature or the addition of an antisolvent during a pre-defined timescale so as to follow a given supersaturation profile in the phase diagram. These profiles are obtained using simple trial-and-error experimentation or more complex model-based or direct-design approaches (Abu Bakar et al., 2009a). Woo et al. proposed an adaptive concentration control strategy that employs the measurement of the number of particle counts per unit time provided by *in situ* laser backscattering in order to detect the onset of nucleation and adapt the operating curve accordingly (Woo et al., 2009). In a move towards new strategies for directly affecting the CSD of a crystallisation process, Abu Baker et al. (Abu Bakar et al., 2009a) used a model-free approach to crystallization control known as 'direct nucleation control' (DNC), in which the number of counts measured by an *in situ* laser probe is directly controlled using a feedback control strategy. An alternative strategy is to chart an oscillating cooling regime within the metastable zone of the crystallization, in which the overall cooling trend drives crystal growth while the temperature fluctuations provided by the oscillating cool allows partial dissolution (Abu Bakar et al., 2009b). The result should be an equalizing of crystal size and shape. Both of these approaches were utilized in the present study. Phenacetin was selected as the subject for this study as it is an Active Pharmaceutical Ingredient - now withdrawn (Prescott, 1980) - and is not known to be crystal polymorphic (Hansen, 2006).

This chapter will present work on phenacetin crystallization using two forms of modified cooling regimes in an attempt to directly affect the product CSD of the crystallization process. We begin with a cooling regime using turbidity counts as a termination clause for each cooling and heating step. We then move to a more in-depth study in Section 3.2, where the cyclic cooling paths that had predefined temperature parameters were used, ensuring that a cyclic temperature profile was maintained within the limits of the metastable zone width (MSZW). This temperature profile was coupled with a seeding regime where each experiment was seeded with crystals that were sieved to give specific size ranges of less than 150 μm (fine particles), between 150-300 μm (medium particles) and over 500 μm (coarse particles). The objective of both of these types of modified cooling regimes is to provide a narrower range of crystal sizes and shapes than would otherwise be obtained by a simple linear cooling regime.

2. Experimental methods

Phenacetin was purchased from Sigma-Aldrich. GPR grade absolute ethanol was purchased from Carbon Group Chemical Ltd. (Cork, Ireland) and was used in all of the experiments. Ethanol was chosen as a solvent for these experiments because the solubility of phenacetin in ethanol varies significantly over the temperature range 10 °C to 30 °C, allowing for the development of a practical crystallization method. The determination of phenacetin solubility and metastable zone width is described below. Phenacetin crystallizes cleanly from ethanol to give well-formed needles which are largely free from the inclusions of other defects.

A HEL Autolab jacketed reactor vessel equipped with a HEL Lasertrack *in situ* particle sizing laser probe was used to monitor the presence and size of phenacetin crystals on a 1 L scale. The laser wavelength was 795 nm. It is possible to get readings of particles as small as 0.5 μm. A PTFE PT100 thermocouple gave *in situ* temperature measurements of the crystallization medium. The temperature of the jacket fluid (Huber DW-Therm thermal fluid, operating range −90 °C to 200 °C) was controlled by a Huber Unistat 815 circulation thermostat. The system was entirely controlled from one PC using HEL WinISO software, allowing the control of stirring rates, the addition of material (the pumping rate and mass pumped), and the heating parameters (heating or cooling rates, reactor vessel temperature set point, jacket fluid temperature set point). The software allowed the pre-programming of crystallization regimes containing as many separate steps as required, e.g., multiple heat/cool steps, solvent addition steps, etc. The steps could be programmed to terminate and move to a next step upon reaching a specified set point, such as the reactor temperature or the number of particles observed.

Crystal habits were observed using a Nikon Polarizing Microscope Eclipse 50i POL and photomicrographs were taken on a Nikon Digital Sight DS-Fi1 digital camera. Figure 1 shows a schematic representation of the experimental setup for phenacetin crystallization experiments.

In the experiments described below, in which seeding was used, crystals obtained from water were used as seed crystals because they typically have a plate-like habit. The plate-like habit provides a greater surface area for growth and so is more suitable for seeds than the needle-like habit obtained from ethanol. The crystals formed from water also have rougher surfaces than the ethanol-derived needles, again assisting crystal growth. In addition, any differences in habit after the crystallization of phenacetin from ethanol using these seed crystals would assist in the determination of whether growth or nucleation dominated these processes. The seed crystals were sieved to give the size ranges described below using a set of U.S.A. standard testing sieves (Avantech Manufacturing) in mesh sizes of 150 μm, 300 μm and 500 μm. The seeds were added to crystallisations in an ethanol slurry using seed crystals with a weight in grams equal to 20% of the total batch concentration.

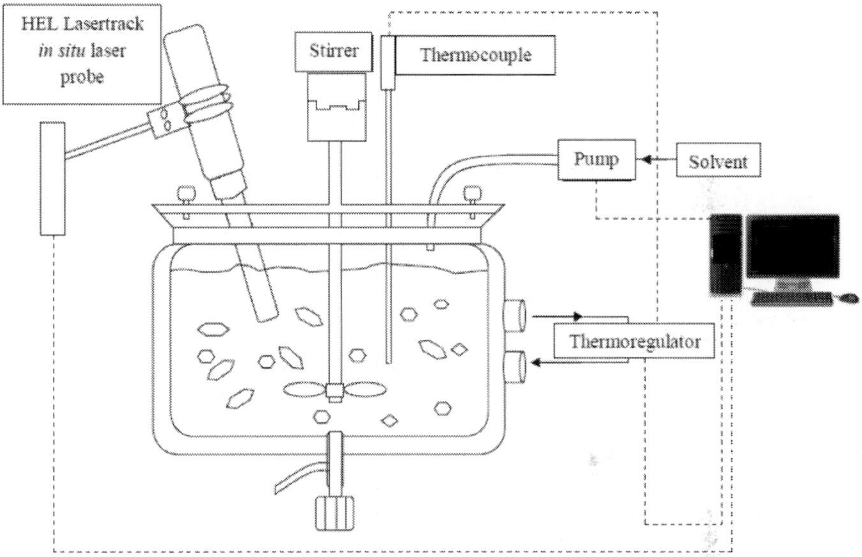

Fig. 1. A schematic representation of the experimental setup for phenacetin crystallization experiments.

Only one crystalline form of phenacetin has been reported to date (Hansen, 2006). To check that no new polymorphs or solvates of phenacetin were obtained during these experiments, powder X-ray diffraction (PXRD) patterns were recorded for all of the samples and compared with the theoretical pattern generated from the structure reported by Hansen et al. (Hansen, 2006). PXRD was performed at an ambient temperature using a Stoe Stadi MP diffractometer operating in transmission mode with a linear PSD detector, with an anode current of 40 mA, an accelerating voltage of 40 kV and Cu $K\alpha_1$ X-radiation (λ = 1.5406 Å), typically over a scan range of 3.5° to 60° 2θ, scanning in steps of 2° for 90 s per step. The samples were held between acetate foils. The theoretical patterns were generated from the crystallographic information file downloaded from the Cambridge Structural Database (CSD) - CSD reference code PYRAZB21 - using the THEO function on the Stoe WinX^{POW} software package. Figure 2 shows a typical example of a PXRD pattern of a phenacetin batch with the theoretical pattern overlaid, with excellent correspondence between the observed and theoretical patterns.

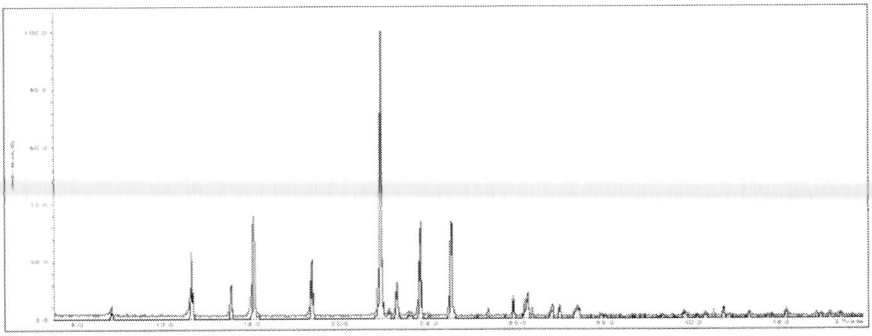

Fig. 2. Comparison of the theoretical PXRD pattern of phenacetin (black) generated from single crystal data (Hansen, 2006; CSD reference code PYRAZB21) overlaid with the experimental pattern (red) of a batch of phenacetin crystallized from ethanol.

3. Results and discussion

3.1 Cooling with feedback control

The metastable zone width (MSZW) of phenacetin from ethanol was first found by a heat / cool experiment in which the concentration was diluted after each heat /cool step until enough data points were collected to plot a solubility/super-solubility diagram. Figure 3 is the experiment progression diagram, illustrating the alternate heating and cooling cyclic programme from which the data points were collected, and Figure 4 is the solubility/super-solubility diagram of phenacetin in ethanol.

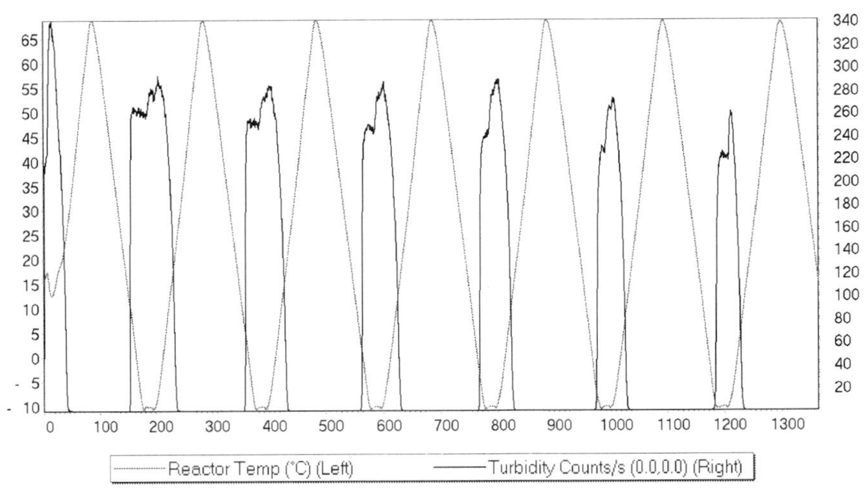

Fig. 3. The experiment progression diagram for measuring the metastable zone width (MSZW) of phenacetin in ethanol.

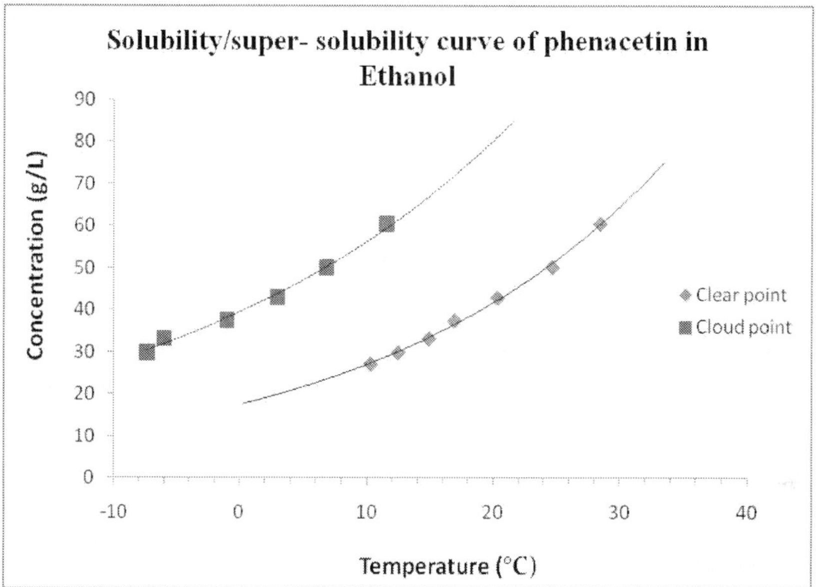

Fig. 4. The solubility/super-solubility diagram of phenacetin from ethanol.

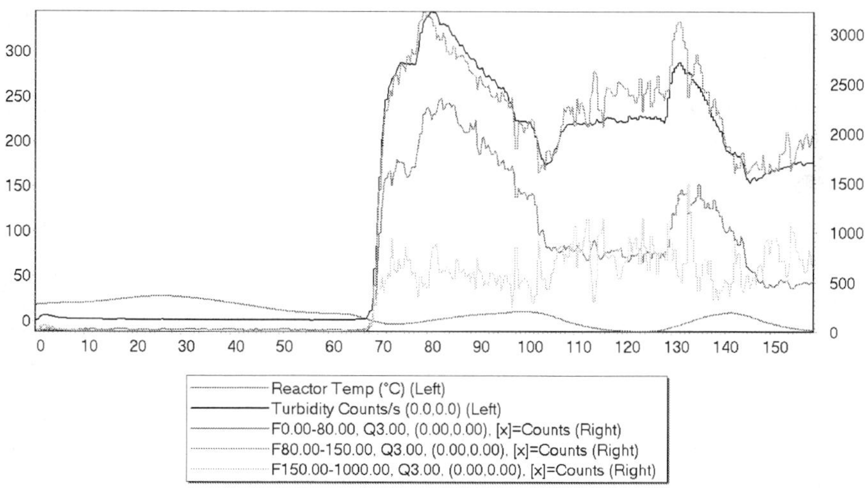

Fig. 5. The progression diagram of a heat/cool experiment of phenacetin in ethanol using turbidity counts as a termination clause (F is the number of particles inside the given size range - e.g., the size range 0.00-80.00 is the content between 0 - 80 µm – and Qz(x) is the less-than-size for parameter x and weight z, where x is counts and z = 3 is interpreted as the volume equivalent distribution).

Our first experiments were undertaken using the phenacetin crystallizations in ethanol involved controlling the crystallization using turbidity counts as a termination clause for each heating and cooling step. The first experiment involved a batch without seeding. Phenacetin, at a concentration of 50 g L^{-1}, was dissolved in ethanol, stirred at 180 rpm and followed by a slow cool at -0.5 °C / min until the phenacetin crystallized out of solution. The termination clause was set on the cooling step so that the cooling was terminated when the turbidity counts exceeded a set number. In this case, the programme was set to cool until the crystal population reached a turbidity of 300 counts / s. Once the turbidity counts reached the pre-set level, the experiment began to heat again until the turbidity counts dropped below another pre-set number. In this experiment, the lower limit was set at 290 counts / s, thereby terminating the step and beginning the cooling process once again. Ideally, this cyclic heating and cooling should be continued until a steady state is reached, which should result in a crystallization process where the turbidity counts are maintained at the pre-set number. Figure 5 gives this experiment's progression diagram of phenacetin in ethanol under this regime.

Using turbidity counts as the terminating clause for controlling the temperature in crystallization batches that were not seeded proved difficult. The temperature response to this terminating clause was slow, causing large fluctuations in the turbidity counts and the fine and medium total particle counts. The coarse particle counts remained relatively stable. The crystals collected at the end of the experiment were needles of varying size. PXRD analysis proved that no new forms of phenacetin were present. Since the experiment was carried out in ethanol, the crystallization of needle shaped crystals was expected. Figure 6 is a photomicrograph of the crystals collected at the end of the experiment.

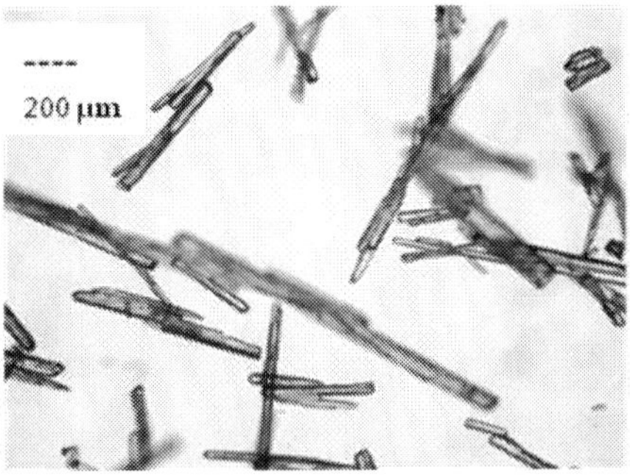

Fig. 6. Photomicrograph of crystals at the end of the experiment from a batch of a phenacetin crystallization without seeding.

It was then proposed to seed a phenacetin crystallization experiment, using the same conditions as before and with seeds of varying sizes, in order to see if better control could be achieved. It was decided to use phenacetin crystals obtained from water as the seed crystals because they typically have a plate-like habit. Any differences in habit after the crystallization of phenacetin from ethanol using these seed crystals would assist the determination of whether growth or nucleation was dominating these processes. Figure 7 is an experimental design showing the proposed heating and cooling path in the seeded experiment and Figure 8 is the experiment progression diagram of a seeded phenacetin crystallization in ethanol using turbidity counts as the termination clause.

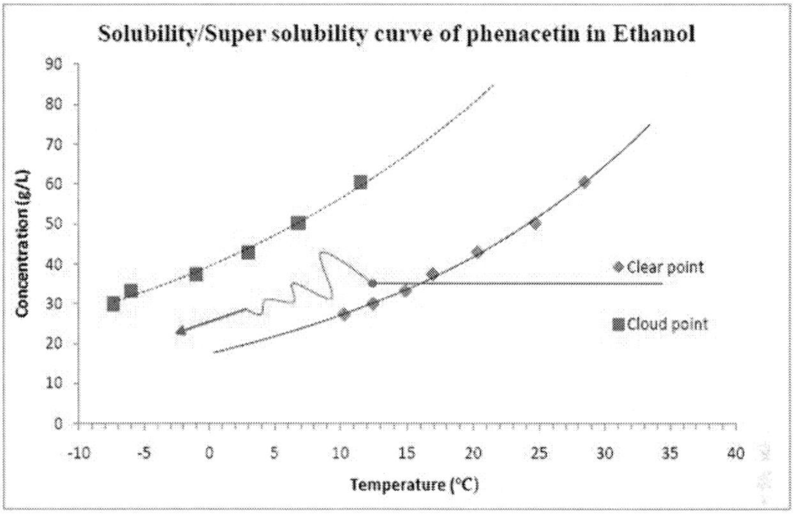

Fig. 7. Design of an experiment showing the proposed heating and cooling path using turbidity counts as the termination clause for each step (the red dot indicates the point of seeding).

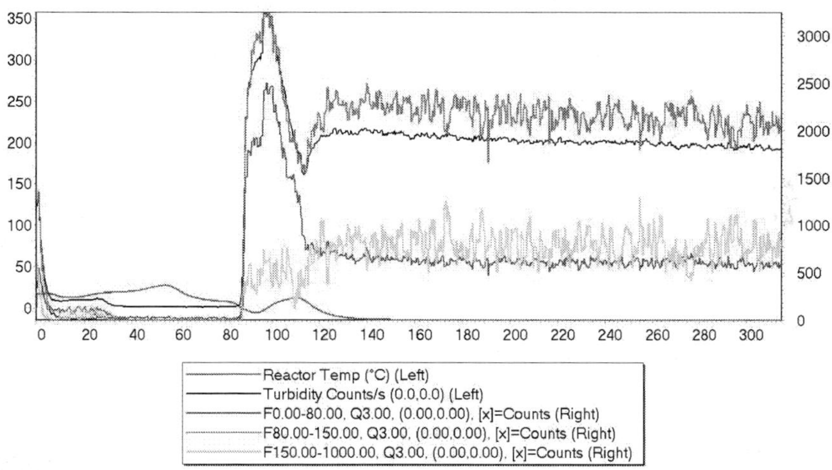

Fig. 8. The experiment progression diagram of a seeded phenacetin crystallization in ethanol using turbidity counts as a termination clause (F is the number of particles inside the given size range - e.g., the size range 0.00-80.00 is the content between 0 - 80 µm - and Qz(x) is the less-than-size for parameter x and weight z, where x is counts and z = 3 is interpreted as the volume equivalent distribution)

The use of turbidity counts as the terminating step to control the temperature in seeded phenacetin crystallizations resulted in a more controlled experiment. The crystallization was slowly cooled to within the solubility limit at a cooling rate of -0.5 °C / min. The batch was seeded with phenacetin crystals of various sizes and the crystallization was rapidly cooled at a rate of -5 °C / min. When the turbidity counts exceeded the stated level - which in this experiment was a turbidity of 200 counts / s - the process began to heat, causing the dissolution of fines and medium particles. Once the turbidity counts had reached the desired level (190 counts / s), cooling began again. The coarse and medium particles began to rise (indicating growth) while the fines remained steady. Figure 9 shows photomicrographs of the seed crystals, which are of varying size, and the end of experiment crystals, which show a more uniform crystal size and are no longer needle-like. The final crystals have a regular prismatic habit, unlike than the needles obtained from ethanol without seeding or the plates obtained from water. This reflects the operation of crystal growth from ethanol on plate-like seeds obtained from water, generating a new and more desirable prismatic habit.

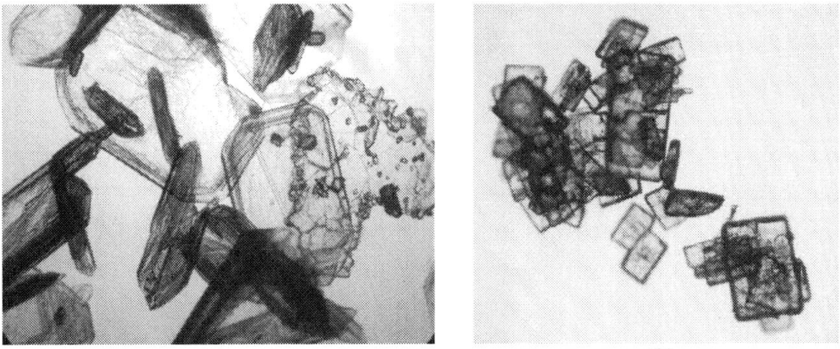

(a) (b)

Fig. 9. Photomicrographs of (a) the seed crystals of phenacetin of varying sizes and (b) end of experiment crystals.

3.2 Cyclic cooling

The next series of experiments used cyclic cooling paths that had predefined temperature parameters, ensuring that a cyclic temperature profile was maintained within the MSZW limits. Rather than using the turbidity counts as a termination clause, these experiments followed plans that were programmed before the experiment began, using the data from the MSZW diagram to find the temperature limits. Figure 10 shows the design of the experiment for this crystallization.

In order to achieve the best control, it was decided to carry out a series of phenacetin crystallizations from ethanol, keeping the crystallization conditions constant throughout but varying the seed crystals. Each experiment was seeded with crystals that were sieved to give a specific size range. The three size ranges used were seed crystals under 150 µm, between 150-300 µm and over 500 µm. All of the crystallization experiments were seeded with crystals grown from water and added to an ethanol slurry using seed crystals with a weight in grams equal to 20% of the total batch concentration. Figure 11 gives the experiment progression diagram of the phenacetin crystallization from ethanol that was seeded with crystals grown from water within the size range 150-300 µm.

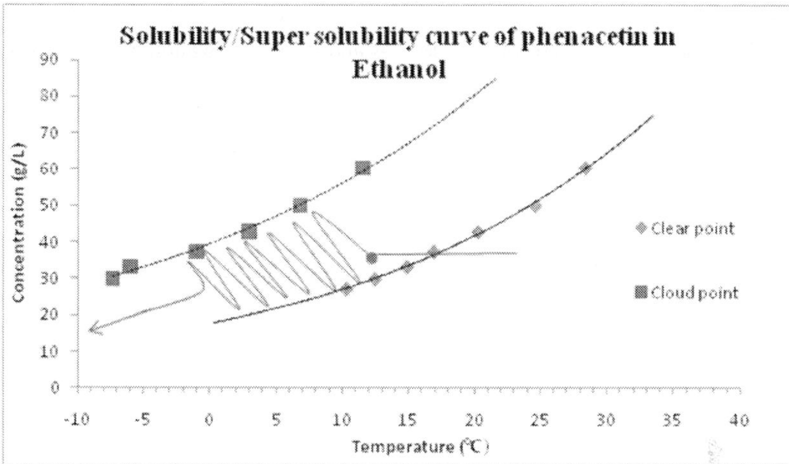

Fig. 10. An experiment design illustrating the pre-set path of the cyclic heating and cooling programme for phenacetin crystallizations in ethanol.

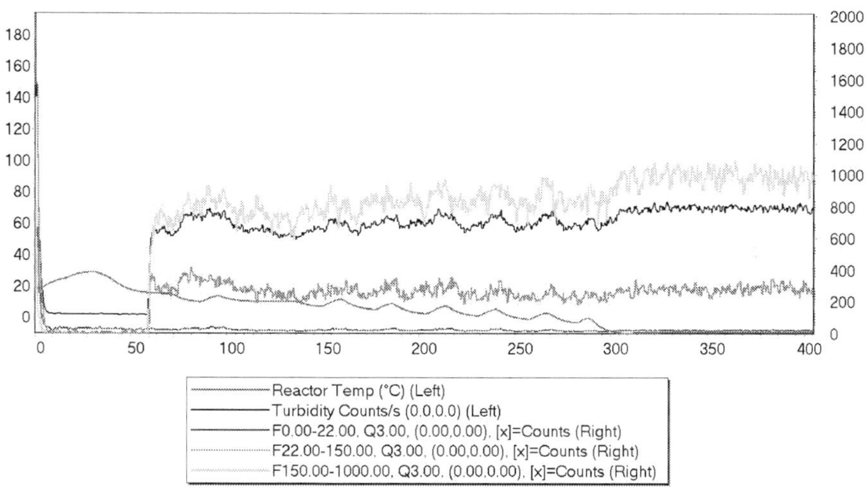

Fig. 11. Experiment progression diagram of the seeded (crystals between 150-300 µm) phenacetin crystallization from ethanol using a pre-set cyclic plan (F is the number of particles inside the given size range - e.g., the size range 0.00-80.00 is the content between 0 - 80 µm - and Qz(x) is the less-than-size for parameter x and weight z, where x is counts and z = 3 is interpreted as the volume equivalent distribution).

The experiment progression diagram shows the point of seeding as the point where there is a simultaneous increase in the turbidity counts (the black line), the medium-sized particles (the pink line indicating particles between 80-150 µm) and the coarse particles (light blue line indicating particles between 150-1000 μm). It should be noted that at this point there is no increase in the fine particles (the dark blue line indicating particles between 0-80 μm). This is expected as the crystallization is seeded when the temperature just reaches the limit of the solubility curve, indicating a saturated solution and that spontaneous nucleation has not occurred. Therefore, the only particles in the solution are the seed crystals. Since these crystals are within the size range of 150-300 μm, no fine particles are present at this stage. As the solution is cooled in a cyclic heat/cool manner, any nucleation will register a rise in the counts for the fraction, indicating fine particles (the dark blue line). It can be seen from the above diagram that the dark blue line stays close to the baseline throughout the experiment, with some small intermittent increases, indicating that a small amount of nucleation occurs when the crystallization is cooling and that these fine particles are being dissolved during heating. This illustrates that this cyclic heating and cooling program is effective in controlling the nucleation process and riding the sample of fine particles, which tend to be problematic when filtering large scale crystallizations. This experiment progression diagram shows that throughout the crystallization, the medium-sized particles (the pink line) are maintained at a steady level with a slight increase in the coarse particles (the light blue) indicating growth. Figure 12 presents photomicrographs of the seed crystals grown from water and the end of experiment crystals.

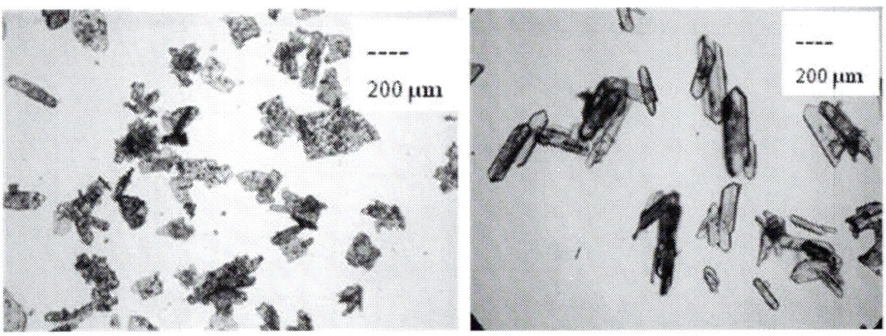

(a) (b)

Fig. 12. Photomicrographs of (a) seed crystals between 150-300 μm and (b) end of experiment crystals.

The experiment was seeded with crystals grown from water which were plate-like in shape, and the end of experiment crystals were elongated plate-like needles. If nucleation occurs after the experiment was seeded, then these crystals will be very needle-like in shape. If the crystallization predominantly favours growth, then those seed crystals that are plate-like to begin with would have a distorted shape due to the solvent effects on the phenacetin crystals. In theory, if very fine particles precipitate during the cooling process, indicating nucleation, then these would be dissolved on the heating step. This continual heating and cooling would ensure that enough time is given to the growth of the seed crystals, while at the same time ridding the crystallization process of fine particles. The heating cycle would also ensure that the very large crystals are dissolved slightly - thereby reducing their size - the result of which would likely be a crystal batch where there is a good crystal size distribution, with the majority of the crystals falling under a mean crystal size. Figure 13 gives the particle size distributions of the crystals at the beginning and at the end of the experiment.

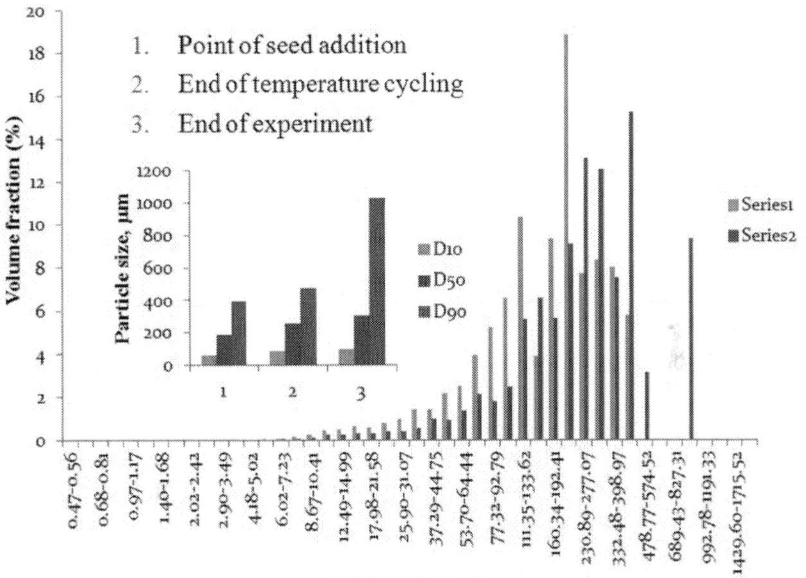

Fig. 13. Crystal size distributions of crystals shortly after the experiment is seeded (series 1) and at the end of the experiment (series 2), showing an increase in the particle size distribution and indicating growth (inset is a bar chart of the D10, D50 and D90 distributions at three points throughout the experiment).

The crystal size distributions in Figure 13 show an upward shift in the particle size from the beginning of the experiment (series 1) to the end of the experiment (series 2), which is indicative of growth. Also shown in Figure 13 are the distributions D10 (meaning the size under which 10% of the crystal population falls), D50 (the size under which 50% of the crystal population falls) and D90 (the size under which 90% of the crystal population falls). The inset of Figure 13 gives these size distributions for the crystal population at the beginning of the experiment, after the cyclic heat/cool program and at the end of the experiment. It agrees with the particle size distributions and also indicates growth. Figure 14 displays photomicrographs of a seed crystal and an end of experiment crystal.

The photographs of the crystals at the beginning and the end of the experiment show the growth of those seed crystals that began plate-like and then grew to elongated plate-like needles. The difference in size also indicates growth. When we take the seed crystal into consideration, we can see that the crystals are flat and plate-like in shape, with round edges and a rough surface. In contrast, the end of experiment crystals tend to be elongated plate-like needles with a rigid, defined shape and a smooth surface.

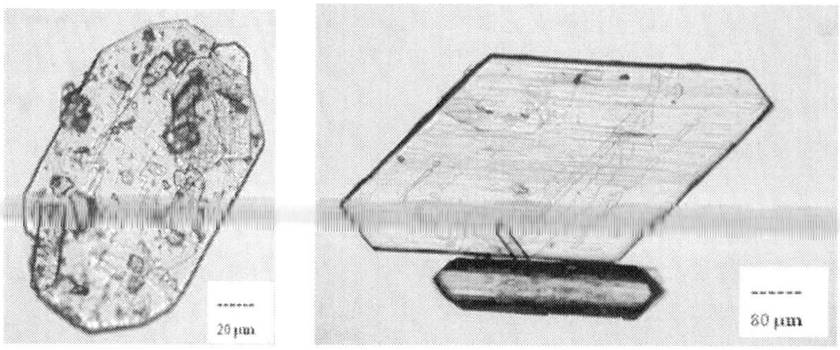

Fig. 14. Photomicrographs illustrating the difference between the seed crystals and the end of experiment crystals.

This experiment was repeated using the same experimental conditions and seeding with crystals (grown from water) under 150 µm. Figure 15 gives the experiment progression diagram of this crystallization. The experiment progression diagram in Figure 15 shows the point of seeding at the point where the turbidity counts (the black line) and the fine (the dark blue line), medium (the pink line) and coarse (the light blue line) particles increase simultaneously. During the cyclic heating and cooling, there is a steady decrease in the fine particle counts (the dark blue line) and a steady increase in the coarse particle counts (the light blue line), both trends indicating growth. The medium particle counts (pink line) remained steady. Figure 16 displays photomicrographs of the seed crystals under 150 µm and the end of experiment crystals. Once again, the crystals grow to form plate-like needles at the end of the experiment.

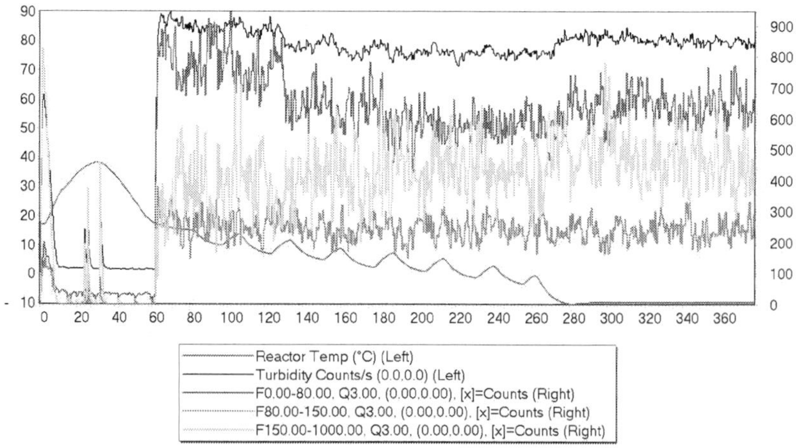

Fig. 15. The experiment progression diagram of a pre-set cyclic heating and cooling crystallization that was seeded with crystals under 150 µm, grown from water (F is the number of particles inside the given size range - e.g., the size range 0.00-80.00 is the content between 0 - 80 µm - and Qz(x) is the less-than-size for parameter x and weight z, where x is counts and z = 3 is interpreted as the volume equivalent distribution).

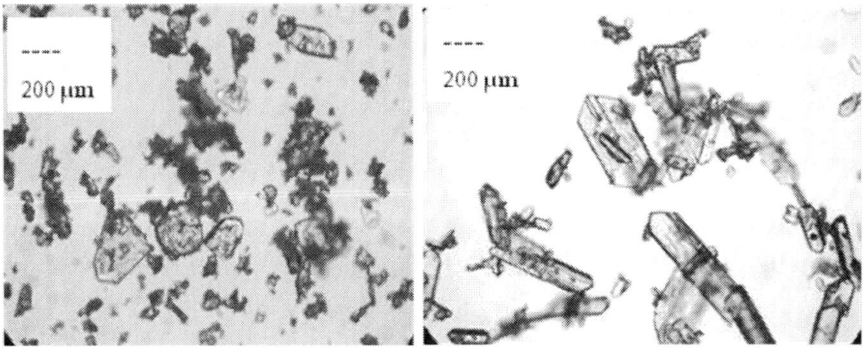

Fig. 16. Photomicrographs of (left) seed crystals grown from water under 150 μm and (right) end of experiment crystals, which are plate-like needles.

Figure 17 is a bar chart of the D10, D50 and D90 particle sizes at various points in the crystallization process, which indicates the growth of the crystal. Figure 18 shows particle size distributions of the crystal population at various points in the phenacetin crystallizations seeded with crystals under 150 μm, with the cumulative distribution in the inset. Both of these distributions show an increase in particle size throughout the experiment, indicating growth.

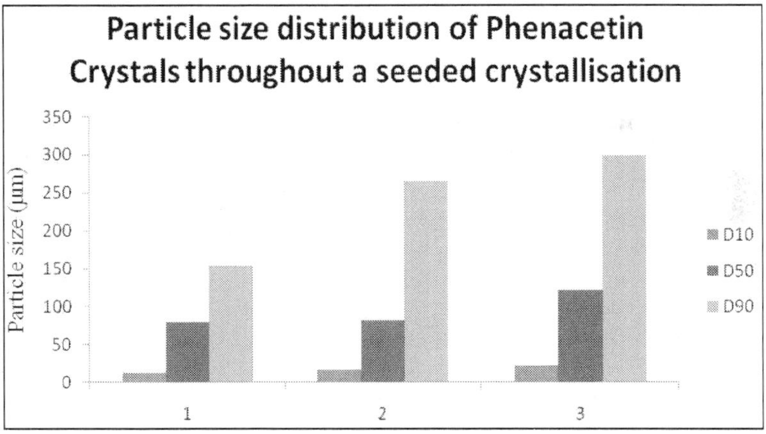

Fig. 17. A bar chart of the D10, D50 and D90 particle sizes of phenacetin crystals throughout the seeded experiments, with the seed crystals grown from water and under 150 μm.

The final experiment in this series used phenacetin crystallizations seeded with particles over 500 μm. Figure 19 is the experiment progression diagram of this crystallization. In this experiment progression diagram, the point of seeding is when the turbidity counts (the black line) and the coarse particle counts (the light blue line) increase simultaneously. Throughout the crystallization, the fine particle counts (the dark blue line) and the medium-sized particle counts (the pink line) pepper the base line, showing control over nucleation

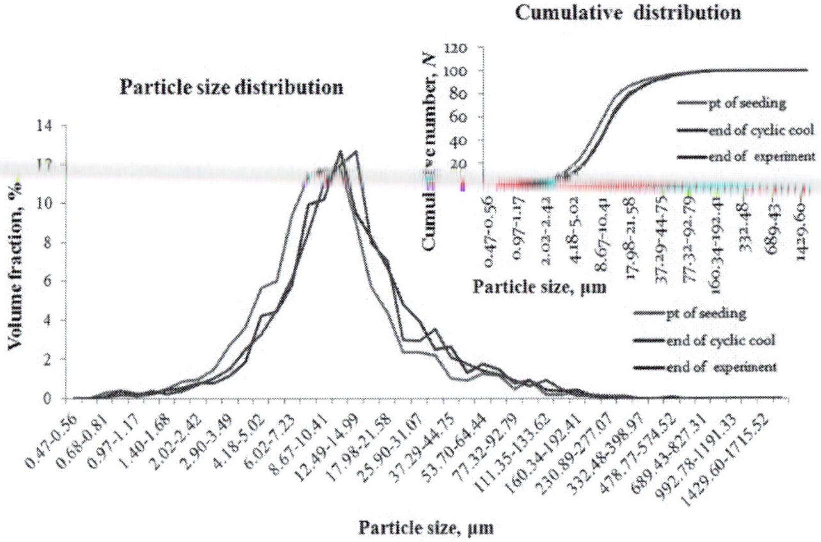

Fig. 18. Particle size distributions taken throughout a seed crystallization seeded with crystals under 150 µm and grown from water.

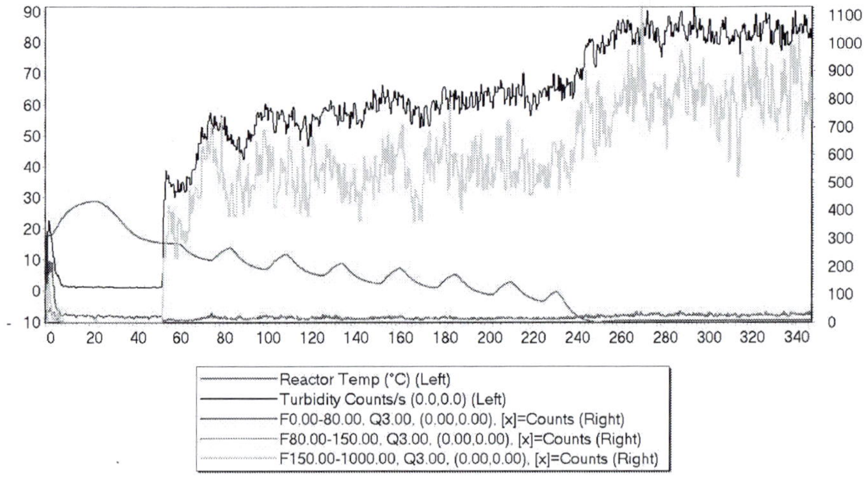

Fig. 19. Experiment progression diagram of a cyclic heating and cooling phenacetin crystallization seeded with crystals over 500 µm, grown from water (F is the number of particles inside the given size range - e.g. the size range 0.00-80.00 is the content between 0 - 80 µm - and Qz(x) is the less-than-size for parameter x and weight z, where x is counts and z = 3 is interpreted as the volume equivalent distribution).

and which is maintained at a very low level. Figure 20 shows photomicrographs of seed crystals and end of experiment crystals showing that little growth has occurred. The plate-like crystals from the water become sharper around the edges, showing the change in shape from the plates to plate-like needles due to the solvent interaction.

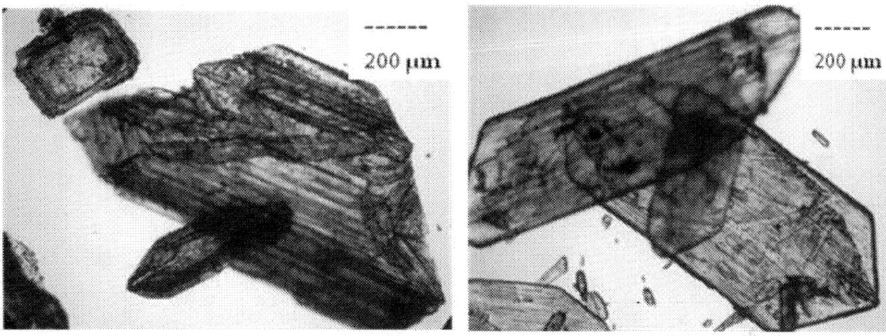

(a) (b)

Fig. 20. Photomicrographs of (a) seed crystals over 500 µm, grown from water and (b) plate-like needles at the end of experiment.

Figure 21 displays particle size distributions of phenacetin crystallizations seeded with crystals over 500 µm, indicating that little growth occurs throughout the crystallization with relatively little change in particle size distribution. The cumulative distribution in the inset also indicates this lack of growth.

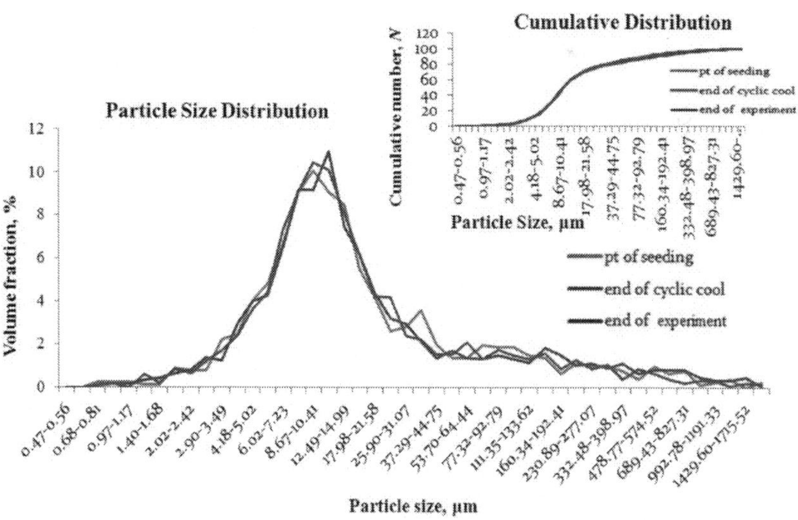

Fig. 21. Particle size distribution of the crystal population at the beginning, after cyclic heat/cool and at the end of the experiment, seeded with crystals over 500 µm (inset is the cumulative distribution at the same points, indicating that little growth is occurring).

The crystal size distributions indicate that little or no growth is taking place in this crystallization, which is not too surprising as the crystals were quite large initially and may have been at their optimum size.

The series of experiments of phenacetin crystallizations using seed crystals grown from water resulted in particle size distributions that were easy to interpret. We now wanted to see how these particle size distributions looked when they were of a crystal population which contained crystals with different dimensions, such as needles. The experiment involved phenacetin crystallizations in ethanol, using seed crystals grown from ethanol. The experiment progression diagram is given in Figure 22.

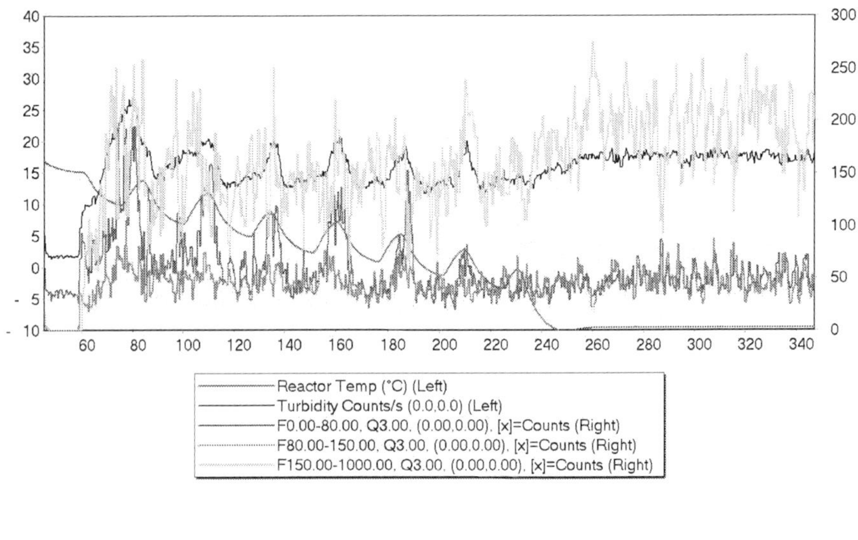

Fig. 22. Experiment progression diagram of phenacetin crystallization seeded with crystals grown from ethanol (F is the number of particles inside the given size range - e.g., the size range 0.00-80.00 is the content between 0 - 80 µm - and Qz(x) is the less-than-size for parameter x and weight z, where x is counts and z = 3 is interpreted as the volume equivalent distribution).

The experiment progression diagram shows a general trend where the fine particle counts (the dark blue line) and medium-sized particle counts (the pink line) maintain a low level and where there is a rise in the coarse particle counts (the light blue line). In Figure 22, the reason for the significant noise in the measurement is likely to be due to the shape of the needles and how they are interpreted as they pass the window of the laser probe. The longest face along the length of a needle may be interpreted as a large particle, while the narrow face at either end of the needle may be interpreted as a fine particle. Thus, every time this needle moves past the window of the laser probe, depending on its orientation, it may be assigned a different particle size leading to significant noise in the measurement. Figure 23 displays photomicrographs of the seed crystals and the end of experiment crystals.

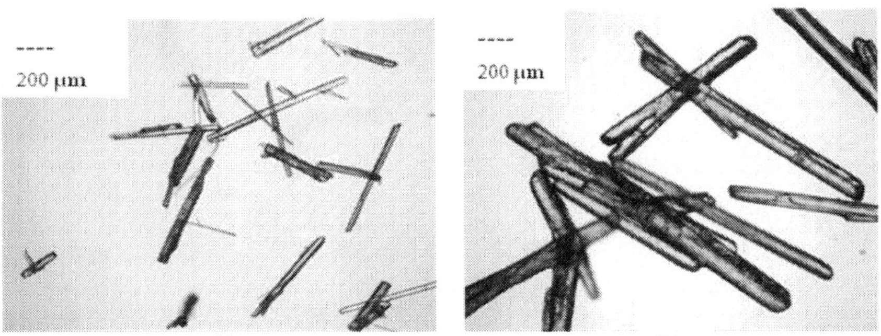

(a) (b)

Fig. 23. Photomicrographs of (a) seed crystals grown from ethanol and (b) end of experiment crystals.

These photomicrographs indicate growth with the crystals maintaining their needle-like shape, as would be expected. Figure 24 shows particle size distributions of the crystal population at various points throughout the crystallization.

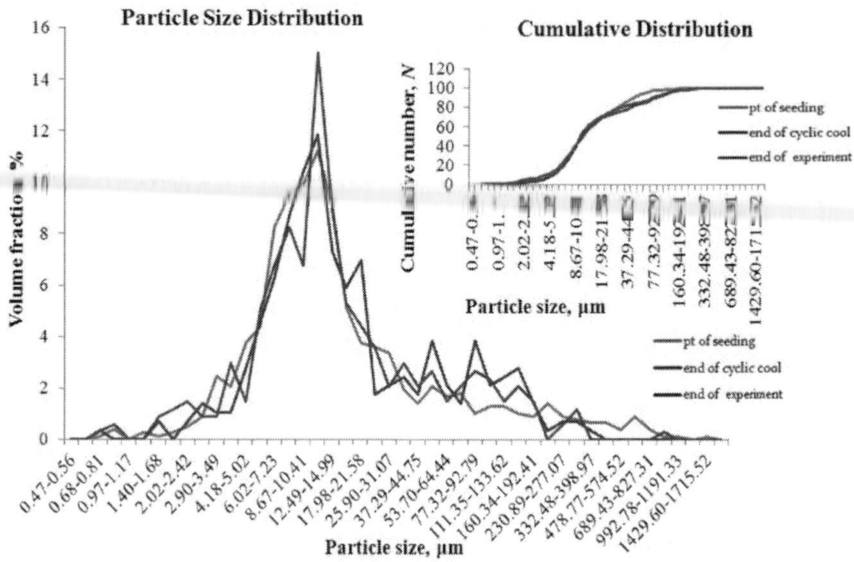

Fig. 24. Particle size distributions of phenacetin crystal populations at various points in a crystallization seeded with crystals grown from ethanol (inset is the cumulative crystallization at the same points).

The particle size distributions indicate growth, as there is an upward shift in particle size throughout the crystallization. These distributions also indicate an environment where there are many differing crystals sizes, as can be seen from the multimodal nature of the trend lines. This is most likely explained by differing orientations of the needle-like crystals as they pass the window of the laser probe. The laser sees the crystal as a two dimensional shape and measures the longest chord length. When the crystal shape is needle-like, the population of crystals passing the sensor can appear as one of the crystals with many varying sizes, depending upon the orientation of the crystals (Barthe, 2008). This is a negative aspect of *in situ* laser probe technology and makes the particle size distributions more difficult to interpret.

The use of seeded batch cooling crystallizations with temperature cycling may be preferable for the control of nucleation and growth, enhancing the size uniformity of the crystals. Cyclic temperature cooling programmes may be preferable to linear temperature cooling programs for the size uniformity of crystals (Lindenberg et al., 2009). Single crystal analysis and microscopy can be used to follow the direction of the growth of seed crystals during a crystallization process.

4. Conclusions

Crystallizations of phenacetin from ethanol were carried out in a 1 L HEL Autolab vessel fitted with a HEL Lasertrack *in situ* particle sizing laser probe. Two forms of modified cooling regimes were examined, each of which was intended to provide a narrower range of crystal sizes and shapes rather than a straightforward linear cool. The first cooling regime involved using turbidity counts as a termination clause for each heating and cooling step. The termination clause was set on the cooling step, so that the cooling was terminated when the turbidity counts exceeded a set number. Once the turbidity counts reached the pre-set level, the experiment began to heat again until the turbidity counts dropped below another pre-set number. Using this approach for crystallization batches that were not seeded proved difficult. The temperature response to this terminating clause was slow, causing large fluctuations in the turbidity counts and the fine and medium total particle counts. The use of turbidity counts as the terminating step for controlling temperature in seeded phenacetin crystallizations resulted in a more controlled experiment, giving more uniform crystals. It was also possible to generate much more prismatic crystals in this way, in contrast to the highly needle-like crystals obtained without seeding.

The second approach used cyclic cooling paths that had predefined temperature parameters, ensuring that a cyclic temperature profile was maintained within the MSZW limits. Each experiment was seeded with crystals that were sieved to give specific size ranges of less than 150 µm, between 150-300 µm and over 500 µm. The use of the seeds in the 150-300 µm range gave relatively uniform elongated plates of good quality. The use of seeds under 150 µm gave rise to elongated plates of reasonable uniformity. Those seeds over 500 µm grew the least during the cooling process and were the least effective in generating uniformity of size and shape, possibly due to these seed crystals' being so large that they had potentially already reached their optimum size. Overall, the pre-programmed cyclic cooling regime was found to be easier to operate and provided good quality crystal batches when used in conjunction with sieved seeds.

5. Acknowledgements

The authors are grateful for financial support from the Solid State Pharmaceutical Cluster (Science Foundation Ireland grant 07/SRC/B1158).

6. References

Abu Baker, M. R.; Nagy, Z. K.; Tan, R. B. H.; Saleemi, A. N. & Rielly, C. D. (2009) The Impact of Direct Nucleation Control on Crystal Size Distribution in Pharmaceutical Crystallization Processes. *Crystal Growth and Design*, 9, 1378-1384

Abu Bakar, M. R.; Nagy, Z. K. & Rielly, C. D. (2009) Seeded Batch Cooling Crystallization with Temperature Cycling for the Control of Size Uniformity and Polymorphic Purity of Sulfathiazole Crystals. *Organic Process Research and Development*, 13, 1343-1356

Barthe, S. C. (2008) Investigation and modeling of the mechanisms involved in batch cooling crystallization and polymorphism through efficient use of the FBRM. PhD Thesis, Georgia Institute of Technology, 119-138

Garside, J.; Gaska, C. & Mullin, J. W. (1972) Crystal Growth Rate Studies with Potassium Sulphate in a Fluidized Bed Crystallizer. *Journal of Crystal Growth*, 13/14, 510-516

Griffiths, H. (1925) Mechanical Crystallization. *J. Soc. Chem. Ind. Trans.*, 44, 7T-18T

Hansen, L. K.; Perlovich, G. L. & Bauer-Brandl, A. (2006) Redetermination of *p*-ethoxyacetanilide (phenacetin). *Acta Crystallographica*, E62, o2712-o2713

Lindenberg, C.; Krattli, M.; Cornel, J.; Mazzotti, M. & Brozio, J. (2009) Design and Optimization of a Combined Cooling/Antisolvent Crystallization Process. *Crystal Growth and Design*, 9, 1124-1136

Prescott, L. F. (1980) Kinetics and metabolism of paracetamol and phenacetin. *British Journal of Clinical Pharmacology*, 10, 291-298

Woo, Xing Yi; Nagy, Z. K.; Tan, R. B. H. & Braatz, R. D. (2009) Adaptive Concentration Control of Cooling and Antisolvent Crystallization with Laser Backscattering Measurement. *Crystal Growth and Design*, 9, 10[illegible]

13

Macromolecular Crystallization Controlled by Colloidal Interactions: The Case of Urate Oxidase

Françoise Bonneté
Institut des Biomolécules Max Mousseron,
Université d'Avignon et des Pays de Vaucluse
France

1. Introduction

Crystallization is a natural or artificial process involving the physical transformation of a fluid or a gas into a regularly organized solid form, the crystal (Fig.1). It occurs in many fields such as health sciences, geosciences, microelectronics, industrial chemical processes.

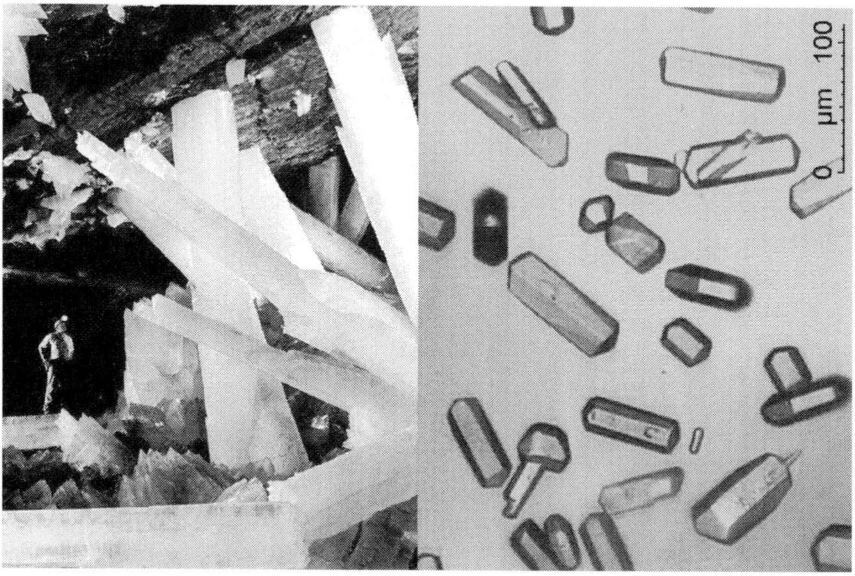

Fig. 1. (Left) Giant Crystal Cave's Mystery Solved (Lovgren, 2007); (Right) Micrometric protein crystals in batch of a therapeutic enzyme, Urate oxidase.

In health sciences, crystals can grow *in vivo* or *in vitro*. *In vivo* this can be due to pathologies (Pande et al., 2001); *in vitro*, crystallization is mainly used to decipher 3D atomic structures of biological macromolecules and understand their structure-function relationship. Originally, crystallization was a method of purification (Sumner, 1926), but now chromatography has replaced protein crystallization in the protein purification process. Today, crystallization is still used for purification and formulation steps in the biotechnology and pharmaceutical industries (Weber et al., 2008). It is a powerful and economical protein purification method, since high-purity proteins can be obtained in a single step operation. However, crystallization remains an empirical process, still based on trial-and-error methods using commercial crystallization cocktails. Crystallization for protein purification is not always practical, since it requires good knowledge of the phase diagram and a substantial quantity of the protein to be crystallized. Although proteins are composed of amino acids of limited types, their structural diversity makes their crystallization conditions difficult to predict. Crystallization conditions include a wide spectrum of parameters such as pH and buffer type, ionic strength, type and concentration of precipitant, temperature, presence and concentration of surfactant molecules and other additives (e.g. cofactors, inhibitors) and, above all, protein concentration. Finding an efficient method to easily crystallize any protein has been, and can be expected to continue to be, a major challenge. Many methods of protein crystallization have been described (Bergfors, 2009). Since the advent of structural genomic programs, a lot of work has been done on automation and miniaturization of crystallization methods in view of the limited amount of protein material available, particularly membrane proteins. Other methods that limit convection effects on crystal growth have shown their efficacy on diffraction quality: crystallization in microgravity (McPherson et al., 1999), in gels (Ng et al., 2003), in microfluidics (Zheng et al., 2005) and other unconventional methods (Sazaki et al., 2004). All these general crystallization methods have recently been thoroughly reviewed (Sauter et al., 2011). But crystallization is not restricted to certain methodologies. Crystallization is a physical process, which consists of two major distinct but inseparable events, nucleation and growth, governed by both thermodynamic and kinetic factors. A pre-nucleation step can be added, creating a supersaturation state. For a long time it was thought that crystallization obeys no comprehensive theory. In the 80's-90's, rational and physical approaches were developed to understand the fundamentals of nucleation and crystal growth of proteins. The concepts of nucleation and crystal growth were described (Feher & Kam, 1985) and recently reviewed (Chernov, 2003; García-Ruiz, 2003). While growth of macromolecular crystals is well characterized, in particular by direct visualization using atomic force microscopy, there are fewer studies of nucleation, since it requires high supersaturation, that is to say often high concentrations of proteins. Some nucleation studies were performed with a model protein such as lysozyme by dynamic light scattering (Mikol et al., 1989) or small angle X ray (Finet et al., 1998) or neutron scattering (Boué et al., 1993) in order to understand the prenucleation and nucleation steps. This was the beginning of the development of rational approaches based on an understanding of the physical properties of macromolecular solutions. Numerous articles and reviews examined the correlation between crystallization and interactions between macromolecules in solution (Ducruix et al., 1996; Muschol & Rosenberger, 1995), applicable to soluble proteins like membrane proteins (Hitscherich et al., 2000). The measurement of the second virial coefficient, noted indifferently A_2, B_2 or B_{22} depending on the authors, appeared to be a powerful tool for predicting crystallization conditions (George & Wilson, 1994). George & Wilson showed that a restrictive range of values of the second virial coefficient was favorable to crystallization of soluble proteins. This "crystallization slot", which is about [-1x10^{-4}; -8x10^{-4} mol.mL.g^{-2}], corresponds to slight or moderate attractions. Second virial coefficients can be measured using different experimental techniques, including osmotic pressure (Haynes et al., 1992), static light scattering (SLS) (Velev et al., 1998), ultracentrifugation (Behlke & Ristau, 1999), small-angle X-ray scattering (SAXS) (Bonneté et al., 1999), small-angle neutron scattering (SANS) (Gripon et al., 1997), size-exclusion chromatography (Bloustine et al., 2003) and self-interaction chromatography (SIC) (Tessier et al., 2002). Studying different biological macromolecules (protein, virus) of various sizes, molecular masses or net charge (Bonneté & Vivares, 2002), we showed that a dimensionless second virial coefficient normalized to protein excluded volume gave a better representation of the effective pair potential between macromolecules (McQuarrie, 2000) when comparing the effects of physico-chemical parameters leading to crystallization. We found a new dimensionless crystallization slot [-10; 3] in which the Brome mosaic virus (BMV) (Casselyn et al., 2001) was found to crystallize with a slightly positive second virial coefficient a_2 = +2.8. This a_2 value was not inconsistent with attractive interactions since the dimensionless second virial coefficient of a pure hard-sphere potential is equal to 4. Indeed the effective pair potential of biomolecules in solution is the sum of different components including, in particular, hard sphere repulsion, van der Waals attraction, electrostatic effects and depletion attraction. Phase diagrams of biological macromolecules are therefore governed by an appropriate combination of these interaction potentials in solution. Repulsive regimes favor solubility, whereas the presence of attractive potentials may induce a variety of phase transitions, including the desired macromolecular crystallization. These pair potentials are medium range, from a few Å up to tenths of Å, and each of them is under separate control of physico-chemical conditions such as pH, temperature and solvent composition (Hansen & McDonald, 1976; Israelachvili, 1994). The desired fine tuning of the interactions for controlled crystallization requires identifying the individual contributions. In order to do so, theoretical simulations were performed on series of SAXS scattered intensities recorded on hen egg white lysozyme as a function of pH, salt concentration, salt type and temperature, on alpha crystallins (Finet & Tardieu, 2001) and urate oxidase (Vivares et al., 2002) as a function of polyethylene glycol (PEG). In this chapter we will show, using the example of urate oxidase, how to choose physico-chemical conditions, how they control the different interaction forces and how these forces can be modulated to design different phase diagrams for different applications (biocrystallography, or pharmaceutical processes for example).

2. Theoretical and experimental background

It is common practice in the colloid field to calculate phase diagrams from interaction potentials in solution (Asherie et al., 1996). The forces between colloids in solution include hard sphere repulsion plus attraction, whose range may modify the appearance of the phase diagram (Fig.2). In a colloid-polymer mixture, the presence of stable or metastable phase transitions can be determined by the ratio of the attraction range and the colloid size (Lekkerkerker, 1997), the range of the attractive interaction between colloids in solution being controlled by the size of the polymer (Poon, 2002). Predicting the phase diagrams of proteins based on knowledge of interaction forces may therefore be a more effective alternative for controlled crystallization than trial-and-error screenings.

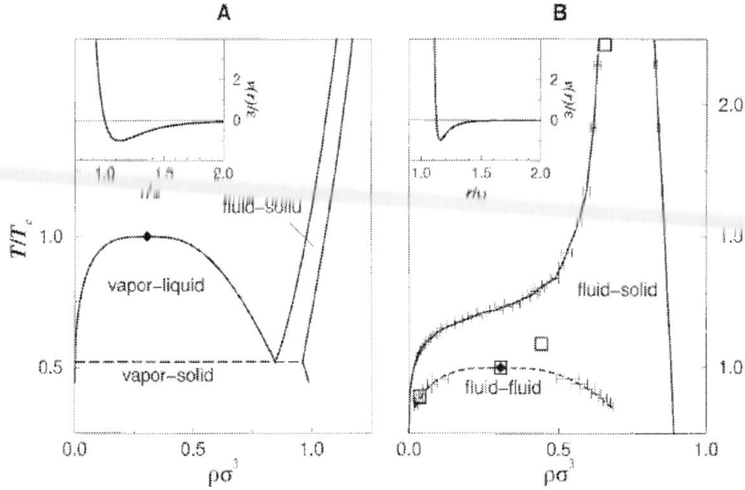

Fig. 2. Typical phase diagram of colloids A) with long-range attraction; B) with short- range attraction (ten Wolde & Frenkel, 1997)

Thus, a thorough understanding of the interaction potentials that govern protein phase diagrams will allow better control of crystal growth. The good news is that this can be applied successfully to biomacromolecule crystallization. The phase diagram of biomacromolecules, usually represented by the solubility curve, is governed by an appropriate combination of interaction potentials (Tardieu et al., 1999). Repulsion favors solubility, whereas attractions induced either by salt or polymer addition favor phase transitions, including crystallization. Ideally, to grow crystals, interaction potentials would be calculated in solution from knowledge of the macromolecule's characteristics, i.e. sequence, molecular mass, charge and isoelectric point as a function of physico-chemical environment (pH, temperature, ionic strength, etc.), leading to its phase diagram. Alas, this is not yet possible. However, using knowledge of the physico-chemical and biochemical characteristics of macromolecules in solution, we can choose relevant parameters to modulate potential forces and control nucleation rate and crystal growth. While it is not yet possible to calculate interaction potentials from macromolecule characteristics, it is possible experimentally to measure the resulting interactions in solution, either repulsive or attractive, through second virial coefficient (A_2) measurements, and then to simulate the underlying pair potentials. A_2 can be measured from the macromolecule concentration dependence of different experimental measurements, such as osmotic pressure, analytical ultracentrifugation, self-interaction chromatography, static light scattering, small angle X-ray or neutron scattering (SAXS or SANS). We used SAXS because it offers additional advantages. By combining SAXS measurements and numerical simulations, we can also analyze forces present in macromolecule solutions, that is to say the different repulsive and attractive components, and study them as a function of common physico-chemical parameters. This type of approach has already been used in the case of colloid-polymer mixtures (Lutterbach et al., 1999a; Lutterbach et al., 1999b; Ye et al., 1996) or biopolymers (Malfois et al., 1996; Vérétout et al., 1989).

2.1 Small Angle X-ray scattering and numerical calculation

To analyze the interaction potentials that control phase diagrams, we combined SAXS and numerical simulations. The total normalized intensity I(c,q), scattered by a solution of monodisperse spherical particles at a scattering angle 2θ, can be expressed as a function of the particle concentration c and the modulus of the scattering vector q, $q = 4\pi\lambda^{-1}\sin\theta$ (note that $q = 2\pi s^{*}$) by:

$$I(c,q) = I(0,q) \times S(c,q) \quad (1)$$

I(0,q) is the intensity scattered by one particle and is usually called the particle form factor. Experimentally, the form factor is generally obtained from curves recorded at low concentrations to avoid interaction effects. The form factor gives information on the particle shape and its oligomeric state. At low angles, the form factor of an ideal solution (c→0, without interaction) can be written:

$$I(c \to 0,q) = I(c \to 0,0).\exp\left(-\frac{R_g^2}{3}q^2\right) \quad (2)$$

Similarly, for non-ideal solutions, the low angle part of the intensity curves recorded as a function of c is written:

$$I(c,q) = I(c,0).\exp\left(-\frac{R_g^2}{3}q^2\right) \quad (3)$$

Therefore a « Guinier plot », i.e. a Ln I(q) plot versus q^2 (Guinier & Fournet, 1955), provides, with intensity at the origin, I(0,0) or I(c,0), and with the structure factor at the origin, S(c,0), since S(c,0)=I(c,0)/I(0,0). Indeed, with interacting spherical particles, departure from ideality can be accounted for by the interference term, S(c,q), usually called the solution structure factor. The value of S(c,q) at zero-q angle gives information on the nature of interactions between particles. With repulsive interactions, the particles are evenly distributed and S(c,0) is lower than 1 (Example Fig. 3A). With attractive interactions, fluctuations in particle distribution are observed and S(c,0) is larger than 1. The nature of the net interactions, either attractive or repulsive, can easily be determined by the plot of the structure factor at the origin, S(c,0), as a function of particle concentration, since it is related to the osmotic with pressure Π by:

$$S(c,0) = \frac{RT}{M}\left(\frac{\partial \Pi}{\partial c}\right)^{-1} \quad (4)$$

With

$$\frac{\Pi}{cRT} = \frac{1}{M} + A_2 c + A_3 c^2 + \ldots \quad (5)$$

the concentration c being expressed in $g.cm^{-3}$.

---

*Note: Depending on SAXS beamlines used for experiments, the scattered intensity was expressed either as a function of $q = 4\pi\lambda^{-1}\sin\theta$ or $s = 2\lambda^{-1}\sin\theta$ in $Å^{-1}$ or nm^{-1}

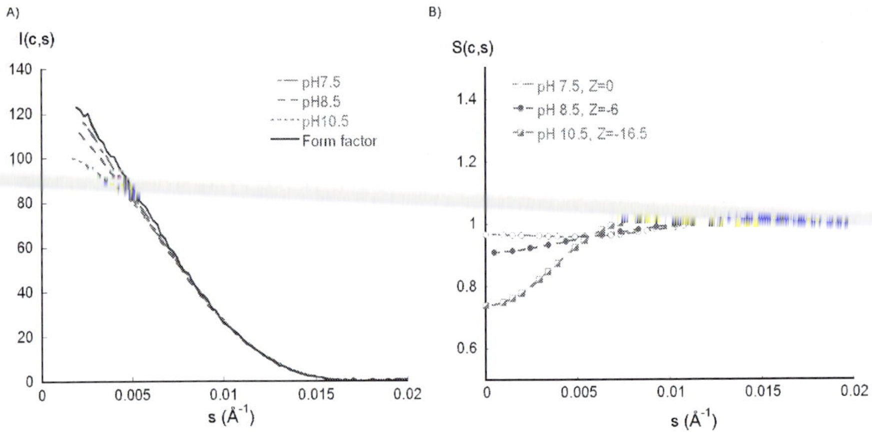

Fig. 3. A) Scattering intensities of urate oxidase as a function of pH. B) Experimental (dots) and fitted (lines) repulsive structure factor of urate oxidase at different pH corresponding to repulsive interactions, S(c,0) < 1.

Therefore, the second virial coefficient can be obtained by the expression:

$$S(c,0) = \frac{I(c,0)}{I(0,0)} = \frac{1}{1 + 2.M.A_2.c + ...} \tag{6}$$

Experimentally the structure factor can be obtained by intensity extrapolation from the expression:

$$S(c,0) = \frac{\lim\limits_{q \to 0} I(c,q)}{\lim\limits_{c \to 0} I(c,0)} \tag{7}$$

The slope of the linear fit gives the coefficient A_2 in mol.ml.g^{-2}.

$$S(c,0) \approx 1 - 2.M.A_2.c \tag{8}$$

If A_2 is positive, the overall interactions are repulsive (Example Fig. 3B); if A_2 is negative, the interactions are attractive.

The solution can be described mathematically as the convolution product of a particle shape and a particle distribution. *S(c,q)* is the Fourier transform of the spherically averaged auto-correlation function *g*(*r*) of the particle distribution or pair-distribution function:

$$S(c,q) = 1 + \rho \int 4\pi r^2 (g(r) - 1) \frac{\sin rq}{rq} dr \tag{9}$$

where $\rho = cN_a/M$ is the number density of particles, i.e. the number of particles per unit volume and *c* is the particle concentration in g.cm^{-3}. Models and numerical methods based on statistical mechanics are extended to proteins in solution (Lomakin et al., 1996; Malfois, et al., 1996). Calculation of structure factors is based on models of pair particle potential $U(r)$, from which a particle distribution $g(r)$ at equilibrium is inferred. The calculation of $g(r)$ is based on the Ornstein-Zernike (OZ) equation and on the hypernetted chain (HNC) integral equation and uses an iterative method (Belloni, 1985; 1988). For one-component fluids, the OZ relationship between total, $h(r) = g(r) - 1$, and direct, $c(r)$, correlation functions is written in Fourier space (where FT indicates a Fourier transform, normalized by the density ρ):

$$S(c, q) = 1 + \mathrm{FT}h(r) = 1/(1/\mathrm{FT} - c(r)) \tag{10}$$

If we introduce an auxiliary function $\gamma(r) = h(r) - c(r)$, the integral HNC equation is:

$$g(r) = \exp[-\beta U(r) + \gamma(r)] \tag{11}$$

where $U(r)$ is the interaction pair potential. To numerically solve the OZ equation with the above closure relation, iterations are used. The structure factor $S(c,q)$, and/or macroscopic properties and thermodynamic variables, are then calculated from the Fourier transform and integrals of pair distribution function $g(r)$ (Hansen & McDonald, 1976). The calculated structure factor is then compared to the experimental structure factor as shown in an example Fig. 3B.

For the simplest "one component" model, only the interaction pair potentials between macromolecules, which interact through solvent and ions, are explicitly considered in the numerical simulations. Since in such calculations the exact potential shape is not critical, we describe the potentials, either attractive or repulsive, in the mathematical form of a Yukawa potential, which is a function of three parameters, hard sphere diameter, σ, depth (strength), J, and range, d, according to:

$$U(r)/k_BT = J\,(\sigma/r)\,\exp[-(r-\sigma)/d] \tag{12}$$

In the case of binary mixtures of macromolecules (mac) and polymers (pol) i.e. of "two component" systems, the total scattered intensity, $I(c_ic_j,q)$, can be expressed as a function of component concentrations c_i by (Belloni, 1991):

$$\begin{aligned} I\left(c_ic_j,q\right) &= \Sigma_i\Sigma_j \sqrt{c_ic_j}\sqrt{I_i(0,q)I_j(0,q)}S_{ij}(q) \\ &= I_{mac}\left(c_{mac},q\right) + I_{mac-pol}\left(c_{mac},c_{pol},q\right) + I_{pol}\left(c_{pol},q\right) \end{aligned} \tag{13}$$

Because of the presence of a cross term, only SAXS intensities and not structure factors can be compared. The polymer form factor $I_{pol}(0,q)$ can be taken as equal to the Debye form factor, which is valid for a Gaussian coil and has already been used successfully with PEG (Debye, 1946):

$$I_{pol}(0,q) = I_{pol}(0,0)\,(2/x^2)\,(\exp(-x) - 1 + x) \tag{14}$$

where $x = q^2R_g^2$, and R_g is the polymer radius of gyration. Partial structure factors $S_{ij}(q)$ are related to the Fourier transform of partial pair distribution function $g_{ij}(r)$:

$$S_{ij}(q) = \delta_{ij} + \sqrt{c_ic_j}\int (g_{ij}(r) - 1)\exp(irq)\,dr \tag{15}$$

where i,j = 1,2, δ_{ij} is the Kronecker symbol (δ_{ij}=0 when i≠j and δ_{ij}=1 when i=j) and r the interparticle distance. To determine the pair distribution function, the usual procedure, as for one component systems, is to use the Ornstein-Zernike equation to link the total, $h_{ij}(r) = g_{ij}(r) - 1$, and the direct, correlation $c_{ij}(r)$ functions. The closure equation is once again the HNC equation. From a set of the three direct potentials, $U_{pol\text{-}pol}(r)$, $U_{mac\text{-}pol}(r)$ and $U_{mac\text{-}mac}(r)$, the theoretical scattered intensity from the binary mixture can be calculated and compared to the experimental scattering curve (Vivares, et al., 2002).

2.2 Direct pair potentials

2.2.1 The DLVO model

The direct protein-protein potential $U_{uox\text{-}uox}(r)$ was chosen equal to the DLVO potential. We had already successfully applied the DLVO model to different proteins at low-salt concentrations (Malfois, et al., 1996; Tardieu, et al., 1999). The DLVO potential is the sum of three potentials: a hard-sphere potential, an electrostatic repulsion and a van der Waals attraction. The hard-sphere potential reflects the fact that proteins cannot interpenetrate, the repulsive coulombic potential is due to the fact that each protein holds the same net charge and the van der Waals attractive potential is the resulting dispersion interaction between proteins. For the sake of simplicity, we chose a Yukawa shape for the coulombic and the van der Waals potentials, a Yukawa shape van der Waals potential having been shown to be sufficient to describe the attractive protein-protein interaction in aqueous solutions.

Respective mathematical expressions of the three potentials are described:

- Hard-sphere potential:

$$U^{HS}_{UOX-UOX}(r) = +\infty \quad r \le \sigma \\ \qquad = 0 \quad r > \sigma \tag{16}$$

with σ the protein diameter

- Repulsive coulombic potential:

$$U^{coul}_{UOX-UOX}(r) = Z^2 L_B / [\sigma(1+0.5\sigma/\lambda_D)^2].(\sigma/r).\exp(-(r-\sigma)/\lambda_D) \quad r > \sigma \tag{17}$$

with Z the effective protein charge, L_B the Bjerrum length (equal to $e^2/(4\pi\varepsilon_0\varepsilon_s k_B T) = 7.31\text{Å}$ at T = 293.15K with $\varepsilon_s = \varepsilon_{H2O} = 80$) and λ_D the Debye length ($\lambda_D(\text{Å}) = 3/\sqrt{I}$ at 293.15 K where I is the ionic strength expressed in mol/l). The potential is expressed in k_BT units.

Repulsive interactions are expected to vary with pH and to be screened with ionic strength (addition of alcohol, which reduces the water dielectric constant, can also reduce the repulsion).

- van der Waals potential:

$$U^{VdW}_{UOX-UOX}(r) = -J_{VdW}.(\sigma/r).\exp(-(r-\sigma)/d) \quad r > \sigma \tag{18}$$

with J_{VdW} (in k_BT units) and d (in Å), respectively, the depth and the range of the potential.

In practice, the van der Waals component is determined at or close to the pI. With small compact proteins, the best fit parameter for depth with a 3Å range was indeed found to be close to the calculated values, 2-3 k_BT (Tardieu, et al., 1999) and the attraction was found to increase with decreasing temperature.

In the DLVO potential, hard-sphere and electrostatic interactions have a repulsive effect which favors solubility. Except possibly at pI, the van der Waals forces are weaker than the coulombic interactions (or even disappear). However, it is clear that the basic interactions considered in the DLVO potential model are generally unable to provide the attraction necessary for macromolecular nucleation and crystal growth. Fortunately, other forces can play a role.

2.2.2 The Hofmeister effect

Salt has long been known to act as a crystallizing agent (Arakawa & Timasheff, 1985). A number of phase diagrams were measured, e.g. by the Ducruix group (Carbonnaux et al., 1995; Guilloteau et al., 1992) and showed that solubility varies with the type of monovalent salt, following the direct/reverse order of the Hofmeister series according to whether the particles are studied at a pH higher/lower than the pI, respectively.

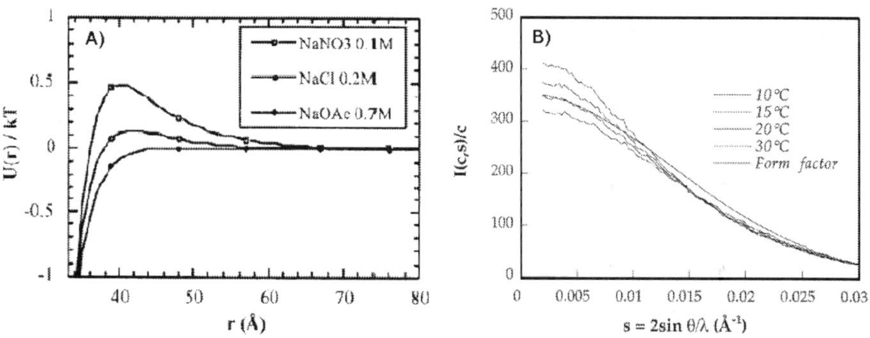

Fig. 4. A) Interaction potential of lysozyme in different salts. B) Normalized scattering intensity of Lysozyme in NaCl 200 mM pH 4.5 at different temperatures (curves in the same order as in legend for both figures).

When the effect of monovalent salts on protein interactions in solution was analyzed (Muschol & Rosenberger 1995, Tardieu *et al.*, 1999), whatever the particle size , at medium ionic strength (> 0.2 M), monovalent anions were observed not only to screen the charges, but to induce an additional attraction, specific to salt type (Finet et al., 2004). This attraction is short-range, about 3Å, and increases with decreasing temperature.

2.2.3 Depletion attraction

The addition of neutral non-interacting polymers to colloidal solutions has long been known to induce a depletion attraction (Asakura & Oosawa, 1954). This depletion interaction can be explained in a simplified manner for an ideal polymer solution.

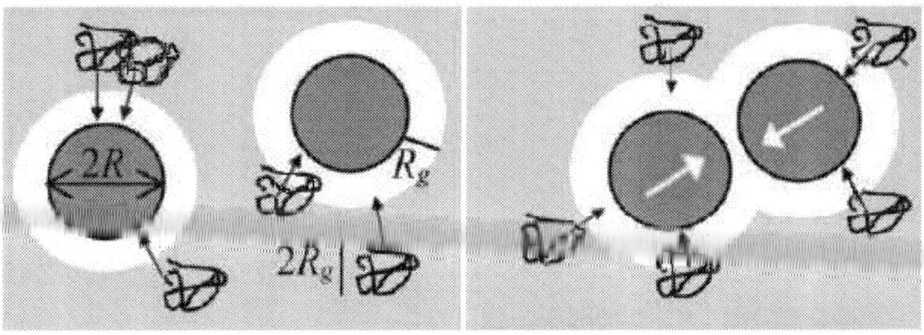

Fig. 5. Schematic illustration of the depletion effect in a colloid-polymer mixture

Molecules of polymer - characterized by their radius of gyration R_g - and colloids cannot mutually interpenetrate, and furthermore, the center of polymers is excluded from a region of thickness R_g around each colloidal particle. This excluded volume is called the depletion zone. When two colloid particles come sufficiently close to each other, their depletion zones overlap and the free volume accessible to the polymer molecules increases, leading to a gain in entropy of the system. Thermodynamically, it is therefore more favorable for the polymer when colloidal particles approach each other, i.e. when there is an attractive interaction between them. This model remains valid as long as polymer molecules do not overlap. To describe the direct polymer-polymer potential $U_{pol\text{-}pol}$ (r), we used an approach where the effective potential is finite for all distances between two polymer molecule centers of mass. Molecules of polymer were therefore considered as "soft colloids". The mathematical form chosen for the polymer-polymer potential was a Gaussian form (Bolhuis et al., 2001):

$$U_{pol-pol}(r) = J_{pol}.\exp(-(r / R_{pol})^2) \quad (19)$$

Where J_{pol} (in k_BT units) and R_{pol} (in Å) are the prefactor parameter and the range of the Gaussian potential respectively.

For the direct protein-polymer potential $U_{pol\text{-}col}(r)$, we chose a Yukawa form, which depends on only two parameters:

$$\begin{aligned} U_{pol-col}(r) &= +\infty \quad r \le \sigma / 2 \\ &= J_{pol-col}.\frac{\sigma}{2r}.\exp(-(r-\sigma/2)/d_{pol-col}) \quad r > \sigma / 2 \end{aligned} \quad (20)$$

with $J_{pol\text{-}col}$ (in k_BT units) and $d_{pol\text{-}col}$ (in Å) the intensity and the range of the potential respectively.

Thus, the numerical simulations provide us with an effective macromolecule-macromolecule potential, $U^{eff}_{col\text{-}col}(r)$, which may be written as the sum of the macromolecular interaction potential in the absence of polymer, $U_{uox\text{-}uox}(r)$, and of the depletion term, $U_{depletion}$ (r):

$$U^{eff}_{col-col}(r) = U_{col-col}(r) + U_{depletion}(r) \quad (21)$$

The depletion potential is therefore obtained from equation 21. From numerical simulations performed on the binary mixture, urate oxidase-polyethylene glycol (Vivares, et al., 2002), we found that the depth and the range of the effective potential increases with the addition of polymer.

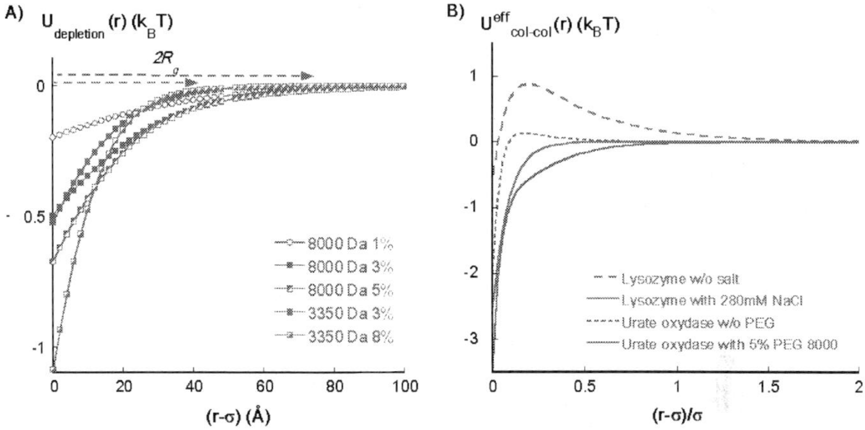

Fig. 6. A) Resulting depletion potential at different PEG concentrations for PEG 8000Da and PEG 3350Da; B) Comparison of the attractive potential induced by salt in lysozyme solutions and by PEG in urate oxidase solutions normalized to the particle diameter.

While depletion potential depth increases with polymer concentration whatever the polymer size, depletion range is approximately equal to $2R_g$ for each polymer, whatever the polymer concentration (Fig. 6A). Crystallization in polyethylene glycol is thus induced by a medium-range attractive potential between macromolecules, in addition to the short-range van der Waals. This result appears in contrast to that observed with small proteins like lysozyme, where crystallization was found to be induced by short-range attraction. The attractive potential range necessary for protein crystallization increases with the size of the macromolecule. In fact, it is interesting to note that, when normalized to the particle diameter, the salt- and PEG-induced interaction potentials that promote the crystallization of lysozyme (14300 Da) and urate oxidase (135000 Da) respectively are quite similar (Fig. 6B). Finally, since salt induces a short-range attraction and since polymer-induced potential range varies both with polymer size and with polymer concentration, by choosing salt in the Hofmeister series as well as polymer concentration and size it is possible to modify the phase diagram and thus control protein crystallization.

Nucleation and crystal growth are inseparable events, both of which depend on supersaturation. The nucleation step controls the structure of the crystalline phase and the number of crystals. Recently, it has been shown that interaction potentials play a key role in the determination of nucleation kinetic parameters (Bhamidi et al., 2005). Depending on the application (biocrystallography, powder diffraction, purification, for example), it appears important to control the nucleation step, supersaturation and, therefore, solubility.

We have seen that, whatever the salt or polymer used, macromolecular crystallization occurs in the presence of an attractive pair potential. We have analyzed how attraction and repulsion can be varied by changes in the environment: pH, ionic strength, type of salt, polymers, and temperature (Tardieu, et al., 1999; Vivares, et al., 2002). With low molecular weight proteins, a coulombic, pH-dependent, repulsive potential plus a short-range, possibly van der Waals, attraction are sufficient to account for the behavior observed at low ionic strength. At higher ionic strengths, salt-specific effects following the (direct or reverse) order of the Hofmeister series correspond to an additional short-range salt-specific attraction. With increasing protein size, the van der Waals contribution becomes negligible. Adding polymers like polyethylene glycol (PEG) induces a depletion attraction. The way forward in growing crystals is therefore a fairly simple extension of these observed effects. While this path is not guaranteed to lead to success, it may well enable us to reduce the time and effort spent on trial-and-error methods.

3. A protein of pharmaceutical interest: Urate oxidase

Lysozyme has long been the model protein most often chosen to study nucleation and crystal growth mechanisms (Drenth, 2005; Gavira & Garcia-Ruiz, 2009; Liu et al., 2010; Vekilov, 2010), since it is easily available at low cost. We used urate oxidase from *Aspergillus flavus* as a new model system to explore the crystallogenesis of proteins in general and the crystallographic structure of urate oxidase in particular. Urate oxidase is used as a protein drug to reduce toxic uric acid accumulation and to treat the hyperuricemic disorders occurring during chemotherapy. Urate oxidase *from Aspergillus flavus* is produced, purified and made commercially available by Sanofi-Aventis (France). Urate oxidase is normally purified using multiple steps of concentration and chromatography (McGrath & Walsh, 2005). Like other proteins (Jacobsen et al., 1998), it could, however, be purified by crystallization (Giffard et al., 2011). Crystallization has the inherent advantages of providing higher final purity yields, not denaturing the protein of interest and often providing some stabilization effects, but it requires a good knowledge of the phase diagram and a substantial amount of the protein to be crystallized. The structure of the urate oxidase from *A. flavus* has been solved in the absence (Retailleau et al., 2004) as well as in the presence of different inhibitors (Retailleau et al., 2005). It is a homotetrameric enzyme of 135kDa with a subunit consisting of 301 amino acids. Although different urate oxidase structures have been determined, its catalytic mechanism is still poorly understood. One possible mechanism for the oxidation of uric acid could be revealed by the precise protonation state of the substrate during the reaction. Neutron crystallography can provide such information and make possible direct determination of the protonation states of the active site residues and substrate analogues (Oksanen et al., 2009). For this purpose, it is necessary to grow large, well ordered, deuterated crystals; here too, therefore, a good knowledge of the phase diagram is required. Finally, whether crystallization is aimed at urate oxidase purification or at growing highly diffracting crystals, a good knowledge of the phase diagram and thus of crystallization conditions are required. We used urate oxidase, then, as a model system both from a fundamental and an applied point of view, in suitable biological and physicochemical conditions, to characterize, simulate and modulate pair potentials present in solution and design appropriate phase diagrams for specific applications. We will show in particular, how macromolecule properties - such as sequence, molecular mass, isoelectric point, stability - can provide the parameters to be screened to obtain adequate phase diagrams.

4. Urate oxidase interactions in solution and implications for crystallization

The stability of urate oxidase from *Aspergillus flavus* was studied by differential scanning calorimetry (DSC) (Bayol et al., 1995) in conjunction with enzymatic activity measurement and size exclusion chromatography. The recombinant urate oxidase is not stable below pH 6 and shows maximum stability between pH 7.25 and 9.5 up to 35 °C. Studies of crystallization were therefore performed at a pH above 7.2 and a temperature below 35 ° C. pH 8 was identified as the value around which the enzyme stabilization activity is optimized (Aleman et al., 1998). The buffer selected for use is a sodium phosphate buffer at pH 8 at a concentration of between 5 mM and 100 mM. It was in this range of urate oxidase stability that different parameters (pH, nature of buffer, temperature, ionic strength, additives) were studied for their effects on interaction forces and their influence on phase diagrams.

4.1 pH and Salt effects

Urate oxidase is a large tetrameric enzyme of 135 kDa stable at pH above 7. Its theoretical isoelectric point (pI), calculated from its primary structure and the pKa of each charged amino acid, is 7.5 (Fig 7), consistent with IEF experiments.

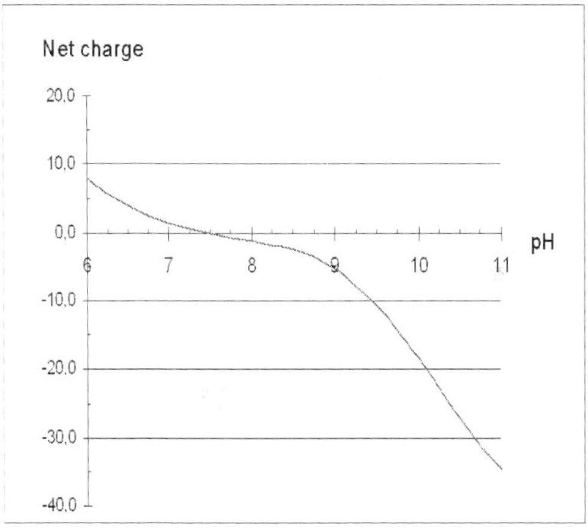

Fig. 7. Urate oxidase net charge as a function of pH

By definition, at pI without any other crystallizing agents, the protein net charge is zero. Thus, there is no electrostatic repulsion. The van der Waals attraction may therefore induce instability and aggregation of proteins. Usually protein solubility is minimum at pI (Riès-Kautt & Ducruix, 1999). At pH above pI, the urate oxidase negative net charge increases as pH increases. We therefore expect urate oxidase electrostatic repulsions to increase as the net charge increases.

A compilation of second virial coefficient (A_2) values of urate oxidase as a function of pH, determined from previous SAXS experiments (Giffard et al., 2008; Vivares & Bonneté, 2002) (Fig. 8A), shows that A_2 is positive whatever the pH above pI, as expected for repulsive interactions, and increases as both pH and protein net charge increase. However, as pH approaches pI, A_2 remains positive, which is not consistent with pure van der Waals attraction at pI. This may be due to residual charges on the protein. In contrast, for a negative second virial coefficient measured in Tris buffer pH 8 close to pI without addition of salt (Fig. 8B), the second virial coefficient increases, becoming positive as salt concentration increases.

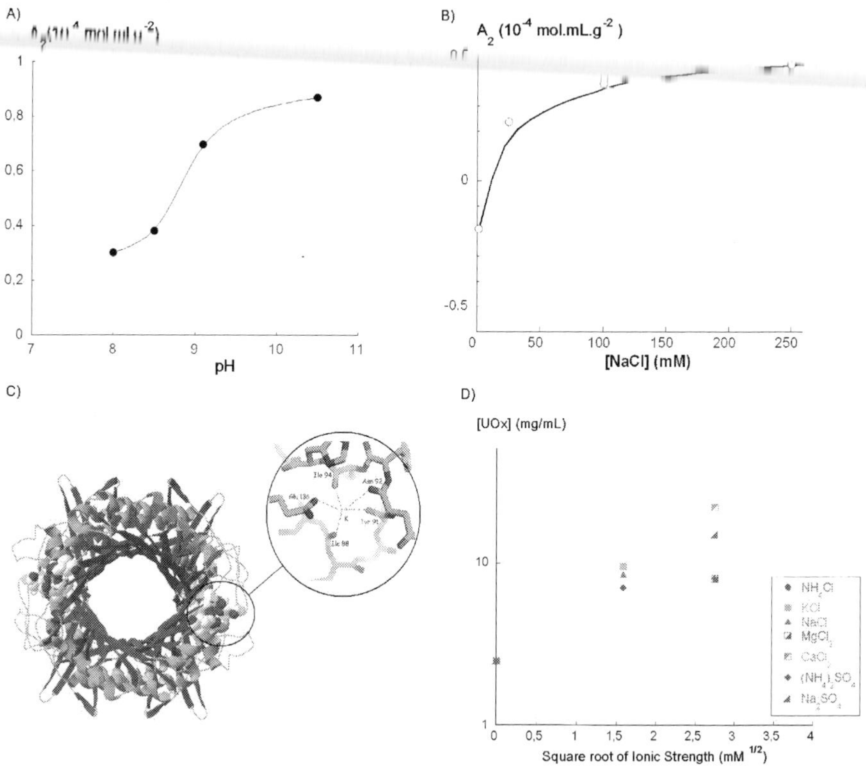

Fig. 8. A) Variation of second virial coefficient (A_2) as a function of pH at 20 °C: pH 8.0 with sodium phosphate buffer; pH 9.1 sodium borate buffer; pH 10.5 sodium carbonate buffer. B) Variation of A_2 in Tris buffer pH with addition of sodium chloride. C) Cationic binding site on the urate oxidase surface (courtesy of G. Marassio). D) Solubility of urate oxidase as a function of different salt concentrations.

This surprising result suggests that the addition of sodium salt, in particular the addition of sodium cation, induces an increase in repulsion ($A_2 > 0$). Most articles report that addition of salt decreases solubility, which is correlated with the fact that salt induces attractive interactions through charge screening. This effect, known as salting-out (Arakawa & Timasheff, 1985), is generally observed at medium and high salt concentrations. At low ionic strength, the opposite effect is expected, i.e. salting-in, where solubility increases with addition of salt. However, salting-in has only sporadically been reported (Faber et al., 2007). Our result is explained by X-ray crystallography (Fig. 8C), which shows that the salting-in effect observed with urate oxidase at pH 8 in Tris buffer results from the direct binding of cations to specific sites on the surface of the protein. A similar effect was recently reported (Gibb & Gibb, 2011), which suggests that salting-in is induced by ion binding. This salting-in effect is not specific to sodium cation, since it was observed with different cations such as K^+, NH_4^+, Mg^{++}, Ca^{++} (Fig. 8D). In all cases, the addition of salt induces positive A_2 values, an indicator of repulsive interactions with rasburicase, leading to an increase in solubility. Finally, the positive A_2 values of urate oxidase in sodium phosphate pH 8 or Tris buffers with 10mM salt are consistent with the high solubility of the protein in these buffers, and the negative A_2 value found in tris buffer pH 8 without salt is consistent with attraction inducing crystallization of urate oxidase (Fig. 9A). Obtaining crystals thus made solubility measurements possible (Fig. 9B).

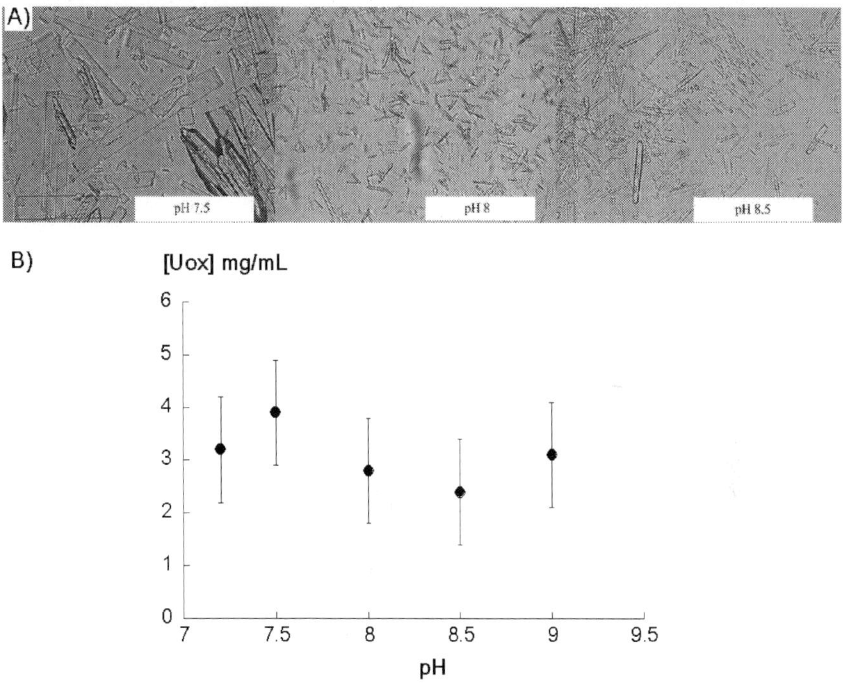

Fig. 9. A) Optical microscopy images of microcrystalline urate oxidase in Tris buffer at different pH values (from Collings et al., 2010). B) Solubility of urate oxidase as a function of pH in Tris buffer w/o salt added.

Even though these crystals were not of sufficient quality for high resolution crystallography, they were perfectly suitable for powder diffraction for space group determination (Collings et al., 2010). Attractive interactions with second virial coefficients in the "crystallization slot" defined by George & Wilson are undoubtedly necessary conditions to obtain crystals; however these interactions must be controlled upstream of the crystallization process to design adequate phase diagrams for appropriate crystallography.

4.2 Buffer type and temperature effects

pH and salt effects observed on second virial coefficient variations are easily explained by a change in the coulombic repulsive interactions, either 1/ due to a decrease in the protein net charge, either by shifting pH towards pI or by screening the repulsion with salt at pH far from pI, or 2/ due to an increase in the protein net charge, by ion binding near the pI. The first assumption was verified by characterizing the different underlying interaction potentials (Vivarès & Bonneté, 2002) and will not be discussed here. The ion binding effect underlines the importance of choice of solubilization buffer, especially since the objective is to test temperature effect. In previous work, a slight temperature effect was observed on the scattered intensity at small angles at pH 10.5 between 283 and 293 K, and this was attributed to a weak van der Waals attraction. In order to characterize the temperature effect on interaction forces and its implications for crystallization, SAXS experiments were performed over a larger range of temperature from 278 to 303 K at a constant pH closer to the isoelectric point, where the repulsive component is weaker. Since Tris buffer is very sensitive to temperature variations (dpKa/dT = -0.028), we used a borate buffer whose pH is known to be less temperature-dependent (dpKa/dT = -0.008). The scattered intensities of urate oxidase in Tris buffer and in borate buffer at pH 8 as a function of temperature are shown in Fig. 10.

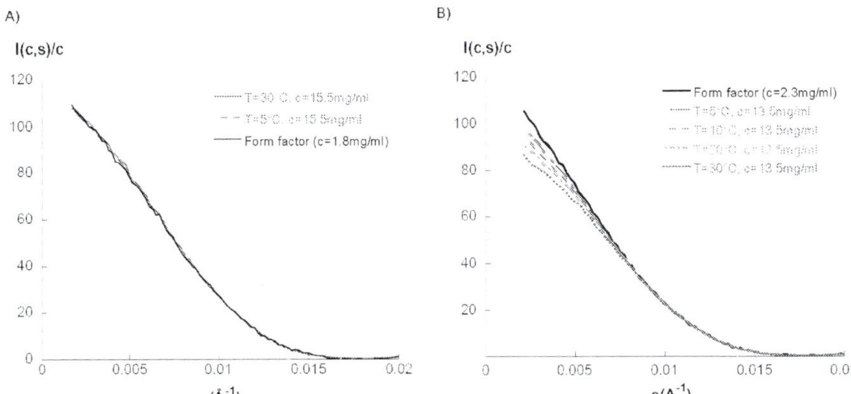

Fig. 10. Variation of scattered intensity of urate oxidase as a function of temperature: A) in tris buffer pH 8; B) in borate buffer pH 8.

It is clear that buffer type influences interactions in urate oxidase in solution. Whereas no variation in intensity is observed in Tris buffer, in borate buffer intensity increases as temperature decreases, probably due to an increase in van der Waals attraction, as previously observed and modelled with lysozyme (Tardieu, et al., 1999).

The second virial coefficient remains positive, probably due to the presence of sodium cations in borate buffer, but decreases as pH decreases (Fig.11A). Since pH in borate buffer does not vary with temperature, the net charge and the repulsive interactions of urate oxidase does not vary. The overall decrease in A_2 can only be induced by an increase in attractive interactions, the van der Waals attraction. In contrast, in Tris buffer, pH increases as the temperature decreases, inducing an increase in the net charge at pH above pI and thus an increase in repulsive interactions, matching the increase in attraction with temperature. This effect can induce the dissolution of crystals obtained at room temperature when stored in cold rooms, for example (data not shown but observed with urate oxidase), even though solubility is direct (Fig. 11B), as in the case of urate oxidase. When care is taken to keep pH constant whatever the temperature, and not to add cations to the buffer, it is therefore possible to obtain crystals and measure solubility (Fig. 11B). Temperature is an important parameter to modify attractive interactions and induce and control crystallization. Neththeless, the appropriate buffer still needs to be chosen and a significant change in solubility needs to be measured if large crystals are to be grown (Budayova-Spano et al., 2007).

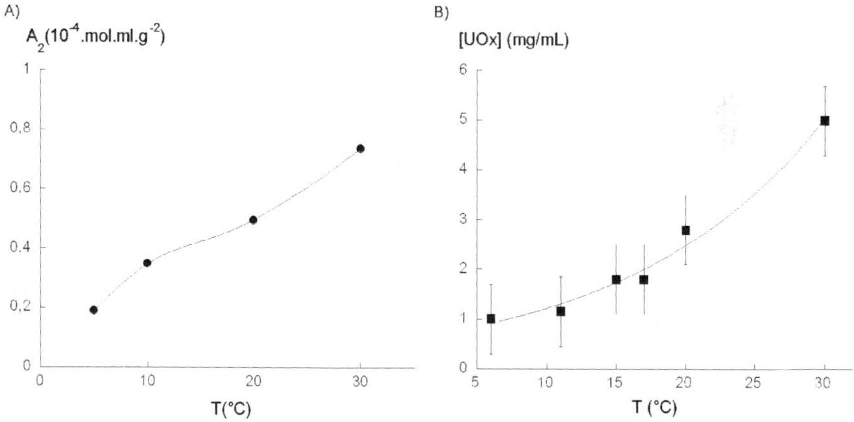

Fig. 11. A) Variation of urate oxidase A_2 with temperature in Na borate buffer pH 8; B) Solubility of urate oxidase at pH 8 in tris buffer w/o cations as a function of temperature

4.3 Usual and unusual polymers

Salts are known to act as crystallizing agents for many proteins, often small compact proteins, by inducing a short-range attraction. Ions, and in particular anions, were observed, at medium ionic strength, not only to screen charges but to induce an attraction, specific to the salt type (Ducruix, et al., 1996). We have seen here with urate oxidase, a large protein, that specific ion binding can induce solubilization of the protein by inducing a repulsive interaction, preventing its crystallization with salts. When salts are not effective in crystallizing proteins, neutral polymers can do the job (Bonneté et al., 2001; Budayova et al., 1999; Hitscherich, et al., 2000). Polyethylene glycol (PEG), a non-adsorbing neutral polymer, has long been used for protein crystallization (McPherson, 1976). The solubility (actually the precipitation) of proteins in solution containing PEG was studied (Atha & Ingham, 1981). A theoretical model was proposed by Asakura-Oosawa to explain the attraction between colloids due to addition of these polymers (§2.2.3). The effectiveness of polymer in precipitating proteins increases with polymer size and concentration, since it increases the depletion zone. By studying the crystallization of urate oxidase, our group greatly contributed to a better understanding of the mechanism of depletion attraction in the presence of different polyethylene glycols. We characterized the attractive interactions between urate oxidase induced by addition of three different PEGs (Fig. 12A), obtaining crystals (Fig. 12B) in conditions where A_2 was found negative in the "crystallization slot". Compared to salt, a major advantage of PEG in crystallization is that the depth and range of the attraction can be varied almost at will, simply by changing polymer size and concentration. However, it may be appropriate to couple the effects of salts (screening of charges or ion binding) and/or of pH with the effects of polymers to modulate the overall interactions and control the phase diagram for [illegible].

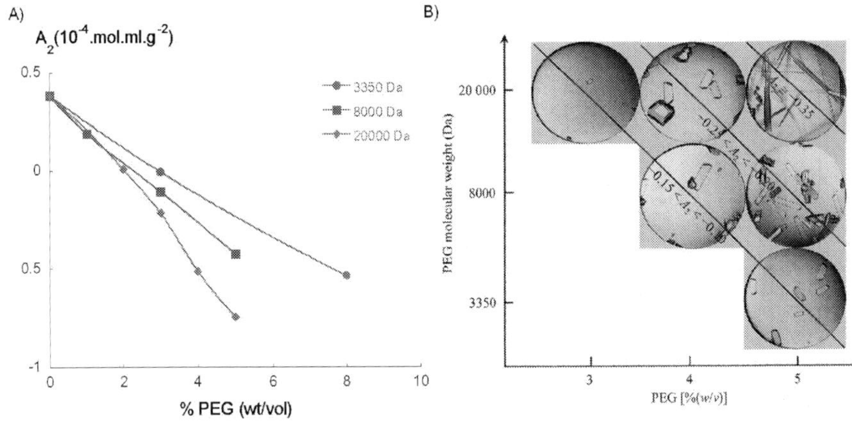

Fig. 12. A) Second virial coefficient urate oxidase in different PEG solutions in Tris pH8. B) Crystals of urate oxidase obtained in attractive ($A_2 < 0$) conditions.

However, as a function of |pH-pI|, the addition of PEG may be insufficient to induce an overall attraction, as shown in figure 13A. Thus, at pH 10, urate oxidase presents repulsive

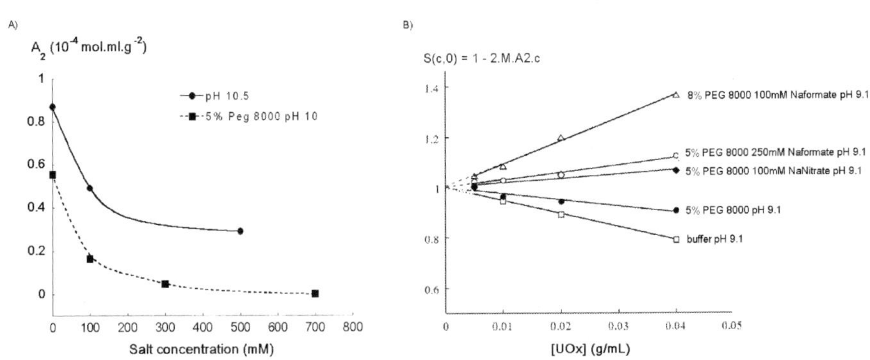

Fig. 13. A) Second virial coefficient of urate oxidase at pH 10 as a function of salt addition far from pI with and w/o PEG. B) Zero-angle structure factor in different physico-chemical conditions as a function of urate oxidase concentration; the slope measures the second virial coefficient (the slope is positive, A_2 is negative, attraction and vice versa)

interactions due to the protein negative net charge (see Fig 7). The depletion attraction induced by 5% PEG 8000 is not sufficient to change repulsive interactions into attractive interactions. To induce an overall attraction by addition of PEG, it is necessary to screen charges by addition of salt (Fig13B) or to be close to pI (Fig. 12A), where electrostatic repulsion is lowest.

At pI, regardless of whether salt is present in the buffer (Tris pH8 in our case), the addition of PEG to urate oxidase solutions induces attraction and leads to a decrease in solubility (Fig. 14). However, at pH 8 the solubility of urate oxidase without salt in the buffer is lower than the solubility of urate oxidase with salt. Since nucleation and crystal growth are driven by the supersaturation β, with β the ratio [initial concentration] / [solubility], a limited variation in solubility and therefore in supersaturation will be unfavorable to the growth of large crystals, and will rather favor the nucleation of small crystals.

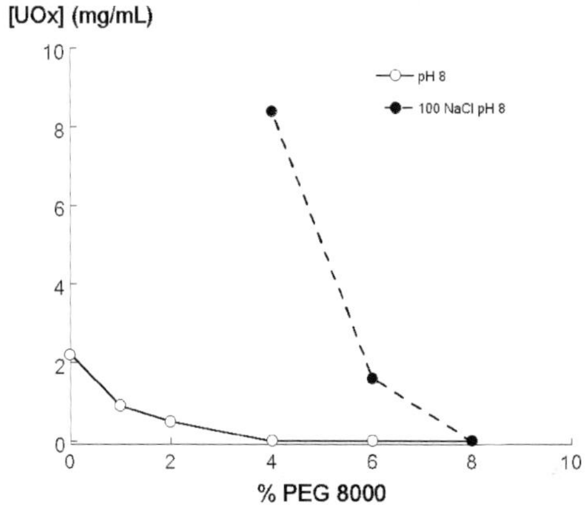

Fig. 14. Solubility of urate oxidase as a function of % of PEG with and w/o salt

As soon as crystals can be obtained, it is therefore possible to characterize the phase diagram with one or several solubility curves depending on the presence of polymorphs and the metastable liquid-liquid phase separation (Fig. 15).

In some cases, salts and neutral polymers are not effective for crystallization or cannot be used in pharmaceutical processes. Other crystallizing agents that can be used include organic compounds such as ethanol (Boyer et al., 1999), isopropanol, methanol, 2-methyl 2,4-pentanediol (Costenaro et al., 2001). Polymers such as amphiphilic multi-block polymers are used in the cosmetic and pharmaceutical industries, more often as emulsifiers, solubilizers, dispersing and wetting agents in the preparation of solid dispersions than as crystallizing agents in biocrystallography or in crystallization processes. We characterized a new class of crystallizing agent for soluble protein crystallization, compatible with both pharmaceutical processes and high-resolution structure determination in bio-crystallography. Poloxamer P188 is a nonionic co-polymer surfactant with a tri-block structure, composed of two hydrophilic segments, poly(oxyethylene) (PEO), and a central hydrophobic segment, poly(oxypropylene) (PPO), linked by ether bonds. The resulting construct can be represented as $HO(C_2H_4O)a(C_3H_6O)b(C_2H_4O)aH$, where a is about 75 and b is about 31 (Takáts et al., 2001) (Fig. 16). Its average molecular weight is around 8400 $g.mol^{-1}$

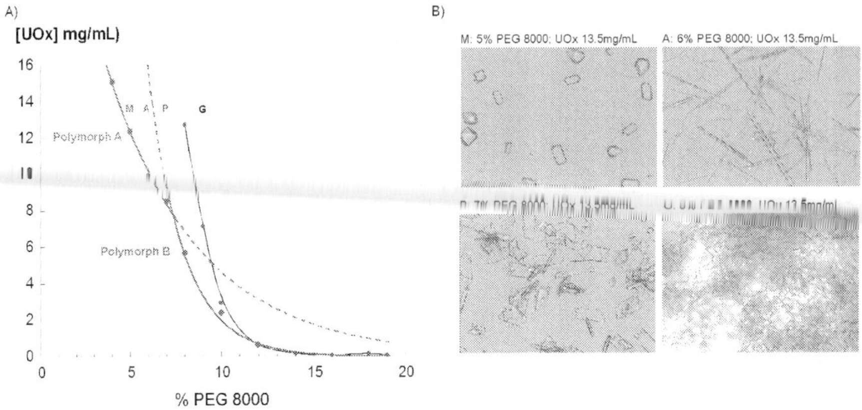

Fig. 15. Phase diagram of urate oxidase in presence of polyethylene glycol 8000 and optical microscopy images of different conditions in this phase diagram (M, A are polymorph A, P polymorph B and G a liquid-liquid phase transition) (Vivares et al., 2006).

and its cmc (critical micellar concentration) is about 0.1 % w/v (Schmolka, 1977). Poloxamer is similar to the usual crystallizing agent for urate oxidase, PEG 8000, which is a linear hydrophilic polymer consisting of approximately n = 180 poly(oxyethylene) units and has a molecular weight of about 8000 g.mol^{-1}. Because of their amphiphilic structure, poloxamers have surfactant properties that make them useful in pharmaceutical applications. They can be used to increase water solubility of hydrophobic, oily substances as well as to increase the miscibility of two substances with different hydrophobicities. They are also used as model systems for drug delivery (Adams et al., 2003) applications. Recently, they have been shown to function as artificial chaperones to facilitate refolding of denatured proteins in solution or to suppress aggregation. In general, all these applications require low concentrations of poloxamer, typically below its cmc and involving monomeric poloxamer in solution. Below the cmc, the hydrophobic segment of polymer can non-specifically interact with exposed hydrophobic domains, preventing aggregation and aiding in the refolding of proteins (Lee et al., 2006).

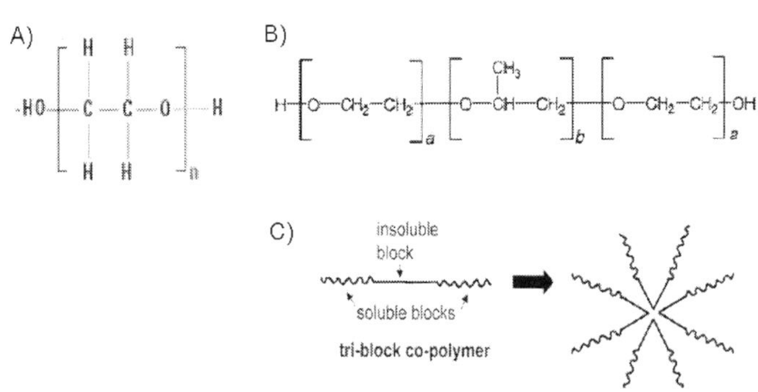

Fig. 16. Chemical structure of poloxamer P188 (A), PEG (B); C) Scheme of micellization of poloxamer at c > cmc (~ 0.1%)

However, high concentrations of poloxamer have also been reported to induce protein aggregation (Garcia, 1975). The addition of 12 to 16 % of the block copolymer to plasma or serum induces the precipitation of high molecular weight proteins including antibodies. This suggests that it would be possible to use high concentrations of poloxamer, thus probably in its micelle form, to induce protein crystallization by depletion, as observed with PEG. To this end, we explored the interactions and crystallization of urate oxidase by addition of poloxamer P188. The second virial coefficient is positive without poloxamer in Tris with 30mM KCl, as expected from previous studies, and increases as the concentration of poloxamer increases up to its cmc (about 0.1%). Above cmc, the A_2 decreases and becomes negative at concentrations higher than 1%. Solubility of urate oxidase was therefore measured in the same conditions, as a function of concentration of poloxamer P188. As with PEG 8000, solubility decreases as the concentration of poloxamer increases and is higher with poloxamer P188 than with PEG 8000 for concentrations lower than 6 % (Fig. 17). Thus, the surfacting or crystallizing nature of poloxamer P188 depends on its concentration. Below 0.1 %, i.e. its cmc, poloxamer is monomeric. In this range of concentrations, it is possible that the hydrophobic central PPO block non-specifically interacts with solvent-exposed hydrophobic patches at the protein's surface, while the hydrophilic surfactant chains remain exposed to the aqueous phase, increasing the solubility and the repulsion of urate oxidase. Above 0.1 % poloxamer, micelles can form. The attractive effect of poloxamer at concentrations above 1 % can be explained by a depletion effect driven by poloxamer micelles. The same effect has been suggested with nanoemulsions whose micelles, after surface saturation by surfactant, do not adsorb on the surface but rather cause attraction by a depletion mechanism (Wulff-Pérez et al., 2009).

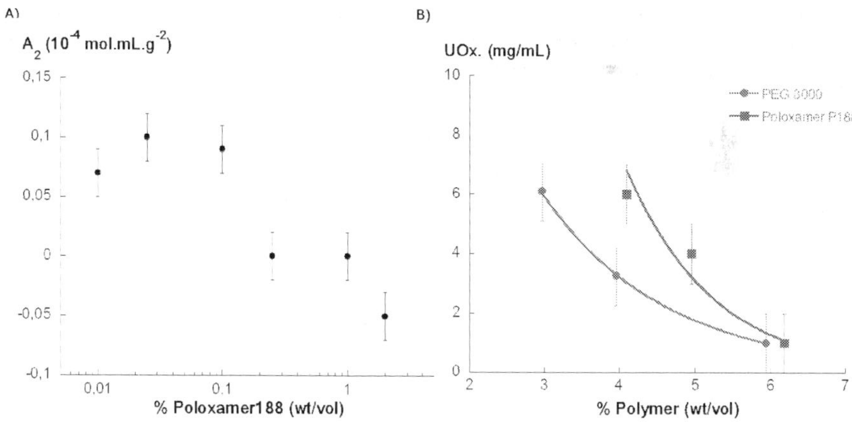

Fig. 17. A) Second virial coefficient of urate oxidase in solution as a function of % Poloxamer P188. B) Comparison between solubility of urate oxidase in PEG 8000 and Poloxamer P188 in 30 mM KCl, 50 mM Tris pH 7.5 at 20°C.

Poloxamer P188 can thus be used as a crystallizing agent for urate oxidase. When other amphiphilic surfactants such as poloxamer P407, polysorbate 20 and polysorbate 80 were studied, poloxamer P407 was found to have a similar structure and molecular formula to poloxamer P188, where a and b are 100 and 60 respectively. The molecular weight of the hydrophobic core is 4000 g.mol-1, which represents 30 % of the total mass of the polymer. P407 average molecular weight is about 13300 g.mol-1. As shown in Fig. 18A, urate oxidase solubility in poloxamer P407 is higher than in either poloxamer P188 or PEG 8000, when the remaining solution components are kept fixed (i.e. 50 mM Tris pH 7.5, 30 mM KCl). Unlike PEG, urate oxidase solubility increases with the mass of the poloxamer used. This result suggests that poloxamer has a solubilizing effect proportional to its hydrophobic core content, and this compensates for its precipitating by depletion. The effect of polysorbates, another class of emulsifiers used in the preparation of pharmaceuticals and food, was also tested. Polysorbates are oily liquids derived from PEGylated sorbitan (a sorbitol derivative) esterified with fatty acids. They all have a hydrophilic moiety characterized by twenty oxyethylene-(CH_2CH_2O)- groups, while the hydrophobic segments vary according to the polysorbate compound. Polysorbates 20 and 80 effectively induce crystallization of urate oxidase (Fig. 18B). Urate oxidase solubility is higher with polysorbates 20 and 80 than with PEG 8000 and poloxamers P188 and P407. As with poloxamers, it appears that the longer the polysorbate hydrophobic chain, the higher the solubility of urate oxidase. Nevertheless, the modification observed in the habit of the urate oxidase crystals with 16% polysorbate 80 suggests that this concentration of surfactant could affect crystal growth, and possibly protein structure.

A)

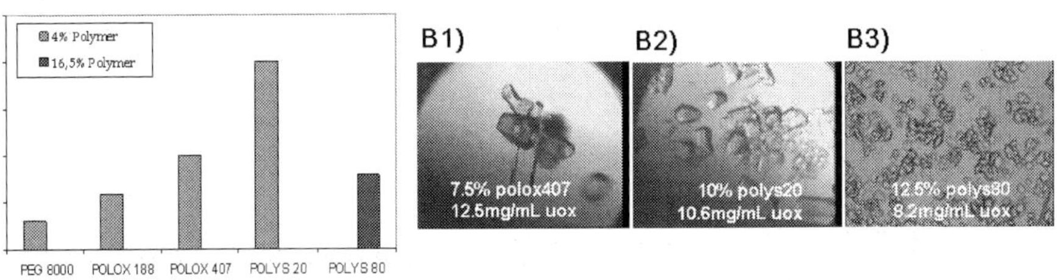

Fig. 18. A) Urate oxidase solubility with various polymers (in 30 mM KCl, 50 mM Tris pH7.5, 20°C). B) Crystals of urate oxidase in 30 mM KCl, 50 mM Tris pH7.5, 20°C: B1) 12.5 mg.mL^{-1} uox with 7.5 % poloxamer P407; B2) 10.6 mg.mL^{-1} uox with 10 % polysorbate 20; B3) 8.2 mg.mL^{-1} uox with 12.5 % polysorbate 80.

4.4 Urate oxidase crystal design for high pressure crystallography

Macromolecules were long supposed to have their own crystallization conditions, since they have a unique sequence. However, the fact is that macromolecules in solution interact through different weak interaction forces. They can be considered as charged colloids under the influence of attraction forces (van der Waals, depletion, Hofmeister) and repulsion forces (hard sphere, electrostatic). These forces are medium-range and each of them is under the separate control of physico-chemical conditions such as pH, temperature and solvent composition. These forces govern macromolecule properties in solution such as solubility, phase separation (liquid-liquid or liquid-solid). The fine tuning of these interaction forces makes the design of phase diagrams for specific applications possible. The overall attractive potential leading to crystallization can be obtained in different ways. We used urate oxidase as a model system, since it was available in large quantities, to explore a wide spectrum of physico-chemical conditions and to see how these parameters influence pair potentials and phase diagrams. For example, the salting-in effect by increasing repulsion allows us to modulate the position of the solubility curve in the urate oxidase phase diagram, leading to a more or less large metastable zone, in order to grow either large single crystals or numerous small crystals. Another example is the use of amphiphilic polymer below or above its cmc (critical micellar concentration), allowing us either to favor the solubility of proteins or to induce their crystallization. In the following example, we describe a strategy for growing protein crystals suitable for macromolecular crystallography under high pressure (HPMX). Studying proteins under high pressure encompasses a wide range of objectives, from understanding the physical chemistry of protein interactions with water to practical applications in food processing. Many proteins are studied using a variety of techniques applied under high pressure, such as spectroscopic techniques, NMR spectroscopy, as well as a wide range of scattering techniques including static and dynamic light scattering, neutron scattering and X-ray crystallography. Such high-pressure approaches are generally used to solve problems involving macromolecule structural changes. For High Pressure Macromolecular Crystallography (HPMX), obtaining the right protein crystal is a challenge. First, as for any crystallographic study, it is necessary to obtain a reasonably large crystal preferentially in a high symmetry space group. Secondly, the crystal has to fit the pressure cell as well as possible, and be grown by a method which allows the crystal to be picked out easily, for example the hanging drop technique. Finally, the crystal has to remain stable in its mother liquor under pressure. Indeed, the effect of pressure on protein nucleation and crystallization can vary widely. In some cases, increased pressure increases the nucleation rate (Visuri et al., 1990), while in other cases, the protein solubility increases, decreasing the nucleation rate (Lorber et al., 1996). Pressure can affect solubility and therefore lead either to dissolution of a crystal or to secondary nucleation. To circumvent the pressure effect on secondary crystal nucleation in the cell, the crystal must be placed in a mother liquor containing no protein and a high concentration of crystallizing agent, since variations in solubility caused by pressure are known to be reduced by increasing the crystallizing agent concentration.

To grow a suitable crystal for HPMX, we first need a low supersaturation β which will lead to a small number of large crystals rather than a large number of small crystals, finally yielding a crystal in equilibrium with a very low protein solution, i.e. at high crystallizing agent concentration. To optimize urate oxidase crystallization conditions, we first characterize the phase diagram without salt, i.e. the conditions where solubility is lowest, as a function of PEG percentage (Fig. 19A) and explore different crystal growth conditions by using the Microbatch technique (Chayen et al., 1992).

The rasburicase crystals obtained as a function of PEG 8000 are of two different shapes: tabular crystals when the percentage of PEG is below 4%, and plate-like crystals when PEG percentage is above 4% (Fig. 19B). Unfortunately, crystals obtained at the lowest solubility, i.e. when percentages of PEG are greatest, are not suitable for crystal diffraction studies, being too numerous and poorly faceted. A suitable tabular crystal grown at a lower PEG percentage is picked out and placed in a glass cell. According to the solubility curve in 50mM Tris pH8, rasburicase is totally insoluble in a solution of more than 10% PEG 8000. To control the stability of the crystal, protein-free mother liquor (15% PEG 8000, 50mM tris pH 8.0) is added to the crystal. The solubility, i.e. the concentration of the protein solution Δc remaining around the picked-out crystal, being too high, this leads to a secondary nucleation around the crystal (Fig. 19C). Indeed, the remaining protein solution around the crystal diluted in the protein-free mother liquor is supersatured and induces a new nucleation. The crystallization conditions must therefore be a compromise between solubility which is sufficient to grow large crystals but not too high, so as to avoid secondary nucleation when the mother liquor is added.

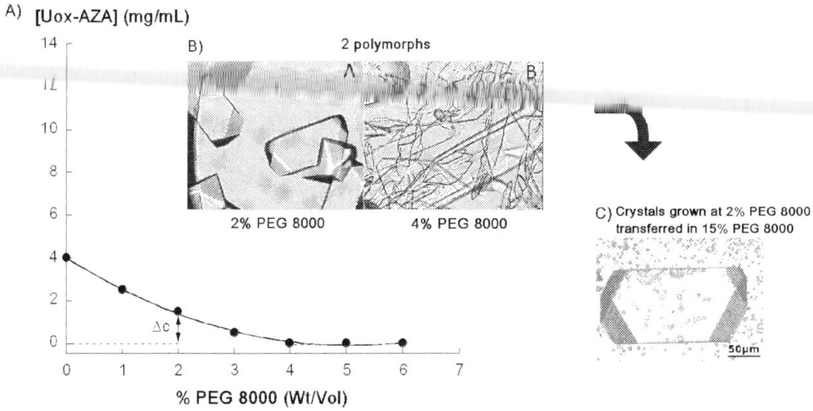

Fig. 19. A) Solubility of urate oxidase complexed with 8-azaxanthine, in Tris buffer pH 8, without salt, as a function of PEG 8000; B) Micrographs of urate oxidase crystals grown in microbatch at two PEG percentages; C) Micrograph of urate oxidase crystals transferred in 15% PEG 8000.

Therefore, to obtain a massive-habit crystal at high PEG percentage, we added 100mM NaCl, despite the salting-in effect which increases solubility, and determined the new solubility curve as a function of PEG 8000 (Fig. 20A).

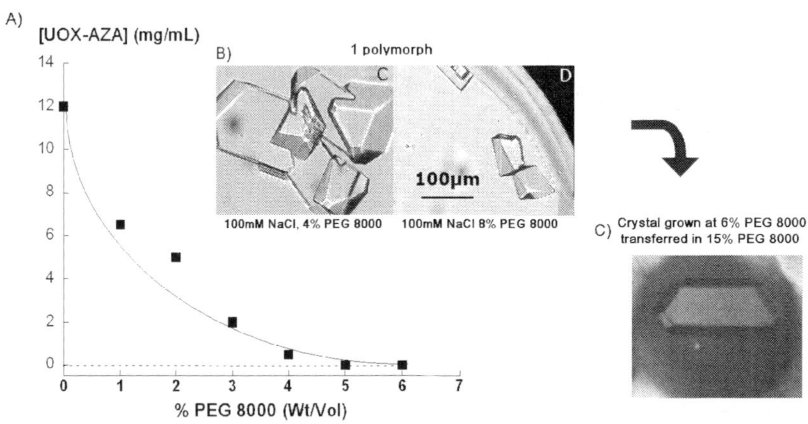

Fig. 20. A) Solubility of urate oxidase complexed with 8-azaxanthine, in Tris buffer pH 8, with 100mM NaCl, as a function of PEG 8000; B) Micrographs of urate oxidase crystals grown in microbatch at two PEG percentages; C) Micrograph of urate oxidase crystal transferred in 15% PEG 8000.

Crystal habits become more suitable for crystal diffraction studies (Fig. 20B) with sizes compatible with the pressure cell dimensions. We successfully transfer a crystal grown in 3.2mg/mL of urate oxidase 6% PEG 100mM NaCl Tris 50mM pH8 into a solution of 15% PEG 8000, 100mM NaCl, 50mM tris pH 8.0. It remains stable for more than one week without secondary nucleation appearing (Fig. 20C). At this concentration of protein, no micro-crystals are generated from the solution surrounding the crystal, whereas at lower PEG and higher protein concentrations in the surrounding solution, many micro-crystals grow.

5. Conclusion

Distribution of macromolecules in solution, phase diagrams and the crystallization process are governed by an appropriate combination of interaction potentials in solution. Three types of parameters or additives appear to play a crucial role: pH, salt and polymers. Whatever the diversity of protein sequences, it is possible, via second virial coefficient measurements, to limit the number of trials for a first screening of crystallization conditions and to rationalize crystallization. The design of small crystals or large diffracting crystals then results from a subtle mix of strong or weak repulsions and attractions, which control the position of the solubility curve and the metastable zone in the phase diagram.

6. Acknowledgment

This work was carried out by Denis Vivarès (1998-2003) and Marion Giffard (2006-2009) during their PhD theses. We gratefully acknowledge the generous support of Bertrand Castro, Mohamed El Hajji and François Ragot from Sanofi-Aventis (France) for this project on fundamental crystallization of urate oxidase since 1998, and their subsequent PhD financial support (2006-2009) enabling us to study the crystallization of urate oxidase for industrial applications.

7. References

Adams, M. L.; Lavasanifar, A. & Kwon, G. S. (2003). Amphiphilic block copolymers for drug delivery. *Journal of Pharmaceutical Sciences*, Vol. 92, No. 7, pp. 1343-1355,

Aleman, C.; Bayol, A.; Breul, T. & Dupin, P. (1998). Stable liquid composition containing urate oxidase and lyophilized composition for its preparation. Sanofi, Paris. United States

Arakawa, T. & Timasheff, S. (1985). Theory of protein solubility. *Methods in Enzymology*, Vol. 114, No. pp. 49-77, ISSN: 0076-6879

Asakura, S. & Oosawa, F. (1954). On the interaction between two bodies immersed in a solution of macromolecules. *J. Chem. Phys.*, Vol. 22, No. pp. 1255-1256,

Asherie, N.; Lomakin, A. & Benedek, G. B. (1996). Phase diagram of colloidal solutions. *Physical Review Letters*, Vol. 77, No. 23, pp. 4832-4835, ISSN: 0031-9007

Atha, D. H. & Ingham, K. C. (1981). Mechanism of precipitation of proteins by polyethylene glycols. Analysis in terms of excluded volume. *The Journal of biological chemistry*, Vol. 256, No. 23, pp. ISSN: 0021-9258

Bayol, A.; Dupin, P.; Boe, J. F.; Claudy, P. & Létoffé, J. M. (1995). Study of pH and temperature-induced transitions in urate oxidase (Uox-EC1.7.3.3) by microcalorimetry (DSC), size exclusion chromatography (SEC) and enzymatic activity experiments. *Biophysical Chemistry*, Vol. 54, No. 3, pp. 229-235, ISSN: 0301-4622

Behlke, J. & Ristau, O. (1999). Analysis of the thermodynamic non-ideality of proteins by sedimentation equilibrium experiments. *Biophysical Chemistry*, Vol. 76, No. 1, pp. 13-23, ISBN 0301-4622

Belloni, L. (1985). A hypernetted chain study of highly asymmetrical polyelectrolytes. *Chemical Physics*, Vol. 99, No. 1, pp. 43-54,

Belloni, L. (1988). Self-consistent integral equation applied to the highly charged primitive model. *J. Chem. Phys.*, Vol. 88, No. 8, pp. 5143-5148,

Belloni, L. (1991). Interacting monodisperse and polydisperse spheres, In *Neutron, X-ray and light scattering*, Series, N.-H.D., pp. 135-155, Elsevier Science Publishers B.V., Amsterdam-Oxford-NewYork-Tokyo

Bergfors, T. (2009). Protein Crystallization ISBN-13: 978-0972077446,

Bhamidi, V.; Varanasi, S. & Schall, C. A. (2005). Protein Crystal Nucleation: Is the Pair Interaction Potential the Primary Determinant of Kinetics? *Langmuir*, Vol. 21, No. 20, pp. 9044-9050, ISBN 0743-7463

Bloustine, J.; Berejnov, V. & Fraden, S. (2003). Measurements of Protein-Protein Interactions by Size Exclusion Chromatography. *Biophysical Journal*, Vol. 85, No. 4, pp. 2619-2623, ISSN: 0006-3495

Bolhuis, P. G.; Louis, A. A.; Hansen, J. P. & Meijer, E. J. (2001). Accurate effective pair potential for polymer solutions. *J.Chem.Phys.*, Vol. 114, No. 9, pp. 4296-4311,

Bonneté, F. & Vivares, D. (2002). Interest of the normalized second virial coefficient and interaction potentials for crystallizing large macromolecules. *Acta Cryst*, Vol. D58, No. pp. 1571-1575, ISSN: 0907-4449

Bonneté, F.; Finet, S. & Tardieu, A. (1999). Second virial coefficient: variations with lysozyme crystallization conditions. *J. Crystal Growth*, Vol. 196, No. pp. 403-414, ISSN: 0022-0248

Bonneté, F.; Vivares, D.; Robert, C. & Colloc'h, N. (2001). Interactions in solution and crystallization of *Aspergillus flavus* urate oxidase. *J. Crystal Growth*, Vol. 232, No. pp. 330-339, ISSN: 0022-0248

Boué, F.; Lefaucheux, F.; Robert, M. C. & Rosenman, I. (1993). Small angle neutron scattering study of lysozyme solutions. *Journal of Crystal Growth*, Vol. 133, No. 3-4, pp. 246-254, ISBN 0022-0248

Boyer, M.; Roy, M.-O.; Jullien, M.; Bonnete, F. & Tardieu, A. (1999). Protein interactions in concentrated ribonuclease solutions. *J. Crystal Growth*, Vol. 196, No. 2-4, pp. 185-192,

Budayova, M.; Bonneté, F.; Tardieu, A. & Vachette, P. (1999). Interactions in solution of large oligomeric protein. *J. Crystal Growth*, Vol. 196, No. pp. 210-219, ISSN: 0022-0248

Budayova-Spano, M.; Dauvergne, F.; Audiffren, M.; Bactivelane, T. & Cusack, S. (2007). A methodology and an instrument for the temperature-controlled optimization of crystal growth. *Acta Crystallographica Section D-Biological Crystallography*, Vol. 63, No. pp. 339-347, ISSN: 0907-4449

Carbonnaux, C.; Riès-Kautt, M. & Ducruix, A. (1995). Relative effectiveness of various anions on the solubility of acidic Hypoderma lineatum collagenase at pH 7.2. *Protein Science*, Vol. 4, No. 10, pp. 2123-2128, ISSN: 0961-8368

Casselyn, M.; Perez, J.; Tardieu, A.; Vachette, P.; Witz, J. & Delacroix, H. (2001). Spherical plant viruses: interactions in solution, phase diagrams and crystallization of brome mosaic virus. *Acta Crystallographica Section D-Biological Crystallography*, Vol. 57, No. pp. 1799-1812, ISSN: 0907-4449

Chayen, N. E.; Stewart, P. D. S. & Blow, D. M. (1992). Microbatch Crystallization under Oil - a New Technique Allowing Many Small-Volume Crystallization Trials. *Journal of Crystal Growth*, Vol. 122, No. 1-4, pp. 176-180, ISSN: 0022-0248

Chernov, A. A. (2003). Protein crystals and their growth. *Journal of Structural Biology, Macromolecular crystallization in the structural genomics era*, Vol. 142, No. 1, pp. 3-21, ISSN: 1047-8477

Collings, I.; Watier, Y.; Giffard, M.; Dagogo, S.; Kahn, R.; Bonnete, F.; Wright, J. P.; Fitch, A. N. & Margiolaki, I. (2010). Polymorphism of microcrystalline urate oxidase from Aspergillus flavus. *Acta Crystallographica Section D-Biological Crystallography*, Vol. 66, No. pp. 539-548, ISSN: 0907-4449

Costenaro, L.; Zaccai, G. & Ebel, C. (2001). Understanding the crystallisation of an acidic protein by dilution in the ternary NaCl-2-methyl-2,4-pentanediol-H2O system. *J. Crystal Growth*, Vol. 232, No. pp. 102-113,

Debye, P. (1946). The Intrinsic Viscosity of Polymer Solutions. *J.Chem.Phys.*, Vol. 14, No. pp. 636-646,

Drenth, J. (2005). The nucleation of lysozyme from a fluctuation point of view. *Crystal Growth & Design*, Vol. 5, No. 3, pp. 1125-1127, ISSN: 1528-7483

Ducruix, A.; Guilloteau, J. P.; Riès-Kautt, M. & Tardieu, A. (1996). Protein interactions as seen by solution X-ray scattering prior to crystallogenesis. *J. Crystal Growth*, Vol. 168, No. pp. 28-39, ISSN 0022-0248

Faber, C.; Hobley, T.; Mollerup, J.; Thomas, O. R. T. & Kaasgaard, S. (2007). Study of the Solubility of a Modified Bacillus licheniformis r-Amylase around the Isoelectric Point. *Journal of Chemical and Engineering Data*, Vol. 52, No. pp. 707-713, ISSN: 0021-9568

Feher, G. & Kam, Z. (1985). Nucleation and growth of protein crystals: general principles and assays. *Method Enzymol.*, Vol. 114, No. pp. 77–112, ISSN: 00766879

Finet, S. & Tardieu, A. (2001). alpha-cristallin interaction forces studied by small angle X-ray scattering and numerical simulations. *J. Crystal Growth*, Vol. 232, No. 1-4, pp. 40-49, ISSN: 0022-0248

Finet, S.; Bonneté, F.; Frouin, J.; Provost, K. & Tardieu, A. (1998). Lysozyme crystal growth, as observed by small angle X-ray scattering, proceeds without crystallization intermediates. *European Biophysics Journal*, Vol. 27, No. 3, pp. 263-271, ISBN 0175-7571

Finet, S.; Skouri-Panet, F.; Casselyn, M.; Bonneté, F. & Tardieu, A. (2004). The Hofmeister effect as seen by SAXS in protein solutions. *Current Opinion in Colloid & Interface Science*, Vol. 9, No. 1-2, pp. 112-116, ISSN: 1359-0294

Garcia, L. A. (1975). Production of antisera comprising fractionating plasma or serum with an ethylene oxide-polyoxypropylene block copolymer. Baxter Laboratories, Inc. United States of America

García-Ruiz, J.-M. (2003). Nucleation of protein crystals. *Journal of Structural Biology, Macromolecular crystallization in the structural genomics era*, Vol. 142, No. 1, pp. 22-31, ISSN: 1047-8477

Gavira, J. A. & Garcia-Ruiz, J. M. (2009). Effects of a Magnetic Field on Lysozyme Crystal Nucleation and Growth in a Diffusive Environment. *Crystal Growth & Design*, Vol. 9, No. 6, pp. 2610-2615, ISSN: 1528-7483

George, A. & Wilson, W. W. (1994). Predicting protein crystallization from a dilute solution property. *Acta Crystallographica*, Vol. D50, No. 4, pp. 361-365, ISSN: 0907-4449

Gibb, C. L. D. & Gibb, B. C. (2011). Anion Binding to Hydrophobic Concavity Is Central to the Salting-in Effects of Hofmeister Chaotropes. *Journal of the American Chemical Society J. Am. Chem. Soc.*, Vol. 133, No. 19, pp. 7344-7347, ISSN: 0002-7863

Giffard, M.; Colloc'h, N.; Ferté, N.; Castro, C. & Bonneté, F. (2008). Salting-in effect on urate oxidase crystal design. *Crystal Growth and Design*, Vol. 8, No. pp. 4220-4226, ISSN: 1932-6203

Giffard, M.; Ferté, N.; Ragot, F.; El Hajji, M.; Castro, B. & Bonneté, F. (2011). Urate Oxidase Purification by Salting-in Crystallization: Towards an Alternative to Chromatography. *Plos One*, Vol. 6, No. 5, pp. WOS:000290483600005

Gripon, C.; Legrand, L.; Rosenman, I.; Vidal, O.; Robert, M. C. & Boué, F. (1997). Lysozyme-lysozyme interactions in under- and super-saturated solutions: a simple relation between the second virial coefficients in H2O and D2O. *Journal of Crystal Growth*, Vol. 178, No. 4, pp. 575-584, ISBN 0022-0248

Guilloteau, J.-P.; Rie`s-Kautt, M. M. & Ducruix, A. F. (1992). Variation of lysozyme solubility as a function of temperature in the presence of organic and inorganic salts. *Journal of Crystal Growth*, Vol. 122, No. 1-4, pp. 223-230, ISSN: 0022-0248

Guinier, A. & Fournet, G. (1955). Small angle scattering of X-rays Wiley, New York

Hansen, J. P. & McDonald, I. R. (1976). Theory of simple liquids Academic Press, London - New York - San Francisco

Haynes, C. A.; Tamura, K.; Korfer, H. R.; Blanch, H. W. & Prausnitz, J. M. (1992). Thermodynamic properties of aqueous .alpha.-chymotrypsin solution from membrane osmometry measurements. *The Journal of Physical Chemistry*, Vol. 96, No. 2, pp. 905-912, ISSN: 0022-3654

Hitscherich, C. J.; Kaplan, J.; Allaman, M.; Wiencek, J. & Loll, P. J. (2000). Static light scattering studies of OmpF porin : implications for integral membrane protein crystallization. *Protein Sci.*, Vol. 9, No. pp. 1559-1566, ISSN: 0961-8368

Israelachvili, J. (1994). Intermolecular and surface forces New York

Jacobsen, C.; Garside, J. & Hoare, M. (1998). Nucleation and growth of microbial lipase crystals from clarified concentrated fermentation broths. *Biotechnology and Bioengineering*, Vol. 57, No. 6, pp. 666-675, ISSN: 0006-3592

Lee, R. C.; Despa, F.; Guo, L.; Betala, P.; Kuo, A. & Thiyagarajan, P. (2006). Surfactant copolymers prevent aggregation of heat denatured lysozyme. *Annals of Biomedical Engineering*, Vol. 34, No. 7, pp. 1190-1200, ISI:000239085800012

Lekkerkerker, H. N. W. (1997). Strong, weak and metastable liquids. *Physica A*, Vol. 244, No. pp. 227-237, ISSN: 0378-4371

Liu, Y. X.; Wang, X. J. & Ching, C. B. (2010). Toward Further Understanding of Lysozyme Crystallization: Phase Diagram, Protein-Protein Interaction, Nucleation Kinetics, and Growth Kinetics. *Crystal Growth & Design*, Vol. 10, No. 2, pp. 548-558, ISSN: 1528-7483

Lomakin, A.; Asherie, N. & Benedek, G. B. (1996). Monte Carlo study of phase separation in aqueous protein solutions. *Journal of Chemical Physics*, Vol. 104, No. 4, pp. 1646-1656, ISSN: 0021-9606

Lorber, B.; Jenner, G. & Giégé, R. (1996). Effect of high hydrostatic pressure on nucleation and growth of protein crystals. *J. Crystal Growth*, Vol. 158, No. 1-2, pp. 103-117, ISSN: 0022-0248

Lovgren, S. (2007). Giant Crystal Cave's Mystery Solved. *National Geographic News*

Lutterbach, N.; Versmold, H.; Reus, V.; Belloni, L. & Zemb, T. (1999a). Charge-stabilized liquidlike ordered binary colloidal suspensions. 1. Ultra-Small-Angle X-ray Scattering characterization. *Langmuir*, Vol. 15, No. pp. 337-344, ISSN: 0743-7463

Lutterbach, N.; Versmold, H.; Reus, V.; Belloni, L.; Zemb, T. & Lindner, P. (1999b). Charge-stabilized liquidlike ordered binary colloidal suspensions. 2. Partial structure factors determined by Small-Angle Neutron Scattering. *Langmuir*, Vol. 15, No. pp. 345-352, ISSN: 0743-7463

Malfois, M.; Bonneté, F.; Belloni, L. & Tardieu, A. (1996). A model of attractive interactions to account for liquid-liquid phase separation of protein solutions. *J. Chem. Phys.*, Vol. 105, No. B, pp. 3290-3300, ISSN: 0021-9606

McGrath, B. & Walsh, G. (2005). Directory of Therapeutic Enzymes CRC Press Inc, ISBN-10: 0849327148,

McPherson, A. (1976). Crystallization of proteins from polyethylene glycol. *The Journal of biological chemistry*, Vol. 251, No. 20, pp. 6300-6303, ISSN: 0021-9258

McPherson, A.; Malkin, J. A.; Kuznetsov, Y. G.; Koszelak, S.; Wells, M.; Jenkins, G.; Howard, J. & Lawson, G. (1999). The effects of microgravity on protein crystallization: evidence for concentration gradients around growing crystals. *Journal of Crystal Growth*, Vol. 196, No. 2-4, pp. 572-586, IBSN 0022-0248

McQuarrie, D. A. (2000). Statistical mechanics Sausalito, CA

Mikol, V.; Hirsch, E. & Giegé, R. (1989). Monitoring protein crystallization by dynamic light scattering. *FEBS Letters*, Vol. 258, No. 1, pp. 63-66, ISBN 0014-5793

Muschol, M. & Rosenberger, F. (1995). Interactions in under- and supersaturated lysozyme solutions. Static and dynamic light scattering results. *J. Chem. Phys.*, Vol. 103, No. 24, pp. 10424-10432, ISSN: 0021-9606

Ng, J. D.; Gavira, J. A. & García-Ruíz, J. M. (2003). Protein crystallization by capillary counterdiffusion for applied crystallographic structure determination. *Journal of Structural Biology, Macromolecular crystallization in the structural genomics era*, Vol. 142, No. 1, pp. 218-231, ISBN 1047-8477

Oksanen, E.; Blakeley, M. P.; Bonnete, F.; Dauvergne, M. T.; Dauvergne, F. & Budayova-Spano, M. (2009). Large crystal growth by thermal control allows combined X-ray and neutron crystallographic studies to elucidate the protonation states in Aspergillus flavus urate oxidase. *Journal of the Royal Society Interface*, Vol. 6, No. pp. S599-S610, ISSN: 1742-5689

Pande, A.; Pande, J.; Asherie, N.; Lomakin, A.; Ogun, O.; King, J. & Benedek, G. B. (2001). Crystal cataracts: human genetic cataract caused by protein crystallization. *PNAS*, Vol. 98, No. 11, pp. 6116-6120, ISBN 0027-8424

Poon, W. C. K. (2002). The physics of a model colloid-polymer mixture. *Journal of Physics-Condensed Matter*, Vol. 14, No. 33, pp. R859-R880, ISSN: 0953-8984

Retailleau, P.; Colloc'h, N.; Vivares, D.; Bonneté, F.; Castro, B.; El Hajji, M.; Mornon, J. P.; Monard, G. & Prangé, T. (2004). Complexed and ligand-free high resolution structures of Urate oxidase (Uox) from *Aspergillus flavus*: a re-assignation of the active site binding mode. *Acta Cryst*, Vol. D60, No. pp. 453-462, ISSN: 0907-4449

Retailleau, P.; Colloc'h, N.; Vivares, D.; Bonneté, F.; Castro, B.; El Hajji, M. & Prangé, T. (2005). Urate oxidase from Aspergillus flavus: new crystal-packing contacts in relation to the content of the active site. *Acta Cryst.*, Vol. D61, No. pp. 218-229, ISSN: 0907-4449

Riès-Kautt, M. & Ducruix, A. (1999). From solution to crystals with a physico-chemical aspect, In: *Crystallization of nucleic acids and proteins - A practical approach,* Giegé, R., pp., Oxford University Press, Oxford, New York, Tokyo

Sauter, C.; Lorber, B.; McPherson, A. & Giegé, R. (2011). Crystallization, In: *General methods In International Tables of Crystallography; Crystallography of Biological Macromolecules,* pp. In press., Chichester

Sazaki, G.; Moreno, A. & Nakajima, K. (2004). Novel coupling effects of the magnetic and electric fields on protein crystallization. *Journal of Crystal Growth*, Vol. 262, No. 1-4, pp. 499-502, ISBN 0022-0248

Schmolka, I. (1977). A review of block polymer surfactants. *Journal of the American Oil Chemists' Society*, Vol. 54, No. 3, pp. 110-116,

Sumner, J. (1926). The isolation and crystallization of the enzyme urease. Preliminary paper. *Journal of Biological Chemistry*, Vol. 69, No. 2, pp. 435-441,

Takáts, Z.; Vékey, K. & Hegedüs, L. (2001). Qualitative and quantitative determination of poloxamer surfactants by mass spectrometry. *Rapid Communications in Mass Spectrometry*, Vol. 15, No. 10, pp. 805-810,

Tardieu, A.; Le Verge, A.; Riès-Kautt, M.; Malfois, M.; Bonneté, F.; Finet, S. & Belloni, L. (1999). Proteins in solution : from X-ray scattering intensities to interaction potentials. *J. Crystal Growth*, Vol. 196, No. pp. 193-203, ISSN: 0022-0248

ten Wolde, P. R. & Frenkel, D. (1997). Enhancement of protein crystal nucleation by critical density fluctuations. *Science*, Vol. 277, No. pp. 1975-1978,

Tessier, P. M.; Vandrey, S. D.; Berger, B. W.; Pazhianur, R.; Sandler, S. I. & Lenhoff, A. M. (2002). Self-interaction chromatography: a novel screening method for rational protein crystallization. *Acta Cryst.*, Vol. D58, No. 10 Part 1, pp. 1531-1535, ISSN: 0907-4449

Vekilov, P. G. (2010). Nucleation. *Crystal Growth & Design*, Vol. 10, No. 12, pp. 5007-5019, WOS:000284675100001

Velev, O. D.; Kaler, E. W. & Lenhoff, A. M. (1998). Protein interactions in solution characterized by light and neutron scattering: comparison of lysozyme and chymotrypsinogen. *Biophys. J.*, Vol. 75, No. 6, pp. 2682-2697, ISSN: 0006-3495

Vérétout, F.; Delaye, M. & Tardieu, A. (1989). Molecular basis of eye lens transparency. Osmotic pressure and X-ray analysis of alpha-crystallin solutions. *J. Mol. Biol.*, Vol. 205, No. 4, pp. 713-728, ISSN: 0022-2836

Visuri, K.; Kaipainen, E.; Kivimaki, J.; Niemi, H.; Leisola, M. & Palosaari, S. (1990). A New Method for Protein Crystallization Using High Pressure. *Nature Biotechnology*, Vol. 8, No. 6, pp. 547-549,

Vivares, D. & Bonneté, F. (2002). X-ray scattering studies of Aspergillus flavus urate oxidase: towards a better understanding of PEG effects on the crystallization of large proteins. *Acta Cryst.*, Vol. D58, No. pp. 472-479, ISSN: 0907-4449

Vivares, D.; Astier, J.-P.; Veesler, S. & Bonnete, F. (2006). Polymorphism of urate oxidase in PEG solutions. *Crystal Growth & Design*, Vol. 6, No. pp. 287-292, ISSN: 1528-7483

Vivares, D.; Belloni, L.; Tardieu, A. & Bonneté, F. (2002). Catching the PEG-induced attractive interaction between proteins. *Eur. Phys. J. E.*, Vol. 9, No. pp. 15-25,

Weber, M.; Jones, M. J. & Ulrich, J. (2008). Crystallization as a Purification Method for Jack Bean Urease: On the Suitability of Poly(Ethylene Glycol), Li2SO4, and NaCl as Precipitants. *Cryst. Growth Des.*, Vol. 8, No. 2, pp. 711-716, ISBN 1528-7483

Wulff-Pérez, M.; Torcello-Gómez, A.; Gálvez-Ruíz, M. J. & Martín-Rodríguez, A. (2009). Stability of emulsions for parenteral feeding: Preparation and characterization of o/w nanoemulsions with natural oils and Pluronic f68 as surfactant. *Food Hydrocolloids*, Vol. 23, No. 4, pp. 1096-1102,

Ye, X.; Narayanan, T.; Tong, P.; Huang, J. S.; Lin, M. Y.; Carvalho, B. L. & Fetters, L. J. (1996). Depletion interactions in colloid-polymer mixtures. *Phys. Rev. E*, Vol. 54, No. 6, pp. 6500-6510, ISSN: 1063-651X

Zheng, B.; Gerdts, C. J. & Ismagilov, R. F. (2005). Using nanoliter plugs in microfluidics to facilitate and understand protein crystallization. *Current Opinion in Structural Biology*, Vol. 15, No. pp. 548–555, ISSN: 0959-440X

14

Crystallization by Antisolvent Addition and Cooling

Marco Giulietti* and André Bernardo
Chemical Engineering Department Federal University of São Carlos UFSCar
Brasil

1. Introduction

Crystallization is the second most important separation process in chemical industry after distillation. Crystallization consists of a solid disperse phase formation into a continuous medium, which usually is a liquid solution in industrial systems. This solid phase formation occurs in two main steps: the appearance of transition structures between solid and fluid phase, or nucleation; and the growth of these structures into solid particles, crystals. The solution concentration must be higher than the equilibrium concentration at that temperature (solubility) in order to nucleation and crystal growth occur. The difference between actual concentration and equilibrium concentration is called supersaturation and is the driving force of crystallization. Supersaturation can be generated in the system by cooling, solvent evaporation, or changing of medium – addition of an antisolvent which reduces the solute solubility in the resultant system, or changing the solute by chemical reaction producing another substance with much lower solubility. Frequently other secondary processes occur, as agglomeration and breakage of those particles, which affect the final product (crystal) size distribution.

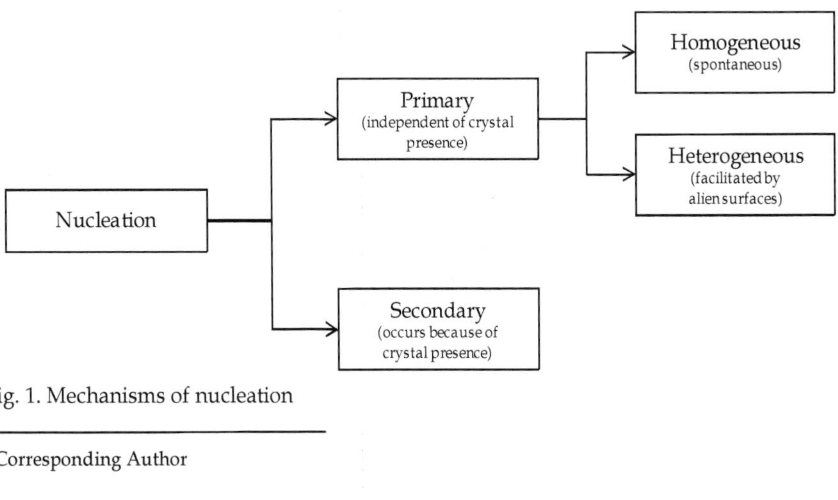

Fig. 1. Mechanisms of nucleation

* Corresponding Author

Nucleation may occur by different mechanisms. It may be primary, when is independent of crystal presence in solution, or secondary, when occurs as result of crystal presence in solution. When nucleation is primary, it may occur as a homogeneous mechanism, i.e. spontaneous depending only of supersaturation degree, or heterogeneous when is facilitated by alien surfaces (reactor or agitator wall, for instance). These different mechanisms are shown in Figure 1.

Primary homogeneous nucleation is the result of successive equilibrium of solute clustering (Figure 2). When a stable cluster is formed, it remains in solution as a transition structure from which the solid dispersed phase may appear. This stable cluster is called nucleus (Mullin, 2001).

$$\begin{aligned} & A + A \leftrightarrow A_2 \\ & A_2 + A \leftrightarrow A_3 \\ & \vdots \\ & A_{n-1} + A \leftrightarrow A_n \end{aligned}$$

Fig. 2. Solute clustering

The excess Gibbs energy is result of a resistance for the formation of a new surface, or the product of the surface tension and the cluster area, and a tendency to reach equilibrium, product of chemical potential and cluster volume:

$$\Delta G = \Delta G_S + \Delta G_V = \gamma \cdot A_{cluster} - \Delta\mu \cdot V_{cluster} = \gamma \cdot 4\pi r^2 - \Delta\mu \cdot \frac{4}{3}\pi r^3 \tag{1}$$

where γ is the surface energy of the system solute-solvent, $\Delta\mu$ is the chemical potential, $A_{cluster}$ is the cluster surface area, $V_{cluster}$ is the cluster volume, and r is the cluster radius (therefore, there is an implicit hypothesis considering cluster spherical)

Because the clusters start to grow from very small sizes (theoretically from the solute volume), it is possible to infer that initially the surface term of the excess Gibbs energy is larger than the volume term. As clusters continue to grow, they reach a critical size which corresponds to the maximum value of the excess Gibbs energy (the critical Gibbs energy). From this point on, Gibbs energy starts to decrease, and crystals grow from the existent stables clusters (nuclei). Mathematically, the critical size is calculated deriving equation (1) and equating to zero the value:

$$r_{crit} = \frac{2\gamma}{\Delta\mu} \tag{2}$$

The above equations as well as Gibbs-Thomson equation for non-electrolytes (Mullin, 2001) may be manipulated to obtain an expression for the chemical potential:

$$\ln\frac{c(r)}{c^*} = \ln S = \frac{2M\gamma}{RT\rho r} \tag{3}$$

where M is the molecular weight of solute, S is the supersaturation ratio defined as c/c^*, R is the gas constant, T the absolute temperature, ρ the crystal density and r the particle radius.

Manipulating equation (3) and considering equation (2):

$$\ln S = \frac{2M\gamma}{RT\rho r} = \frac{2\gamma\upsilon}{\upsilon\frac{R}{V}T\frac{m}{M}r} = \frac{2\gamma\upsilon}{\frac{Rn}{N_A}Tr} = \frac{2\gamma\upsilon}{kTr} \Rightarrow \Delta\mu = \frac{kT\ln S}{\upsilon} \quad (4)$$

where υ is the molecular volume, N_A is the Avogadro number, n is the number of mols, and k is the constant of Boltzmann.

Considering the homogeneous nucleus radius definition (2) and the obtained expression for the chemical potential (4), the critical Gibbs energy for homogeneous nucleation is:

$$\Delta G_{crit} = \frac{16\pi\gamma^3\upsilon^2}{3(kT\ln S)^2} \quad (5)$$

The critical Gibbs energy may be interpreted as an energy barrier to be transposed to the appearance of the nucleus, as primary nucleation depends on ordered clustering of solute until a critical size (or critical radius if cluster and nucleus are considered spherical). The practical consequences of this barrier are the metastability and the induction time. Metaestability is the state of a clear supersaturated solution, i.e. there is driving force for crystallization but there is no crystal in the system yet. The metastable state may persist for different induction times - time interval between the supersaturation imposition and the appearance of first nucleus - depending on supersaturation degree. As this supersaturation degree increases, smaller induction times are observed until the limit of instantaneous nucleation. At this point, the metastable zone width (abbreviated as MZW), i.e. a region between equilibrium concentration and actual solution concentration, was reached.

In primary heterogeneous nucleation, an alien surface (dust particle, reactor wall etc.) facilitates nucleation (Figure 3):

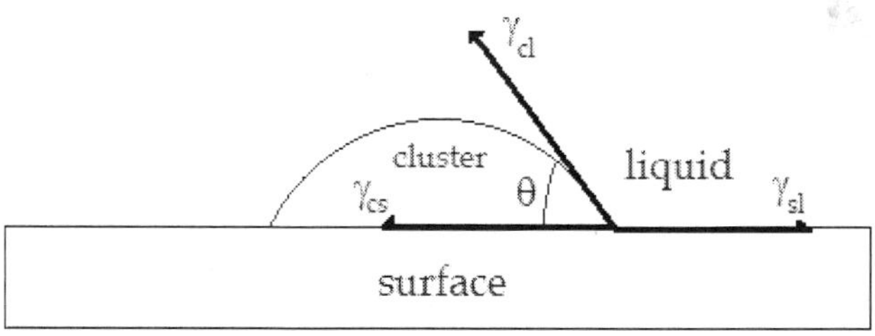

Fig. 3. Surface energies in heterogeneous nucleation. γ_{cs}, γ_{sl}, γ_{cl} are, respectively surface energies between cluster and surface, surface and liquid, and cluster and liquid.

Therefore, in heterogeneous nucleation there is a contact angle θ and a factor Φ which express the affinity between the cluster and the surface (Mullin, 2001):

$$\cos\theta = \frac{\gamma_{sl} - \gamma_{cs}}{\gamma_{cl}} \quad (6)$$

$$\Phi = \frac{(2+\cos\theta)(1-\cos\theta)^2}{4} \tag{7}$$

The factor Φ varies from 0 (total affinity between cluster and surface) to 1 (no affinity). The primary nucleation rate, J, is the rate of appearance of nucleus in a given volume:

$$J = \frac{dN}{Vdt} \tag{8}$$

where N is the number of nucleus, t is time and V is the system volume.

For its characteristics, primary nucleation may be modeled as an Arrhenius thermally activated rate:

$$J = A\exp\left(-\frac{\Delta G}{kT}\right) \tag{9}$$

Equations 5, 7 and 9 allow determining a primary nucleation rate:

$$J = A\exp\left(-\frac{16f\gamma^3\upsilon^2}{3k^3T^3(\ln S)^2}\right) \tag{10}$$

Where f is a factor that equals 1 for homogeneous nucleation or is lower than 1 for heterogeneous nucleation (Bernardo et al, 2004). Consequently, it should be expected that in industrial systems heterogeneous nucleation occurs before homogeneous, and the metastability or the induction times be smaller than in homogeneous nucleation.

It is reported that melts frequently demonstrate abnormal nucleation characteristics – nucleation rate follows the expected exponential curve as supersaturation is imposed to the system, but reaches a maximum and decreases for higher supersaturation values. It was suggested that the viscosity of melts increase starkly with cooling and restricts molecular movement inhibiting the formation of ordered structures. It was modeled by a modification in equation 10, including a 'viscosity' term (Mullin, 2001):

$$J = A\exp\left(-\frac{16f\gamma^3\upsilon^2}{3k^3T^3(\ln S)^2} + \frac{\Delta G'}{kT}\right) \tag{11}$$

where ΔG′ is the activation energy for molecular motion across the system, and is very large for viscous liquids and glasses.

When there are crystals suspended in solution, they may be shorn by relative movement of liquid or broken by collision with other crystals or with crystallizer or impellers surfaces. The consequence of these mechanical processes is the appearance in suspension of small embryos which allow the growth of new crystals, i.e. secondary nucleation. As it exists because of physical interactions between crystals and the system, it cannot be modeled by thermodynamic equations, as primary nucleation. The wide-spread solution is to relate secondary nucleation rate B_0 with process variables which may cause secondary nucleation, as supersaturation ΔC (the difference between C and C*), impeller rotation W, and concentration of solids M_T in a power law equation (Myerson and Ginde, 2002):

$$B^0 = k_N W^i M_T^j \Delta C^n \quad (12)$$

Seeding is a common practice in industry. It consists of adding a small quantity of crystal in the supersaturated solution that will facilitate the crystal growth by the existence of a surface. If the process of crystallization is seeded, it is expected that secondary nucleation be the dominant mechanism of nucleation.

The crystal growth may be defined as the variation in time of the characteristic size of the crystal:

$$G = \frac{dL}{dt} \quad (13)$$

where G is the crystal growth rate and L is the characteristic size of the crystal.

There are many theories to explain and model crystal growth. Briefly explaining, these theories may be grouped in three sets: surface energy theories, that postulate that shape of growing crystals search the minimum energy condition, are limited to molecular modeling studies; diffusion theories, which states that crystal growth is limited by the diffusion of the solute to the crystal surface; adsorption theories, which states that the integration of the solute molecule to the crystal surface is the rate-determining step. For the crystallization of melts, the crystal growth may be limited by heat release. It seems quite obvious that crystal growth may be limited by diffusion of solute to the surface, integration of the solute in the surface, or by the heat release depending on the system composition or the process conditions (Mulin, 2001). In the engineering practice, it is common to describe crystal growth rate as a power law (Myerson and Ginde, 2002):

$$G = k_G \Delta C^g \quad (14)$$

where k_G is a constant which may vary with temperature according to an Arrhenius-like equation, and g is the crystal growth order, generally a number between 1 and 2.

1.1 Antisolvent crystallization

As already mentioned, supersaturation may be generated by changing the solubility of the system by the addition of an antisolvent – a liquid miscible with the solvent which reduces solute solubility in this new mixed solvent. An advantage of the antisolvent crystallization is that the process can be carried out at temperatures near the ambient temperature. It is quite convenient for heat-sensitive substances. Also, the process would demand less energy than a solvent evaporation process. However, the solvent-antisolvent mixture must be separated in order to recover and recycle one or both solvents. Another advantage of antisolvent crystallization is that the change in solvent composition may favor one crystalline structure in those cases where the solute may crystallize in two or more crystalline phases (what is called polymorphism), and only one of them is desired for product application. Because of these characteristics, antisolvent crystallization has been widely used to crystallize pharmaceutical products, which are generally sensitive to degradation by heating and frequently have polymorphism occurrence.

Takiyama et al. (2010) utilized the antisolvent crystallization to control the formation of two possible polymorphs of indomethacin (alpha and gamma forms).To obtain only the desired stable polymorph, it is required to avoid the precipitation of meta-stable polymorph crystals. They postulate that in the antisolvent crystallization, the solubility profiles are essential data for crystallization operation design to selectively isolate the target polymorph. In antisolvent crystallization, indomethacin was dissolved in acetone and heptane as used as antisolvent. Agitation speed and rate of addition control, as well as seeding gamma crystals were done to obtain the desired gamma form. The experiments were done at 288 and 313 K.

Granberg et al. (1999) investigated the influence of solvent composition on the antisolvent crystallization of paracetamol in acetone-water mixtures where extra water was added as antisolvent and concluded that supersaturation degree and not solvent composition defines induction time. They noticed increasing nucleation and agglomeration rate with increasing initial supersaturation, but at a given initial supersaturation, the solvent composition has no clear influence on the crystalline product characteristics. Crystal growth rate showed good relationship with solubility. Their work concluded that antisolvent modifies solubility and crystal shape, but has low influence in crystallization kinetics, governed by supersaturation degree.

Analyzing data of benzoic acid crystallization in water-ethanol solution by the water addition, Kubota (2008) concluded that solvent composition has no effect on induction time or primary nucleation rate, which could be modeled considering only the supersaturation imposed to the system as water as added.

Antisolvent crystallization may be combined with cooling strategies to enhance crystallization. Sheikhzadeh et al. (2008) implemented an adaptive MIMO neuro-fuzzy logic control for crystallization of paracetamol in isopropanol-water system in which water was added as antisolvent and temperature was varied from 40 to 10°C. When seeds were added, product yield reached 99%, while unseeded experiments reached 95% product yield. Seeding allowed to significantly reduce batch time without reduction in crystal mean size.

In combined cooling antisolvent crystallization, it seems that when antisolvent is added before cooling, the results are better than the opposite. Studying crystallization of paracetamol in isopropanol-water system in which water was added as antisolvent, Knox et al. (2009) increased the yield from 78.4% to 93.5% when antisolvent was added before cooling.

Nagy et al. (2006, 2008) utilized the method of moments to model the combined cooling and antisolvent crystallization in order to obtain the optimum recipe for crystallizing lovastatin (a hypolipidemic agent in drugs) in acetone/water mixture and achieve a maximized crystal size. Compared to cooling-only strategy, antisolvent-only strategy improved the product mean size in 15%, while combined strategy improved the mean size in 22%. The width of particle size distribution was lowered in 17 and 23% when only antisolvent and combined cooling antisolvent was used, respectively, compared to cooling-only strategy.

2. Crystallization of sugars

The major amount of crystallization studies for sugars, sucrose is almost an exception, were made concerning the undesirable crystallization or the need to impose the crystallization of these substances in food formulation (Hartel, 2001; Hartel et al., 2011). Very little research concerns industrial crystallization processes, and even for sucrose, industrial processes remain based on practical operational practice. Fundamental studies on sugars solutions behavior have been being done, almost always concerning the behavior of food formulation.

In fact, the most used sugars in industrial formulations - sucrose, glucose, fructose, lactose - seem to have specific characteristics that become the study of their crystallization quite complex. These substances are all very soluble in water, which implies that their industrial processes of crystallization will have very high initial (feed) concentration. Figure 4 shows the solubility of sucrose, glucose, fructose, lactose in water.

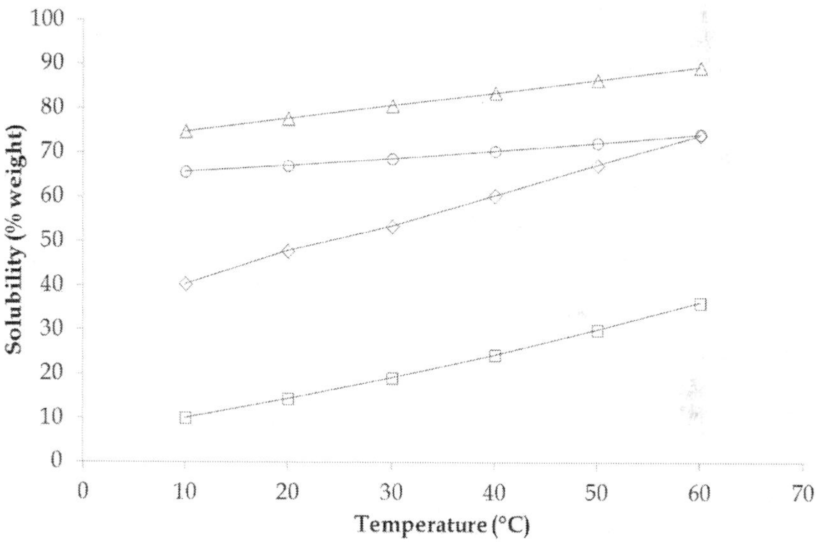

Fig. 4. Solubility in water of fructose △ (Silva, 2010), sucrose ○ (Ouiazzane et al., 2008), glucose ◇ (Alves et al., 2007), lactose □ (Brito, 2007).

It may be observed in Figure 1 that despite of lactose, all the other sugar have solubilities larger than 40 % weight. Even lactose have its solubility larger than 30 % weight at 60°C, where the others haves solubilities larger than 60 % weight.

Other important physical properties to be considered are solution viscosity and glass transition temperature T_g. Chirife and Buera (1997) presented a simple equation to predict the viscosity of sugar solutions at 20°C. Table 1 utilized this equation to calculate viscosities for selected sugars in a 60% weight solution. Hartel et al. (2011) compiled T_g values for different sugars (Table 2). For temperature values around or below T_g, the solution will be expected to exhibit a very low mobility.

Sugar	Viscosity (cP)
Glucose	36,4
Fructose	32,9
Sucrose	59,4
Lactose	65,7

Table 1. Viscosity at 20°C for 60% weight solution of selected sugars

Sugar	T_g (°C)
Glucose	31
Fructose	5-10
Sucrose	62-70
Lactose	101

Table 2. Glass transition temperatures of selected sugars

Considering a sugar crystallization process, if no seed is added, primary nucleation rate will occurs as predict equation (11), i. e. nucleation rate will grow exponentially until a maximum and decrease for higher supersaturation. Crystal growth rate will be limited by diffusion of solute to crystal face, which implies that even seeding policies has limited efficiency.

Further, metastable zone width (MZW) may have reach 30°C as reported in several references (Gharsallaoui et al. (2008); Brito (2007); Silva (2010)). The combination of large MZW, flat solubility curves (Figure 1) and high viscosity makes simple cooling crystallization practically unfeasible. Mathlouthi and Genotelle (1998) compare sucrose crystallization to a 'hurdle race', where viscosity seems to be a minor hurdle and the disassociation of hydration water a major one.

2.1 The role of water affinity in sugar crystallization

Mathlouthi and Genotelle (1998) considered two steps in sucrose crystallization: diffusion of sucrose molecules from the bulk solution to the interface crystal/solution and the incorporation of these molecules to the crystal after releasing their hydration water. Utilizing X-ray diffraction and laser-Raman spectroscopy, they concluded that hydrogen bonds between sucrose molecules in concentrated solutions is so strong that it hinders completely the free diffusion of molecules, and that diffusion in concentrated solutions is not due to viscous flow, but to the transfer of water molecules from one sucrose molecule to another by rotation of these sugar molecules. Consequently, water would diffuse in the concentrated solution and sucrose molecules would remain immobile, becoming the migration of hydration water from the crystal surface to the bulk solution very likely to be the controlling step in sucrose crystal growth.

Gharsallaoui et al. (2008) studied the interactions between water and disaccharides (sucrose, maltitol, and trehalose) in saturated solution and in crystallization conditions. According to them, narrowest metastable zone width was observed for maltitol and the largest for trehalose, because of the higher affinity of trehalose for water. They conclude that the crystallization of anhydrous disaccharides in aqueous solution necessitates that hydration water be removed and evacuated from crystal integration surface to the bulk solution to allow the growth of crystals. It seems to occur because for disaccharides in dilute or concentrated aqueous solutions, folding around the glycosidic linkage and hydrogen bonding influences very much the solution behavior. Specifically in the case of properties such as solubility, viscosity and molecular arrangements that take place before crystallization. As a general rule, high T_g sugars exhibit a greater degree of freedom to rearrange hydrogen bonds during changes in temperature than low T_g sugars.

Molinero et al. (2004) utilizing atomistic simulations investigated the nature of combination of water and glucose in supercooled solutions and concluded that there is a concentration limit not to have water freezing and keep a glassy state for all system.

Bensoussi et al. (2010) compared the metastable zone width of aqueous solutions of sucrose, maltitol, mannitol and xylitol, and attributed the observed differences to the interactions between water and solute molecules, as well as the conformation of solute molecule in solution. Further, they concluded that these factors are at the origin of solution properties like viscosity, diffusivity and surface tension, which interfere in nucleation and crystal growth. They concluded that nucleation of sugars is affected by their solubility, as it affects viscosity and the consequent solution diffusivity. Besides, the stability of bonds established with water may also affect nucleation. Sucrose and xylitol, for instance, have high potential of forming stable hydrogen bonds with water, as well as more favourable water-sugar interactions than to sugar-sugar interaction, which implies in large MZW, low capacity to form spontaneous nuclei, and high hydrophilic behavior. Mannitol and maltitol have similar MZW despite the difference in their solubility in water, because of, in the case of maltitol, its high viscosity at saturation and flexibility of glucitol moiety which decreases the stability of water-maltitol interactions. On the other hand, mannitol has low affinity for water and low viscosity, which favors the conditions to form spontaneous nuclei; its rigid conformation explains the ease of nucleation and the narrow metastable zone.

Shortly, as general rule, sugars have high affinity with water, which frequently implies in highly viscous, highly soluble solutions with large MZW, low ease of nucleation and small crystal growth rates.

2.2 Antisolvent cooling crystallization of sugars

Sugars are very polar compounds, which explain their affinity with water. The dielectric constant of water is 78.54 at 25°C. A 50 weight percent of sucrose aqueous solution has its dielectric constant equals to 60.19; a similar dextrose solution has a dielectric constant of 63.39 (Malmberg and Maryott, 1950). Dielectric constants of ethanol and acetone at 25°C are 24.55 and 20.7, respectively. As dielectric constant provides a good measure of a system polarity, it is obvious that aqueous solutions of sugars are much more polar that the common organic solvents. The solubility of a solute in aqueous solution should be decreased by the addition of an organic solvent with a dielectric constant lower than that of water. Another factor which contributes to precipitation by organic solvents is the redistribution of water and the organic solvent around solute molecule (Arakawa and Timasheff, 1985). In fact, water-organic solvent mixture cannot be regarded as a continuous medium in the vicinity of a sugar molecule, since the sugar surface may be a mosaic of regions with different polarities and different affinities for the solvent components. Furthermore, large organic molecules as sugars may also be excluded by steric hindrance.

If an organic solvent is mixed to aqueous solution of sugar as antisolvent, it is possible to suppose that it would surround the hydrophobic moieties of sugar molecule surface, dehydrating sugar molecule by a steric hindrance mechanism. This dehydration of sugar molecule, exposed to a less polar medium in which it has little affinity to solvent, would decrease viscosity (increasing solute mobility) and facilitate sugar-sugar interaction. The result would be higher ease of nucleation and higher crystal growth rate.

Following, it is presented results of antisolvent cooling crystallization of fructose utilizing ethanol as antisolvent, and for lactose utilizing acetone, ethanol, and iso-propyl alcohol. Crystallization was evaluated utilizing Nýlvlt's method to calculate crystallization kinetics (Nývlt et al., 2001).

Nýlvlt's method utilizes a set of at least nine experiments of no-seeded batch cooling crystallization, with three different cooling rates and three initial concentrations, to determine the apparent average crystal growth rate, expressed as a power-law equation similar to equation (13), and the apparent average nucleation rate $\dot{N}$, expressed as a power-law equation:

$$\dot{N} = k_N M_T^c \Delta C^n \tag{15}$$

where c value allows to comprehend the nucleation mechanism - c = 0 means true primary or secondary nucleation; c = 1 means that crystal-crystal interaction provokes nucleation; c = 2 means that friction between crystals provokes nucleation.

For all experiments, the metastable zone width ΔT_{max} is measured, and from equation:

$$\log \Delta T_{max} = \frac{1-m}{m}\log\frac{dC^*}{dT} - \frac{1}{m}\log k_N + \frac{1}{m}\log\frac{dT}{dt} \tag{16}$$

where m is the apparent nucleation order and C* is the solubility, it is possible to obtain the values of k_N and m by a multiple linear regression.

The cumulative mass distributions of crystals M(L) may be described as the function:

$$M(L) = 100\left(1 + z + \frac{z^2}{2} + \frac{z^3}{6}\right)\exp(-z) \tag{17}$$

where $z = L/(G \cdot t_{batch})$, is the crystal dimensionless size. It is possible to calculate the z values iteratively, and the average crystal growth rate for that experiment may be calculated from the relation between z and L. Crystal mean size L_m is the L value for z = 3, and

$$G = \frac{L_m}{3t_{batch}} \tag{18}$$

The linear coefficient of z-L relation is called z_n and

$$f(z_n) = 100\left(1 + z_n + \frac{z_n^{\ 2}}{2} + \frac{z_n^{\ 3}}{6}\right)\exp(-z_n) \tag{19}$$

The nucleation rate be calculated by the following equation

$$\dot{N} = \frac{27M_TG}{2k_v\rho_Cf(z_n)(L_m - L_n)^4} \quad (20)$$

A multi-linear regression of $\dot{N}$ as function of G and M_T, gives:

$$\ln\dot{N} = \ln\left(\frac{k_N}{k_g^{n/g}}\right) + c\ln(M_T) + \frac{n}{g}\ln G \quad (21)$$

Nucleation order may be calculated:

$$n = \frac{4(m-1)}{3\frac{g}{n}+1} \quad (22)$$

2.3 Antisolvent cooling crystallization of fructose

Silva (2010) studied the antisolvent cooling crystallization of fructose utilizing ethanol as antisolvent. She varied the initial concentration of the aqueous solution of fructose, the quantity of added ethanol expressed as ratio ethanol/water (E/S) and the cooling rate. The agitation rate was 500 rpm and the final temperature was 30°C for all experiments. The results of MZW are shown in table 3.

As it was expected, MZW decreases with added ethanol quantity. The crystals yield, not shown in table, was more than 93% of available fructose quantity for all experiments. The obtained crystals had cubic habit, and agglomeration occurred in all experiments. The crystal mean size, and the crystallization kinetics, calculated by Nývlt's method had no significant difference with the quantity of ethanol added (Table 4).

Initial concentration (% weight)	Ethanol (E/S)	Saturation temperature (°C)	Nucleation temperature (°C)	Cooling rate (°C/min)
86,88	1.5	50.5	30 after 40 min	0.60
86,88	4.0	50.5	30 after 30 min	0.55
86,88	6.0	50.5	42	0.55
86,88	9.0	50.5	47	0.58
88,10	1.5	55	30 after 130 min	0.50
88,10	4.0	55	30 after 20 min	0.55
88,10	6.0	55	38.5	0.60
88,10	9.0	55	42	0.55
89,36	4.0	60	38	0.22
89,36	6.0	60	40.5	0.65
89,36	9.0	60	46.5	0.65

Table 3. MZW of antisolvent cooling crystallization of fructose with ethanol as antisolvent

Flood et al. (2000) studying the same system also concluded that ethanol quantity and temperature did not affect significantly the crystal growth rate. They cited other studies that concluded the same.

Initial concentration (% weight)	Ethanol (E/S)	Product mean size (μm)	G (10^{6}·m/s)	$\dot{N}$ (10^{-11}·#/m^3s)
86,88	1.5	42.50	1.28[1]	202.1[1]
86,88	4.0	109.7	3.14[1]	12.36[1]
86,88	6.0	44.07	1.01	133.5
86,88	9.0	56.75	1.27	49.54
88,10	1.5	51.64	2.82[1]	192.5[1]
88,10	4.0	42.06	1.15[1]	224.1[1]
88,10	6.0	44.02	1.04	156.1
88,10	9.0	63.82	1.46	35.43
89,36	4.0	55.87	1.22	62.32
89,36	6.0	43.95	1.03	177.0
89,36	9.0	60.07	1.36	56.62

[1]As in these experiments nucleation occurred after cooling cessation, calculated kinetic parameters G and $\dot{N}$ must be considered cautiously.

Table 4. Crystallization kinetics of fructose calculated by Nývlt's method

2.4 Antisolvent cooling crystallization of lactose

Brito (2007) studied the antisolvent cooling crystallization of lactose utilizing ethanol (at different pH), isopropanol, and acetone as antisolvents. She varied the initial concentration of the aqueous solution of lactose, the final temperature and the cooling rate; the quantity of added antisolvent was always the same quantity of water in solution (E/S = 1), and the agitation rate was 350 rpm for all experiments. The results of MZW are shown in tables 5, 6 and 7.

C_i (% weight)	pH	Saturation (°C)	Final (°C)	Nucleation (°C)	Cooling rate (°C/min)
25.54	4.00	60	25	52	0.35
36.95	7.00	60	25	55	0.52
33.24	12.41	60	25	53	0.58

Table 5. MZW of antisolvent cooling crystallization of lactose with ethanol as antisolvent

Tables 5, 6 and 7 show that adding the same quantity of antisolvent of water in solution MZW almost disappear for all studied conditions. Tables 8, 9 and 10 present the calculated kinetic parameters (Nývlt's method).

Results presented in tables 8, 9 and 10 allow to conclude that pH has an important role in ethanol cooling crystallization of lactose – crystal growth rate and yield increase with pH, and nucleation rate decreases. For isopropanol, an increase in batch time almost always implies in higher yields and larger mean sizes due to the increase in growth rate. Also, for the same temperature variation, higher initial concentration (that means higher average supersaturation) implies in larger crystal sizes.

C_i (% weight)	Saturation (°C)	Final (°C)	Nucleation (°C)	Cooling rate (°C/min)
42.86	70	20	68	1.00
42.86	70	30	69	0.34
42.86	70	40	68.5	0.53
35.48	60	20	57	1.10
35.48	60	30	57	0.59
35.48	60	40	58	0.33
31.03	52	10	50	0.33
31.03	52	20	49	0.30
31.03	52	30	51.5	0.42

Table 6. MZW of antisolvent cooling crystallization of lactose with isopropanol as antisolvent

C_i (% weight)	Saturation (°C)	Final (°C)	Nucleation (°C)	Cooling rate (°C/min)
35.06	50	20	50	0.15
35.00	50	20	50	0.17
35.07	50	20	48	0.21
35.07	50	20	50	0.27
35.00	50	20	50	0.51
35.07	50	25	50	0.52
35.07	50	35	50	0.45
33.32	45	20	43	0.61
33.32	45	25	43	1.00
33.32	45	30	43.5	0.44
29.88	40	10	39	0.66
30.00	40	20	40	1.19
29.98	40	25	40	0.50

Table 7. MZW of antisolvent cooling crystallization of lactose with acetone as antisolvent

For acetone, larger concentration experiments had a yield of about 90%, increasing cooling rate (which means to increase average supersaturation) caused nucleation rate to increase and mean crystal size to decrease. For lower initial concentration, increasing cooling rate seems to decrease yield and mean size. Intermediate concentration experiments had an increase of mean size with cooling rate.

The presented results corroborated the expectation that the addition of an organic antisolvent eases the crystallization of sugar, as the antisolvent would decrease sugar-water interaction increasing solute mobility (Miranda et al., 2009). An indirect measurement of this effect is the sugar solubility in the solvent mixture. Figure 5 presents lactose solubility in different pH values and in a mixture of 50 percent weight of water and ethanol. Figure 6 presents solubility of lactose in water and in mixture of water with different solvents.

C_i (%weight)	pH	Cooling rate (°C/min)	L_m (μm)	G (10^5·m/s)	$\dot{N}$ (10^{-11}·#/m³s)	Yield (%)
25.54	4.00	0.35	121.18	0.734	5.683	62.4
36.95	7.00	0.52	115.77	3.473	5.308	46.04
33.24	12.41	0.58	168.37	5.051	2.313	82.46

Table 8. Crystallization kinetics of lactose (antisolvent ethanol) calculated by Nývlt's method

C_i (% weight)	Cooling rate (°C/min)	L_m (μm)	G (10^5·m/s)	$\dot{N}$ (10^{-11}·#/m³s)	Yield (%)
42.86	1	103.5	0.6029	3.6751	86.23
42.86	0.34	91.74	5.5047	7.4673	85.72
42.86	0.53	68.09	7.1675	13.769	80.42
35.48	1.1	81.44	6.1077	11.027	49.60
35.48	0.59	53.27	0.2948	18.195	75.65
35.48	0.33	89.98	6.8336	6.2648	49.20
31.03	0.33	47.00	3.1688	28.797	61.01
31.03	0.3	56.47	5.2936	21.703	73.74
31.03	0.42	80.70	5.6968	8.3657	43.85

Table 9. Crystallization kinetics of lactose (antisolvent isopropanol) calculated by Nývlt's method

C_i (% weight)	Cooling rate (°C/min)	L_m (μm)	G (10^5·m/s)	$\dot{N}$ (10^{-11}·#/m³s)	Yield (%)
35.06	0.15	118.71	0.3458	0.7797	88.74
35.00	0.17	99.04	0.3396	1.6071	92.63
35.07	0.21	53.98	0.2722	21.347	89.81
35.07	0.27	77.31	0.4179	4.0916	91.55
35.00	0.51	70.11	0.7130	11.599	86.60
35.07	0.52	93.24	1.1655	6.2562	47.45
35.07	0.45	106.43	1.9351	5.2757	41.55
33.32	0.61	113.84	1.3940	3.3583	73.74
33.32	1.00	72.96	1.7510	17.515	43.85
33.32	0.44	73.94	0.9859	10.476	82.40
29.88	0.66	114.01	1.2906	2.0553	55.72
30.00	1.19	82.07	2.3449	1.2720	40.91
29.98	0.50	149.77	2.2466	1.5606	58.08

Table 10. Crystallization kinetics of lactose (antisolvent acetone) calculated by Nývlt's method

Figure 5 explains the variation in the obtained yields in experiments of table 8, as solubility of lactose has a maximum value in neutral pH and also decreases with ethanol addition. However solubility curves do not explain the yields themselves. In systems where lactose has very low solubility, it would be expected that lactose has low interaction with solvent system and, therefore, high mobility of lactose molecule. A probable reason why lactose crystal yields are about 40% even in system with low solubility could be mutarotation (Miranda et al., 2009). Lactose molecule has two conformations (anomers), α and β forms, and below 93.5°C α-form is the constituent of stable crystals. It is known that mutarotation is affected by temperature, pH and solution impurities. Further, α-form crystallization rate may be faster than mutarotation rate, causing mutarotation to be the rate-determining step for crystallization (McLeod, 2007).

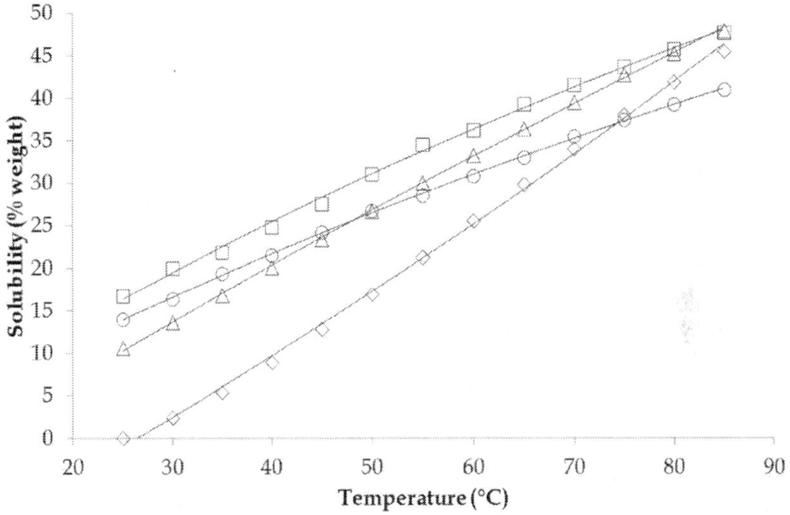

Fig. 5. Solubility of lactose in water at pH 4.0 (◇), water pH 7.0 (□), water pH 12.41 (△), and a mixture 50% weight ethanol water (○).

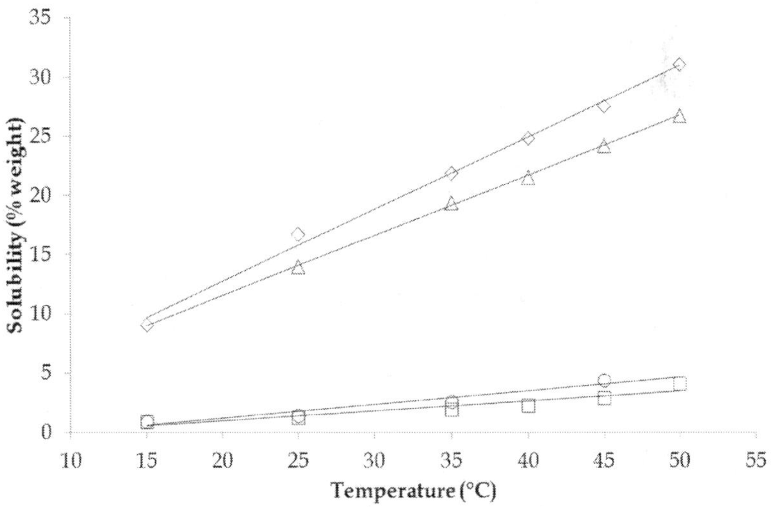

Fig. 6. Solubility of lactose in different solvents: water (◇), solution 50% weight acetone-water (□), solution 50% weight ethanol-water (△), solution 50% weight isopropanol-water (○).

Flood at al. (2000) describes fructose mutarotation issue. According to them, fructose interconverts naturally in solution in five tautomeric forms by mutarotation, but only the β-D-fructopyranose form crystallizes. In aqueous solutions the mutarotation rates would be higher than the crystallization kinetics, but in aqueous ethanolic solutions mutarotation would be sufficiently slow to move the tautomeric equilibrium away from the equilibrium β-D-fructopyranose. So, it would be important express supersaturation in terms of the tautomer, β-D-fructopyranose. In experiments driven by Silva (2010), crystal yield was always higher than 93%, but lower than 100%. In lactose crystallization experiments of Brito (2007), higher yield were about 90%. Therefore, it seems that mutarotation may reduce crystallization yields for lactose and fructose.

From an industrial perspective, in which the maximum crystal yield is an aim, it is important to emphasize that cooling and antisolvent addition must be combined. Figure 7 shows fructose solubility in weight percentage as function of water content in solvent (a mixture of ethanol and water) for the temperatures of 20°C and 60°C.

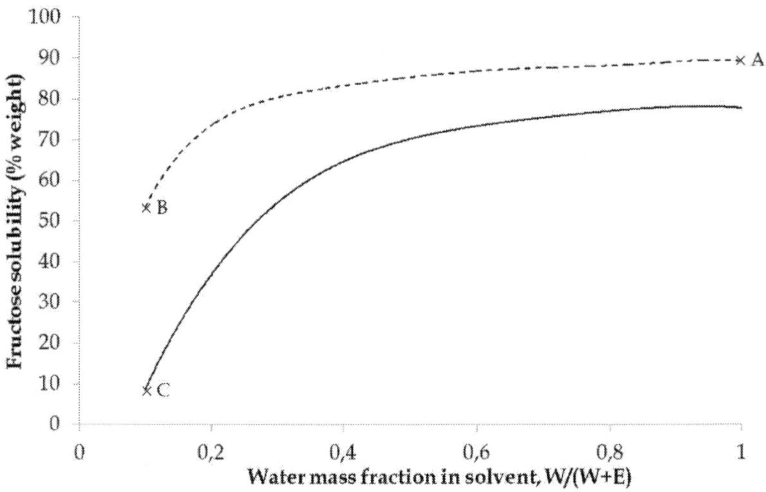

Fig. 7. Fructose solubility in a mixture of ethanol and water at 20°C (continuous line) and at 60°C (dashed line).

The data showed in Figure 7 is based on Silva (2010) work. Figure 7 allows to understand that despite the addition of antisolvent eases crystallization, crystal yield is strongly dependent on final temperature. For 10% of water in solvent (thus 90% ethanol), fructose solubility varies from 53,13% at 60°C to 9,75% at 20°C. For instance, 1000 kg of fructose aqueous solution at 60°C (point A in Figure 7) would contain 106,9 kg of water and 893,1 kg of fructose. Adding 962,1 kg of ethanol to this system, total solvent content would be 1069 kg, with 10% weight of water. In this solvent system, more than 1200 kg of fructose could be dissolved at 60°C (point B in Figure 7) - more than initial quantity with no crystal produced - but only 115,5 kg of fructose would be soluble at 20°C (point C in Figure 7) - giving a theoretical crystal yield of 87% of total fructose, desconsidering possible mutarotation effects, or 81% of total fructose, considering that in Silva's experiments, yield was always more than 93% of available fructose.

3. Conclusion

Crystallization of sugars may be improved by adding an organic liquid antisolvent (as alcohol or ketone) and cooling the system. This addition shuts nucleation hindrance off, as it decreases system viscosity. Simultaneously, the antisolvent competes with solute for water of hydration, throwing solute out of the solution: promoting crystallization. As the solubility of sugars in the mixture water-organic solvent is much lower than in water only, antisolvent addition increases the crystallization rate. Cooling the system maximizes the drowning-out effect.

Sugars may present complex structures which interconvert in solution by mutarotation. As generally only one anomer crystallizes, mutarotation may decrease crystals yield. Mutarotaton may be affected by pH, temperature, and by solvent composition. However, for the studied cases of fructose and lactose, antisolvent cooling crystallization showed to be advantageous even considering mutarotation occurrence.

Fructose was studied utilizing ethanol, and lactose was studied utilizing ethanol, acetone and isopropanol. It is possible to vary antisolvent addition rate and cooling rate simultaneously (Nagy et al., 2008), allowing to optimize crystal quality in industrial operations. The presented kinetic data as well as shape and size distribution measurement for antisolvent cooling crystallization corroborate its utilization in industrial operation.

Combined antisolvent and cooling crystallization is an important technique for obtaining products that are difficult to crystalize due to inherent solution properties like high viscosities, large metastable zone width, low kinetic of nucleation and growth, like sugars and others materials. The good choice of antisolvent must be done carefully with preliminary experiments that can allow to obtain high yields and easiness of solvent recovery. Optimal path to combine the two techniques, antisolvent and cooling, to obtain good crystal size distribution must be evaluated for each particular system, taking into account the couples solvent-antisolvent, solute-mixed solvent. The phase diagram of this ternary system is very important to evaluate that path and possible yield.

4. References

Alves, L. A., Almeida e Silva, J. B., Giulietti, M. (2007). Solubility of D-Glucose in Water and Ethanol/Water Mixtures, *J. Chem. Eng. Data*, 52, 2166-2170.

Arakawa, T., Timasheff, S. N. (1985). Theory of Protein Solubility, in: *Methods of Enzymology 114*, edited by: Wyckoff, H. W., Hirs, C. H. W., Timasheff, S. N., Academic Press.

Bensouissi, A., Roge, B., Mathlouthi, M. (2010). Effect of conformation and water interactions of sucrose, maltitol, mannitol and xylitol on their metastable zone width and ease of nucleation, *Food Chemistry* 122 443–446

Bernardo, A., Calmanovici, C. E., Miranda, E. A. (2004). Induction Time as an Instrument to Enhance Comprehension of Protein Crystallization, *Crystal Growth and Design*, 4 799-805.

Brito, A. B. N. (2007). *Study of lactose crystallization in different solvents*, PhD Thesis (in portuguese), Federal University of São Carlos, São Carlos.

Chirife, J., Buera, M. P. (1997). A Simple Model for Predicting The Viscosity of Sugar and Oligosaccharide Solutions, *Journal of Food Engineering* 33 221-226.

Flood, A. E., Johns, M. R., White, E. T., Crystal Growth Rates and Dispersion for D-Fructose from Aqueous Ethanol, *AIChE Journal*, Vol. 46, No. 2, 239-246 (2000).

Gharsallaoui, A., Roge, B., Mathlouthi, M. (2008). Water-disaccharides interactions in saturated solution and the crystallisation conditions, *Food Chemistry* 106 1329–1339

Granberg, R. A., Bloch, D. G., Rasmuson, A. C. (1999). Crystallization of paracetamol in acetone-water mixtures, *Journal of Crystal Growth* 198/199 1287-1293

Hartel, R. W., Crystallization in foods, in: *Handbook of Industrial Crystallization*, edited by Myerson, A. S., Elsevier, 2001.

Hartel, R. W., Ergun, R., Vogel, S., Phase/State Transitions of Confectionery Sweeteners: Thermodynamic and Kinetic Aspects, *Comprehensive Reviews in Food Science and Food Safety*, 10, 17-32 (2011)

Knox, M., Trifkovic, M., Rohani, S., Combining antisolvent and cooling crystallization: Effect of solvent composition on yield and metastable zone width, *Chemical Engineering Science* 64 (2009) 3555 – 3563

Kubota, N., An interpretation of the metastable zone width concerning primary nucleation in antisolvent crystallization, *Journal of Crystal Growth* 310 (2008) 4647–4651.

Malmberg, C. G., Maryott, A. A., Dielectric Constants of Aqueous Solutions of Dextrose and Sucrose, *Journal of Research of the National Bureau of Standards* Vol. 45, No. 4, October 1950, 299-303.

Mathlouthi, M., Genotelle, J., Role of water in sucrose crystallization, *Carbohydrate Polymers* 37 (1998) 335–342.

McLeod, J., *Nucleation and Growth of Alpha Lactose Monohydrate*, PhD Thesis, Massey University, 2007.

Miranda, E. A., Bernardo, A., Hirata, G.A.M., Giulietti, M,. Crystallization of lactose and whey protein. In: Jane Célia dos Reis Coimbra, Jane S. R.; Teixeira, J. A.. (Org.). *Engineering aspects of milk and dairy products*. 1 ed. Boca Raton, FL, EUA: CRC Press Taylor & Francis Group, 2009, v. 1, p. 121-154.

Molinero, V., Çagın,T., Goddard, III, W.A., Mechanisms of Nonexponential Relaxation in Supercooled Glucose Solutions: the Role of Water Facilitation, *J. Phys. Chem. A* 2004, 108, 3699-3712

Mullin, J. W., Crystallization, 4th Edition, Butterworth-Heineman, London, 2001.

Myerson, A. S.; Ginde, R., Crystals, crystal growth and nucleation, in: *Handbook of Industrial Crystallization*, edited by Allan S. Myerson, 2nd edition, Butterworth, Woburn, 2002.

Nagy, Z.K., Fujiwara, M., Braatz, R.D. Optimal control of combined cooling and antisolvent pharmaceutical crystallization, *Proceedings of BIWIC 2006 13th International Workshop in Industrial Crystallization*, Delft, The Netherlands, 2006.

Nagy, Z.K., Fujiwara, M., Braatz, R.D., Modelling and control of combined cooling and antisolvent crystallization processes, *Journal of Process Control*, 18, 856–864, 2008.

Nývlt, J., Hostomský, J., Giulietti, M., *Cristalização*, EdUFSCar, São Carlos, 2001 (in portuguese).

Ouiazzane, S., Messnaoui, B. Abderafi, S. Wouters, J., Bounahmidi, T., Estimation of sucrose crystallization kinetics from batch crystallizer data, *Journal of Crystal Growth* 310 (2008) 798–803.

Sheikhzadeh, M., Trifkovic, M., Rohani, S. (2008). *Chemical Engineering Science* 63 1261 – 1272

Silva, A. T. C. R. (2010). *Study of fructose crystallization in different media*, MSc. Dissertassion (in portuguese), Federal University of São Carlos, São Carlos.

Takiyama, H., Minamisono, T., Osada, Y., Matsuoka, M. (2010). Operation design for controlling polymorphism in the antisolvent crystallization by using ternary phase diagram, *Chemical Engineering Research and Design*, 88, 1242–1247.

15

Thin Film Growth Through Sputtering Technique and Its Applications

Edgar Alfonso, Jairo Olaya and Gloria Cubillos
Universidad Nacional de Colombia
Colombia

1. Introduction

During the last decade the dc and rf sputtering techniques have been used extensively in their two configurations – balanced and unbalanced magnetron. The main applications have been in the fields of industry and research. Examples of industrial applications are: decorative thin films (Raymond & Baham, 1999), hard wear-resistant thin films (Rodil & Olaya, 2006), low-friction thin films (Heimberg *et al.*, 2001) corrosion-resistant thin films (Flores *et al.*, 2006), and thin films used as a protective optical system (Stefan *et al.*, 2008), as well as maybe the most interesting applications, thin films used in the electronic industry (Monroy *et al.*, 2011). In the research field, the investigation has been oriented toward understanding the main physical mechanisms, such as: interaction between charged particles and the surface of the target material, adherence between the substrate and the deposited material, and chemical reactions near the substrate, as well as the influence of the deposit parameters (substrate temperature, working pressure, density power applied to the target). This research has produced thin films with a high degree of crystallinity and with the possibility of various industrial applications.

Moreover, researchers have made an effort to improve the system of operation. These efforts have been initiated through the so-called conventional or balanced magnetron sputtering in the early 1970s (Waits R, 1978), followed by the development of unbalanced systems in the late 1980s (Window, 1986) and its incorporation into multi-source "closed-field" systems in the early 1990s (Teer, 1989). Finally, the sputtering technique can increase the rate of deposition and ion energy by applying a unipolar high power pulse of low frequency and low duty cycle to the cathode target, referred to as high-power impulse magnetron sputtering (HiPIMS) or high-power pulsed magnetron sputtering (HPPMS). Common to all highly ionized techniques is very high density plasma. Implementing these discharges in sputter deposition technology modifies the surface of components, bringing improvements in mechanical, chemical, optical, electronic, and many other properties of the material. High-current glows are transient discharges operating at simultaneously high voltage (> 300 V) and high current density (> 100mAcm^{-2}). They have recently proven successful for the deposition of thin-film materials. These developments have made it possible to have an exceptionally versatile technique, suitable for the deposition of high-quality, well-adhered films of a wide range of materials with high rates of deposition. Table 1 show the main applications obtained in the last decade with the magnetron sputtering (balanced and unbalanced) rf and dc versions.

Technique	Coating	Reference
Balanced magnetron	Nano-composites of NC-TiC wear applications	(Pavel, *et al.*, 2011)
Balanced magnetron	Optical properties of AlSiN nano-composites	(Liu, *et al.*, 2009)
Balanced magnetron	Hard coatings to decorative applications	(Raymond & Baham, 1999)
Balanced magnetron	Nd-Fe-B Film for magnetic applications	(Liu, *et al.*, 20007)
Balanced magnetron	Optical applications.	(Stefan, *et al.*, 2008)
Unbalanced magnetron	Hard films for corrosion and wear applications	(Rodil & Olaya, 2006).
Unbalanced magnetron	Nb Films for biological applications	(Olivares, *et al.*,2011)
Unbalanced magnetron	NbO films for biological applications	(Ramirez, *et al.*, 2011)
Unbalanced magnetron	Electrical applications of NbN films.	(Olaya, *et al.*, 2008,)
Unbalanced magnetron	Diamond-like carbon films for infrared transmission enhancement	(Lingxia & Xu, 2000).
High power pulsed magnetron sputtering	Thin films for automotive engineering	(Bewilogua, *et al.*, 2009)
High power pulsed magnetron sputtering	TiAlCN/VCN films for tribological applications	(Kamath, *et al.*, 2011)

Table 1. Technological applications of thin films obtained with magnetron sputtering

In sputtering there are two means of operation: dc (diode and triode) and ac (radiofrequency), which also function in two configurations: magnetron dc (balanced and unbalanced) and magnetron ac (balanced and unbalanced).

In dc (diode) discharge, the cathode electrode is the sputtering target and the substrate is placed on the anode, which is often at ground potential (Vossen &Cuomo, 1978). The applied potential appears across a region very near the cathode, and the plasma generation region is near the cathode surface. The cathode in dc discharge must be an electrical conductor, since an insulating surface will develop a surface charge that will prevent ion bombardment of the surface. This condition implies that dc sputtering must be used to sputter simple electrically conductive materials such as metals, although the process is rather slow and expensive compared to vacuum deposition. An advantage of dc sputtering is that the plasma can be established uniformly over a large area, so that a solid large-area vaporization source can be established.

On the other hand, in dc sputtering the electrons that are ejected from the cathode are accelerated away from the cathode and are not efficiently used for sustaining the discharge. To avoid this effect, a magnetic field is added to the dc sputtering system that can deflect the electrons to near the target surface, and with appropriate arrangement of the magnets, the electrons can be made to circulate on a closed path on the target surface. This high current of electrons creates high-density plasma, from which ions can be extracted to sputter the target material, producing a magnetron sputter configuration (Penfold, 1995). A disadvantage of the magnetron sputtering configuration is that the plasma is confined near the cathode and is not available to active reactive gases in the plasma near the substrate for reactive sputter deposition. This difficulty can be overcome using an unbalanced magnetron configuration (see Fig. 1), where the magnetic field is such that some electrons can escape from the cathode region (Windows & Savvides, 1986). A disadvantage of the unbalanced magnetron is that the current of escaping electrons is not uniform, and the plasma generated is not uniform.

In ac sputtering, working at frequencies below about 50 kHz, the potential on the target is periodically reversed, and the ions have enough mobility so that a dc diode-like discharge, where the total potential drop is near the cathode, can be formed alternately on each electrode. The substrate chamber walls can be used as the counterelectrode. At frequencies above 50 kHz, the ions do not have enough mobility to allow establishing a dc-diode-like discharge and the applied potential is felt throughout the space between electrodes. The electrons acquire sufficient energy to cause ionizing collisions in the space between the electrodes. When an rf potential with a large peak-to-peak voltage is capacitively coupled to an electrode, an alternating positive-negative potential appears on the surface. During part of each half-cycle, the potential is such that ions are accelerated to the surface with enough energy to cause sputtering, while in alternate half-cycles, electrons reach the surface and prevent any charge buildup. Rf sputtering can be used to sputter insulating material, although the sputtering rate is low. A major disadvantage of rf sputtering of dielectric targets is that most insulating materials have poor thermal conductivity and high coefficients of thermal expansion, and are usually brittle materials. Since most of the bombarding energy produces heat, this means that large thermal gradients can be generated that result in fracturing the target if high power levels are used.

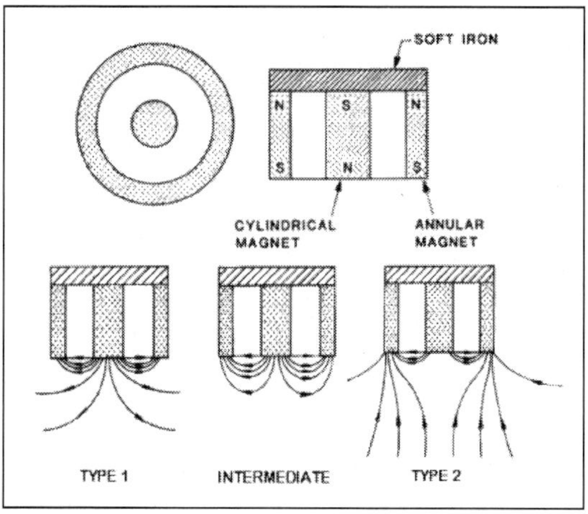

Fig. 1. Schematic configuration of the a- balanced magnetron (intermediate) and b- unbalanced magnetron (type 1and type 2 (B. Window & N. Savvides, 1986)

Rf sputtering can be used with a magnetic field in balanced and unbalanced configurations to obtain a result similar to dc-like diode discharge. In fig 2 the different configurations of dc and rf sputtering are shown.

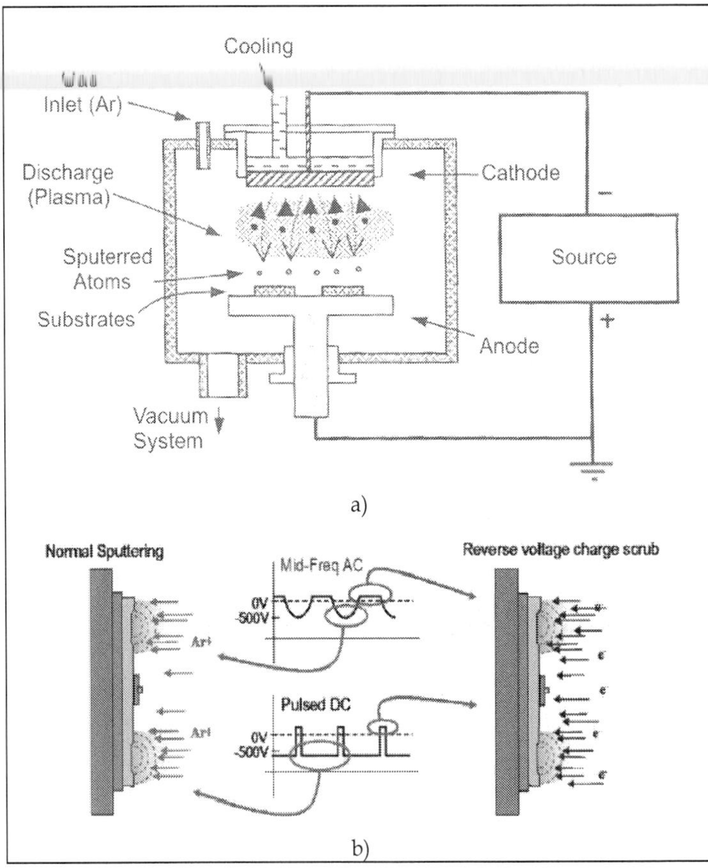

Fig. 2. Sputtering configurations a) dc sputtering, b) rf sputtering

It is important to state that in all the cases discussed above, the target and the substrate were facing (on- axis sputtering). In this configuration, the highly energetic electrons irradiate the substrates and/or the growing surface of the thin films during deposition. Off-axis sputtering reduces the effects of the irradiation of the high-energy particles. In off-axis sputtering, the substrates are settled at the outside of the discharge plasma. The thickness distribution of thin films deposited by off-axis sputtering will be larger than that for on-axis sputtering. A rotating substrate holder with a metal shadow mask is used for the reduction of the thickness distribution of the off-axis sputtering. Under a suitable design, the thickness distribution is less than 2% for substrates of 100 × 100 mm in an rf sputtering system using a 5-inch target (Shibahara *et al.*, 1987).

In this chapter we will present the physical parameters involved in the growth of thin films; also discussed will be the influence that the growth parameters have on the degree crystallinity of the films, the chemical characterization, and the optical characterization of the films; and finally, we will discuss the residual stress, hardness, and corrosion and wear resistance of thin films.

2. Physical sputtering

The main physical phenomenon involved in the sputtering technique is the momentum transfer between energetic atomic-sized particles (usually ions of noble gases) and the atoms of the surface of the material (target). During the interchange of momentum, many effects can be produced on the elastic and inelastic collisions; in the first kind of collision, mainly reflected particles can be found (neutrals, ions of the target and the gas). In the second kind, the collisions can present secondary electrons, UV/visible photons, X-ray and implanted particles; schematically, Fig. 3 shows different processes that may occur during the interaction between charged particles and the surface of the material.

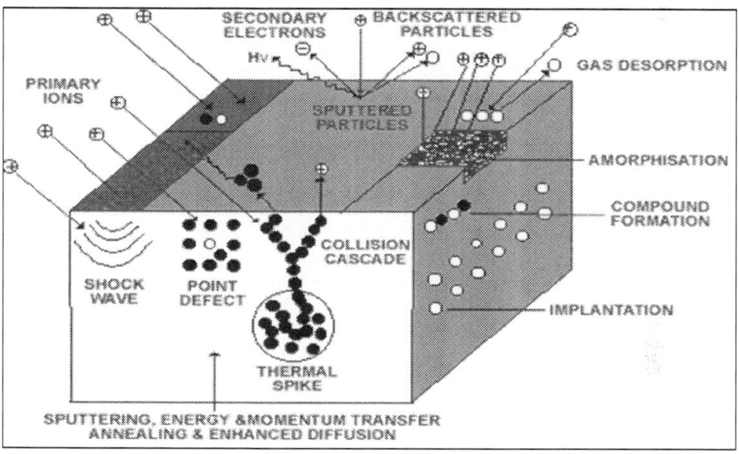

Fig. 3. The main physical process produced in sputtering technique (Weissmantel, 1983).

The momentum-transfer theory for physical sputtering was proposed early on, but was replaced by the "hot-spot" theory, in which the process of thermal vaporization is involved. The confusion about the physical process present in sputtering has only been overcome thanks to the work of Gunthersshulze in the 1920´s and 30´s and Wehner *et al.* in the 1950´s and 60´s, who demonstrated that the effects produced in sputtering could only be explained by the momentum transfer theory. These effects can be summarized as:

1. The sputtering yield (ratio of atoms sputtered to the number of high-energy incident particles) depends on the mass of the bombarding particle as well as their energy. For ion energies from 100 eV to 1000 eV the sputtering yield can be calculated as:

$$Y = \frac{3\alpha}{4\pi^2} \frac{4m_1 m_t}{(m_1 + m_t)^2} \frac{E}{U} \tag{1}$$

where m_i is the atomic mass of the bombarding incident ion, m_t the atomic mass of the target, U the binding energy of the surface atom of the target, E the energy of the incident ion, and α depends on the ratio of the masses of the target atom and the incident ion (monotonically increased with mi/m_t; for a ratio 0.1 α= 0.17 for a ratio 10 α=1.4) (Ochiati, 1986).

The sputtering yield is sensitive to the angle-of-incidence of the bombarding particle.There is a "threshold energy" below which sputtering does not occur no matter how high the bombarding flow.

2. Many sputtering atoms have kinetic energies much higher than those of thermally evaporated atoms.
3. Atoms ejected from single crystals tend to be ejected along the directions of the close-packed planes in the crystal.
4. In a polycrystalline material, some crystallographic planes are sputtered faster than are others.
5. Atoms sputtered from the alloy's surface are deposited in the ratio of the bulk composition, not their relative vapor pressure, as in the case of thermal vaporization.
6. Sputtering yields decrease at very high energies because the ions lose much of their energy far below the surface.
7. The sputtering yield is rather intensive to the temperature of the sputtering target.
8. There is no sputtering by electrons even at very high temperature.
9. The secondary electron emission by ion bombardment is low, whereas high rates from thermo electron emission would be expected if high temperatures were present.

Effects 1 through 7 above are important for the growth of films by sputter deposition. This is particularly true for low pressure (<5 m Torr).

3. Physical models that explain the microstructure of thin film growth through sputtering

The microstructure of thin films is related to the mobility of the adatoms during growth. The energy supply to the atoms is provided by the following mechanism: a- thermal effect, b- ionic bombarding and c- chemical reactions at the substrate. The effects that are produced by these mechanisms in the growth of thin films can be explained by the structure zone model (SZM). The SZM model can determine the morphology and microstructure of the films as a function of the adatoms, regardless of the kind of material. The parameters that the SZM model includes for determining the microstructure of the films are basically the substrate temperature, the final working pressure, the bias voltage applied to the substrate, and the thermal characteristics of the target. For example, in the research of Movchan and Demchishin (Movchan & Demchisshim, 1969), it has been established that the microstructure of the thin films of Ti, Ni, ZrO_2 y Al_2O_3 is related to the normalized temperature, *i. e.* Ts/Tm (Ts is the temperature of the substrate and Tm is the melting temperature of the target). Movchan *et al.* have shown that in metallic films there are three well-defined zones.

The first zone is Ts/Tm <0.3. This zone is formed by small and elongated grains that form a columnar structure with porous morphology and weakly binding grains. The columnar structure is produced by a low diffusion, a low mobility of the atoms adsorbed by the substrate surface, and the atomic shadow effects, which are produced by varying velocity in the growth the columns and the various incidence angles at which the atoms arrive at the surface of the substrate. In the second zone, 0.3≤ Ts/Tm≤0.45, the substrate temperature increasing homogeneous which leads to a higher diffusion of the adatoms, which produce a dense structure with a higher degree of binding among the columns and the borders between columns, with borders of the grain beginning to form. In this zone, the size of the grain can be increased and the grains extended in equiaxed form, from interface substrate-film to film surface. In the third zone, Ts/Tm>0.45, the volumetric diffusion size has a great influence on the morphology of the film, due to the increase in the diffusion into the grains, which produces growth of the grains, formation of the equiaxed grain and re-crystallization. These effects produce a greater crystalline structure.

Thorton (Thorton, 1974) elaborated the zone classification, considering the final working pressure, because this growth parameter can change both the kinetic energy of the ions that arrive at the substrate and the mean free path of the particles, which allows an increase or decrease in the bombardment of the surface of the substrate, which in turn determines the mobility the adatoms in that surface. In the Thorton model, the T zone as a transitional zone between first and second zone discussed above was added. The T zone is formed by grains defined by the limits of the low porosity. The surfaces of the T zone are denser and less rough than the two surfaces around them (see fig 4).

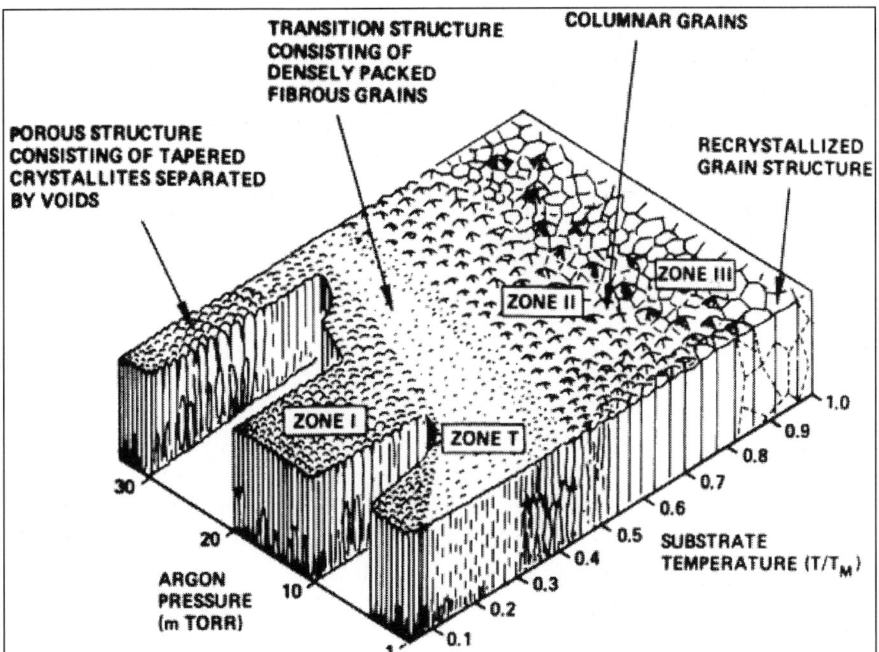

Fig. 4. Thorton zone model (Thonton, 1974)

Moreover, Messier (Messier & Giri, 1984) found that in thin films of TiB_2, BN and SiC there is a non-linear limit between the first zone and T zone, which is a function of the bias voltage applied to the substrate. The bias voltage of the substrate has the same effect on the mobility and adsorption of the atoms as an increase in the temperature, so when it is increased, the T zone increases and the first zone decreases, resulting in denser thin films and with a high degree of crystallinity.

The bias voltage also influences the mechanical properties of thin films; for example, with a voltage (>100V) applied to the substrate, the deformation of the lattice increases, causing high residual efforts and low adherence between the substrate and the film.

A new three-dimensional SZM model has been produced in order to explain the simultaneous influence of the normalized temperature, the bias voltage, and the relation between ion density current (J_i) and density deposited atoms (J_a). In Fig. 5, it can observed that it is possible to obtain the third zone with high density using a combination high-density ionic current, intermediate values of normalized temperature, and low bias voltage (Kelly & Arnell, 1998).

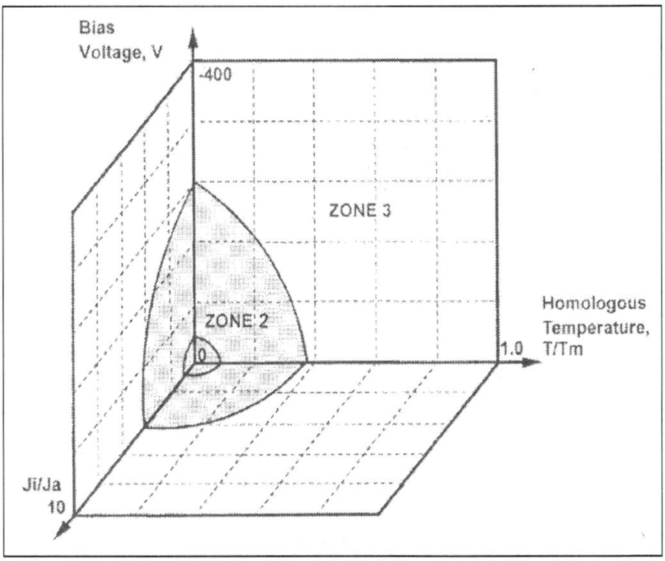

Fig. 5. The zone model including the relation between current density of ions and current density of atoms (Kelly & Arnell, 1999).

Other parameters of growth that affect the microstructure are the power supply to the target and gas flows. For example, changing the nitrogen flow during the growth of TiN and NbN films changes the preferred orientation of the films from the [111] direction to the [200] direction (Alfonso *et al.*, 2010). Explaining the change in the direction of growth of metallic nitride films is a very complex procedure, as is shown in table 2, where it can be observed that any parameter change can influence the direction of growth of the film. There are three models that explain the preferential growth direction: a) the thermodynamic model (Pelleg *et. al.*, 1991); b) the kinetic model (Greene *et al.*, 1995), and c) the atomic model (Petrov *et al.*, 2003).

Material	Parameter	Change of orientation	References
TiN	Ji/Ja	(111) → (200)	(Green *et al.*, 1995)
	N2/Ar	(111) → (200)	(Noda *et al.*, 2002)
	Th	Mixed → (200) → (111)	(Pelleg *et al.*, 1991)
	Vb	(200) → (111) → (220)	(Kobayashi & Doy, 1984)
	W	(200) → (111)	(Je *et al.*, 1997).
	T	(111) → (200)	(Cheng *et al.*, 2002)
AlN	P	axis c → axis a	(Lee & Y. Lee, 1994)
	ds-t	axis c → axis a	(Ishihara *et al.*, 1998)
	λ/ds-t	axis a → axis c	(Ishihara *et al.*, 1998)
	N2/Ar	axis a → axis c	(Lee & Y. Lee, 1994)
	Vb	axis a→ axis c→ axis a	(Wang & Zhao, 1997).
	W	axis a → axis c	(Lee & Y. Lee, 1994)
TaN	N2/Ar	(111) → (200) → (111)	(Nie *et al.*, 2001)
	Vb	(111) → (220)	(Lin & Lee, 2000)
	Ji/Ja	(111) → (200)	(Shi *et al.*, 2002)

Table 2. Relationship between the deposition parameters and film preferential orientation. Ji/Ja: ion–atom flow ratio, N_2/Ar: flow ratios, t h: thickness, Vb: bias voltage, W: power supply, Ts: substrate temperature, P: pressure, ds-t: target-substrate distance y λ: free mean path.

In the thermodynamic model, it has been established that the growth orientation in thin films is produced at the thermodynamic equilibrium, which it reaches when the total energy (W_{hkl}) of the system substrate-film is at a minimum. In this case, W_{hkl} is formed by the addition of surface energy (S_{hkl}) and deformation energy (U_{hkl}), S_{hk} passivation; energy is produced by unsaturated bonding at the surface. S_{hkl} can be calculated by:

$$S_{hkl} = \frac{\Delta H N_{hkl}}{Z}\left[\frac{J}{m^2}\right] \tag{2}$$

where ΔH is the sublimation energy, N_{hkl} is the number of unsaturated bondings per atom at the plane (hkl), and Z is the amount of coordination among neighbors. Equation 2 does not include impurities.

On the other hand, deformation energy is related to intrinsic efforts in the film. Considering only two dimensions, U_{hkl} can be calculated by:

$$U_{hkl} = \epsilon_{hkl}^2 E_{hkl}(1-\mu) \tag{3}$$

where ε_{hkl} is the deformation along the plane (hkl) of the film, E_{hkl} is the elastic module of the plane (hkl), and μ is the Poisson ratio. Research has established that the values of ε_{hkl} and E_{hkl} are different for (111), (220) and (200) planes (Ma *et al.*, 2004), and using equations (2) and (3) an order relation has been obtained: $S_{111}>S_{220}>S_{200}$ and $U_{200}>U_{220}>U_{111}$. These relations imply that the preferential orientation is determined by a configuration of minimum total energy, which results in competition between the (111), with minimum deformation energy, and (200) planes, with minimum surface energy.

Moreover, Pelleg has researched the variation of the total energy (W_{hkl}) for the (111), (220) and (200) planes as a function of the thickness of the TiN film deposited at 373K through the rf sputtering technique. These results have shown that the surface energy does not change with the thickness, the deformation energy increases proportionally with the thickness, and the slope of the straight line represents the deformation per volume.

In the kinetic model, it is proposed that the kinetic process of the ions is involved in the orientation of the growth, which influences the surface substrate, the energy and flow of the ions being the main parameters. A thin film grows in the [111] direction because the adatoms have limited mobility, caused by the low temperature of the substrate and the restricted flow of the ions. Increasing the mobility, either through increasing the substrate temperature or raising the relation J_i/J_a above five, a preferential orientation in the [200] direction is obtained.

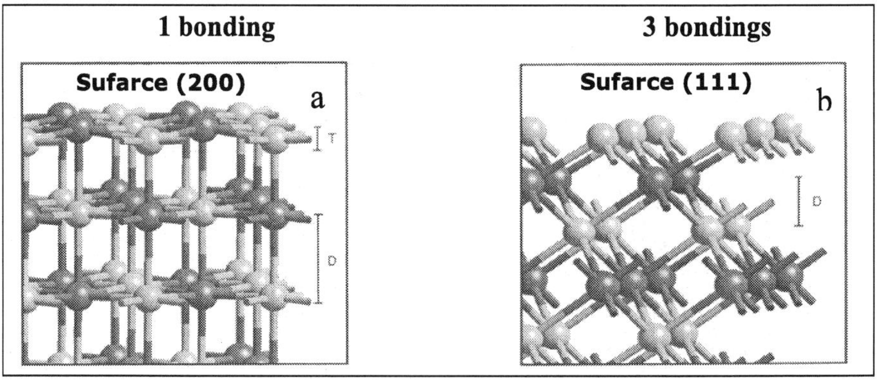

Fig. 6. Surface growth on a- (200) planes and b- (111) planes. The growth on the (200) plane has 1 binding, while the surfaces grown on (111) planes have 3 bandings (M. Marlo & V. Milman, 2000).

Finally, the atomistic model considers that in thin films deposited at high temperatures the thermodynamic parameters control the orientation of the growth, favoring planes with low energy, *i.e.* the (200) planes. This behavior can be understood if adatoms with low diffusion, as in the case of the cation Cl in NaCl, which have three bonds in the (111) surface, and have high diffusion in surface (200), are considered to have only one bond (see Fig. 6), leading to the conclusion that if NaCl thin films are grown at high temperatures, they probably will grow along the [200] direction.

On the other hand, the same model establishes that the degree of ion bombardment is influenced by the flow ratio between the density current of the ions and the density current of the atoms, Ji/Ja, and the ion energy, Ei, and both depend on the pressure, substrate target distance, and substrate bias (Ensinger, 1998; Losbichle & Mitterer, 1997). One parameter that combines both the ion energy and the flow has been suggested by Musil and Kadlec (Musil

& Kadlec, 1990), Ziemann and Kay for Pd coatings (Ziemann and Kay, 1993) and passivation for CBN films (Kulisch, *et al.*,1999). This represents the energy deposited per dense particle. This bombardment-induced mobility parameter, known as the energy parameter Ep, is defined as:

$$E_p = \frac{J_i}{J_a} \times E_i \tag{4}$$

where E_i is incident energy of the ions and is obtained from the plasma potential (V_p) and substrate bias (V_s) and the elementary charge by the expression $E_i = e(V_s - V_p)$.

3.1 Physical mechanisms of thin film growth

So far, we have studied the influence of the growth parameters on the structure and microstructure of thin films; now we discuss the main physical mechanisms involved in the nucleation and crystallization of the films over the substrate. Studies done through X-ray diffraction, optical diffraction and mainly through electron diffraction have allowed establishing that there are three mechanisms of the nucleation and growth of thin films, which depend on the thermodynamic parameters of the deposit and the substrate surface interaction between the adatoms and the substrate material (Green, 1994). The three basic modes are: (i) Volmer-Weber model, (ii) Frank-Van der Merwe model and (iii) Stranski-Krastanov model. A schematic illustration of each of these modes of growth is shown in Fig. 7.

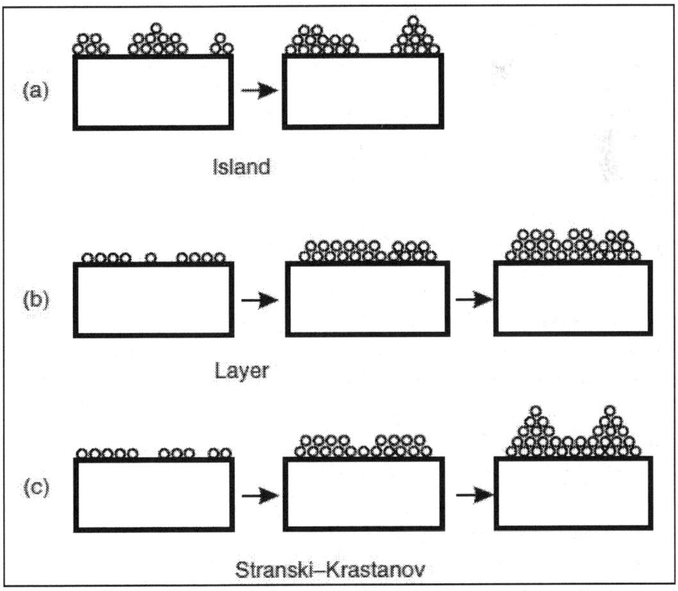

Fig. 7. Modes of growth of film: a- Volmer-Weber island growth b- Frank-vander Merwe layer growth and c- Stranski-Krastanov layer plus island growth (Harsha, 2005).

In the Volmer–Weber model, equilibrium exists in a three-dimensional crystal of the film in contact with the substrate, while the rest of the substrate is devoid of any condensed phase. Nucleation of film occurs in the form of discrete three-dimensional nuclei on the surface of the substrate, for example lead on graphite. Both the number of nuclei and the size of a given nucleus increase. Finally, the nuclei grow in size until they intersperse with each other to form a continuous film.

In the Frank and Vander Merwe model, nucleation occurs in the form of a monolayer island of the deposit. Eventually the monolayer's grow together to form a complete continuous monolayer of the deposit. The process repeats itself so that the deposit grows in a layer-by-layer manner, for example rare gases on graphite. In this growth, the interaction between the substrate and the layer atoms is stronger than that between neighboring layer atoms. Layer-by-layer growth is hindered by elastic constraints at the solid–solid interface.

The Stransky and Krastanov (S-K) model combines the features of layer-by-layer growth and discrete three-dimensional nucleation The S-K nucleation is common with metal-on-metal deposition and at low temperatures where the surface mobility is low (Greene, 1987). The conditions for these types of growth are generally described in terms of thermodynamics and surface energy considerations. In this model, nucleation and growth occurs as in the layer-by-layer mode, so that a finite number of monolayer's is produced. Subsequent formation of film occurs by formation of discrete nuclei. The lattice mismatch between the substrate and the deposit cannot be accommodated when the layer thickness increases, so the three-dimensional growth follows the layer-by-layer growth. Alternatively, symmetry or orientation of the overlayers with respect to the substrate might be responsible for the production of this growth mode.

The foregoing models were summarized by Barna and Adamik (Barna & Adamimik, 1998), who established that the growth of the films has the following evolution: nucleation, island growth, coalescence of islands, formation of polycrystalline islands and channels, development of continuous structure, and thickness growth.

According to Barna, the nucleation starting the growth of individual islands takes place on the substrate surface at the very first stage of the condensation (primary nucleation*)* or later on the bare substrate surface area developing upon liquid-like coalescence (secondary nucleation*)*. A peculiar case of nucleation shows up on the surface of a growing crystal when its growth is blocked by a surface covering layer (SCL) of an impurity phase. This is the repeated nucleation. The primary nucleation starts the condensation and the film growth on the whole substrate surface simultaneously, while the secondary and the repeated nucleation initiates the start of the growth locally in later stages of film formation. It is important to note that on amorphous substrates the nuclei are randomly oriented.

Crystal growth is the fundamental structure-forming phenomenon which incorporates the depositing material into the condensed phase. Two main cases of crystal growth should be considered in the case of polycrystalline thin films: a- the growth of discrete crystals dispersed on the substrate surface, (Fig. 8a and b), and b- the growth of crystals which are parts of a polycrystalline structure (Fig. 8c and d). Crystals growing from the nuclei are randomly oriented due to the random orientation of the nuclei. The complete coalescence of the crystals touching each other produces a grain coarsening, resulting also in the development of discrete single crystals and is connected to some changes in the orientation controlled mainly by the minimization of the substrate–crystal interface energy. The intersection lines of the crystal side faces and the substrate present a specific structural precondition for the growth of these crystals. The intersection lines can be active or passive in the monolayer nucleation on the side crystal faces. In the first case, the movement of the monolayer growth steps proceeds from the intersection line to the top of the crystal, while in the second case, the movement of the growth steps proceeds in the direction of the intersection line. In the presence of impurities, the direction of the movement of the growth steps will be important in determining the location of the developing second phase, *e.g.*, SCL.

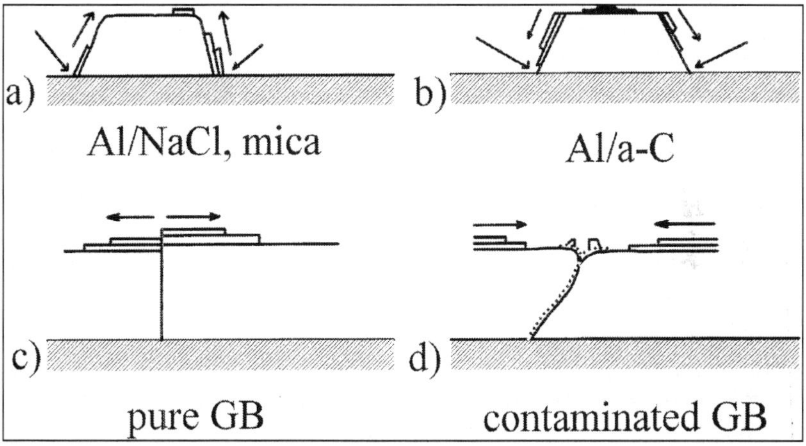

Fig. 8. Types of crystal growth in polycrystalline thin films: a- and b- growth of dispersed individual crystals on the substrate surface, the intersection line between the substrate and the side crystal faces are active (a) and passive (b) in the monolayer nucleation; c- and d- role of grain boundaries in the growth of crystals as parts of a polycrystalline matrix, pure grain boundary active in the monolayer nucleation (a), contaminated grain boundary passive in the monolayer nucleation (b). (Barna & Adamimik)

4. Influence of the normalized temperature (T_s/ T_m) on the structure and microstructure of thin films

The structural behavior, as a function of the normalized temperature of $Ts/T_m<0.3$ is shown in Fig. 9. In general, the patterns recorded from the various films grown present the same reflections as the target (δ-NbN cubic phase), but show important differences in the relative intensities, in particular those corresponding to planes (111) and (200), which increase with an increase of the absolute temperature. Regarding this, it is important to point out that contrary to what occurs in the target and the rest of the film patterns, the XRD pattern of the film grown at 553 K shows a greater intensity for the reflection from plane (111) than from that corresponding to the (200) plane. While the (111) plane in the target pattern has a relative intensity of 100% compared to 86% for the (200) plane, in the 553 K sample the relative intensity of the (200) plane is approximately 6% of the corresponding (111) plane. This result indicates that the supply of additional 22.4 meV to the substrate (which is the energy difference between room temperature and 553 K) does not bring about changes in the polycrystalline character of the materials, but gives place to the growth of films with preferred orientation along the (111) plane.

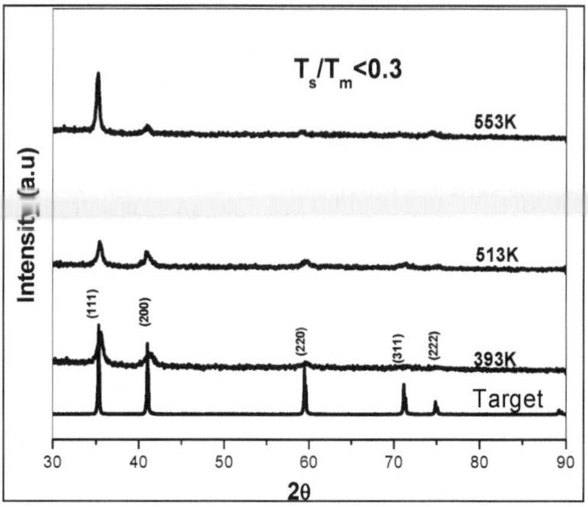

Fig. 9. XRD pattern of NbN thin films deposited at a normalized temperature (T_s/T_m<0.3) at 300W of power.

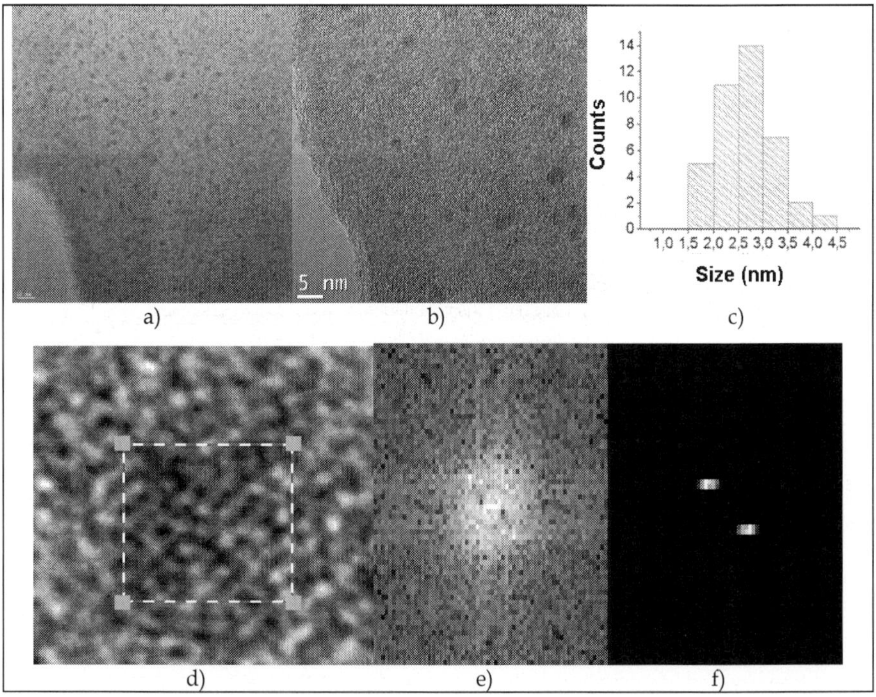

Fig. 10.a) and 10.b) TEM micrographs using multibeam configuration of NbN nanoparticles distributed homogeneously; c) average grain size of NbN nanoparticles; d) HRTEM micrographs shows atomic planes; e) reciprocal space of a NbN nanoparticle; and f) distance between (200) planes.

Fig. 10 shows the micro-structural analysis carried out on the NbN films through transmission ion electron microcopy (TEM). Micrographs 10a and 10b correspond to the image obtained in multibeam configuration, and in them one can observe an amorphous matrix with crystalline nanoparticles with spherical morphology that are homogenously distributed along the growth plane of the film. Using Gatan software, it was determined that the average grain size of the nanoparticles was 2.7±0.6 nm with normal distribution (fig. 10c). When doing magnification of nanoparticles of Fig. 10a, we can observe an interferential pattern produced by the atoms belonging to the NbN film (fig. 9d). Higher magnifications of the nanoparticles visualized in Fig. 10a let us observe atomic planes as a product of the interferential pattern produced by the arrangement of atoms belonging to the NbN film (Fig.10d). Image processing of Fig. 10d, using the Fourier transform, lets us visualize the reciprocal space of one NbN nanoparticle (Fig. 10e). Applying a mask (Fig. 10f) over the corresponding diffracting spots and directly measuring the distance between them (0.224 nm), we can confirm that they correspond to distances between (200) planes of δ-NbN. These results confirm the analysis of XRD, which established that NbN grows preferentially along the (200) plane.

To produce thin films with normalized temperature larger than 0.3 it is necessary to have ultrahigh-vacuum equipment. This condition limits the production the thin films through magnetron sputtering. However, there are studies such as that of Frederick (Frederic & Gall, 2005), who grew CrN thin films on MgO substrate through magnetron sputtering with Ts/tm between 0 .43 to 0.48. These authors found that the films grown at 993 K present complete single-crystal structure with smooth surfaces. The root-mean-square surface roughness for 230-nm-thick layers decreases from 18.8 to 9.3 to 1.1 nm as Ts is raised from 873 to 973 to 1073 K.

4.1 Influence of the gas flow on the structure and microstructure of thin films

In a different set of experiments, nitrogen gas was introduced into the deposition chamber (maintaining the final working pressure constant) in order to study the influence that the addition of this gas during deposition exerts on the structural and micro structural properties of the NbN films. Fig. 11 shows the diffraction patterns recorded from the films grown at 300 W, 553 K and different nitrogen flows. The results obtained make clear that in all cases a preferential growth appears along the (200) plane (Fig.11a). The relative intensity of this diffraction peak is so high that it makes it impossible to distinguish the diffraction peaks corresponding to other planes. To determine the polycrystalline character of the film, we carried out X-ray diffraction experiments at grazing incidence. Fig. 11b shows a representative example. It is clear from Fig. 11 that the films show the same diffraction peaks as the target, confirming their polycrystalline character, although having a preferential orientation along the (200) plane (texture index, 0.65). The grain sizes deduced by Scherer's equation (Cullity, 2001) for the different films grown along the (200) plane vary from 35 nm (Φ= 2 sscm) to 42 nm (Φ = 6 sscm). These results indicate that incorporation of nitrogen during the fabrication process favors the preferential growth of the d-NbN phase along the (200) plane. This effect has also been produced during the growth of TiN films, due to the change of the preferred orientation of the films from the [111] direction to the [200] direction when the gas flow is increased.

The incorporation of nitrogen during the deposition process implies changes in the dynamics of the plasma, since the increase in the number of nitrogen molecules increases the probability of collisions, promoting a larger number of chemical reactions on the substrate surface. These reactions can be explained using a model of low energy (<20 eV) ion bombardment during film growth at a Ts/Tm ratio ranging between 0.1 and 0.3 (a condition which is satisfied in this case) (Petrov & Barna, 2003). According to this model, 25 eV are sufficient to cause collision dissociation of the N_2 ions, providing a continuous source of atomic nitrogen. The nitrogen readily chemisorbs on the (200) planes but not on the N-terminated (111) planes. This in turn reduces the mean free path of the metal cation on the (200) plane due to capture by the nitrogen atoms, and promotes the formation of a NbNi

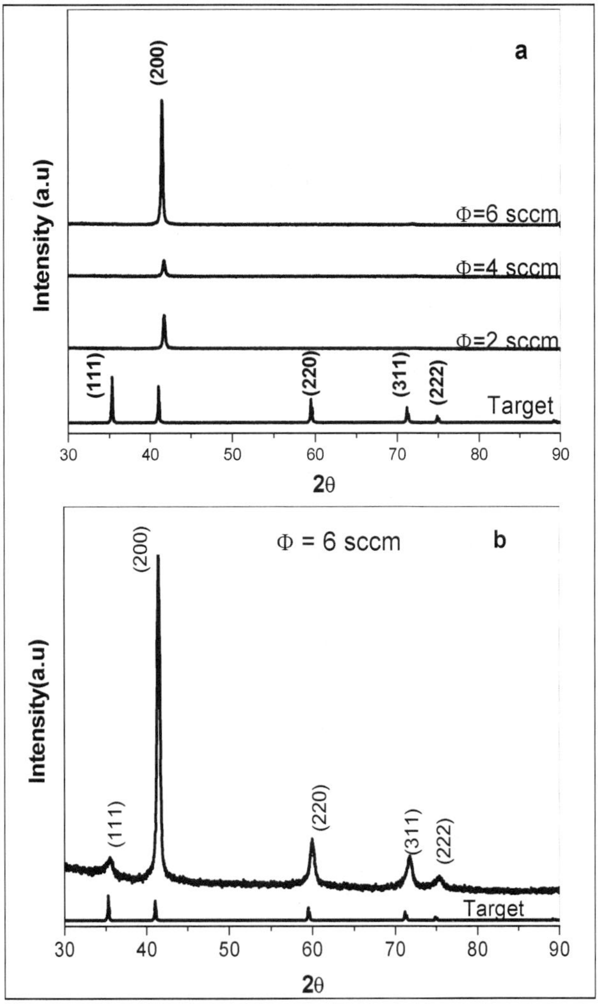

Fig. 11. a) XRD patterns θ-2θ configuration and b) Grazing angle (2°) XRD pattern recorder from the NbN film, grown at 300 W, 553 K, 20 sccm of Ar and 6 sccm of N_2.

(i = 1-4) admolecule or islands of adatoms (Petrov & Barna, 2003). This can be considered as causing an additional decrease in the (200) surface energy relative to that of the (111) plane. Consequently, the presence of the nitrogen atoms reduces the flow of cations from the (200) to the (111) planes, resulting in the orientation of the growth along the [200] direction. In this example, the growth of the films depends on the final working pressure as well as the bombardment energy of the ions.

The SEM study (Fig. 12a and 12b) of film growth as a function of the gas flow indicates that the NbN films present a compact granular structure, with a columnar growth of the type described by Movchan and Demchisshim (Movchan & Demchisshim, 1969), having an average thickness of 0.7 μm, which implies a deposition rate of 20 nm/min.

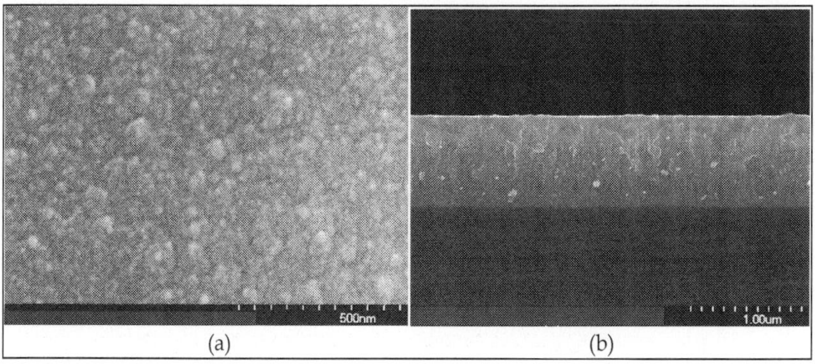

Fig. 12. a) Micrograph of the morphology of NbN thin film. b) Micrograph of the cross-section of NbN film, grown at 300 W and 553 K, 20 sccm of Ar and 6 sccm of N_2.

An interesting example of the growth of thin films in a reactive phase through the rf sputtering technique is ZrN_xO_y thin films, which grow at different flow ratios (N_2/O_2), but with the final working pressure constant (7.4 X10^{-1} Pa). The results of the XRD studies are shown in figures 13a and 13b, where it is possible to observe the influence that the flows have on the crystallographic of the films. Fig. 13a shows the XRD pattern of the films where the nitrogen flow was maintained constant (2.5 sccm) and the oxygen flow was varied, and Fig. 13b shows the XRD pattern of the films where the oxygen flow was maintained constant (2.0 sccm) and the nitrogen flow was varied.

The results obtained allowed establishing that there is an optimum flow ratio of 1.25 in which growth of a film with a high degree of crystallinity is reached. This behavior is very similar to the NbN films discussed above; therefore, the physicochemical mechanisms involved in the growth of thin films are the same for nitrogen molecules as for oxygen molecules, due to the higher reactivity of oxygen as compared to nitrogen (Martin, *et al.*, 2001).

The SEM study (Fig. 14a) of ZrN_xO_y films that were grown at a flow ratio of 1.25 present a highly homogenous and maybe very compact surface in which it is not possible to find contrast, and therefore the growth mechanism is not well defined. The microstructure of the films was evaluated through AFM studies (Fig. 14b). The micrograph reveals that the average size grain was 150 nm with an average rugosity of 5.9 nm.

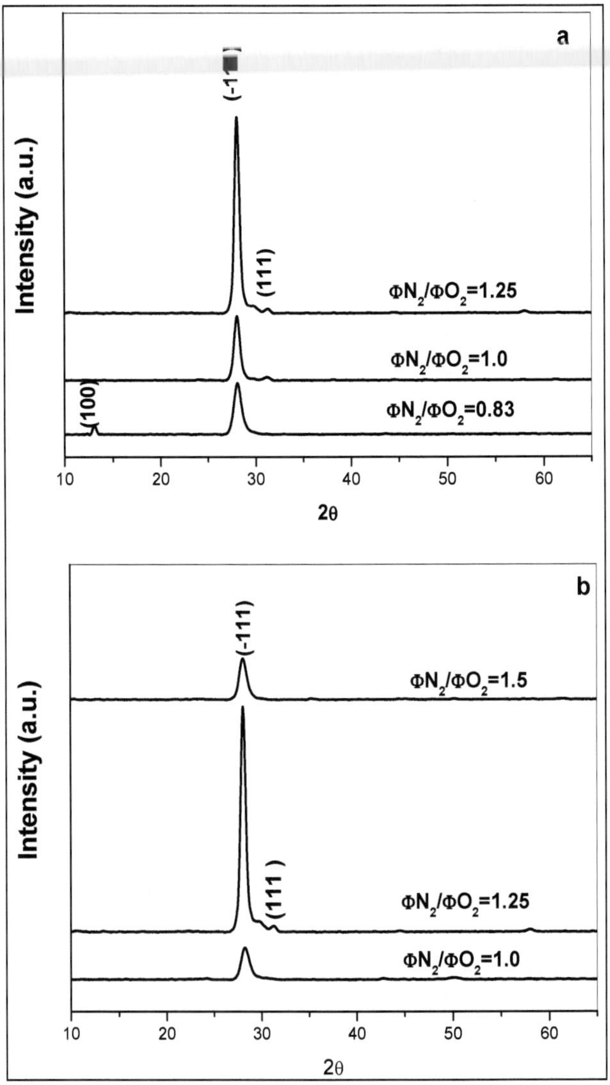

Fig. 13 . XRD Patterns recorded from ZrN_xO_y thin films deposited on common glass at 623K and different ratios flow ($\Phi N_2/\Phi O_2$). a- ΦN_2 constant at 2.5 sccm and b- ΦO_2 constant at 2.0 sccm.

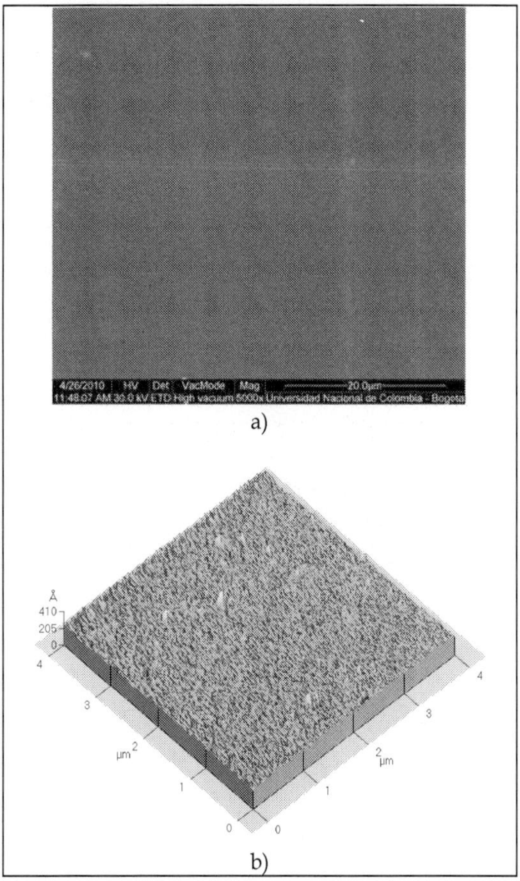

Fig. 14. a- Micrograph of the morphology of ZrN_xO_y thin film. b- AFM micrograph of the ZrN_xO_y film, deposited at 350 W, 623 K and $\Phi N_2/\Phi O_2$= 1.25

4.2 Influence of the power on the structure and microstructure of thin films

The XRD patterns of MgO films (Fig. 15a and 15b) recorded from films grown at room temperature shows the influence which the power supply applied to the target has on the structural behavior of thin films. The films grown within the range from 150 to 200W showed amorphous behavior (not shown in the XRD pattern). On the other hand, starting at 250W, all films showed the (200) plane corresponding to magnesium oxide in the FCC phase. In the XRD pattern it can also be observed that the intensity of the (200) plane is so high that it makes it impossible to distinguish the diffraction peaks corresponding to other planes. To determine the polycrystalline character of the film, we carried out X-ray diffraction experiments at grazing incidence. From Fig. 15b it is clear that the films present two main diffraction peaks, (200) and (220), which belong to the target, confirming their polycrystalline character, although showing a preferential orientation along the (200) plane (texture index, 0.85).

Another example that shows the influence of the power supply applied to the target on the crystallization of the films is the growth of ZrN_xO_y thin films; the films were grown from the Zr target in the reactive phase in an atmosphere of nitrogen and oxygen (Φ_N= 2.5 sccm, Φ_O=2.0 sccm). In the XRD pattern of Fig. 16, it is possible to determine that there is a threshold power (250W) for producing the crystallization of the film on the substrate. At this power, the film grows with a high degree of crystallization along the (111) plane. This behavior can be explained by the energy model, since increasing the power on the target implies increasing the energy of the ions that are bombarding the substrate and therefore improving the mobility of the adatoms, which produce chemical reactions and atomic grouping along the planes with the higher surface energy, which generates films with a high degree of crystallinity.

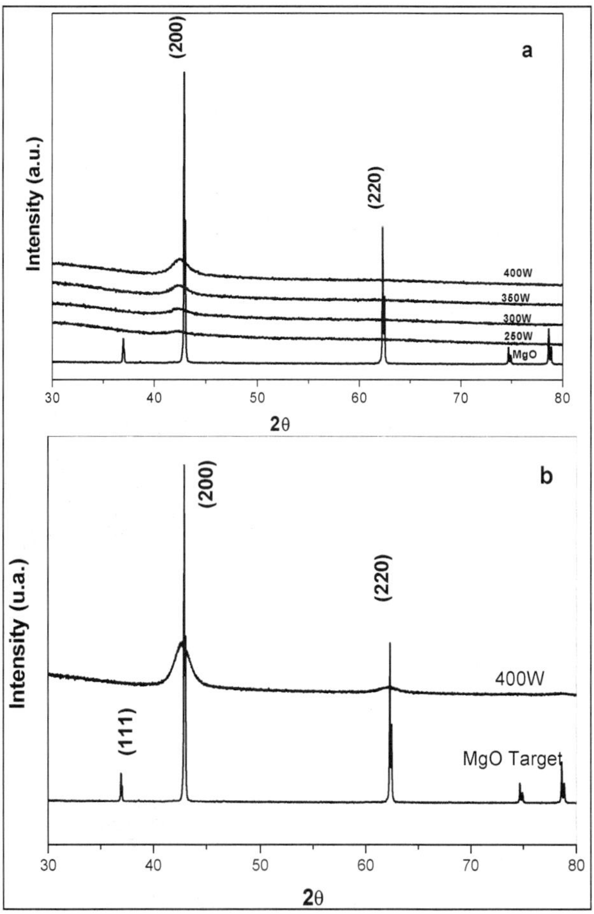

Fig. 15. a- XRD patterns recorded from MgO films deposited at 293 K and different power supplies applied to the target b- Grazing angle (3°) XRD pattern recorded from a MgO film grown at 400 W. The target XRD pattern is included as reference.

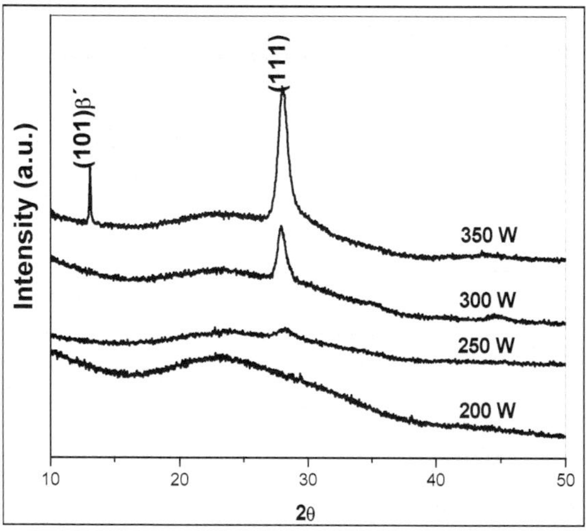

Fig. 16. XRD Patterns recorded from ZrN_xO_y thin films deposited on common glass at 623K at different powers applied to the target.

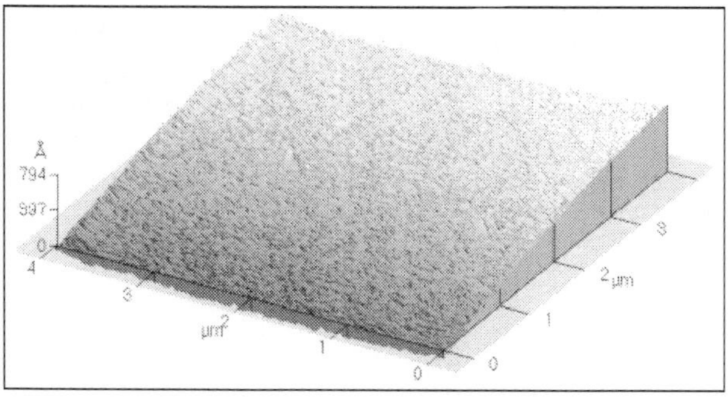

Fig. 17. Atomic force micrographs the MgO thin films deposited at 400 W.

The study of the microstructure of thin films was carried out using atomic force microscopy (AFM). The micrograph of Fig. 17 shows that in the area swept by the cantilever point (16μm²) the film has an average roughness of 20 nm and an average grain size on the order of 110 nm. These results established that there is a threshold of power for the growth of nano-structured MgO thin films with high texture without intentional heating of the substrate.

4.3 Influence of the substrate bias voltage (V_s) on the structure and microstructure of thin films

Fig. 18 shows the influence that the substrate bias voltage has on the growth of the Ti alloy thin films deposited on steel and glass substrates. The films grown on steel at -100V showed, in addition to the diffraction peaks of the substrate material (note that in all the XRD patterns recorded from the films deposited onto steel diffraction peaks from the substrate are still visible), a quite broad peak, which corresponds to the (002) plane of the α- phase of Ti. The XRD pattern corresponding to the films deposited at –120 V present two well-defined peaks at 35.1° (100) and 38.4° (110), which we associate with the (α/β)-alloy phase. The film deposited at a bias voltage of -160 V shows a peak at 38.4°, which can be associated with the (110) plane of the β- phase of Ti. When the bias voltage increases to -200 V, the XRD pattern shows only one high-intensity peak at 38.4°(110), which can be associated with the β- phase of Ti. It can also be observed that the width of the diffraction peaks is narrower for the films deposited at higher bias voltages, especially for that deposited at -200 V, which suggests that increasing the bias voltage increases the grain size of the deposited

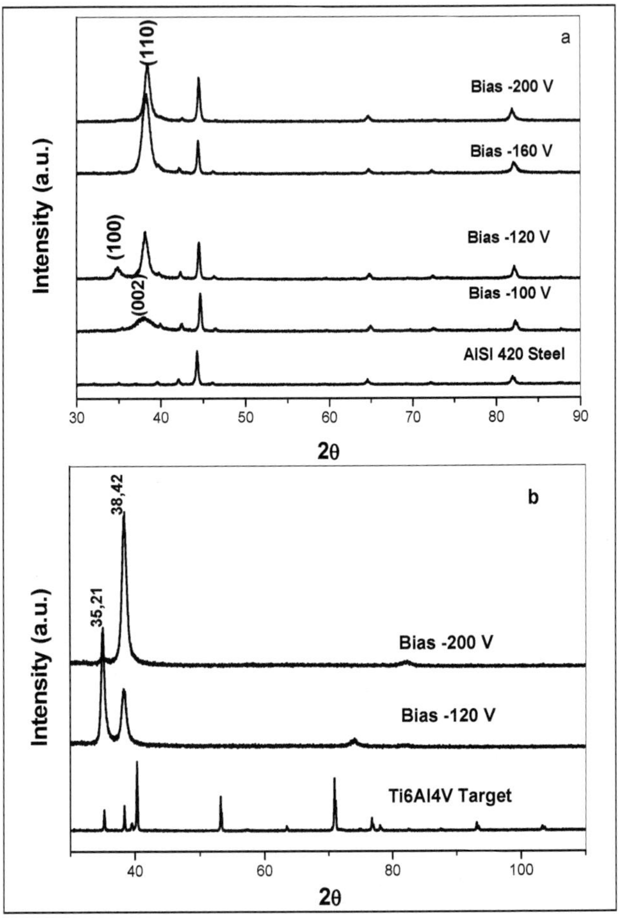

Fig. 18. (a). XRD pattern of the Ti6A14V films deposited by rf magnetron on steel substrates. The substrate appears as a reference. (b) XRD pattern of the films growth on glass and XRD pattern of theTi6A14V in bulk.

films. The films grown on glass at 120V and -200V show a similar behavior, although there are two differences: the two peaks appearing in the XRD pattern of the film grown at -120 V show a different intensity ratio from those of the film grown on steel, and the XRD pattern of the film grown at -200 V shows an additional low-intensity peak at 82.3°, which we associate with the (220) reflection of the β-Ti phase.

In summary, it follows from the XRD data that the substrate bias voltage (Vs) has a great influence on the structure of the deposited films: an increase of the bias voltage promotes the growth of different phases of titanium having well-defined different preferential crystallographic orientations. The increase of V_s is also reflected in a better crystallinity of the deposited films. The data also show that the crystalline phases formed at different V_s are the same in both substrates (see fig 18b). These results are in accord with that established in the foregoing section, where it was indicated that the increasing Vs improves the density of the film, which in turn allows obtaining films with a high degree of crystallinity.

Fig. 19. a) Micrograph of a Ti6A14V film grown on steel with bias voltage of - 200 V. (b) Atomic force Micrograph recorded from Ti6Al4V grown on steel at bias voltage of - 200V.

The analysis of the microstructure of the Ti6Al4V was made through scenic electron microscopy (SEM) and an atomic force microscope (AFM). Fig. 19a shows the SEM results; the recorded micrograph of the film grown at -200V reveals a film with excellent texture, high homogeneity and denser microstructure with grain refinement produced under the enhanced plasma bombardment, which is induced by the substrate bias voltage. Through Scherer's equation (Cullity, 2001) and using the broadening of the (110) and (110) planes of the Ti β-phase, it was found that the average grain size was 13 nm for the (100) plane and 16.5 nm for the (110) plane (Alfonso *et al.*, 2005). Fig. 19b shows the results of the AFM, which indicated that the films grown at -200 V possess a roughness of 20 nm for a scan area of 3µmx3µm.

4.4 Influence of the energy parameter (E_p) on the structure and microstructure of thin films

The X-ray diffraction patterns of the CrNx films deposited at different Ep values are shown in Fig. 20. It was necessary to use two different scales, because as the Ep increased, the intensity of the CrN (200) peak increased by two orders of magnitude and became narrower. At Ep values lower than 30 eV/atom, we observed that the films mainly showed a [200] orientation with traces of the CrN[111] and CrN[220] orientations, while at higher Ep values only the CrN[200] orientation was detected. An estimation of the ⟨200⟩ texture was obtained comparing the (200) peak intensity (I200) relative to the intensity of all the orientations (Ii) appearing in the XRD pattern. Table 3 summarizes some experimental conditions of the deposition of the CrN films.

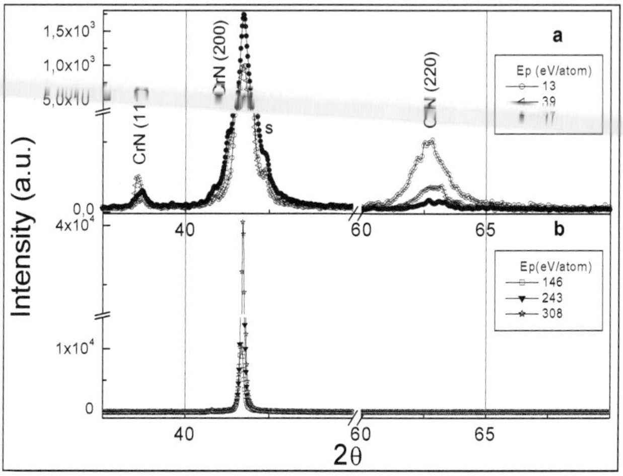

Fig. 20. X-ray diffraction patterns of CrNx coatings: (a) low Ep values, (b) high Ep values.

Sample	d s-t (cm)	Vs (V)	R (μm/min)	Ei (eV)	Ji/Ja	Ep (eV/atom)	$F_{<200>}$
CrN-3	3	0	0.10	10	1.3	13	0.80
CrN-4	4	0	0.064	16	2	32	0.89
CrN-5	5	0	0.051	12	2.2	26	0.90
CrN-6	6	0	0.035	13.5	2.6	35	0.92
CrN-7	7	0	0.026	14.5	2.7	39	0.93
CrN-8	8	0	0.019	17	2.7	46	0.92
CrN-3-100	3	-100	0.10	110	1.3	143	0.99
CrN-4-100	4	-100	0.064	115.6	2.0	231	0.99
CrN-5-100	5	-100	0.05	112.3	2.2	247	1.0
CrN-6-100	6	-100	0.034	113.5	2.6	295	1.0
CrN-7-100	7	-100	0.026	114.5	2.7	310	1.0
CrN-8-100	8	-100	0.019	117	2.7	316	1.0

Table 3. Summary of experimental growth conditions of CrN thin films and some of the results: deposition distance ds–t, r.f. bias voltage Vs, deposition rate R, ion energy Ei, ion–atom flow Ji/Ja, energy parameter Ep and texture factor $f_{<200>}$.

The results of the microstructure behavior of the CrN thin films are in good agreement with that of the theory of Petrov *et al.,* which was explained above. In the case of CrN thin films, increasing the ion energy by applying a higher substrate bias and keeping low ion–atom flows resulted in a film densification and a change in the preferred orientation from [111] to [002] with a subsequent increase in the levels of stress and the incorporation of argon ions. Moreover, by increasing the energy to around 20 eV and working at higher ion–atom ratios, the residual stress remained low, but the texture of the films changed abruptly and completely from the [111] to the [002] direction. According to the Petrov Model, the key factor was that the energy of the N_2 ions should be near 20 eV to promote the dissociation of the ions through collision with the film surface, in this way providing a supply of atomic nitrogen which can chemisorb on [002] oriented grains and later capture metal atoms, resulting in the development of the ⟨002⟩ texture. In CrN films, it was observed that at the lowest Ep values (Ep<50 eV/atom) with $f_{(200)}$ ~80%, the film texture was equivalent to the transition regime in Petrov *et al.*'s research. The ion energy and ion-atom flows were not sufficiently low to produce pure ⟨111⟩ texture, as explained above, but were also not sufficiently high to induce 100% ⟨002⟩. In this transition regime, a competitive growth between the high-trapping (111) plane and the low surface energy (002) plane was established, since as the ds-t increased, the ion energy of the N_2 ions approached 20 eV (see Fig. 20). As the Ep value increased, the conditions favored the formation of the ⟨002⟩ texture. Thus we considered Petrov's theory to be a rather good model to explain the microstructure evolution of the texture in metal nitride films. The competition of the different orientations during growth also affected the microstructure, as observed in Fig. 21a-d. As the ion bombardment increased, the microstructure changed from a non-ordered columnar growth to well-oriented grains that look like fibers coming out from the substrate, *i.e.* equiaxed grains.

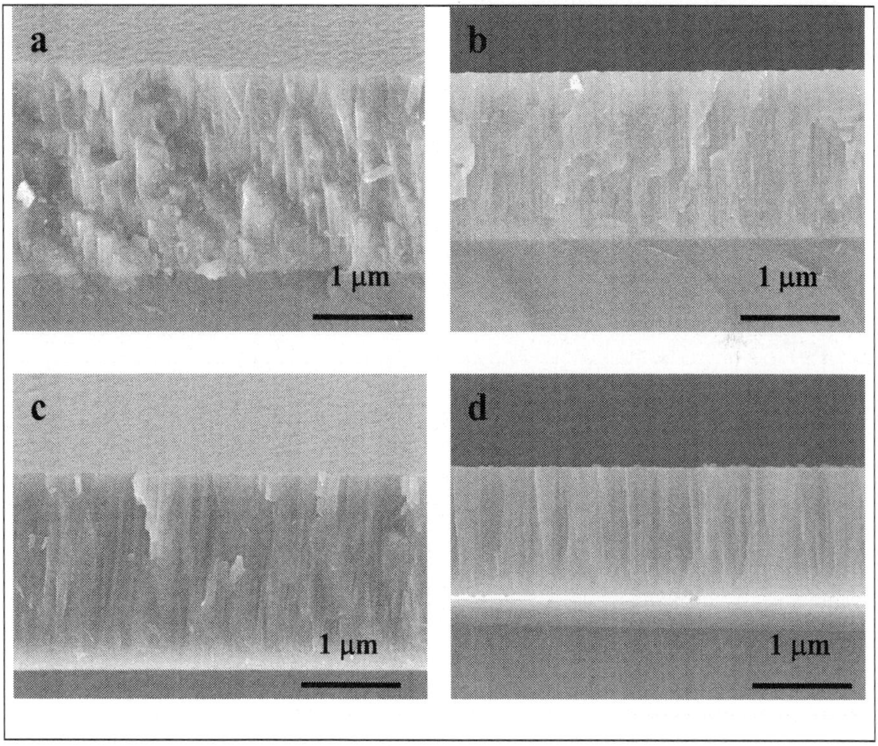

Fig. 21. Cross-section SEM images of samples deposited at increasing Ep values (a) 13, (b) 46, (c) 143 (d) 316 eV/atom, respectively.

5. Chemical characterization of thin films

5.1 X-ray photoelectron spectroscopy (XPS) analysis

The last 10 nm of the surface of thin films can be analyzed using X-ray photoelectron spectroscopy (XPS). As an example, we show the results of XPS on Ti6Al4V thin films. The spontaneously passive film formed on the deposited Ti6Al4V films upon exposure to the air was studied through XPS. Fig. 4 shows high resolution narrow-scan spectra recorded from the Ti 2p, Al 2p, V 2p and O 1s spectra recorded from one of the samples. All the narrow-scan spectra recorded from the rest of the samples are almost identical to those presented in Fig. 22.

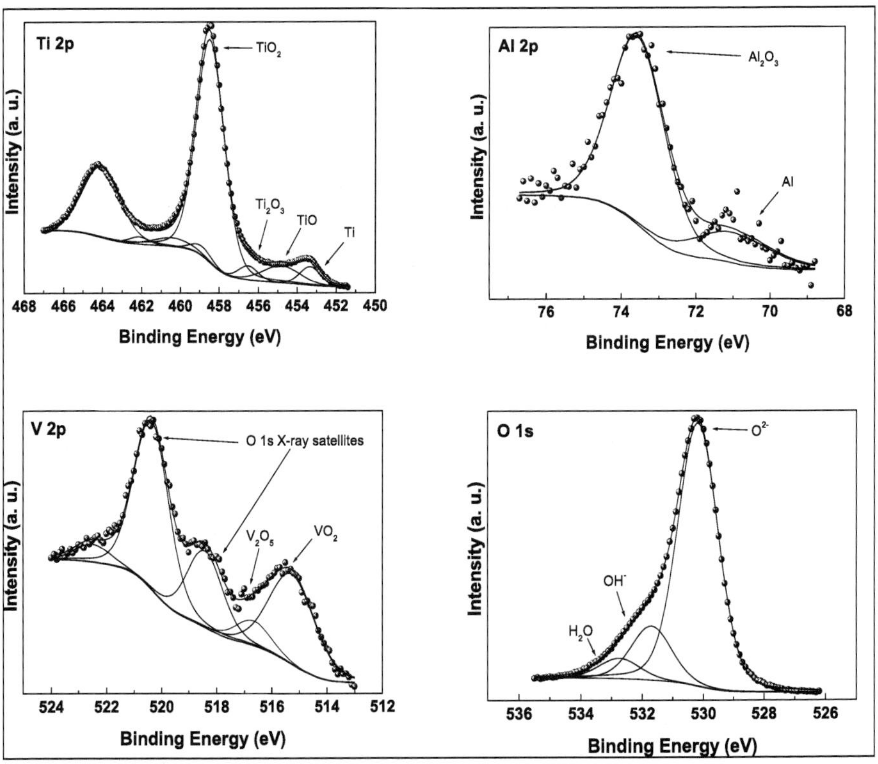

Fig. 22. 2p, Al 2p, V 2p and O 1s XPS narrow-scan spectra recorded from a film deposited at a bias voltage of -140 V.

The Ti 2p spectrum shows several contributions, and is similar to that reported previously by other authors for the spontaneously-formed passive oxide layer on bulk Ti6Al4V alloy. The spectrum is dominated by a major TiO_2 contribution (BE Ti $2p_{3/2}$=458.4 eV, 82%), and shows smaller Ti (BE Ti $2p_{3/2}$=453.3 eV, 5%), TiO (BE Ti $2p_{3/2}$=454.8 eV, 9%) and Ti_2O_3 (BE Ti $2p_{3/2}$=456.5 eV, 4%) contributions. The Al 2p spectrum shows a major Al_2O_3 contribution at 73.6 eV and a minor Al^0 contribution at 71.0 eV. Again this spectrum is similar to that observed for the passive oxide layer on a Ti6Al4V alloy. The O 1s peak shows three different contributions: a main one at 530.2 eV that can be associated with metal-oxygen bonds, and two much less intense contributions at 531.7 eV and 532.8 eV, which correspond to the presence of OH^- groups and chemisorbed water, respectively. The fit of the V 2p spectrum is complicated by the fact that the V peaks overlap strongly with the O 1s $K\alpha_{3,4}$ X-ray satellite peaks. Therefore, the whole spectrum was refined considering the presence of these satellites and the presence of vanadium peaks. The results show the presence of an intense peak at 515.4 eV, which we associate with the presence of VO_2, and a less intense vanadium contribution at 516.7 eV, which corresponds to V_2O_5. This finding contrasts with those of Milosev (Milosev *et al.*, 2000), who did not find oxidized vanadium in the passive layer formed on bulk Ti6Al4V (Alfonso *et al.*, 2006).

These results show that the films that have been grown through rf magnetron sputtering reproduced the stiochoimetry of the target, although on the surface of the thin films chemical reactions will be produced, forming hydroxides and passivation layers.

5.2 Energy dispersive X-ray (EDX) analysis

The elemental composition of thin films is determined through energy dispersive X-ray analysis (EDX), although is important to indicate that elements such as oxygen and nitrogen are not possible to assess because the binding energies are very near to the binding energy of the carbon. As an example, the EDX results for NbN nanoparticle are shown. Fig. 23 shows the elements that constitute the nanoparticles of NbN (N and Nb) that formed the NbN studied in section 4, and the elements of the substrate (common glass) on which the thin films were grown. The Cu belongs to the sampling.

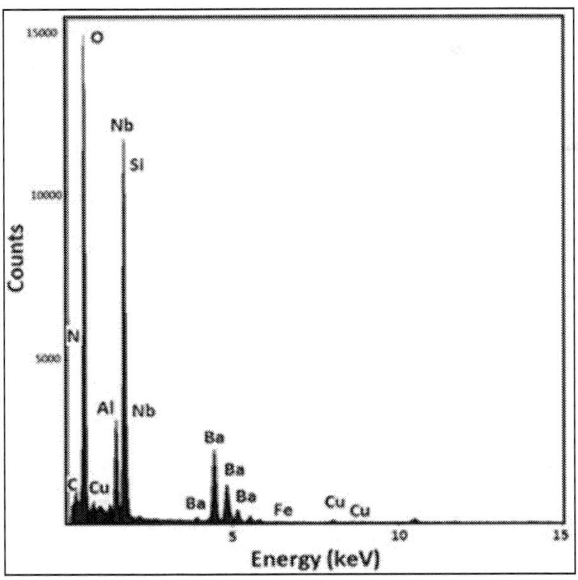

Fig. 23. EDX spectrum of NbN nanoparticle that integrates the NbN thin films.

6. Optical characterization of thin films

An important characteristic of thin films is the optical behavior, since this determines possible industrial applications, which range from transparent coatings to optical filters; optical studies are carried out through measurements of transmittance, absorbance and reflectance. Fig. 24 shows the transmittance percent as a function of the wavelength on MgO thin films that have been grown through rf sputtering with different power supplies applied to the target.

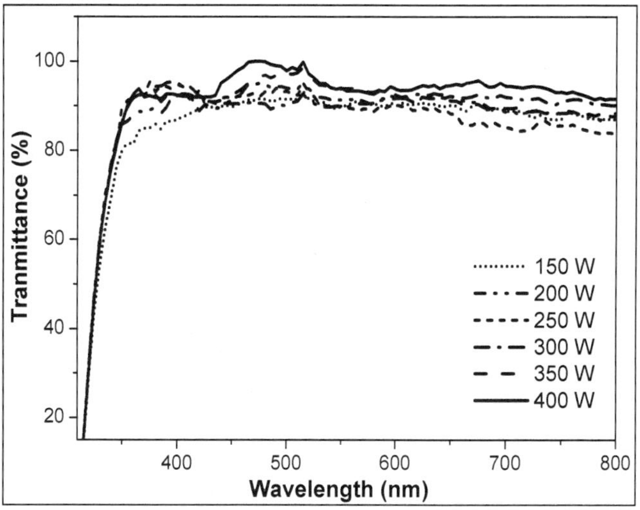

Fig. 24. Transmission spectrum vs wavelength the films the MgO with various strengths of power supply to the target.

The results show (see Fig. 24) that the films have percentages of transparency that range from 84% to 95% in films grown at 150 and 400 W, respectively. These results indicate that films that were grown at a higher power have a high value of transmittance. This behavior is due to the fact that the films are denser and possibly present lower diffraction.

7. Applications of thin films

Surface modification by means of thin film deposition is an important industrial process used to protect basic materials against wear, fatigue, corrosion and many other surface-related damage phenomena (Vyas *et al.*, 2010; Yang *et al.*, 2008; Wesley, 2001; Li *et al.*, 2009; Kumar, A. & Kaur, 2009). The modern methods of plasma-assisted physical vapor deposition techniques provide great flexibility for designing films with specific chemistry and microstructure, leading to coatings with unique properties. Among these, ceramic coatings deposited on metallic substrates have shown excellent improvement of the surface properties, such as a low friction coefficient and a high degree of hardness with associated good wear resistance and also corrosion resistance to aggressive environments; (Yang, *et al.*, 2008; Wesley, 2012).

Films of CrN, TiN, ZrN, TaN and NbN were deposited using an unbalanced magnetron sputtering system with different energy parameters to investigate its effect on some film properties. Fig. 23 shows the main results obtained in these films. In general, it may be observed that the residual compressive stresses, determined using the curvature method and Stoney's equation (Stoney, 1909), and increased with the energy parameter, although for some films the variation was very small.

Fig. 25a shows the microhardness of the substrate-film system. Hardness measurements were made on samples deposited on AISI M2 tool steel. The hardness of the substrate was 800 kg mm^{-2}. The hardness of the films was 2-3 times higher than the substrate hardness. Group IV nitrides generally are harder than those of group V (Hofmann, 1990). Moreover, the data showed that the hardness increased as the energy parameter increased. This could be attributed to the increase in ion bombardment on the substrate surface, increasing adatom mobility and producing denser films, and the effect on hardness of compressive stresses present in these films (Olaya *et al.*, 2009).

Fig. 25b presents the wear coefficient, Kwear, which was evaluated using a ball cratering system, which is a micro scale abrasion test. Two-body grooving abrasion was the wear mechanism observed for all coatings (not shown), probably due to the micro-cutting action of abrasive particles that were dragged across the ball, basically remaining fixed to the ball surface during the test (Adachi, K. & Hutchings, 2003). The wear coefficients were all in the 10^{-6} mm^3 $N^{-1}m^{-1}$ range, and the lowest values were obtained for NbN and TaN films. It may be seen that there was a slight decrease in k wear as the energy parameter increased, but the variation was too small to be considered important. In addition, no clear trend was observed as the unbalance coefficient was increased. This might be a consequence of variations in other film properties, such as the coefficient of friction or the roughness, parameters known to exert a strong influence on the wear performance of the surfaces.

Ceramic films like metallic nitrides on a metal substrate are commonly believed to be immune to corrosion. Fig. 25 c-d shows the potentiodynamic polarization curves for the films deposited on AISI 304 substrate at the two energy parameters and the AISI 304 substrate in an electrolyte of 0.5 M H_2SO_4 + 0.05 M KSCN. The corrosion resistance of a material in the polarization curve is determined by its ability to retain low current densities as the electric potential increases. Quantitatively, Tafel analysis was used to determine the corrosion potential, Ecorr, while the critical passivation current density, Icrit, was estimated from the maximum anodic dissolution current before passivation. The corrosion behavior of PVD ceramic-coated steels in aqueous solutions has been increasingly investigated in recent years (Vyas *et al.*, 2010; Yang *et al.*, 2008; Wesley, 2001; Li *et al.*, 2009; Kumar, A. & Kaur, 2009). One of the major drawbacks is the presence of defects, which are associated with the growth process in PVD ceramic coatings. These coating defects (*e.g.* pores) are particularly deleterious for corrosion protection, since they provide direct paths through which the electrolyte can reach the coating/substrate interface, where localized galvanic corrosion occurs due to the difference in the corrosion potential between the coating and the steel. A lot of research has been undertaken to deal with this problem of coating defects (Stansbury& Buchnan, 2000; Kaciulis *et al.*, 2000; Lang & Yu, 2001; Lee *et al.*, 2009; Chou & Huang, 2003) particularly for transition metal nitride coatings, which in general have excellent wear and corrosion resistance and therefore are widely used in industry.

The critical passivation current density shows that in contrast to the other properties of the films, there was a decrease in the corrosion resistance of the films as the energy parameter increased with Ep. The critical passivation current density is proportional to the exposed area of the substrate due to the existence of pores or pinholes in the film (Olaya *et al.*, 2005). The value of I_{crit} depends on the grain limits that join the columns due to the fact that they can contain vacancies, micropores, pinholes, and possibly microcracks, allowing the diffusion of electrolytes of the corrosive solution toward the substrate and increasing degradation of the coatings.

Therefore, the results reported in Fig. 25 c-d suggest that as the energy parameter increased there was more substrate area exposed at the bottom of the pinholes. These pinholes or defects are usually localized at the grain boundaries, which are defined by the crystal growth process, which consequently models the final film structure. The rise in the corrosion current with Ep was very small for the NbN and TiN films, but significantly higher for TaN, ZrN and CrN. However, the higher ion bombardment induced by the degree of magnetic field created more defects, increasing the residual stresses and therefore deteriorating the film-substrate adhesion and consequently the response of the coatings to the corrosion products, as was visually observed at the end of the corrosion test for the TaN film. This phenomenon can be explained by galvanic coupling produced by a difference in the corrosion potential between coated and uncoated specimens. The potential difference is characterized by anodic dissolution of the substrate material with a high anodic current density at the defect site, leading to an adhesion failure of the coating.

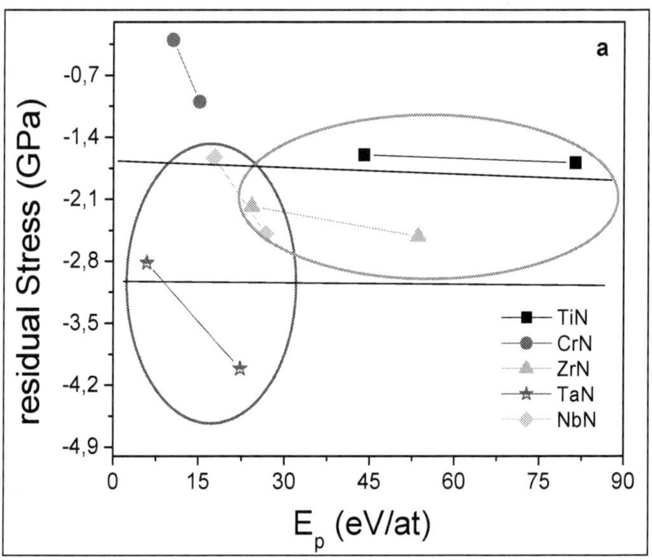

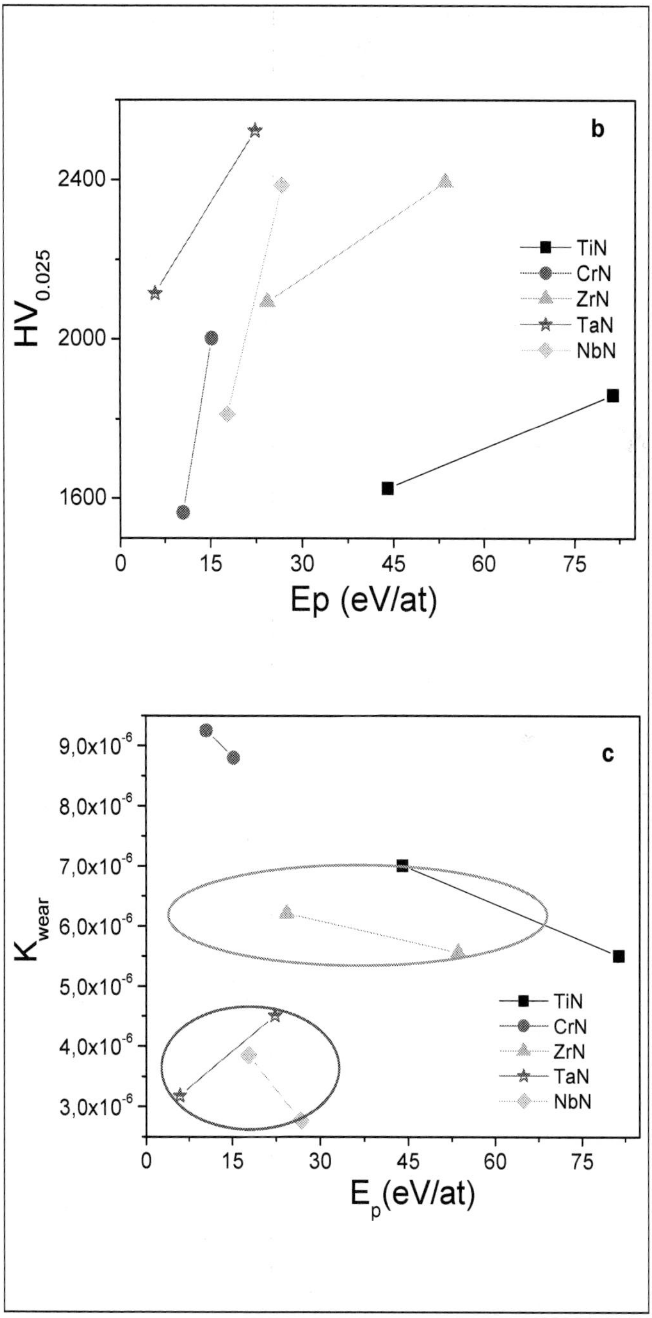
b
2400
2000
1600
$HV_{0.025}$
0
15
30
45
60
75
Ep (eV/at)
TiN
CrN
ZrN
TaN
NbN
c
9,0x10^{-6}
8,0x10^{-6}
7,0x10^{-6}
6,0x10^{-6}
5,0x10^{-6}
4,0x10^{-6}
3,0x10^{-6}
K_{wear}
E_p(eV/at)

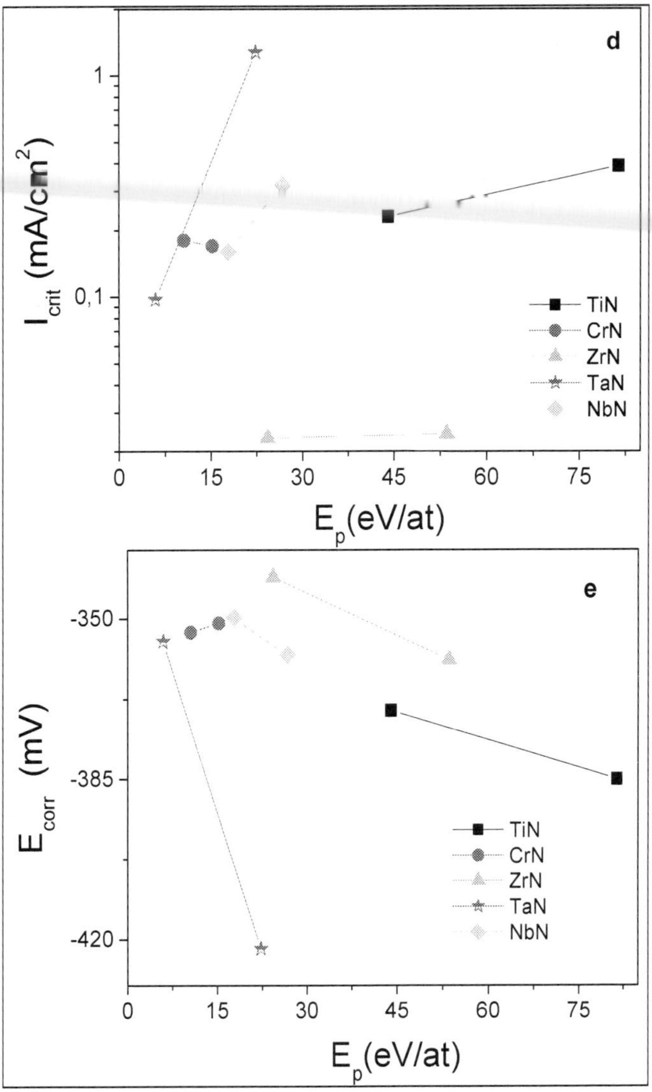

Fig. 25. Summary of mechanical and chemical properties of films deposited through sputtering. a) residual stress b) hardness c) wear coefficient d) critical passivation current and d) corrosion potential

8. Conclusion

In this chapter, the main physical-chemical models utilized to explain the growth of thin films through the rf sputtering technique have been presented, and the influence that the growth parameters have on the crystallinity and the micro-structure of the thin films has been discussed, and based on the described models, examples have been provided of the thin films' growth under each of the growth parameters. The fundamental idea of presenting these basic theories is to introduce the science of materials to young researchers in the world, helping them to understand by means of examples the basic concepts that apply to the growth of thin films, with the hope that a better understanding of these theories will produce new and better applications.

9. References

Adachi, K. & Hutchings, I., Wear-mode mapping for the micro-scale abrasion test. *Wear*, Vol 255, issues 1-6, (august-september 2003) 23-29.

B. A. Movchan and A. V. Demchishin, Study of the Structure and Properties of Dioxide thin, Fiz. Met. Metalloved. Vol. 28, (1969) 83-90.

B. D. Cullity, s: R. Stock, Elements of X-ray diffraction, Prentice Hall, 2001.

B. Rauschenbach, J. W. Gerlach, Texture Development in Titanium Nitride Films Grown by Low-Energy Ion Assisted Deposition, Cryst. Res. Technol. Vol. 35, (2000) 6-7.

B. Wang, Y.N. Zhao, Z. He, The effects of deposition parameters on the crystallographic orientation of AIN films prepared by RF reactive sputtering, Vacuum. Vol. 48, (1997) 427-429.

B. Window & N. Savvides, Charged particle fluxes from planar magnetron sputtering sources, J. Vac. Sci. Technol. A4 (1986) 196–202

Barna, P. , Adamik, M., Fundamental structure forming phenomena of polycrystalline films and the structure zone models. Thin Solid Films, Vol. 317, (April 1998) 27-33.

C. Lin, C. Lee, Growth of Tantalum Nitride Films on Si by Radio Frequency Reactive Sputtering of Ta in N2/Ar Gas Mixtures: Effect of Bias, J. Electrochem. Soc. Vol. 147, (2000) 713-718.

C. S. Shin, D. Gall, Y. W. Kim, N. Hellgren, I. Petrov, J.E. Greene, Development of preferred orientation in polycrystalline NaCl-structure δ-TaN layers grown by reactive magnetron sputtering: Role of low-energy ion surface interactions , J. Appl. Phys. 92, (2002) 5084.

C.R. Weissmantel, Deposition of Metastable Films by Ion Beam and Plasma Techniques, *Proc. 9th Int. Vacuum Congress and 5th Int. Conf. on Solid Surfaces* (Madrid), 26 Sept-1 Oct 1983, p 229-308

Chou, W., Yu, G. & Huang, J. Corrosion resistance of ZrN films on AISI 304 stainless steel substrate. *Surface and Coatings Technology*. Vol. 167, issue 1, (april 2003) 59-67.

Emery, C., Chourasia,A., Yashar, P., A study of CrNx thin films by X-ray photoelectron spectroscopy J. Elect. Spectros. Relat. Phenom. Vol. 104, issue 1-3, (July 1999) 91-97.

Flores, J. F.; Olaya, J. J.; Colás, R.; Rodil, S. E.; Valdez, B. S.; Fuente, I. G. Corrosion behaviour of TaN thin PVD films on steels. Corrosion Engineering, Science and Tecjnology, Vol. 4, issue 2, (June 2006) 168-176.

G. G. Stoney, The measurement of stress in layered structures, in: Proc. Roy. Soc. London Vol. A82, (1909) 172.

G. Kamath , A.P. Ehiasarian, Y. Purandare, P.Eh. Hovsepian, Tribological and oxidation behaviour of TiAlCN/VCN nanoscale multilayer coating deposited by the combined HIPIMS/(HIPIMS-UBM) technique, Surface & Coatings Technology 205 (2011) 2823–2829

Goldfarb, I., Pelleg, J., Zevin, L. and Croitoru, N. (1991). Lattice distortion in thin films of IVB metal (Ti, Zr, Hr) Nitrides, *Thin Solid Film*, Vol. 200, issue 1, (May 1991) 117-127., (May 1991), 117-127.

Greene, J. E., Low Energy Ion Bombardment during Film Deposition from the Vapor Phase: Effects on Microstructure and Microchemistry, *Solid State Technol.*, 30,(4), (1987)115.

Greene, J. E., Nucleation, Film Growth, and Microstructural Evolution, *Deposition Processes for Films and Coating*, 2nd edition, (R. Bunshah,ed.), Ch. 13, Noyes Publications (1994).

H. Liu, W. Tang, D. Hui, L. Hei, F. Lu. Characterization of (Al, Si)N films deposited by balanced magnetron sputtering, Thin Solid Films, Vol. 517, issue 21, 1 (September 2009) 5988-5993.

H.B. Nie, S.Y. Xu, S.J. Wang, L.P. You, Z. Yang, C.K. Ong, J. Li and T.Y.F. Liew, Structural and electrical properties of tantalum nitride thin films fabricated by using reactive radio-frequency magnetron sputtering, Appl. Phys. A. Mater. Sci. process, Vol. 73 (2001) 229- 236.

Hofmann S., Target and substrate surface reaction kinetics in magnetron sputtering of nitride coatings, Thin Solid Films, Vol. 191, (1990) 335-348.

Hwan-Chul Leea, Guen-Hong Kimb, Soon-Ku Hongc, Ki-Young Leea, Yoon-Joong Yonga, Chang-Hwan Chunb, Jai-Young Leea, Influence of sputtering pressure on the microstructure evolution of AlN thin films prepared by reactive sputtering, Thin Solid Films, Vol. 261, issue 1-2, (1 June 1995) 148-153.

I. Milosev, M. Metikos-Hukovic and H.-H Strehblow, Passive film on orthopaedic TiAlV by X-ray photoelectron, Biomaterials, Vol. 21, (2000), 2103.

J. E. Alfonso, Fernando Pacheco, Alvaro Castro P, J. Torres, Influence of the substrate bias voltaje on the crystallographic structure and mechanical properties of Ti6Al4V coatings, deposited by rf magnetro, Phys. Stat. sol. Vol. 10, (2005) 3786-3789.

J. E. Alfonso, J. Torres, and J. F. Marco,. Influence of the Substrate Bias Voltage on the Crystallographic Structure and Surface Composition of Ti6A14V Thin Films Deposited by rf Magnetron Sputtering, Brazilian Journal of Physics, Vol. 36, issue 3, (September 2006) 994-996.

J. E. Greene, J.-E. Sundgren, L. Hultman, I. Petrov, and D. B. Bergstrom, Development of Preferred Orientation in Polycrystalline TiN Layers Grown by Ultra-High-Vacuum Reactive Magnetron Sputtering, Appl. Phys. Letters, Vol. 67, (1995) 2928.

J. H. Je, D. Y. Noh, H.K. Kim, K.S.Liang, Preferred orientation of TiN films studied by a real time synchrotron x-ray scatteringJ. Appl. Phys. 81, (1997) 6126.

J. Musil, S. Kadlec, Reactive sputtering of TiN films at large substrate to target distances, Vacuum, Vol. 40 (1990) 435.

J. Pelleg, L. Z. Zevin, S. Lungo, Lattice distortion in thin films of IV B metal (Ti, Zr, Hr) nitrides Thin Solid Film, Vol. 197, (1991) 117.

J. R. Frederick, D. Gall, Surface morphological evolution of epitaxial CrN(001) layers, J. Appl. Phys., Vol. 98, (2005) 054906.

J.A. Heimberg, K.J. Wahl, & I. L. Singer, Superlow friction behavior of diamond like carbon coatings: Time and speed effects, Applied Physics Letters, Vol. 79, issue 17, (april 2001) 2449-2451.

J.J. Olaya, L. Huerta, S.E. Rodil, R. Escamilla, Superconducting niobium nitride films deposited by unbalanced magnetron sputtering , Thin Solid Films, Volume Vol. 516, issue 23, (1 october 2008) 8768-8773

K. Aso, M. Hayakawa, H. Matsuda, K. Hayashi, W. Ishikawa, On specimen preparation by radio frequency planar magnetron sputtering , J. Vac. Sci. technol. Vol. A4, issue 1, (1986) 19-25.

K. Bewilogua a, G. Bra¨uer a, A. Dietz a, J. Ga¨bler a, G. Goch (1)b, B. Karpuschewski (1)c, B. Szyszka, Surface technology for automotive engineering, Manufacturing Technology, Vol. 58, (2009) 608–627

K. S. Harsha, Principles of Physical Vapor Deposition of Thin Films, Elsevier, 2005.

K. Shibahara, S. Nishino, and H. Matsunam, Antiphase-domain-free growth of cubic SiC on Si(100), Appl. Phys. Lett., 50, (1987) 1888.

Kaciulis, S., Mezzi, A., Montesperelli, G. Lamastra, F., Rapone, M. Casadei, F., Valente, T. & Gusmano, G. Multi-technique study of corrosion resistant CrN/Cr/CrN and CrN : C coatings. *Surface and Coatings Technology*, Vol. 201, issue 1-2, 12, (6 september 2006) 313-319.

Kelly,P., Arnell, R., Magnetron sputtering: a review of recent developments and applications. *Vacuum*, Vol. 56, (September 1999) 159-152.

Kelly,P., Arnell, R., The influence of magnetron configuration on ion current density and deposition rate in a dual unbalanced magnetron sputtering system. *Surface and Coatings Technology*, Vol. 108–109 (1998) 317–322

Kumar,A. & Kaur, D., Nanoindentation and corrosion studies of TiN/NiTi thin films for biomedical applications. *Surface and Coatings Technology*, Vol. 204, issue 6-7, (14 december 2009), 1132-1136.

L. Monroy, J. Olaya, M. Rivera2, A. Ortiz3, G. Santana. Growth study of Y-Ba-Cu-O on buffer layers and different substrates made by ultrasonic spray pyrolysis. Rev. LatinAm. Metal. Mat., Vol. 32 (October 2011) 21-29

Lang, F. & Yu, Z. Surf. Coat. Technolology, The corrosion resistance and wear resistance of thick TiN coatings deposited by arc ion plating. *Surface and Coatings Technology*, Vol. 145, issue 1-3, (august 2001) 80-87.

Lee, K., Kim, W., Young, J., Won, S. & CHOE, H. Effects of TiN film coating on electrochemical behaviors of nanotube formed Ti-xHf alloys. *Transactions of Nonferrous Metals Society of China*, Vol. 19, issue 4, (august 2009) 857-861.

Lingxia Hang , Y. Yin, Junqi Xu, Optimisation of diamond-like carbon films by unbalanced magnetron sputtering for infrared transmission enhancement , Thin Solid Films, Volume 515, Issue 1, (25 September 2006) 357-361.

M. Ishihara, S. J. Li, H. Yumoto, K. Akashi, Y. Ide, Control of preferential orientation of AlN films prepared by the reactive sputtering method Thin Solid Films, Vol. 316 (1998) 152.

M. Marlo, V. Milman, Density- functional study of bulk and surface properties of titanium nitride using different exchange-correlation functional, Physical Review B, Vol. 63, Num. 4 (January 2000), pp. 2899-2907

Martin, O. Banakh, A.M.E. Santo, S. Springer, R. Sanjines, J. Takadoum, F. Levy, Appl. Surf. Sci. Vol. 185 (2001) 123–133.

Mayrhofer,P., Tischler, G., Mitterer, C., Microstructure and mechanical/thermal properties of Cr-N coatings deposited by reactive unbalanced magnetron sputtering, Surf. Coat. Technol. Vol.142-144, (July 2001), 78-84.

Messier, R., Giri, A.P., Roy, R.A., Revised structure zone model for thin film physical structure. *Journal of Vacuum Science and Technology, A: Vacuum, Surfaces, and Films*, 2(2, Pt. 1), 500-3 .

Mitsunori Kobayashi, Yoshihiko Doi, TiN and TiC coating on cemented carbides by ion plating, Thin Solid Films, Vol. 54, Issue 1, 2, (October 1978) 67-74

Olaya, J., Rodil, S., Muhl, S. & Sánchez, E. Comparative study of chromium nitride coatings deposited by unbalanced and balanced magnetron sputtering. Thin Solid Films, Vol. 474, issues 1-2, (march 2005), 119-126.

Olaya, J., Rodil,S., Marulanda, D & B. Bhushan, Propiedades mecánicas de nitruros metálicos depositados con UBM: tecnología eficiente y ambientalmente limpia. *Revista Mexicana de física.* , Vol.55, issue 6 (november 2009), 425-431.

P. Losbichler, C. Mitterer, Non-reactively sputtered TiN and TiB2 films: influence of activation energy on film growth, Surf. Coat. Technol. Vol. 97, (1997) 567.

P. Ziemann, E. Kay, Correlation between the ion bombardment during film growth of films and their structure and electrical properties J. Vac. Sci. Technol., A, Vac. Surf. Films, 1 (1983) 512.

P.B. Barna, M. Adamik, in: F.C. Matacotta, G. Ottaviani Eds., Science and Technology of Thin Films World Scientific Publ., Singapore, 1995.

P.B. Barna, M. Adamik,Fundamental structure forming phenomena of polycrystalline films and the structure zone models., Thin Solid Films, 317, (208) 27-33.

Pavel Souček, Tereza Schmidtová, Evaluation of composition, mechanical properties and structure of NC-TiC/a-C:H coatings prepared by balanced magnetron sputtering, Surface and Coatings Technology, In Press, Accepted Manuscript, Available online 16 September 2011.

Petrov,I, Barna, P, Hultman,L, Greene,J., Microstructural evolution during film growth. *J. Vac. Sci. Technol. A,* Vol 21, No 5, (September 2003), 117-128.

Raymond Constantin, Performance of hard coatings, made by balanced and unbalanced magnetron sputtering, for decorative applications Baham Miremad, Surface and Coatings Technology, Vol. 120-121, (November 1999) 728-733.

René Olivares-Navarrete , Jhon Jairo Olaya, Claudia Ramirez and Sandra Elizabeth Rodil, Coatings 1 (2011) 72-87 G. Ramírez, S.E. Rodil, H. Arzate, S. Muhl, J.J. Olaya Biocompatibility of Niobium coatings, Applied Surface Science, Vol. 257, (2011) 2555-2559

S.E. Rodil and J. J Olaya, Unbalanced magnetic field configuration: plasma and film properties, Journal of Physics: Condensed Matter, Vol. 18,. 1703-1720, 2006.

Siegwart, R. (2001), Indirect Manipulation of a Sphere on a Flat Disk Using Force Information. *International Journal of Advanced Robotic Systems,* Vol.6, No.4, (December 2009),12-16.

Stansbury E. and Buchnan R. *Fundamentals of Electrochemical Corrosion.* First edition. ASM International). ISBN 0-87170-676-8, United States of America(2000).

Stefan Wrehde, Marion Quaas, Robert Bogdanowicz, Hartmut Steffen, Harm Wulff, Rainer Hippler, Optical and chemical characterization of thin TiNx films deposited by DC-magnetron sputtering, Vacuum, Vol. 82, (2008) 1115–1119

T. Q. Li, S. Noda, Y. Tsuji, T. Osawa, and H. Komiyama, J. Vac. Sci. Technol. Vol. A 20 (3), (2002) 583-588.

Teer D.G. A , magnetron sputter ion-plating system, Surf Coat Technol, Vol. 39-40, (1989) 565.

Thornton, J., Influence of apparatus geometry and deposition conditions on the structure and topography of thick sputtered coating, J. Vac. Sci. Technol., Vol. 11, (1974) 666-670.

W. Ensinger, Tribological behavior of duplex coating improved by ion implantation, Surf. Coat. Technol. Vol. 99, (1998) 1

W.F. Liu, S. Suzuki, K. Machida, Magnetic properties of Nd–Fe–B film magnets prepared by RF sputtering, Journal of Magnetism and Magnetic Materials, Vol. 308, Issue 1, (January 2007) 126-130.

Waits R.K. J Planar magnetron sputtering, Vac. Sci. Technol. Vol. 15, Issue 2, (1978) 179-87.

Window B, Savvides N, Unbalanced dc magnetron as source high ion flux, J Vac. Sci. Technol. Vol. A 4, Issue 2 (1986) 453-456.

Y. Ochiai, K. Aso, M. Hayakawa, H. Matsuda, K. Hayashi, W. Ishikawa, On specimen preparation by radio frequency planar magnetron sputtering , J. Vac. Sci. technol. Vol. A4, Issue 1, (1986) 19-25

Y.H. Cheng, B.K. Tay, S.P. Lan, Influence of deposition temperature on the structure and internal stress of TiN films deposited by filtered cathodic vacuum arc, J. Vac. Sci. Technol. Vol. A 20, (2002) 1270.

Crystallization of Ge:Sb:Te Thin Films for Phase Change Memory Application

J. J. Gervacio Arciniega, E. Prokhorov, F. J. Espinoza Beltran and G. Trapaga
CINVESTAV, Unidad Queretaro, Juriquilla, Querétaro, Mexico

1. Introduction

Chalcogenide glasses are a chemical compound consisting of at least one chalcogen element, sulphur, selenium, or tellurium, in combination with other elements. These glasses obtained great attention after discovery between 1962 and 1969 by Kolomiets, Eaton, Ovshinsky and Pearson of the S-shape current-voltage characteristic in chalcogenide glasses and the switching phenomenon from high to low resistivity states (Popescu, 2005). In 1968 S. R. Ovshinsky demonstrated very short (about of 10^{-10} seconds) reversible electrical switching phenomena in $Te_{81}Ge_{15}Sb_2S_2$ thin films due to amorphous-crystalline phase transition (Ovshinsky, 1968). This work opened a new area of phase change technology, which now is one of the most important technologies for memory devices and applications in computers, CD, DVD, phase-change random access memories, etc.

The first electrical phase-change memory devices used various binary, ternary and quaternary Te-based films with compositions made up of Ge:Te, Si:As:Te, Ge:As:Si:Te, etc., systems (Stand, 2005). The early phase change materials used in optical storage comprised simple alloys based primarily on compositions in the vicinity of the tellurium-germanium eutectic. Antimony was primarily used, although other elements including selenium, arsenic and bismuth were all shown to have beneficial effects. Based on the results obtained from films with $Ge_{15}Sb_4Te_{81}$ composition (Ovshinsky, 1971) the first application of Ge:Sb:Te alloys was reported for phase change rewritable optical disks.

In the early 1990s, a second generation of high speed phase change materials based on Ge:Sb:Te alloys was reported by several optical memory research groups (Ohta et al 1989, Yamada et al, 1991, Gonzalez-Hernandez et al, 1992). These alloys have stoichiometric compositions along the $GeTe$-Sb_2Te_3 pseudobinary line of phase diagram such as $Ge_2Sb_2Te_5$, $Ge_1Sb_2Te_4$ and $Ge_1Sb_4Te_7$.

Ge:Sb:Te stoichiometric alloys have three phases: one amorphous and two crystal structures. The first crystalline phase is the rock salt NaCl-like and the second more stable phase is hexagonal. When the amorphous films are heated, the transition from amorphous to rock salt-like structure occurs at around 120-170^{0}C; subsequent heating transforms this phase into a stable hexagonal structure at temperatures around 200-250^{0}C. The exact transition temperature depends on the composition of the film. The hexagonal structure remains stable over a wide temperature range from 200-250^{0}C to around the melting point (593-630 ^{0}C, depending on composition) (Gonzalez-Hernandez et al, 1992). The principle of phase-change memory operation is based on a reversible phase-transformation from the amorphous (high resistance/low reflectivity state) to crystalline NaCl-type (low resistance/high reflectivity state) under short laser or electrical pulses. It is necessary to note that for phase-change memory applications only the amorphous-NaCl-type transition has been used, probably because the heating time produced by the laser or electrical pulse is too short to form the stable hexagonal structure (Yamada, 1991). This transformation is always accompanied by abrupt changes in reflection (about 20-30 %) and resistivity (about 3 orders of magnitude) in chalcogenide films. The reversible transformation from crystalline to the amorphous structurally disordered state can be obtained by increasing the local temperature of the Ge:Sb:Te layer above its melting point by short intense laser, or electrical pulses and the subsequent quenching with a cooling rate about 10^{10} deg/s (Yamada et al, 1991).

The phase transition between the amorphous and NaCl-type crystalline phase in stoichiometric materials is fast because atoms in the amorphous state do not need to travel long distances to take their position in the crystal lattice. On the other hand, non-stoichiometric compounds require long-range diffusion when they crystallize from an amorphous state (Yamada et al, 1991, Yamada 1996), which in general slows down the crystallization process. For example, 30 at. % of the deviation from $Ge_2Sb_2Te_5$ stoichiometric composition increased the crystallization time from 220 to 500 ns (Kyrsta et al., 2001).

For several years, Ge:Sb:Te alloys along the pseudobinary GeTe–Sb_2Te_3 line have been used as the main material for optical phase-change data storage devices and are currently being investigated for nonvolatile electronic storage purposes (which utilize the difference in the electrical resistance of the two phases) due to the high reflectivity and resistivity contrast between the amorphous and crystalline phases. Many investigations have been carried out on this type of phase-change materials with the purpose of improving their properties and to increase its storage capacity, stability, speed and versatility. These investigations have shown that the crystallization of the phase-change (or amorphous-NaCl-type transition) can be considered as a rate-limiting process to obtain a fast data transfer. That is why there are a considerable number of experimental and theoretical studies investigating the amorphous-to-crystalline NaCl-type phase transformation. In spite of the large number of publications, the activation energy for crystallization of Ge:Sb:Te materials reported in the literature shows a large discrepancy; the activation energy of crystallization for $Ge_2Sb_2Te_5$ reported in the literature varied between 0.8 and 2.9 eV, for example. Such dispersion in the kinetic parameters might depend on the differences in the deposition methods, type of substrate, dielectric cover layer, film thickness and/or parameters of deposition processes, which lead to differences in the crystallization temperature, activation energy, and the like. In addition, experimental data show relatively long incubation times for crystallization of Ge:Sb:Te films and a relatively large amount of crystallized material during this period of time (Morales-Sanchez et al 2010). Incubation time manifests itself as the time necessary to reach critical values of nucleus for the crystallization to occur (Senkader et al, 2004). The existence of a non-negligible amount of crystallite during the incubation time could also be responsible for the dispersion in the kinetic parameters reported in the literature.

Based on such antecedents, we aimed in this chapter at investigating the crystallization kinetics of thin films on a glass substrate obtained by the same deposition process (DC sputtering), with the same thickness (around 200 nm), and composition along the GeTe–

Sb_2Te_3 pseudobinary line ($Ge_2Sb_2Te_5$ $Ge_1Sb_2Te_4$ and $Ge_1Sb_4Te_7$) including GeTe and Sb_2Te_3 (which lie at the end of GeTe-Sb_2Te_3 pseudobinary line), taking into account the processes during the incubation time. It is necessary to note that GeTe and Sb_2Te_3, as well as Ge:Sb:Te, can be used in phase-change data storage. Additionally, the crystallization kinetics of $Ge_4Sb_1Te_5$ films will be present in this chapter. The structure of this alloy is unclear. In contrast to other Ge:Sb:Te ternary compounds, it does not belong to the GeTe - Sb_2Te_3 homologous series, although it lies on (or very close to) the GeTe-Sb_2Te_3 pseudobinary line in the Ge:Sb:Te phase diagram between GeTe and $Ge_2Sb_2Te_5$. It is suggested (Coombs et al., 1995) that this alloy is a solid solution between GeTe and a compound with a composition close to $Ge_4Sb_1Te_5$. This material, in contrast to stoichiometric alloys, demonstrates only one amorphous-NaCl-type phase transition, but it has the largest optical (Kato et al., 1999) and electrical (Morales-Sanchez et al., 2005) contrast between the crystalline and amorphous states, compared with other Ge:Sb:Te ternary alloys.

Necessary to note that as-prepared and melt-quenched Ge:Sb:Te amorphous materials show different crystallization kinetics (Nobukuni et al., 1999, Park et al., 1999, Khulbe et al., 2000, Wei et al., 2003, Kalb et al., 2004, Raoux et al., 2008). It has been proposed in the literature that for melt-quenched amorphous materials may exist: embryos following the condensation and evaporation of embryos to form crystalline clusters (Khulbe et al., 2000), preexisting clusters (Park et al., 1999), crystal nuclei (Wei et al., 2003), sinks and voids after repeated overwriting (Nobukuni et al., 1999), locally ordered regions with structure similar to that of crystalline Sb (Naito et al., 2004) or a crystalline amorphous border (Raoux et al., 2008). In spite of different models, crystallization in the melt-quenched amorphous material can start from these nucleation centers, which decrease incubation time and increase crystallization speed. But in this chapter only crystallization of as-prepared materials will be discussed.

Studies of crystallization kinetics of phase change materials are mostly analyzed using the Johnson-Mehl-Avrami-Kolmogorov (JMAK) model for isothermal annealing (see for example: (Weidenhof et al., 2001, Ruitenberg et al., 2002, Trappe et al., 2000, Morales-Sanchez et al., 2010)), which permitted to determine the activation energy for the crystallization process. This method will be used for investigation of the crystallization kinetics in the compositions mentioned above.

Figure 1 shows in-situ optical reflection (using a laser diode emitting at 650 nm) as a function of temperature with a heating rate of $5^0C/min$ for all investigated films with composition indicated on the graph. An abrupt change in reflection is associated with the onset of phase crystallization. Stoichiometric $Ge_2Sb_2Te_5$, $Ge_1Sb_2Te_4$ and $Ge_1Sb_4Te_7$ materials demonstrate two changes: the first corresponds to amorphous-NaCl-type transition and the second to NaCl-type-hexagonal transformation. In contrast, GeTe, $Ge_4Sb_1Te_5$ and Sb_2Te_3 show only one amorphous-crystalline transition. The highest crystallization temperature (T_c) has been observed in GeTe, the lowest in Sb_2Te_3. The crystallization kinetics of the GeTe-Sb_2Te_3 pseudobinary line will be analyzed according to this Figure from the highest to the lowest crystallization temperature.

2. Deposition of phase change films

An important aspect to develop thin films of phase change materials is the deposition method. It is well known from the literature that deposition technology strongly affects the microstructure and as a results the optical, electrical and crystallization (especially the crystallization temperature) properties of phase-change materials. For example, the crystallization temperature of the films with the same composition but obtained by different deposition methods can differ for more than 20^{0}C, and no suitable explanation has yet been proposed (Morales-Sanchez et al., 2005). That is why, to compare crystallization properties all studied films have to be obtained with the same deposition process (DC sputtering), with the same thickness (around 200 nm), and composition along the GeTe-Sb_2Te_3 pseudobinary line. This method has been chosen because sputtering can produce films with the desired stoichiometry. Compared, for example, with thermal evaporation, in sputtering the thickness and chemical composition of thin films are easier to control.

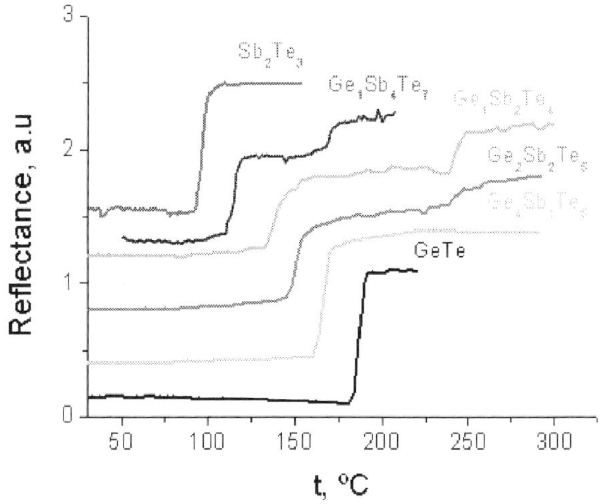

Fig. 1. Dependencies of optical reflection on temperature for films with compositions indicated on the graph.

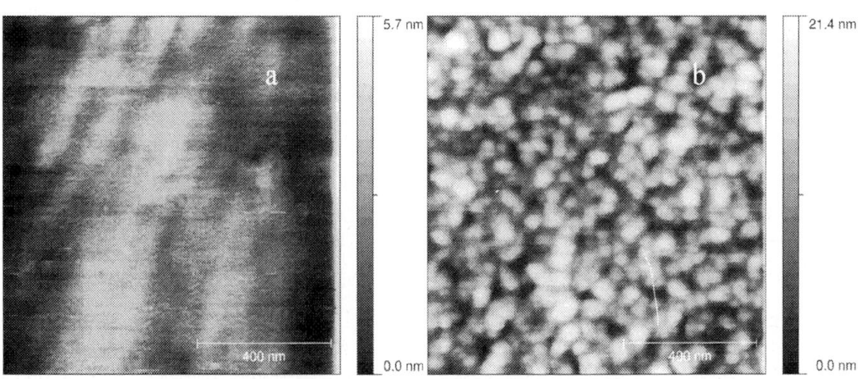

Fig. 2. Surface topography of as-prepare amorphous (a) and crystalline (b) $Ge_2Sb_2Te_5$ films.

The thin films have been prepared on glass and Si substrates by DC magnetron sputtering from one composite target. The deposition conditions were: 4.5×10^{-6} mbar base pressure, 2.0×10^{-5} mbar work pressure, 180 cm^3/min Ar gas input flow, 0.49 W/cm^2 DC power density, and 10 min deposition time. Before films sputtering, the targets were cleaned for 10 minutes by sputtering to remove the oxide from the surface. According to XRD measurements all films were obtained in amorphous state. Energy dispersive spectroscopy has shown that deviations of films compositions from the targets were approximately 2%.

Figure 2a shows typical surface topography obtained from AFM measurements of as-prepared amorphous thin films. All amorphous films demonstrated vary smooth surface with the surface roughness between 1 and 3 nm. In contrast, in the crystalline films (annealed to the temperature of 200^0C) the topography image shows grains with average size between 20-120 nm (Fig. 2b). The dimensions of grains are dependent on the films composition.

3. Crystallization of GeTe

GeTe is a promising candidate for the application of in phase-change technology due to its higher crystallization temperature (onset of crystallization temperature at about 184^0C) when compared to other alloys, since it offers a significant improvement in data-retention at high temperature (Perniola et al., 2010, Fantini et al., 2010). Amorphous films, in contrast to Ge:Sb:Te ternary alloys, crystallize in the rhombohedral phase. This phase can be viewed as a rock salt structure distorted along the [111]-direction (Caravati et al., 2010). Additionally, the central atom is displaced along the [111]-direction from the center of the rhombohedron. Upon crystallization, in a rhombohedral (distorted rock salt) structure appeared about 10% of vacancies occurring on Ge sites (Kolobov 2004). GeTe is a classical ferroelectric material that shows displacive type ferroelectric-paraelectric transition from a rhombohedral ferroelectric phase to a rock salt type structure with paraelectric properties. Two sublattices form this rock salt structure: Ge atoms compose one of the lattices and tellurium atoms the other (Rabe et al., 1987).

The NaCl type crystalline structure of GeTe is an unstable phase. This crystalline structure exists for temperatures above 300°C. The ferroelectric-paraelectric transition occurs in the interval of 327-427^0C depending on various factors, such as exact composition, carrier density, etc. For example, changes in Te composition from 50 to 50.7 at. % shift the transition temperature from 427^0C to 356^0C (Okura, 1992).

Activation energy of amorphous to rhombohedral phase crystallization reported in the literature demonstrates large dispersion: 1.7 eV (Libera et al., 1993), 1.77 eV (Lu et al., 1995), 1.96 eV (Fantini et al., 2010), 2.5 eV (Fan et al., 2004), 3.9 eV (Matsushita et al., 1989). Such dispersion can be dependent on the differences in the deposition methods, type of substrate, method of results interpretation, etc. According to the literature and to our measurements (Fig. 1), GeTe has the highest crystallization temperature for materials on the GeTe-Sb_2Te_3 pseudobinary line and must demonstrate the highest activation energy of crystallization.

Figure 3 shows X-ray patterns of GeTe film measured at the temperature indicated on the graph. At temperatures below 180^0C, the material shows only wide bands, which are characteristic of amorphous materials. At higher temperatures, a rhombohedral GeTe phase appears. At a temperature of 350^0C two phases have been observed: rhombohedral and

NaCl-type, and at higher temperatures the material transforms into a NaCl-type phase. This phase exists only at temperatures above 400^{0}C and in the process of cooling, again transforms into the stable rhombohedral phase.

Crystallization kinetics in GeTe, as in other materials, has been investigated using optical reflection during isothermal measurements. Reflection measurements were made with a 650 nm wavelength laser diode and a PIN diode as detector. In the reflectance measurements, the generally employed assumption has been used; the reflection signals are linearly related to the total transformed crystalline volume fractions x (Weidenhof et al., 2001):

$$x = [R(t) - Ra] / [Rc - Ra]\,, \tag{1}$$

where R(t), the experimental measurement value of reflectance, Ra and Rc are reflectance of amorphous and fully crystalline phases, respectively.

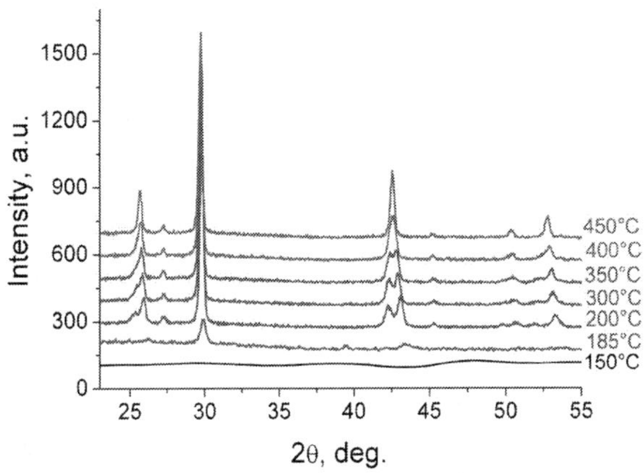

Fig. 3. XRD patterns of GeTe film measured at the temperature indicated on the graph. Blue patterns correspond to rhombohedral phase. Red patterns correspond to NaCl-type phase.

Figure 4 shows the evolution of the total crystalline volume fraction x, for GeTe samples, calculated from reflection measurements for films isothermally annealed at the temperatures indicated on the graph.

In this chapter, crystallization kinetics will be analyzed using the Johnson–Mehl–Avrami-Kolmogorov (JMAK) model for isothermal annealing. According to the classical JMAK model, the transformed volume fraction x can be determined by the following expression:

$$x(t) = 1 - \exp(-Kt^n)\,, \tag{2}$$

where $K=\gamma exp(-E/kT)$, γ is the frequency factor, E is the effective activation energy, t is the annealing time, and n the Avrami exponent, which provides information about the mechanisms of crystallization. The value of n can be evaluated from the slope of the $ln[-ln(1- x)]$ versus $ln(t)$ plot, which in materials with random nucleation and isotropic growth should be linear. The JMAK model is based on several assumptions such as:

homogeneity of the system, random and uniform nucleation, and nucleation rate taking place at the very beginning of the transformation is time independent and applied for isotropic growth at a constant rate.

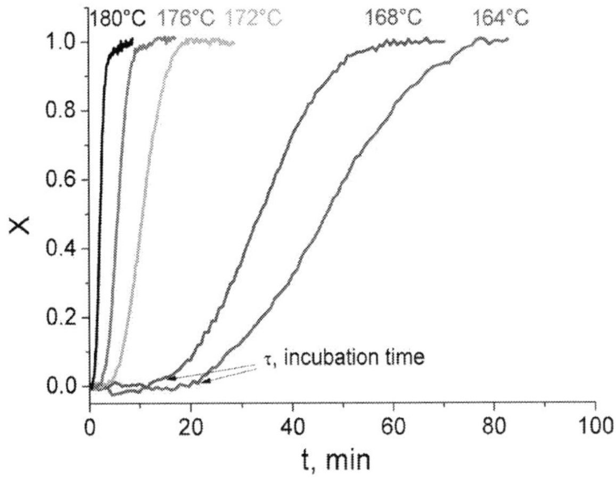

Fig. 4. Dependencies of the volume fraction x versus time obtained for GeTe samples from reflection measurements at the temperature indicated on the graph.

The value of activation energy for GeTe obtained using Equation (2) was very high: close to 12 eV. Such high value of activation energy has been obtained because GeTe, as other Ge:Sb:Te materials, shows long incubation times τ for crystallization, namely, the annealing time required to reach a critical nuclei size or to observe an abrupt increase in the crystalline volume fraction. In this case, the nucleation rate cannot be considered to be time independent for the entire crystallization process as is considered in classical JMAK models (Senkader et al., 2004).

In the case of GeTe, the volume fraction of crystalline phase approximately equals zero during all the incubation time (see Fig. 4). In such cases than the volume fraction of crystalline phase does not demonstrate a substantial increase during the incubation time it is possible to define the beginning of the transformation after the incubation time (Weidenhof et al., 2001). The JMAK equation can now be expressed as:

$$x(t) = 1 - \exp[-K(t-\tau)^n], \quad (3)$$

where τ is the incubation time. According to Equation (3) the plot $ln[-ln(1- x)]$ versus $ln(t-\tau)$ must be a straight line with slope n. Figure 5 shows such modification of an Avrami plot for GeTe. The insert shows the rate constant as a function of the reciprocal temperature.

The calculated activation energy for GeTe as determined from the modification of the Avrami Equation (3) using the Arrhenius type relation for K was 3.98±0.12 eV. The value of the Avrami exponent was about 1.6, which corresponds to diffusion controlled growth from small dimension grains with decreasing nucleation rate (Christian, 1975).

In addition, X-ray measurements have shown that during isothermal annealing from amorphous phase only one rhombohedral phase appeared. As time increases, the volume fraction of amorphous phase decreases (Figure 6); in the same time, the volume fraction of crystalline phase increases. No shift in the position of diffraction peaks was observed. This observation will be important in the analysis of crystallization processes of $Ge_2Sb_2Te_5$ and $Ge_1Sb_2Te_4$ materials.

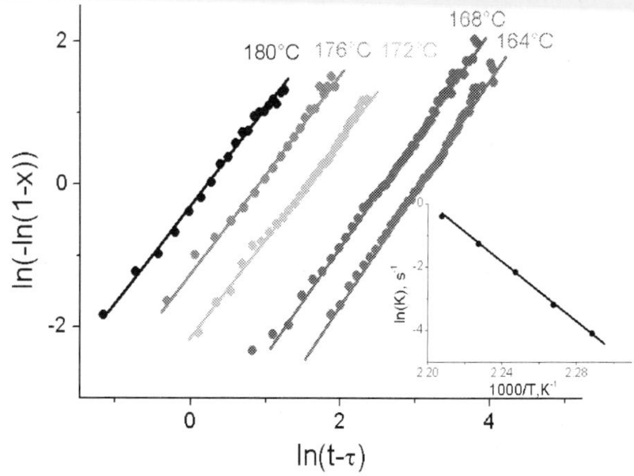

Fig. 5. Avrami plot of *ln[−ln(1−x)]* vs *ln(t−τ)* for GeTe. Points - experiment, lines - result of fitting using JMAK equation (3). The insert shows the rate constant *K* as function of the reciprocal temperature.

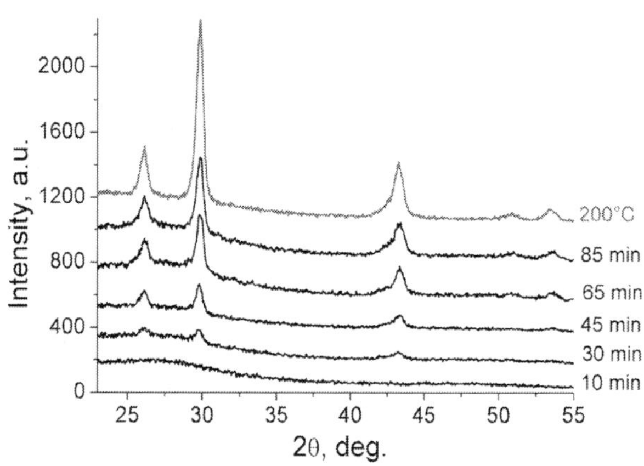

Fig. 6. X-ray diffraction spectra for a GeTe film obtained in the process of isothermal annealing at a temperature of 164°C during the time indicated on the plot. The upper pattern corresponds to film annealing at 200°C.

4. Crystallization of $Ge_4Sb_1Te_5$

As mentioned above, the structure of $Ge_4Sb_1Te_5$ is not known. It lies on or is close to the GeTe-Sb_2Te_3 pseudobinary line in the Ge:Sb:Te phase diagram between GeTe and $Ge_2Sb_2Te_5$. Because of such position, the onset of crystallization temperature is about 160ºC, less than in GeTe but higher than in another Ge:Sb:Te ternary alloys. This alloy crystallizes in NaCl-type structure (Gonzalez-Hernandez et al., 1992, Wamwangi et al., 2002, Ruiz Santos et al., 2010). For $Ge_4Sb_1Te_5$ films the reported values of the activation energy varied in a wide range: 1.13 eV (Kato et al., 1999), 3.09 eV (Morales-Sanchez et al., 2005), 3.48 eV (Wamwangi et al., 2002).

In addition, similarly to GeTe, bulk material and thin $Ge_4Sb_1Te_5$ films demonstrate at a temperature of approximately 327ºC ferroelectric-paraelectric transition (Bahgat et al., 2004, Ruiz Santos et al., 2010). But in contrast to GeTe, in which ferroelectric-paraelectric transition relates to a transformation from a rhombohedral phase to a rock salt type structure, in $Ge_4Sb_1Te_5$ the ferroelectric-paraelectric transition relates to a transformation between two different rock salt type structures.

Figure 7 shows DSC measurements on a crystalline $Ge_4Sb_1Te_5$ bulk alloy and on the amorphous material obtained by removing the as-deposited films from the glass substrates. In the bulk alloys, a nonsymmetrical endothermic peak at a temperature of 329ºC has been observed. In contrast, DSC measurements on amorphous material showed an exothermic peak at 176ºC, which is associated with the crystallization of the sample and a broad endothermic peak, with a maximum at approximately 327ºC, and that corresponds to the same temperature as in the bulk alloy. The observed endothermic peak can be associated with the ferroelectric-paraelectric transition.

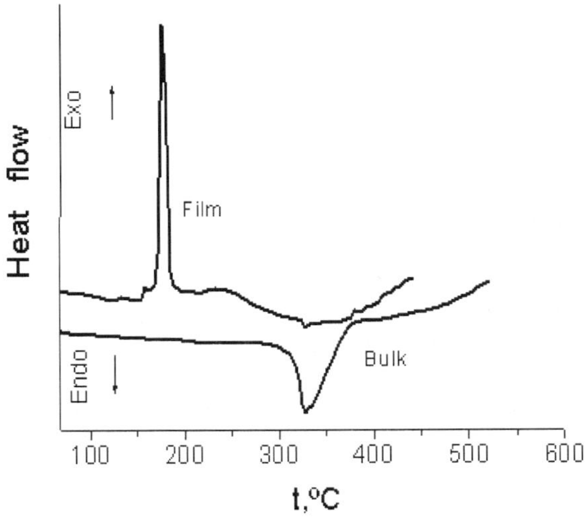

Fig. 7. DSC measurements on a bulk crystalline alloy and on the amorphous material, obtained by removing the as-deposited films from the glass substrates.

Additional confirmation of ferroelectric-paraelectric transition in $Ge_4Sb_1Te_5$ films has been obtained by temperature capacitance measurements (Ruiz Santos et al., 2010). In this work, it has been shown that the reciprocal capacitances (which are proportional to the reciprocal dielectric constant) as a function of temperature show a typical Curie-Weiss behavior for temperatures above 327^{0}C.

Specific data of crystallization process and ferroelectric-paraelectric transition in $Ge_4Sb_1Te_5$ films can be obtained by in situ XRD measurements at different temperatures. Figure 8 shows that below the crystallization temperature the film is in an amorphous state (pattern at 140^{0}C). As the temperature is increased, the material crystallizes in the NaCl type structure. As the temperature continue increasing, the position of all $Ge_4Sb_1Te_5$ peaks shift to lower 2θ values, reaching saturation at a temperature of approximately 327^{0}C at positions close to the NaCl type of GeTe structure (ICSD cart #602124), which demonstrate paraelectric properties. The position of the NaCl type of GeTe is marked on the graph with the horizontal line. This shift can be clearly seen on the insert of Figure 8, which shows the temperature dependence of the (200) diffraction peak in $Ge_4Sb_1Te_5$. The horizontal line on the insert indicates the position of the (200) peak in crystalline GeTe.

Figure 9 shows the evolution of the total crystalline volume fraction x, for $Ge_4Sb_1Te_5$ samples, calculated from reflection measurements for films isothermally annealed at temperatures indicated on the graph. As in the case of GeTe, the volume fraction of crystalline phase of $Ge_4Sb_1Te_5$ is approximately equal to zero during all the incubation time. The same as for GeTe, kinetics parameters for $Ge_4Sb_1Te_5$ can be calculated using Eq. (3).

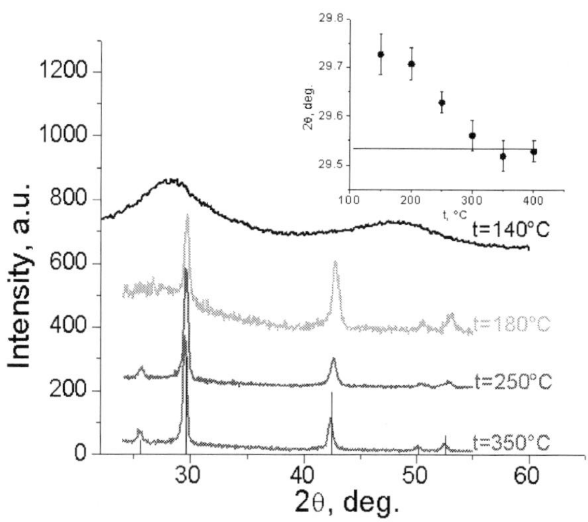

Fig. 8. XRD patterns of $Ge_4Sb_1Te_5$ film measured at the temperature indicated on the graph. The insert shows temperature dependence of (200) diffraction line in $Ge_4Sb_1Te_5$.

Figure 10 shows the modification of an Avrami plot for $Ge_4Sb_1Te_5$. The insert shows the rate constant as a function of the reciprocal temperature. The effective activation energy, determined from the Avrami plot for different temperatures using the Arrhenius type relation for *K*, was 3.46±0.22 eV (insert in Fig. 10). This value is in good agreement with other values reported in the literature (3.48 eV±0.12 eV) obtained using the Kissinger analysis (Wamwangi et al., 2002). The Avrami exponent for $Ge_4Sb_1Te_5$ was close to 1.8, which corresponds to a diffusion controlled growth from small dimension grains with decreasing nucleation rate, an exponent similar to that of GeTe (Christian 1975).

Additional X-ray measurements have shown that during isothermal annealing from amorphous phase only one NaCl-type phase appeared.

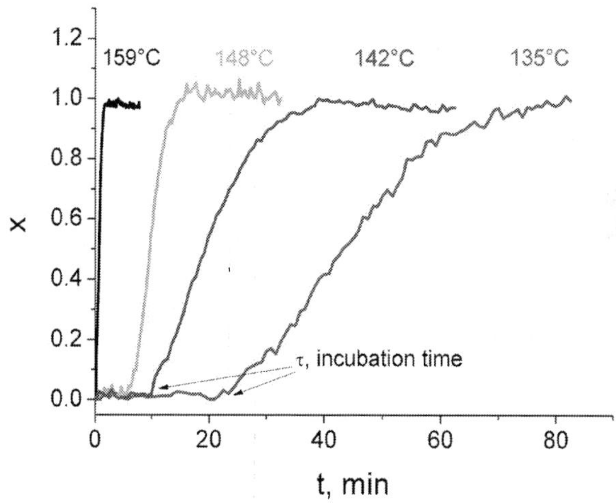

Fig. 9. Dependencies of the volume fraction *x* versus time obtained for $Ge_4Sb_1Te_5$ samples from reflection measurements at the temperature indicated on the graph.

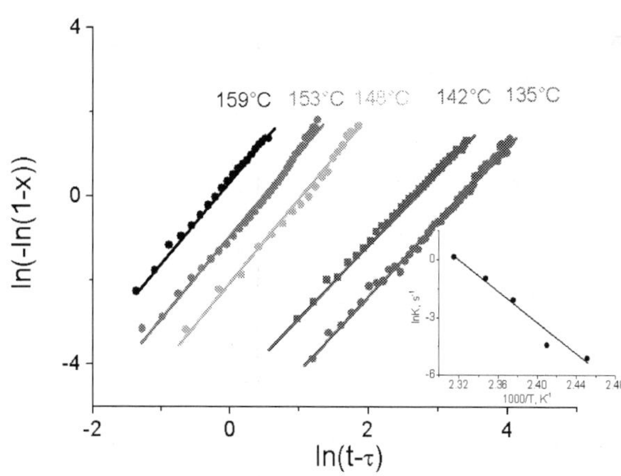

Fig. 10. Avrami plot of $ln[-ln(1-x)]$ vs $ln(t-\tau)$ for $Ge_4Sb_1Te_5$ films. Points - experiment, lines - results of fitting using JMAK equation (3). The insert shows the rate constant *K* as function of the reciprocal temperature.

5. Crystallization of $Ge_2Sb_2Te_5$

$Ge_2Sb_2Te_5$ thin films are the most commonly employed materials for phase-change memory technology application due to its high crystallization speed and relatively high crystallization temperature (but less than in GeTe and $Ge_4Sb_1Te_5$), which lead to high thermal stability. Because of this, many extensive experimental and theoretical studies have been conducted to understand the structure and crystallization phenomena in this material (see for example (Yamada et al, 1991, Weidenhof et al., 2001, Matsunaga et al., 2004, Matsunaga et al., 2006, Paesler et al., 2007, Im et al., 2008, Claudio et al 2006)).

As mentioned above, $Ge_2Sb_2Te_5$ material demonstrates two phase changes: first at a temperature close to 145°C there is an amorphous-NaCl-type transition and second (at about 240°C) a NaCl-type-hexagonal transformation.

Under heating, the amorphous $Ge_2Sb_2Te_5$ films crystallize at around 130-170°C, depending on the preparation method and heating rate, into a phase with a NaCl-type structure ($Fm\bar{3}m$). In this structure, the 4(a) site is fully occupied by Te atoms, whereas the 4(b) site is randomly occupied by Ge and Sb atoms and vacancies. The composition of Ge:Sb:Te ternary system, which lie on the GeTe-Sb_2Te_3 pseudobinary line, can be described as $(GeTe)_x$+ $(Sb_2Te_3)_{1-x}$ ($0<x<1$). In such case the site occupancy of the vacancy varies continuously according to $(1-x)/(3-2x)$ (Matsunaga et al., 2004, Matsunaga et al., 2006). But Ge and Sb atoms deviate from the ideal rock-salt positions in $Ge_2Sb_2Te_5$ not in a random way but in a strongly correlated manner with respect to the neighboring Te atoms (Kolobov et al., 2004, Kolobov et al., 2006). The off-center location of Ge atoms means that there is a net dipole moment and suggests that a NaCl-type phase of the $Ge_2Sb_2Te_5$ is a ferroelectric material (Tominaga et al., 2004). The ferroelectric properties in $Ge_2Sb_2Te_5$ NaCl-type phase have been observed using capacitance-temperature measurements. The temperature dependence of the capacitance shows an abrupt change with a maximum at the temperature that corresponds to the end from a NaCl-type to a hexagonal transition. In addition, the reciprocal capacitance for temperatures above this transition shows the Curie–Weiss dependence, which is typical of ferroelectric materials (Gervacio Arciniega et al., 2010).

In spite of the large number of publications about crystallization phenomena in $Ge_2Sb_2Te_5$, the activation energy of crystallization reported in the literature shows a large discrepancy in the range between 0.8 and 2.9 eV and between 1.2 and 4.4 for the Avrami exponent (Morales-Sanchez et al., 2010, Liu et al., 2009). Such parameter dispersion can be related not only to the difference in the preparation methods but to the specific crystallization process.

Experimental data in $Ge_2Sb_2Te_5$ films, as in all materials along the pseudobinary GeTe-Sb_2Te_3 line, show relatively long incubation times during isothermal crystallization (Weidenhof et al., 2001, Laine et al., 2004, Zhang et al., 2008, Morales-Sanchez et al., 2010). But in GeTe and $Ge_4Sb_1Te_5$ the amount of crystallized material is close to zero during the incubation. In contrast, in $Ge_2Sb_2Te_5$ films large amount of crystallized material has been observed during this period of time (ranging from 50% in Ref. (Zhang et al., 2008), 10% in Ref. (Sian et al., 2008), 9% in Ref. (Laine et al., 2004) and 8% in Ref. (Weidenhof et al., 2001)). It is necessary to note that the amount of crystalline phase material during incubation time depends on the annealing temperature. The existence of a non-negligible amount of crystallite during the incubation time, as will be shown below, is the most important factor responsible for the dispersion in the kinetic parameters reported in the literature.

Figure 11 shows the evolution of the total crystalline volume fraction x, for $Ge_2Sb_2Te_5$ samples, calculated from reflection measurements for films isothermally annealed at the temperatures indicated on the graph. In contrast to GeTe and $Ge_4Sb_1Te_5$, the volume fraction of crystalline phase of $Ge_2Sb_2Te_5$ does not equal zero during incubation time.

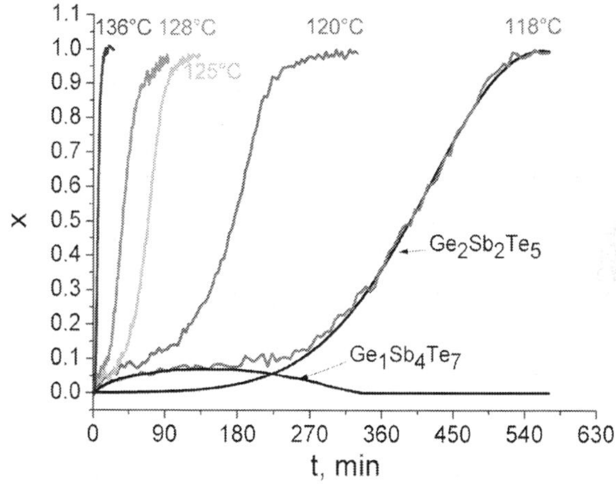

Fig. 11. Dependencies of the volume fraction x versus time obtained for $Ge_2Sb_2Te_5$ samples from reflection measurements at the temperature indicated on the graph. Black curves results from fittings using equation (7).

A typical JMAK plot for $Ge_2Sb_2Te_5$ films is shown in Figure 12. If we evaluate the slope only on a linear behavior, which is observed after incubation time τ, using a classical equation (2), the effective activation energy and the Avrami exponent will be equal to 5.61 eV and 3.3, respectively. The evaluation using a modification of a JMAK equation (3), which takes into account the incubation time, renders values for effective activation energy and an Avrami exponent equal to 2.07 eV and 1.5, respectively. This value of activation energy correlates well with what is reported in the literature (2.0 eV) and is obtained from the same interpretation of the reflection measurements (Weidenhof et al., 2001). Thus what comes into question is the pertinence of using these two approximations.

As has been mentioned above, the JMAK model is based on the following assumptions: the system is homogeneous, nucleation is random and uniform, the nucleation rate is time independent and takes place at the very beginning of the transformation, the isotropic growth maintains a constant rate, etc. (Henderson, 1979). However, nucleation in $Ge_2Sb_2Te_5$ films is not random nor uniform (Senkader et al., 2004) and due to partial crystallization during incubation time the nucleation rate cannot be considered to be time independent for the entire crystallization process as is considered in the classical model (equation 2). To overcome this limitation it has been proposed to avoid using a modification of the JMAK model (equation 3): after incubation time the rate constant can be considered as independent in time (Weidenhof et al., 2001). But in the case of $Ge_2Sb_2Te_5$ a large amount of crystallized material has been observed during incubation time. Moreover, experimental investigation has shown that nuclei, which appeared in $Ge_2Sb_2Te_5$ films during incubation time, have a composition corresponding to $Ge_1Sb_4Te_7$ which differs from the nominal value for the amorphous matrix (Laine et al., 2004, Claudio et al., 2006). Figure 13 shows X-ray diffraction patterns for $Ge_2Sb_2Te_5$ film isothermal annealing at the temperature 118⁰C and times indicated on the graph. The patterns of samples annealed during 50, 100, and 200 minutes show weak peaks, with positions corresponding to $Ge_1Sb_4Te_7$ NaCl-type phase. After annealing for more than 200 min, the positions of the peaks begin to change to those

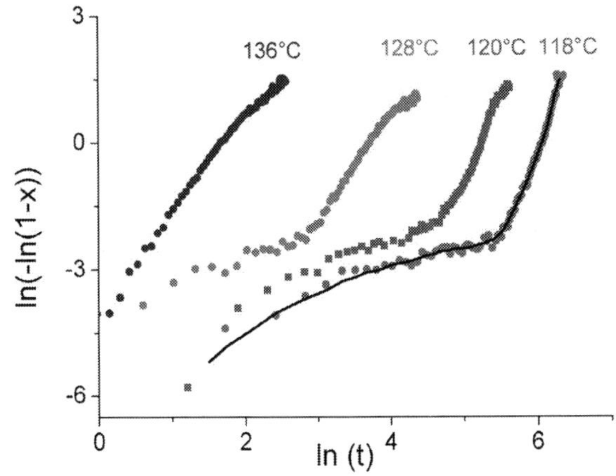

Fig. 12. Avrami plot of *ln[−ln(1−x)]* vs *ln(t)* for $Ge_2Sb_2Te_5$ films. Points - experiment, black line - results of fitting using equation (7).

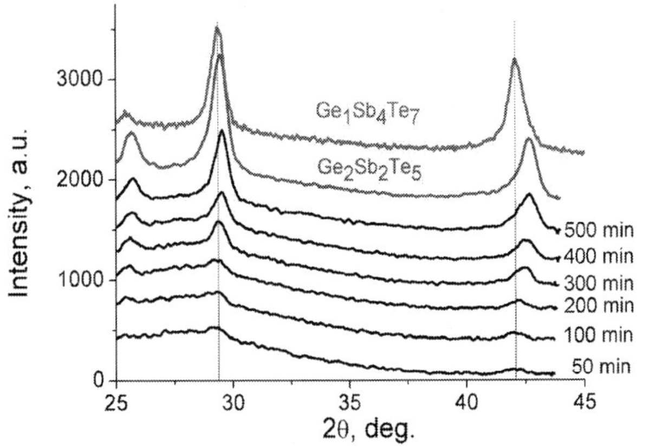

Fig. 13. X-ray diffraction spectra for a $Ge_2Sb_2Te_5$ film obtained in the process of isothermal annealing at a temperature of 118⁰C during the time indicated in the plot. Upper patterns correspond to $Ge_1Sb_4Te_7$ (red curve) and $Ge_2Sb_2Te_5$ (blue curve) taken at 180⁰C. The positions of the $Ge_1Sb_4Te_7$ peaks are indicated with the vertical lines.

corresponding to the NaCl-type phase of $Ge_2Sb_2Te_5$ composition. In this figure the blue pattern corresponds to material annealed at 180⁰C; the upper pattern corresponds to $Ge_1Sb_4Te_7$ film annealed at 180⁰C. Thus, in $Ge_2Sb_2Te_5$ films the formation of a stable rock salt type crystalline phase is preceded by the formation of a metastable $Ge_1Sb_4Te_7$ phase.

On the basis of these results, the crystallization of $Ge_2Sb_2Te_5$ material during isothermal annealing can be considered as a process which takes place in two stages: in the first stage nuclei of a metastable $Ge_1Sb_4Te_7$ phase appear; in the second stage, the nuclei transform into the equilibrium NaCl-type stoichiometric $Ge_2Sb_2Te_5$ structures.

In materials in which phase transformation starts with the formation of metastable phases the isothermal crystallization process cannot be simply described by JMAK theory and the plots of $ln(-ln(1 - x))$ versus $ln(t)$ are not linear, which implies that the Avrami exponent n does not remain constant during the crystallization process.

In ref. (Claudio et al., 2006) an analytical model that can describe the isothermal crystallization process of materials is proposed. In this process, the stable crystalline phase is preceded by the formation of a metastable phase. The model assumes that the volume fraction of metastable phase f_m grows up to a maximum value f_{max} and then stops growing when the stable phase that has nucleated into it overpasses the metastable grain boundaries. The kinetics of the metastable fraction can be represented by a JMAK-type equation subtracting the fraction of stable phase f_{sm} which grows inside the metastable phase:

$$f_m(t) = f_{\max}(1 - \exp(-K_m t^{n_m})) - f_{sm}(t), \tag{4}$$

where K_m is the crystallization rate constant and n_m the Avrami exponent.

The kinetic behavior of the stable phase into a metastable phase can be represented by the modification of classical JMAK formula, which takes into account the incubation time t_{sm} of the stable phase into the metastable phase:

$$f_{sm}(t) = f_{\max}(1 - \exp(-K_{sm}(t - t_{sm})^{n_{sm}})) \quad \text{for } t \ge t_{sm} \quad \text{and} \quad f_{sm}(t) = 0 \text{ for } t < t_{sm} \tag{5}$$

where K_{sm} and n_{sm} are the crystallization rate constant and the Avrami exponent, respectively, which represent the kinetic parameters of the transformation from stable phase into metastable phase.

Furthermore, it is possible to propose that the stable phase can grow into the amorphous phase. The stable phase transformation kinetics in amorphous phase can be described by JMAK-type equation with its specific parameters:

$$f_{sa}(t) = (1 - f_{\max})(1 - \exp(-K_{sa}(t - t_{sa})^{n_{sa}})) \text{ for } t \ge t_{sa} \quad \text{and} \quad f_{sa}(t) = 0 \text{ for } t < t_{sa} \tag{6}$$

where f_{sa} is the fraction of the stable phase in amorphous phase (from 0 to $1-f_{max}$), K_{sa} and n_{sa} are the JMAK parameters and t_{sa} is the incubation time of stable phase in amorphous phase.

The total fraction of transformed material is given by the equation:

$$f_{total}(t) = f_m(t) + f_{sm}(t) + f_{sa}(t) \quad (7)$$

This model allows fitting the experimental transformation data obtained in $Ge_2Sb_2Te_5$ films using a genetic algorithm. The continuous lines on Fig. 11 show the evolution on time of the volume fractions of metastable $Ge_1Sb_4Te_7$ and stable $Ge_2Sb_2Te_5$ phase obtained from fitting using Equation (7). Results of simulations have shown that the volume fraction of metastable $Ge_1Sb_4Te_7$ phase grows up to a maximum value and then decreases and disappears. After some incubation time, the $Ge_2Sb_2Te_5$ phase began to grow and all material is transformed into a stable crystalline phase.

In addition, the model is capable of predicting the three slopes clearly shown in the JMAK plot (Fig. 12) corresponding to three distinguishable stages in the crystallization process observed at dependencies obtained at low annealing temperature: the first one related to the metastable transformation, the second one (low value) with the step between metastable and stable transformation, and the last one with the stable phase nucleation and growth. It is quite evident from these simulated results that the first slope of the JMAK plot can be related to the kinetic behavior of the metastable phase.

The results of the investigation have shown that during the isothermal process of amorphous to crystalline phase transformation in $Ge_2Sb_2Te_5$ film a metastable phase with composition $Ge_1Sb_4Te_7$ and a stable NaCl-type $Ge_2Sb_2Te_5$ phase coexist within a certain time range. The appearance of nuclei of the $Ge_1Sb_4Te_7$ composition could be related to the local fluctuation in the film composition and to the fact that the crystallization temperature of $Ge_1Sb_4Te_7$ is lower than in $Ge_2Sb_2Te_5$ material.

6. Crystallization of $Ge_1Sb_2Te_4$

$Ge_1Sb_2Te_4$, as with $Ge_2Sb_2Te_5$ film, during annealing crystallize at a temperature of approximately 133^{0}C (lower than the crystallization temperature of $Ge_2Sb_2Te_5$) into a NaCl-type structure ($Fm\bar{3}m$) (Matsunaga et al., 2004) and at a higher temperature (about 235^{0}C) into a hexagonal phase. Because of the lower crystallization temperature, if compared with $Ge_2Sb_2Te_5$, this material may possess a low programming current in random access memory.

Crystallization kinetics, the same as in $Ge_2Sb_2Te_5$ material, is sufficiently complicated. Figure 14 shows the dependence of the total crystalline volume fraction x versus time for $Ge_1Sb_2Te_4$ samples calculated from reflection measurements for films isothermally annealed at the temperatures indicated on the graph and Figure 15 shows a typical JMAK plot. If we evaluate the slope only on a linear behavior, which is observed after incubation time τ, using classical equation (2), the effective activation energy will be equal to 5.45 eV. The evaluation using modification of JMAK equation (3), which takes into account incubation time, renders values of effective activation energy equal to 1.77 eV. But in $Ge_1Sb_2Te_4$, the same as in $Ge_2Sb_2Te_5$ films, a large amount of crystallized material has been observed during the incubation time. Moreover, XRD measurements (Figure 16) have shown that nuclei, which appeared in $Ge_1Sb_2Te_4$ films during the incubation period, have a composition corresponding to the $Ge_1Sb_4Te_7$ structure. The upper XRD pattern in this Figure corresponds to a fully crystallized film with the $Ge_1Sb_4Te_7$ composition; the peak positions of this phase are marked with the vertical lines. The pattern mark as $Ge_1Sb_2Te_4$ shows diffraction lines of $Ge_1Sb_2Te_4$ film annealing to 180^{0}C.

The results obtained have shown that in $Ge_1Sb_2Te_4$, the same as in $Ge_2Sb_2Te_5$ film, during the isothermal process of amorphous to crystalline phase transformation, a metastable phase appeared with composition of $Ge_1Sb_4Te_7$. Thus, for the interpretation of the crystallization process in $Ge_1Sb_2Te_4$ an analytical model can be used in which the formation during crystallization of metastable phase is taken into account (Equation 7).

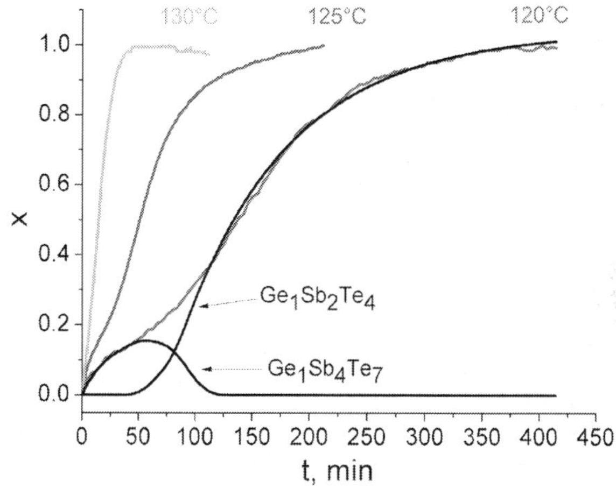

Fig. 14. Dependencies of the volume fraction x versus time obtained for $Ge_1Sb_2Te_4$ samples from reflection measurements at the temperature indicated on the graph. Black curves - results from fittings using equation (7).

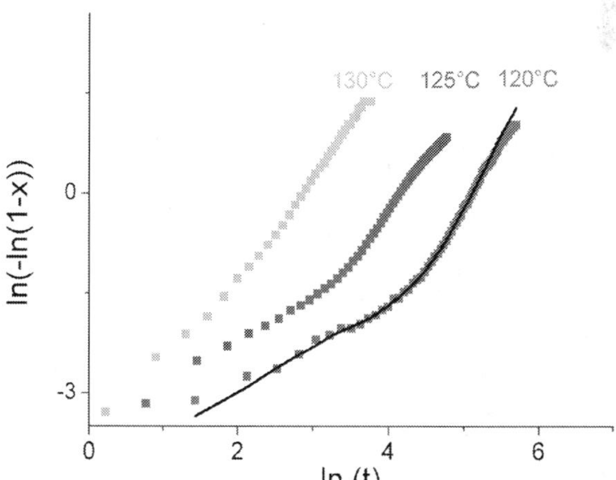

Fig. 15. Avrami plot of $ln[-ln(1-x)]$ vs $ln(t)$ for $Ge_1Sb_2Te_4$ films. Points - experiment, black curve - results from fitting using equation (7).

Continuous black curves on Fig. 14 show the calculated the volume fractions of metastable $Ge_1Sb_4Te_7$ and stable $Ge_1Sb_2Te_4$ phase obtained from fittings using Equation (7). Results of simulation have shown that the same as in $Ge_2Sb_2Te_5$, the volume fraction of metastable $Ge_1Sb_4Te_7$ phase grows up to a maximum value and then decreases and disappears. After some incubation time, $Ge_1Sb_2Te_4$ phase began to growth and all material was transformed into a stable crystalline phase.

Additionally, the model is capable to simulate the JMAK plot of Fig. 15, in which the first slope of the JMAK plot can be related to the kinetic behavior of the metastable $Ge_1Sb_4Te_7$ phase.

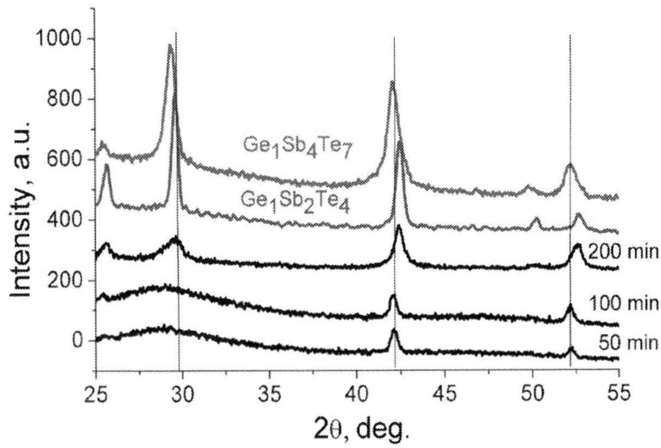

Fig. 16. X-ray diffraction spectra for a $Ge_1Sb_2Te_4$ film obtained in the process of isothermal annealing at a temperature of 120ºC during the time indicated on the plot. Upper patterns correspond to $Ge_1Sb_4Te_7$ (red curve) and $Ge_1Sb_2Te_4$ (blue curve) taken at 180ºC. The position of the $Ge_1Sb_4Te_7$ peaks is indicated with the vertical lines.

7. Crystallization of $Ge_1Sb_4Te_7$

Similarly, as with $Ge_2Sb_2Te_5$ and $Ge_1Sb_2Te_4$, $Ge_1Sb_4Te_7$ demonstrates a two phase transition: amorphous to the rock salt-like structure at a temperature of approximately 110ºC and a second transition to the hexagonal phase at the temperature close to 168ºC. This material demonstrates the fastest phase transformation among Ge:Sb:Te ternary alloys and phase-reversible transformations can be observed in the femtoseconds range (Huang et al., 2006). But because of the low crystallization temperature, $Ge_1Sb_4Te_7$ has lower thermal stability when compared with other ternary alloys (Miao et al., 2006).

Crystallization transformation from amorphous to NaCl-type structure in $Ge_1Sb_4Te_7$ differs from the transformations observed in $Ge_2Sb_2Te_5$ and $Ge_1Sb_2Te_4$ films. In contrast with $Ge_2Sb_2Te_5$ and $Ge_1Sb_2Te_4$, in-situ X-ray diffraction measurements show that amorphous films with the $Ge_1Sb_4Te_7$ composition crystallize in the NaCl-type phase with the same composition as at the beginning of the transformation (Figure 17). Moreover, isothermal reflection measurements (Figure 18) have shown practically zero value of crystalline phase during the much shorter incubation time compared with $Ge_2Sb_2Te_5$ films (Figure 11). In this case, as is for GeTe and $Ge_4Sb_1Te_5$, it is also possible to neglect the amount of crystallized material during τ, and describe the transformation using the modified JMAK equation (3).

Figure 19 shows a plot of $ln[-ln(1-x)]$ versus $ln(t-\tau)$ for a $Ge_1Sb_4Te_7$ film, which demonstrates a linear dependence. This means that the crystallization process in $Ge_1Sb_4Te_7$ material displays random nucleation and isotropic growth with effective activation energy of 1.7±0.27 eV and an Avrami exponent n close to 1.94. According to (Christian, 1975), the values of n in the range $1.5 < n < 2.5$ correspond to a crystallization process dominated by all shapes growing from small dimensions with decreasing nucleation rate.

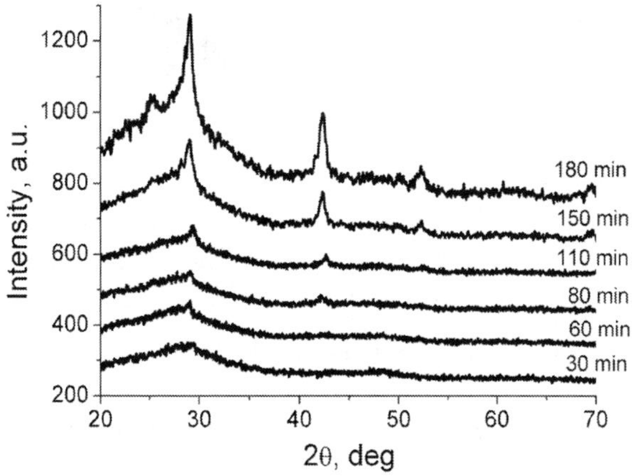

Fig. 17. X-ray diffraction spectra for a $Ge_1Sb_4Te_7$ film obtained in the process of isothermal annealing at a temperature of 90^{0}C during the time indicated on the plot.

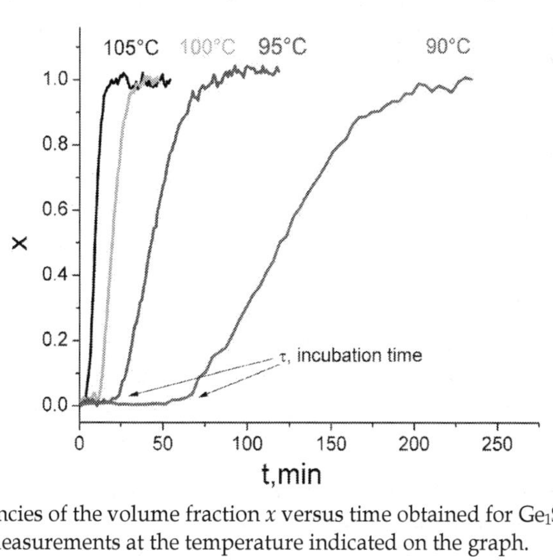

Fig. 18. Dependencies of the volume fraction x versus time obtained for $Ge_1Sb_4Te_7$ samples from reflection measurements at the temperature indicated on the graph.

8. Crystallization of Sb_2Te_3

Sb_2Te_3 lies at the end of the GeTe-Sb_2Te_3 pseudobinary line and has the lowest crystallization temperature but the highest crystallization speed. This material is an attractive candidate for phase change random access memory due to its rapid crystallization speed, but the crystallization temperature of Sb_2Te_3 (about 94°C in material investigated in this chapter and between 90-100°C reported in the literature (Yin et al., 2007, Kim et al., 2010)) is too low for it to be of practical use.

Upon annealing, amorphous Sb_2Te_3 films crystallize in rhombohedral Sb_2Te_3 (Kim et al., 2008, Lv et al., 2010) or fcc structure, which at a temperature above 200°C transforms into an hexagonal phase (Yin et al., 2007, Zhu et al., 2011). In the investigated Sb_2Te_3 films within the temperature range comprised between 95-200°C, only the rhombohedral phase (JCPDS data #15-0874) has been observed.

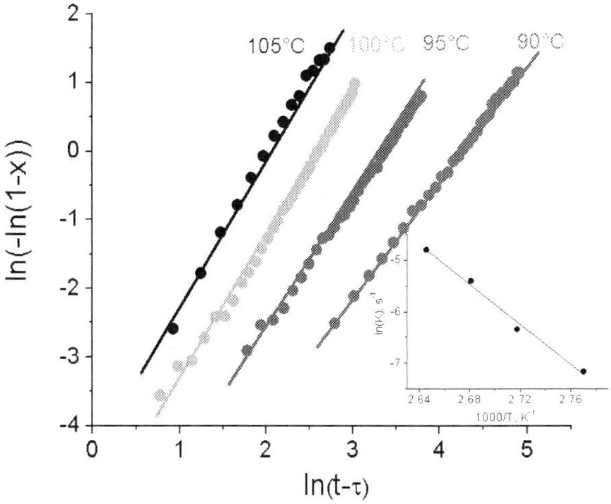

Fig. 19. Avrami plot of *ln[−ln(1−x)]* vs *ln(t−τ)* for $Ge_1Sb_4Te_7$ films. Points - experiment, lines - results from fittings using JMAK equation (3). Insert shows the rate constant *K* as a function of the reciprocal temperature.

Isothermal reflection measurements (Figure 20) have shown the same as for $Ge_1Sb_4Te_7$: practically zero value of crystalline phase during the incubation time. In this case, the same as for GeTe, $Ge_4Sb_1Te_5$ and $Ge_1Sb_4Te_7$ it is possible to neglect the amount of crystallized material during τ, and describe the transformation using the modified JMAK equation (3).

Figure 21 shows a plot of *ln[-ln(1-x)]* versus *ln(t-τ)* for a Sb_2Te_3 film, which demonstrates a linear dependence. This means that the crystallization process in Sb_2Te_3 films, the same as in $Ge_1Sb_4Te_7$ material, display random nucleation and isotropic growth with an effective activation energy of 1.54±0.15 eV and an Avrami exponent n close to 1.1, which according to ref. (Christian, 1975), corresponds to a crystallization growth of particles of appreciable initial volume. It is necessary to note that the values obtained for crystallization activation energy is in good agreement with values reported in the literature (1.51 eV) obtained using the Kissinger analysis in the films deposited on the glass substrates by the same DC magnetron sputtering (Zhai et al., 2009).

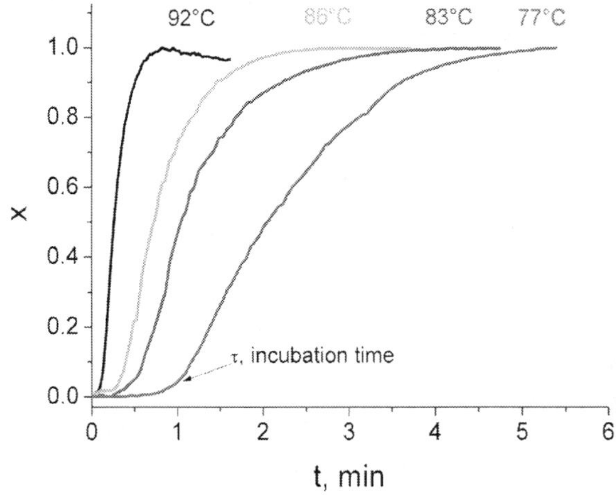

Fig. 20. Dependencies of the volume fraction x versus time obtained for Sb_2Te_3 samples from reflection measurements at the temperatures indicated on the graph.

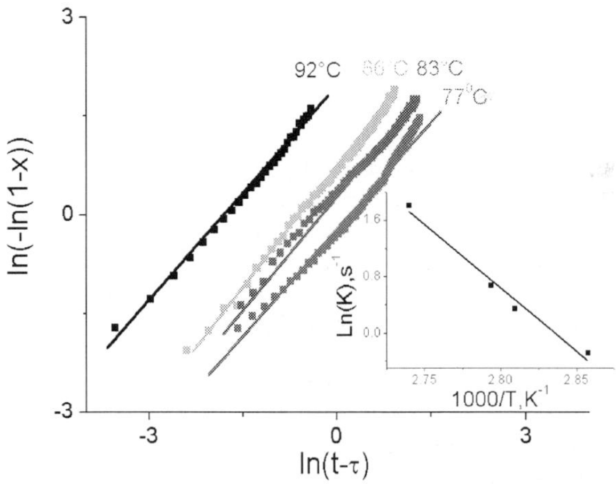

Fig. 21. Avrami plot of $ln[-ln(1-x)]$ vs $ln(t-\tau)$ for Sb_2Te_3 films. Points - experiment, lines - results from fittings using JMAK equation (3). Insert shows the rate constant K as function of the reciprocal temperature.

9. Conclusions

This chapter presents some of the results concerning the crystallization properties of the most commonly employed materials on the GeTe-Sb_2Te_3 pseudo-binary line for phase change memory application. Despite the fact that these alloys have found application in already commercialized optical disc and workable electrical phase change memory devices, their properties are not well known so far. This leads to many contradictory and controversial results in the literature.

As has been shown in this chapter, the crystallization properties of GeTe, $Ge_4Sb_1Te_5$, $Ge_1Sb_4Te_7$, and Sb_2Te_3 cannot be analyzed using the classical Johnson–Mehl–Avrami-Kolmogorov model for isothermal annealing due to the long incubation times. This limitation can be avoided by using a modification of the JMAK model, which takes into account that after incubation time the rate constant of crystallization can be considered as independent from time. Such model gave reasonable values of crystallization energy, which proved compatible with crystallization energy obtained by other methods. The difference in interpretation of isothermal measurements, in addition to the difference in preparation methods, can be responsible for the dispersion of crystallization parameters reported in the literature. But this modified model can be applied if during incubation time non-neglible amounts of crystalline phase material are observed. In the case of $Ge_2Sb_2Te_5$ and $Ge_1Sb_2Te_4$ films during incubation times, the nuclei fraction is of about 10% and has a composition corresponding to $Ge_1Sb_4Te_7$, which is different from the nominal value for the amorphous matrix. That is why the crystallization in $Ge_2Sb_2Te_5$ and $Ge_1Sb_2Te_4$ materials during isothermal annealing can be considered as a process that takes place in two stages: in the first stage, nuclei of a metastable $Ge_1Sb_4Te_7$ phase appear; in the second stage, the nuclei transforms into the equilibrium NaCl-type stoichiometric structures correspondent to the material composition. Such crystallization process can be described by a model in which the stable crystalline phase is preceded by the formation of a metastable phase. The most questionable issue is to explain these crystallization properties.

It is necessary to note that crystallization with phase separation has been observed in materials close to the GeTe-Sb_2Te_3 pseudo-binary line with excess of Ge ($Ge_{2+x}Sb_2Te_5$ x=0.5 (Privitera et al., 2003) and Sb ($Ge_2Sb_{2+x}Te_5$, $0<x<1$, (Yamada et al., 2000), and $Ge_2Sb_{2.3}Te_5$ (Yao et al., 2003)). In all of these materials, separation occurs during the crystallization phase, with segregation of small amounts of the excess elements, which remain in the amorphous state at the grain boundaries. By increasing the annealing temperature, the residual amorphous material can convert into another polycrystalline NaCl-type structure with a slightly lower lattice parameter. As a result, after the formation of the first phase, the crystallization rate is strongly reduced and a further conversion of the film into another crystalline structure that crystallizes at higher temperatures can occur (Privitera et al., 2003).

Furthermore, $Ge_2Sb_2Te_5$, $Ge_1Sb_2Te_4$ and $Ge_1Sb_4Te_7$ materials crystallize into the same NaCl-type ($Fm\bar{3}m$) structure (Matsunaga et al., 2004, Matsunaga et al., 2006) with lattice constants of 6.001, 6.044 and 6.0876 Å respectively (Morales-Sanchez et al., 2005). In materials with excess Sb, as the content of Sb increases, the lattice constant also increases (Yamada et al., 2000). Thus, during crystallization, atoms in the amorphous state must travel less distance to take their position in the $Ge_1Sb_4Te_7$ crystal lattice than in any other ternary alloys. This effect is responsible for the highest crystallization speed of $Ge_1Sb_4Te_7$ when compared with $Ge_2Sb_2Te_5$ and $Ge_1Sb_2Te_4$ materials. Also, $Ge_1Sb_4Te_7$ has the lowest crystallization temperature. All these factors and the possible existence of a local composition fluctuation can be responsible for the appearance of $Ge_1Sb_4Te_7$ nuclei in the process of crystallization in $Ge_2Sb_2Te_5$ and $Ge_1Sb_2Te_4$ materials.

The clarification of crystallization mechanisms in alloys that lie on the GeTe–Sb_2Te_3 pseudo-binary line will allow the production of phase-change materials with better recording properties.

10. References

Bahgat, A. A., Mahmoud, E. A., Abd, Rabo, A. S. & Mahdy, I. A. (2006). Study of the ferroelectric properties of Ge–Sb–Te alloys. *Physica B: Condensed Matter,* 382 No. 1-2, (April 2006) 271-278, ISSN 0921-4526.

Caravati, S., Bernasconi, M. & Parrinello, M. First principles study of the optical contrast in phase change materials. *Journal of Physics: Condensed Matter,* 22, No. 31, (July 2010) 315801, ISSN 0953-8984.

Claudio, D., Gonzalez-Hernandez, J., Licea, O., Laine, B., Prokhorov, E. & Trapaga, G. (2006) An analytical model to represent crystallization kinetics in materials with metastable phase formation. *Journal of Non-Crystalline Solids,* 352, No. 1, (January 2006) 51–55, ISSN 0022-3093.

Christian, J. W. *The Theory of Transformation in Metals and Alloys,* (1975), Pergamon Press, Oxford. ISBN 978-0-08-044019-4, United Kingdom.

Coombs, J. H., Jongenelis, A. P. J. M., Van Es-Spiekman, W. & Jacobs, B. A. (1995). Laser induced crystallization phenomena in GeTe based alloys. II. Composition dependence of nucleation and growth. *Journal of Applied Physics,* 78, No. 8, (October 1995) 4918-4956, ISSN 0021-8979.

Fan, Z., Wang, L. & Laughlin, D. E. (2004). Modeling of Crystallization Activation Energy for GeTe-Sb_2Te_3 Based Phase Change Materials. *Proceedings of SPIE,* ISBN 0-8194-5293-9, Monterey CA. USA, April 2004.

Fantini, A., Sousa, V. Perniola, L., Gourvest, E., Bastien, J. C., Maitrejean, S., Braga, S., Pashkov, N., Bastard, A., Hyot, B., Roule, A., Persico, A., Feldis, H., Jahan, C., Nodin, J. F., Blachier, D., Toffoli, A., Reimbold, G., Fillot, F., Pierre, F., Annunziata, R., Benshael, D., Mazoyer, P., Vallee, C., Billon, T., Hazart, J., De Salvo, B. & Boulanger, F. (2010). N-doped GeTe as Performance Booster for Embedded Phase-Change Memories. *Proceedings of International Electron Device Meeting (IEDM).* ISBN 978-1-4424-7418-5, San Francisco, CA. USA, December 2010.

Gervacio, Arciniega, J. J., Prokhorov, E., Espinoza, Beltran, F. J. & Gonzalez-Hernandez, J. (2010). Ferroelectric properties of $Ge_2Sb_2Te_5$ phase-change films. *Applied Physics Letters,* 97, No. 6, (August 2010) 063504, ISSN 0003-6951.

Gonzalez-Hernandez, J., Chao, B. S., Strand, D., Ovshinsky, S. R., Pawlik, D. & Gasiorowski, P. (1992). Crystallization studies of GeSbTe optical memory materials. *Applied Physics Communications,* 11, No. 4, (April 1992) 557-581, ISSN 0277-9374.

Henderson, D.W. (1979). Thermal analysis of non-isotermal crystallization kinetics in glass forming liquits. *Journal of Non-Crystalline Solids,* 30, No. 3, (July 1979) 301-315, ISSN 0022-3093.

Huang, S. M., Huang, S. Y., Zhao, Z. J. & Sun, Z. (2006). Investigation of phase changes in $Ge_1Sb_4Te_7$ films by single ultra-fast laser pulses. *Applied Physics A*, 82, No. 3, (March 2006) 529–533, ISSN 0947-8396.

Im, J., Eom, J. H., Park, C., Park, K., Suh, D. S., Kim, K., Kang, Y. S., Kim, C., Lee, T. Y., Khang, Y., Yoon, Y. G. & Ihm, J. (2008). Hierarchical structure and phase transition of $(GeTe)_n(Sb_2Te_3)_m$ used for phase-change memory. *Physical Review B*, 78, No. 20, (November 2008) 205205, ISSN 1098-0121.

Kalb, J., Spaepen, F. & Wuttig, M. (2004). Atomic force microscopy measurements of crystal nucleation and growth rates in thin films of amorphous Te alloys. *Applied Physics Letters*, 84, No. 25, (June 2004), 5240–5242, ISSN 0003-6951.

Kato, N., Takeda, Y., Fukano, T., Motohiro, T., Kawai, S. & Kuno, H. (1999). Compositional Dependence of Optical Constants and Microstructures of GeSbTe Thin Films for Compact-Disc-Rewritable (CD-RW) Readable with Conventional CD-ROM Drives *Japanese Journal of Applied Physics*, 38, No. 3B, (March 1999) 1707-1708, ISSN 0021-4922.

Khulbe P. K., Wright, E. M. & Mansuripur, M. (2000). Crystallization behavior of as-deposited, melt quenched, and primed amorphous states of $Ge_2Sb_{2.3}Te_5$ films. *Journal of Applied Physics*, 88, No. 7, (October 2000) 3926-3933, ISSN 0021-8979.

Kim M. S., Cho, S. H., Hong, S. K., Roh, J. S. & Choi. D. J. (2008). Crystallization characteristics of nitrogen-doped Sb_2Te_3 films for PRAM application. *Ceramics International*, 34, No. 4, (May 2008) 1043–1046. ISSN 0272-8842.

Kim, M.Y. & Oh, T. S. (2010). Crystallization behavior and thermoelectric characteristics of the electrodeposited Sb_2Te_3 thin films. *Thin Solid Films*, 518, No. 22, (September 2010) 6550-6553, ISSN 0040-6090.

Kolobov, A. V., Fons, P., Frenkel, A. I., Ankudinov, A. L., Tominaga, J. & Uruga, T. (2004). Understanding the phase-change mechanism of rewritable optical media. *Nature Materials*, 3, (September 2004) 703-708, ISSN 1476-1122.

Kolobov, A. V., Fons, P., Tominaga, J., Ankudinov, A. L., Yannopoulos, S. N. & Andrikopoulos, K. S. (2004). Crystallization-induced short-range order changes in amorphous GeTe. *Journal of Physics: Condensed Matter*, 16, No. 44, (October 2004) S5103–S5108, ISSN 0953-8984.

Kolobov, A. V., Fons, P., Tominaga, J. & Uruga, T. (2006). Why DVDs work the way they do: The nanometer-scale mechanism of phase change in Ge–Sb–Te alloys. *Journal of Non-Crystalline Solids*, 352, No. 9-20, (March 2006) 1612–1615, ISSN 0022-3093.

Kyrsta, S., Cremer, R., Neuschutz, D., Laurenzis, M., Bolivar, P. H. & Kurz, H. (2001). Deposition and characterization of Ge–Sb–Te layers for applications in optical data storage. *Applied Surface Science*, 179, No.1-4, (July 2001) 55-60, ISSN 0169-4332.

Laine, B., Trapaga, G., Prokhorov, E., Morales-Sanchez, E. & Gonzalez-Hernandez, J. (2003). Model for isothermal crystallization kinetics with metastable phase formation. *Applied Physics Letters*, 83, No. 24, (December 2003) 4969–4971, ISSN 0003-6951.

Libera M. & Chen, M. (1993). Time-resolved reflection and transmission studies of amorphous Ge-Te thin-film crystallization. *Journal of Applied Physics*, 73, No. 5, (March 1993) 2272-2282, ISSN 0021-8979.

Liu, D., Xu, L., Liao, Y. B., Dai, M., Zhao, L., ...Xu, J., Wu, L. C., Ma, Z. Y. & Chen, K. J. (2006). Comparison of Thermal Stability and Electrical Characterization between $Ge_2Sb_2Te_5$ and $Ge_1Sb_2Te_4$ Phase-Change Materials. *Japanese Journal of Applied Physics*, 48, No. 12, (December 2009) 121104, ISSN 0021-4922.

Lu Q. M. & Libera, M. (1995). Microstructural measurements of amorphous GeTe crystallization by hot-stage optical microscopy. *Journal of Applied Physics*, 77, No. 2, (January 1995) 517-521, ISSN 0021-8979.

Lv, B., Hu, S., Li, W., Di, X., Feng, L., Zhang, J., Wu, L., Cai, Y., Li, B. & Lei, Z. (2010). Preparation and Characterization of Sb_2Te_3 Thin Films by Coevaporation. *International Journal of Photoenergy*, Article ID 476589 (June 2010), ISSN 1110-662X.

Matsunaga, T., Kojima, R., Yamada, N., Kifune, K., Kubota, Y., Tabata, Y. & Takata, M. (2006). Single Structure Widely Distributed in a GeTe-Sb_2Te_3 Pseudobinary System: A Rock Salt Structure is Retained by Intrinsically Containing an Enormous Number of Vacancies within Its crystal, *Inorganic Chemistry*, 45, No. 5, (February 2006) 2235-2241, ISSN 0020-1669.

Matsunaga, T. & Yamada, N. (2004). Structural investigation of $GeSb_2Te_4$: A high-speed phase-change material. *Physical Rewiew B*, 69, No. 10, (March 2004) 104111, ISSN 1098-0121.

Matsushita, T., Nakau, T., Suzuki, A. & Okuda, M. (1989). Measurements of activation energy in the initialization process of amorphous GeTe films. *Journal of Non-Crystalline Solids*, 112, No. 1-3, (October 1989) 211-214, ISSN 0022-3093.

Miao, X. S., Shi, L. P., Lee, H. K., Li, J. M., Zhao, R., Tan, P. K., Lim, K. G., Yang, H. X. & Chong, T. C. (2006). Temperature Dependence of Phase-Change Random Access Memory Cell. *Japanese Journal of Applied Physics*, 45, No. 5A, (May 2006), 3955–3958, ISSN 0021-4922.

Morales-Sanchez, E., Lain, B., Prokhorov, E., Hernarndez-Landaverde, M. A., Trapaga, G. & Gonzalez-Hernandez, J. (2010). Crystallization process in $Ge_2Sb_2Te_5$ amorphous Films. *Vacuum*, 84, No. 7, (March 2010) 877–881, ISSN 0042-207X.

Morales-Sanchez, E., Prokhorov, E., Gonzalez-Hernandez, J. & Mendoza-Galvan, A. (2005). Structural, electric and kinetic parameters of ternary alloys of GeSbTe. *Thin Solid Films*, 471, No. 1-2, (January 2005) 243-247, ISSN 0040-6090.

Naito, M., Ishimaru, M. & Hirotsu, Y. (2004). Local structure analysis of Ge-Sb-Te phase change materials using high-resolution electron microscopy and nanobeam diffraction. *Journal of Applied Physics*, 95, No. 12, (June 2004) 8130-8135, ISSN 0021-8979.

Nobukuni N., Takashima, M., Ohno, T. & Horiea, M. (1995). Microstructural changes in GeSbTe film during repetitious overwriting in phase change optical recording. *Journal of Applied Physics*, 78, No. 12, (December 1995) 6980-6988, ISSN 0021-8979.

Ohta, T., Inoue, K., Uchida, M., Yoshida, K., Akiyama, T., Furakawa, S., Nagata, K. & Nakamura, S. Phase change disk media having rapid cooling structure. *Japanese Journal of Applied Physics*, 28, Suppl. 28, No. 3, (March 1989) 123-128, ISSN 0021-4922.

Okura, M., Naito, H. & Matsushita, T. (1992). Discussion on the Mechanism of Reversible Phase Change Optical Recording. *Japanese Journal of Applied Physics*, 31, No. 2B, (February 1992) 466-470, ISSN 0021-4922.

Ovshinsky S. R. (1968). Reversible electrical switching phenomena in disordered structures, *Physical Review Letters*, 21, No. 20, (November 1968) 1450–1453, ISSN 0031-9007.

Ovshinsky, S. & P. Klose, Reversible High-Speed High-Resolution Imaging in Amorphous Semiconductors, *Society for Information Display International Symposium Digest of Technical Papers*, First Edition, 58-61 (1971).

Paesler, M. A., Baker, D. A., Lucovsky, G., Edwards, A. E. & Taylor, P.C. (2007). EXAFS study of local order in the amorphous chalcogenide semiconductor $Ge_2Sb_2Te_5$. *Journal of Physics and Chemistry of Solids*, 68, No. 5-6, (March 2007) 873-877, ISSN 0022-3697.

Park, J., Kim, M. R., Choi, W. S., Seo, H. & Yeon, C. (1999). Characterization of Amorphous Phases of $Ge_2Sb_2Te_5$ Phase-Change Optical Recording Material on Their Crystallization Behavior. *Japanese Journal of Applied Physics*, 38, No. 8, (May 1999) 4775–4779, ISSN 0021-4922.

Perniola, L., Sousa, V., Fantini, A., Arbaoui, E., Bastard, A., Armand, M., Fargeix, A., Jahan, C., Nodin, J.-F., Persico, A., Blachier, D., Toffoli, A., Loubriat, S., Gourvest, E., Betti, Beneventi, G., Feldis, H., Maitrejean, S., Lhostis, S., Roule, A., Cueto, O., Reimbold, G., Poupinet, L., Billon, T., De Salvo, B., Bensahel, D., Mazoyer, P., Annunziata, R., Zuliani, P. & Boulanger, F. (2010). Electrical Behavior of Phase-Change Memory Cells Based on GeTe. *Electron Device Letters IEEE*, 31, No. 5, (April 2010) 488-490, ISSN 0741-3106.

Popescu, M. A. Ovonic Materials. (2005). *Journal of Ovonic Research*, 1, No. 6, (December 2005), 69-76, ISSN 1584–9953.

Privitera, S., Rimini, E., Bongiorno, C., Zonca, R., Pirovano, A. & Bez, R. (2003). Crystallization and phase separation. in $Ge_{2+x}Sb_2Te_5$ thin Films. *Journal of Applied Physics*, 94, No. 7, (October 2003) 4409-4413, ISSN 0021-8979.

Rabe, K. & Joannopoulos, J. (1987). Theory of the structural phase transition of GeTe. *Physical Rewiew B*, 36, No. 12, (October 1987) 6631-6639, ISSN 1098-0121.

Raoux, S., Shelby, R. M., Jordan-Sweet, J., Munoz, B., Salinga, M., Chen, Y. C., Shih, Y.H., Lai, E. K. & Lee, M.H. (2008). Phase change materials and their application to random access memory technology. *Microelectronic Engineering*, 85, No. 12, (August 2008) 2330–2333, ISSN 0167-9317.

Reddy, G. B., Dhar, A., Malhotra L. K. & Sharmila E. K. (1992). Comparative study of crystallization processes in Sb,Te, films using laser and thermal annealing techniques. *Thin Solid Films*, 220, No. 1-2 , (1992) 111-115, ISSN 0040-6090.

Ruitenberg, G., Petford-Long, A. K. & Doole, R. C. (2002). Determination of the isothermal nucleation and growth parameters for the crystallization of thin $Ge_2Sb_2Te_5$ films. *Journal of Applied Physics*, 92, No. 6, (September 2002) 3166-3123, ISSN 0021-8979.

Ruiz, Santos, R., Prokhorov, E., Espinoza, Beltran, F. J., Trapaga, Martınez, L. G. & Gonzalez-Hernandez, J. (2010). Crystallization and ferroelectric properties of $Ge_4Sb_1Te_5$ films. *Journal of Non-Crystalline Solids*, 356, No. 52-54, (July 2010) 3026-3031, ISSN 0022-3093.

Senkader S. & Wright, C. D. (2004). Models for phase-change of $Ge_2Sb_2Te_5$ in optical and electrical memory devices. *Journal of Applied Physics*, 95, No. 2, (January 2004) 504-511, ISSN 0021-8979.

Sian, T. S., Mittal, S., Goel, A., Chakraborty, S., Gautam, S., Budhwar, A., Kapil, R., Achar, P. C. & Nyati, G. (2008). Correlations between electrical parameters of rewritable digital versatile disc and optical and thermal properties of phase change alloy: effect of sputter power. *Japanese Journal of Applied Physics*, 47, No. 2, (February 2008) 936–941, ISSN 0021-4922.

Strand, D. (2005). Ovonics: from science to products. *Journal of Optoelectronics and Advanced Materials*, 7, No. 4, (August 2005) 1679–1690, ISSN 1454-4164.

Tominaga, J., Shima, T., Kuwahara, M., Fukaya, T., Kolobov, A. &Nakano, T. (2004). Ferroelectric catastrophe: beyond nanometre-scale optical resolution. *Nanotechnology, 15, No. 5, (January 2004) 411-415, ISSN 0957-4484.*

Trappe, C., Bechvet, B., Hyot, B., Winkler, O., Facso, S. & Kurz, H. (2000). Recrystallization Dynamics of Phase Change Optical Disks with a Nitrogen Interface Layer. *Japanese Journal of Applied Physics*, 39, No. 2B, (February 2000) 766-769, ISSN 0021-4922.

Wamwangi, D., Njoroge, W. K. & Wuttig, M. (2002). Crystallization kinetics of $Ge_4Sb_1Te_5$ films. *Thin Solid Films*, 408, No. 1-2, (March 2002) 310–315, ISSN 0040-6090.

Weidenhof, V., Friedrich, I., Ziegler, S. & Wuttig, M. (2001). Laser induced crystallization of amorphous $Ge_2Sb_2Te_5$ films. *Journal of Applied Physics*, 89, No. 6, (March 2001) 3168–3176, ISSN 0021-8979.

Wei J. & F. Gan. (2003). Theoretical explanation of different crystallization processes between as-deposited and melt-quenched amorphous $Ge_2Sb_2Te_5$ thin films. *Thin Solid Films*, 441, No. 1-2, (September 2003) 292-297, ISSN 0040-6090.

Yamada, N. (1996). Erasable phase-change optical material, *MRS Bulletin*, 21, No. 9, (October 1996) 48-50, ISSN 0883-7694.

Yamada, N. & Matsunaga, T. (2000) Structure of laser-crystallized $Ge_2Sb_{2+x}Te_5$ sputtered thin films for use in optical memory. *Journal of Applied Physics*, 88, No. 12, (December 2000) 7020-7028, ISSN 0021-8979.

Yamada, N., Ohno, E., Nishiuchi, K., Akahira, N. & Takao, M. (1991). Rapid-phase transitions of GeTe-Sb, Te, pseudobinary amorphous thin films for an optical disk memory, *Journal of Applied Physics*, 69, No. 5, (March 1991) 2849-2856, ISSN 0021-8979.

Yao, H. B., Shi, L. P., Chong, T. C., Tan, P. K. & Miao, X. S. (2003). Optical Transition of Chalcogenide Phase-Change Thin Films. *Japanese Journal of Applied Physics*, 42, No. 2B, (February 2003) 828-831, ISSN 0021-4922.

Yin, Y., Sone H. & S. Hosaka. (2007). Characterization of nitrogen-doped Sb_2Te_3 films and their application to phase-change memory. *Journal of Applied Physics*, 102, No. 6, (September 2007) 064503, ISSN 0021-8979.

Zhai, F., Wang, Y., Wu, Y. & Gan, F. (2009). Phase Transition Kinetics of Sb_2Te_3 Phase Change Thin Films. *Proceedings of SPIE*, ISBN 0-8194-5293-9 Wuhan, China, November 2008.

Zhang, T., Cheng, Y., Song. Z., Liu, B., Feng, S., Han, X., Zhang, Z. & Chen, B. (2008). Comparison of the crystallization of Ge–Sb–Te and Si–Sb–Te in a constant-temperature annealing process. *Scripta Materialia*, 58, No. 11, (February 2008) 977–980, ISSN 1359-6462.

Zhu, M., Wu, L., Rao, F., Song, Z., Peng, C., Li, X., Yao, D., Xi, W. & Feng. S. Phase Change Characteristics of SiO_2 Doped Sb_2Te_3 Materials for Phase Change Memory Application. *Electrochemical and Solid-State Letters*, 14, No. 10, (July 2011) H404-H407, ISSN 1099-0062.

17

Metal Induced Crystallization

Ahamad Mohiddon Mahamad and Ghanashyam Krishna Mamidipudi
School of Physics, University of Hyderabad, Hyderabad, India

1. Introduction

Polycrystalline silicon thin films have attracted the attention of semiconductor industries in the past few decades due to their wide applications in thin film transistors, solar cells, display units and sensors (Schropp & Zeman, 1998; Choi et al., 2005; Mahan et al., 2008). Polycrystalline Si thin films are generally fabricated by crystallizing amorphous Si (a-Si) thin films, because these can render larger grains compared to the conventional poly Si film deposition. As a consequence, a variety of methods for lowering the crystallization temperature of a-Si have been developed. Excimer laser annealing is one of the promising ways to achieve large grain size poly Si films at lower substrate temperatures. Its high costs and nonuniform grain size, however, are significant obstacles that prevent its wide use (Parr et al., 2002). The other promising technique is the solid phase crystallization method. But this technique is essentially a high-temperature process and many substrates, including most forms of glass, cannot withstand the thermal processing. In order to achieve lower costs and have a wider range of application, inexpensive materials such as glass and special polymers have to substitute quartz or Pyrex^{TM} substrates. In the case of glass substrates, all of the processing steps need to be limited to temperatures below 550 °C. The other known technique is rapid thermal annealing (RTA). In RTA infrared radiation is used as a heating source, and has the advantage of the high heating speed (up to 60 °C/s) that reduces the crystallization time. In RTA radiation is applied in pulses to heat the sample without heating the glass substrate (which is transparent to the infrared radiation). However, the grain size obtained in the crystallization of a-Si is also in the range of a few micrometers.

In an effort to reduce the crystallization temperature and crystallization time, and to increase the grain size, metal-induced crystallization (MIC) has been investigated as an alternative crystallization process for thin-film device fabrication. The MIC process involves the deposition of a-Si films on top of which a layer of suitable metal is deposited. This bilayer of metal and Si is then annealed in a furnace at temperatures ranging from 150 to 700 °C for durations between one minute to several hours leading to crystallization of the a-Si. As a result the Si nanocrystals grow in the metal/metal silicide/metal oxide matrix and their growth rate depends on the annealing conditions of the bilayer. Thus, to grow the Si nanocrystals, both thermodynamics and kinetics of the process have to be understood in detail. In the MIC process, metals like aluminum, nickel, gold, silver etc are used to decrease the crystallization temperature below 700 °C. Among the metals employed for the study of the MIC (Jin et al.,1999; Kumar et al.,2007; Kumar et al.,2008; Mohiddon & Krishna, 2011), the preferred metal up to date has been nickel (Ni) due to its low residual metal contamination in the poly-Si region (Wang & Wong, 2001). The mobility and the transfer characteristics of the poly Si TFT are improved by reducing the leakage current caused by the metal agents. To achieve this, metal-induced lateral crystallization (MILC) has been introduced. Ni induced MILC TFTs' are now widely used in large display applications. Large number of metals have been investigated to crystallize the a-Si at possible low temperature. These metals are classified in to two groups basing on the mechanism they follow in crystallizing the a-Si matrix. One group of metals forms eutectic with a-Si (follows Layer exchange mechanism) and the other reacts with a-Si and forms silicide. The eutectic phase forming metals can produce extended continuous poly-Si films with large grain size of 10 µm (Oliver & Hartmann, 2000). However, these metals contaminate the so produced *c*-Si and degrade its properties. Thus, the search for metals that would lead to crystallization of *a*-Si thin film at low temperatures and also reduce the contamination is a subject of recent research.

The aim of this review is to present an overview of the current state of understanding of the mechanisms involved in MIC and to discuss some of the recent work of the current authors in this area. The motivation is to crystallize a-Si films using metals that have not been investigated in detail so far, in addition to Ni which is a well studied metal used in the MIC process. The two other metals that were selected for this study are Tin (Sn) and Chromium (Cr). The former is known to form a eutectic alloy with the Si whereas the latter is a silicide/oxide forming metal, similar to Nickel. Furthermore, results on two different types of geometries are presented; the first involves the deposition of metal on top of the a-Si film, while in the second geometry the a-Si film is deposited on top of the metal film. This provides very interesting insights in to the mechanisms that cause crystallization of a-Si.

2. Mechanisms of metal induced crystallization

The main mechanisms of metal induced crystallization are discussed in this section.

2.1 Layer exchange mechanism

The proposed formation of poly-Si films on foreign substrates by eutectic forming metals relies on the overall layer exchange of adjacent Si and metal films during the transformation of amorphous to polycrystalline Si. The phenomenon is named as "layer exchange mechanism". The general driving force behind metal-induced crystallization is the reduction of the free energy of the silicon material during the transformation of the amorphous to the crystalline phase. When *a*-Si is in contact with certain metals, electronic screening of the covalent bonding in the Si material occurs, which weakens the Si bonds, and therefore, facilitates the inter diffusion of the metal and silicon atoms (Hiraki, 1980). The representatives of the eutectic forming metals are Au, Ag and Al. Aluminum-induced crystallization [AIC] is extensively studied metal to form extended continuous poly-Si films of large-grained material on glass, which is desirable for thin film solar cells.

The AIC comprises the diffusion of Si atoms into the Al layer, which occurs due to the fact that adjacent Al and Si layers are not in thermal equilibrium at elevated temperatures. According to the Al/Si phase diagram, up to 1.5 at.% of Si can be dissolved in Al at temperatures below the eutectic temperature of 577 °C (Murray & McAlister, 1984). The initial interaction of the Al and *a*-Si layers involves the partial dissolution of the Al or Si oxide interface layer, depending on the layer sequence, to allow the inter diffusion of the materials. The Si oxide layer is transformed by the Al into a mixture of Al oxide and an Al_xSi phase. This newly formed Al_xSi phase provides a diffusion channel for the Si and Al atoms (Nast & Wenham, 2000). The evaporated Al films are of polycrystalline nature, there are four different diffusion paths available for the diffusion of the Si atoms. 1) Diffusion inside the Al grains, 2) Diffusion along Al grain boundaries, 3) Diffusion along the Al/*a*-Si interfaces, and 4) Diffusion along the glass/Al interface.

The diffused Si atoms start crystallizing near the Al/*a*-Si interface. The Al grain boundaries are preferential nucleation sites for the dissolved Si atoms, since they are the sites of low critical free energy for nucleus formation. According to the random successive nucleation model, at early stage of crystallization, Si nuclei are formed at preferred sites located at the Al/Si interface. The released heat (crystallization energy and strain energy) leads to rise the local temperature in ambient regions. The temperature field propagates fast, since the thermal diffusion is faster than the atomic diffusion, so that the heat flow can stimulate new nuclei appearing randomly in nearby region. These nuclei of the next generation also produce a local temperature rise. This process repeats many time during annealing, resulting in a fractal formation.

Silicon atoms, dissolved in the Al layer, have a high mobility and, hence, diffuse quickly within the film and/or along the interfaces towards the growing Si grains (McCaldin & Sankur, 1971). It is this fast growth within the Al layer compared to any other growth normal to the layer structure that leads to the formation of a continuous poly-Si film. At an early stage of the crystallization process the newly formed Si grains within the Al matrix are far apart and do not influence each other. The nucleation rate strongly depends on the defects and grain boundaries in the Al layer as well as the temperature and corresponding silicon concentration. Si solute depletion occurs up to the effective diffusion distance in the vicinity of the growing grain (Zener, 1949). In this region the possibility of new nucleation decreases with increasing depletion towards the advancing Al/Si grain interface. When the effective diffusion distances of adjacent grains begin to overlap, competition for the available Si atoms dissolved in the Al layer occurs. At this point, the possibility of new nuclei formation decreases. This type of crystallization pattern, where isolated grains start to interfere at an early stage of the process, is different to solid-phase crystallization of amorphous silicon (Spinella et al., 1998).

The grain-size distribution is dependent on the ratio of the grain growth to the nucleation rate. This ratio increases with decreasing annealing temperature (Nast & Wenham, 2000). The decrease in grain growth due to a lower annealing temperature is less than the decrease in nucleation rate. The effective diffusion distance is longer at lower annealing temperatures leading to larger depletion areas around the growing grains, which prevents further nucleation. Therefore, the grains grow to a larger size before impingement occurs. The nucleations as well as the diffusion-controlled growth are thermally activated processes. Once the size of the Si grains is equivalent to the thickness of the Al layer, they solely advance laterally since they are constrained by the substrate and the *a*-Si/Al interface. Throughout the poly-Si growth process the Al layer is gradually displaced. The continuous supply of Si atoms from a-Si layer and gradual displacement of Al atoms leads to the growth of poly Si film with the large grain size. As an alternative member of this group of metals, we have investigated the tin induced a-Si crystallization.

2.2 Mechanism of diffusion assisted crystallization

Two main mechanisms of diffusion assisted crystallization are proposed in literature. One is based on the diffusion and movement of Ni in the *a*-Si matrix causing crystallization of Si around dispersed nanostructured $NiSi_2$ seeds (Ni diffusion assisted MIC); the other is based on the formation of the silicides at the interface between Ni and Si layers, which then moves into the *a*-Si matrix, leading to crystallization of Si (MILC)

According to the first mechanism, the metal (Ni) at first diffuses through *a*-Si matrix and then forms $NiSi_2$ at a sufficient high temperature. This silicide will then act as seed for the crystallization of *a*-Si. The concept of metal diffusion is proposed on the basis of the experimental observations reported in the literature. Park et al. [Park et al., 2001] reported secondary ion mass spectroscopy (SIMS) and transmission electron microscope studies on excimer laser annealed Ni induced crystallization of poly Si. They measured the melting temperatures as a function of Ni content and reported that *a*-Si, just below the Ni layer, melted at lower temperature and the melting point increases with movement away from the Ni rich layer. The change in melting point was attributed to change in Ni concentration. Their explanation for the observed behaviour was that, the Ni diffuses through *a*-Si matrix only when a sufficiently high temperature is reached. Ferri et al. [Ferri et al., 2001] studied diluted metal contaminated *a*-Si system by detailed Raman spectroscopy measurement and came to similar conclusions.

The knowledge of formation, structure and electrical behaviour of a metal-semiconductor interfaces plays an important role in semiconductor technology. The nickel monosilicide (NiSi) is a low resistance silicide in Ni-Si binary system, and is a key material to reduce the contact resistance of gate and source-drain regions. I. H. Hong et al. (Hong et al., 2006) carried out scanning photoelectron spectromicroscopy on the a-Si/Ni films, which were deposited by chemical vapor deposition at ultra high vacuum pressure. They state that when the Si/Ni interface is heated, Ni_2Si, which is an unstable phase with high resistivity, is first formed at 200-300 °C. At around 300 °C or above, Ni_2Si starts transformation into a low resistivity phase NiSi. The Ni_2Si phase disappears at above 400°C. Another high resistivity phase, $NiSi_2$, nucleates at above 750 °C. Another method of enhancing metal diffusion is electric field assisted MIC (FAMIC). In FAMIC, the crystallization of a-Si is enhanced by applying an electric field to the metal/Si binary system.

The two major driving forces for the MIC process are (1) the free-energy gradient between silicon and metal silicides, (2) the concentration gradient caused by atomic diffusion in metal/silicon stacked films. In addition to these driving forces, the applied electrical field introduces an additional driving force to lower the activation energy for crystallization. Lee (Lee et al., 2000) suggested in the study of a Cu/Si FAMIC system that the electrons can charge up at the a-Si surface of the Cu silicide/a-Si interface under an applied voltage. The Cu ions with positive charges could then migrate at a faster speed, resulting in a directional and rapid lateral movement of Cu silicide/a-Si interface and so the crystallized poly-Si can be formed at the enhanced speed. The similar enhanced growth rate is reported in Ni/Si system. It has been postulated that the enhanced growth rate is a result of field-enhanced diffusion of nickel atoms (Park et al., 1999; Yoon et al., 2001). Having an effective charge of -0.3299e in the Si lattice, Ni atoms move toward the positive electrode under an electric field

(Yoon et al., 2001). The other mechanism of FAMIC considers the bombardment of Ni atoms by electrons traveling in an *a*-Si layer. Nickel atoms are pushed toward the positive electrode under a large number of low-momentum collisions (electron wind). Also, energetic electrons are capable of breaking bonds by exciting the bonding electrons through the collision. In fact both field-enhanced diffusion under electric force and electron bombardment effects are responsible for the enhanced crystallization in FAMIC. In both mechanisms, the diffusion of metal is a common argument. Grisenti et al. (Grisenti et al., 2008) reported the Extended x-ray absorption fine structure (EXAFS) analysis on Ni induced *a*-Si produced by co-sputtering. They came to conclusion that, Ni segregates as $NiSi_2$ even at temperatures as low as 200 °C and this $NiSi_2$ enhances the MIC of *a*-Si. The transformation from one kind of silicide to other is well explained by Ni diffusion through Ni/Si interface and this mechanism is widely accepted in the semiconductor technology.

2.3 Metal Induced Lateral Crystallization (MILC)

In case of MILC first metal silicides are formed at the interface between metal and Si layers at the lower annealing temperatures. On further annealing, nucleation of c-Si on metal silicide precipitate starts, which is then followed by migration of the metal silicide precipitates throughout the a-Si thin film, resulted in crystallization of the entire a-Si film. This mechanism is supported by the following experimental observations. Jin et al. (Jin et al, 1998) studied the XPS depth profile of Ni (10 nm) covered a-Si films (70 nm) which was annealed at 500 C for 1 hr. Ar ions with energy of 4 keV were used for sputtering during depth profiling. They reported the concentration variation of selected elements (Ni, O and Si) across the 100 nm stack. From their results it is found that the Ni concentration first decreased with Ar sputtering then stabilized to about 4% in the bulk of the film and finally increased to about 10% near the interface of the film. No Si was detected on the surface of the stack before the Ar sputtering, indicating the diffusion of Si in Ni and NiO_x was slow at the heat treatment temperature of 500 °C. The presence of excess metallic Ni at the bottom of the stack was attributed to the diffusion of the $NiSi_2$ nodules. Hayzelden et al. (Hayzelden & Batstone, 1993) reported an in-situ transmission electron microscopy study on Ni induced *a*-Si crystallization and suggested a similar mechanism for the diffusion of $NiSi_2$ in *a*-Si matrix. According to the mechanism proposed by these authors, Si first crystallizes on one of the eight faces of $NiSi_2$ crystalline seeds and then it dissociates as Ni and Si atoms at the *c*-Si/ $NiSi_2$ interface. The Nickel atoms then diffuse and react again with *a*-Si to form $NiSi_2$ at *a*-Si/ $NiSi_2$ interface. The cause of such dissociation, diffusion and reformation of $NiSi_2$ was explained by chemical potential free energy of the reactions at different interfaces. The chemical potential of Ni is lower at $NiSi_2$/a-Si interface, while the chemical potential of Si is lower at the $NiSi_2$/c-Si interface. Thus Ni moves to the $NiSi_2$/a-Si interface and Si atoms are froced to diffuse to the $NiSi_2$/c-Si interface. The consumption of a-Si at the $NiSi_2$/a-Si interface and diffusion of the Ni atoms result in the growth of needle like Si crystallites. The diffusion of Ni in crystalline Si is much faster than that in a-Si. This supports the fast diffusion of Ni from the c-Si interface to the a-Si interface through a $NiSi_2$ crystallite. Hwang et al. (Hwang & Li, 2005) studied the Auger depth profile on Ni induced MIC and came to a similar conclusion. Our recent work on the transmission electron microscopy study of Ni induced MIC of a-Si films also supports the $NiSi_2$ diffusion mechanism. The details of the experiment and result are presented in the section 4.2 of the chapter.

3. Tin induced a-Si crystallization

Tin is a representative of the eutectic forming group along with Al and Au. The Sn–Si alloy has a relatively low eutectic temperature of 232 °C (Jeon et al., 2010) compared other members of the group. Our recent work on Tin induced a-Si revealed that it follows the eutectic forming metal mechanism well below the Al-Si system (Mohiddon & Krishna, 2011). Jeon et al. (Jeon et al., 2010) studied the Sn-Si system for growing Si nanowires and quoted that Sn is a favourable catalyst for low temperature synthesis of Si nanowires.

The experimental details are as follows. Sn films of 500 nm thickness were deposited onto Borosilicate Glass (BSG) substrates by resistive thermal evaporation. Si films of 500 nm thickness were deposited by electron beam evaporation over the Sn layers. Fig. 1 shows the block diagram of experimental deposition setup of electron beam evaporator and block diagram of film annealing setup. The starting materials were granular pure silicon powder (99.999%) and Sn (99.99% pure). A pressure 5×10^{-6} Torr was maintained throughout the depositions. The depositions were carried out at ambient temperature and in all cases the substrate to source distance was kept constant at 10 cm. The thickness of the films was measured after deposition using a surface profilometer (model XP-1 of Ambios Technology, USA). The films were annealed in a furnace atmosphere at different temperatures for 1 hr. X-ray diffraction patterns were recorded on a powder x-ray diffractometer (CPS120 of Inel, France) machine equipped with a Co K*α* x-ray source (wavelength = 0.178896 nm) and gas phase position sensitive detector.

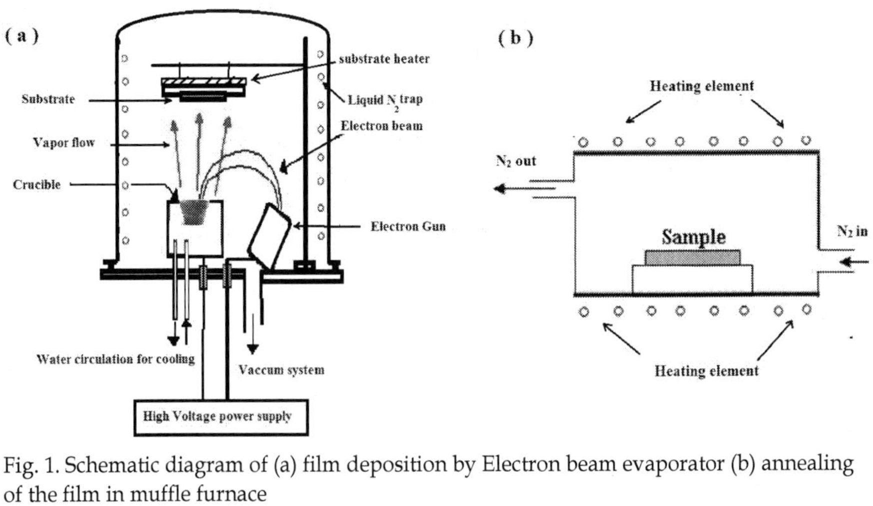

Fig. 1. Schematic diagram of (a) film deposition by Electron beam evaporator (b) annealing of the film in muffle furnace

X-ray diffraction patterns of as deposited BSG/Sn/*a*-Si stacks and stacks post deposition annealed at different temperatures from 300 °C to 500 °C each for 1 hour are presented in Fig. 2. XRD pattern of Sn film deposited on BSG and annealed at 500 °C for 15 hrs is also presented in the same figure. It is evident from the XRD pattern of the annealed Sn film that, it is well crystallized and the pattern can be identified as belonging to the orthorhombic phase of SnO_2 (the peaks were indexed according to the PCPDF file no-781063). The as deposited BSG/Sn/*a*-Si stack, shows sharp diffraction peaks along with a broad diffuse hump around 2θ = 27°, due to the amorphous substrate and *a*-Si film. The peaks can be assigned to the (200), (101), (220), (211), (112) and (312) planes of the tetragonal phase of metallic Sn (the peaks are indexed according to the PCPDF file no- 23456). There is no evidence for the presence of crystalline Si.

On annealing this stack at 300 °C for 1 hr, there is a very interesting transformation in the crystalline behavior of the films. Significantly, all the peaks due to metallic Sn are completely suppressed and the x-ray diffraction pattern resembles that of an amorphous film except for a very diffuse peak at 2θ = 33.58° assigned to the (111) plane of the diamond cubic form of Si. The intensity of this peak increases with increasing annealing temperature up to 500 °C as shown in Fig. 2 indicating that the extent of crystallization of Si is improved by increasing the annealing temperature. However, there is no evidence for crystalline Sn in the films. The reason for this is, most probably, partial oxidation of Sn. The peak at 2θ = 33.58° assigned to the (111) plane of diamond cubic silicon (PCPDF file no- 23345) is shifted by 0.45° which is attributable to strain in the film. Evidently, the onset of crystallization of *a*-Si occurs at 300 °C by contacting with the Sn metal. M Joen et al. have also reported Sn induced crystallization of *a*-Si at 300 °C (Jeon et al., 2010). Significantly, there is no evidence for the formation of silicides in the Sn-Si system even after annealing at 500 °C. Deconvolution of the peak at 2θ = 33.58° provides more insight into the mechanism of crystallization of *a*-Si. The peak was deconvoluted by smoothening and best fitting with a Gaussian function. The data is shown in the inset of Fig.2 for the samples annealed at 300, 400 and 500 °C. Closer observation of the peak corresponding to the sample annealed at 500 °C shows that the peak can be resolved into two peaks centered at 33.7 and 34°. The deconvoluted peaks fitted using Gaussian function is shown in the inset of the Fig 2. The peak centered at 2θ=34° is assigned to the (113) plane of orthorhombic SnO_2 phase (PCPDF file no- 781063). By eliminating the effect of SnO_2 phase in 500 °C annealed samples, the crystallite size is estimated as 18 nm. The average nanocrystal size calculated by Scherrer's equation (Mohiddon & Yadav, 2008) is found to increase from 5 nm to 14 nm with increasing the annealing temperature from 300 to 400 °C. The solid solubility of Sn in Si is known to be nil, *i.e.* they are immiscible in the solid state (Leonard & Koch, 1992). Hence it is not expected to form a silicide. X-ray diffraction data presented earlier indicates that this is the case even in the present samples. The crystallographic evolution of the c-Si phase in the current case is traced by studying the dependence of the crystallite size of Sn and Si on the annealing temperature. This is plotted in Fig. 3, from which it is evident that the crystallite size of Si is zero in the as-deposited film while that of the Sn phase is high. Interestingly, there is a cross-over region below 300 °C when the crystallite sizes of both phases are low. This indicates that at this temperature, the film is essentially amorphous. On increasing the annealing temperature to 400 and 500 °C, there is further increase in the crystallite size of Si while crystallite size of the Sn phase is invariant at zero. This behavior is very similar to that observed in the case of Al induced crystallization (AIC) of a-Si (McCaldin & Sanku, 1971; Zener,1949;Spinella et al.,1998) with one very crucial difference. No peaks corresponding to crystalline Sn are observed on annealing the films unlike the AIC case. This is evidently due to oxidation of Sn during the annealing process as revealed by the deconvolution of the (111) peak of Si in the inset of Fig.2. Thus we conclude from this work that *a*-Si crystallizes when it is brought in contact with Sn metal layer without involving the formation of a silicide, exactly as followed by eutectic forming metals at 300 °C.

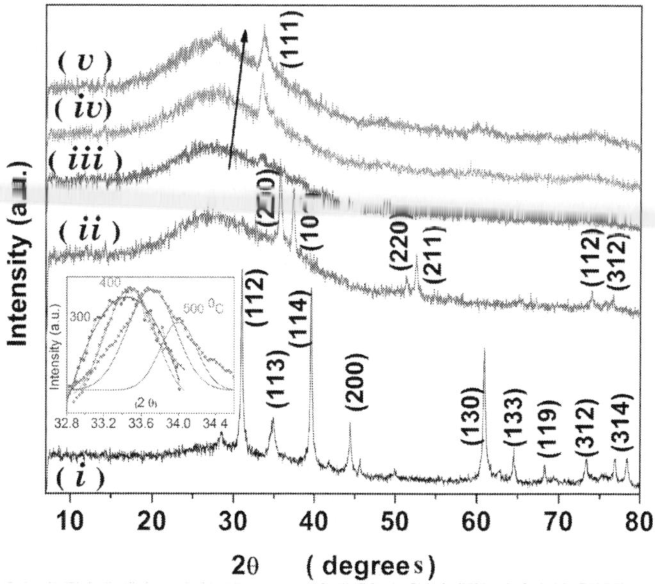

Fig. 2. X-ray diffraction pattern of (*i*) BSG/Sn film annealed at 500 °C for 15 hours. (*ii*) as-deposited BSG/Sn/a-Si stack and BSG/Sn/a-Si stack annealed at (*iii*) 300, (*iv*) 400 and (*v*) 500 °C along with the expanded and smoothened (111) peak of *c*-Si for BSG/Sn/a-Si stack, deconvoluted to show the presence of SnO_2 (Mohiddon et al., 2012).

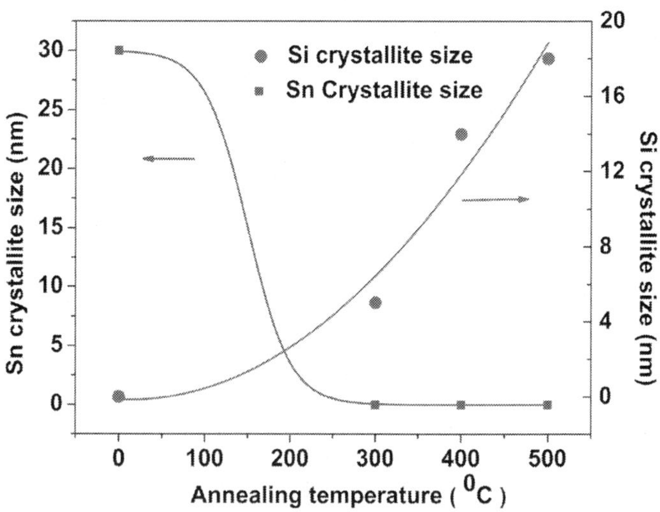

Fig. 3. Variation in crystallite size of Si and Sn phases as a function of annealing temperature (Mohiddon et al., 2012).

4. Nickel and Chromium induced crystallization

The second group of metals used for MIC, reacts with silicon and forms silicides. e.g.: Ni, Cr, Pd, Mo etc. A very small concentration of metals contamination crystallizes large part of *a*-Si and hence they are highly useful in electronic applications. In this section we review our work on Nickel and Chromium induced crystallization of amorphous Si films.

4.1 Nickel induced crystallization

Nickel induced MIC is one of best representatives of this group. To investigate Nickel based MIC, two types of experiments were carried out. In the first case, a 400 nm *a*-Si film is deposited on a fused silica (FS) substrate followed by 200 nm Ni film forming a FS/*a*-Si(400nm)/Ni(200nm) (FSN) stack. The depositions were carried out at different substrate temperatures from 200 to 400 °C. After the deposition is completed, the films were annealed in the vacuum, in the same chamber without removing the vacuum for 30 min at the deposition temperature. The second growth process involves the deposition of 200 nm Ni films on fused silica (FS) substrates followed by deposition of 400 nm amorphous Si films at 200 °C substrate temperature, forming a FS/Ni(200nm)/a-Si(400nm) (FNS) stack.

All films were deposited by the electron beam evaporation technique. The starting materials were granular pure silicon powder (99.999%) and nickel powder (99.99% pure). The vacuum chamber was evacuated using a diffusion-rotary pump combination equipped with a liquid nitrogen trap. A pressure 5×10^{-6} Torr was maintained throughout the depositions. In all cases the substrate to source distance was kept constant at 10 cm. The thickness of the film is measured after deposition using a surface profilometer (model XP-1 Ambios Tech., USA). X-ray diffraction patterns were recorded on a powder x-ray diffractometer (CPS120 of Inel, France) machine equipped with a Co $K\alpha$ = 0.178896nm and gas phase position sensitive detector. XAFS measurements on the FNS stacks were performed at the Ni K-edge at 8333 eV, in fluorescence mode at two different incidence geometries. The two incident angles are 2° and total reflection (TR) geometries. In total reflection geometry only few nanometers of top layer are monitored (10 to 20 nm). At 2 degree incident angle, the entire thickness of the film is under focus. The radiation source was the European Synchrotron Radiation Facility (ESRF, Grenoble, France) and XAS measurements were performed at the BM08 (GILDA) beamline, with an average storage ring current of 180 mA. Data are collected in fluorescence mode using a 13- element hyper pure Ge detector.

Figure 4 shows the X-ray diffraction pattern (XRD) of FS/ *a*-Si / Ni (FSN) stack deposited at different substrate temperatures. The FSN stack deposited at 200 and 300 °C, shows similar XRD pattern. These XRD patterns, indexed according to the PCPDF file no -701849 and refined using PowderX software, are identified as belonging to face centred cubic (FCC) structure of metallic Nickel. The refined FCC unit cell parameter is a=3.53Å. No change in the unit cell parameter is observed for the samples deposited at 200 and 300 °C. In the FSN stack deposited at 400 °C, it is observed that, a new set of XRD peaks are grown along with the metallic Ni XRD peaks. These new set of peaks belong to the face centre cubic structure of $NiSi_2$ indexed according to the PCPDF file no-652974. The FCC unit cell parameter of metallic Ni calculated using (111) peak is found to be 3.52 Å and that of $NiSi_2$ using (111) peak is 5.36 Å. There is no change in the unit cell parameter of metallic Ni for the samples deposited at 300 and 400 °C. The small change of 0.01Å is considered within the range of instrumental measurement error. The disagreement of our result with that of I. H. Hong et al. (Hong et al., 2006) work is expected to be due to the ultra high vacuum pressure deposition conditions, which may facilitate the reaction of the diffused metallic Ni atoms at lower temperature to form different silicide phases.

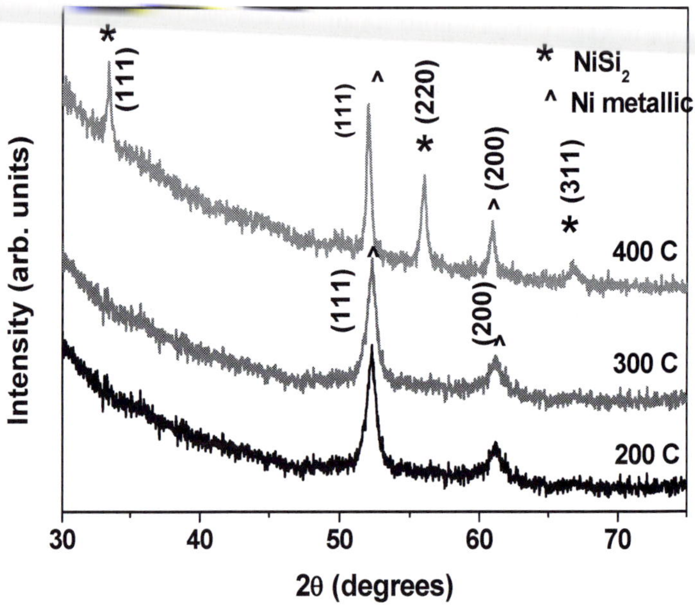

Fig. 4. XRD pattern of FS/Si/Ni stack deposited at different substrate temperature

Figure 5(a) compares the experimental XAFS spectrum of Ni K edge in the FNS stack, which were measured in two different incident angles (TR and 2 degree), along with the reference metal Ni foil spectra. The first derivative near the XANES region of the spectra is presented in the inset of the Fig. 5(a). The observation of the inset figure shows that the XANES spectra of FNS stack deposited at 200 °C and measured in TR and 2 degree has a resemblance with that of the metallic Ni reference spectra. The qualitative analysis of the XAFS spectra shows that the nature of Ni impurity measured in TR and 2 degree are similar and has close resemblance with that of the metallic Ni. Thus at 200 °C annealing temperature the metallic Ni, which is deposited at the bottom, is diffused through 400 nm a-Si layer. The similar observation is found in the EXAFS part of the spectra. Figure 5 (b) compares the Fourier Transform (FT) of the EXAFS signal at the K edge of Ni presented for the FNS stack deposited at 200 °C and measured in two different incident angles, along with the metallic Ni. It shows that the FNS stack deposited at 200 °C and measured in two different incident angles has the features of metallic Ni. The detailed quantitative analyses were carried by FEFF8 and FEFFIT theoretical code and are presented elsewhere (Mohiddon et al., 2011). The conclusions of the work are in good agreement with that of the qualitative analysis presented above. Thus the EXAFS work carried on the FNS spectra support the Ni diffusion assisted MIC mechanism.

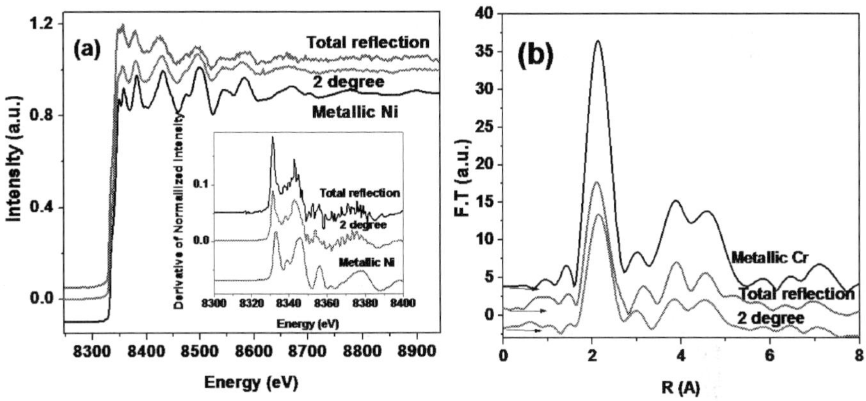

Fig. 5. (a) X-ray absorption spectra along with its derivative near XANES in the inset (b) F.T of EXAFS part of FS/Ni/a-Si stack deposited at 200 °C measured in total reflectance and 2 degree along with metallic Ni reference compound.

4.2 Ni induced Lateral Crystallization (NILC)

As discussed in Sec 2.3, in case of MILC first metal silicides are formed at the interface between metal and Si layers at the lower annealing temperatures. On further annealing, nucleation of c-Si on metal silicide precipitate starts, which is then followed by migration of the metal silicide precipitates throughout the a-Si thin film, resulted in crystallization of the entire a-Si film. We have carried a detailed transmission electron microscope study and found similar observation. The details of the experiment are as follows. The Ni and Si films are deposited using electron beam evaporation at $4x10^{-6}$ Torr and ambient temperature onto Si (311) substrates. The starting materials were granular pure silicon powder (99.999%) and nickel powder (99.99% pure). The (311) oriented Si wafer was well cleaned and subjected to heat treatment of 1000 °C for 2 hours in normal furnace atmosphere to form a thick SiO_2 layer, before the deposition is started. The SiO_2 layer is expected to act as a barrier to stop the diffusion of Ni atom into the Si substrate. X-ray diffraction patterns were recorded in grazing incidence of 0.5 degree on a powder x-ray diffractometer (CPS120 of Inel, France) machine. A 50 nm film of Ni was deposited on the *c*-Si/SiO_2 substrate, followed by the deposition of the 400 nm Si film without breaking vacuum in the deposition chamber. The thickness of the films was measured after deposition using a surface profilometer (model XP-1 Ambios Tech., USA). The films were subsequently annealed in a furnace (in air) at 600 °C for 1 hr. Transmission electron micrographs were obtained by a Tecnai 20 G2 STwin, FEI electron microscope, operated at 200 kV. Electron diffraction patterns (EDPs) were recorded with a Gatan CCD camera. A 10 nm gold film deposited on the grid was used for purpose of camera length calibration. The samples for TEM measurement were prepared by scratching the film and transferring it on to the grid.

Figure 6 is a bright-field transmission electron micrograph obtained from a region of the Ni/Si thin film crystallized by the MIC process. The typical microstructure shown in Fig. 6 consists of dark region with an irregular shape. The microstructure consists of dark dendrite like nanowire structures spread over the bright looking matrix. To find the nature of nanostructures at two different regions of the sample, selected area electron diffraction (SAED) patterns were recorded at both the dark and bright regions of the sample. A typical diffraction pattern from the dark part of the image is shown in the inset of the Fig. 6. The diffraction pattern consists mainly of bright diffraction spots radially positioned on rings. The diffraction spots arise from the crystalline part of the sample, whereas the ring originates from the polycrystalline parts of the sample. The inset of the Fig 6 shows that, the diffraction spots A, B and C lie on the polycrystalline ring of radius 0.314 nm, which is indexed to the (111) plane of the Silicon diamond cubic phase. The spot D has a d=0.185 nm which can be indexed as belonging to the $(2\,\bar{2}\,0)$ plane of the diamond cubic system [JCPDS-892955]. However, an alternative assignment is possible for the same diffraction spots (and the higher order diffraction spots), because they can be equally assigned to the face centred cubic c-$NiSi_2$, with d values of 0.312 nm, 0.191 nm for (111) and $(2\,\bar{2}\,0)$ planes, respectively [JCPDS-652974]. Hence, the diffraction taken in dark region of the sample shown in Fig 6 may belong either to diamond cubic Si or to face centre cubic $NiSi_2$.

The diffraction pattern from the bright area region is shown in other inset of Fig.6. Even at the higher magnifications, the microstructure of this bright region is seen to contain dark dendritic lines. It was thus inpossible to eliminate the presence of the darker regions. Hence, the diffraction pattern shown in second inset of Fig 6 has a combined effect, with a small contribution from dark nanostructures and large contribution from the bright area. The diffraction pattern is slightly different from that of dark region diffraction pattern. It consists of one diffuse diffraction rings with d=0.314 nm and a bright diffraction spots with d = 0.192 nm (indicated by E). There are, however, no diffraction spots disposed on the ring, thus suggesting that the dark region is better crystallized compared to bright region although both the regions show similar diffraction patterns.

To exactly identify the material(s) that are contributing to the diffraction from the bright and dark regions of the samples, Z (atomic number)-contrast scanning transmission electron microscopy (STEM) study was carried out. Fig. 7 is a Z-contrast STEM image of the same selected area that is shown in Fig. 6. The change in contrast in the image is due to change in Z value of Ni (Z= 28) and Si (Z= 14). This observation suggests that the central high contrast region is due to the contribution from Ni component, i.e. $NiSi_2$ and the low contrast region is from Si. Hence, it can be inferred that the $NiSi_2$, after its formation, starts spreading through the matrix of a-Si like a dendrite and leading to its crystallization. The results presented in this part of the study supports the mechanism of movement of $NiSi_2$ dendrites in the a-Si matrix as the cause for crystallization. Similar movement of Cu_3Si dendrites with time of annealing was reported using in-situ bright field transmission electron microscopy, but no diffraction study was reported (Russell et al.,1991). It is evident from the study of SAED and STEM, that when Ni makes contact with the Si layer at high temperature it forms $NiSi_2$. This silicide starts diffusing as dendrites into the a-Si matrix and crystallizes the silicon at the nanoscale. This strongly supports the mechanism of silicide movement as the basis for crystallization of a-Si by the MILC process using silicide forming metals.

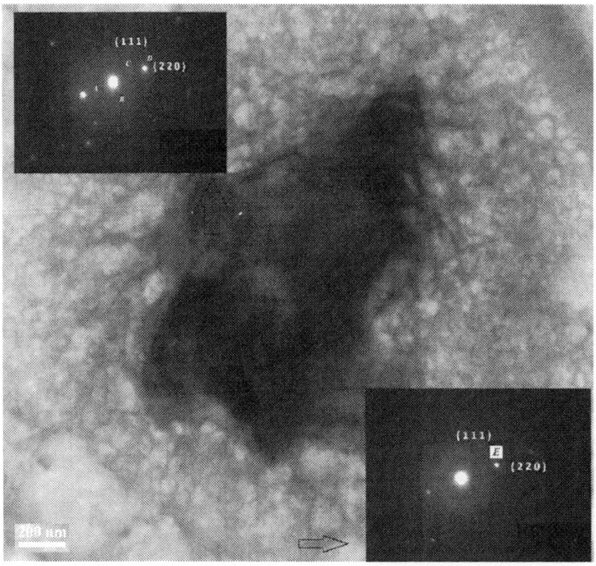

Fig. 6. Selected Bright field image of Ni/Si film annealed at 600 °C for 1 hr along with the diffraction pattern in the dark and bright regions (Mohiddon et al., 2012).

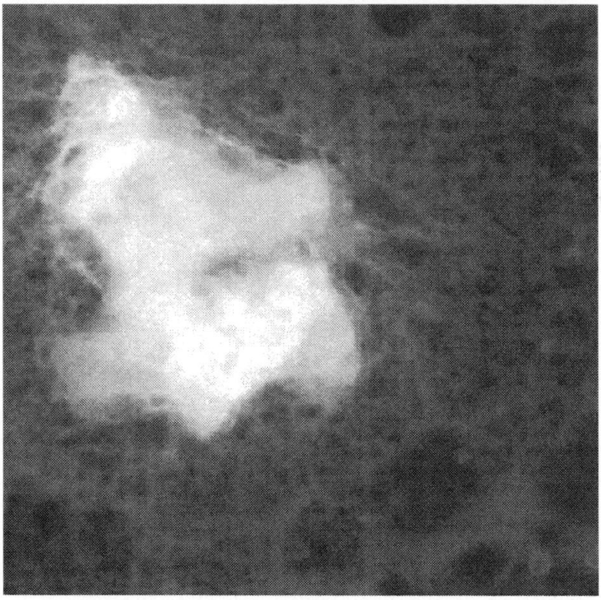

Fig. 7. STEM image of Ni/Si film annealed at 600 °C for 1 hr, selected the same region as that selected for bright field image in Fig. 6 (Mohiddon et al., 2012).

4.3 Stabilization of wurtzite Si

Silicon (Si) usually crystallizes in the cubic diamond structure with fourfold coordinated symmetry. Numerous high-pressure experiments have been performed, revealing no less than 12 different polymorphs of silicon between the well characterized diamond cubic phase and theoretically intractable amorphous phase. With the release of pressure several metastable phases are observed (Wu et al., 2000). For example, a nonmetal to metal transition of cubic-diamond phase, which occur by changing pressure from ambient atmosphere pressure to 10–13 GPa to form a β-tin structure, which further transforms to rhombohedral phase on slow pressure release. The latter transforms reversibly at a pressure of 2 GPa to a body-centered-cubic phase (bc8). Hexagonal-wurtzite silicon (*w*-Si) can be formed by heating the bc8 phase to above 200 °C, or directly from the cubic-diamond phase in the presence of shear stresses at twin intersections, or a nonhydrostatic stress of 8 GPa in indentation experiments (Kailer et al., 1997). *w*-Si material is rarely studied by spectroscopic measurements, because it cannot be obtained in a stable phase. Zhang et al. (Zhang et al., 1999) produced stable *w*-Si phase by laser ablation. Its identification by electron diffraction has been confirmed by micro Raman spectroscopy. Bandet et al. (Bandet et al., 2002) deposited *w*-Si during elaboration of SiO_2 thin films and reported that oxygen seems to play a crucial role in the stabilization of the uncommon metastable structure. Kim et al. (Kim & Lee, 1996) reported that the micrometer-sized diamond cubic silicon (*c*-Si) crystals contain minority part of hexagonal silicon, when *a*-Si films crystallized by a pulsed laser. In the present work we have crystallized the silicon film with a metal contact and found that silicon nanocrystals are in wurtzite phase. The experimental details are as follows.

Nickel and Silicon films were deposited by electron beam evaporation on to BoroSilicate Glass (BSG) substrates in high vacuum of the order of 10^{-6} Torr. First a 50 nm thin nickel blanket bottom layer was deposited on BSG substrates maintained at ambient temperature. This is followed by the deposition of a 700 nm thick Si film, without breaking vacuum, to form a BSG/Ni/Si stack. The thickness of the films is measured after deposition using a surface profilometer (model XP-1 Ambios Tech., USA). The films were annealed in a furnace atmosphere (in air) at 550 °C for 1 hr. Transmission electron micrographs were obtained by a Tecnai 20 G2 STwin, FEI electron microscope, operated at 200 kV. Electron diffraction patterns (EDPs) were recorded with a Gatan CCD camera. A 10 nm gold film deposited on the grid was used for camera length calibration purposes. The Raman spectra were recorded in air using an Nd-YAG 532nm laser in the back scattering geometry in a CRM spectrometer equipped with a confocal microscope and 100× objective (1 μm diameter focal spot size) with a CCD detector (model alpha 300 of WiTec Germany). The phase content with in the samples was investigated in a spectral region 200–1500 cm^{-1}.

Figure 8 shows the electron diffraction pattern along with a bright-field transmission electron micrograph obtained from a region in the silicon thin film crystallized by a MIC process in the inset. The typical microstructure shown in inset of Fig. 8 consists of spherical particles with approximate diameter of 40 nm. The diffraction spots can be identified as belonging to well defined orientations. The calculated interplanar distances of the lattice planes at different spots of the diffraction pattern assigned by symbols A, B and C in Fig. 8 are A=0.256nm, B=0.237nm and C=0.237nm respectively. From the calculated interplanar distances the diffraction spots were indexed as (1 0 0), (1 $\bar{1}$ 1), and (O $\bar{1}$ 1), which belong to the $[1\bar{2}13]$ zone axis. These diffraction spots were diffracted from a grain which has a hexagonal structure and oriented in the $[1\bar{2}13]$ direction and the spots A, B and C were diffracted from the (1 0 0), $(1\ \bar{1}\ 1)$, and (O $\bar{1}$ 1) planes, respectively. The corresponding lattice parameters were calculated as a=0.296nm, c=0.629nm with c/a ratio of 2.13. The angle calculated between these planes using c/a value is found to be exactly matching with the measured values suggesting that Si crystallizes in the hexagonal structure. For diamond cubic Si crystals the interplanar distances of the lattice planes should be 0.33 1 nm, 0.33 1 nm and 0.331 nm, respectively and the angles between transmitted beam and the spots will be 60°, 60° and 60°, with the diffraction pattern showing the expected six fold symmetry. But in our case the angles between the dots are $\angle$AOB=65°, $\angle$BOC=55° and $\angle$AOC=120° respectively, clearly not cubic Si. Thus, the TEM analysis shows that the BSG/Ni/Si stack after heat treatment consists of 40 nm particles of hexagonal Silicon with c/a ratio of 2.11.

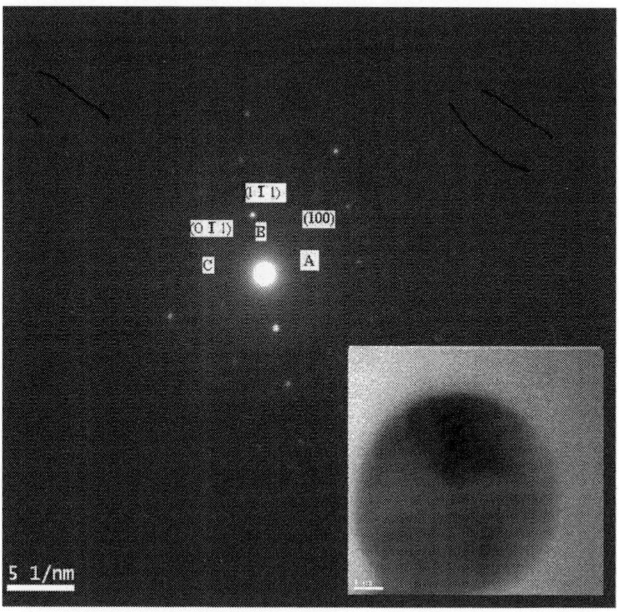

Fig. 8. Electron diffraction pattern along with indexing of selected w-Si part of BSG/Ni/Si stack, inset shows the bright field image of the above diffraction area (Mohiddon et al., 2011).

Additional evidence for the formation of *w*-Si is provided in the form of Raman spectra of the films. It has earlier been used to characterize porous Si (Sui et al., 1992) and Si nanostructures (Kozlowski et al., 1991). The Raman shift and the shape of the Raman peak yield information on the degree of crystallinity achieved in Si by MIC process. Fig. 9 shows Raman spectra from the Silicon film annealed at 550 °C for 1 hrs. A sharp peak at 504 cm^{-1} and weak broad peaks around 300 and 950 cm^{-1}, are associated with optical phonons, two transverse acoustic (2TA) phonons, and two transverse optical (2TO) phonons of *w*-Si (Mohiddon & Kirshna, 2011) respectively. The asymmetric nature of the peak at 504 cm^{-1} can be attributed to the presence of small amounts of *a*-Si, which has its broad peak at 480 cm^{-1}. In general, when Si crystallizes in cubic diamond structure the Raman peak occurs at 520 cm^{-1}. The broad peaks around 283 cm^{-1} in our observation is may belong to NiSi, $NiSi_2$ or to wurtzite TA mode. However, this observation implies that, at 550 °C nickel silicides are present in the samples. Tan et al. (Tan et al., 1981) reported a stress-induced metastable form of hexagonal silicon, which has the wurtzite structure with the c/a ratio close to 1.63, the interplanar distances and angular relationships obtained from our experiment were compared with those of silicon which has the wurtzite structure. However, for silicon with the wurtzite structure, no exact coincidence was obtained. Kim et al. reported a similar observation and suggested a hexagonal phase of Si with *a*=0.382nm, *c*=1.024nm and *c*/*a* ratio of 2.68 (Kim & Lee, 1996). Parson and Hoelke suggested similar hexagonal structure for Ge with *c*/*a* ratio of 2.17 (Parsons & Hoelke, 1983). Zhang *et al.* (Zhang et al., 1994) have reported wurzite silicon by laser ablation, and they observed Raman peaks at 516 and 518 cm^{-1} due to the hexagonal silicon. Kumar et al. observed *w*-Si in chromium induced Si crystallization evidenced by a peak at 495 cm^{-1} (Kumar & Krishna, 2008). Kailer et al. (Kailer et al., 1997) performed indentation Raman investigations and reported that, shear deformation leads to *w*-Si phase formation. They also stated that high pressure phases leads to the formation of *w*-Si at moderate temperatures and to the reversal to the original diamond structure at higher temperatures.

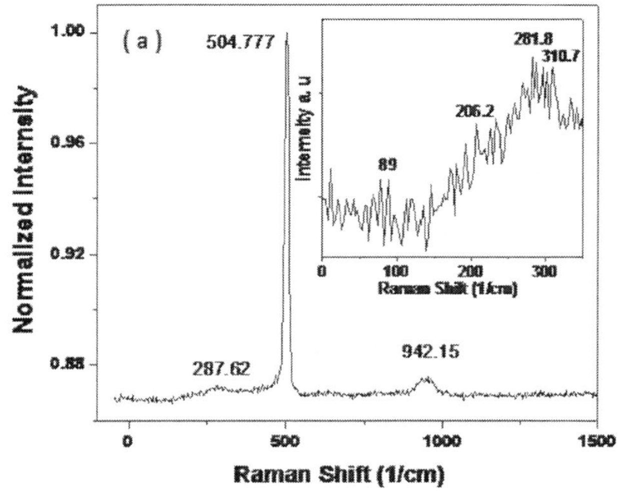

Fig. 9. Raman spectra of BSG/Ni/Si stack annealed at 550 °C for 1 hr (Mohiddon et al., 2011).

4.4 Chromium induced crystallization

The search for new metal in the MIC process to reach the best device needs is still continuing. Chromium is another member of the silicide forming group. Chromium induced MIC has attracted our attention due to its ability to crystallize the *a*-Si into wurtzite phase (Kumar et al., 2008). Our recent work on Cr induced MIC has revealed that Cr deposited at ambient substrate temperature, remains in metallic form up to an annealing temperature of 500 °C (Mohiddon et al.,2012). As an extension of this work we deposited Cr/Si stacks at higher substrate temperature. The details of the experiment are as follows. A Cr(200nm)/Si(400nm) stack was deposited over a fused silica substrate at 400 °C substrate temperature and a similar stack of Cr(50nm)/Si(400nm) was deposited at ambient substrate temperature by electron beam evaporation. We have carried out a detailed EXAFS analysis on these set of samples (Mohiddon et al., 2012) and concluded that, in case of film deposited at ambient substrate temperature, major part of Cr is in metallic form apart from being oxidized. In the case of the stack deposited at 400 °C, the entire Cr is turned into either Cr_2O_3 or $CrSi_2$. After the EXAFS measurements were completed, these samples were heat treated at different temperatures. Samples were annealed at 600, 650 and 700 °C for 1 hr. in excess Nitrogen flux. At each heat treatment, the samples were cooled to room temperature and then the Raman spectra were collected at three to four different position of the each sample. The details of the Raman measurements are discussed in the section 4.3 of the chapter.

Figure 10(a) and 10(b) compares the Raman spectra of the FS/Cr(200nm)/Si(400nm) stack deposited at 400 °C substrate temperature and FS/Cr(50nm)/Si(400nm) stack deposited ambient substrate temperature then annealed at different temperature respectively. Fig. 10(a) shows that the as deposited stack contains a broad diffuse peak around 480 cm^{-1} that can be assigned to the a-Si phase. After the heat treatment at 600 °C, a sharp peak, around 520 cm^{-1}, appears that can be assigned to c-Si. The intensity of the c-Si peak increases with annealing temperature up to 650 °C. No such change in the Raman spectra is observed in the Fig. 10 (b). By combining the EXAFS and Raman results we conclude that the FCS stack deposited at 400°C possesses $CrSi_2$ in the as deposited condition. When this stack is further heat treated the $CrSi_2$ acts as a seed for the Si crystallization. In the FCS stack deposited at minimum substrate temperature, no $CrSi_2$ is observed from the EXAFS analysis. Thus even after heat treatment up to 700 C, no Si crystallization is observed.

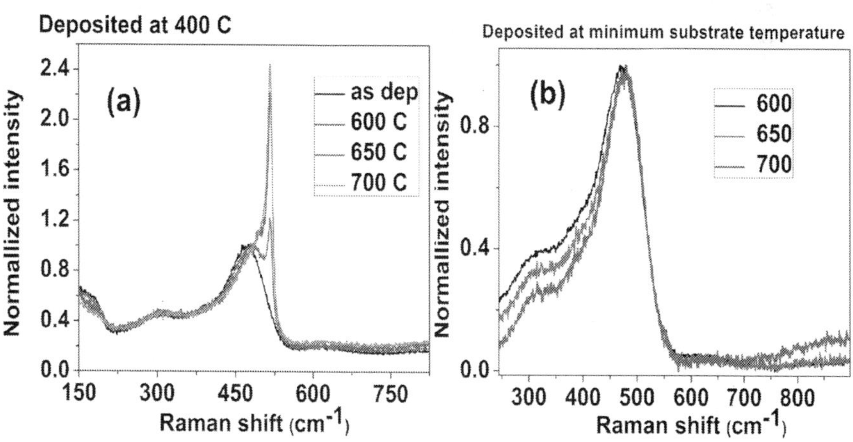

Fig. 10. Raman spectra of (a) FS/Cr(200nm)/Si(400nm) stack deposited at 400 C; (b) FS/Cr(50nm)/Si(400nm) stack deposited at minimum substrate temperature

5. Conclusions

An overview of the metal induced crystallization of amorphous Silicon films has been presented. The possible mechanisms of crystallization have been discussed. Two metals Sn and Cr, which have not been used frequently for MIC, have been shown to be as attractive as Ni for metal induced crystallization of a-Si. Our results clearly show that these are very useful materials for the microelectronics industry. The crystallization occurs at temperatures ranging from 200 to 600 °C, which is much lower than the normal crystallization temperature of Si (1100 °C). The stabilization of the wurtzite form of Si has also been demonstrated.

6. Acknowledgements

The authors acknowledge discussions with Prof. F. Rocca and Prof. G. Dalba of University of Trento and funding for this work through the DST-ITPAR program. Facilities provided by the CAS programme of the School of Physics and DST funded Centre for Nanotechnology at the University of Hyderabad are also acknowledged.

7. References

Choi, J.H., Cheon, J.H., Kim, S.K., & Jang, J. (2005). Giant-grain silicon (GGS) and its application to stable thin film transistor. *Displays* 26, 3, (June 2005), pp. (137-142), ISSN 0141-9382

Bandet, J., Despax, B., & Caumont, M. (2002). Vibrational and electronic properties of stabilized wurtzite like silicon. *Journal of Physics D: Applied Physics*, 35, 3, (January 2002), pp. (234-239), ISSN 1361-6463

Ferri, F.A., Zanatta, A.R., & Chambouleyron, I. (2006). Metal-induced nanocrystalline structures in Ni-containing amorphous silicon thin films. *J. Appl. Phys.*, 100, 9, (November 2006), pp. (094311-094317), ISSN 0021-8979

Grisenti, R., Dalba, G., Fornasini, P., Rocca, F., Kumar, K.U.M., & Krishna, M.G. (2008). XAFS study on Ni surroundings in metal induced crystallization of thin film amorphous silicon. Solid State Comm., 147, 9-10, (September 2008), pp. (401-404) ISSN 0038-1098

Hayzelden, C., & Batstone, J.L. (1993). Silicide formation and silicide-mediated crystallization of nickel-implanted amorphous silicon thin films. J. Appl. Phys., 73, 12, (February 1993), pp. (8279-8289), ISSN 0021-8979

Hiraki, A. (1980). A model on the mechanism of room temperature interfacial intermixing reaction in various metal-semiconductor couples: what triggers the reaction?. *J. Electrochem. Soc.* 127, 12, (December 1980), pp. (2662-2665), ISSN 1945-7111

Hong, I., Hsu, T.C., Yen, S.C., Lin, F.S., Huang, M.L., & Chen, C.H. (2006). Proc. 8th Int. Conf. X-ray Microscopy, IPAP Conf. Series 7 (2006) 270.

Hwang, J.D., & Lin, J.W. (2005). Nanostructural silicon films prepared by metal induced growth using RTCVD system. *J. Electrochem. Soc.*, 152, 1, (Novermber 2005), pp. (G35-G39), ISSN 1945-7111

Jeon, M., Jeong, C., & Kamisako, K. (2010). Tin induced crystallization of hydrogenated amoprphous silico, K., Kowk, H.S., & Wang, M. (1999), The effects of extended heat treatment on Ni induced lateral crystallization of amorphous silicon thin films. *IEEE Trans. Electron Devices*, 46, 1, (January 1999), pp. (78-82), ISSN 0018-9383

Jin, Z., Bhat, G.A., Yeung, M., Kwok, H.S., & Wong, M. (1998). Nickel induced crystallization of amorphous silicon thin film. *J. Appl. Phys.* 84, 1, (April 1998), pp. (194-200) ISSN 0021-8979.

Kailer, A., Gogotsi, Y.G., & Nickel, K.G. (1997). Phase transformation of silicon caused by contact loading. *J. Appl. Phys.*, 81, 7, (January 1997), pp. (3057-3063), ISSN 0021-8979

Kim, J.H., & Lee, J.Y. (1996). Hexagonal silicon formation by pusled laser beam annealing. *Materials Letters* 27, 6, (August 1996), pp. (275-279), ISSN 0167-577X

Kozlowski, F., Petrova-Koch, V., Kux, A., Stadler, W., Fleischmann, A., & Sigmund, H. (1991). Structural instability of Si-nanocrystals in a $(SiH_2)_n$. *J. Non-Cryst. Solids*, 137-138, 1, (January 1991), pp. (91-94), ISSN 0022-3093

Kumar, K.U.M., Brahma, R., Krishna, M.G., Bhatnagar, A.K., & Dalba, G. (2007). An optical study of Ni induced crystallization of a-Si thin films. *J. Phys. Condens. Matter* 19, 49, (December 2007), pp. (496208-1-11), ISSN 1361-648X

Kumar, K.U.M., & Krishna, M.G., (2008). Chromium-induced nanocrystallization of a-Si thin films into the wurtzite structure. *J. Nanomater.* 2008, (February 2008), pp. (736534-1-6)

Lee, C.J., Lee, J.B., Chung, Y.C., & Choi D.K. (2000). Lateral crystallization of amorphous silicon films. *Jpn. J. Appl. Phys.* 39, (2000), pp. (6191-6195), ISSN 1347-4065

Leonard, R.T., & Koch, C.C. (1992). Nanoscale composites of Si/Sn and Ge/Sn synthesized by mechanical attrition. *Nanostructured Materials,* 1, 6, (December 1992), pp. (471-478), ISSN: 0965-9773

Mahan, A.H., Ahrenkiel, S.P., Schropp, R.E.I., Li, H., & Ginley, D.S. (2008). A comparison of grain nucleation and grain growth during crystallization of HWCVD and PECVD a-Si:H films. *Thin Solid Films,* 516, 5, (January 2008), pp. (529-532), ISSN 0040-6090

McCaldin, J. O., & Sankur, H. (1971). Diffusivity and solubility of Si in the Al metallization of integrated circuits. *Appl. Phys. Lett.*, 19, 12, (July 1971), pp. (524-527), ISSN 0003-6951

Mohiddon, M.A., & Krishna, M.G. (2011). Nanocrystalline wurtzite Si-nickel silicide composite thin films with large band gap and high resistivity. *J Mater. Sci.*, 46, 8, (April 2011), pp. (2672-2677), ISSN 1573-4803

Mohiddon, M.A., & Krishna, M.G. (2012) Growth and optical properties of Sn-Si nanocomposite thin films *J. Material Science*, DOI: 10.1007/s10853-012-6647-0

Mohiddon, M.A., & Yadav, K. L. (2008). Effect of 90° domain on ferroelectric properties of alkali modified SBN. *J. Phys. D: Appl. Phys.*, 41, 22, (October 2008), pp. (225406-1-7), ISSN 1361-6463

Mohiddon, M.A., Naidu, K.L., Dalba, G., Rocca, F., & Krishna, M.G. (2012) unpublished

Mohiddon, M.A., Naidu, K.L., Dalba, G., Rocca, F., & Krishna, M.G. (2012) Cr induced nanocrystallization of *a*-Si thin films: its mechanism *Physica Status Solidi C,* 9 (April 2012), pp. (1493–1495)1493–1495

Mohiddon, M.A., Naidu, K.L., Dalba, G., Rocca, F., & Krishna, M.G. (2011) Growth, optical, and electrical properties of silicon films produced by the metal-induced crystallization process, *J Nanopart Res* 13 (June 2011), pp. (5999–6004).

Mohiddon, M.A., Krishna, M.G., Dalba, G. & Rocca, F. (2012) Transmission electron microscopy study of Ni-Si nanocomposite films *Mate. Sci. Eng. B,* DOI: 10.1016/j.mseb.2012.05.018

Murray, J. L., & McAlister, A.J. (1984). Al-Si. *Bull. Alloy Phase Diagram,* 5, 1 (January 1984), pp. (74-84) ISSN 0197-0216.

Nast, O., & Wenham, S.R. (2000). Elucidation of the layer exchange mechanism in the formation of polycrystalline silicon by aluminum induced crystallization. *J. Appl. Physics,* 88, (January 2000), pp. (124-131), ISSN 0021-8979

Oliver, N., & Hartmann, A. J. (2000). Influence of interface and Al structure on layer exchange during aluminum induced crystallization of amorphous silicon. *J. Appl. Phys.* 88 (April 2000), pp. (716-724) ISSN 0021-8979

Park, J.H., Jun, S.I., Song, K.S., Kim, C.K., & Choi, D.K. (1999). Field aided lateral crystallization of amorphous silicon thin film. *Jpn. J. Appl. Phys.,* 38, 2, (January 1999), pp. (L108-L109), ISSN 1347-4065

Park, K.C., Song, I.H., Jeon, J.H., & Han, M.K. (2001). Excimer laser annealing effect on Nickel-induced crystallized polycrystalline silicon film. *J. Electrochem. Soc.,* 148, 10, (August 2001), pp. (G563-G565), ISSN 1945-7111

Parr, A.A., Gill, K., Gardiner, D., Holyland, J., Sands, D., Brunson, K., & Carline, R. (2002). A comparison of laser- and furnace- annealed poly-silicon structure. *Semicond. Sci. Technol.* 17, 1, (January 2002), pp. (47-54), ISSN 1361-6641

Parsons, J.R., & Hoelke, C.W. (1983). Crystallography Of hexagonal germanium. *Nature,* 30, 5901, (1983), pp. (591-592), ISSN 1476-4687

Russell, S. W., Jian, Li., & Mayer, J. W. (1991). In situ observation of fractal growth during a-Si crystallization in a Cu_3Si matrix. *J. Appl. Phys.,* 70, 9, (August 1991), pp. (5153-5155), ISSN 0021-8979

Schropp, R.E.I., & Zeman, M. (1998). *Amorphous and Microcrystalline Silicon Solar Cells:Modeling, Materials and Device Technology,* Kluwer Academic Publishers, ISBN 978-0-7923-8317-8, Boston.

Spinella, C., Lombardo, S., & Priolo, F. (1998). Crystal grain nucleation in amorphous silicon. *J. Appl. Phys.,* 84, 10, (July 1998), pp. (5383-5415), ISSN 0021-8979

Sui, Z., Leong, P.P., Herman, I.P., Higashi, G.S., & Temkin, H. (1992). Raman analysis of light emitting porous silicon. *Appl. Phys. Lett.,* 60, 17, (February 1992), pp. (2086-2088), ISSN 0003-6951

Tan, T.Y., Foll, H., Hu, S.M. (1981). On the diamond-cubic to hexagonal phase transformation in silicon. *Philos. Mag. A,* 44, 1, (July 1981), pp. (127-140), ISSN 1478-6443

Wang, M., & Wong, M. (2001). Characterization of an individual grain boundary in metal-induced laterally crystallized polycrystalline silicon thin-film device. *IEEE Trans. Electron Devices,* 48, 8, (August 2001), pp. (1655-1660), ISSN 0018-9383

Wu, B.R. (2000). First- principles study on the high-pressure behavior of the zone-center modes of lonsdaleite silicon. *Phys. Rev. B,* 61, 1, (January 2000), pp. (5-8), ISSN 1550-235X

Yoon, S. Y., Park, S. J., Kim, K. H., & Jang, J. (2001). Metal induced crystallization of amorphous silicon. *Thin Solid Films,* 383, 1-2, (February 2001), pp. (34-38), ISSN 0040-6090

Zener, C. (1949). Theory of growth of spherical precipitates from solid solution. *J. Appl. Phys.,* 20, 10, (March 1949), pp. (950-953), ISSN 0021-8979

Zhang, Y., Iqbal, Z., Vijayalakshmi, S., & Grebel, H. (1999). Stable hexagonal wurtzite silicon phase by laser ablation. *Appl. Phys. Lett.,* 75, 18, (September 1999), pp. (2758-2760), ISSN 0003-6951

Zhang, S.L., Wang, X., Ho, K., Li, J.J., Diao, P., & Cai, S. (1994). Raman spectra in a broad frequency region of p- type porous silicon. *J. Appl. Phys.* 76, 5, (May 1994), pp. (3016-3019), ISSN 0021-8979

18

ArF Excimer Laser Annealing of Polycrystalline Silicon Thin Film

Noriah Bidin and Siti Noraiza Ab Razak
Laser Research Group, Advanced Photonic Science Institute, Faculty of Science, Universiti Teknologi Malaysia Johor Malaysia, Malaysia

1. Introduction

Crystallization of amorphous silicon (a-Si) using excimer laser annealing (ELA) has been reported since 1994 by Watanabe group. It is known as the best method to fabricate a good poly-silicon because it can heat the film up to the melting point and, at the same time no thermal damage occur into the glass substrate (Carluccio et al., 1997; Matsumura and Oh, 1999). ELA technique is widely used to increase the grain size and changes the microstructure of polysilicon thin film which is the most important characteristics of excellent built-in polysilicon devices (Palani et al., (2008). The principle advantage of excimer lasers is the strong absorption of ultraviolet (UV) light in silicon beside having larger beam size and high energy density than other laser light sources (Watanabe, et al. 1994). Another major advantage of the excimer lasers is it low-temperature polysilicon annealing. Excimer laser with 308 nm wavelength for example has been reported can transform 50-nm-thin layers of amorphous silicon into high-quality polycrystalline silicon with greatly enhanced electron mobility, for use in flat-panel displays for mobile phones and flat-screen televisions (Delmdahl, 2010). In the low-temperature annealing of polysilicon, excimer lasers with UV output energies of over 1 J per pulse and output powers of 600 W are used to manufacture liquid-crystal and organic LED backplanes at a rate of 100 $cm^2\ s^{-1}$. A new VYPER/LB750 line beam annealing system enables volume production of low-temperature polysilicon (LTPS) on large generation 6 glass panels. LTPS is the key material for high-resolution liquid crystal displays (LCDs) and organic light-emitting diode (OLED) displays for smartphones, tablet PCs and TVs.

Today, the most prominent applications of excimer lasers are in semiconductor chip manufacturing. This relies on the deep-UV emission wavelength of 193 nm, with semiconductor chip manufacturing capitalizing on the optical resolution of excimer lasers. For example, high repetition rate line-narrowed excimer laser models with 10 mJ per pulse, pulse frequencies of up to 6 kHz and narrowed bandwidths of 0.35 pm are used in advanced photolithography to produce computer chips with feature sizes of 45 nm.

Ongoing miniaturization in microelectronics and the trend towards thin-film technologies demands increasing lateral resolution and selective machining. Functional structures and active layers are often only tens of nanometres thick and has to be annealed, patterned and removed in a selective manner without damaging underlying layers or substrates. Because excimer lasers provide the shortest wavelength of all laser technologies, they will continue to have a crucial role in many industries over the next decade. In years to come we will see high-power applications of excimer lasers, with output energies reaching 2 J and output power levels of 1,000 W and above (Delmdahl, 2010).

The aim of this chapter is to demonstrate the crystallization of polysilicon thin film by excimer laser annealing. Various aspects of excimer laser annealing process will be covered in this chapter including principle and experimental results. The definition and the history of excimer laser annealing are described in introduction section. Since excimer laser is the key component in this annealing process, the detail regarding excimer laser system is discussed. Mechanism of laser annealing is explained under third section of laser ablation which includes the ablation phenomena and photochemical mechanism. The detail sample preparation using solid phase crystallization technique is described under section 4. Section 5 will give information regarding polycrystalline silicon. The basic annealing process relies on transformation temperature in which three important parameters, comprising critical point, recalescence effect and super lateral growth energy are desired to be identified. The crystallization of silicon thin film is characterized via atomic force microscope analysis. Finally the excimer laser annealing on silicon thin film is summarized in section 8.

2. Excimer laser

2.1 History and application

An excimer laser is a form of ultraviolet laser which is commonly used in the production of microelectronic devices (semiconductor integrated circuits or "chips"), eye surgery, and micromachining. The excimer laser was invented in 1970 by Nikolai Basov, V. A. Danilychev and Yu. M. Popov, at the Lebedev Physical Institute in Moscow, using a xenon dimer (Xe_2) excited by an electron beam to give stimulated emission at 172 nm wavelength. A later improvement, developed by many groups in 1975 was the use of noble gas halides (originally XeBr).

The excimer laser typically uses a combination of a noble gas (argon, krypton, or xenon) and a reactive gas (fluorine or chlorine). Under the appropriate conditions of electrical stimulation and high pressure, a pseudo-molecule called an excimer (or in the case of noble gas halides, exciplex) is created, which can only exist in an energized state and can give rise to laser light in the ultraviolet range.

Laser action in an excimer molecule occurs because it has a bound (associative) excited state, but a repulsive (dissociative) ground state. This is because noble gases such as xenon and krypton are highly inert and do not usually form chemical compounds. However, when in an excited state (induced by an electrical discharge or high-energy electron beams, which produce high energy pulses), they can form temporarily-bound molecules with themselves (dimers) or with halogens (complexes) such as fluorine and chlorine. The excited compound can give up its excess energy by undergoing spontaneous or stimulated emission, resulting in a strongly repulsive ground state molecule which very quickly (on the order of a picosecond) dissociates back into two unbound atoms. This forms a population inversion. Excimer lasers are usually operated with a pulse repetition rate of around 100 Hz and a pulse duration of ~10 ns, although some operate at pulse repetition rates as high as 8 kHz and some have pulsewidth as large as 30 ns.

Excimer lasers are widely used in high-resolution photolithography machines, one of the critical technologies required for microelectronic chip manufacturing. Current state-of-the-art lithography tools use deep ultraviolet (DUV) light from the KrF and ArF excimer lasers with wavelengths of 248 and 193 nanometers (the dominant lithography technology today is thus also called "excimer laser lithography), which has enabled transistor feature sizes to shrink below 45 nanometers. Excimer laser lithography has thus played a critical role in the continued advance of the so-called Moore's law for the last 20 years.

The most widespread industrial application of excimer lasers has been in deep-ultraviolet photolithography, a critical technology used in the manufacturing of microelectronic devices (i.e., semiconductor integrated circuits or "chips"). Historically, from the early 1960s through the mid-1980s, mercury-xenon lamps had been used in lithography for their spectral lines at 436, 405 and 365 nm wavelengths. However, with the semiconductor industry's need for both higher resolution (to produce denser and faster chips) and higher throughput (for lower costs), the lamp-based lithography tools were no longer able to meet the industry's requirements. This challenge was overcome when in a pioneering development in 1982, deep-UV excimer laser lithography was proposed and demonstrated at I.B.M. by Kanti Jain. With phenomenal advances made in equipment technology in the last two decades, and today microelectronic devices fabricated using excimer laser lithography totaling $400 billion in annual production, it is the semiconductor industry view that excimer laser lithography has been a crucial factor in the continued advance of Moore's law, enabling minimum features sizes in chip manufacturing to shrink from 0.5 micrometer in 1990 to 32 nanometers in 2010. This trend is expected to continue into this decade for even denser chips, with minimum features approaching 10 nanometers. From an even broader scientific and technological perspective, since the invention of the laser in 1960, the development of excimer laser lithography has been highlighted as one of the major milestones in the 50-year history of the laser.

For micro- and nano-technologies, laser micromachining is currently used in a large number of R&D and industrial applications. The range of applications to which laser methods are applied is continuously expanding, supported also by the development of novel processing techniques. Over the last decades the excimer laser has obtained the key position among lasers in various sectors of micromachining. Excimer lasers have developed into powerful manufacturing tools mainly because of two reasons: (i) The short wavelengths of the excimer laser offers excellent quality of machining and a great versatility in features which can be produced. (ii) The progress in basic excimer laser technology has made the excimer lasers to reliable machines suitable for the industrial environment.

There are certainly not many types of lasers which have found such broad markets as the excimer laser. Over the last years the main growth results from increasing industrial use followed by medical applications while new sales into R&D applications stay nearly constant. Today the largest known industrial applications of excimer lasers are (i) based on micromaching of different materials as polymers, ceramics and glasses, applied for example in the production of ink jet cartridges by drilling the nozzles, (ii) excimer laser radiation is being used for changing the structure and properties of materials as oxides, silicon or glass in bulk or thin films, as applied for the production of active matrix LCD monitors, fiber Bragg gratings in telecommunication, and high temperature superconducting films, (iii) employing the excimer laser as "short wavelength light bulb" in optical microlithography for the production of computer chips with critical dimensions below 0.25 μm (the largest homogeneous market for excimer lasers).

All the widespread applications of excimer lasers in micromachining and medicine are based on the early use of excimer lasers leading to the discovery of the ablation of materials under intense illumination with ultraviolet laser pulses by R. Srinivasan.

2.2 Crystallization source

The source of energy used for crystallization in this work is an argon fluoride excimer laser. The excimer laser generates deep ultraviolet light measuring 193 nm when high voltage energy is discharged into a cavity containing a mixture of a rare gas argon and a halogen (i.e., fluoride) Almost, 90% or more of the mixture contain buffer rare gas (typically helium) that does not take part in the reaction. This gas mixture is pre-ionized by a set of electrodes before a high-voltage current (about 30,000 eV is applied) resulting in formation of highly unstable rare-gas halide molecules, which rapidly dissociate, emitting UV light whose wavelength is determined by the particular gas mixture chosen. Excimer lasers use in this particular project is comprised of a mixture of argon and fluoride gas. The term excimer is derived from the two words "excited" and "dimer" which are used to describe the reaction in which the laser transfer energy through an ultraviolet beam of light (Schneider, 1998). All emitted powerful pulses lasting in 10 nanoseconds at wavelength of ultraviolet region of 193 nm.

2.3 Excimer laser structure

The gain of excimer lasers is extremely high, so the output is superradiant. A single rear mirror is employed and an output coupler of 4 to 8% which transmit in the region of interest (which is in UV region) is used (Figure 1). Divergence of the beam is reduced when a full optical cavity is used, and alignment is easy since the laser operates even when cavity mirrors are completely misaligned. The beam is rectangular profile. Quartz cannot be used as an output coupler for argon fluoride (ArF) Excimer laser since fluorine attacks the material, hence magnesium fluoride are used to replace quartz as it absorbs less UV at most wavelength compare to quartz (Csele, 2004). In discharge excitation, electric current flows through the laser medium, typically ranging from a kilovolt (kV) to over tens of kilovolts deliver energy to the laser gas (Hecht, 1992).

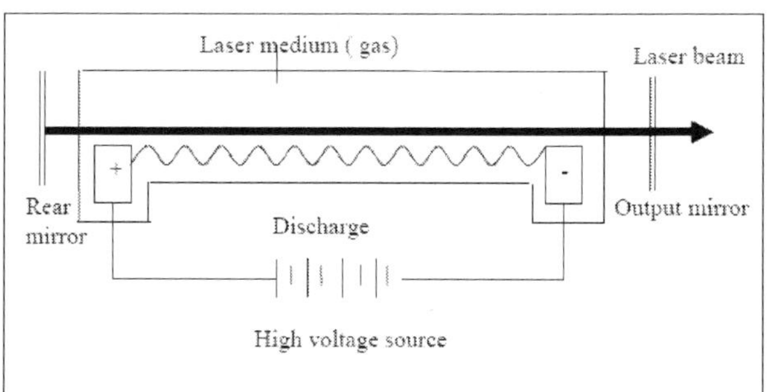

Fig. 1. An electrical discharge exciting a gas laser (Hecht, 1992)

Excimer laser is excited by passing a short, intense electrical pulse through a mixture of gases containing the desired rare gas and halogen. However, the molecule were found more stable when it is on excited state and become less stable on ground state and this properties can be found on argon fluoride (ArF), krypton fluoride (KrF_2) and xenon chloride ($XeCl_2$). This phenomenon could be explained in detail by referring on Figure 2.

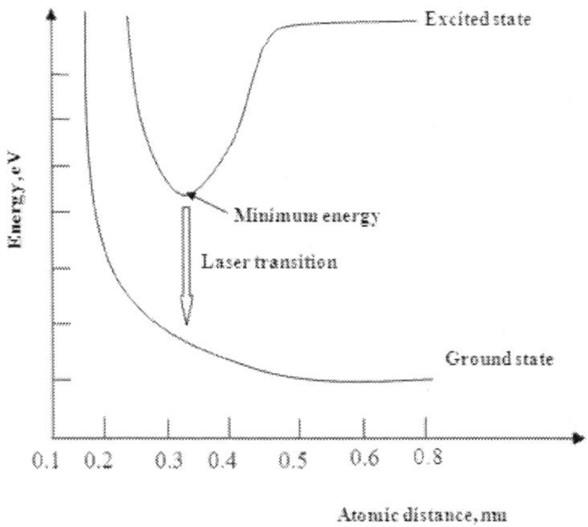

Fig. 2. Excimer laser potential energy curve.

Figure 2 shows a graph of potential energy curve as a function of atomic distance between the molecules. Generally, the potential energy is minimized at equilibrium molecule distance. However, for ArF, KrF_2 and $XeCl_2$ material, these properties did not occur. They are found not stable even when they are in their ground state. Thus, excitation process cannot directly be done on ground state. Hence, indirect excitation in discharge system is employed. For example, in ArF excitation process, electron reaction process is as followed,

$$e + F^2 \rightarrow F^- + F \quad (1)$$

The negative ion produced will combine with positive ion to produce excited molecule

$$Ar^+ + F^- \rightarrow (ArF)^* \quad (2)$$

This reaction can produce an efficient excited dimer molecule. The unstable molecules at ground state leads to a small number of population. Hence, it causes the population inversion to easily happen and laser transition will trigger. The nature of dimer material that is not stable on their ground state causes the atom to easily breaks-up (Shah, 2009). Because of the ground state essentially does not exist, there is a population inversion as long as there are molecules in the excited state. This process is performed again and again; this is how the pulse excimer laser is trigger out.

2.4 External triggering

An argon fluoride ArF excimer laser, manufactured by GAMLaser model EX5/200-110, is employed as a source of ablation. The wavelength of the laser is 193 nm with pulse duration of 10 ns and repetition rate which can be varied from 20 Hz to 200 Hz. The beam size of the laser output when it hits the sample is 4 mm x 6 mm. The size of the beam also depends on the output energy of the laser beam (GAMLaser Inc., 2003).

The excimer laser is internally operated via the aids of computer using 32 bit Windows software (GAMLaser Inc., 2003). The laser parameters including the frequency, high voltage and number of pulse can be commendable according to the requirement of the experiment. The output energy of the laser can be varied by manipulating the capacitor voltage from 10 kV to 15 kV. The dose number of laser exposure is not limited; however, the minimum number of pulses can be set at 100 pulses. However, to trigger the number of pulse as minimum as 1 pulse, a Sony Tektronix arbitrary function generator model AFG 310 is employed. Figure 3 shows photograph of the excimer laser, which coupled to an arbitrary function generator in the real field. In this particular experiment, only a few pulses of laser exposure are required to obtain the desired optical surface treatment.

Fig. 3. An excimer laser connected to a function generator for external triggering.

The function generator can be used to control the number of pulses lower than 100 pulses. As an external trigger unit, the function generator is set up at '5 Volts', operated at 'pulse' output function, and with 'burst' mode (Sony Tektronix Inc., 2003). The frequency of the function generator is synchronized with the excimer laser system.

2.5 Excimer laser annealing

A GAM LASER Argon Fluoride ArF excimer laser is used to anneal the hydrogenerated silicon thin film. The ArF excimer laser generates ultraviolets laser light at 193 nm with 10 ns pulse duration. The lasers parameters such as number of pulses, repetition rates and pumping voltages can be controlled via a computer program. In this experiment, the laser capacitor voltage is kept constant at 12 kV with repeatition rate of 20 Hz, operating at room temperature. The gas pressure is set at 3200 Torr.

The excimer laser annealing experimental setup is shown in Figure 4. The laser energy (in mJ) for each laser pulses is measured using 13PEM001 Broadband power and energy Meter.

The laser is controlled externally using a AFG310 function generator for obtaining laser pulse below than 100 pulses. In this present works, the number of pulse is varied from one to ten pulses. The operational voltage of the function generator is set at 5 Volts 'pulse' output function, and with 'burst' mode. The laser pulse is generated after 2.5 µs upon application of the trigger pulse rising edge with DC charging is operating in external trigger mode. The frequency of the function generator and the excimer laser system is fixed at 20.0 Hz. The silicon copper thin film placed on x and y translational stage in and vertical position with respect to the excimer laser beam axis. The number of pulses of excimer laser beam exposed on the thin film target is, controlled by the external trigger unit.

Maximum laser performance can be achieved by aligning the laser mirrors. The mirrors are alligned via the guidance of a visible of helium-neon He-Ne laser. The He-Ne laser is aligned to coaxis with excimer laser beam. The excimer laser is optimized by adjusting the output coupler via the aids of power/energymeter. A bright and rectangular dimension of beam spot is recorded permanently using beam profiler. The images are arranged in order of 3D, 2D and the bright beam spot in the real field such as shown in Figure 5 a, b and c respectively. The dimension of excimer laser beam spot is 4 × 6 mm^2. The a-Si films is recrystalized at different energy densities. Finally, nanostructure of the crystallized silicon is examined via atomic force microscope AFM.

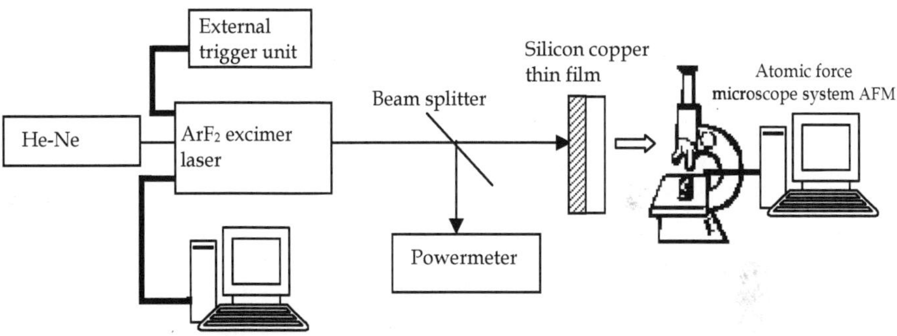

Fig. 4. The schematic diagram of excimer laser annealing process.

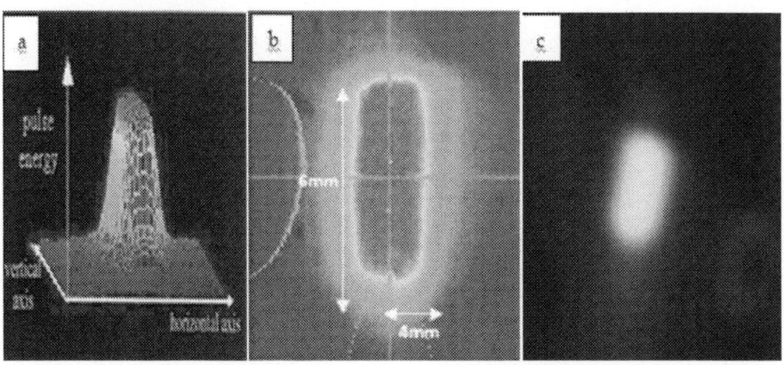

Fig. 5. Beam profile and beam spot of excimer laser. a. 3D beam profile, b. 2D beam profile, and c. The beam spot in the real field.

3. Laser ablation

In laser annealing process the main key for its successful work is the ablation process between optical material and UV light from argon fluoride (ArF) excimer laser. In order to understand the laser-matter interaction, it is better to understand the fundamental mechanism of laser ablation and optical material behavior with respect to the ultraviolet (UV) light illumination. In this section, the photochemical process occurred during ablation process will be discussed.

Ablation is usually described in terms of physical mechanism such as vaporization and shock effects. It is often performed in vacuum or air. Examples of ablation application include polymer ablation, removal of thin metal films from an insulating substrate, and deposition of high-temperature superconducting thin films by ablation of bulk targets (Brannon, 1997).

The laser ablation process requires an intense UV light (λ = 193 nm- 351 nm) produced by an excimer laser. Because of the high absorption of UV light and relatively poor conductivity of many materials, particularly silicon, the energy is deposited in a very thin layer. Whenever the energy density or fluence exceeds the ablation threshold value for the material, chemical bonds are broken, fracturing the material into energetic fragments. The fragments are atoms, groups of atoms, ions and electrons. Because the fragments leave the reaction zone as an energetic gas and solid debris, the ablation process resembles explosive evaporation of the material (Speidell, 1997).

Generally, there are two classes of laser ablation mechanisms: thermal and electronic (non-thermal). Thermal processes rely on an intense laser pulse to very rapidly heat a surface at rates of the order of 10^{10} K/sec. These processes can produce expansion and vaporization from solid and melted regions.

Laser ablation of given material can involve both thermal and non-thermal mechanisms. For example, during the interaction of the leading edge of the laser pulse with the solid, non-thermal processes may predominate because significant heating has not yet occurred. As the rest of the pulse is absorbed, the temperature rise rapidly and thermally activated mechanism may commence. It should be noted that laser excitation of a strongly absorbing solid will always result in substantial heating regardless of the ablation mechanism. Although non-thermal process may occur, there is no way to "turnoff "the thermal processes. The dominant process is ultimately determined by the detailed chemistry and physics of the solid.

3.1 Photochemical

In the photochemical processes the photon absorption event break a chemical bonding within a molecule forming molecules with smaller number of atoms. The resulting photoproducts occupy a large volume and create pressure inside the irradiated volume that can then convert to the translational energy of ablation (Sato et al., 2001). During this process, thermal and mechanical damage to the surrounding material is minimized, therefore achieving more precise control over the ablated region.

There has been much un-certainty and debate over the fundamental ablation mechanisms in material. Several photochemicals and photothermals models have been developed to explain the ablation mechanisms. In photochemical mechanisms, the photon break the chemical bonds of the material directly where as in photothermal mechanism the material is ablated by heating, melting and vaporizing the material (Chen et al., 1995). It is found that the material removal by laser ablation approximately obeys the Beer lamberts Law at lower laser fluences (Srinivasan, 1982). For photochemical ablation to occur, energy of the photons at that wavelength should overcome the intermolecular bond energies of the target material. The relation between the photon energy of light and laser wavelength is given by (Sauerbrey, 1989),

$$E = 1.245 / \lambda \quad (3)$$

Where λ is wavelength of light (193 nm), E is energy of photon (eV). The photon energy of the lights depends on the wavelength of the light and as the wavelength increases, the energy decreases. An UV laser with a wavelength of 193 nm has photon energy of 6.45 eV. It can be seen that for photochemical ablation to occur in thin film the photon energy of the light should be greater than the bond energy of the material. While excimer laser does not produce as high average powers as longer wavelength infrared lasers such as Nd:YAG and CO_2, their shorter wavelength carried high photon energy. As for example, argon fluoride (ArF) excimer laser with wavelength of 193 nm have photon energy of 6.45 eV, while Nd:YAG with wavelength of 1064 nm have photon energy of 0.18 eV. Hence, only laser with short wavelength can initiate photochemical reaction while the laser with longer wavelength may only lead to heating the material only.

Apparently, the advantage of photochemical reaction, is that the mechanism can change the molecule structure as well as the refractive index of the material, but not involve in external damage unless the energy deliver is too high. As a result the photochemical mechanism is important effect in the nucleation and the growing of grain size or in the crystallization process.

4. Sample preparation

Target used to be annealed is an amorphous silicon. In this works silicon is chosen to be the main material and deposited on the glass substrate as the first layer of thin film. Silicon is widely used in semiconductors industry. It can stand at higher temperatures compared to germanium. Furthermore its native oxide is easily grown in a furnace. Plus it can form a better dielectric interface rather than any other materials.

An ordinary glass substrate is conducted to make it possible for low annealing temperature. The glass substrate used in this research has maximum operation temperature in the range of 350°C to 400°C. The thickness of the glass substrate is in between 1.0 mm to 1.2 mm. The glass substrates are cut into small pieces using diamond cutter. Each of them having dimension of 10×10 mm^2. All glass substrates are cleaned using Ultrasonic Branson 3210 cleaning machine to ensure the glass surface is free from any dust, oil, or contaminations.

4.1 Thin film fabrication

Solid phase crystalization (SPC) is prepared by using a vacuum chamber technology. An Edwards 360 thermal evaporation is a vacuum chamber used to fabricate the silicon thin film. Silicon is deposited on a glass substrate. Initially, the source material to be deposited or evaporant is loaded into a molybdenum boat. The glass substrate is placed 10 cm above the evaporant. The chamber is set at a vacuum condition until the pressure achieved down to 10^{-6} mbar. The current is increased slowly until the evaporant on the molybdenum boat begin to melt. The thickness of film is detected by quartz crystal thickness detector inside the vacuum chamber. The thickness is measured via an Edwards FTM 7 which connected to the vacuum system. To deposit the silicon for about 100 nm thick, 0.3 g of silicon powder is needed. A schematic diagram of the thin film is shown in Figure 6.

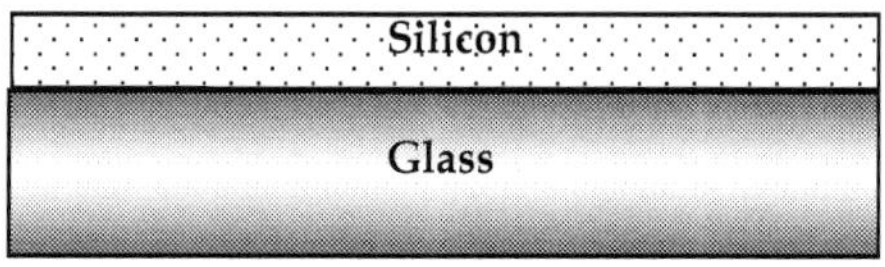

Fig. 6. Silicon thin film on glass substrate

4.2 Annealing process

The annealing of silicon thin film is a combination from heat treatment and excimer exposure. Immediately after fabricated the thin film, it is heated via conventional Tube Furnace type F21100. This is important procedure in order to dry out the water inside the thin film so that the hydrogen percentage can be lowered. The hydrogenated thin film is then annealed by using an argon fluoride (ArF) excimer laser.

4.2.1 Heat treatment

The prepared solid phase crystallization of silicon thin film is initially under-went heat treatment. As mention earlier the aim is to extract the water contained in the film during deposition. An encapsulated annealing technique is utilized for this heat treatment in order to avoid vaporization of thin film material. The silicon thin film is placed in a carbon block (both the container and its cover are made from the same element). The capsulated thin film is then placed in the tube furnace. The annealing process is carried out under atmospheric pressure for four hours in air. The annealing is set at a fixed temperature of 350 °C. This furnace has a single set point control which enables the end user to control the furnace up to a preset temperature and remained constant at the set temperature.

4.2.2 Laser annealing

Nearly all optical materials show moderate to intense absorption in the ultraviolet region. This absorption is usually ascribed to electronic transition from a ground singlet to the first excited state. The unique feature in the UV laser ablation of optical material is encountered only in those wavelength region in which such electronic absorption exist (Srinivasan et al., 1988). Material ablation depends on two key conditions, as shown in Figure 7. First, the material must absorb light strongly at the laser's wavelength. Clean, precise ablation usually requires linear absorption coefficients of at least 10^4 cm^{-1}. Secondly, ablation occurs only after the material has absorbed a minimum energy per unit volume, i.e., the laser intensity must exceed a threshold value (Brannon et al., 1985).

The laser is absorbed in amorphous silicon thin film surface without heating the substrate. A homogenized excimer laser beam spot is exposed on the thin film surface. Within the laser pulse duration the silicon layer is rapidly heated and melted as shown in Figure 7(a, b). As it cools down the crystallization into poly-silicon occurs (Figure 7c, d). In the process of excimer laser annealing, the amorphous silicon thin film is exposed by multiple pulses of laser. The exposed area is partially melted. The period for heating as well as for freezing is the same as the pulse duration of the excimer laser. The heat release in high speed freezing is more compared to the slow rate once. As a result during cooling the heat released by convection force-liquid flow which break the planar symmetry so that the crystal grow develops along the columns or finger such as shown in Figure 7c and d. In Figure 7c, there is partially melted and partially crystallized growth. The totally crystal growth is illustrated in Figure 7d. The crystal grows in the finger formation or poly-crystallization.

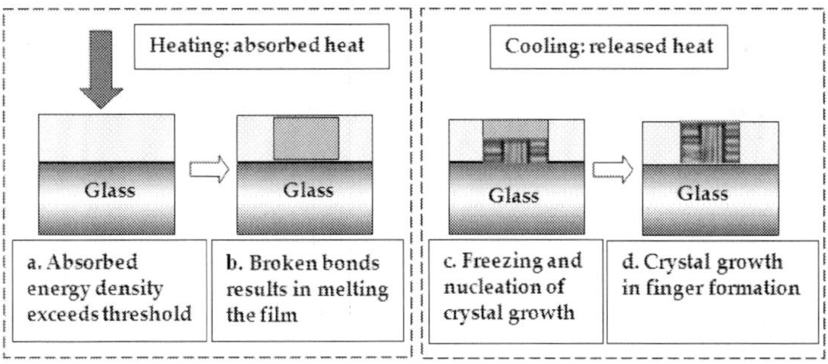

Fig. 7. Schematic diagram of excimer laser annealing on amorphous silicon copper thin film.

5. Polycrystalline silicon

Silicon films are fabricated with a microstructure tailored to the application depending on economic and performance requirements. Silicon film is typically divided into three categories: monocrystalline, amorphous, and polycrystalline.

Monocrystalline silicon is a form in which the crystal structure is homogenous throughout the material; the orientation, lattice parameter, and electronic properties are constant throughout the material (Green, 2004). Dopant atoms such as phosphorus and boron are often incorporated into the film to make the silicon n-type or p-type respectively. Monocrystalline silicon is fabricated in the form of silicon wafers, usually by the Czochralski Growth method, and can be quite expensive depending on the radial size of the desired single crystal wafer (Green, 2004). This monocrystalline material is one of the chief expenses where approximately 40% of the final price of the product is attributable to the cost of the starting silicon wafer used in cell fabrication (Campbell, 2001).

Amorphous silicon has no long-range periodic order. The application of amorphous silicon as a standalone material is somewhat limited by its inferior electronic properties (Streetman and Banerjee, 2000). When paired with microcrystalline silicon in tandem and triple-junction solar cells, however, higher efficiency can be attained than with single-junction solar cells (Shah et al., 2003). This tandem assembly of solar cells allows one to obtain a thin-film material with a bandgap of around 1.12 eV (the same as single-crystal silicon) compared to the bandgap of amorphous silicon of 1.7-1.8 eV bandgap (Shah et al., 2003). Tandem solar cells are then attractive since they can be fabricated with a bandgap similar to single-crystal silicon but with the ease of amorphous silicon.

Polycrystalline silicon is composed of many smaller silicon grains of varied crystallographic orientation (Figure 8). This material can be synthesized easily by allowing liquid silicon to cool using a seed crystal of the desired crystal structure. Additionally, other methods for crystallizing amorphous silicon to form polycrystalline exist such as high temperature chemical vapor deposition (CVD). Each grain is crystalline over the width of the grain. The grain boundary separates the grains where the adjoining grain is at a different orientation than its neighbor. The grain boundary separates regions of different crystal structure thus serving as a center for recombination. '*d*' here is a characteristic grain size, which should be maximized for maximum film efficiency. Typical values of *d* are about 1 micrometer.

Presently, polysilicon is commonly used for the conducting gate materials in semiconductor devices such as MOSFETs; however, it has potential for large-scale photovoltaic devices (Streetman and Banerjee, 2000; Ghosh et al., 1980). The abundance, stability, and low toxicity of silicon, combined with the low cost of polysilicon relative to single crystals makes this variety of material attractive for photovoltaic production (Ghosh et al. 1980). Grain size has been shown to have an effect on the efficiency of polycrystalline solar cells. Solar cell efficiency increases with grain size. This effect is due to reduce recombination in the solar cell. Recombination, which is a limiting factor for current in a solar cell, occurs more prevalently at grain boundaries as shown in Figure 8 (Ghosh et al., 1980).

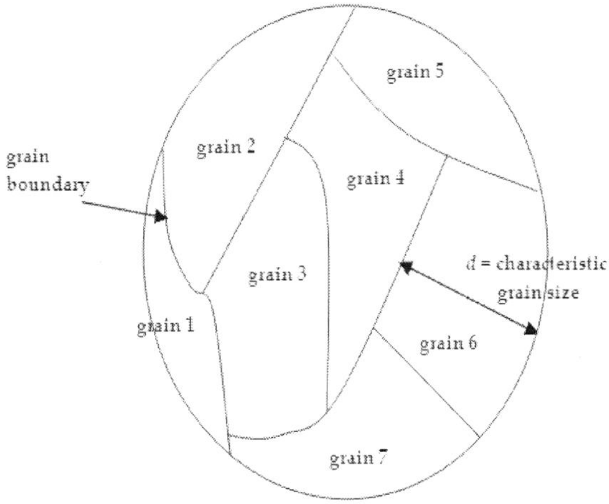

Fig. 8. Grain boundaries of polysilicon

The resistivity, mobility, and free-carrier concentration in monocrystalline silicon vary with doping concentration of the single crystal silicon. Whereas the doping of polycrystalline silicon does have an effect on the resistivity, mobility, and free-carrier concentration, these properties strongly depend on the polycrystalline grain size, which is a physical parameter that the material scientist can manipulate (Ghosh et al., 1980). Through the methods of crystallization to form polycrystalline silicon, a scientist can control the size of the polycrystalline grains which will vary the physical properties of the material.

5.1 Novel ideas for polycrystalline silicon

The use of polycrystalline silicon in the production of solar cells requires less material and therefore provides for higher profits and increased manufacturing throughput. Polycrystalline silicon does not need to be deposited on a silicon wafer to form a solar cell, rather it can be deposited on other-cheaper materials, thus reducing the cost. Not requiring a silicon wafer alleviates the silicon shortages occasionally faced by the microelectronics industry. An example of not using a silicon wafer is crystalline silicon on glass (CSG) materials (Basore, 2006).

A primary concern in the photovoltaics industry is cell efficiency. However, sufficient cost savings from cell manufacturing can be suitable to offset reduced efficiency in the field, such as the use of larger solar cell arrays compared with more compact or higher efficiency designs. Designs such as CSG are attractive because of a low cost of production even with reduced efficiency (Basore, 2006). Higher efficiency devices yield modules that occupy less space and are more compacts however the 5-10 % efficiency of typical CSG devices still makes them attractive for installation in large central-service stations, such as a power station (Basore, 2006). The issue of efficiency versus cost is a value decision of whether one requires an "energy dense" solar cell or sufficient area is available for the installation of less expensive alternatives. For instance, a solar cell used for power generation in a remote location might require a more highly efficient solar cell than one used for low-power applications, such as solar accent lighting or pocket calculators, or near established power grids.

Thin film silicon photovoltaics are typically produced by chemical vapor deposition processes yielding an amorphous, polycrystalline, or nanocrystalline film. Conventionally, amorphous silicon thin films are most common. Silicon is usually deposited on glass, plastic, or metallic substrates coated with a transparent conducting oxide material. While chalcogenide-based Cadmium-Tellurium (CdTe) and Copper-Indium-Selenium (CIS) polycrystalline thin films cells have been developed in the lab with great success, there is still industry interest in silicon-based thin film cells. Silicon-based devices exhibit fewer problems than their CdTe and CIS counterparts such as toxicity and humidity issues with CdTe cells and low manufacturing yields of CIS due to material complexity. Additionally, due to political resistance to the use non-"green" materials in solar energy production, there is no stigma in the use of standard silicon. Three major silicon-based module designs dominate: amorphous silicon cells, amorphous or microcrystalline tandem cells, and thin-film polycrystalline silicon on glass (Green, 2003).

Amorphous or microcrystalline silicon consists of a mixed phase of small crystalline regions surrounded by amorphous material. This material typically behaves more like crystalline silicon than the amorphous variety. A 3-month field study has shown that hybrid amorphous or microcrystalline cells degrade roughly to the same degree as triple-junction amorphous cells while maintaining higher conversion efficiencies (7.0% versus 5.0% as measured at the conclusion of the study). This result suggests hybrid designs of this type may supplant traditional amorphous-based modules (Green, 2003).

A new attempt to fuse the advantages of bulk silicon with those of thin-film devices is thin film polycrystalline silicon on glass. These modules are produced by depositing an antireflection coating and doped silicon onto textured glass substrates using plasma-enhanced chemical vapor deposition (PECVD). The texture in the glass enhances the efficiency of the cell by approximately 3% by reducing the amount of incident light reflecting from the solar cell and trapping light inside the solar cell. The silicon film is crystallized by an annealing step, temperatures of 400 – 600 °C, resulting in polycrystalline silicon.

These new devices show energy conversion efficiencies of 8% and high manufacturing yields of >90%. Crystalline silicon on glass (CSG), where the polycrystalline silicon is 1-2 micrometers, is noted for its stability and durability; the use of thin film techniques also contributes to a cost savings over bulk photovoltaics. These modules do not require the presence of a transparent conducting oxide layer. This simplifies the production process twofold; not only can this step be skipped, but the absence of this layer makes the process of constructing a contact scheme much simpler. Both of these simplifications further reduce the cost of production. Despite the numerous advantages over alternative design, production cost estimations on a per unit area basis show that these devices are comparable in cost to single-junction amorphous thin film cells (Green, 2003).

5.2 Low temperature induced crystallization of amorphous silicon

Amorphous silicon can be transformed to crystalline silicon using well-understood and widely implemented high-temperature annealing processes. This typical method is the typical method used in industry but requires high-temperature compatible materials, such as special high temperature glass that is expensive to produce. However, there are many applications for which this is an inherently unattractive production method. Flexible solar cells have been a topic of interest for less conspicuous-integrated power generation than solar power farms. These modules may be placed in areas where traditional cells would not be feasible, such as wrapped around a telephone pole or cell phone tower. In this application a photovoltaic material may be applied to a flexible substrate, often a polymer. Such substrates cannot survive the high temperatures experienced during traditional annealing. Instead, novel methods of crystallizing the silicon without disturbing the underlying substrate have been studied extensively. Aluminum-induced crystallization (AIC) and local laser crystallization are common in the literature, however not extensively used in industry. In both of these methods, amorphous silicon (a-Si or a-Si:H) is grown using traditional techniques such as plasma-enhanced chemical vapor deposition (PECVD). The crystallization methods diverge during post-deposition processing.

Another method of achieving the same result is the use of a laser to heat the silicon locally without heating the underlying substrate beyond some upper temperature limit. An excimer laser or, alternatively, green lasers such as a frequency-doubled Nd:YAG laser is used to heat the amorphous silicon, supplying energy necessary to nucleate grain growth. The laser fluence must be carefully controlled in order to induce crystallization without causing widespread melting. Crystallization of the film occurs as a very small portion of the silicon film is melted and allowed to cool. Ideally, the laser should melt the silicon film through its entire thickness, but not damage the substrate. Toward this end, a layer of silicon dioxide is sometimes added to act as a thermal barrier (Yuan et al., 2009). This allows the use of substrates that cannot be exposed to the high temperatures of standard annealing, polymers for instance. Polymer-backed solar cells are of interest for seamlessly integrated power production schemes that involve placing photovoltaics on everyday surfaces.

A third method for crystallizing amorphous silicon is the use of thermal plasma jet. This strategy is an attempt to alleviate some of the problems associated with laser processing - namely the small region of crystallization and the high cost of the process on a production scale. The plasma torch is a simple piece of equipment that is used to thermally anneal the amorphous silicon. Compared to the laser method, this technique is simpler and more cost effective (Lee et al., 2009).

Plasma torch annealing is attractive because the process parameters and equipment dimension can be changed easily to yield varying levels of performance. A high level of crystallization (~90%) can be obtained with this method. Disadvantages include difficulty achieving uniformity in the crystallization of the film. While this method is applied frequently to silicon on a glass substrate, processing temperatures may be too high for polymers.

6. Transformation temperature

Laser annealing on the thin film is a high-speed process. It depends on the pulse duration of the excimer laser normally in nanoseconds region and the threshold energy to nucleate and optimum energy to stop crystallization. Such short duration of annealing process is referred as transient crystallization of amorphous silicon film (Viatella and Singh, 1997). Within transient duration, the annealing process still needs to pass through several critical points. In the next section, such phase transformation point will be described in detail.

6.1 Critical points

In order to figure-out the physical change involve in laser annealing it is better to understand the basic concept of phase transformation during heating and cooling of hardening process. There are several important stages in heating and cooling of material. The crucial parameter is temperature at several critical points (Figure 9). The "critical points" are the temperatures at which certain changes in the chemical composition of the material take place, during both heating and cooling. Steel for example at normal temperatures has its carbon (which is its chief hardening element) in a certain form called *pearlite* carbon, and if the steel is heated to a certain temperature, a change occurs and the pearlite becomes *martensite* or hardening carbon. If the steel is allowed to cool slowly, the hardening carbon changes back to pearlite. The points at which these changes occur are the decalescence and recalescence or critical points, and the effect of these molecular changes is as follows: When a piece of steel is heated to a certain point, it continues to absorb heat without appreciably rising in temperature, although its immediate surroundings may be hotter than the steel. This is the *decalescence* point. Similarly, steel cooling slowly from a high heat will, at a certain temperature, actually increase in temperature, although its surroundings may be colder. This takes place at the *recalescence* point. The recalescence point is lower than the decalescence point, and the lower of these points does not manifest itself unless the higher one has first been fully passed. These critical points have a direct relation to the hardening of steel. Unless a temperature sufficient to reach the decalescence point is obtained, so that the pearlite carbon is changed into a hardening carbon, no hardening action can take place; and unless the steel is cooled suddenly before it reaches the recalescence point, thus preventing the changing back again from hardening to pearlite carbon, no hardening can take place. The critical points vary for different kinds of material, and must be determined by tests in each case.

Similar phase transformation also experience by other materials during heating and cooling process. The different may be arisen in the aspect of temperature and the time taken to complete the same process. If the steel is replaced with other material like any amorphous silicon thin film and take into account the duration involves in nanosecond time domain then the whole event is considered as a high-speed phenomenon. This includes the rate of absorption and release of heat which occurs in transient time and enough energy to cause the phase change. In case excimer laser ablation which posses in nanoseconds pulse duration and fluence more than 500 J/cm^2, having high potential to cause the target thin film experience the same process as hardening steel. This implies excimer laser annealing on thin film should passing through the decalescence and recalescence points to ensure the formation of crystallization.

In general when amorphous silicon thin film is heated and cooling, normal phase change is occurred including the melting and re-solidification transformation. The phenomenon of this phase transformation is similar as illustrated in Figure 9. The red line is indicated the heating process, while the blue line represents the cooling process. In heating process, there are two important points that are decalescence and critical points. Decalescence is a phenomenon that occurs when a material is being heated and there is a sudden slowing in the rate of temperature increase. The slowing rate is due to the change in the internal crystal structure of the material. Meanwhile during cooling down or re-solidification, the phase change is passing through a recalescence point. A sudden spontaneous increase in the temperature of cooling resulting from an exothermic change in crystal structure occurs. This is the critical state in cooling process whereby the formation of crystallization is taken place.

Furthermore, a single pulse of the excimer laser, carried out temperature of 6.42 eV which is higher than threshold to break the bonding of silicon molecules. Once the excimer laser pulse is exposed on the film, it must passing through the decalescence and recalescence points in order to form crystallization. If a lower energy than threshold material or too high energy achieve up to the peak temperature in the transformation process, no crystallization will be formed on silicon film.

In both temperatures, there is latent heat of fusion is liberated from the solid during heating and from liquid during cooling. The latent heat is liberated during both processes to delay either the process of heating or cooling. The coexistent both solid and liquid associated in heating and cooling, resistant the rate of transformation. The advantage of this drawback is given an opportunity to nucleate and grow the crystallization in the silicon.

If the heating is over than optimum energy by exposing with higher number of pulses, the thin film is totally melted. No solid form is left to be as a seed for crystallization. In this case the treatment is achieved up to the peak temperature, passing over the decalescence point. Subsequently re-solidification occurs faster without passing through the recalescence point. Hence no formation of crystallization and end up with homogenous or no change in crystal structure in the excimer laser annealing process.

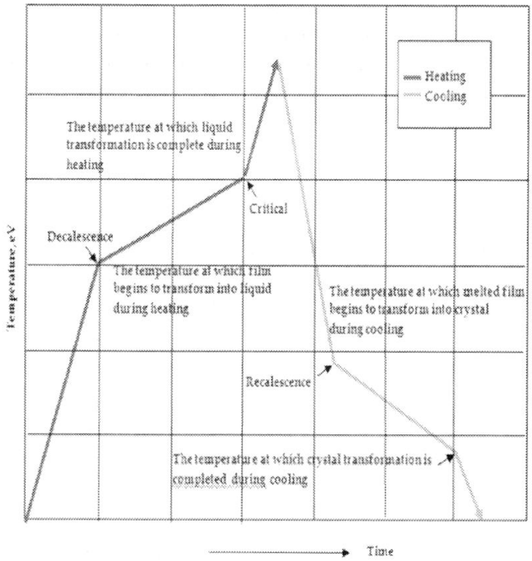

Fig. 9. Transformation temperature

6.2 Recalescence effect

There are several researches on the recalescence phenomenon have been reported. Among them are Zhang and Atren, (1992), Armengol et al., (2003) and Mariucci et al., (2003). Zhang and Atren, (1992) have studied the recalescence characteristics in rapid solidification of copper by using a thermokinetic model. They have investigated the effects of the heat transfer coefficient, the melt thickness and the nucleation temperature. Their results had shown that the lower nucleation temperature and thinner melt lead to a longer recalescence effect while larger heat transfer coefficient results in a weaker recalescence effect. A dimensionless number was derived to measure the extent of recalescence.

Mariuci group (2003) had claimed that crystallization less than complete melting point will experience the recalescence effect. Structural properties of thin polycrystalline silicon films, crystallized by single shot excimer laser annealing at different laser energy densities, have been investigated by various researchers (Armengol, et al., 2003,; Mariucci et al., 2003; Razak and Bidin, 2010; Bidin and Razak, 2011a, b). Formation of disk structures has been observed in a wide range of energy densities, from complete melting down to 180 mJ/cm^2. These structures have been correlated to the lateral growth of grains starting from the small grains present in the central regions of the disks. They had proposed a new crystallization scenario for energy densities below the complete melting. In this framework, the recalescence effect plays an important role while the super lateral growth-regime is no longer a particular crystallization condition but simply represents the upper energy density limit of partial melt crystallization regime.

Similar recalescence effect after bulk solidification in germanium films melted by nanosecond laser pulses had also reported by Armengol, et al., (2003). Rapid solidification dynamics in amorphous germanium films melted by nanosecond laser pulses has been analyzed by means of single-shot subnanosecond time resolved reflectivity measurements using a streak camera based setup. The results showed that once a minimum melt depth is induced, a bulk solidification process followed by the release of the solidification enthalpy dominates the solidification process. Moreover, the laser-melted material solidifies completely before being remelted as a consequence of the solidification enthalpy release, something only observed, up to date, upon irradiation with picosecond laser pulses (Armengol et al., 2003).

Bossuyt and Greer, (2004) have been investigated the effects of positive feedback on crystallization kinetics and recalescence. They described that in a series of alloys in the Cu-Ni-Ti-Zr system, with compositions close to the bulk glass forming alloy Cu47Ni8Ti34Zr11 (Vit101), the nucleation density has been shown to be spatially inhomogeneous; in an amorphous matrix there are spherical clusters with a high density of nanocrystals. The implied positive feedback in the nucleation rate was analyzed in terms of recalescence instability, where the latent heat released upon crystallization causes the nucleation rate to increase locally, if the nucleation rate is an increasing function of temperature and thermal diffusivity is low enough to avoid distributing the heat evenly over the sample. In deeply under cooled liquids, the first of these requirements is satisfied, but not for the second requirement.

Bossuyt and Greer have used numerical technique to trace the cooling rate with and without the feedback heat in crystallization formation. They identified a critical cooling rate for recalescence to occur. At slightly faster cooling rates, the heat released by crystallization causes the actual cooling rate to drop to nearly zero, but the heat output is never sufficient to actually raise the temperature. When that does occur, there is an abrupt transition from approximately 40% crystallization at cooling rates just above the critical cooling rate, to full crystallization below the critical cooling rate. Furthermore they claimed that disabling the heat of crystallization eliminates the recalescence effect from the temperature traces, and the transition to full crystallization is now much more gradual, without a clearly defined critical cooling rate.

Other feedback mechanisms have effects similar to those of recalescence: there is always a threshold condition where the acceleration due to feedback offsets the quenching effect of decreasing temperature. At cooling rates above this threshold, the crystallization kinetics did not deviate much from those in the absence of feedback. The effect of feedback becomes noticeable over a relatively narrow range of cooling rates just above the threshold, and then dominant below the threshold, abruptly increasing the degree of crystallization (of at least one of the phases). The decalescence and recalescence effect are shown in Figure 9. Further explanation about the critical points will be discussed on section 6.5.

However not many researchers have been realized about the decalescence effect in the re-crystallization process. Without determination the decalescence effect during heating process the recalescence point cannot be achieved. Therefore in order to crystallization of polycrystalline silicon film the emphasized is focusing on these both phenomena (decalescence and recalescence effects). In this project, silicon thin film is crystallization by controlling the number of laser pulses.

In most cases, silicon is deposited onto a corning glass substrate and annealed with high energy excimer laser which requires high fabrication cost for industrial purposes. In this attempt, low energy ArF (wavelength of 193 nm) excimer laser was employed together with conventional annealing method to anneal the silicon thin film. As reported by previous researchers (Carluccio et. al., 1997, Han et. al., 2003) combination annealing techniques between conventional and laser annealing would produce a better poly-si thin film. Figure 10 shows a different in terms of grain size structure between polysilicon thin film PSTF annealed by (a) conventional annealing and (b) combination of conventional and excimer laser annealing. Polycrystalline silicon thin film (PSTF) has been an interest currently due to high potential as electrical interconnections in microelectronics and microsensors.

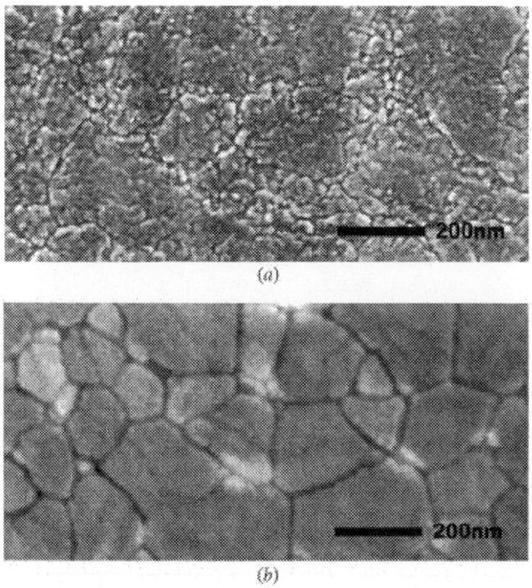

Fig. 10. FESEM images of poly-Si (a) annealed at 600 °C for 24 h (*a*) and the combination annealing with ELA at 470 mJ cm^{-2} (Han et. al., 2003)

6.3 Super lateral growth energy

In order to optimize the annealing process, a super lateral growth SLG energy density needs to be identified. The SLG point is known as the critical energy density or laser fluence desires to generate the maximum crystallization. Greater than this energy normally the crystallization will be reduced and soon eliminated. In other word the SLG energy density is equivalent with the critical point in the temperature transformation curve (Figure 9) during heating process. Beyond this energy, the silicon film will totally melt and cause elimination of spontaneous crystallized formation. This implies that calibration needs to be done in determining this crucial energy.

There are two parameters commonly used to quantify the degree of crystallization that are, grain size and surface roughness. The later parameter is due to the melting effect. Both parameters are utilized for measurement of crystallization. If the boundaries of the grain size is expanding or the surface roughness is increasing after being exposed by a series of excimer laser pulses that means the annealing process is within the decalescence and recalescence region. Once the optimum energy density that is super lateral growth energy is achieved that is at the critical point in the temperature transformation curve, no more number of pulses can be added or no more laser energy density can be increased.

Otherwise, the excess energy exposed that would cause the film totally melted and no solid form leaves at the interface between substrate and film to seed the crystallization.

Hence in excimer laser annealing the most crucial process is to identify the SLG energy. Under this particular SLG energy, the heating and the cooling are associated with the decalescence and recalescence effects. This implies that, the phase change is accompanied with the coexistence of liquid and solid forms. This is illustrated in Figure 11. At the interface between film and substrate, there are some particles remain in solid form whereas the rest in liquid phase. The left over the particles are used to seeding the crystallization. This region is similar as the condition achieved at recalescence point in the transformation temperature curve during quenching or cooling period.

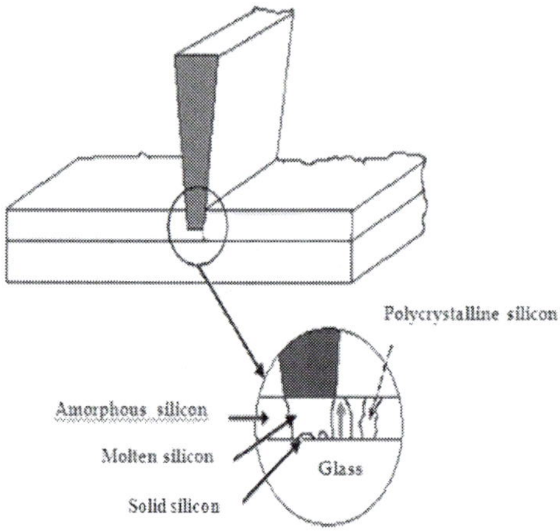

Fig. 11. Nearly complete melt of amorphous silicon layer. The liquid/solid interface seeds the crystallization process

It is well known that laser ablation involves photothermal and photochemical process, depending on the nature of the material used and the experimental conditions, for example laser fluence, wavelength, and pulse duration (Sato et al., 2001; Serafetinides et al., 2001; Hahn et al., 1999). A 193 nm wavelength radiation can cause a neat and clean etch region because at this wavelength the molar extinction coefficient of the material is high (~ 10^4 L mol^{-1} cm^{-1}) so that 95% of the light is absorbed in the first 300 nm of the substrate. On the other hand, a laser with longer wavelength (~ 500 nm) light melts and damage nearby regions of the sample as a longer wavelength light vibrationally excites the irradiation sites. This process can eventually cause material to desorb from the heated region (Barbara et al., 1984).

7. Atomic force microscope analysis

The crystallization result is quantified based on metallurgical technique. The annealed silicon thin film is examined under the atomic force microscope. The scan area for each sample is fixed to 5000 nm X 5000 nm and zoomed until 1000 nm X 1000 nm to get clearer image of the thin film surface. The typical of the nucleation and the growth of silicon thin film is shown in Figure 12. The frames of crystallization are arranged in the increasing order of the energy density. Prior annealing, the grain size of the amorphous silicon thin film is measured. The average nanocrystal size, G_{av} of the amorphous silicon film before doping and heating as shown in Figure 12(a) is approximately 17 nm. After experienced four hours heat treament via conventional annealing, a new crystal structure is realized. The crystallization pattern almost uniform with an increment up to 56 nm as depicted in Figure 12(b). Immediately after crystalization with a single shot of excimer laser corresponding to energy density of 65.50 $mJcm^{-2}$, another new pattern nanostructure is realized. The crystallization of the silicon film is accelerated further to a grain size of 75 nm as illustrated in Figure 12(c). The crystallization of thin film is continuously increasing after received the sequential number of pulses from argon fluoride ArF excimer laser. The maximum crystallization of silicon film is notified after received five number of pulses as depicted in Figure 12(g). The enlargement achieved approximately to 143 nm corresponding to the super lateral growth energy of 345 $mJcm^{-2}$. Beyond the critical energy as shown in Figure 12(h) the crystallization reduced drastically.

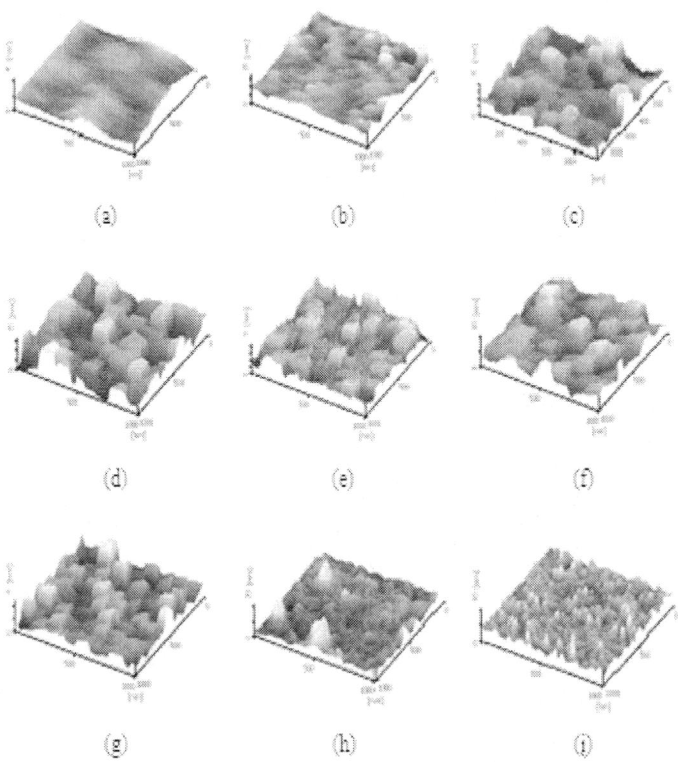

Fig. 12. AFM images of silicon thin film with magnification area of 1000 nm x 1000 nm at different energy density.

The stress developed during solidification due to the melting effect of silicon thin film is quantified based on surface roughness measurement. A typical example of the surface roughness of the silicon copper film is shown in Figure 13. Figure 13(a) indicates the roughness before annealing process. The measurement of surface roughness for the amorphous silicon thin film is 0.3 nm. After annealing the intergration of crystalization formation comprised of the agglomeration and coalesence of grain. Simialr results have also reported by He et al., (2007). The surface roughness of the annealed silicon thin film then increased up to 3.0 nm. The profile of the grain sized and surface roughness measurement are presented in Figure 14 and 15 respectively. However the maximum roughness is not overlapping with maximum grain size. The possible reason for such different may be due to the measured surface roughness is not matched with measured area of the grain size.

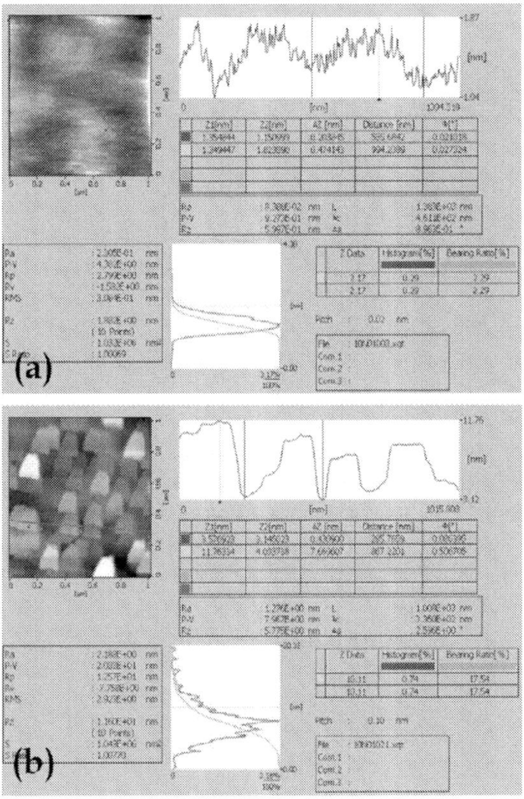

Fig. 13. Surface roughness of silicon at different number of pulses: a. o pulses, b. 3 pulses.

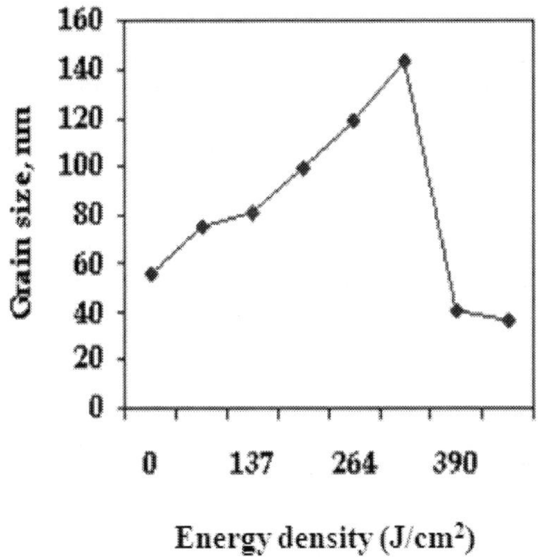

Fig. 14. Grain size as function of energy density

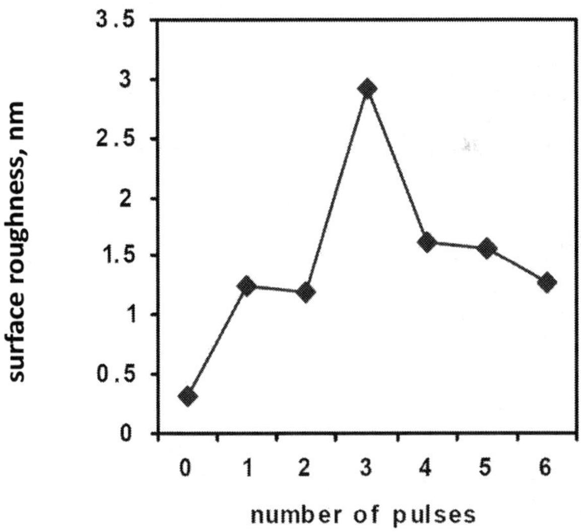

Fig. 15. Surface roughness of silicon thin film with respect to number of pulses

8. Conclusion

In general, the excimer laser annealing is capable to crystallize the silicon thin film based on the increment of the grain size as well as the surface roughness. In the process of annealing, the film experience transient heating and freezing. The UV light of excimer laser carried out energy of 6.42 eV which is equivalent with 7.5×10^4 K that is more than enough to melt the thin film. Considering the temporal length of pulse duration each of laser pulse is only 10 ns, this means the silicon copper thin film is experience an ultra-speed heating and freezing. The energetic excimer is enough to control the phase transformation with the decalesence point in which the melted is associated with some solid form to nucleate the crystal growth. The decalescence effect is controlled via the number of pulses. In contrast during rapid re-solidification the liquid form is associated with recalescence effect which allows the formation of crystallization. The latent heat is released to delay the transformation from liquid to solid. Inherently the recalescence effect responsible to spontaneously crystallize the silicon film. When the energy density greater than the end limit of decalescence point which equivalent with the super lateral growth energy, the heating achieved the peak temperature where the film totally melted. Consequently no recalescence effect occur which result homogenous solidification without crystallization formation during cooling. Thus, in annealing it is crucial to determine the super lateral energy in order to ensure the process is within the decalescence and recalescence points to allow the formation of crystallization.

In addition, laser annealing process of silicon thin film promising well-arranged and large poly-Si grains. As far as we can see, laser annealing has become the most technique used to fabricate a good quality of poly-Si film for ultra-large scale integrated circuit devices, thin film transistors, active matrix liquid crystal displays, solar cells and other optoelectronic devices for industrial purposes. The employment of excimer laser for low temperature poly-Si (LTPS) thin film fabrication makes it possible to replace glass substrate with plastic substrate. This would certainly reduce the cost for large scale productions. Furthermore the high speed crystallization process will shorten the time to anneal as compare to the conventional method. The combination of time for annealing and the low energy consumption will give advantage for the production of high quality silicon thin film.

9. Acknowledgement

This work has been supported from the government of Malaysia via FRGS grant vote 4F001 and GUP vote 00H10. We would like to acknowledge the support given by all technicians during experimental performances.

10. References

Armengol, J.; Vega, F.; Chaoui, N.; Solis, J.; Afonso, C. N. (2003). Recalescence after bulk solidification in germanium film melt by ns laser pulses. J. of Appl. Phys. 93(3):1505

Barbara, J. and Srinivasan, R., (1984). Microscopic Model for the Ablative Photodecomposition of Polymers by Far-Ultraviolet Light Radiation, Appl. Phys. Lett. 44(9):849

Basore, P. A. (2006). "CSG-2: Expanding the production of a new polycrystalline silicon PV technology", Proc. of the 21st European Photovoltaic Solar Energy Conference, http://www.csgsolar.com/downloads/CSG-2%20Basore.pdf.

Bidin N. and Razak S.N. (2011a). Crystallization of Poly-Silicon Film by Different Annealing Techniques. Proc. of Saudi Int. Electronic, communication and Photonic conference SIECPC-2011 Riyadh Saudi Arabia, 24-26 April, 2011

Bidin N. and Razak S.N. (2011b). Nucleation and Growth of Cu Thin Films on Silicon Base Induced By Excimer Laser Annealing (Invited paper). Proc. of International symposium on Functional materials, Bali Indonesia, April 27-28, 2011

Bossuyt, A. S. and Greer L. (2004). Effects of Positive Feedback on Crystallization Kinetics and Recalescence. Mat. Res.Soc. Symp. Proc. Vol. 806, Materials Research Society

Brendel K.; Nickel N. H.; Lengsfeld P.; Schöpke A.; Sieber I.; Nerding M.; Strunk H. P. and Fuhs W. (2003). Excimer laser crystallization of amorphous silicon on metal coated glass substrates. Thin Solid Films. 427(1-2): 86

Brannon, J.; Lankard, J.R.; Baise, A.; Burns I. and Kaufman F. (1985). Excimer Laser Etching of Polymide. J.Appl. Phys. 58:2036

Brannon, J. (1997). Excimer Laser Ablation and Etching: Making a Mark in Many Applications. *Circuit & Devices*, 97: 11

Campbell, S. A. (2001).The Science and Engineering of Microelectronic Fabrication (2nd ed.), New York: Oxford University Press, ISBN 0195136055

Chen,Y.H.; Zheng, H.Y.; Wong, K.S. (1995). Excimer Laser Drilling of Polymers, Proc. SPIE , 3184:202

Csele, M. (2004). Fundamental of Light Sources and Lasers. New Jersey: Wiley-Intersciece

Carluccio R.; Cina S.; Fortunato G.; Friligkos S.; and Stoemenos J. (1997). Structure of Poly-Si Films Obtained By Excimer Annealing. Thin Solid Films. 296: 57

Delmdahl, R. (2010). The excimer laser; precision engineering. Nature photonic. 4: 286

GAMLaser Inc. (2003). EX5 Excimer Laser Manual. Florida: User Manual

Ghosh, A. K.; Fishman, C. and Feng, T. (1980). Theory of the electrical and photovoltaic properties of polycrystalline Silicon. Journal of Applied Physics. 51 (1): 446. doi: 10.1063/1.327342

Green, M. A. (2003). Crystalline and thin-film silicon solar cells: state of the art and future potential. Solar Energy 74 (3): 181. doi: 10.1016/S0038-092X (03)00187-7

Green, M. A. (2004). Recent Developments in Photovoltaics. Solar Energy 76 (1–3): 3. doi:10.1016/S0038-092X(03)00065-3

Hahn, C. H.; Lippert, T. and Wokaun, A. (1999). Comparison of the Ablation Behaviour of Polymer Films in the IR and UV with Nanosecond and Picosecond pulse. J. Phys. Chem. B. 103: 1287

Han, W. H.;Kang, C.J. and Yong, S.K (2003). A Novel Structured Polycrystalline Silicon Thin Film Transistors tha increases the On/Off Current ratio. Semicond. Sci. Technol. 18 (845-849).

He M.; Ishihara R.; Metselaar W. and Beenakker K. (2007). Agglomeration of amorphous silicon film with high energy density excimer laser irradiation. Thin Solid Films. 515(5): 2872

Hecht, J. (1992). Understanding Laser: An Entry -Level Guide. New York: IEEE Press

Lee, H. S.; Choi, S.; Kim, S. W.; Hong, S. H. (2009). Crystallization of Amorphous Silicon Thin Film by Using a Thermal Plasma Jet. Thin Solid Films 517 (14): 4070. doi: 10.1016/j.tsf.2009.01.138

Mariucci L.; Pecora A.; Fortunato G.; Spinella C. and Bongiorno C. (2003). Crystallization mechanisms in laser irradiated thin amorphous silicon films. Thin Solid Films. 427(1-2): 91

Matsumura M. and Oh C. H. (1999). Advanced Excimer-Laser Annealing Process for Quasi Single-Crystal Silicon Thin-Film Devices. Thin Solid Films. 337: 123

Palani I. A., Vasa N. J., and Singaperumal M. (2008). Crystallization and Ablation in Annealing of Amorphous-Si Thin Film On Glass and Crystalline-Si Substrates Irradiated By Third Harmonics of Nd^{3+}:YAG Laser, Materials Science in Semiconductor Processing 11: 107

Razak S.N and Bidin N. (2010). Effect of Multipulses Exposure On Excimer Laser Annealed Polycrystalline Silicon Thin Film. Proc. of 2010 International Conference on Enabling Science and Nanotechnology. Escienano2010, 1-3 December, 2010, KLCC, Malaysia

Sato H. and Nishio S. (2001). Polymer laser photochemistry, ablation, reconstruction, and polymerization, J. Photochem. Photobiol. 2:139

Sauerbrey, R. and Petit, G. H. (1989). Theory of Etching Organics Materials by Ultraviolet Laser Pulses. Applied Physics Letter, 55(5): 421

Schneider, T. L. (1998). Refractive Surgery and the Excimer Laser. Jacksonville Medicine

Serafetinides, A. A.; Makropoulou M. I.; Skordoulis, C. D; and Kar, A. K. (2001). Ultrashort Pulsed Laser Ablation of Polymers. Appl. Surf. Sci. 180: 42

Shah, A.; Meier, J.; Vallat-Sauvain, E.; Wyrsch N.; Kroll U.; Droz C.; Graf U. (2003). Material and solar cell research in microcrystalline silicon", Solar Energy Materials and Solar Cells. 78 (1–4): 469. doi: 10.1016/S0927-0248(02)00448-8

Shah, Z. A. (2009). Surface Treatment of Polymethylmethacrylate by Excimer Laser Ablation. MSc Thesis. Universiti Teknologi Malaysia 2009

Speidell, J. L.; Pulaski D.P.; Patel R.S. (1997). Mask for laser ablation technology: New requirement and challenges, IBM Journal and research development. 8618-8646.

Srinivasan, R. and Mayne-Banton, V. (1982). Self Developing Photoetching of Poly(ethyleneterephtalate) Film by Far Ultraviolet Excimer Laser Radiaion.Appl. Phys. Lett. 41(6): 576

Streetman, B. G. & Banerjee, S. (2000). Solid State Electronic Devices (5th ed.), New Jersey: Prentice Hall. ISBN 0130255386

Takaoka G. H.; Fujita K.; Ishikawa J., and Takagi T. (1989). Grain Boundary Diffusion Control of Cu Films Prepared By The ICB Technique. Nuclear Instruments and Methods in Physics Research B37/38: 882

Viatella J. and Singh R. K. (1997). Transient recrystallization of amorphous silicon films. Materials Science and Engineering B. 47(1): 78

Watanabe, H.; Miki H.; Sugai, S.; Kawasaki K.; Kioka, T. (1994). Crystallization process of polycrstalline silicon by KrF_2 Excimer laser annealing. Jpn. J. Appl. Phys. 33(8): 4491

Yuan, Z.; Lou, Q.; Zhou, J.; Dong, J.; Wei, Y.; Wang, Z.; Zhao, H.g; Wu, G. (2009). Numerical and experimental analysis on green laser crystallization of amorphous silicon thin films. Optics & Laser Technology 41 (4): 380. doi: 10.1016 /j.optlastec.2008.09.003

Zhang X and Atrens A. (1992), Rapid solidification characteristics in melt spinning. Materials Science and Engineering: A. 159(2): 243. doi:10.1016/0921-5093(92)90295.c

19

Oriented Lateral Growth and Defects in Polycrystalline-Silicon Thin Films on Glass Substrates

Kuninori Kitahara[1] and Akito Hara[2]
[1]*Shimane University*
[2]*Tohoku Gakuin University*
Japan

1. Introduction

Silicon (Si) thin films on glass substrates have been extensively developed as a semiconductor material for electronic devices. This material is especially useful for large-area panel devices such as thin-film transistors (TFTs) on active-matrix flat panel displays. The most widely used Si films are hydrogenated amorphous Si (a-Si:H), which can be deposited at temperatures lower than the strain point of the substrate. However, improved electronic properties are required to achieve higher device performance. Using polycrystalline Si (poly-Si) films instead of a-Si:H films enhances carrier mobility by two or three orders of magnitude; thus, driver circuits can be incorporated into display panels, as shown in Fig. 1. The application of poly-Si will be extended to mobile displays with large pixel density, microprocessor–display combined panels, and thin-film solar cells.

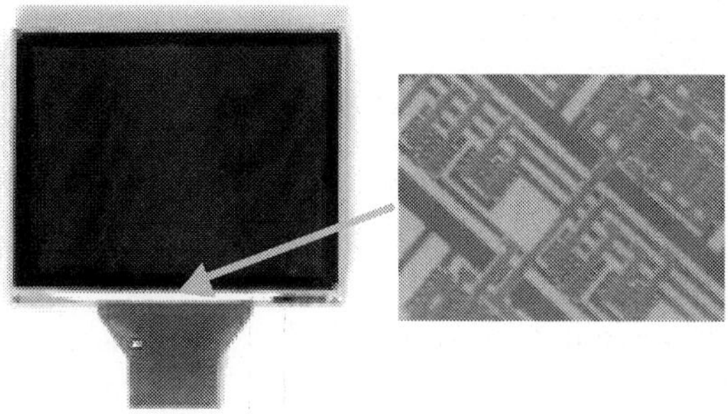

Fig. 1. Low-temperature poly-Si liquid crystal display (2 in. diagonal). Driver circuits are integrated at the periphery of the panel.

The poly-Si used for TFT must be ≤ 50 nm thick to ensure the desired device performance. Furthermore, the crystalline fraction should be almost 100%. Such thin films cannot be deposited directly on glass; they must be formed by recrystallization of a-Si precursor films. For this purpose, manufactures have employed solid-phase crystallization (SPC) and excimer-laser crystallization (ELC) techniques, as shown in Fig. 2 (Hayashi et al., 1984; Sameshima et al., 1986).

The grain boundaries in poly-Si thin films typically feature random configurations. However, random grain boundaries were reported to severely degrade device performance (Blake et al., 1997). Therefore, grain enlargement is desirable. To achieve this, lateral growth during crystallization is necessary. The lateral growth described in this manuscript corresponds to oriented overgrowth of the crystalline film on a non-crystalline layer (i.e., the glass substrate), which is regarded as a type of epitaxy (Givargizov, 1991). Epitaxy proceeds so as to minimize the free energy and is expected to effectively reduce the defect density.

In this work, we investigate the relationship between lateral growth and defects using various characterization techniques. First, we give an overview the crystallization process on non-crystalline substrates and the characterization of defects in poly-Si thin films. Next, we describe previous studies and our work on SPC, ELC, and flow-shaped growth of poly-Si on glass from the perspective of lateral growth. Finally, the growth of a quasi-single-crystal SiGe thin film on a glass substrate, that is super-lateral growth, is demonstrated.

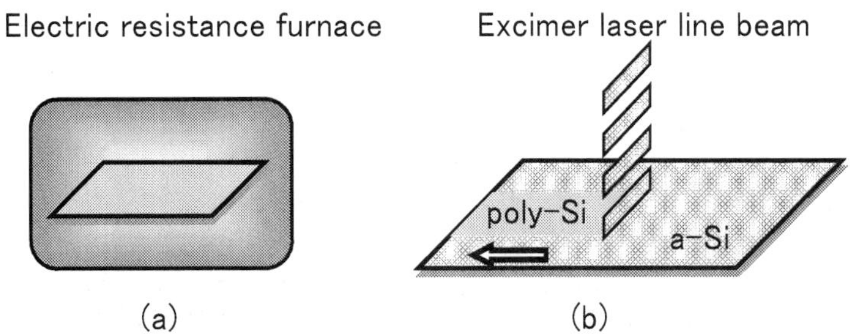

Fig. 2. Techniques for recrystallization of a-Si films on glass substrates using (a) solid-phase crystallization and (b) excimer-laser crystallization.

2. Overview of lateral crystallization on non-crystalline layer

The basic technology for poly-Si formation on a glass substrate is Si on insulator (SOI), which was developed for applications such as low power consumption large-scale integrated circuits (LSIs), three-dimensional LSIs, radiation-hardened electronics and solar cells. SOI technologies have been reviewed in detail elsewhere (Givargizov, 1991; Colinge, 2004). The substrate used for SOI is typically single-crystalline Si (c-Si) coated with a dielectric film such as SiO_2 or Si_3N_4. Although SOI technology includes a wide range of methods, we focus on the recrystallization of a-Si thin films. The principle SOI techniques used for this purpose are zone-melting recrystallization and lateral SPC.

Zone-melting recrystallization is performed by applying a heat source to an a-Si film, which induces melting followed by solidification of the film. Common heat sources are line-shaped electric resistors, electron beams, and continuous-wave (CW) Ar ion lasers. Lateral SPC is also performed by zone heating; its advantages include low T_s (500–700 °C) and consistent surface smoothness.

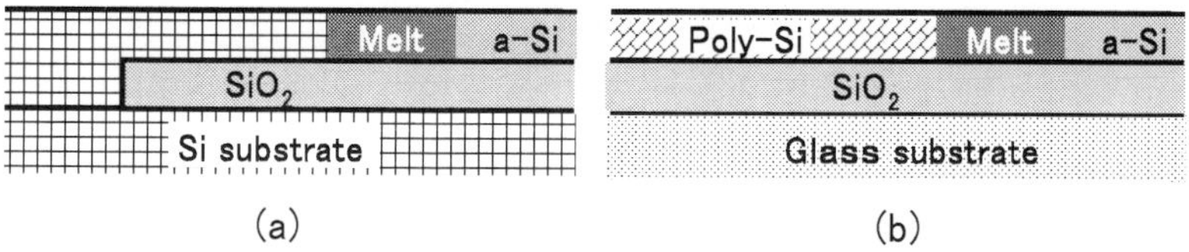

Fig. 3. Structures of Si layers on non-crystalline layers: (a) Si on insulator, where Si substrate is coated by SiO_2 film with windows for seed areas: (b) excimer-laser crystallized Si, where glass substrate is fully coated by SiO_2 film.

Control of nucleation during the initial stage of lateral growth and crystalline orientation is essential for the growth of single or quasi-single crystals. To achieve this control, the non-crystalline underlayer is partially opened to reveal the crystalline Si substrate, which acts as the seed area, as shown in Fig. 3(a). Alternatively, the fundamental layers are trenched in order to form a grating-like shape to guide the crystalline orientation; this is known as artificial epitaxy or graphoepitaxy. However, the structures thus created require additional photolithographic processing. In contrast, the laser crystallization process enables control of the crystalline orientation without seeding or photolithography; however, the obtained crystalline quality is not always adequate for the submicron-scale processes required for LSI.

For applications in electronic displays, the substrate should be glass or plastic rather than c-Si. On these substrates, Si thin films become essentially polycrystalline because of the seedless growth, as shown in Fig. 3(b). Poly-Si on glass has been applied in devices called giant microelectronics or large-area electronics (Kuriyama et al., 1992; Sameshima, 2009), which have areas several orders of magnitude larger than those of SOI devices. The use of glass or plastic substrates strongly restricts the upper limit of T_s. Technologies are classified as high-temperature poly-Si or low-temperature poly-Si depending on T_s (Blake, 1997). The border between them is determined by the strain point of a low alkali glass substrate at ~590 °C (Corning's technical glass catalog).

SPC requires a T_s that is higher than 600 °C. Thus, fused quartz glass is used as the substrate because its strain point is as high as 990 °C. SPC poly-Si is used mainly in 1-in. diagonal or smaller liquid crystal displays (LCDs) for multimedia projectors.

Laser crystallization technology decreases T_s to below the strain point of low alkali glass substrates and in many cases to room temperature (RT) owing to selective heating of the Si layer. Progress in laser annealing has been reviewed elsewhere (Sameshima, 2009). Poly-Si films for medium-sized LCD panels suitable for mobile information terminal applications have been formed.

There are two opposing requirements for laser crystallization: ensuring uniformity in order to integrate TFTs, and eliminating grain boundaries in order to improve the performance of individual TFTs. The former and latter requirements are met by the growth of grains that are notably smaller and larger than the TFT channels, as shown in Fig. 4(b) and 4(c), respectively. Those opposing needs are expected to be reconciled by the growth of flow-shaped grains, as shown in Fig. 4(d).

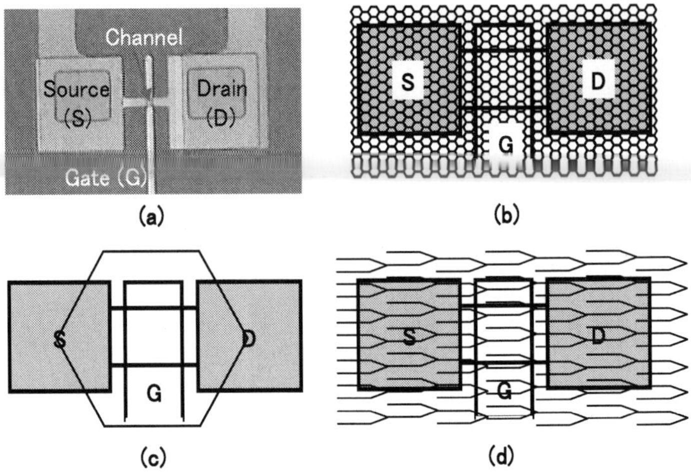

Fig. 4. Structures of thin-film transistors (TFTs) on glass substrates: (a) test device for characterizing electrical performance; geometry of grains and TFT channel for (b) grains notably smaller than the channel region, (c) a large grain containing the channel region, and (d) flow-shaped grains running parallel to the drain current.

3. Characterization of poly-Si films on glass

The principal terms for the characterization of poly-Si thin films on glass substrates are summarized in Table 1. The dangling bonds in a-Si and poly-Si films were qualitatively analyzed via the spin density of unpaired electrons using electron-spin resonance. The spin density was found to be reduced by hydrogenation (Nickel et al., 1997; Spinella et al., 1998), which causes termination of the dangling bonds with H atoms. The geometry and crystal orientation of the grains were investigated using images and diffraction patterns obtained by scanning electron microscopy (SEM), transmission electron microscopy (TEM), and electron backscattering diffraction (EBSD). When growth accompanies melting, the surface morphology can vary with the displacement of the melt. The surface morphology is observed using atomic force microscopy (AFM) or cross-sectional TEM. The electronic characteristics were estimated by fabricating test n-channel and p-channel TFT devices, as shown in Fig. 4(a). The devices exhibit characteristics such as the field-effect electron and hole mobilities (μ_{FEn} and μ_{FEp}, respectively), threshold voltage, subthreshold voltage, and off-current.

The characterization techniques described above require the specimen be processed. Moreover, they are not simple enough to be used to characterize large numbers of films. In contrast, optical techniques are suitable for simple, non-destructive characterization. We analyzed the defects and stress in a poly-Si/glass system by Raman scattering spectroscopy (Kitahara et al., 2002, 2003, 2011a). Micro-Raman spectroscopy was conducted by using a Renishaw System 1000 in the backscattering geometry under excitation with a 514.5 nm Ar ion laser. Details of the setup and band component analysis were described elsewhere (Frost & Shurvell, 1997). The full width at half maximum (FWHM) values of spectra obtained for poly-Si (described below) were calibrated according to the spectral resolution of the setup. The peak frequency of the spectra was exactly determined by fitting to a Lorentzian curve.

Typical first-order Raman scattering for c-Si, Si on sapphire (SOS), and poly-Si on glass are shown in Fig. 5. In the crystalline phase, the Raman spectrum exhibits an isolated band corresponding to the degeneration of a single transverse-optical (TO) phonon mode and two longitudinal-optical (LO) phonon modes; the c-Si band exhibits a linewidth as narrow as 2.5–3.5 cm^{-1} at RT (Temple & Hathaway, 1973; Menendez & Cardona, 1984) owing to the $q = 0$ (q: wave vector) selection rule. In the amorphous phase, four continuous phonon bands are observed; they are broad because the $q = 0$ selection rule is loosened as a result of a decrease in the phonon correlation length.

The features of the optical phonon mode (OPM) of the crystalline phase are represented by the peak frequency (ω), FWHM, intensity, and in some cases asymmetry. When the grains are smaller than a few tens of nanometers, the spatial correlation of the phonon decreases, which causes the peaks to shift to a lower frequency ($\Delta\omega$) and increases the FWHM (Richter et al., 1981). Compressive and tensile stress in the films cause the peak to shift to higher and lower frequencies, respectively. Moreover, $\Delta\omega$ is proportional to the magnitude of the stress (Englet, 1980). Therefore, the stress can be quantitatively estimated only when the impact of the other factors is negligible. Note, however, that the intensity of the stress depends on the crystallization technique and conditions used. Therefore, an analysis of individual crystallization techniques should be developed by comparing the Raman spectroscopy results with those obtained by SEM and TEM.

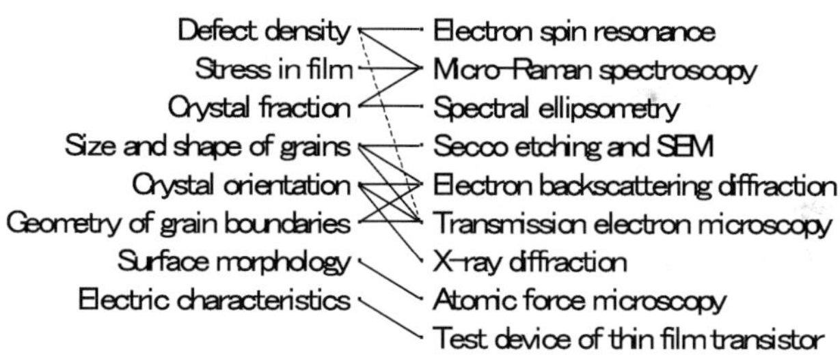

Table 1. Principal terms for characterization of poly-Si thin films on glass substrates.

Defects in poly-Si are frequently observed in SEM images. Grain boundaries and some defect clusters are clearly revealed by Secco etching (Secco d'Aragona, 1972). These defects are evidently electrochemically active, which is confirmed by the fact that they disappear upon hydrogenation before etching. On the other hand, inactive defects such as twin boundaries are observed only as shallow contrasts and are independent of hydrogenation (Kitahara et al., 2009a).

To summarize, various electron microscopies provide direct knowledge of defects and the geometric configuration of grains in poly-Si/glass systems. However, defects cannot always be detected by those methods. Raman microscopy is useful for easy macroscopic (submicron-scale) defect analysis and quantitative analysis of stress in films. However, semi-empirical analysis aided by other techniques is required to simultaneously characterize the defect density and stress in films by Raman spectroscopy.

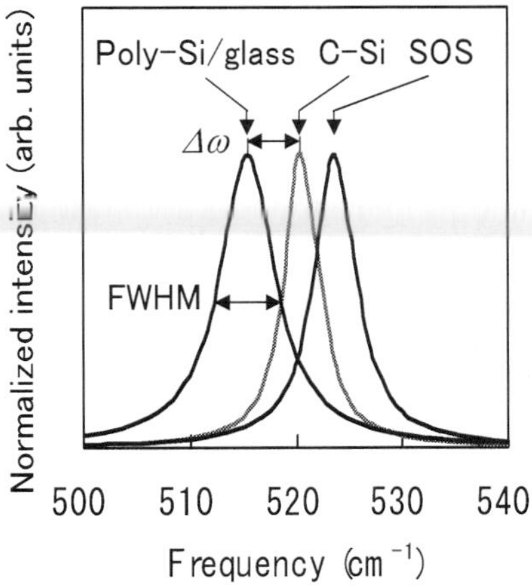

Fig. 5. Typical optical-phonon mode observed in the Raman scattering spectra for Si on sapphire (SOS), single crystalline Si (c-Si) bulk containing no stress, and poly-Si thin film on glass. Intensity is normalized to the individual peak values. Peaks are shifted to higher and lower frequencies by compressive and tensile stresses, respectively.

4. Solid-phase crystallization

The heat source for SPC typically consists of an electric resistance furnace and an infrared lamp annealing furnace. The force that drives crystallization of a-Si in the metastable state is the lowering of the Gibbs free energy through the crystalline phase change. SPC poly-Si technologies are reviewed in detail elsewhere (Hatalis & Greve, 1988; Spinella et al., 1998).

The typical grain size of solid-phase crystallized poly-Si for TFTs is a few hundred nanometers, which is at least several times larger than the film thickness. This suggests that SPC on non-crystalline substrates is dominated by lateral growth rather than columnar growth which is frequently seen in chemical vapor deposition.

SPC is initiated by homogeneous nucleation after incubation. In pure Si, the activation energy of nucleation is larger than that of growth by 0.25 eV (Spinella et al., 1998). Therefore, growth proceeds immediately after nucleation. The lower limit of the nucleation temperature depends on the quality of the precursor film and is ~560 °C for highly pure a-Si. The grain size decreases with increasing crystallization temperature (T_c), which is associated with increasing nucleation frequency. Lateral grain growth is first interrupted by collisions among grains. Then, the grain size increases due to secondary grain growth; the secondary growth rate increases with the grain boundary energy, surface energy anisotropy, grain boundary mobility, and T_c (Thompson, 1985). Furthermore, twin formation at the early stage of crystallization preferentially accelerates the lateral growth rate in the <112> direction, producing grains with an ellipsoidal outline (Nakamura et al., 1989). Growth along the minor axis of the ellipsoid, i.e., in the <111> direction, proceeds with simultaneous formation of many micro-twins (Drosed & Washburn, 1980). Although solid-phase crystallized poly-Si films have a considerable density of defects, hydrogenation of TFTs effectively inactivates those defects, producing a μ_{FEn} of 24 cm^2/Vs (Little et al., 2001).

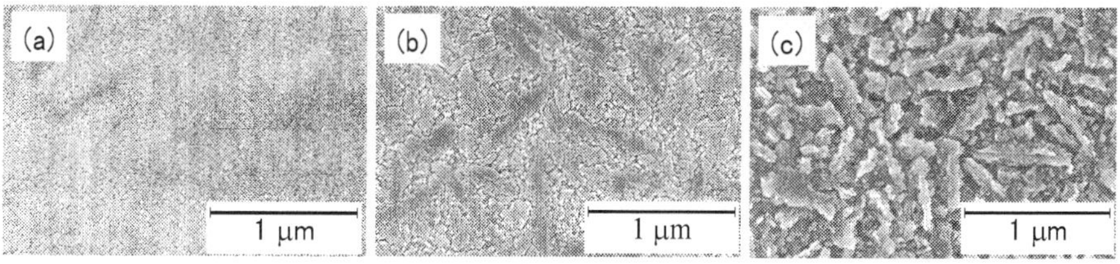

Fig. 6. Scanning electron microscopy images for solid-phase crystallized poly-Si heated at 900 °C for 1.5 h. Images were taken (a) before and after Secco etching for (b) 20 s and (c) 60 s.

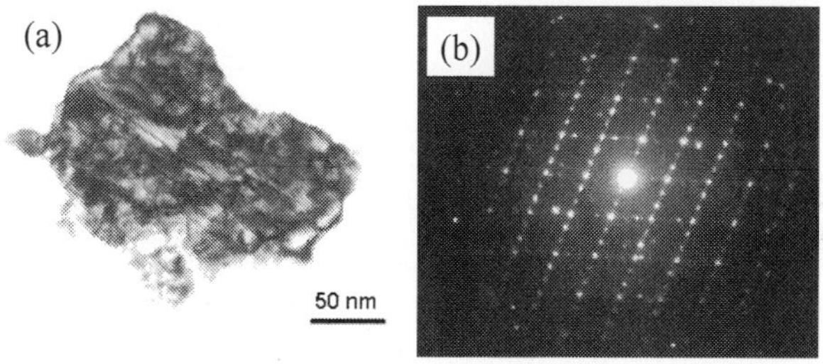

Fig. 7. Transmission electron microscopy image and diffraction pattern for solid-phase crystallized film heated at 900 °C for 3 h.

We investigated the microscopic geometry of the grains and defects by SEM and TEM. We then studied the phase variation at the initial stage macroscopically using Raman spectroscopy and spectroscopic ellipsometry. SPC was performed in a N_2 flow at atmospheric pressure. Figure 6 shows SEM images before and after Secco etching after SPC at T_c = 900 °C for 1.5 h. The surface of the as-crystallized film is relatively smooth. Grain boundaries were not clearly revealed by the short etching time (20 s) owing to the fine structures at the grain boundaries and the overlapping of grains (Spinella et al., 1998). The longer etching time (60 s) revealed feather-like grains lying parallel to the substrate. In addition, major twin boundaries appeared in the individual grains as shallow lines. The plan-view TEM image shown in Fig. 7 indicates that the twin boundaries extend along the major axis. Furthermore, a large density of micro-twins is distributed in the grains. The subspots and streaks evident in the diffraction pattern also indicate the presence of twins and {111} microfacets in the grains.

Figure 8 shows the OPMs in the Raman spectra during the initial stage of SPC at T_c = 580 °C. Only the amorphous phase is detected at 1 h. The crystal component begins to appear in the spectrum after a latent time of ~3 h. Figure 9 shows the crystal fraction in the 580 °C SPC film as a function of the heating time. In Raman spectroscopy, the volume fraction of the crystal was estimated by the area-intensity ratio of the amorphous (I_a) and crystal (I_c) components, i.e., $I_c/(I_c + I_a)$, which yields a relative value. In ellipsometry, the volume fraction of the crystal was analyzed using the Tauc-Lorentz model (Jellison, 1998), which yields an absolute value. Although no crystal component was found by Raman spectroscopy until 3 h had passed, the volume fraction estimated by ellipsometry began to increase at 1 h. The spectra of the imaginary dielectric function observed by ellipsometry show that the E1 and E2 bands were somewhat visible even in the latent time. This suggests that atomic reconstruction toward ordering before the appearance of the crystal was detected by ellipsometry.

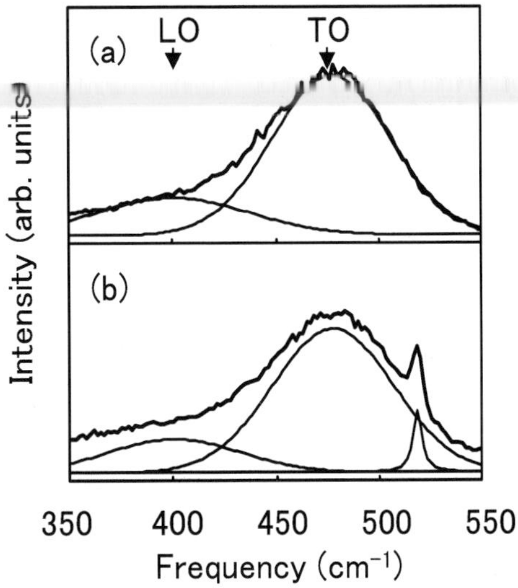

Fig. 8. Optical phonon modes in the Raman spectra for the initial stage of solid-phase crystallized film heated at 580 °C for (a) 1 h and (b) 3 h. Thick lines and thin lines are experimental values and fitting curves, respectively.

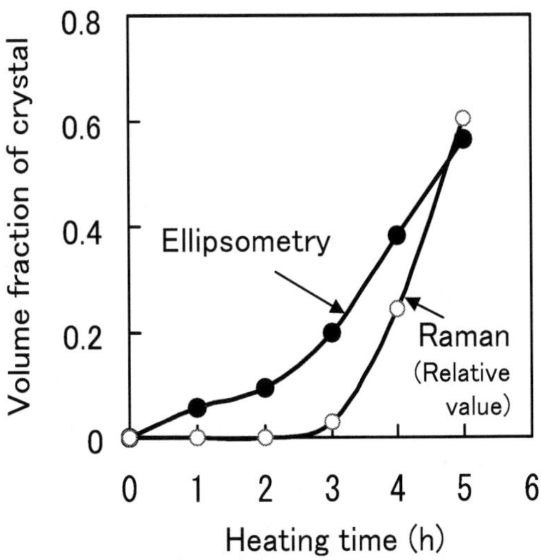

Fig. 9. Relationship between crystal fraction and heating time for solid-phase crystallized film heated at 580 °C. Crystal fraction was estimated using spectral ellipsometry and Raman spectroscopy.

Figure 10 shows the relationship between the FWHM of the OPM and T_c in the range of 625–900 °C with annealing times of 3–8 h; the times were varied in order to achieve sufficient crystallization. The FWHM evidently decreases with increasing T_c. Thus, the defect density decreases with increasing T_c, which is a universal tendency for crystal growth under essentially stable thermal equilibrium. The plots for metal-induced lateral crystallization (MILC) will be described in the next section.

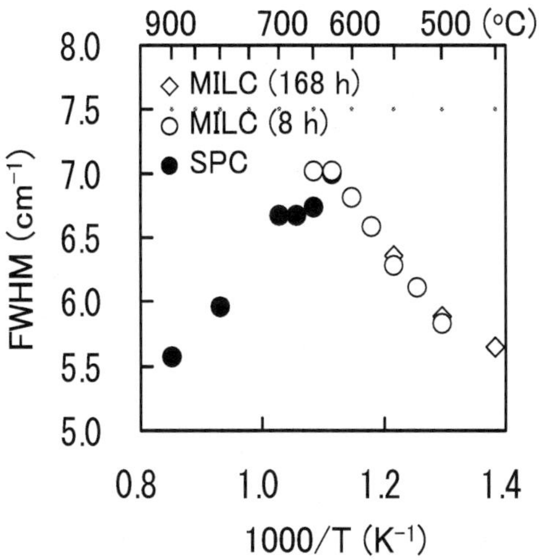

Fig. 10. Relationship between the full width at half maximum (FWHM) of the optical phonon mode and the crystallization temperature (T_c). Plots for poly-Si films fabricated by metal-induced crystallization (MILC) and solid-phase crystallization (SPC) are shown. Note that the dependence on T_c of MILC is opposite to that of SPC.

To summarize, Si atoms cause reconstruction toward ordering in the amorphous phase in the initial stage of annealing. SPC begins at homogeneous nucleations after the incubation time. Growth begins immediately after nucleation. The grain size increases with acceleration of lateral growth by the major twin boundaries formed at the early stage and the coalescence of grains by secondary grain growth. Growth proceeds with simultaneous formation of many micro-twins, resulting in many defects. Increasing T_c decreases the defect density in grains as the grain size is reduced.

5. Metal-induced lateral crystallization

The nucleation temperature of a-Si is decreased by the addition of a catalytic metal; the process is referred to as MILC and catalyst-assisted SPC (Kawatsu et al., 1990; Cammarata & Thompson, 1990; Takayama et al., 2000). Although metals such as Al, Mo, Ni, Pd, and Ti have been applied to cause MILC, Ni is the most frequently used catalyst. Ni can be supplied to the seed area of an a-Si film by evaporation, sputtering, or application of an acetate solution. Annealing of the film causes lateral diffusion of Ni atoms into the amorphous area before the silicide reaction occurs. $NiSi_2$ begins to segregate at $T_c > 420$ °C. Although $NiSi_2$ has the fluorite crystal structure, which is not the same as that of Si, the mismatch of their lattice constants can be as small as 0.4%. Thus, $NiSi_2$ operates as an excellent nucleus for Si growth. $NiSi_2$ seeds then migrate laterally, leaving needle-like Si crystals in the a-Si. The activation energies reported for the growth rate of Ni-MILC, which range from 1.3–2.3 eV (Makihara et al., 2003; Kitahara et al., 2009b), are smaller than those for SPC (3.4 eV) and solid-phase epitaxy (2.7 eV) (Roth et al., 1990; Spinella et al., 1998). Lower values of T_c lead to straighter crystallization. Annealing at 450 °C reportedly results in crystals 10 μm in length and 160 nm wide (Makihara et al., 2003). Needle-like crystals coalesce and form larger grain when the remaining a-Si in the interstitial area is crystallized via SPC.

The application of MILC to devices is not easy because of the long processing time required and the existence of silicide in the crystallized region. However, TFTs have been fabricated by combining MILC with ELC. In this method, the silicide was gettered to ion-implanted regions corresponding to the source and drain electrodes and achieved a μ_{FEn} value of 320 cm^2/Vs (Mizuki et al., 2004).

The density and geometry of the residual defects in MILC films are expected to differ from those in SPC films because MILC is more dominated by lateral growth than SPC is. We investigated the difference in defects between SPC and MILC (Kitahara et al., 2009b). MILC was performed by evaporating Ni onto an a-Si/SiO_2/fused quartz glass substrate through a metal mask with rectangular windows. The thicknesses of Ni and a-Si were 13 and 110 nm, respectively. The plan-view TEM image after MILC at 450 °C for 168 h is shown in Fig. 11(a). The image exhibits needle-like crystals with a {110} plane extending toward <111>. Note that the grains exhibited little contrast in the strain corresponding to defects. The coexistence of sharp diffraction spots and a halo in Fig. 11(a') indicate that the grain is a single crystal with a surrounding amorphous component. The TEM image after MILC at 500 °C for 168 h [Fig. 11(b)] shows accumulated needle-shaped grains with strain contrast due to defects. Small subspots in the diffraction pattern shown in Fig. 11(b') also indicate the presence of defects in the grains. However, the streaks corresponding to microfacets that were seen for the SPC films are not observed here. These results suggest that the number of defects in the grains decreases with decreasing T_c, which is opposite to the trend for SPC.

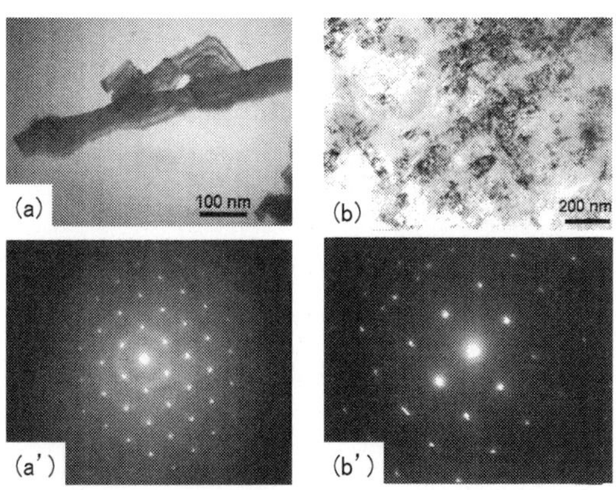

Fig. 11. Transmission electron microscopy images for (a) metal-induced crystallization (MILC) at 450 °C for 168 h and (b) MILC at 500 °C for 168 h. Individual diffraction patterns are also shown in (a') and (b').

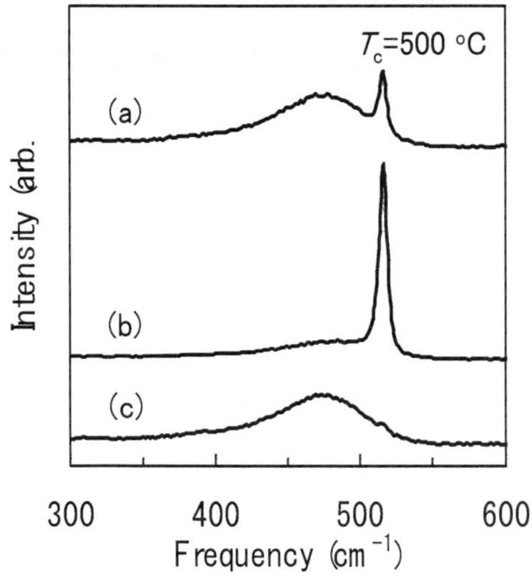

Fig. 12. Typical Raman spectra for (a) Ni-evaporated source region, (b) lateral growth region, and (c) uncrystallized region.

Typical Raman spectra for a Ni-evaporated source region, lateral growth region, and uncrystallized region are shown in Figs. 12(a)-(c). Here, MILC was conducted at 500 °C for 8 h. The spectrum for the Ni-evaporated region indicates the presence of a large amorphous fraction. In this area, the Ni density is likely too high to effectively induce lateral crystallization. In contrast, in the majority of the lateral growth region, the OPM is accompanied only by a weak amorphous-like mode. The uncrystallized region naturally remains amorphous. These results suggest that the amorphous-to-polycrystalline transition during MILC requires both a Ni supply and the lateral growth process.

The relationship between T_c and the FWHM of the OPM for the lateral growth region is shown in Fig. 10. The FWHM decreases with decreasing T_c and reaches 5.7 cm^{-1} at 450 °C, which is nearly equal to the FWHM of SPC at 900 °C, 5.6 cm^{-1}. A factor other than thermal energy is responsible for the T_c dependence of MILC because the dependence on T_c is opposite to that of SPC. In MILC, the Ni atoms exhibit a lateral gradation of density in a-Si due to diffusion from the source region. In particular, at T_c < 500 °C, the observed lateral growth length is smaller than the calculated Ni diffusion length in a-Si.

We concluded that the lateral growth at relatively low T_c is directed by the gradation in Ni density. Furthermore, the growth at T_c as low as 450 °C is dominated by needle-like lateral epitaxy and results in a low defect density. Increasing T_c restricts the length of straight growth and also enhances random growth due to SPC; these two effects increase the defect density despite the increase in T_c.

6. Excimer-laser crystallization

Crystallization is performed primarily by KrF or XeCl excimer lasers, which supply intense pulsed light with durations of ~30 ns and wavelengths of 249 and 308 nm, respectively. The luminous flux is typically shaped as a linear beam. Overlapping irradiation by the scanning pulsed light enables uniform crystallization over a wide area. The laser energy density is adjusted to be slightly lower than that at which complete melting of the Si film occurs. In practical use, T_s is maintained at around RT during crystallization; this forms grains a few hundred nanometers in size with high reproducibility.

Because of the short pulse duration, T_s does not almost increase. However, rapid cooling after laser irradiation is an obstacle to lateral growth. Some techniques, including double beam irradiation, have been proposed to decrease the cooling rate. Applying multiple laser irradiations at $T_s < 400$ °C reportedly increases the grain size to 4.5 μm and strongly aligns the surface orientation to the <111> direction (Kurimiya et al., 1993).

One of the features of ELC is a rough surface morphology containing a large density of hillocks. The hillocks are generated by a positive feedback effect between the beam-induced periodic surface roughness pattern and the interference in subsequent pulses (McCulloch & Brotherton, 1995). Another feature of ELC is that μ_{FEn} is considerably higher than that of SPC. The μ_{FEn} value of an excimer-laser crystallized poly-Si TFT increases with increasing grain size and reaches 320 cm^2/Vs at an average grain size of 700 nm, at which the dominant factor determining μ_{FEn} varies from grain boundary scattering to lattice scattering (Hara et al., 2002a). Furthermore, a μ_{FEn} value of 914 cm^2/Vs was reportedly obtained by position-controlled large grain growth (Mitani et al., 2008).

The large μ_{FEn} value of excimer-laser crystallized poly-Si has been attributed to a low defect density in grains owing to a regrowth procedure through the liquid phase. In practice, TEM images typically exhibit few defects in the grains. The average dislocation density was reportedly 8×10^6 cm^{-2} for large (2–4 μm) grains and 10^8–10^9 cm^{-2} even for defective grains (Christiansen et al., 2001). However, the cooling velocity of ELC can be as high as ~10^{10} K/s (Sameshima & Usui, 1993). Therefore, recrystallization is expected to proceed under significant deviations from thermal equilibrium; consequently, large numbers of defects are frozen in the grains. In defects in bulk c-Si, impurity complexes generated at high temperature were shown to be frozen by quenching at a rate of ~10^3 K/s (Takahashi et al., 1995). The cooling rate of ELC is far larger than that in the quenching experiment. Hence, in ELC, defects undetectable by TEM are presumed to reside in the grains.

We studied defects in excimer-laser crystallized poly-Si (Kitahara et al., 2007, 2009a). We irradiated a-Si films (50 nm thick) on SiO_2-coated fused quartz substrates with a XeCl excimer laser with 95% overlap. Figure 13 shows the surface of the excimer-laser crystallized poly-Si observed by SEM after Secco etching. Many hillocks appear as bright spots, as shown in Fig. 13(a); they are located at the junctions of grain boundaries. The hillock interval under the optimum conditions for ELC is nearly equal to the wavelength of the laser. AFM images indicated that ridges also lay along grain boundaries.

The grain boundaries were more evident after a longer etching time, as shown in Fig. 13(b). In contrast, grain boundaries did not appear in the hydrogenated film even after a long etching time, as shown in Fig. 13(c). This result implies that the grain boundaries were electrochemically inactivated by hydrogenation. On the other hand, structures other than grain boundaries were detected as shallow lines independently of hydrogenation, which could be due to the lack of electrochemically active dangling bonds.

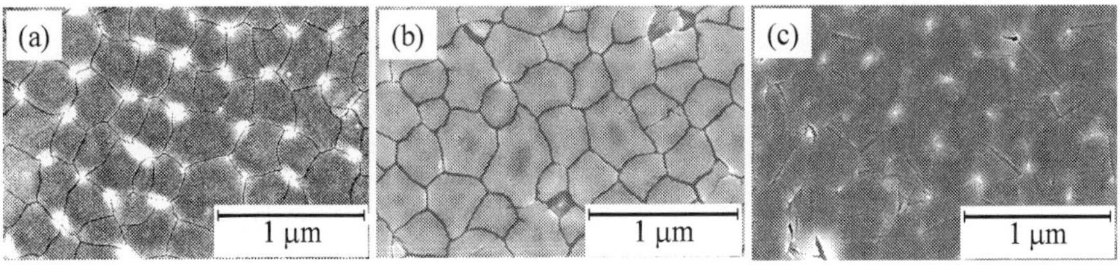

Fig. 13. Scanning electron microscopy images for (a, b) non-hydrogenated and (c) hydrogenated excimer-laser annealed poly-Si films. Secco etching was performed for (a) 15 s and (b, c) 40 s. Hydrogenation interfered with the appearance of grain boundaries after etching.

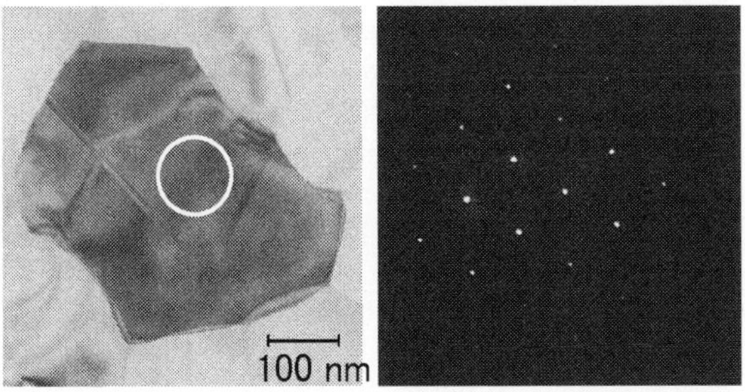

Fig. 14. Typical transmission electron microscopy image and diffraction pattern for a grain of excimer-laser crystallized poly-Si; diffraction pattern is that in the circled region..

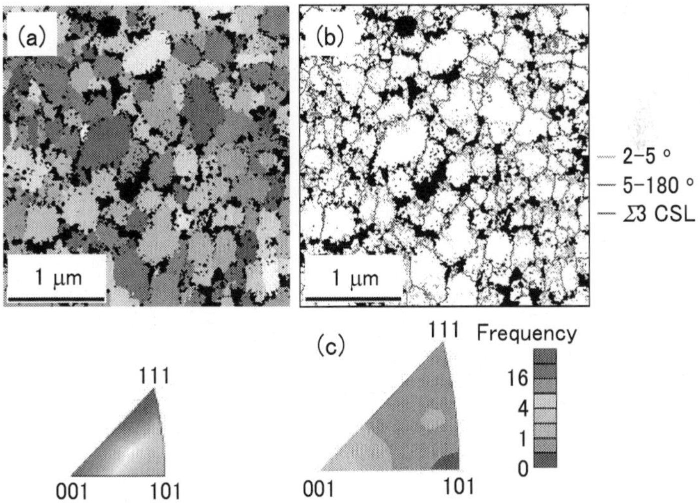

Fig. 15. (a) Crystalline orientation map in terms of the normal direction, (b) grain boundary map, and (c) inverse pole diagrams of the excimer-laser crystallized poly-Si, as observed by electron backscattering diffraction. Surface exhibits weak orientation to the {001} plane. Twin boundaries appear for some grains.

A typical plan-view TEM image and diffraction pattern of a grain of excimer-laser crystallized poly-Si are shown in Fig. 14. The major twin boundary lies across the grain. However, unlike the results for SPC, no micro-twin was found in the grain. Dark contrast due to dislocation, defect clusters, and stacking faults tends to appear at the periphery of the grain.

The EBSD pattern of the surface normal direction (ND) is shown in Fig. 15. Although the surface orientation is scattered over a wide range, the {001} orientation has the highest frequency. A low index plane tends to exist owing to the small growth rate. The grain boundary map indicates that most grain boundaries are random and a quarter of the grains contain twin boundaries corresponding to the Σ3 coincident-site lattice (CSL). The twin boundaries in other grains might be eliminated by grain coalescence due to secondary grain growth.

Raman spectroscopy was applied in order to obtain macroscopic and microscopic characterization of excimer-laser crystallized poly-Si. Figure 16 shows the variation in the FWHM and $\Delta\omega$ of the OPM with the energy density of the laser used for crystallization. The space correlation model (Richter et al., 1981) is useful for analysis of Raman spectra in order to deduce the decrease in the regularity of the crystalline structure, i.e., the presence of nanocrystals and large density defects. The relationship between the FWHM and $\Delta\omega$ of the OPM is plotted in Fig. 17, which also shows a line calculated using the space correlation model. The experimental plots exhibit the same tendency as the calculated values but are shifted to a lower frequency by ~3 cm^{-1}. The magnitude of that deviation is close to the frequency shift caused by tensile stress in the film, which is induced by shrinkage of the film during solidification. The large values of both the FWHM and $\Delta\omega$ at low laser energies are due to the formation of nanosized crystals. At higher laser energies, the space correlation is determined by the defect density because the grains are sufficiently large. This suggests that the defect density is greater than 10^{12} cm^{-2}, which is estimated on the basis of the space correlation length deduced from the FWHM.

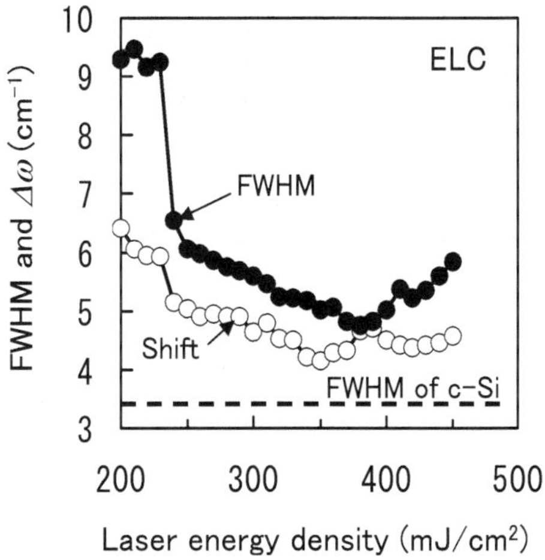

Fig. 16. Full width at half maximum (FWHM) and peak shift ($\Delta\omega$) of the optical-phonon mode as a function of the laser energy density used for crystallization.

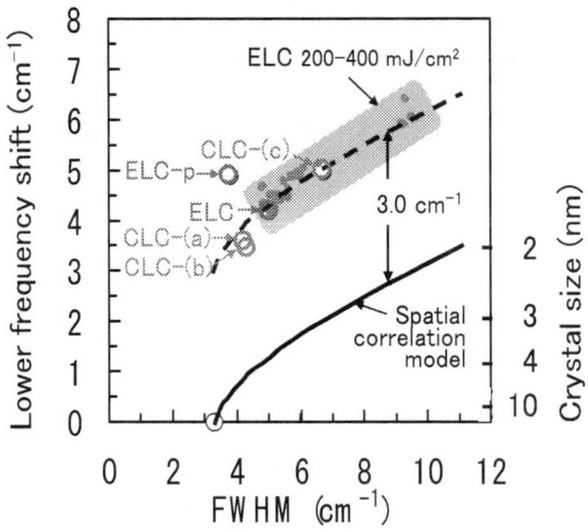

Fig. 17. Relationship between full width at half maximum (FWHM) and lower frequency shift ($\Delta\omega$) for excimer-laser crystallized poly-Si films irradiated at various energy densities. Plots for the film crystallized under the optimum conditions (excimer-laser crystallization, ELC) and that polished to remove hillocks and ridges on the surface (ELC-p) are also shown. Solid line is calculated from the space correlation model; broken lines are values shifted to larger $\Delta\omega$ by 3 cm^{-1}. Plots for continuous-wave laser lateral crystallization (CLC) will be described in the next section.

The Raman spectra for excimer-laser crystallized poly-Si exhibit remarkable features. For example, the intensity of Raman scattering is more than ten times that for SPC films with the same thickness and even that of bulk c-Si. This enhanced Raman scattering has been reported for a roughened semiconductor surface, Si nanocones, and Si nanotubes (Sridharan et al., 2003; Jayavel et al., 2006; Cao et al., 2006). Figure 18 shows a two-dimensional map of the OPM intensity for excimer-laser crystallized poly-Si. The observed area consists of a few micrometer-size grains enlarged by super-lateral growth (SLG) and nanocrystals adjacent to the SLG region (Im & Kim, 1993). The OPM intensity is clearly enhanced at grain boundaries. Thus, the large intensity for the excimer-laser crystallized films is attributed to the enhancement of Raman scattering by hillocks and ridges around the grain boundaries. Therefore, the Raman spectra of excimer-laser crystallized poly-Si reflect mainly the situation around grain boundaries. Accordingly, polishing the sample to remove the hillocks and ridges decreases the intensity to nearly one-tenth. Then, the Raman spectra begin to reflect the region inside the grain. The FWHM changes from 5.0 to 3.8 cm^{-1} after polishing. This supports the conclusion that Raman spectra of unpolished specimens reflect mainly the situation around grain boundaries, where the high defect density increases the FWHM.

The other feature of ELC is the hydrogenation effect. Excimer-laser crystallized poly-Si films with and without hydrogenation were examined in terms of the OPM. Figure 19 shows the variation in $\Delta\omega$ with Secco etching time; the results of SPC are also shown for comparison. The non-hydrogenated film exhibits relaxation of the tensile stress after a short etching period; the relaxation is due to penetration of the etching solution through clusters of defects in the grains to the poly-Si/SiO_2 interface. Defects inside the grain tend to accumulate rather than remain as point defects. In contrast, the hydrogenated excimer-laser crystallized film exhibits almost no stress relaxation. Therefore, a large number of electrochemically active defects are assumed to be present in the grains even though they were not detected by TEM. In the solid-phase crystallized films and excimer-laser crystallized films annealed at 1000 °C, the reduction in stress with etching was not as abrupt, which implies that the thermal treatment reduced the number of dangling bonds in the grains.

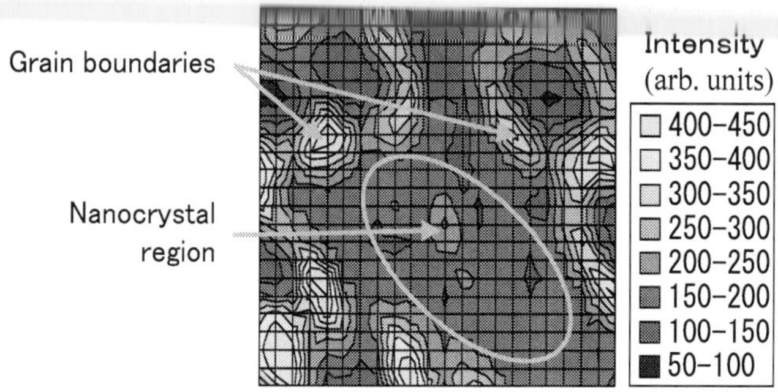

Fig. 18. Two-dimensional map of optical-phonon mode intensity for excimer-laser crystallized poly-Si consisting of a few micrometer-sized grains enlarged by super-lateral growth. View area is 4.2 × 4.2 μm². The intensity is apparently large at grain boundaries.

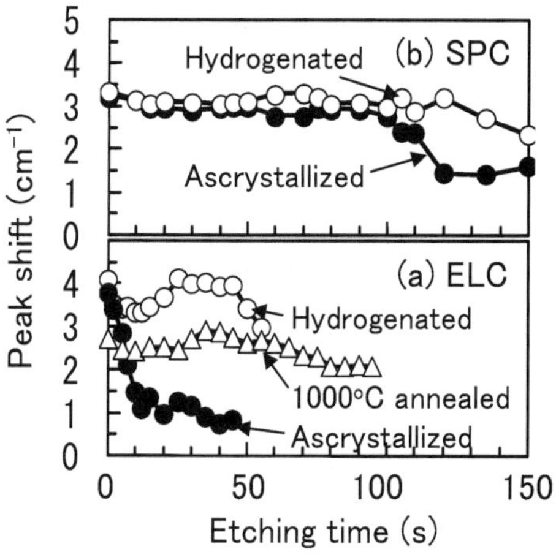

Fig. 19. Variation in peak shift with Secco etching time observed for (a) excimer-laser crystallization (ELC) and (b) solid-phase crystallization (SPC). Only the as-crystallized excimer-laser crystallized film exhibits a rapid decrease in the peak shift, which implies stress release in the film.

On the basis of the results shown above, ELC is considered to progress as follows. Excimer-laser irradiation under optimum conditions melts the film, leaving an adequate seed density. Because solidification occurs extremely quickly, a large number of defect clusters are frozen in the grains. Multiple irradiations affect the merging of relatively high-energy grains with relatively low-energy grains. At a practical grain size of a few hundred nanometers, the lateral growth of individual grains at random orientation progresses, forming high-energy grain boundaries. A large number of the dangling bonds at grain boundaries and the accumulated defects in the grains are electrochemically and probably electronically active.

7. Flow-shaped growth

Free carriers in the channels of poly-Si TFTs are frequently scattered by grain boundaries. The presence of flow-shaped crystals aligned parallel to the channel current is expected to improve the electrical performance. The advantage of this configuration is that it does not require photolithography or complicated optical laser setup.

Flow-shaped growth of Si films has been achieved via techniques such as sequential-lateral solidification (SLS), selectively enlarging laser crystallization (SELAX), and CW laser lateral crystallization (CLC) (Im et al., 1997; Crowder et al., 2000; Hatano et al., 2002; Hara et al., 2002b; Fujii et al., 2007). SLS and SELAX employ pulsed excimer lasers and pulse-modulated diode-pumped solid-state (DPSS) CW lasers (a Nd:YVO_4 laser with a wavelength of 532 nm), respectively, as heat sources. In these methods, the region crystallized by the previous laser pulse acts as the seed for the next laser irradiation. Grain boundaries and subgrain boundaries run approximately parallel to the motion of the solidification interface. The general pattern consists of a branching river-like or wishbone structure (Crowder et al., 2000). The CLC procedure is based on the laser annealing method used in SOI technology. A compact CW DPSS laser or blue-ray semiconductor laser diode (Noguchi et al., 2010) is used as the heat source. Flow-shaped lateral growth is achieved by adjusting the laser scanning velocity and output power. TFTs fabricated by SELAX and CLC exhibit μ_{Fn} values of 440 and 566 cm^2/Vs, respectively; these values are evidently larger than those of TFTs fabricated by conventional SPC and ELC.

We investigated the grain geometry and defects in grains produced by CLC (Kitahara et al., 2009a, 2011b). CLC was performed on 150 nm thick a-Si. Figure 20 shows SEM images of CW laser lateral crystallized poly-Si after Secco etching. The growth geometry varies from the center to the periphery of the laser beam and includes flow-shaped crystals, granular crystals, and leaf-shaped crystals extending to the periphery. The grain boundaries are distinctly visible as sharp etched lines. In the flow-shaped region, most of the etched lines were generated in the crystal, and the number of lines was diminished by coalescence with other lines. Lateral growth processes repeated the generation and reduction of boundaries, as in SLS. Note that the outlines of grains in the leaf-shaped region resemble those of the solid-phase crystallized poly-Si except for the length, which extends toward the outside of the beam. It is likely that the temperature gradient enhances lateral growth in SPC. Figure 20(a′)–(c′) show the surface of a hydrogenated specimen after Secco etching. Most of the grain boundaries in all regions became undetectable, which implies that they were electrochemically active.

Figure 21 shows the crystalline orientation maps and inverse pole diagrams of the Si film in terms of the ND and reference direction (RD) determined by EBSD; the RD is parallel to the laser scanning direction. The ND is distributed and tends to exhibit a high index, whereas the RD tends to be oriented toward <101> or <100>. This suggests that the growth direction is stabilized by the liquid–solid interface rather than the surface or film–substrate interface. Many of the grains began to exhibit ND orientation to <100> with increasing film thickness

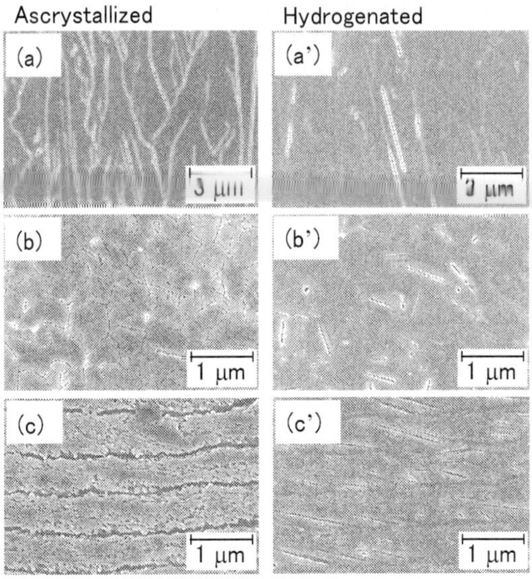

Fig. 20. Scanning electron microscopy images of continuous-wave laser lateral crystallized poly-Si. Secco etching was performed for 25 s. Observations were made in areas from the center to the periphery of the laser beam, as follows: (a) flow-shaped along the laser scanning direction, (b) granular, and (c) leaf-shaped extending to the periphery. (a′)-(c′) Images of the etched film after hydrogenation.

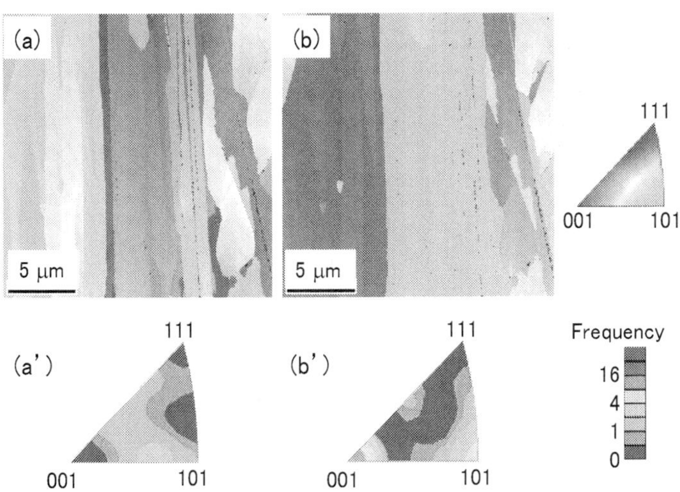

Fig. 21. Crystalline orientation maps for (a) normal direction and (b) reference direction and (a′, b′) individual inverse pole diagrams. Poly-Si film was formed by continuous-wave laser lateral crystallization.

(Hara et al., 2002b). The grain boundary map in Fig. 22(a) indicates that the Σ3 CSL corresponding to the first-order twin boundary lies between the high-angle grain boundaries and extends in the laser scanning direction. Figure 22(b) shows the number fractions of the boundaries in terms of their misorientation angles. The Σ3 CSL with a tilt angle of 60° occupies a fraction of ~0.4. First-order twin boundaries reportedly are not electrically active, whereas second-order twin boundaries act as strong recombination centers (Cunningham et al., 1982). Therefore, the present Σ3 CSL is regarded as electrically inactive. The remaining boundaries, which occupy a fraction of ~0.6, are distributed over a wide range of angles.

Figure 17 shows the relationship between the FWHM and $\Delta\omega$ for three regions in the CW laser lateral crystallized film: (a) flow-shaped growth, (b) granular growth, and (c) SPC-like growth. Regions (a) and (b) exhibit very similar FWHM and $\Delta\omega$ values, which are somewhat smaller than those obtained by ELC. The difference between CLC and ELC can be attributed to the cooling rate. That of CLC is three to four orders of magnitude smaller than that of ELC; the cooling rate of ELC is determined by the scan speed and energy gradation at the edge of the laser beam.

As described above, flow-shaped crystals were successfully grown by CLC. The crystalline boundaries in the resulting films, except for the Σ3 CSL, are electrochemically active, like those of excimer-laser crystallized films. The defect cluster density in the grains of the CW laser lateral crystallized films is smaller than that of the excimer-laser crystallized films; this is explained by the fact that the cooling rate of CLC is several orders of magnitude smaller than that of ELC.

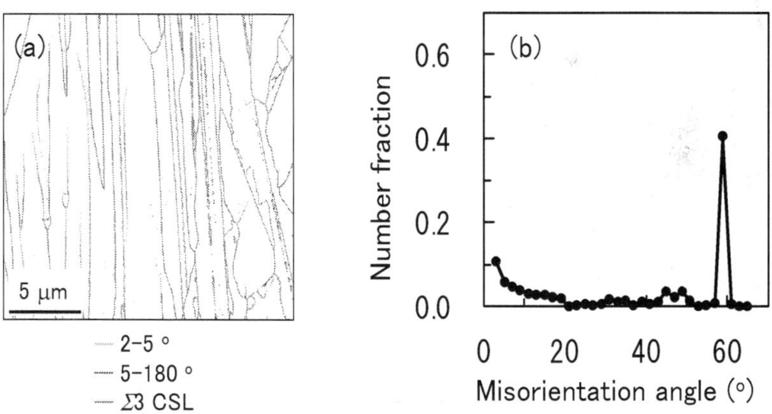

Fig. 22. (a) grain boundary maps and (b) number fractions of the boundaries in terms of the misorientation angle for the poly-Si films shown in Fig. 21. Σ3 coincident site lattice dominates the boundaries.

8. Growth of quasi-single crystalline SiGe thin films

Compounding Si with germanium (Ge) not only extends the application of band engineering but also reduces the melting point, which is advantageous for growth on glass substrates. Rapid melt growth on sapphire substrates and ELC on glass substrates have been reported as techniques for the growth of SiGe thin films (Sameshima et al., 2005; Weizman et al., 2005; Koh et al., 2010; Tanaka et al., 2010) Strong lateral segregation reportedly caused the local Ge content to differ by as much as 40% from the average value (Weizman et al., 2005). Therefore, it is necessary to examine the segregation of Ge and the appropriate growth mode for alloys in order to form crystalline SiGe on glass substrates.

We investigated the crystal configuration, Ge segregation, and growth modes specific to the alloy (Kitahara et al., 2011b). SiGe thin films were crystallized on glass substrates using CLC with a DPSS CW laser. The precursor film was amorphous $Si_{0.7}Ge_{0.3}$ with a thickness of 100 nm.

A typical Raman spectrum of the $Si_{0.7}Ge_{0.3}$ film is shown in Fig. 23. The crystallization of the SiGe films was confirmed by the appearance of sharp optical phonon modes at ~300 cm^{-1} for the Ge-Ge bond, ~400 cm^{-1} for the Si-Ge bond, and ~500 cm-1 for the Si-Si bond.

Figure 24 shows a reflection electron image. In the region corresponding to the central position of the laser beam, multiple areas of bright contrast that extend for more than 100 µm run parallel to the laser scanning direction at intervals of ~1 µm. The coalescence and generation of grain boundaries observed in pure Si films were not evident here, which indicates a type of super-lateral growth. Because the reflection electron intensity depends on the atomic number, the contrast suggests strong segregation of Ge along the bright lines.

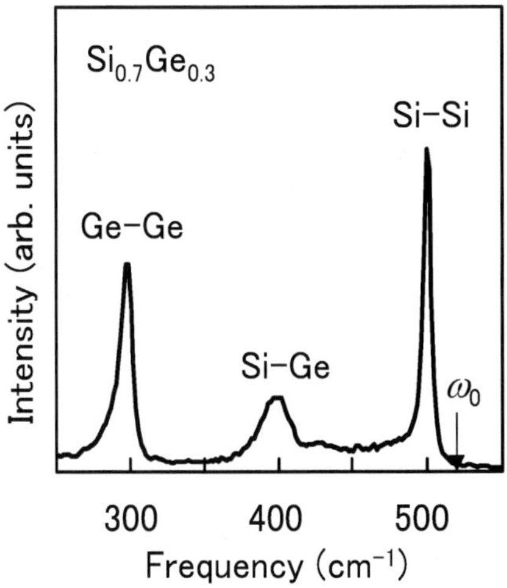

Fig. 23. Typical Raman spectrum for the optical phonon modes of Ge-Ge, Si-Ge, and Si-Si bonds in a crystallized $Si_{0.7}Ge_{0.3}$ film. Peak position for unstrained Si is indicated by ω_0. The lower shift of the Si-Si band from ω_0 depends primarily on the Ge content.

Figure 25 shows crystalline orientation maps and inverse pole diagrams of the film determined by EBSD. "ND" indicates that the crystals are strongly aligned with the <111> plane. The RD, which corresponds to the laser scanning direction, tends to align with the <101> plane. Figure 26 shows the grain boundary map and number fractions of boundaries in terms of their misorientation angles. The map shows that there were very few high-angle grain boundaries. Twin islands were aligned parallel to the laser scanning direction and were outlined by the Σ3 CSL. Although low-angle grain boundaries with a misorientation of <5° appeared, their population was relatively small. Thus, these findings indicate that a quasi-single crystal was formed in the film.

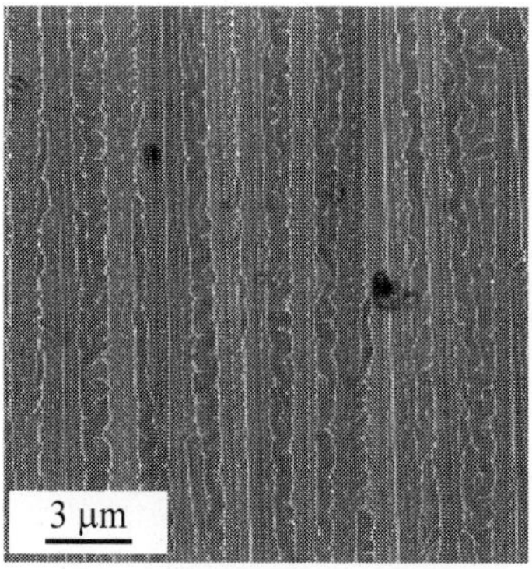

Fig. 24. Reflection electron image of the surface of the $Si_{0.7}Ge_{0.3}$ film. Bright areas corresponds to regions of Ge accumulation.

Figure 27 shows a TEM image and an energy-dispersive X-ray (EDX) profile of Si and Ge. Multiple areas of dark contrast run nearly parallel to the laser-scanning direction in the TEM image. The EDX profile indicates that Ge is strongly segregated in the dark regions of the TEM image. This finding is consistent with the Ge segregation estimated from the reflection electron image in Fig. 24. Thus, Ge segregation was found to occur at many Σ3 CSLs and at both low- and high-angle grain boundaries.

The Σ9 CSL for pure Si reportedly can construct a dangling-bond-free stable structure through the formation of 5- and 7-membered rings (Kohyama & Yamamoto, 1994). In SiGe alloys, the energy at the boundaries can be further reduced by locally varying the Ge content and thus adjusting the lattice constant.

The alloy's characteristics should be considered when discussing the proper growth of SiGe. In pure materials, the solid–liquid interface under a positive temperature gradient is essentially flat during solidification. In contrast, the solid phase in alloys extends into the melt owing to the instability of the solid–liquid interface, resulting in oriented cellular growth; this corresponds to the constitutional undercooling model (Chalmers, 1964). Si and Ge form a complete solid solution. From the phase diagram and the calculation of the equilibrium partition coefficient, the undercooled area was estimated to be on the order of submicrometers in front of the growth region (Kitahara et al., 2011b).

As described above, the growth mode of CLC was changed from flow-shaped growth to super-lateral growth by compounding Ge with Si. This result was attributed to the constitutional undercooling that is peculiar to alloys. At crystalline boundaries running along the laser scanning direction, the energy was lowered by Ge segregation, resulting in the suppression of high-angle grain boundaries. As a result, quasi-single-crystalline films were formed on glass substrates.

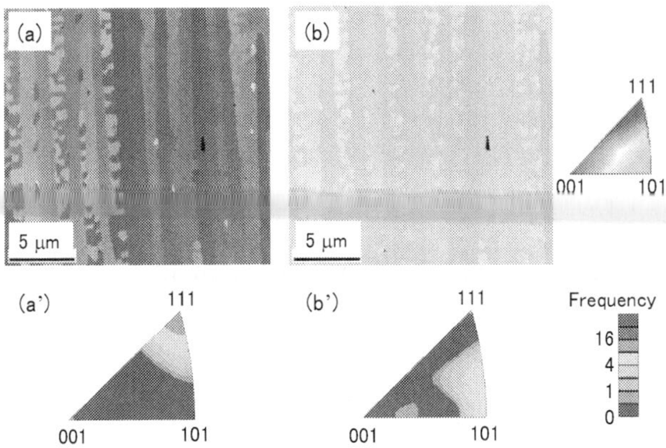

Fig. 25. Crystal orientation maps and inverse pole diagrams for (a, a') normal direction and (b, b') reference direction of $Si_{0.7}Ge_{0.3}$ film. Surface is strongly oriented to <111>.

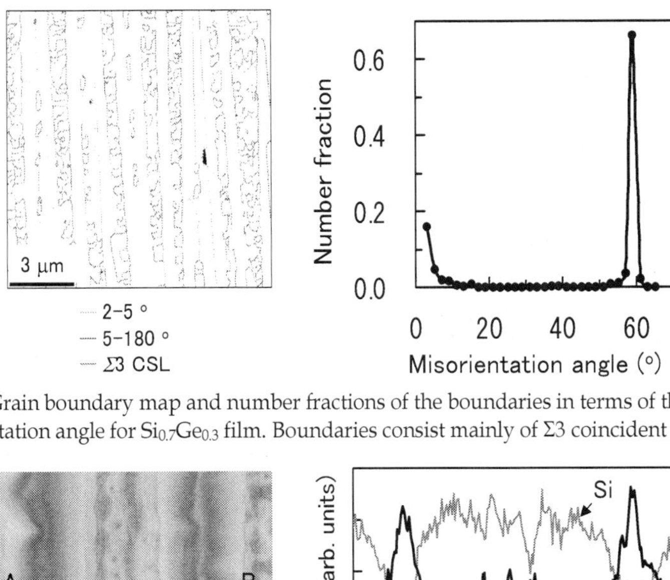

Fig. 26. Grain boundary map and number fractions of the boundaries in terms of the misorientation angle for $Si_{0.7}Ge_{0.3}$ film. Boundaries consist mainly of Σ3 coincident site lattice.

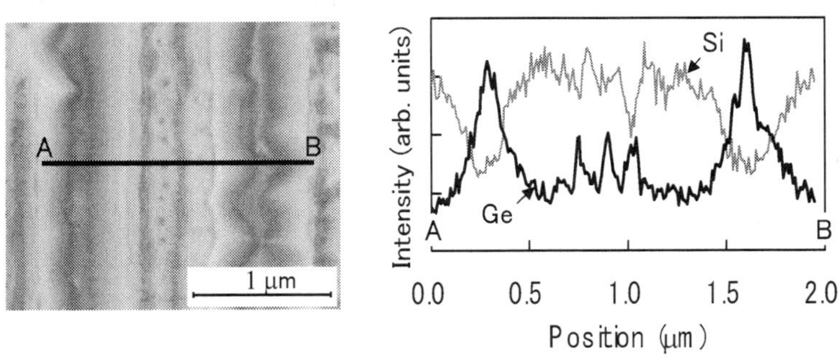

Fig. 27. Transmission electron microscopy image and electron dispersion X-ray profile of Si and Ge atoms in crystallized $Si_{0.7}Ge_{0.3}$ film. Profiles were taken between A and B in the image.

9. Conclusions

Lateral crystallization is indispensable for the growth of poly-Si films on glass substrates for application in large-area devices such as electronic displays and solar cells. In this paper, an outline of the various crystallization techniques was introduced, and our study of lateral growth and defects was described.

When SPC was used, lateral growth proceeded along the major twin boundary. However, a large density of micro-twins appeared in the grains. Oriented lateral growth is expected to decrease the number of defects. In practice, MILC exhibited a decreased defect density with lowered growth temperature, which is attributed to enhanced directivity of needle-like growth.

In ELC, growth is accompanied by melting and recrystallization of the film. TEM revealed only low-density defects in grains. Those defects consisted primarily of dislocations. However, Raman spectroscopy and chemical etching indicated the presence of a considerable density of defects in the grains and at the grain boundaries. The existence of defects in the grains was attributed to the high cooling velocity during recrystallization.

CLC is advantageous for its low cooling velocity and ability to produce oriented lateral growth. Relatively small defect densities were observed in pure Si films. Lateral growth proceeds by generation and disappearance of sub-grain boundaries along the laser scanning direction. Lateral growth was significantly enhanced by compounding Ge with Si, which results in a type of super-lateral growth. Such growth was attributed to the constitutional undercooling effect that is characteristics of alloys. The undercooling effect and the segregation of Ge decrease the energy at the boundaries running parallel to the laser scanning direction, resulting in the growth of quasi-single-crystalline films on glass.

Not only Si and Si-related films but also metal oxides and organic films are beginning to be extensively developed as materials for electronic devices on glass or plastic films. Si will share this role in the development of high-performance devices. One of the remarkable achievements of Si technology is the growth of single-crystalline films on large-area substrates, although such films suffer from cracking, which must be avoided in heterostructures. Both the advance of laser technologies and the control of the physical properties of materials, such as alloying and stacking of different layers, are more promising for the growth of single-crystal films on non-crystalline substrates.

10. Acknowledgements

The authors would like to thank Prof. Tsuda H. in Osaka Prefecture Univ. for TEM observation of SPC and MILC poly-Si. This study was funded in part by the Japan Society for the Promotion of Science [Grants-in-Aid for Scientific Research, (B) 19360165 and (C) 21560329].

11. References

Blake, J. G.; King, M. C.; Stevens III, J. D. & Young R. (1997). Low-Temperature Polysilicon Reshapes FPD Procudtuin, *Solid State Technology*, Vol, 40, No. 151, pp. 151-161

Cammarata, R. C.; Thompson, C. V.; Hayzelden, C & Tu, K. N. (1990). Silicide Precipitation and Silicon Crystallization in Nickel Implanted Amorphous Silicon Thin Films, *Journal of Materials Research*, Vo. 5, No. 10, (October 1990), pp. 2133-2138

Cao, L.; Nabet, B. & Spanier, J. E. (2006), Enhanced Raman Scattering from Individual Semiconductor Nanocones and Nanowires, *Physical Review Letters*, Vol. 96 (April 2006), pp. 157402-1-4.

Chalmers, B. (1964). *Principles of Solidification*, John Wiley and Sons Inc., New York

Christiansen, S.; Lengsfeld, P.; Krinke, J.; Nerding, M.; Nickel, N. H. & Strunk, H. P. (2001). Nature of Grain Boundaries in Laser Crystallized Polycrystalline Silicon Thin Films, *Journal of Applied Physics*, Vol. 89, No. 10 (May 2001), pp. 5348-5354

Crowder, M. A.; Limanov, A. B. & Im J. S. (2000). Sub-Grain Boundary Spacing in Directionally Crystallized Si Films Obtained via Sequential Lateral Solidification, *Materials Research Society Symposium Proceedings*, Boston, 2000, Vol. 621, pp. Q9.6.1-Q9.6.6

Colinge, J.-P. (2004). *Silicon-on-Insulator Technology: Materials to VLSI*, Springer, New York, ISBN: 1-4020-7773-4

Cunningham, B.; Strunk, H. & Ast, D. G. (1982). First and Second Order Twin Boundaries in Edge Defined Film Growth Silicon Ribbon, *Applied Physics Letters*, Vol. 40, No. 3 (February 1982), pp. 237-239

Drosed, R. & Washburn, J. (1980). A New Technique for Observing the Amorphous to Crystalline Transformation in Thin Surface Layers on Silicon Wafers, *Journal of Applied Physics*, Vol. 51, No. 8 (August 1980), pp. 4108-4110, ISSN 0021-8979

Englert, Th.; Abstrerrer, G. A.; & Pontcharra, J. (1980). Determination of Existing Stress in Silicon Films on Sapphire Substrate Using Raman Spectroscopy, *Solid-State Electronics*, Vol. 23, pp. 31-33

Frost R. L. & Shurvell H. F. (1997). Raman Microprobe Spectroscopy of Halloysite, *Clays and Clay Minerals*, Vol. 45, No. 1, pp. 58-72

Fujii S., Kuroki S., Kotani K. & Ito T. (2007). Enlargement of Crystal Grains in Thin Silicon Films by Continuous-Wave Laser Irradiation, *Japanese Journal of Applied Physics*, Vol. 46, No. 4B (April 2007), pp. 2501-2504

Givargizov, E. I. (1991). *Oriented Crystallization on Amorphous Substrate*, Premium Press, New York, ISBN: 0-306-43122-X

Hara, A.; Takeuchi, F. & Sasaki, N. (2002a). Mobility Enhancement Limit of Excimer-Laser-Crystallized Polycrystalline Silicon Thin Film Transistors, *Journal of Applied Physics*, Vol. 91, No. 2, (January 2002) pp. 708-714, ISSN 0021-8979

Hara, A.; Takeuchi, F.; Takei, M.; Suga, K.; Yoshino, K.; Chida, M.; Sano, Y. & Sasaki, N. (2002b). High Performance Polycrystalline Silicon Thin Film Transistors on Non-Alkali Glass Produced Using Continuous Wave Laser Lateral Crystallization, *Japanese Journal of Applied Physics*, Vol. 41, Part 2, No. 3B, (March 2002), pp. L311-L313

Hatalis, M. K. & Greve, D. W. (1988). Large Grain Policrystalline Silicon by Low-Temperature Annealing of Low-Pressure Chemical Vapor Deposited Amorphous Silicon Films, *Journal of Applied Physics*, Vol. 63, No. 7, (April 1988), pp. 2260-2266, ISSN 0021-8979

Hatano, M.; Shiba, T. & Ohkura, M. (2002). Selectively Enlarging Laser Crystallization Technology for High and Uniform Performance Poly-Si TFTs, *SID Symposium Digest of Technical Papers*, Boston, May 2002, Vol. 33, Issue 1, pp. 158-161

Hayashi, H.; Noguchi, T. & Oshima, T. (1984). Polysilicon Super-Thin Film Transistor (SFT), *Japanese Journal of Applied Physics*, Vol. 23, No. 11, (November 1984), L819-L820

Im, J. S. & Kim, H. J. (1993). Phase Transformation Mechanisms Involved in Excimer Laser Crystallization of Amorphous Silicon Films, *Applied Physics Letters*, Vol. 63, No. 14, (October 1993), pp. 1969-1971, ISSN 0003-6951

Im, J. S.; Sposili, R. S. & Crowder, M. A. (1997). Single-Crystal Si Films for Thin-Film Transistor Devices, *Applied Physics Letters*, Vol. 70, No. 25, (June 1997), pp. 3434-3436, ISSN 0003-6951

Jayavel, P. ; Nakamura, S.; Kesavamoorsly, R.; Srivastava, G. P.; Tomoda, W.; Koyama, T. & Hayakawa, Y. (2006). Surface Morphology Effects on the Optical Phonon Modes in $InAs_xSb_{1-x}$ Epilayers on GaAs(001), *Physica Status Solidi* (b), Vol. 243, No. 4, (February 2006), pp. R19-R21

Jellison, Jr. G. E. (1998). Spectroscopic Ellipsometry Data Analysis: Measure Versus Calculated Quantities, *Thin Solid Films*, Vol. 313-314, (February 1998), pp. 33-39, ISSN 0040-6090

Kawazu, T.; Kudo, H.; Onari, S. & Arai, T. (1990). Low-Temperature Crystallization of Hydrogenated Amorphous Silicon Induced by Nickel Silicide Formation, *Japanese Journal of Applied Physics*, Vol. 29, No. 12, (December 1990), pp. 2698-2704

Kitahara, K.; Yamazaki, R.; Kurosawa, T.; Nakajima, K. & Moritani, A. (2002). Analysis of Stress in Laser-Crystallized Polysilicon Thin Films by Raman Scattering Spectroscopy, *Japanese Journal of Applied Physics*, Vo. 41, Part 1, No. 8, (August 2002), pp. 5055-5059

Kitahara, K.; Ohnishi, K.; Katoh, Y.; Yamazaki, R. & Kurosawa, T. (2003). Analysis of Defects in Polycrystalline Silicon Thin Films Using Raman Scattering Spectroscopy, *Japanese Journal of Applied Physics*, Vol. 11, Part 1, No. 11, (November 2003), pp. 6742-6747

Kitahara, K.; Ohashi, Y. & Matsumoto, S. (2007). Hydrogenation Effects on Chemical Activity of Defects in Excimer-Laser-Annealed Polycrystalline Silicon Thin Films, *Japanese Journal of Applied Physics*, Vol. 46, No. 19, (May 2007), pp. L448-L450

Kitahara, K.; Ohashi, Y.; Yamamoto, K. & Sasaki, N. (2009a). Electrochemical and Raman-Scattering Characterizations of Defects in Polycrystalline Silicon Thin Films Formed by Excimer-Laser Annealing, Solid-Phase Crystallization, and Continuous-Wave Lateral Crystallization, *Japanese Journal of Applied Physics*, Vol. 48, (February 2009), pp. 021205 (6 pages)

Kitahara, K.; Kambara, J.; Kobata, M. & Tsuda, H. (2009b). Suppression of Defects during Metal-Induced Lateral Crystallization of Polycrystalline-Silicon Thin Films by Directed Lateral Growth, *Japanese Journal of Applied Physics*, Vol. 48, pp. 091203 (5 pages)

Kitahara, K.; Ishii, T.; Suzuki, J.; Bessyo, T.; Watanabe, N. (2011a). Characterization of Defects and Stress in Polycrystalline Silicon Thin Films on Glass Substrates by Raman Microscopy, *International Journal of Spectroscopy*, Vol. 2011, (August 2011), pp. 632139 (14 pages), doi:10.1155/2011/632139

Kitahara, K.; Hirose, K.; Suzuki, J.; Kondo, K. & Hara, A. (2011b) Growth of Quasi-Single-Crystal Silicon-Germanium Thin Films on Glass Substrates by Continuous Wave

Laser Lateral Crystallization , *Japanese Journal of Applied Physics*, Vol. 50, (October 2011), pp, 115501 (6 pages)

Koh, H.-Y. S.; Chen, S.-L.; Griffin, P. B. & Plummer, J. D. (2010). High Quality Single-Crystal Laterally Graded SiGe on Insulator by Rapid Melt Growth, *Electrochemical and Solid State Letters*, Vol. 13, Issue 8, (May 2010), pp. H281-H283

Kohyama, M. & Yamamoto, R. (1994). Theoretical Study of Grain Boundaries in Si: Effects of Structural Disorder on the Local Electronic Structure and the Origin of Band Tails, *Physical Review* B, Vol. 50, No. 12, (September 1994), pp. 8502-8522, ISSN 0163-1829

Kuriyama, H.; Kuwahara, T.; Ishida, S.; Nohda, I.; Sano, K. & Iwata, H. (1992). Improving the Uniformity of Poly-Si Films Using a New Excimer Laser Annealing Method for Giant- Microelectronics, *Japanese Journal of Applied Physics*, Vol. 31, Part 1, No. 12B, (December 1992), pp. 4550-4554

Kuriyama, H.; Nohda, T.; Ishida, S.; Kuwahara, T.; Noguchi, S.; Kiyama, S.; Tsuda, S. & Nakano, S. (1993). Lateral Grain Growth of Poly-Si Films with a Specific Orientation by an Excimer Laser Annealing Method, *Japanese Journal of Applied Physics*, Vol. 32, Part 1, No. 12B, (December 1993), pp. 6190-6195

Little, T. W.; Takahashi, K.; Koide, H.; Nakazawa, T.; Yudasaka, I. & Ohshima, H. (2001). Low Temperature Poly-Si TFTs Using Solid Phase Crystallization of Very Thin Films and an Electron Cyclotron Resonance Chemical Vapor Deposition Gate Insulator, *Japanese Journal of Applied Physics*, Vol. 30. No. 12B, (December 1991), pp. 3724-3728

Makihara, K.; Nozaki, H.; Asano, T. & Miyasaka, M. (2003). Metal-Induced Lateral Crystallization of Amorphous Silicon under Reduced Nickel Supply, *Solid State Phenomena*, Vol. 93, pp. 207-212

McCulloch, D. J. & Brotherton, S. D. (1995). Surface Roughness Effects in Laser Crystallized Polycrystalline Silicon, *Applied Physics Letters*, Vol. 66, No. 16, (April 1995), pp. 2060-2062, ISSN 0003-6951

Menendez, J. & Cardona, M. (1984). Temperature Dependence of the First-Order Raman Scattering by Phonons in Si, Ge, and a-Sn: Anharmonic Effects, *Physical Review* B, Vol. 29, No. 4, (February 1984), pp. 2051-2059

Mitani, M.; Endo, T.; Taniguchi, Y.; Katou, T.; Shimoto, S.; Ohno, T.; Tsuboi, S.; Okada, T.; Azuma, K.; Kawachi, G. & Matsumura, M. (2008). Ultrahigh-Performance Polycrystalline Silicon Thin-Film Transistors on Excimer-Laser-Processed Pseudo-Single Crystal Films, *Japanese Journal of Applied Physics*, Vol. 47, No. 12, (December 2008), pp. 8707-8713

Mizuki, T.; Matsuda, J. S.; Nakamura, Y.; Takagi, J. & Yoshida, T. (2004). Large Domains of Continuous Grain Silicon on Glass Substrate for High-Performance TFTs, *IEEE Transactions on Electron Devices*, Vol. 51, No. 2, (February 2004), pp. 204-211, ISSN 0018-9383

Nakamura,A.; Emoto, F.; Fujii, E.; Yamamoto, A.; Uemoto, Y.; Senda, K. & Kano, G. (1989). Analysis of Solid Phase Crystallization in Amorphized Polycrystalline Si Films on Quartz Substrates, *Journal of Applied Physics*, Vo. 66, No. 9, (November 1989), pp. 4248-4251, ISSN 0021-8979

Nickel, N. H.; Anderson, G. B. & Johnson R. I. (1997). Grain-Boundary Defects in Laser-Crystallized Polycrystalline Silicon, *Physical Review* B, Vol. 56, No. 19, (November 1997), pp. 12062-12068, ISSN 0163-1829

Noguchi, T.; Chen, Y.; Miyahira, T.; Mugiraneza, J. D.; Ogino, Y.; Iida, Y.; Sahota, E. & Terao, M. (2010). Advanced Micro-Polycrystalline Silicon Films Formed by Blue-Multi-Laser-Diode Annealing, *Japanese Journal of Applied Physics*, Vol. 49, (March 2010), pp. 03CA10 (3 pages)

Richter, H.; Wang, Z. P. & Ley, L. (1981).The One Phonon Raman Spectrum in Microcrystalline Silicon, *Solid State Communications*, Vol. 39, pp. 625-629, ISSN 0038-1098

Roth, J. A.; Olsen, G. L.; Jacobson, D. C. & Poate, J. M. (1990). Kinetics of Solid Phase Epitaxy in Thick Amorphous Si Layers Formed by MeV Ion Implantation, *Applied Physics Letters*, Vol. 57, issue 13, pp. 1340-1342

Samesima, T.; Usui, S. & Sekiya, M. (1986). XeCl Excimer Laser Annealing Used in the Fabrication of Poly-Si TFT's, *IEEE Electron Device Letters*, Vol. EDL-7, No. 5, (May 1986), pp. 276-278, ISSN 0741-3106

Sameshima, S. & Usui, S. (1993). Pulsed Laser-Induced Melting Followed by Quenching of Silicon Films, *Journal of Applied Physics*, Vol. 74, No. 11, (December 1993, pp. 6592-6598, ISSN 0021-8979

Sameshima, T.; Watakabe, H.; Kanno, H.; Sadoh, T. & Miyao, M. (2005). Pulsed Laser Crystallization of Silicon-Germanium Films, *Thin Solid Films*, Vol. 487, pp. 67-71, ISSN 0040-6090

Sameshima, T. (2009). Laser Crystallization for Large-Area Electronics, *Applied Physics* A, Vol. 96, (April 2009), pp. 137-144

Secco d' Aragona, F. (1972). Dislocation Etch for (100) Planes in Silicon, *Journal of Electrochemical Societies: Solid State Science and Technology*, Vol. 119, No. 7, (July 1972), pp. 948-951

Sridharan, M.; Narayandass, Sa. K.; Mangalaraj, D. & Lee, H. C. (2003). Effect of Boron Ion Implantation on the Structural and Optical Properties of Polycrystalline $Cd_{0.96}Zn_{0.04}Te$ Thin Films, *Nuclear Instruments and Methods in Physics Research* B, Vol. 201, pp. 465-474

Spinella, C.; Lombard, S. & Prioro, F. (1998). Crystal Grain Nucleation in Amorphous Silicon, *Journal of Applied Physics*, Vol. 84, No. 10, (November 1998), pp. 5383-5413, ISSN 0021-8979

Takahashi, H.; Suezawa, M. & Sumino, K. (1995). Generation and Dissociation of Complexes of Iron and Phosphorus Atoms in Silicon, *Journal of Applied Physics*, Vol. 78, No. 5, (September 1995), pp. 3078-3082, ISSN 0021-8979

Takayama, T.; Ohtani, H.; Miyanaga, A.; Mitsuki, T.; Ohnuma, H.; Nakajima, S. & Yamazaki, S. (2000). Continuous Grain Silicon Technology and Its Application for Active Matrix Display, *Digest of Technical Papers: AM-LCD 2000*, Tokyo, July 2000, pp. 25-28

Tanaka, T.; Toko, K.; Sadoh, T. & Miyao, M. (2010). High Quality Single-Crystalline Ge-Rich SiGe on Insulator Structures by Si-Doping Controlled Rapid Melting Growth, *Applied Physics Express*, Vol. 3, (February 2010), pp. 031301 (3 pages)

Temple, P. A. & Hathaway, C. E. (1973). Multiphonon Raman Spectrum of Silicon, *Physical Review* B, Vol, 7, No. 8, (April 1973), pp. 3685-3697

Thompson, C. V. (1985). Secondary Grain Growth in Thin Films of Semiconductors: Theoretical Aspects, *Journal of Applied Physics*, Vol. 58, No. 2, (July 1985), pp. 763-772, ISSN 0021-8979

Weizman, M.; Nickel, N.H.; Sieber, I.; Bohne, W.; Rohrich, J.; Strub, E. & Yan, B. (2005). Phase Segregation in Laser Crystallized Polycrystalline SiGe Thin Films, *Thin Solid Films*, Vol. 487, (March 2005), pp. 72-76, ISSN 0040-6090

20

Crystallization, Fractionation and Solidification of Co-Magmatic Alkaline Series Sequentially Emplaced in the Carbonatite Complex of Tiruppattur, Tamil Nadu, India

R. Ramasamy
Ocean Engineering, IITM, Chennai, India

1. Introduction

A mineral is naturally occurring substance with a characteristic crystal structure and chemical composition. It occurs in rocks. Frequently, it has homogeneous crystalline structure in which one or more types of atoms or molecules may be partially substituted for the original atoms and molecules without changing the structure. Thus, most rock-forming minerals are formed in solid solutions. Aggregates of minerals produce characteristic rock-textures which shed light on their trends of magmatic evolution. They crystallize from molten rock, called magma. Magmas are a natural material of hot silicate / carbonate /oxide / phosphate/ sulphide and sulphur melts from which igneous rocks form. They develop by the partial melting of deep-seated rocks at depth. Primary melts are composed of suspended crystals and dissolved gases. The crystallization of a primary melt and the isolation of the rest liquid freed from suspended solids are called the parent magma, which then produces a series of residual magmas of a secondary nature with varying compositions through differentiation and fractionation. The molten material is necessarily forced to the surface, either by hydrostatic head in the mantle where the encasing rock is denser than the melt or else via gas by propulsion. Pressure enhances the solubility of H_2O and other volatiles, which lower the equilibrium temperature of solidification. These volatile substances may largely escape during the course of the ascension of the magma as well as during the course of crystallization. Rising toward the Earth's surface, the magma enters zones of lower temperatures and pressures. Decreasing temperatures tend to bring about crystallization, which produces solid crystals suspended in the liquid. Other solid fragments are incorporated from the walls and roof of the conduit through which the magma is rising and the magma changes its composition by assimilation. A closed magmatic chamber is formed by the filling up of magma in cavities at intermittent stopping places. Over the course of the magmatic crystallization of the parent magma in a closed magmatic chamber, volatiles may be concentrated and interacted with minerals during slow cooling. Magma is generally composed of eight major oxides SiO_2, Al_2O_3, Fe_2O_3, FeO, MgO, CaO, Na_2O and K_2O, as well as from lesser proportions of TiO_2, P_2O_5, H_2O, CO_2, S and other volatiles. In addition to these there are other substances constituting various gases and trace elements. As crystallization progresses, the volatiles and the more soluble silicate components are concentrated in the remaining liquid. They increase the fluidity of the magma. Moreover, they do not enter appreciably into the earlier minerals forming from the magma, but are instead finally concentrated in an aqueous solution as hydrothermal fluid. Minerals initially crystallized from the magma are denser than the magma and buoyancy forces lead to gravity settling the crystals. In some rare and denser iron-rich or carbonatite magmas, feldspar crystals may float. The viscosity of silicate magma generally increases with a decrease in the temperature with decrease in dissolved H_2O, with an increase in crystal fraction and an increase in the degree of SiO_2 content.

2. Polymerization

The polymerization of silicate melt takes place by the reaction of a silicate melt to form three dimensional networks of SiO_4 chains. This structure of a silicate melt provides information on its structure and physical, chemical and thermal properties. From this information, it is possible to understand the conditions of crystallization and the evolution of the silicate minerals from the melt /magma. In magmatic processes, phase equilibria in melt-crystal-vapour systems, diffusion in melts and the thermodynamic, electrical and rheological properties of magma systems are of particular importance. Silicate minerals essentially contain silicon and oxygen in the form of a network of tetrahedra in which a Si ion is surrounded by 4 oxygen ions. The tetrahedra occur as isolated $(SiO_4)^{4-}$ or else joined together in various ways, such as $(Si_2O_7)^{6-}$ and rings $(Si_6O_{18})^{12-}$. Ortho-silicates have isolated $(SiO_4)^{4-}$ tetrahedra that are connected only by interstitial cations. Sorosilicates have isolated double tetrahedral groups with $(Si_2O_7)^{8-}$. Cyclosilicates have linked tetrahedra with $(Si_xO_{3x})^{2x-}$. Inosilicates have interlocking chains of silicate tetrahedra with either SiO_3 for single chains or Si_4O_{11} for double chains. Sheet-silicates form parallel sheets of silicate tetrahedra with Si_2O_5. Tectosilicates have a three-dimensional framework of silicate tetrahedra with SiO_2. As a silicate melt cools, minerals crystallize that are in equilibrium with the melt. Ortho-silicates like olivine generally crystallize from a silicate melt at a very high temperature in a disordered state; on the other hand, quartz and alkali feldspar crystallize from the melt at a relatively low temperature state. At the ends of the scale, the structural states of silicate melts vary widely. Similarly, within the crystallizing minerals, the olivine crystallizing at the early stage of magmatic evolution and the quartz crystallizing at very high temperature states at the late magmatic stage are deficient in Si ions in their tetrahedral co-ordinations. Thus, the structural states of co-existing minerals and the silicate melt from which they crystallize vary widely. In studying such variations, it is possible to see the trend of magmatic evolution. Magmatic evolution may reveal the cooling history of minerals and their accompanying rocks. Volcanic rock which has been suddenly quenched also reveals genetic relationships between phenocrysts and their groundmass matrix, which represents a quenched silicatemelt. There exists a genetic relationship between the structure of a silicate melt and the structure of the silicate mineral crystallizing stemming from it. For example, under an increase of alkalis and alkaline earth elements in the silicate melt, tetrahedrally co-ordinated Al^{3+} displays a strong preference for three-dimensional network units in the form of feldspars in the silicate melts with an increase of the Al/(Al+Si) of the melt and a decrease of SiO_4^{4-} units in the melt (Mysen et al., 1985). The structure of a silicate melt depends upon the viscosity of the melt and the subsolidus structure of the crystallizing silicate mineral at a high temperature state. Under these favourable conditions, the crystal-melt equilibrium of the partition of one element between them sets. Magmatic evolution can be traced by studying the changes of the structural state of crystallizing minerals and the changes in structure of residual melts in spatial and temporal conditions. Slow cooling permits extensive large crystal growth under plutonic magmatic crystallization. As melting and crystallization are reversible processes, different types of phase diagrams are needed to understand how melts crystallize. Magmatic liquids have a sub-solidus structure which is created largely by varying the degrees of bonding among SiO_4 tetrahedra (polymerization). The difference between a silicate liquid and a silicate mineral is that the mineral has a definite long-range order structure that is the same throughout while a silicate melt shows different types of the short-range order of polymerization throughout the melt. Additionally, the degree of the polymerization of the melt controls the viscosity of the melt. Mafic magmas that have relatively low Si contents have depolymerized melts that produce depolymerized crystals - like olivine or pyroxene - and these melts tend to have a high temperature and a low viscosity. In contrast, felsic magmas have abundant Si and Al, whereby the melts are highly polymerized owing to their high viscosity such that they crystallize as sheet and framework silicates. Because of these consequences, Mg-Fe-rich basaltic magmas erupt easily and the lavas flow several kilometres from their vents, while the Si-rich rhyolite lava erupts explosively without escaping out of its inherent gas bubbles. Volatiles are elements that dissolve in magmas but transform to gas when magma crystallizes or because of a sudden decrease in pressure. The eolution of magmatic rocks may be traced out based on the concept of "Bowen's Reaction Series". The order of crystallization of common plagioclase feldspars from cooling magma is evolved from Ca-rich to more Na-rich plagioclase feldspars as a continuous series during decreasing temperature. The high temperature olivine would react with residual magma and change to the next mineral pyroxene in the series. Pyroxene continues to cooling, it would convert to amphibole and then biotite by adjusting the mineral crystalline lattice to achieve stability of different temperatures as discrete minerals. At lower temperatures both continuous and discrete series merge together and orthoclase, muscovite and quartz tend to crystallize. The difference in crystallization temperatures for different kinds of minerals plays a major role in the differentiation of rock composition as magma cools. Thus, various types of magmatic rocks from the early formed ultamafic peridotites to granitic rocks at the late magmatic stages are evolved (Bowen, 1956). Bowen's reaction series show that one homogeneous body of magma can form more than one kind of igneous rock. It reveals the relationship between cooling magma and the formation of minerals that make up igneous rocks.

Some generalizations:

1. The compositional paths followed by both the melts and the growing crystal depend on the initial bulk composition of the melt.
2. The composition of both the melt and the crystal at any point in the crystallization process is dependent on the extent to which the crystallization process has been completed.
3. For minerals that exhibit a solid solution, crystallization proceeds in a continuous manner with the composition of the mineral changing along with that of the magmatic liquid.
4. For minerals that do not exhibit a solid solution, melting or crystallization will proceed in discrete discontinuous steps. As a melt cools, such discrete minerals will appear suddenly when the melt reaches the appropriate temperature.
5. Minerals that crystallize at high temperatures are rich in Mg and Fe and relatively poor in Si.

6. Minerals that crystallize at the lowest temperatures are rich in Si and Al.
7. Intermediate minerals that crystallize at intermediate temperatures are intermediate in relation to Si.
8. Another important observation in that those minerals which crystallize at high temperatures show less polymerization of the Si-tetrahedra than those which crystallize at lower temperatures.
9. The discrete mineral sequences starts with orthosilicates (olivine), single chain (pyroxene), double chain (amphibole) and sheet silicates (biotite), and exhibit a high degree of silica polymerization with incorporations of H_2O, F and other volatiles. The crystallization of plagioclase in a continuous series also indicates an increase of silica constituents, even though they crystallize in framework silicates.
10. Highly polymerized silica-rich tectosilicates are more stable than silica-poor tectosilicates. The degree of polymerization increases from orthosilicates to tectosilicates.
11. The initial crystallization of olivine, the residual magma, is enriched in silica.
12. The initial crystallization of pyroxene under dry conditions, the residual liquid, is deficient silica.
13. The initial crystallization of alkali pyroxene, alkali amphibole, biotite or magnetite, from a silica undersaturated volatile-rich basic alkali magma, residual magma slowly gets enriched with silica and finally at the late magmatic stages free quartz crystallizes in significant amount as an end product.
14. Rock textures are controlled by conditions of crystallization and cooling. Studying rock textures and the mode of the occurrence of rock types in the field, it is possible to correlate rocks and minerals.

The most common magma which is rich in silica is highly polymerized and has a low activity of oxygen ions and a basic magma rich in MgO and CaO has a high activity of oxygen ions. During the course of crystallization, the most abundant, least soluble mineral grows first followed by the less abundant, more soluble substances from the magma. During the course of crystallization, silicate magma displays progressive enrichments of both soda relative to lime and ferrous iron relative to magnesia in crystals and residual melts. Slowly ascending very hot magmas may cool before reaching the surface.

The crystallization of minerals from magma is a complex process because of changes in the temperature, pressure and composition (TPX). The TPX can have dramatic effects on the order of crystallization and the order-disorder structural state of various minerals in a sequence. The addition or loss of water, CO_2, H_2 and O_2 changes the course of magmatic differentiation. The initial crystallization and separation of early-formed minerals from the residual liquid plays a critical role in the trend of magmatic evolution. The general order of crystallization from the magma is as follows: a) low silica containing basic minerals, such as olivine and calcic plagioclase, together with non-silicate minerals, b) medium-silica-bearing minerals, such as clinopyroxene or labradorite, c) high silica-bearing minerals, such as orthoclase and quartz. Depending on the initial chemical composition of the common basaltic magma, either plagioclase or pyroxene may start to crystallize first, changing the composition of the residual magma until simultaneous crystallization occurs. Slight differences in the composition of the parent basaltic magma may result in the production of an end-magma which is either over-saturated or under-saturated in silica (Barth, 1962). The early crystallization and fractionation of olivine / plagioclase / diopside, the respective residual liquid, is progressively enriched in silica over-saturated rocks (Bowen, 1956) or else alumina poor or silica under-saturated rocks (Barth, 1962) or silica under-saturated alkali-aluminium silicates (Schairer and Yoder, 1960). If the crystal's settling and fractionation are irregular, the composition of the residual magma continuously changes over the course of time. The early crystallized and fractionated silicate or non-silicate mineral from the magma controls the chemical composition of the residual melt and its subsequent trend of crystallization. The composition of the parent magma and its volatile constituents decide which of the minerals start the initial crystallization. Only by studying the field relationship, petrographic features and compositions of the minerals in the rocks and their mode of occurrence and geological setting, along with experimental results, is it possible to trace the magmatic evolution of co-magmatic series of rocks.

3. Magmatic melt structure

The kinetic properties of magma depend upon the atomic mobility within variably polymerized silicate melts. Most polymerized silicic crystal fractions affect the properties of melts by resisting shear deformations. They provide bases for polymerized links to attach in melts' rich fluids and by providing porous networks for the melts and the chemical species within them to flow or diffuse from one location to another. Crystals with large interfacial angles (>60 degrees) represent a near collapse of the grain-supported structure, which can act to either press out or lock melt fractions from one another within a crystal - a melt mush at lower temperatures. Therefore, it may also be impossible to initiate the types of grain boundary movements continuously needed for both minerals in new solutions. This may have an impact on grain compositions, resulting in the zoning of the crystals that form silicic melts becoming more viscous. The magmatic melt structure is viewed as linkages of Si^{4+} and Al^{3+} ions in tetrahedral coordination with O^{2-}ions. They are polymerized in the more Si-Al rich melts. Dissolved water breaks the Si-O chains and reduces the degree of polymerization and makes it less viscous. The presence of CO_2 rich fluid inclusions in calcites, feldspars, pyroxenes and olivines (Ramasamy and Shapenko, 1980) could be considered to be vapour trapped phases during the growth of these minerals. In addition to H_2O, the vapour phases of CO_2, SO_3, and P_2O_5 play a critical role in the viscosity of the melt.

The formation of crystal nucleus and crystal growth or the accretion of atoms onto the nucleus may take place in two or more stages, depending upon the kinetic conditions of the magma. Microlites are relatively associated with higher energies of surface tension and surface energies than their associated phenocrysts. The rates of growth increase with the increasing of under-cooling up to a maximum value and then diminish. High nuclei populations yield fine-grained rocks. The close associations of both fine-grained aplitic syenites and coarse-grained pegmatitic syenites with subvolcanic miarolitic crystallizations indicate wide variations in the textural pattern of the alkaline rocks in the area under study. The drastic cooling of highly viscous melts produces amorphous glass. Vitrophyric textures are produced by rapid cooling while emplacement occurs in some dolerites in this area. Complex minerals with solid solution relationships often grow when their growth rates are faster than their diffusion rates. The diffusion of chemical constituents is due to variations in the thermal energy in the magmatic chamber. Zoned plagioclases, clinopyroxenes, zoned alkaline rocks and zoned carbonatites are evolved in this manner. An increase of volatile proportions in late magmatic melts produces boiling, exsolution and / or vesiculation in the residual rocks. The sizes of comagmatic bodies and their volumes during fractionation in sequences and subsequent emplacements also play vital roles in evolution of melt structures at various levels of emplacements as well as in residual magmas. In studying the order / disorder relationships and partitioning elements, it is possible to trace the cooling history of the magma.

4. Carbonatite complex

The carbonatite complex of Tiruppattur (12° 00′00″-12°30′00″N and 78°25′00″-78°35′00″E) Tamil Nadu India is an ideal area to trace magmatic evolution and its magmatic zonal variations (Fig. 1). It is unique in its geological settings, in the fractional crystallization of various minerals crystallized during the course of a prolonged period of magmatic differentiation, and in the emplacement of co-magmatic zoned alkaline rocks and the immiscible separation of carbonatites and alkali syenites from common parent magma. The geological field settings present in this area indicate the occurrences of co-magmatic sequences of both silica undersaturated series of rocks and silica oversaturated series of rocks together. Both these series of rocks have co-magmatic relationships and continuous compositional variations among mineral assemblages. Progressive enrichment of alkali, silica and volatile constituents plays a critical role during the course of magmatic evolution from a highly silica undersaturated shonkinite magma to silica oversaturated alkali syenite or granite magma under favourable tectonic environment..

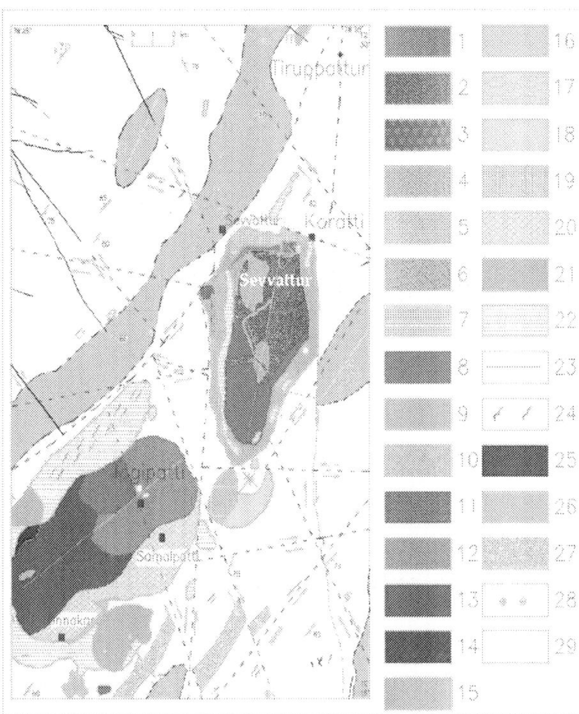

1. Carbonatite,
2. Shonkinite,
3. Vogesite,
4. Oligoclasite,
5. Hornblende syenite,
6. Hornblende porphyritic syenite,
7. Augite syenite,
8. Porphyritic augite syenite,
9. Foliated syenite,
10. Foliated porphyritic syenite
11. Hornblede biotite syenite,
12. Riebeckite syenite,
13. Garnetiferous syenite,
14. Microcline garnet syenite,
15. Wollastonite syenite,
16. Skarn,
17. Granite
18. Granodiorite,
19. Biotite oligoclase pyroxenite
20. Biotite pyroxenite,
21. Ultrabasics
22. Ultramafics,
23. Dolerite,
24. Pegmatite-aplite
25. Albitite
26. Charnockite,
27. Pyroxene granulite,
28. Norite,
29. Granite gneiss

Fig. 1. a. Geological map of the carbonatite complex of Tiruppattur, India

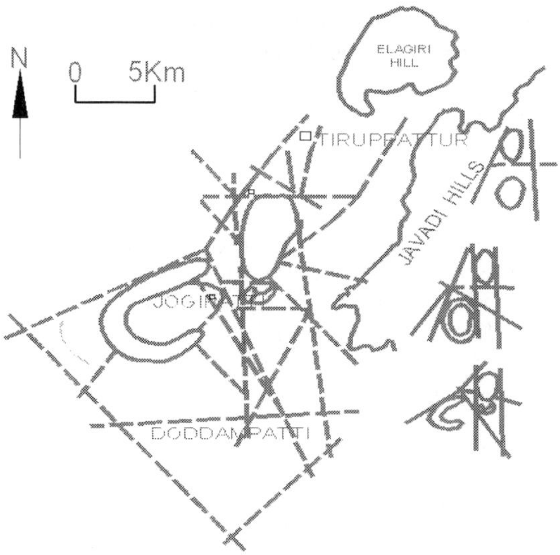

Fig. 1. b. Structural sketch map

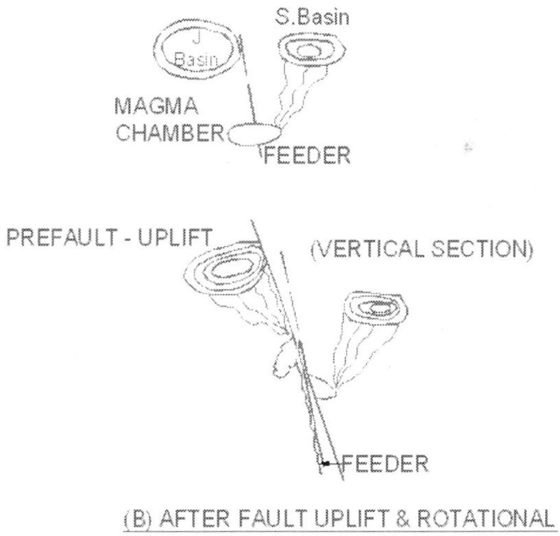

Fig. 1. c. Parent magma ascending from a deep-seated source filled in a magmatic chamber

5. Methodology

Several fieldtrips and observations were made during the course of geological mapping. Thin sections were prepared for various rocks collected in the field and studied under a universal stage (Naidu, 1958) attached polarizing microscope for the identification and determination of the volume proportions of the minerals.

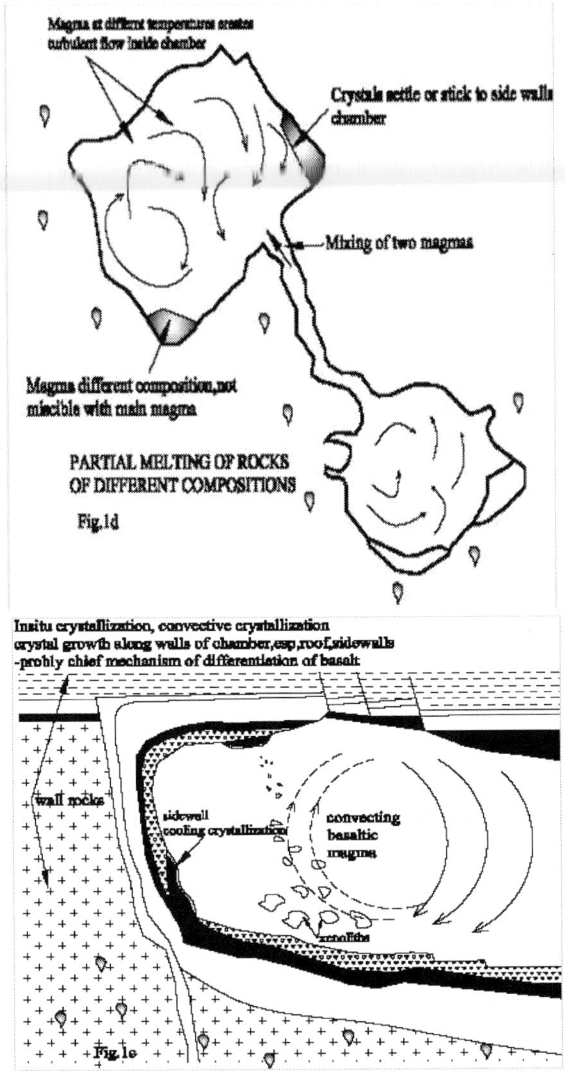

Fig. 1. d. The partial melting of rocks of different compositions concentrates in some large cavities of magmatic chambers. Hot magma rising from the feeder chamber mixes with the already existing magma at a relatively lower temperature and turbulently circulates inside the magmatic chamber. Some of early-formed crystals stick to the walls. The pre-existing magma may dissolve with the incoming magma and mixes with it. The undissolved immiscible magma of different compositions may settle at certain portions. Fig. 1e Magma ascending through fractures breaks the wall rock into pieces of rock fragments which are assimilated by the magma while the magma changes its composition.

i. Differentiation and fractionation of the parent magma for a quiescent prolonged period
ii. Emplacements of a series of soda-rich comagmatic syenites and carbonatites in the Sevvattur Basin
iii. Shifting of the magmatic emplacements of potash rich co-magmatic alkali series and carbonatites into the continuously uplifted and rotated Jogipatti Basin towards the SW
iv. The two basins are enclosed with ultrabasic rocks

By measuring the relationships between the crystallographic and optic axes of the chemical compositions of the minerals and their order-disorder, the structural features are estimated. The textural and structural features of the mineral assemblages, the order of crystallization of the crystallization of the minerals and their magmatic evolutions are traced. The composition of feldspar grains, Fe-Mg ratios in olivine, pyroxene, amphiboles and biotite were calculated by measuring the refractive indices and by finding the optical properties and using the appropriate nomograms given in the texts (Winchell, 1945; Deer et al., 1965). X-ray powder diffraction analyses were made to study the order-disorder relationship by measuring the triclinicity of the feldspars and the M2, M1 and T sites in the clinopyroxenes and the M_4 sites in the amphiboles. Staining techniques were adopted to distinguish calcite and ferro carbonates from dolomite, both in the laboratory as well as in the field (Dickinson, 1965). The chemical compositions of rocks were given by the conventional gravimetric analyses (author in the Geology Dept. Presidency College Chennai and Petrochemical Laboratory, Geology Faculty, Moscow State University) XRF and EDAX under a high resolution scanning electron microscope (IITM, Chennai) . The trace elements contents were determined by Atomic Absorption Spectrometry, Quartz -Grating Spectrograph and HR EDAX. These instrumental facilities were availed by the Department of Geology, Presidency College, Chennai, the Petrography Department, Faculty of Geology, Moscow State University, Russia, the Petrology laboratory, Department of Geology, Government of Tamil Nadu, Chennai and the Material Science Laboratory, Indian Institute of Technology Madras, Chennai.

6. Structure

The carbonatite complex is located in Tamil Nadu, India in a NE-SW trending rift valley formed between the Javadi Hills and the Elagiri Hills (Fig. 1 b), which are located within 5 km of one another. According to Grady (1971), the rift valley, bounded by fault planes, belongs to the category of the deep main faults of Peninsular India, extending more than 200 km in distance. Before rifting, the region was folded, faulted and uplifted by major tectonic deformations, trending with fold-axial plunges in the NW, N, NE, ENE, E and ESE directions. Owing to the number of re-worked fault systems occurring, since the Archaeans, the area has been dissected into several segments of block faults with rotational movements. Hence, it is very hard to analyse the age relationships of different fold styles. However, the regional pattern of fault systems present is seen around the carbonatite complex of Tiruppattur. The culmination of a NE-SW fold system caused the coeval and collinear formation of the Sevvattur and Jogipatti basins by the refolding of ultramafic, basic granulite, charnockite and granite gneiss (Fig. 1b & 1c). The regional NE-SW and N-S fault systems caused formation of a graben structure (Grady, 1971). Due to block faulting, the Jogipatti block was moved towards the west and then towards the north. The melting of the mantle rocks occurred at depth due to E-W crustal shear, which later provides an access route for the rising magma from a mantle source with intermittent stopping in a cavity, forming a closed magmatic chamber at depth. After a prolonged period of magmatic crystallization, differentiation and fractionation with the immiscible separation of carbonatite magmas, the sequences of magmatic rocks were emplaced, first, in the Sevvattur basin and the block was then down warped. A forceful southern movement of the Sevvattur block abutting over the Jogipatti block caused an uplift and a rotational effect over which the rotational axis remained normal to the fault plane trending E-W and resting on the NE-SW fault plane. By the geometrical analysis of this fault system (Ramasamy, 1982), it was estimated that the Jogipatti basin was uplifted by more than 1000 m and, apparently, rotated by about 10° towards the east, and that it caused the emplacements of continuous sequences of late co-magmatic rocks. The magmatic activities were shifted to the latter basin from the original feeding source at depth (Fig. 1 b and 1 c). The generalized features of the magmatic activities are described in Fig. 1 d and 1 e.

7. Field relationship, mineralogical and textural variations in the alkaline complex

The carbonatite complex of Tiruppattur (Borodin et al., 1971; Saravanan and Ramasamy, 1971; Ramasamy, 1982) belongs to the Proterozoic Period, being around 800 Ma old (Schieicher et al., 1998). It is emplaced in two adjacent structural basins amidst Pre-Cambrian granite gneisses and charnockites (Fig. 1). Both of the basins are bounded by ultramafic rocks. The outer shell of the northern Sevvattur body is composed of a fine-grained speckled hornblende syenite with chilled margins and sharp contact with ultramafics and granite gneisses. Towards the south, the specked hornblende syenite imperceptibly grades into mottled augite syenite. The syenite complex exceeds 30 km^2 in extent. These syenites imperceptibly grade inwards towards the centre into their porphyritic syenite variants. The mottled porphyritic syenite is composed of large plates of feldspars containing smaller grains of feldspars in a matrix with accessories of aegirine augite, ferro-hastingsite, hornblende and magnetite. Large plates of plagioclase exhibit normal zoning from andesine to sodic oligoclase. The anorthite contents of the plagioclases decrease from the portions from the north to the south and from the peripheral portions towards the centre. The normal zoning and twinning of the plagioclase lamellae can be clearly seen in large platelets, even in hand-specimens. These rocks are essentially composed of alkali feldspars measuring up to 10 cm x 7 cm x 1 cm. The accessory mafic minerals are diopside, hornblende, augite and ferro-hastingsite. Feldspathic phenocrysts are present in the primary foliation of the rock. However, those mafic minerals occurring in the south are slightly coarse-grained, subhedral to euhedral in form, and are characteristically enriched in alkali constituents and potash feldspars. In places, cumulate feldspar phenocrysts with primary flow orientations are seen in some syenite bodies. The plagioclase and potash feldspars are composed of inclusions of plagioclases with complex twinning lamellae belonging to two or three generations and oriented in different directions. Microcline micro-perthite is commonly present in mottled porphyritic syenite. Albite lamellae, as blebs, orient along crystallographic directions. The plagioclase laths are increasingly plate-like and inform towards the centre of the basin. Similarly, the sizes of the mafic minerals increase and their compositions change with increases in alkali and ferric iron contents. Some places among the syenites - owing to the concentration of mafic minerals in felsic rocks or else the mush of feldspar phenocrysts in hypo-melanocratic syenites - form cumulate textures. The flow textures of feldspar phenocrysts and amphiboles are common in most of the syenites and they orient in the same directions. The rocks of the inner portions are successively younger than the rocks of the peripheral portions. The core portions are occupied by highly differentiated magmatic layers derived from deeper levels. Small veins of fine-grained acmite-syenite are composed by a significant amount of magnetite and quartz and large plates of acmite are surrounded by the released products of magnetites. Abrupt variations from a porphyritic syenite to a non-porphyritic syenite are also seen. The impregnation of cumulates of feldspar phenocrysts into a syenite body is also seen. Many melanocratic streaks, thin slivers and patches of pyroxenites, are strewn as disseminated or as scattered material in the syenite body as mafic cumulates. The dissemination of mafic minerals is often crowded around mafic xenoliths. It seems to be the case that melanocratic rocks rich in mafic minerals are developed in this manner. Foliated fine-grained syenite occurs as a linear band in the porphyritic syenites in the Sevvattur basin. It exhibits crude foliation by the presence of abundant mica. In thin sections, it displays a xenomorphic granular texture. It is composed of oligoclase, anorthoclase, microcline, biotite, margarite and calcite. The grain size of the feldspar ranges from 0.1 mm to 0.3 mm. Discrete grains of microcline are formed as an interstitial to biotite. The replacement of oligoclase by potash feldspar is found along the cleavage planes. The development of patchy perthite is obvious: it imperceptibly grades into a pink coloured porphyritic syenite with a trachitoid texture. It occurs as a pale pink coloured rock with large plates of potash feldspar and oligoclase set in a fine-grained feldspathic matrix. Most of the feldspar platelets are square shaped with up to 5 cm sides. The breadth of the plates varies between 2 mm to 4 mm. In thin sections, feldspar displays an inequigranular poikilitic texture. Hornblende needles are present as inclusions within the plates of the feldspars and the needles orient in parallel towards the foliation planes of the rock. It is essentially composed of microcline, oligoclase and orthoclase, with accessories of chlorite, biotite, hornblende and epidote. The microcline and oligoclase are formed in two generations. Often, the feldspars are unoriented. They carry inclusions of fine needles of mafic minerals (with a helicitic texture) and show characteristic foliation. The core portion of the Sevvattur syenite body is occupied by oligoclasiteand albitite. The amphibole in the biotite-oligoclasite is altered into epidote granules and the amphibole is impregnated with felsic minerals producing a sieve texture. In some oligoclasites oligoclase and biotite, flakes exhibit a sheath texture with radiating platelets orienting towards a common centre. Along the peripheral portion of the carbonatite, albitite is exposed. In the albitite, the feldspar plates have inclusions of magnetite. The modal composition of potash feldspars varies widely and their content increases towards the core of the basin. Again, the volume percentage of the potash feldspars increases towards the south. Albitite, oligoclasite and biotite-bearing oligoclasites occur adjacent to the carbonatite body. A fine-grained monomineralic rock composed of more than 90% of albite in volume with accessories of magnetite and others exhibits the hypidiomorphic granular texture. Post magmatic growth of lueshite (sodium pyrochlore) in this rock produces expansion cracks in the surrounding albite. .

Syenite pegmatites and aplites are common in the carbonatite complex. They also occur as sheets along the fracture planes present in the syenites. Numerous anastomosing veins of pegmatites and aplites are seen in the syenites. Small bodies of pegmatites and aplites carry coarse-grained crystals along their borders. The mafic minerals grow towards it as shooting grains from the wall rocks and the core is only composed of felsic minerals. Massive bodies of pegmatites and aplites occupy over 50 km^2 in the Jogipatti basin. The northern portion of the syenite complex of this basin is composed of riebeckite-anorthoclase syenite while its southern portion is composed of orthoclase-bearing garnetiferous pegmatitic syenite. Garnetiferous syenite exhibits a hypidiomorphic granular texture. Euhedral to subhedral grains of garnets are embedded in the potash feldspar. The grossularite-andradite garnet gradually transforms into melanite towards the southwest of this region. Melanite-microcline syenite, wollastonite syenite, wollastonite-scapolite syenite and wollastonite-melanite carbonatite are some of the aplitic veins found in this complex. Fine-grained microcline melanite syenite displays a xenomorphic granular texture. Micro-grains of microcline are surrounded by accessories of anhedral albite and melanite. These microclines and anorthoclases are highly disordered feldspars, having significant trace amounts of ferric iron. The biotite-syenite has a xenomorphic granular texture with euhedral to subhedral biotite. The biotite syenite porphyry is composed of large phenocrysts of anorthoclase, displaying zonal variation from the core to the periphery and set in a micro-granular feldspathic matrix. Such syenite porphyries are also composed of potash feldspar phenocrysts with a cryptoperthitic texture at the core and with homogeneous peripheral portions. The feldspars belong to two or more generations and they occur as large plates of phenocrysts as well as very small anhedral grains in the matrix. Younger aplites trend in N85° crosscuts while older aplites trend in the N45° direction. Small pockets of zircon- and magnetite-bearing syenites are seen in this basin. The miarolitic texture is a commonly seen in riebeckite pegmatites and aplites. This texture is produced by a crystal growth mechanism that was initially governed under the influence of surface-controlled kinetics. Larger crystals tend to grow by absorbing the required amount of water from the pore fluids present in the surrounding fine-grained matrix. Therefore, the fine-grained grains in the matrix are unable to grow larger due to a lack of water. Moreover, owing to their rapid cooling, larger phenocrysts formed at high temperatures exhibit lognormal rates of growth and in sizes relative to the small grains present in the matrix, forming syenite porphyries or porphyritic rocks under sub-volcanic conditions. The sudden increase of the viscosity of the residual magma also plays a critical role in the crystallization process of felsic and mafic minerals of two or more generations. Large crystals enclosing inclusions of euhedral feldspars oriented in two or three different directions are seen in mottled augite porphyritic syenite and this feature indicates slow cooling of the syenite magma followed by rapid ascending and consolidation of the magma. On the other hand, the anorthoclase-bearing miarolitic syenitic aplites and pegmatites carry abundant miarolitic cavities in which radiating needles of riebeckite are grown towards the centre of the cavities. Cross-cutting anorthoclase-bearing aplites are seen in mottled porphyritic syenites. Two generations of magnetite, apatite and zircon crystallize, both as early-formed minerals as well as very late-formed minerals. The late-formed minerals tend to have comparatively high ratios of surface area to volume. The large feldspar phenocryst platelets increasingly thin in speckled and mottled porphyritic syenites. Moreover, feldspars which crystallize very late in the mottled porphyritic syenite appear to be formed as early crystallized mineral in some of the agpaitic syenite pegmatites and aplites in the Jogipatti basin. The concentration of garnet, aegirine-augite or magnetite along a particular zone produces a banded structure. Radiating bundles and prisms of dark green riebeckite are seen in the aplitic riebeckite syenite and in the carbonatites of Jogipatti. All these varieties of syenites are more or less interrelated to each other in their mineralogy and chemical composition. The syenites of the Sevvattur basin exhibit magmatic continuity with the syenites of the Jogipatti basin. They exhibit a co-magmatic relationship and gradational, zonal, spatial and temporal variations and continuities.

An arc-like vogesite outcrop with a lamprophyric pandiomorphic texture with black needles of katophorite set in a feldspathic matrix of two generations occurs in the Sevvattur basin with close proximity to carbonatite exposures. The vogesite imperceptibly grades to ferro-hastingsite syenite, speckled hornblende syenite and hornblende-biotite oligoclasite. An acmite-bearing syenite grades with an enrichment of cumulate needles of katophorite into lamprophyric vogesite, displaying a pandiomorphic texture. The shonkinite occurring in Jogipatti basin is a coarse-grained inequigranular rock composed of equal proportions of sanidine / anorthoclase and augite, with accessories of olivine, hastingsite, phlogopite, apatite and magnetite and exhibiting a lamprophyric pandiomorphic texture. In the field, it occurs as xenoliths or nodules in ultramafic rocks. The ultramafic rock which is the host rock of the shonkinite appears to be possessed of kimberlitic affinity (Ramasamy et al., 2010). Mineralogical gradation exists between the shonkinite and garnetiferous syenite. An abrupt mineralogical and compositional gradation exists between the vogesite and shonkinite. Similarly, there exists such gradation between vogesite and hornblende-biotite oligoclasites. Again, the shonkinite exhibits such a type of gradation between the garnetiferous-orthoclase syenite and the mottled augite syenite. On the whole, one syenitic member has genetic relationship with any other member in this zoned complex, either in mineralogical or compositional gradations. In a detailed petrographic and mineralogical study, it is shown that the shonkinite is considered as a parent magma for this carbonatite alkali-syenite complex (Saravanan and Ramasamy, 1995; Ramasamy et al., 2010). Dykes, ring dykes, sills, veins, cavity fillings, cone sheets and plugs of syenites and carbonatites show sharp contacts with chilled margins with their host rocks. Curvilinear tensional cracks are seen around the exposures of the syenites and carbonatites. Late magmatic aplites and pegmatites filled these cracks and show co-magmatic relationships with the adjoining plutons.

All of the carbonatites occurring in the three places in Sevvattur, Jogipatti and Onnakarai grade with zonal variation from ferro-carbonatite, beforsite and sovite. Calcite, dolomite and ankerite constitute the essential minerals in these rocks and biotite, phlogopite, magnetite, riebeckite, aegirine augite, acmite, wollastonite, garnet and the rare-earth minerals of pyrochlore, zircon and niobian rutile constitute the notable accessory minerals in these rocks. Two generations of magnetites, both in the form of opaque dusts and anhedral, euhedral and twinned crystals, are present in some of the carbonatites and in the shonkinite. The Fe- and Mg-rich carbonatites appear to be emplaced from relatively deeper horizons. By their characteristic minerals, the zoned cone-sheets of the carbonatites in Sevvattur exhibit various sheets and layers as hybrid sovite, biotite sovite, phlogopite sovite, aegirine sovite, pyrochlore-bearing para-ankeritic sovite, pyrochlore-rich para-ankeritic beforsite, magnetite beforsite and apatite soviet. The elongated rings and veins of the riebeckite-carbonatites are seen in the riebeckite syenites of the Jogipatti basin. The hybrid carbonatites of the Jogipatti basin are enriched with richterite, magnesio-arfvedsonite, acmite, wollastonite and scapolite. The carbonatites of Onnakarai appear to be younger than the Jogipatti carbonatite, which is intermediate and younger itself than the Sevvattur carbonatite (Ramasamy, 1973). Again, the zoned alkali syenites of Sevvattur are older than those in the Jogipatti basin. A carbonatite body crops out in arcuate cone-sheets with a sharp contact along the meeting between the ultramafic rocks and the syenites in the northern periphery of the Sevvattur body. Mineralogical variations occur as zones which are almost conformable with one another, starting from the convex side of the arc and proceeding to the concave side in the following order: a) micaceous zone, b) ankerite zone, c) beforsite zone and d) sovite zone. About 400 m west of the carbonatite exposure is evidence of the forceful injection of carbonatite, as manifested in a syenitic outcrop which has been thoroughly brecciated with feldspar grains which were twisted and ragged with the emplacement of fine-grained carbonatitic cementing media. Along the contacts of the carbonatite, the host rocks have been carbonatized by the intrusion of a thin vein in which lots of carbonates and replacements of silicate minerals into calcites are observed. Iron-enriched phlogopite is found at the contact of carbonatite with the ultramafics. Schiller-type inclusions are common and abundant along the cleavages and partings of the augite crystals. Hornblende wraps round the augite and is pleochroic, from green to yellow. Small prisms of biotite are found along the margins of the pyroxene and hornblende. An economical deposit of vermiculite has been formed at the contact between the ultramafics and the carbonatites. The carbonatite exhibits a well-developed concentric flow structure with linear crystals of apatite, alkali-amphibole and magnetite or else a schlieren of hornblende and biotite and streaky patches of ferromagnesian minerals. The carbonatite imperceptibly grades into sovite, beforsite and ankerite, depending upon an increase of dolomite or ferro-carbonates. In the ferro-carbonatites, which are relatively emplaced from deep-seated sources, pyrochlore, ilmenite and ilmenorutile concentrated in the Sevvattur, Jogipatti and Onnakarai villages, respectively. The broad spectrum of the evolution of the carbonatite complex is manifested in: a) introduction of the calcite into the host rocks, b) the formation of calcium-bearing minerals like grossularite-andradite, melanite, wollastonite, richterite paragonite, margarite, sphene and epidote, c) the alkalinization of clinopyroxene and amphiboles, d) phlogopitization and e) feldspathization. The carbonatite exhibits a xenomorphic granular texture by the presence of anhedral calcite or dolomite grains. The oligoclase and potash feldspars in the plates of the calcites of carbonatites are often corroded with an embayed outline. The micro-fractures in the host rocks are filled with calcites and some mafic minerals are transformed into margarite. The potash feldspar is found as inclusion along the contact of carbonatites and syenites and the host rock is enriched with turbid granular potash feldspar. Lumps of granular clinopyroxenes are developed in the host rock of ultramafics along the point of contact with the carbonatites. The grain size of the granular clinopyroxene has been transformed into well-developed clinopyroxene with larger dimensions. Some of them grow to over 10 cm in their dimensions as single crystals. Apatite is the more common accessory mineral in this contact zone. Clinopyroxene and amphiboles are converted into biotite. Wollastonite, garnet (melanite) or scapolite are formed as small prisms interstitial to acmite and appear to be formed by magmatic crystallization (Eckermann, 1966).

Both the basins are surrounded by ultramafic rocks. They include dunite, peridotites, kimberlites, pyroxenite, biotite pyroxenite, massive amphibolites and biotite. It is not possible to classify these rocks in the field owing to their scarce exposure on the surface and so they are together classified as ultramafic rocks. Magnesite and kankar are found on the weathered surfaces of the ultramafics. Dunite and peridotites are altered to talc-steatite and calcite-bearing rocks. Massive dunite exhibits a saccharoidal texture. However, the grain sizes of these rocks vary widely. Porphyritic pyroxenite with kankar veins is also seen in some well sections. Ultramafics rich in biotite and amphibole are very common. Veins and selvages of calcite-bearing riebeckite showing a schistose structure fill the cracks developed within the ultramafics. The ultramafics are essentially comprised of pyroxenites and peridotites. Thin veins of oligoclasites, albitites and carbonatites intrude into the ultramafics, which are transformed into biotite-pyroxenite and hornblende-oligoclase-calcite-bearing pyroxenite. Many varieties of hybrid carbonatites are developed at the point of contact of the ultramafic rocks with the carbonatites. Alkali pyroxenes like aegirine-augite and augite have developed in the ultramafic rocks and in the syenites. Alkali amphiboles like magnesio-riebeckite, magnesio-arfvedsonite and richterite have developed extensively in the aplitic syenites. Grossularite, melanite, scapolite, calcite and wollastonite were developed at the point of contact with the alkali syenites-carbonatite-ultramafics. Feldspathic selvages in ultramafics often display the development of biotite. At the contact between the garnetiferous syenites and the ultramafics, a skarn rock has been developed at around 2000 m length and with a width ranging from 10 to 200 m. It is comprised of calcite, wollastonite, grossularite and epidote as major minerals. In these rocks, no accessory minerals characteristic of carbonatites, such as apatite, magnetite, zircon, alkali pyroxenes, amphiboles and REE minerals, are seen. The individual minerals in the skarn rock do not have any mineral lineation steeply plunging towards the centre of the basin. However, evidence such as the partial transformation of diopside into calcite, garnet or wollastonite is present for the carbonate metasomatism of one mineral into another. The rock has been extensively folded, sheared and deformed, carrying ultrabasic nodules of varying dimensions from less than 1 cm to over 40 cm. These nodules are partially carbonatized and they exist discontinuously along the folded axial planes. The intricate folding pattern with layers and veins of calcite is even seen in hand specimens. Granulitic and gneissic layers occur within it as contorted and elongated lenses. These knotty inclusions stand out boldly on the weathered surfaces. The skarn rock exhibits a characteristic ribbed structure due to weathering. The ribs are parallel to the foliation direction and are developed along the trend of the elongation of the direction of the relics of mobilized ultramafic nodules towards the N 45° direction. Specks of native copper, bornite, covellite, chalcopyrite, pyrite, galena pyrrhotite are disseminated into the nodules. The modes of the occurrence of these nodules indicate that part of the ultramafic rock occurring along the contact of the garnetiferous syenite has been transformed into a skarn rock during the emplacement of the syenite body along its contacts. Moreover, along these contacts, riebeckite-sovites, acmite and riebeckite-bearing ferro-carbonatites, veins of ferro-carbonatite breccias are seen. The ferro-carbonatite breccia carrying angular fragments of ferro-carbonatite and riebeckite syenite, quartz carrying inclusions of riebeckite needles and euhedral magnetite crystals sets in very fine-grained carbonate matrix. Besides breccias, there are exposures of monazite bearing riebeckite syenite, benstonite carbonatite, barite veins, veins of galena and sulphide hosted aplites and pegmatites are emplaced just inside and on the western outside of the skarn rock exposures. The ultramafic body exposed to the west of the skarn rock is seen with the extensive development of biotite and with specks of sulphide minerals aligned characteristically along the foliation or schistose planes of the rock. It is subjected to vermiculitization with the emplacement of carbonatitic veins. In places, veins of ilmenorutile, barite, apatite-ilmenorutile rock, ankeritic carbonatite and riebeckite are seen in it. The biotite is extensively developed in the ultramafic rock along its contact with garnetiferous syenite. Small pockets of massive biotite-hornblende granites with a sharp point of contact with the country rocks of granite-gneisses occur in the north and south of the carbonatite complex. An arcuate fault-bounded aegirine granite exposure imperceptibly grades towards the east as the hornblende granodiorite in the east is seen in between the two syenite basins. The other granites occurring in this complex also have a gradational variation in their mineralogy and in their chemical compositions, and they also have a genetic relationship with that of the alkali syenites.

8. Zoned carbonatite-alkali complex

The mineral assemblages in the alkaline rocks occurring in the two adjacent basins exhibit continuous chemical variations from one end to the other in various members of co-magmatic alkaline rocks. Larger syenite bodies show imperceptible gradation to the adjacent syenite bodies. The syenites and carbonatites are composed of both early-formed and late-formed minerals (Saravanan and Ramasamy, 1995). All of these rocks are classified based on their wide variations in their textural features and mineralogical compositions. The mapping was carried out based on these features in the field. The anorthite content in plagioclase feldspars in the syenites varies between An28% and An35%, and often oligoclase is the predominant in the speckled syenite and it decreases to An4% in the places from the mottled porphyritic syenite which imperceptibly grade from that speckled porphyritic syenite with an oligoclase of An18%. Plagioclase growth twins are evident in the syenites. Carlsbad (twin axis {001}, penetration twin), acline(composition plane {001}) Baveno (contact twin, twin plane {021}) and Manebach (contact twin, twin plane {001} twins as well as periclines (twin axis {010}) and albite (twin plane {010}) twins are common in the feldspars of the syenites. Normal zoning (progressively sodic towards the rim) is common and reverse zoning (progressively more calcic towards the rim) is also found, occasionally, in the syenites. In places, the oscillatory normal zonings ranging between An34% and An24% are seen in the speckled syenites. Patchy zonings are found in the biotite syenite porphyries. The anorthite contents of the phenocrysts in the porphyritic syenites, vogesites and the syenite porphyries are slightly higher than those of the plagioclases in the matrix. The potash feldspars have the following mean variations of orthoclase and albite components: speckled hornblende syenite (Or_{77} Ab_{23}), speckled hornblende porphyritic syenite ($Or_{87}Ab_{13}$), mottled augite syenite (Or_{50} Ab_{50}), mottled augite porphyritic syenite (Or_{87} Ab_{13}), shonkinite (Or_{61} Ab_{39}), garnetiferous syenite (Or_{62} Ab_{38}) and riebeckite syenite (Or_{43} Ab_{57}). Microcline cryptoperthite is commonly found in mottled porphyritic syenite and in syenite porphyries. The plot of the percentages of or-ab-an in a trilinear diagram shows that most of alkali feldspars fall within the field of sanidine, anorthoclase, albite and oligoclase. The Al/Si ratios (0.11 in albitite, -0.49microcline-melanite-syenite) in the alkali feldspar appear to be related to the history of the unmixing of the alkali feldspar (Deer et al., 1965). These components are plotted in a diagram constructed for the experimental data under 5000 bars H_2O pressure; they indicate that the individual components fall within the range of 800°C to 950°C (Yoder et al., 1957). Microcline, microperthite, anorthoclase, sanidine and orthoclase are present in various syenite members and show a high degree of triclinicity (ranging between 0.6 and 1.0), indicating its disorder states and their high temperature formations and the emplacements (Ramasamy, 1986) of the syenite bodies. The microcline-microperthite, potassic host ranges from 48% to 67%, while the sodic guest ranges from 33% to 52%, indicating their high disorder relationships and high temperature formations. The feldspars display clouding owing to the presence of very fine dusty inclusions of iron oxides in some varieties of shonkinite and syenites. Such types of clouding are present in calcite, dolomite and in ankerite, with varieties of textural patterns in the carbonatites of Sevvattur. It is interpreted that dissociation of ankerite into calcite and magnetite during the course of the ascension of the magma during successive emplacements of cone sheets of carbonatites from deep-seated sources through the centre of the magmatic body at the late magmatic stages would have produced such textural variations (Ramasamy et al., 2001).

The diopside in pyroxenite and feldspar syenites adjacent to the carbonatite exposures are transformed into calcite. The ultramafic body in the Jogipatti basin is extensively deformed and carbonatized. the proportions of plagioclase content decrease towards the centre of the mottled porphyritic syenite. The colour index (Rittmann, 1973) for the syenites varies widely in the alkaline rocks, depending upon the degree of fractionation and the separation of the clinopyroxenes or mafic minerals during the course of magmatic differentiation. According to the volume proportion and mineral structure of the mafic minerals present in the various alkaline members, they are subjected to transformations of clinopyroxenes into amphibole, biotite or magnetite. There exists continuity in the compositional variations of the mafic minerals from one syenite member to another (Ramasamy, 1986a). The clinopyroxenes from the youngest alkali syenites fall into the fields of ferrosilite and the calcium-rich hedenbergite end. The clinopyroxene from the melanite-orthoclase syenite contains a high content of CaO owing to the partial transformation of the clinopyroxene into grossularite-andradite garnet (Saravanan and Ramasamy, 1995). During late magmatic stages, due to increase of agpaitic coefficient {(Na+K)/Al} with progressive depletion of Al, orthoclase transforms into highly disordered microcline incorporating ferric iron in its lattice. Release of Ti^{iv} and Fe^{iii} in the appropriate sites of orthoclase and grossularite-andradite garnet as per the exchange of $Ti^{iv}Fe^{iii} \leftrightarrow Ti^{lv}Al^{iii}$ from subsolidus garnetiferous orthoclasite transforms to melanite-microcline syentite (Ramasamy, 1986). Owing to increasing crystallization and the transformation of mafic minerals into alkaline mafic minerals, it is also dependent on the mineral proportion of mafic minerals present in the individual alkaline rocks. Replacing Ca ions from the amphiboles of $Ca_{76}Na_{15}K_9$ by the substitution of (Na+K) in the M4 sites of the amphiboles to $Ca_{27}Na_{65}K_8$ in the younger generation of syenites caused the crystallization of the riebeckite. Furthermore, the ratio of 100 Mg/(Mg+Fe+Mn) of the calciferous amphiboles decreases from 85 to 30% because of the changing composition from pargasite to ferrohastingsite in the younger syenites. The katophorite and richterite are restricted to vogesite and hybrid carbonatites respectively. The eckermannite, magnesio-arfvedsonite, arfvedsonite, magnesioriebeckite and riebeckite are widely distributed in potash rich syenites and carbonatites. Both the clinopyroxenes and amphiboles are deficient in Si in their tetrahedral sites and are compensated by Al and Ti ions. However, under increasing P_{H2O} in the late magmatic syenites, Si is saturated and Na replaces Ca with increasing silica activity. The "m" value of biotite Mg/(Mg+Fe) decreases from 0.91 to 0.29 in the late magmatic syenites. The Cs (4-10 ppm), Rb (30-345 ppm) and Li (1.5-7.5 ppm) contents increase as K increases in this complex.

The emplacement of younger alkali syenite bodies at the contacts of older syenite units plays a critical role in the transformation and metasomatism of mafic minerals into alkali-rich minerals. Again, feldsic minerals are also subjected to metasomatism, such as the transformation of oligoclase into potash feldspars. The oligoclase surrounded by the rims of reverse zoning in the mottled porphyritic syenite indicates that it is ascending to higher levels. The presence of peristerites formed by the sub-microscopic intergrowth of sodium-rich and calcium-rich phases is observed in the mica syenite porphyries. The compositional zoning indicates that with a decrease of the temperature, the sequence solidified from the margin inwards. However, such metasomatism appears to take place at sub-solidus phases of the phenocrysts through the escape of volatile constituents carrying the required ions in the vapour or liquid phases. Sheets and plugs emplaced at the inner portions appear to be derived from a deep-seated magmatic column. The solidification was interrupted repeatedly by surges of fluid core magma. The emplacements of the various plutons are structurally controlled. The isoclinal folding of the flow bands of the carbonatitic or syenitic layers were developed, indicating the stretching and spreading of the carbonatitic or syenitic magma along the narrow zones of the emplacements. The residual magmas formed through the differentiation and fractionation of a shonkinitic parent magma resulted in decreasing silica activity and steadily increasing oxygen fugacity and agpaicity due to the progressive enrichment of alkalis, ferric iron and volatile constituents in a closed magmatic chamber (Saravanan and Ramasamy, 1995). The relative rate of the development of agpaicity in the residual liquid and the crystallization of femic and felsic minerals also played a critical role in the alkaline magma's evolution. The silica deficiency developed in the residual magma due to the early crystallization and fractionation of the clinopyroxene and alkali feldspar, which may have been compensated for through the formation of amphibole, biotite, carbonates, phosphates and magnetites (Yagi, 1953). Therefore, all of these magmatic bodies exhibit compositional zonings. The compositional zoning indicates that, with a limited decrease in the temperature, the sequence solidified from the margin inwards. Sheets and plugs emplaced at the inner portions appear to be derived from a deep-seated magmatic column. The emplacements of the various plutons are structurally controlled.

The early-formed clinopyroxene in some of the syenites reacted with the residual magma and transformed into calcic-rich diopside and then into augite and aegirine-augite to acmite. Similarly, the acmite transformed into garnet which, again, transformed into melanite with an enrichment of titanium. the calcite reacted with silica-forming wollastonite in the garnetiferous syenite and it reacted with plagioclase forming scapolite, taking the required amount of SO_3 from the volatile phase enriched with this constituent (Tables 2 and 3), as revealed from the chemical analyses by the enrichment of SO_3 in bulk rock compositions. The quartz content decreases from the rocks outcropping at the outer shell to those rocks occurring towards the centre of the Sevvattur basin. On the other hand, such variation is remarkable given the formation of riebeckite, biotite, magnetite, wollastonite, garnet, melanite, scapolite and calcite by the heteromorphic transformation of mineral assemblages to compensate for the silica deficiency of the residual magma impoverished in the silica owing to the crystallization of sodaclase anorthoclase, sanidine and high temperature microclines in the syenites of the Jogipatti basin. The enrichment of volatile constituents like H_2O, CO_2, SO_3, P_2O_5 and F in the late magmatic residual magmas developed different varieties of oxide minerals, like magnetite, rutile, ilmenorutile, perovskite and zircon, carbonate minerals like calcite, dolomite, para-ankerite, benstonite, sulphate minerals like barite, gypsum and scapolite, phosphate minerals like apatite and monazite, and sulphide minerals like galena, chalcopyrite, pyrite and pyrrhotite. The very presence of these minerals indicates that the residual magma was enriched in these volatile constituents, which compensates for any silica under-saturation owing to the extensive development of feldspathic constituents. Most of the rocks in this complex have insufficient alumina to form adequate alkali feldspars. The emplacements of carbonatites and its co-magmatic alkali syenites are in a sequential order due to a series of pulses occurring during intermittent tectonic disturbances.

Olivine is present in significant proportions in ultramafic rocks, shonkinite and carbonatites. The compositions of olivine in the ultrabasic rock represent an intermediate position in the course of the magmatic evolution of the olivine occurring in the shonkinite. The olivine in the shonkinite exhibits peripheral zonal variation augite, aegirine-augite and jadeite, indicating a prolonged period of crystallization under the liquid stage. The high concentration of Ca-Na-K-Al in the olivine from the ultramafics indicates its kimberlitic affinity. These features indicate that shonkinite magma is the parent magma for the ultramafics. The early fractionated olivine from the parental shonkinite magma reacts with the residual magma and the peripheral portion of the olivine is transformed into diopside on slow cooling with release of volatile constituents into the residual magma, inducing magmatic pressure by incorporating CaO. The crystallization of the clinopyroxene in the place of the olivine, the residual magma, is depleted in the silica. Under increasing magmatic pressure and in order to meet the silica deficiency, more Na_2O and Al_2O_3 are incorporated with the development of aegirine augite. With the impoverishment of SiO_2 and Fe_2O_3, jadeite crystallizes from the residual magma along the peripheral portion of the olivine or the clinopyroxene. The jadeite develops along the peripheral portions of the olivine within the temperature range of 600 to 800°C with a pressure ranging from 10 to 20 kbar under progressive enrichments of $Al_2 O_3$ and Na_2O and with decreasing silica activity in the residual magma (Ramasamy et al., 2010). The crystallization of jadeite (jd) in the place of albite (Ab) or nepheline (Ne) compensates the silica deficiency in the residual magma.

$$2jd \leftrightarrow Ne + Ab$$

The further crystallization of jadeite - the residual magma - is enriched with K_2O, Al_2O_3 and SiO_2 (Yagi, 1953).

Thus, the development of jadeite produces as end products silica under-saturated nepheline-free shonkinite with the crystallization of potassium feldspars from the residual magma. With an increase of the vapour pressure during the late magmatic processes, particularly under high P_{CO2}, the mafic cumulates become unstable and release more CaO into the melt. The Ca-enriched residual magma immiscibly separates into silicate and carbonate magmas under high P_{CO2}. The further differentiation of these magmas produces a series of alkali syenites and carbonatites, successively emplaced in a sequential order, first in the Sevvattur basin and then in the Jogipatti basin, and resulting in the formation of the zoned alkali carbonatite complex of Tiruppattur, Tamil Nadu.

Normally, a carbonatite complex is associated with alkaline rocks accompanied with significant amounts of alkali-rich ferro-magnesium minerals. Often, a fenitized aureole is present around the carbonatite complex owing to the alteration of country rocks by its reaction with alkali fluids escaping from the carbonatite body. The country rock may be changed into silica-impoverished nepheline-bearing fenites. In this complex, such a development of fenites is absent. As magmatic differentiation takes place within a closed magmatic chamber for a prolonged period of crystallization and differentiation, the alkaline fluid evolves during the course of magmatic evolution, reacting with the minerals and crystallizing from the magma and increasing the alkali constituents up to the development of ultrapotassic syenites. The transformation of diopside into acmite and jadeite during the course of magmatic evolution is adjusted with the development of acmite (Fig. 2) and the ensuing silica deficiency is compensated for by the reaction of albite with silica under-saturated residual magma forming a jadeite component. Under high oxygen fugacity and a volatile concentration, biotite and magnetite form in the place of jadeite. The distribution of volume proportions of alkali feldspar against quartz (Fig. 3a) and alkali feldspar against plagioclase is calculated on the basis of the Rittmann norm using the chemical analyses listed in Tables 1 and 2 for the plot in the Q-A-P-F (Rittmann, 1973) diagram, indicating a linear differentiation trend for co-magmatic alkali series of rocks (Fig. 4 a & b). In Fig. 5a-j, it is indicated that the distribution of major elements in binary variations indicate overlapping smooth linear trends for syenites, ultramafics, carbonatites and dolerites. The dolerites in this area also show a trend of alkali-enrichment during its magmatic evolution.

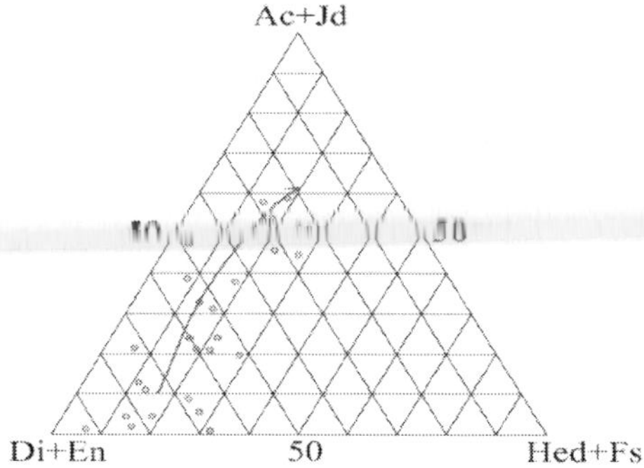

Fig. 2. The chemical compositions of alkaline rocks calculated on the basis of Acmite+Jadeite, Diopside+Enstatite and Hedenbergite+Ferrosalitse compositions, indicating a magmatic differentiation trend moving towards the Ac+Jd end members. The diagram indicates the enrichment of alkalis and iron during the course of differentiation

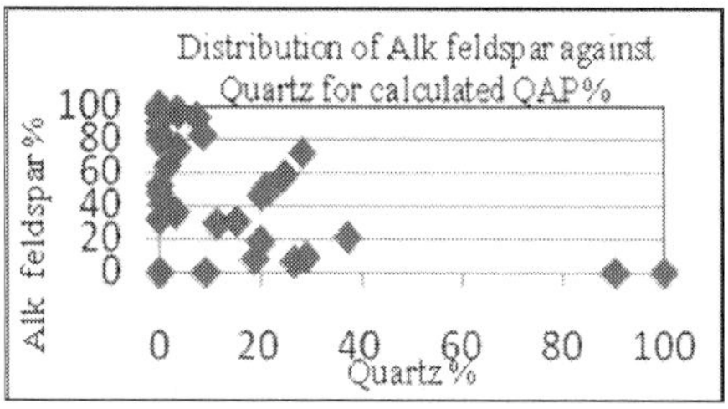

a)

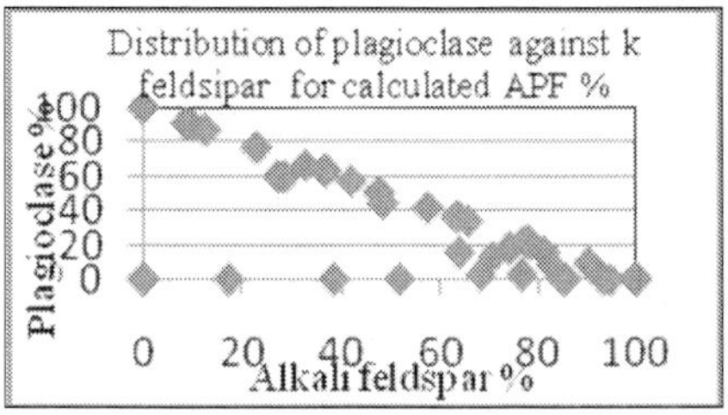

b)

Fig. 3. a. & b. Streckeisen (1967) QAPF double triangle plots are presented in Qz-alk-felds and Alk-felds - Pl variation diagrams showing linear variations in co-magmatic series (The C.I. for syenites varies from 5-30; for hybrid rocks 30-70; for ultramafics 70-100).

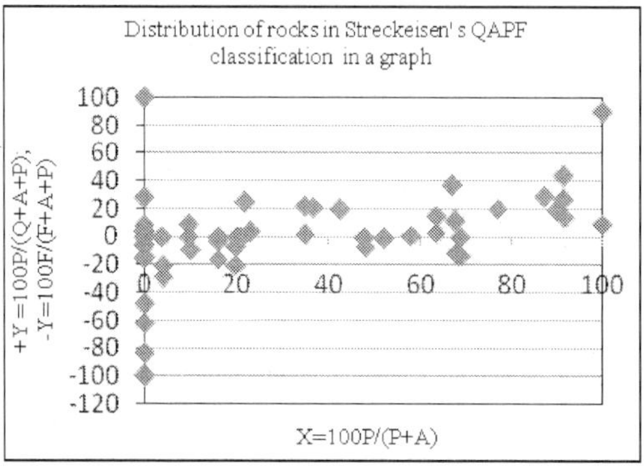

a)

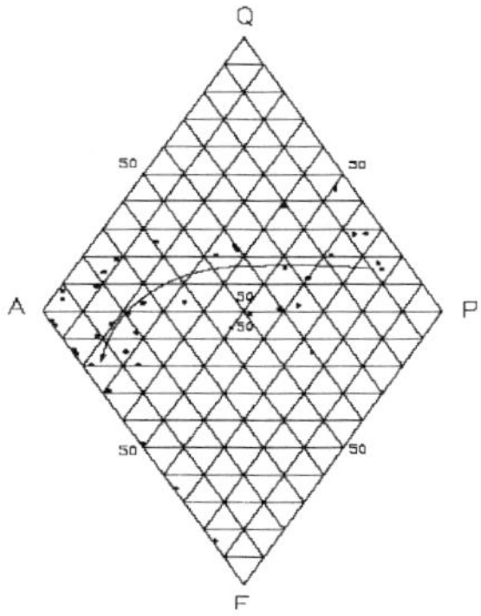

b)

Fig. 4. a., b. Streckeisen graph and double triangle showing the linear trends of the magmatic evolution of alkaline rocks in the Tiruppattur carbonatite complex

Rock	No	SiO2	Al2O3	Fe2O3	FeO	MnO	MgO	CaO	Na2O	K2O	TiO2	P2O5	CO2	SO3	H2O
Bio oligoclasite	723	48.35	12.97	3.78	3.26		2.68	12.91	3.54	5.00	0.75	1.06	2.76		3.15
Bio fol aplite	1122	50.28	23.25	5.54	2.98	0.36	3.82	5.60	2.08	1.80	0.90	0.60	0.60		
Bihbpor syenite	428	51.53	18.22	7.53	2.79		4.00	7.50	3.85	2.70	0.65	0.22	0.39		0.44
Vogesite	733	53.07	13.03	4.28	5.09		3.75	11.75	4.00	3.93	0.69	0.48			0.36
Aughb syemite	464	53.17	14.06	5.32	5.42	0.35	3.62	9.06	3.80	2.25	1.30	1.50			0.68
Ac albitite	468	54.37	5.30	19.84	1.95	0.52	0.98	2.50	11.75	0.80	1.50	0.36			0.46
Bihbpor syenite	370	54.44	17.80	4.63	3.82	0.34	2.71	5.90	3.80	3.60	0.65	0.85			0.95
Augpor syenite	459	57.72	18.18	1.75	2.24		2.38	7.90	4.15	4.10	0.28	0.46			0.34
Augpor syenite	476	58.16	17.98	2.88	2.70		1.87	5.98	3.69	4.70	0.65	0.82			0.39
kato syenite	18	58.18	17.19	2.64	3.19	0.17	1.36	6.36	5.40	4.08	0.69	0.29			0.20
Augpor syenite	485	58.58	11.14	3.47	3.01	0.30	2.97	8.37	4.84	3.50	0.70	1.48	0.21		1.20
S.concave syenite	34	59.59	20.67	0.74	0.28		0.01	0.01	4.86	8.66	0.01	0.01	0.29		0.07
Albitite	461	60.87	14.09	0.32	2.62	0.40	2.51	4.14	8.00	2.57	0.20	1.81	0.51		1.58
Sev gar peg syenite	43	61.25	20.76	1.78	0.28		0.01	1.98	10.62	0.65	0.01	0.01	1.08		0.07
Kakangarai syenite	7	61.26	20.91	0.89	0.53		0.01	1.10	4.67	8.84	0.01	0.01	0.25		0.10
Sevvattur syenite	10	62.22	20.98	1.14	0.28		0.01	1.44	4.54	9.07	0.01	0.01	0.48		0.08
SevNE syeite	36	62.68	19.41	1.17	0.13		0.01	0.93	9.07	3.15	0.01	0.01	0.62		0.03
Kunnattur syenite	3	63.05	22.39	0.43	0.28		0.01	0.67	8.55	4.00	0.01	0.01	0.44		0.05
Koratti syenite	44	63.53	20.60	0.59	0.27		0.01	0.40	7.77	6.96	0.01	0.01	0.40		0.02
Karapattu syenite	6	63.54	22.27	0.74	0.27		0.01	0.75	8.55	2.86	0.01	0.01	0.30		0.09
PinkBihbporsy	463	64.24	14.74	2.84	2.02	0.83	1.13	2.42	4.93	3.34	0.62	0.52	0.44		1.43
Bi-ab-peg syenite	635	67.01	14.06	3.99	1.98		0.52	2.38	6.77	1.40	0.15	0.81	0.34		0.20
Bi hb syenite	369	70.30	13.35	1.94	2.63	0.20	1.17	4.05	3.36	1.58	0.45	0.62			
Richteite ultramafic	321	21.02	1.64	5.48	2.87	0.63	6.39	31.68	1.96	1.44	0.45		23.65	0.69	0.65
pl px ultramafic	319	41.26	3.65	10.18	8.26	0.29	9.67	20.18	0.41	1.20	1.48	0.73	0.41	0.34	1.15
bi-fels px ultramafic	320	44.12	9.10	3.64	1.26	0.06	14.20	21.78	1.15	1.24	0.88	0.10	1.12		0.60
gar pegmatite	572	44.84	11.35	12.14	0.50	0.40	0.70	19.01	3.95	4.98	0.40	1.20			0.40
px hb bi ultramafic	203	45.82	15.53	1.50	5.59	0.19	3.95	9.62	5.27	5.16	0.66		4.66	0.14	1.80
hb oligo ultramafic	204	49.16	21.03	2.57	5.16	0.08	5.43	12.37	2.43		0.83		0.14	0.16	0.30
Shonkinite	561	49.20	7.61	6.40	5.58	0.36	10.43	10.03	3.44	4.29	0.76	0.82			0.80
Olaipatti syenite	78	51.94	22.27	3.53	0.41	0.14	0.01	4.50	0.92	13.54	0.01	0.01	0.93		
gar aplite	39	52.28	19.59	2.82	1.73	0.13	0.25	5.15	0.14	17.16	0.56	0.13			0.44
bi px ultramafic	318	52.92	12.98	5.31	1.80	0.11	3.21	6.18	2.03	11.76	1.09	0.46	2.24	0.16	0.44
Olaipatti syenite	75	53.43	19.31	6.94	0.28		0.01	7.58	2.35	7.72	0.01	0.01	0.93		
Olaipatti syenite	77	54.25	22.99	2.83	0.56		0.01	2.12	2.44	12.05	0.01	0.01	1.04		
Olaipatti syenite	79	54.60	23.21	4.05	0.55	0.16	0.01	5.70	1.93	8.84	0.01	0.01	1.04		
Bi-mus oligoglacite	219	55.02	18.75	2.17	0.36	0.03	1.24	4.47	3.31	10.32	0.28		2.54	1.30	1.30
hb syeite	361	56.84	15.69	2.37	3.02	0.17	1.73	7.73	3.24	6.96	0.50	0.47	0.44	0.76	0.82
aplite	201	58.50	18.65	3.11	0.85	0.07	0.49	2.75	6.48	6.24	0.44		0.30	0.13	1.00
Jogipatti syenite	51	58.92	19.14	1.89	0.56		0.01	1.20	1.09	14.68	0.01	0.01	1.00		
anorthoclasite	349	58.98	16.20	2.51	2.72	0.12	2.22	5.15	6.75	3.48	0.56	0.17	0.22	0.74	0.50
Rieb gar pegmatite	360a	59.78	18.27	1.32	1.14	0.07	1.21	5.61	5.40	6.78	0.20	0.17			0.10
Bi pegmatite	340	61.82	16.90	0.61	0.34	0.02	0.49	2.06	0.27	16.80	0.09		0.42	0.25	0.32
Garigaipalli syenite	85	62.69	20.15	0.21	0.14		0.01	0.01	1.43	15.06	0.01	0.01	0.19		
bihb oligoclasite	206	63.18	15.02	2.73	4.39	0.14	2.72	5.50	3.98	1.32	0.66		0.30		0.38
rieb pegmatite	301	63.24	17.74	0.88	0.72	0.03	0.49	2.41	4.73	10.12	0.28			0.15	
rieb pegmatite	85	63.75	13.01	1.98	0.65		2.59	0.89	5.21	9.98	0.10	0.40			1.10
rieb aplite	541	64.01	15.44	2.44	0.25	0.40	1.62	0.21	4.04	9.15	0.14	0.22			1.58
Rieb diorite	80	64.97	16.75	2.61	1.31	0.40	0.53	3.75	5.04	2.70	0.30	0.48			0.82
Rieb sy pegmatite	36a	65.01	17.66	1.25	0.57	0.03	0.49	2.41	7.02	4.92			0.16	0.05	0.20
Pink granite	12	69.72	14.92	1.40	0.71	0.07	0.71	3.15	4.79	4.32	0.31				0.48
Pink granite	1	70.20	14.42	1.94	0.57	0.04	0.47	3.09	4.19	4.74	0.40	0.16			0.62
Rieb granite	78	71.40	13.60	2.53	0.63	0.50	0.72	2.67	3.80	2.93	0.15	0.32			0.52
ap-mt rock	498	0.48	0.73	54.17	18.07	0.49	0.73	6.66	0.14	0.42	13.20			0.77	
barite rock	452	0.50		0.10			0.05	0.50	0.05				0.53	33.40	0.80
apatite rock	450	2.06	6.91	2.89	2.06	0.07	1.46	41.36	0.47	0.10	10.01	31.77			0.26
Crocidolite rock	60	6.50	1.09	0.49	0.26	0.09	0.85	50.84	0.01	0.01	1.68	0.02	37.86		0.00
Ultramafic	460	18.84	0.91	28.94	17.60	0.19	5.42	15.42	0.06	0.14	3.15	3.00	4.50	0.30	1.30
Feore rock	13	20.36	0.07	27.60	37.85	0.46	0.01	11.16	0.71	0.34	1.57	0.04			
Ultramafic	492	27.72	0.80	16.25	7.15	0.78	6.37	19.09	4.04	0.28	0.60	0.21	15.26	0.23	0.36

Sevvattur syenites No723-369; Jogipatti syenites No 321-36a; Granites No12-78; Ultramafics No 498-47; Carbonatites No 310-495; Dolerites No 1176-1224

Table 1. Chemical composition of the alkaline rocks in the carbonatite complex of Tiruppattur

Rock	No	SiO2	Al2O3	Fe2O3	FeO	MnO	MgO	CaO	Na2O	K2O	TiO2	P2O5	CO2	SO3	H2O
Peridotite	31	30.09	2.00	19.09	6.95	0.79	9.91	15.68	3.67	0.01	0.63	1.29	8.93		
Biofels pyroxinite	9	36.88	1.47	15.49	10.72	0.10	8.00	17.76	1.53	1.88	1.74	1.97	1.10		
Kimberlite	[illegible]	37.71	0.10	[illegible]	[illegible]	[illegible]	0.11	21.10	0.32	2.93	0.73	0.09	16.60		
Samalpatti ultramafics	S3	38.09	5.17	1.05	14.10	0.18	14.18	15.52	0.23	3.44	1.66	3.69			1.90
Feore rock	21	39.30	0.07	49.84	10.36	0.02	0.01	0.01	0.16	0.17	0.33	0.14			
Kimberlite	7a	40.85	0.74	10.33	0.57	0.13	19.08	7.00	0.40	0.01	0.01	0.01	20.65		
Riebcc ultramafic	491	42.36	1.27	7.54	1.98	0.19	12.97	15.65	4.59	0.90	0.44	0.10	9.75		2.15
Olaugoligo ultramafic	40	42.58	4.62	6.32	5.25	0.15	10.62	25.08	0.34	0.06	0.93	3.10	0.60	0.33	0.60
apatite rock	451	42.82	1.27	21.86	2.51	0.04		14.76	1.95	0.06	1.56	7.90	2.46	1.65	0.24
Samalpatti ultramafic	S4	45.21	1.28	0.90	4.32	0.09	14.82	20.00	2.34	0.28	0.23	0.05	9.68		1.10
Ultramafic nodule	497	45.92	6.12	5.52	5.33	0.18	9.60	22.46	1.08	0.18	1.26	0.19	1.02		0.50
Rieb ultramafic	402	46.40	10.41	9.31	2.87	0.22	8.32	12.42	3.04	4.38	0.84	0.44	0.17	0.55	
Sam ultramafics	S5	46.41	2.09	1.64	2.90	0.10	14.46	18.90	2.55	0.74	0.17	0.74	8.73		1.08
Sev Ultramafics	S1	46.77	2.17	3.43	12.58	0.13	14.02	19.76	0.19	0.35	0.66	0.33			
wollastonite rock	491	49.66	1.22	2.01	0.63	0.67	1.52	41.86	0.34	0.30	0.07	1.15	0.30	0.30	0.20
Biopx ultramafic	736	52.40	12.26	2.99	6.75	0.02	6.01	8.48	3.11	4.26	0.99	1.53	0.49		0.00
Sam ultramafics	S2	54.65	1.01	0.85	4.12	0.15	17.17	21.37	0.44	0.12	0.22		0.02		
Fe ore ultramafics	47	62.14	0.17	33.58	1.22	0.01	0.01	1.07	0.01	0.34	1.30	0.01	0.39		
Sev carbonatite	310	0.20	0.18	3.60	2.97	0.55	17.40	29.54	0.10	0.60	0.10	0.05	43.56		0.50
Icelandspar	499	0.28	0.08		0.14		0.25	55.03	0.21	0.12	0.14		43.04		0.50
Sev.carbonatite	11	0.30		0.12		0.06	3.84	52.12	0.10	0.05			43.10		0.20
Sev. Carbonatite	59	0.50	0.29	1.25	1.38	0.48	5.00	45.59	0.46	0.07	0.40	1.30	41.45		1.10
Sev carbonatite	389	0.60	0.79	1.23	3.29		15.28	32.28	0.50	0.10	0.10	2.33	42.35		0.40
Sev carbonatite	250	1.00	0.70	4.08	0.60	0.06	0.45	50.46	0.34	0.22	0.20	0.82	39.28		0.50
Sev.carbonatite	410	1.26	0.42	5.00	1.33		4.52	45.46	0.68	0.43	0.15	1.80	39.10		0.28
Jogi. carbonatite	313	1.88	0.65	3.55	2.36	0.62	11.25	36.04	0.20	0.09	0.15	0.02	40.02	1.66	0.40
Jogi.carboatite	315	2.06	0.60	3.90	2.93	0.60	13.23	32.51	0.37	0.33	0.10	0.04	40.24	1.35	0.45
Jogi.carbonatiate	311	2.64	0.50	1.10	0.42	0.50	2.45	48.28	0.20	0.13	0.20	0.04	40.46	0.57	0.32
Jogi. carbonatite	990	3.24	0.94	2.30	0.41		1.06	48.43	2.30	0.50	0.20	1.48	38.52		0.52
Sev.carbonatite	369	4.90	2.95	4.62	0.95		12.29	32.24	0.73	0.08	0.20	1.66	38.14		0.44
Sev. carbonatite	20	9.97	5.06	0.88	2.35		3.12	46.90	1.10	0.57	0.30	2.01	26.32		1.00
Onna Bens carbonatite	500	12.51	5.73	2.05	0.13	0.16		24.39	1.05	0.98			27.17		0.59
Jogi carbonatite	321	21.02	1.64	5.48	2.87	0.63	6.39	31.68	1.96	1.44	0.45		23.65	0.69	0.65
Jogi. carbonatite	1010	24.70	2.01	2.54	2.00	0.30	7.06	33.51	2.92	0.90	0.20	3.68	19.45		1.20
Onna carbonatite	496	27.26	1.92	16.11	7.09	2.46	4.91	17.16	3.58	0.24	1.23	1.26	16.34		0.95
Onna.carbonatite	492	27.72	0.80	16.25	7.15	0.78	6.37	19.09	4.04	0.28	0.60	0.21	15.56	0.23	0.36
Onna.carbonatitae	495	41.36	1.45	11.16	5.30	0.32	8.32	13.61	5.20	0.24	0.98	0.52	10.95		0.20
dolorite	1176	47.22	18.44	1.30	9.76	0.38	7.19	11.09	1.68	0.68	0.75	0.40			0.80
dolorite	1155	48.08	19.06	2.30	9.41	0.15	5.42	10.01	2.52	0.90	1.25	0.50			0.38
dolorite	1159	48.58	15.49	1.60	12.80	0.40	4.68	7.80	2.92	0.90	2.60	0.93			0.84
dolorite	33	49.50	15.25	1.67	9.01		7.23	11.85	2.60	0.20	0.53	0.68	0.33		0.62
dolorite	1186	54.18	12.76	1.80	8.80	0.15	8.96	8.20	2.08	0.90	0.85	0.23			0.63
dolorite	1187	55.13	8.46	3.32	9.98	0.30	7.21	9.04	3.36	1.33	1.65	0.12			0.20
dolorite	1224	56.04	8.26	3.21	9.62	0.20	4.61	8.41	6.02	1.58	1.10	0.31			0.55

Table 1. Chemical composition of the alkaline rocks in the carbonatite complex of Tiruppattur (continued)

9. Geological setting and tectonomagmatic evolution

The location of the Precambrian carbonatite complex of Tiruppattur in the Indian peninsula along the Eastern Ghats Paleo-rift System (Ramasamy, 1982, 1987), extending over a stretch of 3000 km x 200 km from Cape Comorin / Palghat Gap to the Brahmaputra valley (Eastern Syntaxes of the Himalayas), is favourable site for a low degree of the partial melting of alkali-enriched upper mantle rocks from a low velocity zone (Schleicher et al., 1998). The rift system is a superimposed structure over which a series of older block-faulted horsts and graben structures are comprised of a number of magmatic emplacements of charnockites, anorthosites, alkaline rocks, carbonatites and volcanic effusives. It is also favourable for the genesis of under-saturated alkaline magma charged with relatively more anhydrous volatiles (such as CO_2, SO_3, P_2O_5, F, Cl and CH_4 etc.) with respect to H_2O vapour, which is the main constituent in the volatile phase derived from a deep-seated source. The $Mg/(Mg+Fe^t)$ ratios of the magmatic melts are frequently used as an indicator of whether a melt could be a partial melting product of a mantle material. The high mg values between 0.76 and 0.86 in this area depend on the TPX and the volatile constituents by which they were formed; moreover, these ratios of the starting material of the partial melts from the mantle may fall between the ratios of 0.9-0.3 (Mysen, 1975). Accordingly, all of the rocks in this complex can be derived from the partial melts from the mantle horizons. The propagation and penetration of the Eastern Ghats Paleo-rift system by reactivations during the subsequent tectonic episodes from the Early Proterozoic Period may extend to different depth levels, causing certain low degrees of partial melting which have been attained only after some prolonged period of tectonic deformation. According to Ramasamy (1982), a span of charnockitic activities extends from 3100 Ma to 2600 Ma, with anorthositic events extending from 2000 Ma to 1100 Ma and alkali syenitic- carbonatitic activities commencing from 1200 Ma onwards (Ramasamy, 1981). It seems that successively younger magmatic episodes were formed under more anhydrous- and alkali-rich environments and at deeper levels than the older ones, owing to the deeper penetration of the rifted continental plates into the mantle. The emplacement of most of the carbonatites from the Proterozoic to recent periods occurring in various parts of the world is restricted along deep crustal fractures that were controlled by regional structures and tectonics. According to Macintyre (1975), the ages of carbonatite complexes younger than 200 Ma indicate that many of them are intimately related with major changes in plate motion which were globally synchronous. The continental plate formed by the propagation and penetration of the Eastern Ghats Paleo-rift system led to the later separation of Gondwana Land from Peninsular India. The wide-spread Deccan Trap volcanic activity covered more than 500000 km^2 and the thickness of the lava flows exceeds 2000 m in some places (Krishnan, 1962), with younger eruptions of olivine-tephrite, soda-trachyte and carbonatite eruptions at Kudangulam near Cape Comorin indicating that Peninsular India is prone to repeated volcanic and magmatic activities (Ramasamy, 1987) and are also bearing evidence of Indian plate movement towards the north after the break of Antarctica from Peninsular India. The ascension of magma and its rate of cooling, pressure and the volume of volatile constituents played a critical role in the magmatic evolution of the residual magma under a specific geological setting and tectonic movement. The influence of local variations in the TPX conditions during the course of the crystallization of minerals creates complexities in tracing the trends of magmatic evolution in a spatial order. There exists a compositional relationship between these new discrete minerals in younger plutons with the minerals in older plutons in this area. The distribution of incompatible HFSE of K, Ti, P, Zr, Nb, Ba, Sr (Table 4) in these rocks and high ratios of HFSE such as Ti/P, K/(K+Na), Nb/Ta, Zr/Hf, Sr/Ba, LREE/HREE and the presence of minerals like allanite, zircon, apatite, monazite, pyrochlore, niobian rutile, magnetite, galena and feldspars, indicate that the low degree of partial melts from the mantle horizon and the parent magma separated from the melt were derived from HFSE-enriched portions of the mantle during the folding and up-arching of the mantle rock at a depth there by enriching the above characteristic elements just prior to the initial stage of the rifting of the diverging continental plates of the Eastern Ghats Mobile Belt.

Mineral	SiO2	Al2O3	Fe2O3	FeO	MnO	MgO	CaO	Na2O	K2O	TiO2	BaO	SrO	P2O5	CO2	SO3	F	H2O	H20	Total
Wollastonite	49.66	1.22	2.01	0.63	0.67	1.52	41.86	0.34	0.30	0.07	0.00	0.00	1.15	0.20	0.00	0.00	0.00	0.00	99.63
Epidote	37.72	17.61	4.67	0.18	0.06	2.29	34.34	0.41	0.84	0.96	0.00	0.00	0.07	0.58	0.00	0.00	0.00	0.00	99.73
Apatite1	0.05	0.18	0.00	0.00	0.00	0.52	54.25	0.30	0.00	0.00	0.00	0.80	39.80	0.00	0.00	0.00	0.00	0.00	95.90
Apatite 2	0.22	0.06	0.00	0.00	0.00	0.17	54.15	0.29	0.00	0.00	0.00	1.51	40.03	0.00	0.00	0.00	0.00	0.00	96.43
apatite 1a	0.00	0.00	0.00	0.00	0.00	1.51	55.73	0.00	0.00	0.00	0.00	0.00	40.95	0.00	0.00	1.56	0.00	0.00	99.75
apatite2a	0.00	0.00	0.00	0.00	0.00	1.51	53.28	0.00	0.00	0.00	0.00	0.00	37.96	0.00	0.00	0.80	0.00	0.00	93.55
apatite 450	2.06	6.91	2.89	2.06	0.07	1.46	41.36	0.47	0.00	0.00	0.00	0.00	41.50	0.00	0.00	1.70	0.26	0.00	100.74
Magnetite 498I	0.10	0.18	67.64	22.57	0.39	0.25	1.05	0.00	0.00	8.10	0.00	0.00	0.00	0.00	0.00	0.00	0.00	0.00	100.28
magnetite 498ii	0.72	0.73	54.17	18.07	0.49	0.73	6.66	0.14	0.42	0.00	0.00	0.00	0.00	0.44	0.77	0.00	0.22	0.00	83.56
Iceladspar	0.36	0.00	0.14	0.00	0.00	0.25	55.03	0.21	0.12	0.14	0.00	0.00	0.00	43.04	0.34	0.00	0.00	0.00	99.63
calcite cc250	2.00	0.00	5.40	0.50	0.06	0.25	51.36	0.24	0.12	0.00	0.00	0.00	0.00	40.28	0.00	0.00	0.00	0.00	100.21
dolomite cc310	0.38	0.00	3.80	2.87	0.55	18.40	30.54	0.10	0.60	0.00	0.00	0.00	0.00	46.06	0.00	0.00	0.00	0.00	103.30
calcite cc311	3.74	0.00	1.40	0.32	0.50	2.25	48.52	0.10	0.03	0.00	0.00	0.00	0.00	40.46	0.57	0.00	0.00	0.00	97.89
Carbonatite cc3	36.40	0.00	2.00	0.25	0.37	1.50	31.93	0.17	0.18	0.00	0.00	0.00	0.00	26.62	0.41	0.00	0.00	0.00	99.83
Mg calcite 313	2.78	0.00	4.05	2.16	0.62	11.65	36.44	0.10	0.09	0.00	0.00	0.00	0.00	41.02	1.66	0.00	0.00	0.00	100.57
carbonatite 314	35.30	0.00	1.70	0.36	0.49	1.12	32.72	0.07	0.11	0.00	0.00	0.00	0.00	26.34	0.67	0.00	0.00	0.00	98.88
ferrocarbonate c	3.26	0.00	4.30	2.73	0.60	13.63	33.31	0.27	0.23	0.00	0.00	0.00	0.00	41.84	1.35	0.00	0.00	0.00	101.52
ferrocarbonate 4	53.02	0.00	6.25	1.15	0.20	3.37	19.09	0.04	0.08	0.00	0.00	0.00	0.00	15.56	0.29	0.00	0.00	0.00	99.05
ferrocarbonatec	49.74	0.00	2.25	0.00	0.42	1.00	23.25	0.04	0.11	0.00	0.00	0.00	0.00	19.92	0.60	0.00	0.00	0.00	97.33
ferrocarboate49	49.82	0.00	2.50	0.00	0.23	1.25	23.95	0.04	0.11	0.00	0.00	0.00	0.00	20.78	0.39	0.00	0.00	0.00	99.07
Benstonite	12.51	5.73	2.03	0.13	0.16	0.00	24.39	1.08	0.98	0.00	20.69	4.33	0.00	27.17	0.00	0.00	0.59	0.00	99.79
Biotite	32.96	15.64	15.09	0.00	0.05	16.10	3.51	0.41	6.42	3.42	0.00	0.00	0.32	0.00	0.00	1.68	5.96	0.00	101.56
vermiculite1	30.92	8.17	23.56	0.00	0.00	10.15	6.60	4.95	4.14	2.88	0.00	0.00	0.00	0.00	0.00	0.00	7.54	1.09	100.00
vermiculite2	29.88	7.33	23.15	0.00	0.00	11.18	8.50	4.93	3.25	3.17	0.00	0.00	0.00	0.00	0.00	0.00	8.04	0.57	100.00
vermiculite3	32.45	24.47	4.09	0.00	0.00	9.96	11.20	1.27	4.08	2.19	0.00	0.00	0.00	0.00	0.00	0.00	7.97	1.61	99.29
vermiculite4	31.58	14.24	24.75	0.00	0.00	6.61	3.15	3.19	4.11	2.41	0.00	0.00	0.00	0.00	0.00	0.00	7.94	2.02	100.00
vermiculite5	31.64	10.47	21.36	0.00	0.00	12.26	3.15	2.39	5.05	3.17	0.00	0.00	0.00	0.00	0.00	0.00	8.18	2.33	100.00
vermiculite6	30.14	11.64	13.00	0.00	0.00	12.92	8.69	0.20	1.33	2.27	0.00	0.00	0.00	0.00	0.00	0.00	11.50	7.05	98.74
vermiculite7	22.44	13.99	3.42	0.00	0.00	9.02	23.97	2.39	0.88	1.22	0.00	0.00	0.00	0.00	0.00	0.00	19.93	2.82	100.08
vermiculite8	26.25	20.76	3.63	0.00	0.00	12.53	15.38	2.54	0.88	1.33	0.00	0.00	0.00	0.00	0.00	0.00	14.54	2.04	99.88
vermiculite9	26.41	20.48	3.76	0.00	0.00	8.24	20.46	2.27	0.82	1.61	0.00	0.00	0.00	0.00	0.00	0.00	13.72	1.88	99.65
vermiculite10	34.10	25.49	4.51	0.00	0.00	12.27	4.97	3.27	1.15	1.97	0.00	0.00	0.00	0.00	0.00	0.00	9.30	2.93	99.96
vermiculite11	21.82	13.70	3.68	0.00	0.00	7.39	24.93	1.41	0.60	0.99	0.00	0.00	0.00	0.00	0.00	0.00	22.01	3.83	100.36
vermiculite12	30.92	18.20	17.16	0.00	0.00	13.65	4.88	0.00	0.00	1.04	0.00	0.00	0.00	0.00	0.00	0.00	14.47	0.00	100.32
vermiculite13	32.15	21.58	12.51	0.00	0.23	4.36	4.89	0.18	1.52	3.01	0.00	0.00	0.00	0.00	0.00	0.00	18.12	0.00	98.55
Barite452	0.50	0.00	0.10	0.00	0.00	0.05	0.60	0.05	0.00	0.00	62.54	1.30	0.00	0.56	33.40	0.00	0.70	0.00	99.80
BariteAlan	4.56	0.17	0.06	0.00	0.00	0.00	1.00	0.00	0.00	0.00	58.74	4.54	0.00	0.05	30.66	0.00	0.08	0.00	99.86

Wo wollastonite from wollastonite-calcite-garnet syenite Epi Epidote from skarn rock from Garigaipalli Aps1, Aps2 and ap1
apatite from Sevvattur carbonatites; ap2,ap450 apatite from Onnakarai carbonatites mt498 magnetitae from Sevvattur carbonatite
Ice Icelandspar from skarn rock of Garigaipalli Cc250 Calcite from carbonaitte of Sevvattur Cc310, cc311. cc312, cc313, cc314,
Carbonates from carbonaites of Jogipatti Cc492, cc493, cc494 carbonates from Carbonatites of Onnakarai cc315
Ben Benstonite from Benstonite carbonatite from Onnakarai Bi Biotite from Biotite pyroxenite verm 1-13 Vermiculite from
Sevvattur carbonatites Ba452 Barite from Onnakarai BaAlan Barite from Alangayam

The chemical compositions of the rocks are listed in Tables 1 and 2 on the basis of rock types and their places of occurrence. The chemical compositions of these rocks vary widely in their silica, alkalis and carbonate contents. CO_2-rich rock carbonatites, Alkali syenites vary in SiO_2 between 35% and 65%, Ultramafic rocks vary in SiO_2 between 40% and 52%. Hybrid rocks have both SiO_2 and CO_2 in an intermediate position.

Table 2. Compositions of various minerals occurring in the carbonatite complex of Tiruppattur

Rock	No	Ba	Sr	Nb	Zr	La	Ce	Nd	Y	Th	U	V	Cu	Zn	Cr	Ni	Co	Pb	Sc
Sev sovite	Sub21		5225		5180	2690	2930	1900	45	140		450							
Sev beforsite	Sub22	3570	9700	220		350	1500	450	90	10	110								
Sev ferrocarbonaite	Sub23		4250	635	40	180	1860	570	130	30	40								
Jogi sovite	Sub24	35	300		10	25	40	10		10	10	10							
Gari skarn	Sub25	220	350		20	15	30			10	10	10							
Onna ferrocarbontite	Sub26	1159	7795	60	50	290	760	240	105	10	25	40							
Sev syenite	Sub11		270	740	120	200	650	430	200	50	10		20	20	15			120	
Sev syenite	Sub12	2780	1120	15	120	60		30	10				25	104	15			100	
Sev syenite	Sub13		955	35	260	70	120	45	30	15	10		20	50	25			170	
Sev ultramafic	Sub1		375		12	10	10	10	10			510							90
Jogi ultramafic	Sub2		275		15	10	25	10	10	10		40							55
Jogi ultramafic	Sub3		900		30	80	180	50	35										
Gari ultramafic	Sub4		217		20	10	10					90							40
Sev granite gneiss	Sub32	2321	1424	60	155	55			25	10	10		45	70	10			20	
Jogi syenite	Sub33	1726	900	60	170	55			30	30	10		20	50	20			35	
Jogi syenite	Sub34	1945	1120	35	240	60			20	15	10		20	60	15			180	
Jogi albitite	Sub35	220	30	180	350	20			10	250	100		10	15	35			10	
Jogi syenite	Sub36		470		80	10	25	50	50		10		20	20	10			880	
Jogi shonkinite	8	1000	2000		600				15			2000	1000	1000	1000	300	200	10	300
Jogi Riebeckite	R	600		60	30				10			200	9	100	2000	30	10	10	20
Jogi hy carbonatite	319	200	200		10				8			100	100	100	8	20	30		10
Jogi hy carbonatite	320	200			20							60	60	200	8	8	8		8
Jogi hy carbonatite	321	2000	2000		8				8			60	30			8	8	8	20
Jogi Rieb carbonatite	401	200	300	60	8				8			30	8	200	600	30	20	20	20
Jogi Rieb carbonatite	402	600	100		20							6010	8		100	30	30		10
Jogi Rieb carbonatite	490	100			10				8			10	8		30	10	10		
Jogi Wollastonite rock	491	300	300		8							60	300	100			8		
Jogi ultramafic nodule	497	2000			10							300	100	300	100	60	30		10
Jogi ultramafic	40	200	100		20				30			60	300	300	600	100	60	8	60
Sev ultramafic	201	3000	300	50	60				10			60	20	100		8	8	100	
Sev ultramafic	203	6000	2000	100	50				60			100	8	200	100	30	20	100	
Sev hy carbonatite	204		300		30				20			200	30	200	200	200	60		20
Sev hy carbonatite	206	600	600		60				60			100	100	100	600	100	60	20	60
Sev oligoclasite	219	1000	1000	60	100				10			300	60		30	10	10	20	
Jogi carbonatite	318	4000	1000	20	60				60			100	100	100	200	30	30	30	30
Apatite magnetite rock	460	100	100		8							200	300	100	300	30	30		8
Sev pyroxenite	789	300			400				10				100	60		60	60		60
Jogi ultramafic	493		790			170	145	90	950			50			100	175	75		675
Jogi ultrapot syenite	39	475	2300			880	380	790	2630			315	280	670	320	375	560	260	390
Jogi melanite sovite	990	120	4913		1115	250	165	123	733			78	73	110	73	73	160	133	610
Jogi shonkinite	561	322	5960		708	136	142	76	534			76	24	64	76	70	50	262	74

Table 3. Trace elements' distribution in carbonatites and associated alkaline rocks

10. Conclusion

The dolerites are hypabyssal equivalents of gabbro and basalts, which are common products of the partial melting of mantle / crustal rocks at depth. The overlapping petrographic

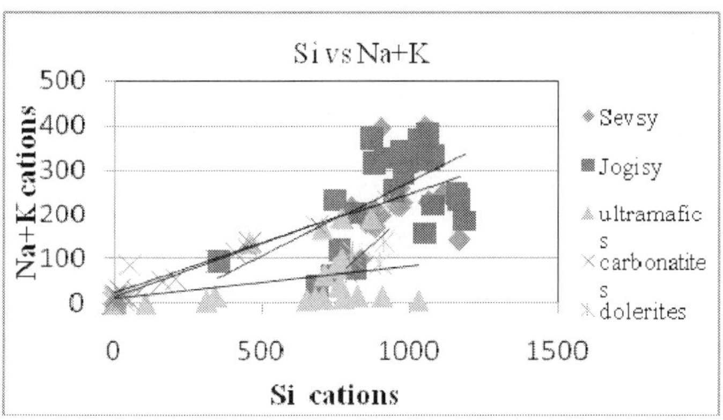

a)

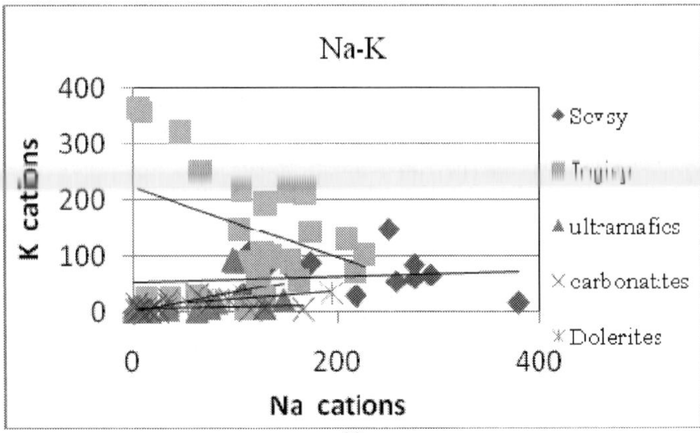
Na-K
K cations
Na cations
0
100
200
300
400
0
200
400
ultramafics
carbonatites
Dolerites
b)

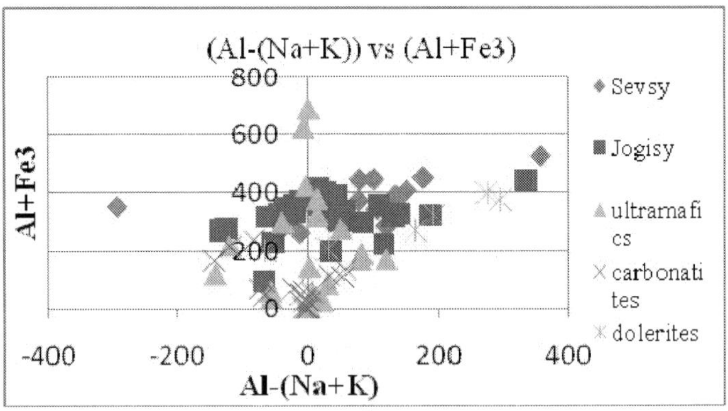
(Al-(Na+K)) vs (Al+Fe3)
Al+Fe3
Al-(Na+K)
0
200
400
600
800
-400
-200
0
200
400
Sevsy
Jogisy
ultramafics
carbonatites
dolerites
c)

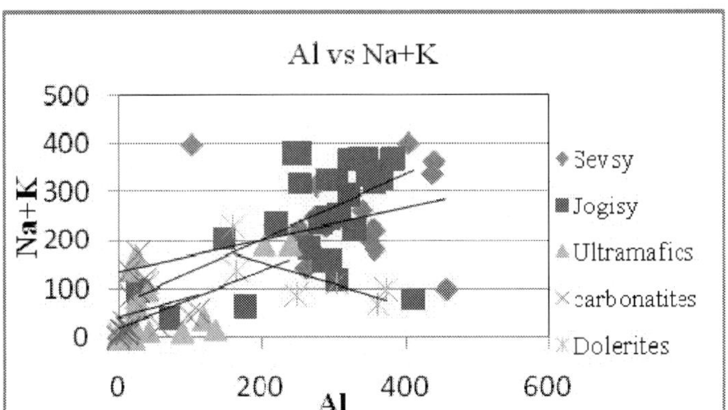
Al vs Na+K
Na+K
Al
0
100
200
300
400
500
0
200
400
600
Sevsy
Jogisy
Ultramafics
carbonatites
Dolerites
d)

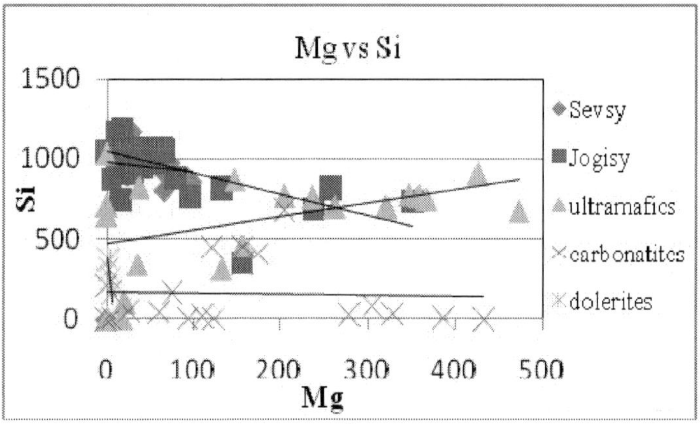
Mg vs Si
1500
1000
500
0
Si
0
100
200
300
400
500
Mg
Sevsy
Jogisy
ultramafics
carbonatites
dolerites

e)

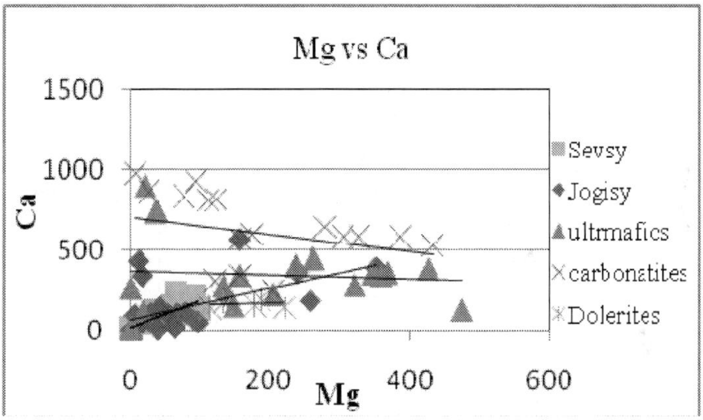
Mg vs Ca
1500
1000
500
0
Ca
0
200
400
600
Mg
Sevsy
Jogisy
ultrmafics
carbonatites
Dolerites

f)

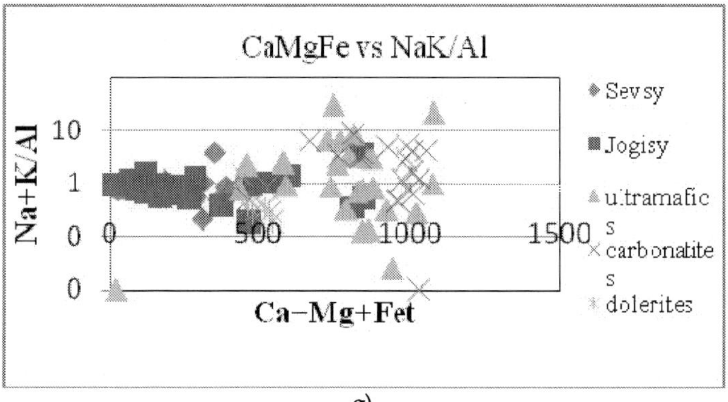
CaMgFe vs NaK/Al
10
1
0
0
Na+K/Al
0
500
1000
1500
Ca−Mg+Fet
Sevsy
Jogisy
ultramafics
carbonatites
dolerites

g)

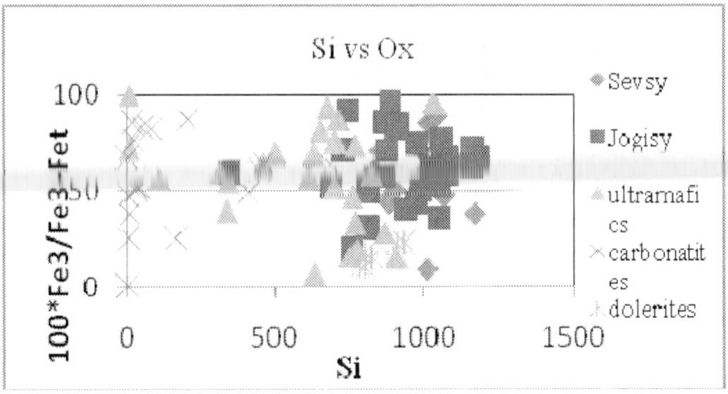

h)

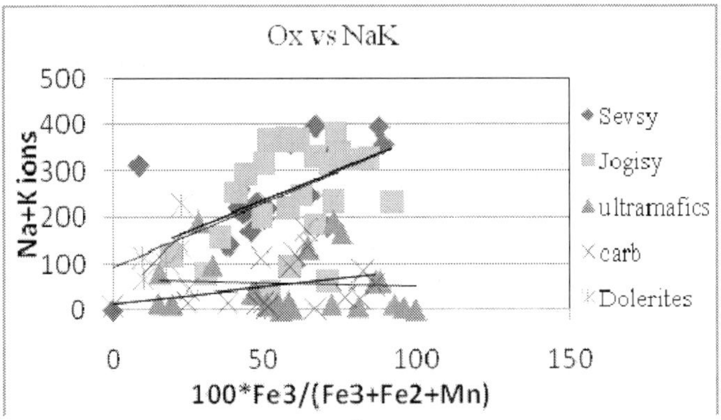

i)

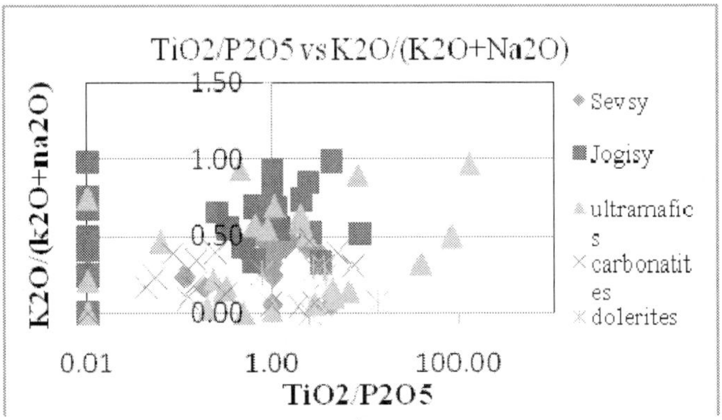

j)

Fig. 5. a-j show linear trends of the magmatic evolution of different types of rocks from common parent magma variations of all these rocks indicate that they are all derived from the same mantle source by different degrees of partial melting. The composition of the parent magma for this complex appears to be very close in composition to that of shonkinite magma, which might have been derived by liquid fractionation and separation from the low degree of the partial melt of the mantle material. The partial melt might have initially been a crystallized olivine and calcic plagioclase; therefore, the residual shonkinitic parent magma so evolved is impoverished in alumina and in silica and acts as a primary magma in a closed magmatic chamber under volatile enriched conditions. Later on, with early crystallizations of calcium-rich clinopyroxenes, the residual magma is impoverished in silica with the enrichment of volatile constituents such as H_2O and CO_2. In an ascending convection current which motivated a low viscous and high temperature shonkinite magma, clinopyroxenes are crystallized with the enrichment of Ca, Mg, Fe, Ti^{iv} and Al^{iv} and a depletion in Si, Al^{vi}, Na and K in low pressure zones towards the top of the magmatic column. On the other hand, from the top of the magmatic column, sinking clinopyroxenes liberate Ca, Mg, Fe, Ti^{iv} and Al^{iv} and accumulate Na and Al^{vi} in their subsolidus crystal lattices during descending convection currents and towards the bottom of the chamber with the crystallization of subsolidus aegirine. Thus, both salic and mafic constituents in the form of ions are concentrated at the top of the magmatic column in different portions. In this closed magmatic chamber under high P_{H2O} and P_{CO2}, an immiscible separation of camafic carbonate liquids and alkali silicate liquids are derived at depth (Saravanan and Ramasamy, 1995). The separated alkali silicate magmas are emplaced in sequences of co-magmatic bodies of syenites, first in the Sevvattur Basin and subsequently emplaced in the Jogipatti basin followed by carbonate magmas in both (Fig. 1 a-d). It is quite a complex matter to study the impact of all these processes by tracing the history of the magmatic evolution inscribed on the rocks over the course of millions of years with the help of detailed petrographic and field investigations.

11. References

Barth, T.F.W. (1952) Theoretical Igneous Petrology, Ms Book and Mineral Company, 2nd Edn, New York, p. 416

Borodin, I.S., Gopal, V., Maralev, V.M. and Subramanian, V. (1971) Precambrian carbonatites of Tamil Nadu, South India, Journal of the Geological Society of India, v. 12, pp. 101-112

Bowen N.L. (1956) The Evolution of Igneous Rocks, Dover, New York, 2nd Edn, p. 332

Deer, W.A., Howie, R.A. and Zussman, J. (1965) Rock Forming Minerals, v. 1 to v. 5, Longmans, London

Dickson,J.A.D. (1965) A modified staining technique for carbonates in thin sections, Nature, v. 205, p. 587

Grady, J.C. (1971) Deep main faults in South India, Journal of the Geological Society of India, v. 12, pp. 56-62

von Eckermann, H. (1966) Wollastonite in carbonatite rocks, Geochimica et Cosmochimica Acta, v. 31 (11), pp. 2253-2254

Schieicher, H., Kramm, U., Pericka, E., Schidlowski, M., Schmidt, F., Subramanian, V., Todt, W. and Viladkar, S.G. (1998) Enriched sub-continental upper mantle beneath Southern India: Evidence from Pb, Nd, Sr and C-O Isotopic studies on Tamil Nadu carbonatites, Journal of Petrology, v. 39, (10)m, pp. 1765-85

Krishnan, M.S. (1962) Geology of India and Burma, Higgin Bothams, Chennai

Macintyre, R. (1975) Age and significance of carbonatites and alkaline intrusive, Mem. Noticicas. Mus. Lab. Min. Geol. Univ. Coimbra, v. 80, pp. 99-103

Mysen, B. (1975) Partitioning of iron and magnesium between crystals and partial melts in Peridotites Upper Mantle, Contrib, Mineral Petrol., v. 52, pp. 69-76

Mysen, B.O., Virgo, D. and Seifert, F.A. (1985) Relationships between properties and structure of aluminosilicate melts, Am. Mineralogist, v. 70, pp. 88-105

Naidu, P.R.J. (1958) 4-Axes Universal Stage, Dept. of Geology, Madras University

Ramasamy, R. (1973) Geology of the area South west of Tiruppattur, Madras State (Tamil Nadu), India, Ph.D Thesis, p. 226, University of Madras

Ramasamy, R. and Shapenko, V. (1980) Fluid inclusion studies in carbonatites of Tiruppattur, Tamil Nadu , India (Abst) Int. Geol. Cong. Paris , v. 1, p. 79

Ramasamy, R. (1981) Tectonomagmatic Episodes of the Peninsular India, Sympet, Jaipur, Geol. Surv. India, Abstract L2-L3.

Ramasamy, R. (1982a) A possible paleo-rift system of the eastern Ghats in the Peninsular India, J. Moscow State Univ., Geology, Ser. 4, No. 2, pp. 32-37 (in Russian)

Ramasamy, R. (1982) Structure and tectonics of carbonatite complex of Tiruppattur, Tamil Nadu In: Bhattacharya, A./K. (Ed.) Current Trends in Geology, IV Indian Geological Congress, Today and Tomorrows Publishers, New Delhi, pp. 119-136

Ramasamy, R. (1986) Titanium-bearing garnets from alkaline rocks of carbonatite complex of Tiruppattur, Tamil Nadu, Current Science, v. 55, pp. 1026-29

Ramasamy, R. (1986a) Ca-rich pyroxenes from the carbonatite complex of Tiruppattur, Tamil Nadu, Current Science, v. 55, pp. 981-984

Ramasamy, R. (1987) Reactivation of the Eastern Ghats Paleorift system during Tertiary and other periods, Proc. National, Seminar Tertiary Orogeny, pp. 109-127.

Ramasamy, R., Gwalani, L.G. and Subramanian, S.P. (2001) A note on the occurrence and formation of magnetite in the carbonatites of Sevvattur, North Arcot district, Tamil Nadu, Southern India Journal of Asian Earth Sciences, v. 19, pp. 297-304

Ramasamy, R., Subramanian, S.P. and Sundaravadivelu, R. (2010) Compositional variations of olivine in shonkinite and its associate ultrabasic rock from the carbonatite complex of Tiruppattur, Tamil Nadu, Current Science, v. 99(10), pp. 1428-1433

Rittmann, A. (1973) Stable Mineral Assemblage of Igneous Rocks, Springer, Berlin, p. 251

Saravanan, S. and Ramasamy, R. (1971) A magnesio-riebeckite from Samalpatti area, Tamil Nadu, India, Mineralogical Magazine, v. 38, pp. 376-377

Saravanan, S. and Ramasamy, R. (1995) Geochemistry and petrogenesis of shonkinite and associated alkaline rocks of Tiruppattur carbonatite complex, Tamil Nadu, Journal of the Geological Society of India, v. 46, pp. 235-243

Schairer, J.F. and Yoder, H.S.J.R., (1960) The nature of residual liquids from crystallization data on the system Nepheline-diopside-silica, American Journal of Science, Bradley Volume, v. 258A, pp. 273-83

Udas, G.R. and Krishnamurthy, P. (1970) Carbonatites of Sevvattur and Jogipatti, Madras State, India, Proc. Of the Indian National Academy of Science, 36A, pp. 331-343

Verwoerd, W.J. (1966) South African Carbonatites and their probable mode of origin, Ph.D. Thesis, Univ. of Stellenboch

Viladkar, S.G. and Subramanian, V. (1995) Mineralogy and geochemistry of the Sevvattur and Samalpatti complexes, Tamil Nadu, Journal of the Geological Society of India, v. 45, pp. 505-518

Winchell, A. N. (1945) Elements of Optical Mineralogy, part II Optical properties of Minerals

Yagi, K. (1953) Petrochemical studies on the alkaline rocks of the Morotu District, Sakhalin, GSA Bull, v. 64, 769-810

Yoder, H.S., Stewart, D.B. and Smith, J.R. (1957) Ternary feldspars, Carnegie Inst. Washington Year Book, pp. 206-214